		IIIA	IVA	VA	VIA	VIIA	2 He 4.00260
		5 B 10.81	6 C 12.011	7 N 14.0067	8 O 15.9994	9 F 18.998403	10 Ne 20.179
IB	IIB	13 Al 26.98154	14 Si 28.0855	15 P 30.97376	16 S 32.06	17 Cl 35.453	18 Ar 39.948

28 Ni 58.69	29 Cu 63.546	30 Zn 65.38	31 Ga 69.72	32 Ge 72.59	33 As 74.9216	34 Se 78.96	35 Br 79.904	36 Kr 83.80
46 Pd 106.42	47 Ag 107.868	48 Cd 112.41	49 In 114.82	50 Sn 118.69	51 Sb 121.75	52 Te 127.60	53 I 126.9045	54 Xe 131.29
78 Pt 195.08	79 Au 196.9665	80 Hg 200.59	81 Tl 204.383	82 Pb 207.2	83 Bi 208.9804	84 Po (209)	85 At (210)	86 Rn (222)

63 Eu 151.96	64 Gd 157.25	65 Tb 158.9254	66 Dy 162.50	67 Ho 164.9304	68 Er 167.26	69 Tm 168.9342	70 Yb 173.04	71 Lu 174.967
95 Am (243)	96 Cm (247)	97 Bk (247)	98 Cf (251)	99 Es (252)	100 Fm (257)	101 Md (258)	102 No (259)	103 Lr (260)

INTRODUCTION TO
General, Organic, AND Biological
CHEMISTRY

INTRODUCTION TO
General, Organic,
AND
Biological
CHEMISTRY

Robert J. Ouellette

The Ohio State University

Macmillan Publishing Company

NEW YORK

Collier Macmillan Publishers

LONDON

Earlier editions, entitled *Introductory Chemistry*, copyright © 1970
and 1975 by Robert J. Ouellette.

Macmillan Publishing Co., Inc.
866 Third Avenue, New York, New York 10022

Collier Macmillan Canada, Inc.

Library of Congress Cataloging in Publication Data

Ouellette, Robert J., 1938–
  Introduction to general, organic, and
biological chemistry.

  Includes index.
  1. Chemistry. I. Title.
QD31.2.O84 1984   540   83-976
ISBN 0-02-389880-1

Printing: 2 3 4 5 6 7 8   Year: 4 5 6 7 8 9 0 1

ISBN 0-02-389880-1

Preface

This textbook has evolved from the experiences of two decades of teaching a two-quarter course sequence covering general, organic and biological chemistry. The students in the course are interested in agriculture, health education, home economics, medical technology, nursing, nutrition, physical therapy and other fields generally described as life or health sciences. For most of these students, their first chemistry course will be their last, but the applications of chemistry will permeate a number of subjects that they will study as part of their major.

This book is written with the assumption that the students have had no previous course in chemistry and that their mathematical background is limited. Emphasis is placed on essential chemical concepts rather than giving a cursory treatment to a large number of topics related to each of the diverse interests of the students.

A chemistry course must first teach chemistry and then point to selected applications to justify the study of chemistry as a science which is basic to many other fields. The detailed applications of chemistry must remain for other courses. A chemistry course cannot discuss all of the potential applications for students who wish to become exercise physiologists, radiation technicians or dieticians, for example.

An author, like any teacher, must present the complex structured science of chemistry at a level suitable for students. Oversimplifications and analogies may result in misinterpretations and even incorrect conclusions in extensions of the acquired knowledge. Throughout the text the author has tried to carefully define the limits to which the theories may apply without introducing excessive exclusionary comments.

Discussions are qualitative in many areas of this book. However, since chemistry is a quantitative science, an appreciation of chemical concepts can be best achieved by working quantitative problems in areas such as stoichiometry or gas laws. The students need essentially only arithmetic skills. All problems are solved using the factor-unit method, which is emphasized from the very start of this book. In order to provide a variety of problems, this book contains 200 solved in-chapter examples and 1600 end-of-chapter questions—many of which have multiple parts. References to similar end-of-chapter problems are given in the chapter examples.

Although every author feels that his book provides the best approach to teaching the subject, the successful book acknowledges the variety of instructors, students and goals of the many courses offered in this country. Accordingly, care has been used to write sections of each chapter to be as free-standing as possible. In general, the more difficult concepts or areas that may not be covered are placed at the end of the chapter.

The unifying theme of this book is the chemistry of the human body. Thus, in the general chemistry portion of the book, there are references to both subsequent organic and biological chemistry. Similarly the organic section contains references to biological chemistry. As a consequence, information is not segmented, but flows from one subject area to the next. If the students can view chemistry as a broad science rather than a conglomeration of specialties, then an educational goal has been achieved.

Each new term is introduced carefully with a clear definition in each chapter. The New Terms section at the end of the chapter contains these definitions or modifications thereof, and the terms are used in a chapter summary written in narrative form. Learning objectives at the end of the chapter also emphasize the importance of most of the new terms introduced in the chapter.

Small essays, called "An Aside," are set off from the main sections of the chapters. These "asides" vary in content and style and are not directly related to the general flow of knowledge to be gained from the book. However, because of the interesting aspects of the "asides" and their relationships to material or applications in the students' fields of study, most students will find it worthwhile to read them.

The author wishes to thank the many reviewers selected by Macmillan: Thomas Berke, Brookdale Community College; Robert W. Coley, Montgomery College; Gary L. Gard, Portland State University; James T. Hoobler, Chemeketa Community College; James E. House, Illinois State University; James J. Leary, James Madison University; Cortlandt G. Pierpont, University of Colorado; Alan J. Pribula, Towson State University; Alan R. Price, University of Michigan; Thomas I. Pynadath, Kent State University; Daniel J. Smith, The University of Akron; Gerald Swanson, Daytona Beach Community College; Vaughan Vandegrift, Murray State University; and Albert Zabady, Montclair State College.

The author acknowledges the valuable contributions of the Chemistry Editor, Thomas Vance. Whereas the final manuscript may have been deemed suitable for publication one year earlier by other publishers, the decision to take additional time and to rewrite the entire manuscript and obtain another round of reviews gave the author the opportunity to, in effect, prepare a second edition without a published first edition. In addition, class testing of prepublication material for another year aided in a better integration of the various components of the book.

It was a pleasure to have Elisabeth Belfer as the Production Supervisor for this book. She not only provided control on consistency of style and terminology, but carefully scrutinized the chemistry. Her carefully phrased questions about chemistry content were immensely helpful and allowed the author to make substantial changes to the copy-edited manuscript.

The majority of this text was typed in its several versions by Joy Snyder. Her services under deadline pressures were provided with good humor.

One special person, who doesn't want her name mentioned, read every page of manuscript and galleys. Her professional editorial skills in chemistry were invaluable and are acknowledged with love.

R. J. O.

Contents

1 Chemistry and Measurements
1

2 Classification of Matter
23

3 The Structure of Matter
47

7

Gases
133

8

Liquids and Solids
153

9

Water and Solutions
170

13

Introduction to Organic Chemistry
267

14

Saturated Hydrocarbons
286

15

Unsaturated Hydrocarbons
311

19
Amines and Amides
411

20
Introduction to Biochemistry
437

21
Carbohydrates
460

25

Biochemical Energy
581

26

Metabolism of Carbohydrates
609

27

Metabolism of Lipids
629

Chemistry and Measurements

This type of scale, formerly used to weigh produce, has been replaced by more accurate electronic scales.

1.1
What Is Chemistry?

Chemistry *is a science of the composition, structure, and reactions of matter.* This statement gives you a feeling for the term chemistry only if you understand the words used. Let's assume you know what science means, since it is part of your vocabulary. But do you know what matter is? **Matter** *is anything that has mass and occupies space.* A colloquial synonym for matter is "stuff." Anything around you that you would call "stuff" is matter. Examples of matter include the oxygen you breathe, the water you drink, and the sugar you eat.

Mass is not a common term that you use. **Mass** *is a quantity of matter.* Normally you describe a quantity of a material in terms of its weight. There is a subtle difference between mass and weight. **Weight** *is the result of the gravitational attraction of the earth for the object.* Without gravity, an object would not have weight, but it would still have mass. For example, astronauts are weightless in space but still have mass. The force of gravity is about the same anywhere on the earth and at any location, two objects that have the same mass have the same weight. Therefore, mass and weight are often used interchangeably.

In chemistry the word **composition** *means the identity and amount of the components of matter.* The nursery rhyme tells us that little girls are made of (composed of) sugar and spice and all things nice. This whimsical expression of composition gives the components of little girls but not the amounts. An extraterrestrial being might describe a little girl in part as consisting of two hands, two arms, one trunk, one head, and so on. Such a description gives a more detailed composition of little girls, since it gives the amounts as well as the identities of the components.

Composition can be stated in terms other than the number of items. Mass can also be used. You often purchase produce by weight rather than by number of items. Composition can be based on this weight. Consider a bag of groceries containing one grapefruit and two oranges. The composition of the bag can be stated in terms of the weight of the fruit. If the weight of an orange is 7 oz and the weight of a grapefruit is 14 oz, then the total weight is 28 oz.

$$
\begin{aligned}
2 \text{ oranges} &= 2 \times (7\text{ oz}) \\
1 \text{ grapefruit} &= \underline{1 \times (14\text{ oz})} \\
\text{total weight} &= 28\text{ oz}
\end{aligned}
$$

The percent composition of the fruit based on weight is 50% oranges and 50% grapefruit.

$$
\begin{aligned}
\text{percent} &= \frac{\text{weight oranges}}{\text{total weight}} \times 100 \\
&= \frac{2 \times (7\text{ oz})}{28\text{ oz}} \times 100 = 50\%
\end{aligned}
$$

Example 1.1

A 110 lb woman may eat food consisting of 10 oz of carbohydrates, 2 oz of fat, and 18 oz of protein on an average day. What is the percent of carbohydrate in her diet?

Solution. The total food intake is 30 oz. Since percent is the ratio of a part divided by the total with the result multiplied by 100, we may write

$$\frac{10 \text{ oz}}{10 \text{ oz} + 2 \text{ oz} + 18 \text{ oz}} \times 100 = 33\%$$

The composition of her diet is 33% carbohydrate.

[Additional examples may be found in **1.3–1.5** at the end of the chapter.]

Chemists describe the composition of matter in terms of atoms and molecules. Atoms are the simplest units of matter, whereas molecules are combinations of atoms. Helium gas consists of helium atoms. Water consists of molecules that are composed of two hydrogen atoms and one oxygen atom. In addition to knowing the number and type of the components of a substance, such as water, chemists use mass to describe the composition of matter. Hydrogen is the lightest atom and has a mass of 1 atomic mass unit (amu). The oxygen atom is 16 times heavier than the hydrogen atom and has a mass of 16 amu. A water molecule then has a mass of 18 atomic mass units.

mass of 2 hydrogen atoms = 2 × (1 amu)
mass of 1 oxygen atom   = 1 × (16 amu)
_____
mass of 1 water molecule = 18 amu

In terms of mass, the water molecule is then 11% hydrogen and 89% oxygen.

$$\% \text{ composition} = \frac{\text{part}}{\text{whole}} \times 100$$

$$\% \text{ hydrogen} = \frac{2 \text{ amu}}{18 \text{ amu}} \times 100 = 11\%$$

$$\% \text{ oxygen} = \frac{16 \text{ amu}}{18 \text{ amu}} \times 100 = 89\%$$

Example 1.2

Carbon dioxide consists of molecules that have one carbon atom and two oxygen atoms. The masses of carbon and oxygen atoms are 12 and 16 amu, respectively. Calculate the composition of carbon dioxide in terms of the mass of the atoms.

Solution. The total mass of a carbon dioxide molecule is 44 amu.

mass of 1 carbon atom   = 1 × (12 amu)
mass of 2 oxygen atoms   = 2 × (16 amu)
_____
mass of 1 carbon dioxide molecule   44 amu

The percent of carbon in the carbon dioxide molecule based on mass is

$$\% \text{ carbon} = \frac{12 \text{ amu}}{44 \text{ amu}} \times 100 = 27\%$$

$$\% \text{ oxygen} = 100 - 27 = 73\%$$

[Additional examples may be found in **1.6–1.8** at the end of the chapter.]

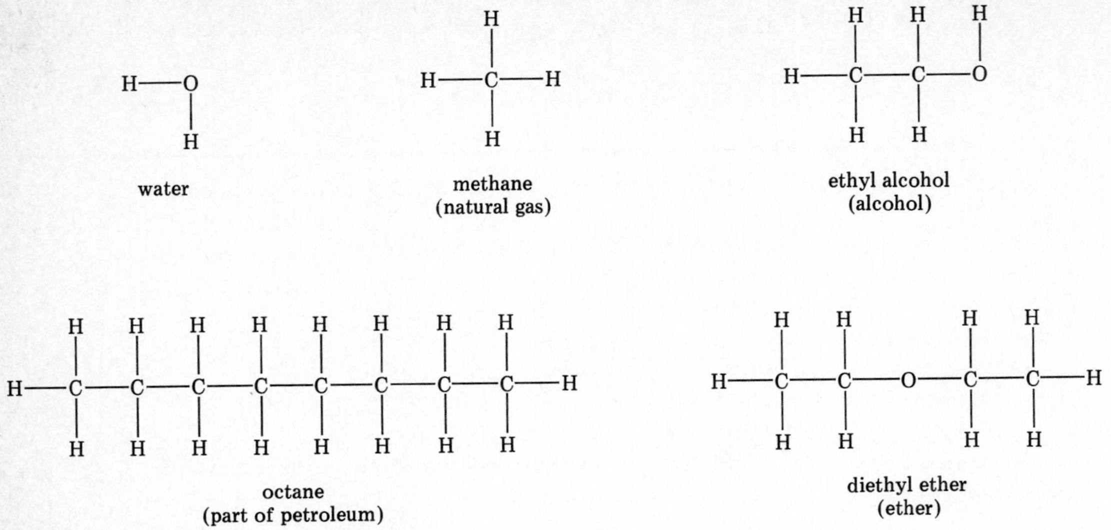

H—O
|
H

water

H
|
H—C—H
|
H

methane
(natural gas)

H   H   H
| | |
H—C—C—O
| |
H   H

ethyl alcohol
(alcohol)

H   H   H   H   H   H   H   H
| | | | | | | |
H—C—C—C—C—C—C—C—C—H
| | | | | | | |
H   H   H   H   H   H   H   H

octane
(part of petroleum)

H   H   H   H
| | | |
H—C—C—O—C—C—H
| | | |
H   H   H   H

diethyl ether
(ether)

C is the symbol for the carbon atom.
O is the symbol for the oxygen atom.
H is the symbol for the hydrogen atom.
The dashes between the symbols represent bonds that hold the molecule together and create the structure.

Figure 1.1
Structures of Some Common Substances.

Composition by itself does not tell us all that we really need to know about matter. For example, would the inhabitants of another world be able to picture you if they were given only your composition? The answer is no, since they need to know how the parts of your body go together.

In chemistry **structure** *means the arrangement of the components of a substance.* In other words, how things are put together. As the science of chemistry is developed in this book, you will learn that atoms in molecules are bonded to each other. **Bonding** *means held or fastened together.* The manner in which atoms are bonded to each other then creates a structure. For example, in water the two hydrogen atoms are bonded to a central oxygen atom but not with each other (Figure 1.1). The capital letters H and O are used to represent (or symbolize) the hydrogen and oxygen atoms, respectively. The dashes between the symbols for the atoms represent the bonds. Structures for some other common substances are also shown in Figure 1.1. More details on composition and structure will be given in later chapters.

Example 1.3 ───────────────────────

Describe the composition of ethyl alcohol in terms of the atoms in the molecule.

Solution. The representation of ethyl alcohol in Figure 1.1 tells us that the molecule has two carbon atoms, six hydrogen atoms, and one oxygen atom.

Example 1.4

Describe the structure of methane.

Solution. The representation of methane in Figure 1.1 tells us that methane has four hydrogen atoms bonded to a central carbon atom.

Example 1.5

Describe the structure of ethyl alcohol centering your discussion on the carbon atoms.

Solution. From the structure in Figure 1.1, one can see that the two carbon atoms are bonded to each other. One carbon atom is also bonded to three hydrogen atoms. The other carbon atom is bonded to two hydrogen atoms and one oxygen atom. The oxygen atom is bonded to a carbon atom and a hydrogen atom.

[Additional examples may be found in **1.9–1.11** at the end of the chapter.]

A **reaction** *involves changes in the composition and structure of matter.* There are many examples of chemical reactions occurring in your everyday life. Examples include sight perception, muscle movement, and the metabolism of food. Although you might think of these as physiological processes, they all involve chemical reactions. The metabolism of alcohol converts it into carbon dioxide and water. The body uses oxygen in this reaction. Chemists represent the reaction with an equation.

$$\text{alcohol} + \text{oxygen} \longrightarrow \text{carbon dioxide} + \text{water}$$

The words used to the left of the arrow are the **reactants,** *the materials that react.* The words to the right of the arrow are the **products,** *the substances produced in a chemical reaction.* The arrow means "is converted into." In this chemical reaction substances are transformed or converted into other substances that have different compositions and structures.

1.2 Who Studies Chemistry and Why?

The field of chemistry is so broad that many specialists are included under the term chemist. At one time lines could be drawn between the various sciences. However, chemistry merges in one direction into physics and in another into biology. There are both chemical physicists and physical chemists who are interested in the structure of the atom and the energy that can be obtained by "splitting" the atom in a nuclear reactor. Similarly, there are molecular biologists and biochemists interested in the same research areas with only small differences in emphasis. Studies of the chemistry of DNA (which carries the genetic code) and the workings of the nerves, muscles, and even the brain are part of the research areas of biochemists and molecular biologists.

Chemistry forms the basis for understanding applied sciences such as dietetics, inhalation therapy, medical technology, nursing, nutrition, physiology, and x-ray technology. Chemistry courses are usually required in the curricula of these applied sciences. Chemistry is also necessary for students studying to enter professional fields such as medicine, dentistry, or veterinary medicine.

Because of the chemical nature of the body, you should be interested in chemistry so that you know how your body works. Deficiencies in some chemicals, the presence of improper chemicals, or the breakdown of the body's system of working with chemicals can result in illness or death. A background in chemistry thus will help you understand illness and medical treatment.

1.3
The Scientific Method

Observation and Laws

All sciences depend on observations. *The method of making observations, of cataloging them, and of using them in the solution of a problem is called the* **scientific method.**

Accumulation of quantitative observations of certain regularities in nature leads to the statement of laws of nature. *A* **law** *is simply the conclusion or an explicit statement of fact that is already inherent in the information obtained by observations.* Nothing new in understanding nature is obtained by stating a law; the law merely summarizes what has been observed.

You could state a "law of the morning." The law would be that upon leaving your bed in the morning your feet will touch the floor. Your statement or law is based on observations made many times. Of course, this isn't a law of science found in scholarly texts.

An example of a scientific law is the law of conservation of mass. This law states that in a chemical reaction the mass of products formed is equal to the mass of reactants used. In other words, the mass has been conserved. This law is important to the study of chemistry and will be discussed later. For now it is stated to inform you that chemists have made observations on the mass of reactants and products in many chemical reactions. In every case mass is conserved.

Models

Science goes beyond the observation of facts and the statement of laws. There is also an interest in how and why matter behaves as it does. Scientists use models to describe nature. *The term* **model** *refers to an idea that may correspond to what is responsible for the natural phenomenon.* Picture models are easy to understand and will be used frequently in this text. These models must not be regarded as more important than the experimental facts. The experimental facts reflect nature; the model is a human creation. There are limitations to models, for they may not describe all of the more complex aspects of nature. A scientific model is not a small-scale working version of a larger entity, as model planes are miniature representations of actual planes. Scientific models are attempts to picture that which we cannot see. These models are imagined; they are not photographs of the real thing.

The structure of matter is submicroscopic in scale and cannot be seen by optical microscopes. However, the chemist has developed models of matter by using indirect means. An analogy is that of the chemist as a detective seeking to unravel the mystery of structure by accumulating circumstantial evidence. We cannot directly "see" the atoms and molecules, but there is an overwhelming body of circumstantial evidence on chemical structure. However, as in a mystery, the case is never closed. Each new bit of evidence is used to recheck our model.

The terms hypothesis and theory are often used in science. *A* **hypothesis**

AN ASIDE _____

The scientific method is a process that may be used in many parts of our everyday life. Consider that you drop your calculator on the way back to the dorm after your chemistry class. The next day, a friend borrows your calculator and finds that it doesn't give the lighted display when the ON button is pushed. Why doesn't the calculator work?

If your friend knew that you dropped the calculator, he might think that the calculator does not work as a result of the accident. That would be a reasonable hypothesis. However, without knowledge of the accident, he could make some hypotheses that could lead to a solution of the problem and perhaps make the calculator work properly.

One hypothesis would be that there is no battery in the calculator. This hypothesis is easily checked by opening the calculator. If there is a battery in the calculator, another hypothesis is necessary. The second hypothesis might be that the battery connections are not firmly in place. That hypothesis can also be easily tested. Finally, he might hypothesize that the battery is dead. A way to test this hypothesis is to borrow a battery from someone else's calculator to see if the calculator will now work.

If none of the hypotheses is correct, then there could be some physical damage to some of the components of the calculator. The ON–OFF button could be defective, or one of the chips within the calculator may be damaged.

Your friend would then have to return the calculator to you and explain that for some reason that he doesn't understand your calculator does not work. Based on his tests, you might consider purchasing a new calculator or finding a repair service. However, prior to these costly possibilities, it would be scientifically proper to repeat the tests performed by your friend. If the tests were done properly, you should obtain the same results. If in any of the steps, the calculator begins to work, then you might conclude that your friend did not do all of the tests properly. However, there is a possibility that some loose connection or short exists in the calculator that can return to give you trouble at a later date. That date may be the day of an important examination. Obviously, you need to consider further how to ensure that your calculator will work when you need it.

_____

refers to a tentative model, whereas a **theory** *describes a model that has been tested many times.* The dividing line between the two is arbitrary and cannot be precisely defined. When a substantial majority of scientists accept a model, it is called a theory.

The English chemist John Dalton (1766–1844) proposed a model to account for the then known laws of chemistry. As a result of many years of testing his model is called Dalton's atomic theory. The theory, which forms part of the foundation of chemistry, will be discussed in later chapters. The theory is stated in five parts.

1. There are simple substances called elements that are composed of tiny indivisible particles called atoms.
2. Atoms of the same element are identical, but differ from those of any other element. Each type of atom has a definite mass that is different from that of any other atom.

3. Atoms of different elements combine only in certain whole number combinations to produce molecules.

4. Complex substances or compounds are composed of molecules. The composition of all molecules in a compound is identical.

5. In chemical reactions atoms are transferred or rearranged to produce different substances. The identity of each atom is unchanged, and its mass is unchanged.

1.4
What Is a Measurement?

We all use measurements. The pound of hamburger in a supermarket display case is measured by a scale that is set to a standard of weight we have come to accept socially. The prescription of the quantity of a drug to be administered to a patient by a nurse involves accepted standard units.

Science depends on measurements. From the simplest to the most exacting experiments, **measurements** *are made by use of or comparison to a standard measuring device.* In the U.S.A., the English system of measurement, with its inconvenient relationships between different units, is the standard for measurements. A mile is 1760 yards; a yard is 3 feet; a foot is 12 inches.

It is often necessary to convert one measured quantity into another equivalent quantity with different units. This is accomplished by a **conversion factor,** which is *a multiplier having two or more units associated with it.* The approach is called the factor unit method, because the factor is numerically equivalent to one. When a quantity is multiplied by the factor, it is changed into an alternate but equivalent number having different units.

The factor unit method is widely used to solve problems. An outline of this problem-solving technique follows.

1. Examine the data given and note the units associated with all numbers.
2. Determine what is asked for, that is, what units are desired.
3. Write down the vital data given, along with the units, to the left of an equality sign.
4. Write down a symbol for the desired unknown, along with the units, to the right of the equality sign.
5. Develop the conversion factors, with their units, that, when multiplied by the known data, will give the desired unknown.
6. Check your work to see that the units are equal on both sides of the equation.
7. Carry out the arithmetic, and check the answer for mathematical reasonableness.

To illustrate the factor unit method, let us calculate the length in feet of a 100 yard football field. While you know the answer without the factor unit method, the use of the factor unit method with familiar quantities is the best way to learn the method.

The number of feet in the field is the unknown quantity. Your given quantity is 100 yards. Therefore, write

100 yards × factor = ? feet

In addition, we know that one yard is equal to three feet.

1 yard = 3 feet

Dividing both sides of this statement by 1 yard gives a factor, F_1, that is equal to one.

$$F_1 = \frac{1 \text{ yard}}{1 \text{ yard}} = \frac{3 \text{ feet}}{1 \text{ yard}} = 1$$

This factor, 3 feet divided by 1 yard, is just a mathematical way of saying that 3 feet equals 1 yard.

Alternatively, dividing both sides of the original statement by 3 feet gives factor F_2, which is also equal to 1.

$$F_2 = \frac{1 \text{ yard}}{3 \text{ feet}} = \frac{3 \text{ feet}}{3 \text{ feet}} = 1$$

This factor is another mathematical way of saying that 1 yard equals 3 feet.

Both F_1 and F_2, then, express the same information, but their focus is different; F_1 focuses on feet, and F_2 focuses on yards. Since you know the length of the field in yards and you are interested in its length in feet, F_1 has the proper focus for you. Indeed, multiplication of 100 yards by F_1 gives an answer in terms of feet since the yard units cancel.

$$100 \text{ yards} \times F_1 = 100 \text{ yards} \times \frac{3 \text{ feet}}{1 \text{ yard}} = 300 \text{ feet}$$

Example 1.6

The phosphorus content of a cow manure used as a natural fertilizer is 2 pounds per ton of manure. Calculate the amount of phosphorus in a pound of manure.

Solution. The given datum is 2 lb phosphorus per ton or, in a ratio,

$$\frac{2 \text{ lb phosphorus}}{1 \text{ ton manure}}$$

In order to calculate the desired quantity, you need to convert to pounds of manure. Writing down the given and desired quantities on the proper sides of an equal sign, we have

$$\frac{2 \text{ lb phosphorus}}{1 \text{ ton manure}} = \frac{? \text{ lb phosphorus}}{1 \text{ lb manure}}$$

The possible conversion factors to use both involve the quantities 2000 lb and 1 ton.

$$F_1 = \frac{1 \text{ ton}}{2000 \text{ lb}} \quad \text{or} \quad F_2 = \frac{2000 \text{ lb}}{1 \text{ ton}}$$

Only the first of these two factors will give the proper units when multiplied by the given quantity.

$$\frac{2 \text{ lb phosphorus}}{1 \text{ ton manure}} \times \frac{1 \text{ ton manure}}{2000 \text{ lb manure}} = \frac{0.001 \text{ lb phosphorus}}{1 \text{ lb manure}}$$

[Additional examples may be found in **1.17–1.20** at the end of the chapter.]

1.5
The Metric System

In the sciences, as well as in most nations of the world, the metric system is the standard of measurement. The metric equivalents of some of the English measures are given in Table 1.1

Table 1.1

A Comparison of Metric and English Measurements

Dimension	Metric unit	English unit
length	1.00 meter	39.4 inches
	2.54 centimeters	1.00 inch
	1 kilometer	0.6 mile
volume	1.00 liter	1.06 quarts
	0.94 liter	1.0 quart
mass	454 grams	1.00 pound
	1.0 kilogram	2.2 pounds

The metric system is simple and convenient to use, since all units are based on multiples of 10. Conversions are easy, because they involve moving only the decimal point as the units are changed. The metric system was established in France in 1790. A revised system called the International System of Units, known as SI, was adopted by international agreement in 1960. While there are some differences in units between the SI and metric systems, they are seldom important in the subjects to be covered in this text. Metric units will be used.

The standard units of the metric system are the second (s) for time, the meter (m) for length, the gram (g) for mass, and the liter (L) for volume. Fractions and multiples of the standard units of the metric system use prefixes to indicate the size of the unit relative to the standard unit. A list of prefixes appears in Table 1.2.

Table 1.2

Prefixes Used in the Metric and SI Systems
(The common prefixes and symbols are given in color.)

Multiplier	Prefix	Symbol
$1\ 000\ 000\ 000\ 000\ 000\ 000 = 10^{18}$	exa	E
$1\ 000\ 000\ 000\ 000\ 000 = 10^{15}$	peta	P
$1\ 000\ 000\ 000\ 000 = 10^{12}$	tera	T
$1\ 000\ 000\ 000 = 10^{9}$	giga	G
$1\ 000\ 000 = 10^{6}$	mega	M
$1\ 000 = 10^{3}$	kilo	k
$100 = 10^{2}$	hecto	h
$10 = 10^{1}$	deka	da
$1 = 10^{0}$		
$0.1 = 10^{-1}$	deci	d
$0.01 = 10^{-2}$	centi	c
$0.001 = 10^{-3}$	milli	m
$0.000\ 001 = 10^{-6}$	micro	μ
$0.000\ 000\ 001 = 10^{-9}$	nano	n
$0.000\ 000\ 000\ 001 = 10^{-12}$	pico	p
$0.000\ 000\ 000\ 000\ 001 = 10^{-15}$	femto	f
$0.000\ 000\ 000\ 000\ 000\ 001 = 10^{-18}$	atto	a

You will use these units and their equivalents in applying the factor unit method of solving problems. The abbreviations for the prefixes kilo, deci, centi, milli, and micro should be memorized.

We are familiar with a kind of metric system in our currency. For example, we know that $1.50 is the same thing as 150 cents or that $0.25 is 25 cents. A cent is a hundredth of a dollar, just as a centimeter is a hundredth of a meter.

Example 1.7

How many cents are there in $2.27?

Solution. Although you may automatically move the decimal point two places to the right to obtain the correct answer, 227 cents, the factor unit method may be used in preparation for problems involving less familiar units.

$$\$2.27 \times factor = ? \text{ cents}$$

The factors possible are

$$F_1 = \frac{\$1.00}{100 \text{ cents}} \quad \text{and} \quad F_2 = \frac{100 \text{ cents}}{\$1.00}$$

The solution requires F_2 in order to cancel the units and convert $ into cents.

$$\$2.27 \times \frac{100 \text{ cents}}{\$1.00} = 227 \text{ cents}$$

The Meter

The meter (m) is the basic unit of length in the metric system. However, in chemistry, medicine, and the health sciences there is seldom a need to measure length in meters. A meter rule is divided into 100 equal parts called centimeters (cm). This quantity is more commonly used. Each centimeter consists of ten smaller units called millimeters (mm). There are 1000 millimeters in one meter. A list of metric units for length is given in Table 1.3.

Table 1.3

Metric Units of Length, Volume, and Mass

Prefix	Relation to basic unit	Length	Volume	Mass
kilo	$\times 1000$ (10^3)	kilometer (km)	kiloliter (kL)[a]	kilogram (kg)
(none)	$\times 1$ (10^0)	meter (m)	liter (L)	gram (g)
deci	$\times 0.1$ (10^{-1})	decimeter (dm)	deciliter (dL)[a]	decigram (dg)[a]
centi	$\times 0.01$ (10^{-2})	centimeter (cm)	centiliter (cL)[a]	centigram (cg)
milli	$\times 0.001$ (10^{-3})	millimeter (mm)	milliliter (mL)[b]	milligram (mg)
micro	$\times 0.000001$ (10^{-6})	micrometer (μm)	microliter (μL)	microgram (μg)

[a]These units are not widely used.
[b]The milliliter occupies a volume of one cubic centimeter (cc). Both mL and cc (or cm³) may be used interchangeably.

Example 1.8

Convert 6721 mm into meters.

$$6721 \text{ mm} \times \text{factor} = ? \text{ m}$$

Solution. The factors interrelating meter and millimeter are derived from the fact that 1 m = 1000 mm.

$$F_1 = \frac{1 \text{ m}}{1000 \text{ mm}} \quad \text{and} \quad F_2 = \frac{1000 \text{ mm}}{1 \text{ m}}$$

The solution to the conversion requires the use of F_1 in order to cancel the units and convert millimeters to meters.

$$6721 \text{ \cancel{mm}} \times \frac{1 \text{ m}}{1000 \text{ \cancel{mm}}} = 6.721 \text{ m}$$

[Additional examples may be found in **1.31–1.38** at the end of the chapter.]

Some idea of the size of the units of length can be given in common examples. The kilometer is approximately 0.6 mile. A centimeter is approximately $\frac{5}{16}$ inch (in.), or in another, more exact way there are 2.54 cm in 1 in. (Figure 1.2).

The Liter

Volume refers to the amount of space occupied by matter. Because the meter is the standard for length, the cubic meter (m^3) could be a standard unit of volume, but it is large and unwieldy. A volume of a cubic centimeter (cm^3 or cc) offers an alternative standard that is often used in laboratory work.

The liter has been chosen as a volume standard intermediate between cubic meters and cubic centimeters. (A liter is the volume of a cube that is 10 cm or

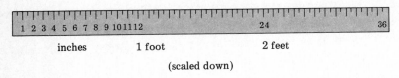

yardstick

inches   1 foot   2 feet

(scaled down)

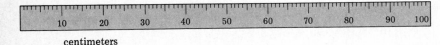

meterstick

centimeters

(scaled down)

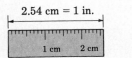

2.54 cm = 1 in.

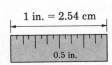

1 in. = 2.54 cm

Figure 1.2
The Meter and a Comparison of Length with the English System.

Figure 1.3

The Liter contains 1000 Cubic
Centimeters or 1000 Milliliters.

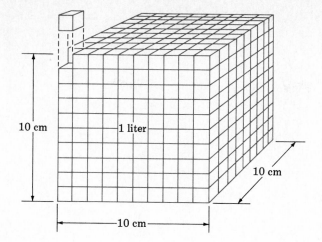

10 cm

1 liter

10 cm

—10 cm—

1 dm on each side.) A liter (L) then has a volume of 1000 cm^3 or 1 dm^3 (Figure 1.3).
Fractions of a liter can be expressed by use of the proper prefix. For example, $\frac{1}{1000}$
liter is 1 milliliter (mL). Note that since a liter is 1000 cm^3, the milliliter and the
cubic centimeter are the same volume. These units are used interchangeably.

Example 1.9 ——————————————————————————————

An adult of average activity inhales 10,000 L of air a day. What is the equivalent volume in milliliters?

Solution. In order to convert liters into milliliters the conversion factor
1000 mL/L is needed.

$$10,000 \; \cancel{L} \times \frac{1000 \; \text{mL}}{1 \; \cancel{L}} = 10,000,000 \; \text{mL}$$

Note that the units of liters cancel by proper use of the conversion factor. In
addition, the number of milliliters contained in a volume is far greater than the
number of liters, as required by the relative sizes of the units.

——

There are several devices for measuring volumes of liquids (Figure 1.4). Both
the beaker and the graduated cylinder are used to store and dispense approximate volumes in the chemistry laboratory. They are used to measure in much the
same way as a measuring cup is used in cooking.

A pipet or buret is used in accurate work. The pipet contains a definite
amount when filled to its calibration mark. The buret is used to deliver variable
amounts of liquids. Both pipets and burets are available in many sizes.

The volumetric flask is used to contain a specified volume of liquid when it
is filled to the calibration mark. It is often used to store accurately prepared
samples of liquids.

A syringe is similar to the buret since it can dispense a variable volume of a
liquid. Of course, the syringe is used for injecting small liquid volumes.

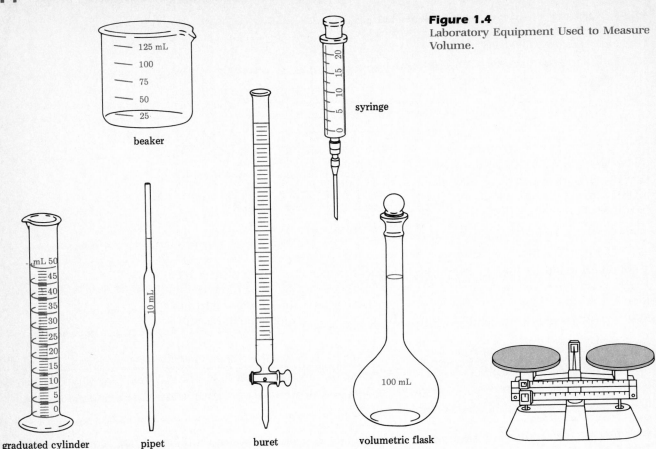

Figure 1.4
Laboratory Equipment Used to Measure Volume.

beaker

syringe

graduated cylinder   pipet   buret   volumetric flask

(a) double pan platform balance

The Gram

The unit of mass most commonly used in chemistry is the gram (g). Smaller quantities of matter used in chemistry are expressed in milligrams (mg). The mg unit is used to designate the contents of drugs in capsules or pills. Micrograms, nanograms, and even picograms are used to measure some highly potent biological chemicals. For example, some hormone levels in the blood are detected in picogram quantities. Body mass is expressed more conveniently in terms of the kilogram. A kilogram is defined as the mass of 1 L of water at 4°C (the Celsius temperature scale will be discussed in Chapter 2). Since 1 L of water contains 1000 mL and has a mass of 1000 g, it follows that 1 mL of water has a mass of 1 g.

A pound is equivalent to 453.6 g. The metric equivalent of mass has been printed for years on cans of vegetables and fruits. A kilogram is equal to approximately 2.2 lb.

(b) single pan balance

Figure 1.5
Laboratory Equipment to Determine the Mass of Matter.

Example 1.10

A linebacker weighs 231 lb. What is his mass in kilograms?

Solution. A pound is a smaller quantity of matter than a kilogram. Several pounds are needed to make a kilogram. Therefore, the mass will be numerically less than 231. The proper conversion factor is 1 kg/2.2 lb.

$$231 \text{ lb} \times \frac{1 \text{ kg}}{2.2 \text{ lb}} = 105 \text{ kg}$$

[Additional examples may be found in **1.21–1.30** at the end of the chapter.]

While you are accustomed to using scales to measure weights, scientists use balances to determine mass. A balance measures the mass of an object by comparing it to the mass of standards. The platform balance (Figure 1.5), in common use in laboratories, compares mass of an unknown object placed on the left side with known masses on the right side. When the two platforms are level, the masses must be identical. More sensitive instruments called single pan analytical balances measure mass electronically.

1.6
Scientific Notation

Many numbers that are used in chemistry are either so large or so small that the number of zeros that must be written is inconvenient. For example, 18.00 g of water contains 602300000000000000000000 molecules of water. Each water molecule then weighs 0.00000000000000000000002988 g.

We can write both large and small numbers by using scientific notation. *In* **scientific notation** *a number is expressed as a product of a coefficient multiplied by 10 raised to a power. The coefficient is a number equal to or greater than 1 but less than 10.* The number of molecules of water in an 18.00 g sample is 6.023×10^{23}. The 6.023 is the proper coefficient; it is larger than 1 but less than 10. Neither 0.6023 nor 60.23 is acceptable. The 10^{23} indicates the number of places required to move the decimal point from its position in the coefficient to obtain the actual number. In this case the decimal point in the actual number is 23 places to the right counting from the decimal point in the coefficient.

The weight of a water molecule is 2.988×10^{-23} g. A negative exponent indicates that the actual number is smaller than the coefficient. To figure out the actual number you must move the decimal point to the left by 23 places. Some additional examples of scientific notation are given in Table 1.4.

Table 1.4
Scientific Notation

Numbers larger than 1		Numbers smaller than 1	
Number	**Scientific notation**	**Number**	**Scientific notation**
1,111,100	1.1111×10^6	0.0000077	7.7×10^{-6}
222,200	2.222×10^5	0.0000666	6.66×10^{-5}
33,300	3.33×10^4	0.0005	5.0×10^{-4}
4,400	4.4×10^3	0.004444	4.444×10^{-3}
555	5.55×10^2	0.033333	3.3333×10^{-2}
66	6.6×10^1	0.22	2.2×10^{-1}
7.7	7.7×10^0		

1.7
Precision and Accuracy

While the terms precision and accuracy may appear to mean the same thing, their meanings are very different in science. **Precision** *means the degree of re-producibility of a measurement.* The precision of a measurement is determined by repeated individual measurements and depends on the skill of the individual making the measurement. For example, you may determine your temperature. Then the thermometer may be shaken down and used to measure your temperature again. The series of readings might be 98.6, 98.5, and 98.7°F. Your temperature is then 98.6°F. The precision is to the nearest 0.1°F.

Accuracy *means the degree to which a measurement represents the true value of what is measured.* Accuracy depends on the quality of the measuring device. For example, if your thermometer is improperly calibrated by the manufacturer, you cannot accurately determine your temperature. If the thermometer is calibrated one degree low, then your temperature is actually 99.6°F and you have a slight fever.

Normally we assume that measuring devices are accurately manufactured. However, it is always good practice to compare one measuring instrument against another. If the instruments are accurate, then we can try to make precise measurements.

All measuring devices have limits to their accuracy. For example, some watches have second hands, while others only record minutes. A second hand is needed to determine a pulse rate accurately. The minute hand is accurate enough to keep an appointment on time at the doctor's office.

1.8
Significant Figures

Consider the significance of any numerical quantity that we observe, measure, or are given. Are all of the digits in a number necessarily reliable or well established? In other words, what is the precision of the measurement? *The number of digits in a number that gives us reliable information is called the number of* **significant figures.** Some numbers we encounter are exact quantities. Thus the number of items in a dozen is exactly 12. However, when a ruler is used in a measurement of a distance, there is an uncertainty in the exactness of the measurement, and the quantity must be stated so as to indicate its reliability. This reliability is given by the number of significant figures.

In all sciences it is necessary to state quantities numerically and simultaneously give the precision of the number. If a sample is said to have a mass of 9 grams (g), the scientist means that to the nearest gram the sample contains 9 g of matter. The mass may be between 8.5 and 9.5 g. If it were less than 8.5 g or more than 9.5 g, the mass would be reported as 8 g or 10 g, respectively. The mass is precise to one significant figure. If the sample were placed on a more accurate balance, the mass might be found to be 8.8 g. This quantity has two significant figures, and we know that the actual mass is less than 8.85 g but more than 8.75 g. In each of the two cases cited, the last significant figure is only approximate. The degree of precision of each number is given by the number of significant figures. When we write down a number, it is necessary to make sure that the number does indeed contain the same number of significant figures as the reliability of the measurement. No measurement should ever be expressed or written as more reliable than the actual measurement.

Zeros are sometimes significant figures and at other times are merely used to place the number on the measurement scale being used. Students often have trouble determining which zeros in a quantity are significant. Suppose two different samples are weighed on a balance that can give the mass accurately to 0.001 g. If the mass of one sample is found to be 5.000 g, the sample mass is between

Table 1.5

Rules for Significant Figures

1. All digits 1 through 9 inclusive are significant. Thus 14.9 contains three significant figures while 125.62 contains five significant figures.
2. Zero is significant if it appears between two nonzero digits. Thus 306, 30.6, 3.06, and 0.306 all contain three significant figures.
3. A terminal zero to the right of a decimal in a number greater than 1 is significant if it is expressing a reliable measurement. Thus 279.0, 27.90, and 2.790 all contain four significant figures.
4. A terminal zero to the right of a decimal point in a number less than 1 is significant if it expresses reliable information. Thus 0.2790 contains four significant figures if indeed the value has been shown to be zero in the ten-thousandth place.
5. A zero that is used only to fix the decimal in a number less than 1 is not significant. Thus 0.456, 0.0456, 0.00456, and 0.000456 all contain only three significant figures.
6. Terminal zeros in an integer are not significant. Thus 450 contains only two significant figures.

4.9995 g and 5.0005 g, and the number of significant figures is four. Thus, the three zeros following the decimal point are significant. If the other sample has a mass of 0.015 g, the first zero after the decimal point is not a significant figure; this zero only gives the magnitude of the number. The sample actually has a mass between 0.0145 g and 0.0155 g, and the number 0.015 contains only two significant figures.

Numbers containing zeros, such as 10,000, are difficult to interpret, and their significance may depend on the source of the quantity. The number might contain five significant figures and so represent a quantity between 9999.5 and 10,000.5. If the quantity refers to the number of people in a crowd, the accuracy of the number may be between 9500 and 10,500. For this reason it is convenient to use scientific notation involving powers of ten. For the quantity 10,000 known to the nearest unit, the number is expressed as 1.0000×10^4. For the number of people in a crowd known only to the nearest thousand, we use 1.0×10^4.

Unless the power of ten notation is used, we normally assume that zeros at the end of a number indicate only the relative magnitude of the number and are not significant figures. Thus 10,000 is viewed as containing only one significant figure. The rules of significant figures are summarized in Table 1.5.

Example 1.11

What is the number of significant figures in each of the following numbers? (Give the rules used for your answer.)

a. 5041	b. 5401	c. 5410
d. 54100	e. 0.5401	f. 0.5410
g. 0.05401	h. 0.05410	i. 54.10

Solution

a. 4 (1, 2)	b. 4 (1, 2)	c. 3 (1, 6)
d. 3 (1, 6)	e. 4 (1, 2)	f. 4 (1, 4)
g. 4 (1, 2, 5)	h. 4 (1, 4, 5)	i. 4 (1, 3)

[Additional examples may be found in **1.41** and **1.42** at the end of the chapter.]

1.9
Mathematical Operations and Significant Figures

Whenever two or more quantities representing measurements and expressed to their proper number of significant figures are added, subtracted, multiplied, or divided, **the resulting answer cannot be expressed to any more reliability than the least significant quantity used.** The rules governing the reliability of numbers derived from mathematical operations are given in the next three subsections.

Addition and Subtraction

When numbers are added or subtracted, **the answer must not contain any significant figures beyond the place common to all of the numbers.** Thus, the sum of the numbers 12.2 and 13.31 is 25.5 and not 25.51. Only the tenths place is common to both numbers. The hundredths place is not given in 12.2 and is not known. The same rule applies to subtraction. The difference of the numbers 13.31 and 12.2 is 1.1 and not 1.11. In each case the number of significant figures reflects the reliability of the resultant number based on the reliability of the quantities used in obtaining it.

Example 1.12

What is the sum of 25.1 + 15 + 14.15? Express the answer to the proper number of significant figures.

Solution. The place common to all numbers is the units place. A line can be placed at this point to avoid using the numbers to the right in the answer.

$$
\begin{array}{r|l}
25 & .1 \\
15 & \\
14 & .15 \\
\hline
54 & .25 \\
\end{array}
$$

The sum is 54.

Example 1.13

What is the difference of 16.29 and 3.168?

Solution

$$
\begin{array}{r|l}
16.29 & \\
-\ \ 3.16 & 8 \\
\hline
13.12 & 2 \\
\end{array}
$$

The answer is 13.12.

Multiplication and Division

In multiplication or division of numbers, the answer must not contain more significant figures than the least number of significant figures used in the operations. Thus the product of 201 × 3 is 603, but the answer can be expressed to only one significant figure since 3 has only one significant figure. The proper answer is 600 in which only the 6 is significant. Similarly, the quotient 603/3 is not 201 but rather 200.

Example 1.14

What is the product of 304 × 11?

Solution

304 × 11 = 3344

However, the answer can be expressed to only two significant figures, 3300 or 3.3×10^3.

Example 1.15

What is the quotient of these quantities?

$$\frac{1760.1}{25}$$

Solution. Division yields the number 70.4. However, since the divisor contains only two significant figures, the quotient must be expressed as 70 or 7.0×10^1.

Rounding Off

In the preceding examples of this section, the nonsignificant figures were discarded. All of the discards were mathematically valid because the examples were carefully chosen so that the nonsignificant figures were less than 5. However, if the nonsignificant figure is 5 or greater than 5, the figure cannot be discarded and must be used in rounding off to the required number of significant figures. Hence if 25.21 must be expressed to three significant figures, the answer is 25.2 as the 1 in the hundredths place is less than 5. However, if the number 25.2 must be expressed to one significant figure, the answer must be rounded off to 30 as the nonsignificant number 5.2 is greater than 5.

If the number to be rounded off is 5, the 5 is dropped and the last significant digit is left the same if it is even and the last significant digit is increased by one if it is odd.

Example 1.16

What is the sum of 10.7 + 17.43 + 3.56?

Solution

$$
\begin{array}{r|l}
10.7 & \\
17.4 & 3 \\
\underline{3.5} & \underline{6} \\
31.6 & 9
\end{array}
$$

Rounding off to the tenths place gives the correct answer, 31.7.

Example 1.17

What is the product of 5.61×21?

Solution. The product is 117.81. Since 21 contains only two significant figures, only two are allowed in the answer. The seven cannot be dropped but must be used to round off to the correct answer, 120 or 1.2×10^2.

Summary

Every material or object we observe consists of matter that has mass and occupies space. Chemistry is a study of matter, its composition, structure, and reactions.

In order to communicate and develop science, the scientific method is used. This approach is based on observation and experiment to obtain facts that are stated as laws. The facts in turn are studied to determine what is responsible for a natural phenomenon. An established explanation for a natural phenomenon is called a theory.

The metric system is used for uniform precise measurements required in chemistry. Measurements in one type of unit may be converted into another unit by using conversion factors.

Every measurement involves an uncertainty, and it is necessary to express a measured quantity so that the uncertainty is clear. The certainty of a measurement is indicated by using the proper number of significant figures. All arithmetic operations on measured quantities must be carried out to result in derived quantities with the proper number of significant figures.

Learning Objectives

As a result of studying Chapter 1 you should be able to
- Distinguish among the terms law, hypothesis, and theory.
- Use factor units in problem solving.
- Interconvert a given measurement in one prefixed metric unit into another prefixed unit.
- Distinguish between accuracy and precision.
- Express numbers in scientific notation.
- Determine the number of significant figures in a number.
- Express, rounding off where appropriate, the results of mathematical operations to the proper number of significant figures.

Key Terms

Accuracy means the degree to which a measurement represents the true value of what is measured.

Bonding describes how the atoms in a compound are held or fastened together.

Composition means the identity and amount of the components of a sample of matter.

A **conversion factor** is a multiplier having two or more units associated with it that is used to convert a quantity in one unit to its equivalent in another unit.

A **hypothesis** is a tentative model to explain a law.

A **law** is an explicit statement of fact obtained by observation and experimentation.

Mass is a quantity of matter.

Matter is anything that occupies space and has mass.

Measurements are comparisons to a standard measuring device.

A **model** is an idea that may correspond to what is responsible for a natural phenomenon.

Precision means the degree of reproducibility of a measurement.

Products are the substances that are produced in a chemical reaction.

Reactants are the substances that enter into a chemical reaction.

A **reaction** involves changes in the composition and structure of matter.

The **scientific method** is a sequence of steps involving observation, experimentation, and the formulation of laws and theories that lead to scientific knowledge.

In **scientific notation** a number is expressed as a product of a coefficient multiplied by 10 raised to a power.

Significant figures are the digits in an experimentally measured quantity that indicate the precision with which the quantity is known.

Structure means the arrangement of the components of a substance.

A **theory** is a concept or model that is used to explain a law.

Weight is a result of the gravitational attraction between an object and the earth.

Questions and Problems

Terminology

1.1 Clearly distinguish between precision and accuracy.

1.2 Explain the difference between a theory and a law.

Percent Composition

1.3 In a class of 80 students there were 8 A, 16 B, 40 C, 12 D, and 4 E grades. What is the percent distribution of each grade assignment?

1.4 A 110 lb woman has a blood volume of 3.6 L. Her plasma volume is 2.3 L, and her red cell volume 1.3 L. What percent of her blood volume is the red cell volume?

1.5 Approximately 18% of human body weight is carbon. How much carbon is there in a 150 lb man?

1.6 Carbon monoxide, a poisonous gas, consists of molecules containing one carbon atom and one oxygen atom. The atomic weights of carbon and oxygen atoms are 12 amu and 16 amu, respectively. Calculate the percent composition of carbon monoxide.

1.7 Methane is natural gas used for cooking in the home. It consists of molecules containing one carbon atom and four hydrogen atoms per molecule. The atomic weights of carbon and hydrogen are 12 and 1 amu, respectively. Calculate the percent composition of methane.

1.8 Ethyl alcohol consists of molecules containing two carbon atoms, six hydrogen atoms, and one oxygen atom. The atomic weights of carbon, hydrogen, and oxygen are 12, 1, and 16 amu, respectively. Calculate the percent composition of ethyl alcohol.

Structure of Molecules

1.9 A foul-smelling gas having the odor of rotten eggs consists of molecules containing two hydrogen atoms and one sulfur atom. The symbol for sulfur is S. The compound consists of a central sulfur atom bonded to two hydrogen atoms. The two hydrogen atoms are not bonded to each other. Draw the molecule.

1.10 The molecule of butane, a substance used in butane cigarette lighters, consists of four carbon and ten hydrogen atoms. The carbon atoms are connected in a chain similar to that shown for octane in Figure 1.1. Given the facts that a carbon atom forms four bonds and a hydrogen atom forms one bond, draw the structure of butane.

1.11 Methyl alcohol, a poisonous alcohol that taken internally in small quantities can cause permanent blindness, consists of molecules containing one carbon atom, four hydrogen atoms, and one oxygen atom. Given the facts that a carbon atom forms four bonds, an oxygen atom forms two bonds, and a hydrogen atom forms one bond, draw a structure of methyl alcohol.

The Scientific Method

1.12 Einstein said "God is subtle, but he is not malicious." Explain how Einstein's statement is related to the study of science.

1.13 You live in an apartment with three other students. Just prior to leaving for your class, you discover that your calculator is "missing." Describe a scientific way to approach this problem.

1.14 It has been said that science progresses most rapidly when scientists can suggest several hypotheses to explain a natural phenomenon. Explain why this statement is valid.

1.15 Explain the reason for repeating experiments several times before developing a law.

1.16 Why is it important that scientific ideas be published?

Miscellaneous Interconversions

1.17 A Tennessee walking horse is 15 hands high. A hand is 4 in. How high is a Tennessee walking horse in feet?

1.18 A heavyweight boxer from Great Britain weighs 13 stones. A stone is 14 lb. What is the weight of the boxer in pounds?

1.19 A cuffisco used in Sicily to measure oil is 5.6 gal. In Tangiers kulas, which equal 4.0 gal, are used for oil. How many kulas are present in 3.0 cuffiscos?

1.20 The barrel used for petroleum is equal to 42 U.S.A. gal. How many barrels are needed to provide 21,000,000 gal of petroleum?

English–Metric Interconversions

1.21 Make the following conversions using conversion factors.

  (a) 12.1 yards into meters
  (b) 5.5 meters into feet
  (c) 41 cm into inches
  (d) 14.5 in. into centimeters
  (e) 245 g into pounds
  (f) 212 lb into kilograms
  (g) 452 cc into quarts
  (h) 2.1 qt into liters

1.22 A U.S.A. sprinter can run 100 yd in 9.1 s. How long should it take the sprinter to run 100 m?

1.23 A French male chauvinist enjoys "objects" of dimensions 91, 66, 91. What is the object if the measurements are in centimeters? Express the measurements in inches.

1.24 A speed sign on the approach to a Mexican town reads 40 km/hr. Should you drive your American-made car at 25 or 65 mph?

1.25 If 20 kg is the baggage allowance on an international air flight, how many pounds are you allowed to carry?

1.26 A Methuselah of champagne has a volume of 1.40 gal. How many milliliters are there in a Methuselah?

1.27 An individual has a mass of 75 kg and is 1.70 m tall. What is the equivalent weight in pounds and height in feet?

1.28 The dust deposited from city air might be 2 tons per square mile per day. Express this dust level in milligrams per square meter per day.

1.29 A medication label reads "Administer 2 mg per kilogram of body weight." How much medication should be given to a 110 lb patient?

1.30 A premature infant has a mass of 2 kg. What is the child's weight in pounds?

Metric Conversions

1.31 Make the following conversions using conversion factors.
(a) 59 mm into centimeters
(b) 153 cm into kilometers
(c) 348 mL into liters
(d) 5.328 L into milliliters
(e) 248 mg into grams
(f) 0.056 kg into grams

1.32 Your stomach releases 2.5 L of gastric juice each day. What is the equivalent volume in milliliters?

1.33 Assume that your heart beats at a rate of 70 beats per minute. If 60 mL of blood is pushed into the aorta by each beat, what volume of blood is circulated each day?

1.34 A 160 lb adult has 5 L of blood. Each milliliter of blood contains 5 million red blood cells. How many red blood cells does the adult have?

1.35 A patient receives intravenous fluids in the amounts of 250, 500, 250, and 100 cm³. How many liters in total has the patient received?

1.36 A drop of blood has a volume of 0.05 mL. How many drops of blood are there in an average adult body, which has 5 L of blood?

1.37 The level of vitamin C in your blood is about 0.2 mg per 100 mL of serum. What is the amount of vitamin C in grams per liter?

1.38 The ozone in a city one day is 100 μg per cubic meter. What is the amount in nanograms per liter?

Scientific Notation

1.39 Convert each of the following quantities into scientific notation.
(a) 147.89   (b) 0.0375   (c) 2146.8
(d) 0.000408   (e) 21.6489   (f) 0.0000039

1.40 Write the numerical equivalent of each of the following numbers expressed in scientific notation.
(a) 1.56×10^{-3} (b) 4.897×10^3
(c) 3.468×10^4 (d) 1.987×10^{-4}
(e) 5.02×10^2 (f) 6.059×10^{-5}

Significant Figures

1.41 How many significant figures does each of the following numbers have?
(a) 3015   (b) 0.00025   (c) 3.015
(d) 14.0   (e) .03015   (f) 0.00104
(g) 0.000003   (h) 1000.12

1.42 How many significant figures does each of the following numbers have?
(a) 5.02×10^4 (b) 1.256×10^3
(c) 3.15×10^{-4} (d) 1.2×10^{-6}
(e) 2.100×10^{-5} (f) 2.01×10^{-30}
(g) 4.50×10^{10} (h) 4.0×10^{40}

1.43 Perform the following calculations and determine the number of significant figures in your answer.
(a) $340 + 3.2 + 0.4589$ (b) $1.345 + 0.6789 + 0.1$
(c) $45.63 - 34.26$ (d) $2.456 - 1.56$
(e) 15.60×3.4567 (f) 1.89×0.3
(g) $7.65/3.2$ (h) $34.5/56.45$

Classification of Matter

2

GOLDEN DELICIOUS APPLES 79¢ LB.

RED DELICIOUS APPLES 79¢ LB.

Apples are one type of fruit. Within the class of apples there are subclasses with individual characteristics. Regardless of the type of apple, you would recognize it as an apple.

2.1
Classification in Science

To facilitate communication and organize knowledge in an intelligent manner, scientists group or classify items of interest according to their similarities. A classification may be simple and consist of large numbers of items in few classes or complex and consist of many classes and subclasses. Regardless of the classification process used, the similarities and differences allow us to make generalizations about the members of each class. The generalizations then provide a focus for scientific hypotheses and theories that help us understand nature.

You are probably familiar with the idea of classification from your high school biology course. The number of living things is enormous and the first classification scheme devised by the early Greeks is overly simple. Aristotle divided the living world into animals and plants: animals gather and ingest food, whereas plants are rooted in soil and make their own food. Animals were classified according to whether they lived in the air, in the sea, or on land. Plants were classified as trees, shrubs, undershrubs and herbs.

Scientists in the eighteenth century developed more extensive classifications based on detailed information gathered from sophisticated observations and experiments. Currently a five kingdom system is used for all life. Each kingdom is subdivided into phyla, which in turn are further subdivided. Man, for example, belongs to the kingdom Animalia, the phylum Chordata, the subphylum Vertebrata, the class Mammalia, subclass Eutheria, Order Primates, suborder Anthropoidea, family Hominidae, genus *Homo*, and species *sapiens*. In order to study biology, an understanding of the method of classification and the meaning of all terms is essential.

In chemistry the process of simplification by classification is important to understand matter, its composition, structure, and reactions. The classification used depends on the objective of the study. For example, in classifying matter one can divide substances into atoms, molecules, and ions. For a study of chemical reactions the classes could be acid–base, oxidation–reduction, and many other types of reactions. In this chapter matter will be classified in broad terms based on composition. A more detailed picture of matter based on structure will be developed in the next chapter. As further details of classification are presented throughout this text, you should be sure that you understand the classification scheme and the terminology used in the classification method.

2.2
Density and Specific Gravity of Matter

One of the ways of discussing matter is to use one of its characteristic properties, its mass. We sometimes talk about one thing being "heavier" than another. Lead is "heavier" than water and will not float, whereas wood is "lighter" than water and will float. However, what we are comparing is not really the weight of the materials but their densities.

If two 1.00 cm³ samples of lead and maple wood are weighed, the mass of lead is 11.3 g, whereas that of the wood is 0.490 g. Such an observation is the basis of our feeling that lead is "heavier" than wood. However, more exactly we should state that for comparable volumes there is more mass of lead than of wood (Figure 2.1). We could obtain equal weights of lead and wood. Thus, 11.3 g of lead and 11.3 g of maple wood have the same mass, but the 11.3 g of maple would occupy 23.0 cm³.

Density

Density *is defined as mass per unit volume.* Density does not depend on the amount of the material. For example, a 1 mL sample of water has a mass of 1 g.

Figure 2.1

Mass, Volume, and Density of Matter.

In experiment (a) the equal volumes of lead and maple wood do not have the same mass. The density of lead is greater than the density of maple wood. In experiment (b) the mass of lead is equal to the mass of maple wood. Since the density of maple wood is less than that of lead, the sample of maple wood has a larger volume.

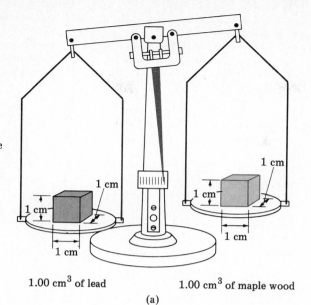

1 cm

1 cm

1 cm

1 cm

1 cm

1 cm

1.00 cm³ of lead   1.00 cm³ of maple wood

(a)

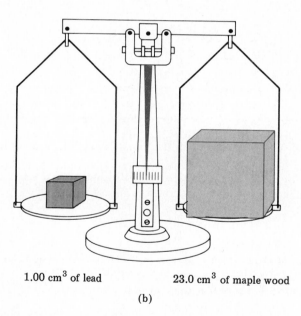

1.00 cm³ of lead   23.0 cm³ of maple wood

(b)

Thus, water has a density of 1 g/mL. A 10 mL sample of water has a mass of 10 g. The ratio of mass to volume is still 1 g/mL.

$$\text{density} = \frac{\text{mass}}{\text{volume}} = \frac{1 \text{ g}}{1 \text{ mL}} = \frac{10 \text{ g}}{10 \text{ mL}} = 1 \text{ g/mL}$$

A variety of units can be used in measuring the volume and mass of a substance. **Usually the densities of liquids and solids are expressed in grams per milliliter (g/mL)** because the resultant values avoid small fractions and extremely large numbers. For the same reason **the densities of gases are expressed in grams per liter (g/L).** A list of the densities of some common substances at 25°C is given in Table 2.1.

Table 2.1

Densities of Some Liquids, Solids, and Gases

Liquid	Density (g/mL)	Solid	Density (g/cm³)	Gas	Density (g/L)
alcohol	0.80	iron	7.86	carbon dioxide	1.80
bromine	3.12	gold	19.3	carbon monoxide	1.14
ether	0.71	lead	11.3	hydrogen	0.08
olive oil	0.92	rock salt	2.2	helium	0.16
turpentine	0.87	sugar	1.59	methane	0.66
water	1.00	uranium	19.0	oxygen	1.31
mercury	13.53	wood, maple	0.49	nitrogen	1.14

Example 2.1

A 1.50 cm³ sample of aspirin has a mass of 1.74 g. What is the density of aspirin?

Solution. Density is defined as mass per unit volume, and the units used for solids are normally grams per cubic centimeter (g/cm³). Therefore, the mathematical operation is

$$\frac{\text{mass}}{\text{volume}} = \frac{1.74 \text{ g}}{1.50 \text{ cm}^3} = 1.16 \text{ g/cm}^3 = \text{density}$$

Example 2.2

Alcohol has a density of 0.80 g/mL. What is the mass of 35 mL of alcohol?

Solution. The mass of the sample may be obtained by multiplying the density by the volume. In this way the units of volume cancel and mass remains.

$$0.80 \, \frac{\text{g}}{\text{mL}} \times 35 \text{ mL} = 28 \text{ g}$$

[Additional examples may be found in **2.3–2.8** at the end of the chapter.]

Specific Gravity

Water, which has a density of 1 g/mL and is widely distributed on earth, has been chosen as a reference material to compare the densities of other materials. **Specific gravity** *is a ratio of the density of a substance to the density of water at the same temperature.*

$$\text{specific gravity} = \frac{\text{density of substance (g/mL)}}{\text{density of water (g/mL)}}$$

Because the specific gravity is a ratio of two numbers having identical units, the specific gravity is without dimensions. Of course, since the density of water is 1 g/mL, the specific gravity of a substance is numerically equal to its density in

Figure 2.2
Examples of a Hydrometer Used to Measure the Specific Gravity of Urine Samples.

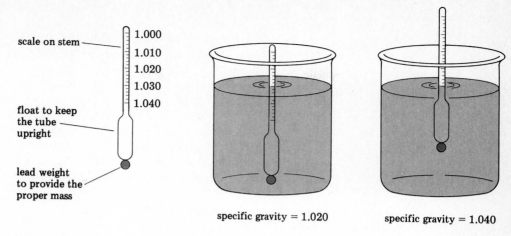

specific gravity = 1.020   specific gravity = 1.040

grams per milliliter. Since the density of alcohol is 0.80 g/mL, its specific gravity is 0.80.

Specific gravity can be measured with a hydrometer. The hydrometer sinks until it displaces a mass of liquid equal to its own mass. Thus the higher the density of the liquid, the higher the hydrometer tube will float.

A common test used in hospitals and clinics is the determination of the specific gravity of urine with a hydrometer called a urinometer. The normal range of specific gravities of urine is 1.009–1.030 with an average of about 1.018. If an abnormal amount of material such as glucose is present in the urine, the specific gravity is higher and the urinometer will ride "higher" in the sample. A urinometer is illustrated in Figure 2.2.

Example 2.3

A urine sample has a density of 1.013 g/mL. What is the mass of a 50.0 mL sample of urine?

Solution. The given units are g/mL and the desired unit is one of mass such as grams. By multiplying by mL the units of mL will cancel and the unit g will remain.

$$1.031 \frac{g}{mL} \times 50.0 \ mL = 51.5 \ g$$

Many properties of matter depend on temperature. For this reason the temperature of a substance must be known before any of its properties can be discussed.

Almost all substances expand when heated and contract when cooled. This property is used in the mercury thermometer to measure temperature. The mercury expands when heated and rises up the stem from the bulb at the base of the thermometer. More accurate electronic thermometers with digital readouts operate on somewhat different principles (Figure 2.3). However, in all cases the tem-

2.3
Temperature Units

perature recording results from a transfer of heat from the object to the thermometer or vice versa. Heat flow occurs from hot objects to cold objects.

The Celsius scale is used most frequently in scientific work. *On the* **Celsius scale** *the freezing point and boiling point of water are defined as* 0 *and* 100°C, *respectively*. There are 100 degrees between these two points. *On the* **Fahrenheit scale,** the one most commonly used in nonscientific applications in the U.S.A., *the freezing point of water is* 32°F, *and the boiling point of water is* 212°F. There are 180 degrees between the freezing point and boiling point of water. The Celsius and Fahrenheit scales are compared in Figure 2.4.

The Kelvin scale, an important temperature scale for certain measurements, is identical to the Celsius scale in the size of the degree intervals, but differs in the value assigned to the reference points previously discussed. *On the* **Kelvin scale,** *which is also called the absolute scale, the zero point is* 273.15 *degrees below the zero point on the Celsius scale.* No temperature lower than this value is possible. The freezing point of water is 273.15 K, and the boiling point is 373.15 K. For all but the most exact work the 0.15 degrees can be disregarded.

Mathematical expressions that interrelate the various scales are given below. Note that the degree symbol is not used with K.

$$°F = (\tfrac{9}{5} × °C) + 32°$$

$$°C = \tfrac{5}{9} × (°F − 32°)$$

$$°C = K − 273°$$

$$K = 273° + °C$$

Figure 2.3
An LED Digital Thermometer. This digital electronic thermometer, which resembles a calculator, displays temperature to 0.1° in either Celsius or Fahrenheit. (Courtesy of Markson Science, Phoenix, AZ.)

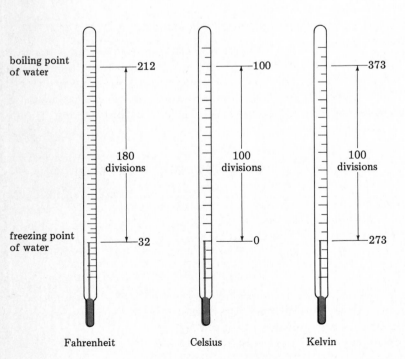

boiling point of water — 212

freezing point of water — 32

180 divisions

Fahrenheit

— 100

100 divisions

— 0

Celsius

— 373

100 divisions

— 273

Kelvin

Figure 2.4
A Comparison of the Fahrenheit, Celsius, and Kelvin Temperature Scales.

Example 2.4 _____

The normal body temperature is 98.6°F. What temperature is considered normal by a European doctor who uses the Celsius scale?

Solution. You may substitute directly into either the first or the second formula, but the following is the most efficient.

$$°C = \tfrac{5}{9} × (°F - 32.0°)$$
$$= \tfrac{5}{9} × (98.6° - 32.0°)$$
$$= \tfrac{5}{9} × (66.6°) = 37.0°$$

You may note in future studies that experiments in many hospitals and biology and biochemistry laboratories are done at 37°C, which is the normal body temperature.

Example 2.5 _____

Many enzymes that are important to our well-being are denatured (lose their biological ability to cause reactions to occur easily) at 45°C. What is this temperature in degrees Fahrenheit.

Solution. You may substitute directly into the following formula and solve for the required degrees Fahrenheit.

$$°F = (\tfrac{9}{5} × °C) + 32°$$
$$= (\tfrac{9}{5} × 45°) + 32°$$
$$= 81° + 32° = 113°$$

Note that this temperature is above even that of a high fever. Nevertheless, biological damage may result from prolonged high body temperatures below this value.

[Additional examples may be found in **2.9–2.17** at the end of the chapter.]

Energy *is the ability to do work.* Various types of energy that you may encounter are heat energy, electrical energy, and light energy. There is another important form of energy called chemical energy.

2.4
Energy

Chemical energy *is the energy stored in substances that can be released during a chemical reaction.* As the subject of chemistry is developed in this text, you will see that the chemical energy stored in various substances determines the reactions that they will undergo.

Interconversion of Energy

Energy may be converted from one form to one or several other forms. For example, electrical energy can be converted into heat energy in a heater. Although it may not be obvious, in this process none of the energy is destroyed. The

AN ASIDE

All of the chemical reactions that occur in living organisms are metabolic reactions, and they are divided into two types, anabolic and catabolic reactions. Anabolic reactions build molecules necessary for the life of the cell and require energy. Catabolic reactions decompose large molecules into smaller molecules and release energy. Part of the released energy is used to supply the necessary energy for anabolic reactions. Energy is also required to circulate blood, breathe, use muscles, and operate various glands as well as a variety of other body functions that are part of the body's basal activities.

When the metabolic rate is increased by work or exercise, the energy released by catabolic reactions exceeds that required for normal body functions. Without a means of dissipating the heat energy, the body temperature would rise and eventually death would result. In order to stay within the temperature range for life, the body releases heat energy by evaporation, conduction, convection, and radiation.

Evaporation of any liquid from the surface of the body results in a cooling of the body. Any liquid requires heat energy to evaporate. Thus as the body supplies heat energy for the evaporation of the liquid, the body is cooled. The evaporation of water from the body is an important means of cooling especially during strenuous work or exercise. Evaporation of water without sweat gland activity is called insensible perspiration. Evaporation as a result of sweat gland activity is called sensible perspiration. It is the latter type of active perspiration that provides the cooling during periods of high metabolic rates.

Evaporation of other liquids placed on the body also causes a cooling effect. An alcohol rub cools the body as the alcohol evaporates. Some liquids are used to spray on injured muscles of athletes. The evaporation process cools the skin and prevents further injury to tissue.

Conduction of heat involves direct contact of two substances with each other. The heat is transferred from the hot body to the cool body. You might have tried to cool your face by placing ice on it. This type of cooling is an example of conduction. Conduction is a method used to cool individuals who have a high temperature condition known as hyperthermia. In hyperthermia, the body cannot dissipate the heat being generated at a sufficiently rapid rate. The condition may be the result of lack of sufficient body fluids to allow normal sensible perspiration. Immersion of the body in an ice bath may be necessary to save the life.

Convection cooling is a result of transfer of heat to surrounding air. Air in contact with your body is heated. As the air is moved away by wind, additional local air is heated. In order to prevent excessive heat loss in cold weather, we wear warm clothing that traps some air in pockets in the fabric. The air is warmed, but is not blown away by the wind. Loss of heat as a result of improper clothing can be very severe especially under windy conditions. The combination of low temperature and high wind velocity is measured as the wind-chill factor.

Heat can also be transferred by radiation. You are most familiar with the radiation of the infrared heat lamp. Even though the air between you and the lamp doesn't feel hot, you do. You can still feel the effect of the radiation on your body even if air is moving about you.

Your body releases a type of infrared radiation. During cold weather the loss of heat by radiation may be as much as 50% of your heat loss. The rate of heat loss is proportional to the difference in temperature between your body and the air. The major source of heat loss is through the head. It is for this reason that it is advisable to wear a hat during cold weather.

The radiation of heat from a body can be detected electronically. There are security devices that can detect the presence of a body in a closed area. This radiation can be used to set off a silent alarm and bring security forces.

Differences in the radiation given off by various parts of the body can be used for medical diagnosis. Areas of high metabolic activity give off much radiation and are "hot" areas. One example is a cancerous growth that grows at a more rapid rate than normal tissue and has a high metabolic activity.

---

only change is in the form of the energy. *The **law of conservation of energy** states that energy may be converted from one form to another but may be neither created nor destroyed.*

In running, an electric motor transforms a quantity of electrical energy into an equivalent amount of energy in other forms. The motor develops energy of motion, kinetic energy, and also some heat energy because of friction. The total of the kinetic energy and the heat energy equals the original electrical energy.

Green plants store chemical energy in complex chemicals produced from carbon dioxide and water. The energy stored is obtained from the radiant energy of light. Energy is released when plants are metabolized by animals. No energy has been created or destroyed. It has only changed its form.

The energy required for or generated by chemical reactions may be in the form of electrical, heat, or light energy. *A reaction that occurs with the evolution of heat is **exothermic**, whereas a reaction in which heat energy is required from the surroundings is **endothermic**.*

The Calorie

The most common unit of heat energy is the calorie (cal). *A **calorie** is the amount of energy required to raise the temperature of 1 g of water 1°C.*

The kilocalorie (kcal), equivalent to 1000 cal, is the unit used in dietary tables, although it is referred to as a Calorie (Cal). Thus the dietetic Calorie is the amount of energy required to raise the temperature of 1 kg of water by 1°C.

Each person has caloric requirements that depend on, among other things, body size, metabolic rate, and physical activity. A rough estimate of a minimum caloric intake to maintain a sedentary patient in a hospital bed is 1.0 Cal/hr for each kilogram of body weight, or $1.0\ \text{Cal kg}^{-1}\ \text{hr}^{-1}$.

Example 2.6

How many Calories are required for a 165 lb patient in a hospital in one day?

Solution. First, the mass equivalent of 165 lb must be determined.

$$165\ \text{lb} \times \frac{1.0\ \text{kg}}{2.2\ \text{lb}} = 75\ \text{kg}$$

Next, the number of Calories required per hour is calculated.

$$75\ \text{kg} \times \frac{1.0\ \text{Cal}}{\text{kg hr}} = 75\ \frac{\text{Cal}}{\text{hr}}$$

Finally, the number of Calories required per day is calculated.

$$75 \, \frac{Cal}{\cancel{hr}} \times 24 \, \frac{\cancel{hr}}{day} = 1800 \, \frac{Cal}{day}$$

[Additional examples may be found in **2.18–2.24** at the end of the chapter.]

2.5
States of Matter

Matter can be classified according to its physical form or state. There are three states: gaseous, liquid, and solid. In our experience almost every substance is thought of in terms of a single state. Air is a gas, oil is a liquid, and iron is a solid. However, most matter can exist in any of these three states under the proper experimental conditions.

Water is one of the few substances that most people have observed in all three states. At normal atmospheric pressure water exists as a solid below 0°C and as a gas above 100°C. Whereas the terms ice and steam are used to describe the solid and gaseous states of water, to the chemist the substance is still water. Regardless of the state in which it exists, the composition of water is 88.81% oxygen and 11.19% hydrogen.

The Gaseous State

A gas has no characteristic shape or volume and completely fills any container in which it is placed. At normal atmospheric pressure and temperature the densities of gases are in the range of 0.00008–0.008 g/mL. Therefore it is more convenient to state the densities of gases in g/L. When gases are heated, they expand and their density decreases compared to surrounding cooler air. Hence, hot air rises.

Gases are highly compressible. Under pressure, the volume of a gas decreases. This compressibility allows a large volume of oxygen to be stored in small cylinders such as those for use in hospitals.

The Liquid State

Liquids have a constant volume at a specified temperature and pressure, but do not have a characteristic shape. They will fill a container from the bottom up.

The densities of liquids are generally less than those of solids and range from 0.5 to 1.5 g/mL. One notable exception is mercury, a liquid metal, with a density of 13.5 g/mL.

Liquids are only slightly compressible but usually more so than are the corresponding solids. The slight compressibility allows them to be used to transmit pressure, as in the case of hydraulic fluids. If hydraulic fluids were very compressible, the pressure exerted by a foot on a brake pedal would be used to compress the liquid rather than stop the car.

The Solid State

Solids are recognized by their definite volume and rigid shape. For all practical purposes, solids are incompressible. For example, the volume of a piece of coal beneath the enormous pressure of the earth is the same as it is on the surface of the earth.

The densities of solids are in the range of 1–20 g/cm^3. Metals such as lead, with a density of 11.3 g/cm^3, and gold, with a density of 19.3 g/cm^3, are among the densest solids.

Most solids will expand slightly when heated. For a temperature increase of 1°C the expansion is usually less than 0.01%. Such a small expansion means that solids such as wood, metals, and even plastics can be used in construction.

Every substance can be identified by a set of properties that is characteristic of that substance. Several substances may have some properties in common, but no two substances have a completely common set. These properties are conveniently divided into physical and chemical properties. Chemical properties will be discussed in Section 2.7.

Physical properties *are characteristics that are determined without altering the chemical composition of the substance.* Some examples of physical properties include boiling point, melting point, density, electrical conductivity, odor, and color. Pure water is a colorless, odorless, and tasteless liquid that freezes at 0°C and boils at 100°C. Both the qualitative adjectives and the quantitative numbers describe physical properties of water. Similarly, chlorine can be described as a yellow-green gas with a suffocating odor and a sharp sour taste, and it boils at −34.5°C and freezes at −101.6°C.

Physical properties are used by chemists to identify substances. For example, aspirin melts at 143°C, whereas sugar melts at 186°C. A chemist can then easily identify a sample as aspirin or sugar by determining the temperature at which it melts.

Matter can change in appearance without changing its chemical composition. For example, the metal filament in a light bulb changes appearance when the electricity is turned on. The glowing filament is still chemically the same, and when the light is turned off, the filament is as it originally appeared (Figure 2.5). The composition of the filament is unchanged. *Any conversion of matter that occurs without a change in the composition of the substance is called a* **physical change.**

The most common examples of physical changes that you have observed are changes of state. The melting of ice and boiling of water are both examples of physical changes (Figure 2.6). Examples of other liquid substances that can form solids at lower temperatures and gases at higher temperatures are alcohol and ether. Ethyl alcohol forms a solid at −117.3°C and is converted into a gas at 78.5°C. Ether freezes at −116.2°C and boils at 34.6°C.

 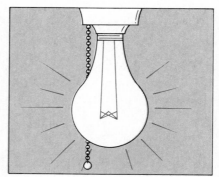 

Figure 2.5
A Physical Change.
When a current is passed through a light bulb, there is a change in its physical appearance. However, no change in composition occurs.

ice (0°C)   water (20°C)   steam (100°C)

Figure 2.6
The States of Water.
Water can exist in three states. When a change in state occurs, there is no change in chemical composition.

Although most substances change from solid to liquid and then to a gas as the temperature is increased, there are exceptions. *The change of a substance from a solid to a gas without passing through the liquid state is called* **sublimation.** Dry ice, which is solid carbon dioxide, sublimes into a gas at $-78.5°C$.

2.7
Chemical Properties

Chemical properties *are the characteristic changes in composition that a substance undergoes during chemical reactions.* For example, water reacts violently with sodium metal to produce hydrogen gas and a caustic substance known as sodium hydroxide. Chlorine reacts with sodium to form sodium chloride, known as table salt. In both examples, the substances formed are of different composition than the initial materials used.

Let's consider a flashbulb (Figure 2.7), and compare it to the light bulb discussed in the previous section. When a small current is passed into a flashbulb, which consists of fine wires of magnesium metal in a gaseous atmosphere of oxygen, a flash results. Is this a physical change similar to that of turning on a light? The answer is no! The bulb will not flash again and the contents of the bulb have been altered. A white powdery substance known as magnesium oxide has

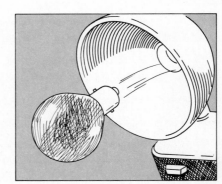

Figure 2.7
A Chemical Change.
When a flashbulb is used, its chemical composition changes.

been formed. This substance has very different physical and chemical properties from the original materials. *A* **chemical change** *(reaction) is a process in which the composition or structure of one or more substances is altered.*

Chemical Equations

As indicated in Chapter 1, chemists use chemical equations to record the chemical reactions of matter. At this point we shall use word equations to describe chemical changes, but later the equations will be altered and made more quantitative. In the case of the flashbulb, the word equation is

magnesium + oxygen $\longrightarrow$ magnesium oxide

Recall that in an equation, the substances to the left of the arrow are reactants. The substances on the right in the equation are products. In the flashbulb experiment, both magnesium and oxygen are reactants, whereas magnesium oxide is a product. There may be one or several reactants and one or more products in a chemical reaction. For example, when methane (natural gas) is burned, there are two products, carbon dioxide and water.

methane + oxygen $\longrightarrow$ carbon dioxide + water

Law of Conservation of Mass

It has been shown experimentally many times that the products of a chemical reaction have the same total mass as the sum of the masses of the starting materials. These observations have led to the **law of conservation of mass:** *there is no experimentally detectable gain or loss of mass during an ordinary chemical reaction.*

The flashbulb provides a good example of the law of conservation of mass since both the reactants and products are sealed in the bulb. If the mass of the bulb is determined both before and after the flash, it can be shown that the chemical reaction occurred without a change in mass (Figure 2.8).

Figure 2.8
Law of Conservation of Mass. The mass of a flashbulb is unchanged after a chemical reaction has occurred.

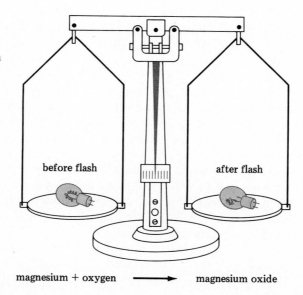

before flash   after flash

magnesium + oxygen $\longrightarrow$ magnesium oxide

Example 2.7 ————————————————————————————————

A 50.0 g sample of lead is heated with excess sulfur to form a compound called lead sulfide. The excess sulfur is burned off as a gas, sulfur dioxide. The amount of lead sulfide formed is 57.8 g. How much sulfur reacted with the lead?

Solution. The gain in mass must be due to the amount of sulfur that reacted with the lead to give lead sulfide.

mass of lead sulfide − mass of lead = mass of sulfur
$$57.8 - 50.0 = 7.8$$

The mass of sulfur is 7.8 g.

Example 2.8 ————————————————————————————————

A 5.00 g sample of a red-colored compound of mercury is heated. The only visible product is 4.63 g of the grey liquid metal mercury. Explain what occurred.

Solution. The decrease in mass of 0.37 g is the result of a chemical reaction. This mass must be accounted for as another product that cannot be seen. The most probable second product is a gas.

[Additional examples may be found in **2.28–2.30** at the end of the chapter.]

2.8
Mixtures and Pure Substances

Mixtures

Most matter we encounter every day is a complex mixture. Air contains essentially four gases, whereas gasoline is a mixture of at least twenty liquids. Practically all food consists of an incredible number of components. The human body contains hundreds of thousands of substances.

A **mixture** *can be separated into pure substances by physical means.* Physical separations include any process by which mixtures are separated without changing the identity of the individual components. For example, a pepperoni and cheese pizza can be seen to be a mixture, and its components can be separated manually.

A mixture of iron filings and sulfur can also easily be seen to contain two components. Iron filings are heavy and hard chips of iron. Sulfur is a light fluffy yellow powder. The mixture can be physically separated using a magnet (Figure 2.9). The iron filings are attracted to the magnet and the yellow sulfur remains.

If you put salt and water in a cooking pot, a mixture results, although by looking at the material you could not tell that it is a mixture. The mixture can still be separated by physical means; that is, you could boil off the water and the salt would remain in the pot. Each of the two components of the mixture is a pure substance. Neither the water nor the salt can be further separated into simple materials by physical means.

All mixtures can be classified into two categories, heterogeneous and homogeneous mixtures. *A* **homogeneous mixture** *has the same physical and chemical properties throughout.* The salt–water example given is a homogeneous mixture.

Figure 2.9
Example of Physical
Changes.

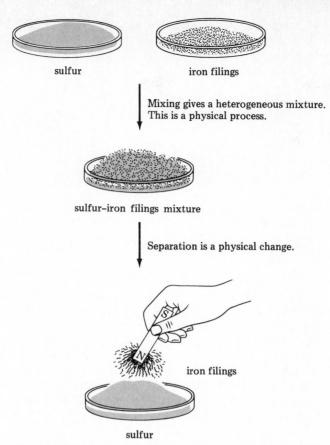

sulfur   iron filings

Mixing gives a heterogeneous mixture.
This is a physical process.

sulfur–iron filings mixture

Separation is a physical change.

iron filings

sulfur

Heterogeneous mixtures *have physical and chemical properties that are not uniform throughout the sample.* A pepperoni pizza is obviously a heterogeneous mixture.

All mixtures can have variable compositions. For example, a salt–water mixture might contain a pinch of salt or a teaspoon or tablespoon of salt. A pepperoni pizza could have a few slices of pepperoni or be generously covered with it. A summary of the characteristics of homogeneous and heterogeneous mixtures is given in Figure 2.10.

Figure 2.10
The Classification of Mixtures.

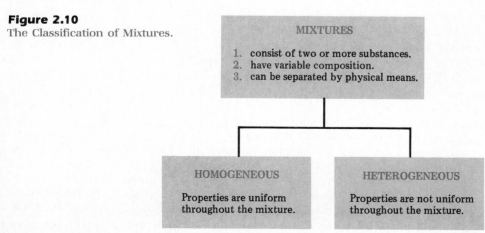

MIXTURES

1. consist of two or more substances.
2. have variable composition.
3. can be separated by physical means.

HOMOGENEOUS

Properties are uniform
throughout the mixture.

HETEROGENEOUS

Properties are not uniform
throughout the mixture.

Homogeneous mixtures are commonly called **solutions.** Solutions of all states of matter are possible. The air we breathe is a gaseous solution. Salt in water is a liquid solution. An alloy, such as brass, which contains zinc and copper, is a solid solution. We will discuss solutions in more detail in Chapter 9.

Pure Substances

There are relatively few materials in common use that are pure, that is, consist of a single substance. Copper used in electrical wiring is pure; however, most other metals, such as steel, are mixtures of several substances. Sugar is pure, but table salt can have additives such as potassium iodide to help prevent goiter. A medicine may consist of a single substance, although mixtures are more common.

Like homogeneous mixtures, pure substances have uniform properties. However, there are two important differences.

1. A pure substance cannot be separated into components by physical means.
2. A pure substance has a definite composition.

Pure substances are divided into two classes—elements and compounds—with elements forming the smaller class. At one time an element was defined as a substance that cannot be made from or decomposed into simpler substances. This definition needed to be changed when it was recognized that elements are composed of even simpler components called electrons, protons, and neutrons. These will be discussed in Chapter 3. As a result, the modified definition is: **elements** *are substances that cannot be constructed from or decomposed into simpler substance(s) by ordinary chemical processes.* Examples of elements include hydrogen and oxygen.

A compound is composed of two or more elements in fixed proportion by mass. A compound can be converted into other compounds or elements only by chemical reactions. Why is a compound different from a homogeneous mixture? Why can't a compound be separated by physical means?

Compounds *are composed of elements joined together by forces called bonds,* which will be discussed in Chapter 6. Bonds are not destroyed by physical changes. Only chemical reactions affect bonds. In a mixture, the components are simply mixed together and are not bonded to each other.

The definite composition of compounds is the result of bonds between elements. Only when the elements are present in the proper amounts and have reacted to form bonds does a compound result. For example, water consists of 88.81% oxygen and 11.19% hydrogen. In water, oxygen and hydrogen atoms are bonded to form a molecule. The elements hydrogen and oxygen may be mixed in all proportions to form many gaseous mixtures. However, none of the gaseous mixtures has the properties of water since there are no chemical bonds between hydrogen and oxygen. Water is the result of a chemical reaction of hydrogen and oxygen; it consists of a special bonded arrangement of hydrogen and oxygen. Water is different from a mixture because it consists of fixed proportions of its elements. A summary of the characteristics of elements and compounds is given in Figure 2.11.

2.9
Elements

About 6 million pure substances have been isolated or prepared by chemists. Of these, however, only 105 cannot be broken down by chemical means. These 105 simple substances are called elements. All of the other pure substances can be

Figure 2.11
The Classification of Homogeneous Matter.

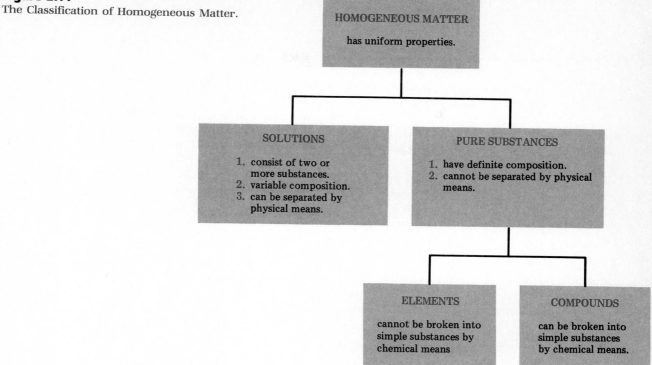

shown to consist of combinations of elements; that is, they are compounds. In the history of chemistry, a number of substances were at one time thought to be elements. However, as new chemical techniques were developed, it was shown that these substances were either mixtures or compounds. Quicklime was believed to be an element until it was eventually decomposed into calcium and oxygen. Therefore, quicklime is actually a compound of calcium and oxygen.

Chemical Symbols

Chemists have long used symbols to represent the elements. *A **chemical symbol** is a chemical shorthand or abbreviation of the name.* Chemical symbols for elements may be one or two letters. **When two letters are used, only the first is capitalized.**

Eleven of the 105 elements have symbols derived from older names and bear no resemblance to the currently used name. These elements and their symbols are listed in Table 2.2. The first letter of the currently accepted name is used as a symbol for twelve of the elements. You should start to learn the symbols for those elements described in this text as they appear. The remaining elements and their symbols are listed inside the back cover of this text.

The majority of the elements have symbols consisting of two letters. The symbols of many elements consist of the first two letters of the name or the first and third letter of the name. However, there are also symbols in which some other letter contained in the name is used. The necessity for using other than the first or first two letters in the name of the element is the result of the similarities in the names of elements. Consider the example of carbon, calcium, cadmium, and californium, whose symbols are C, Ca, Cd, and Cf. A variety of choices of letters is necessary to provide symbols for these four elements whose first two letters are identical.

Example 2.9

Examine the chemical symbols of the two metals magnesium and manganese. Determine how the symbols were chosen.

Solution. The third letter of the name has been chosen as the lowercase letter of the symbol. Thus magnesium has the symbol Mg, whereas manganese has the symbol Mn.

The two symbols are easily confused. For example, examine the second syllable of each name. In magnesium the second syllable starts with n, whereas the second syllable of manganese starts with g. Thus when each element name is pronounced, one might feel that the correct letter for the symbol should be the sound that starts the second syllable. This would give you the wrong chemical symbol.

[Additional examples may be found in **2.41–2.43** at the end of the chapter.]

Element Abundance

Which elements are the most important to us? About 90% of the mass of the universe is hydrogen. It is the nuclear fusion of hydrogen that provides the energy of the sun and other stars. Helium is the second most abundant element, amounting to 9%.

If only the surface of the earth is considered, the abundance of the elements is considerably different. The percent composition by mass listed in Table 2.3 is for the ten-mile-thick shell of the earth, the atmosphere, and the oceans. Only ten elements make up approximately 99% of this small but very important part of the universe. If the entire earth is considered, then the five most abundant elements are iron, oxygen, silicon, magnesium, and nickel.

Table 2.2

Symbols of Elements Derived from Names No Longer Used

Present name	Symbol	Former name
antimony	Sb	stibium
copper	Cu	cuprum
gold	Au	aurum
iron	Fe	ferrum
lead	Pb	plumbum
mercury	Hg	hydrargyrum
potassium	K	kalium
silver	Ag	argentum
sodium	Na	natrium
tin	Sn	stannum
tungsten	W	wolfram

Table 2.3

The Abundance of Elements in Percent by Mass

Element	Earth crust, sea, and air	Total earth	Atmosphere	Human body
hydrogen	0.88		0.00005	10.0
oxygen	49.20	29.5	20.9	65.0
carbon	0.08		0.03	18.0
nitrogen	0.03		78.0	3.0
calcium	3.39	1.1		2.0
potassium	2.40			0.2
silicon	25.70	15.2		
magnesium	1.93	12.7		0.04
phosphorus	0.11			1.1
sulfur	0.06	1.9		0.2
aluminum	7.50	1.1		
sodium	2.64			0.1
iron	4.71	34.6		
titanium	0.58			
chlorine	0.19			0.1
argon			0.93	
boron				
nickel	0.02	2.4		
neon			0.0018	

Another way of deciding which elements are most important is to consider the composition of the human body. About 25 elements are present, but only ten of these occur in quantities greater than 0.1%. The percent composition of the human body is given in Table 2.3. Many elements in trace quantities are vital to life. Cobalt occurs in vitamin B_{12}; zinc, manganese, and magnesium are incorporated in a variety of enzymes that are responsible for regulating many chemical reactions in living systems.

About 6 million compounds have been identified. Another 6000 new compounds are being discovered or prepared by chemists each week. The compounds that contain carbon and hydrogen are far more numerous than the compounds of all the other elements. Because many of them occur naturally in living material, carbon-containing compounds are called organic compounds and are discussed in Chapters 13–19, separate from the consideration of other compounds.

The composition of a compound is independent of its source, providing it has been rigorously purified. The composition of pure water, for example, is always 88.81% oxygen and 11.19% hydrogen, whether it is obtained by the purification of seawater or rainwater. The composition can be determined by chemically breaking the compound into its elements. Alternatively, the elements can be recombined in the proper amounts in a chemical reaction to produce the water. These facts are summarized in the **law of definite proportions:** *When elements combine to form compounds, they do so in definite proportions by mass.* The law is really equivalent to the definition of a compound.

The law of definite proportions can be illustrated with the reaction represented in Figure 2.12. Iron and sulfur can combine to form the compound iron

2.10
Compounds

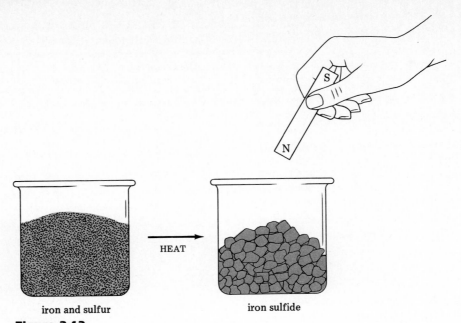

Figure 2.12
Example of Chemical Change.
A chemical reaction of iron and sulfur produces iron sulfide. The iron sulfide cannot be separated by using a magnet.

sulfide. If 55.85 g of iron is heated with 32.06 g of sulfur, then 87.91 g of iron sulfide results. No iron or sulfur remains. If 90.00 g of iron is heated with 32.06 g of sulfur, then 87.91 g of iron sulfide results, but 34.15 g of iron is unreacted. Iron and sulfur can only combine in the ratio of 55.85 g of iron per 32.06 g of sulfur to form iron sulfide. As further verification of the law of definite proportions, consider the experiment in which 55.85 g of iron is heated with 40.00 g of sulfur. Again 87.91 g of iron sulfide results, but in this case 7.94 g of sulfur remains unreacted.

In Section 2.8 you were told that a mixture of iron filings and sulfur can be separated using a magnet. When iron and sulfur react to form a compound, neither the iron nor the sulfur remains as an element. Thus a magnet cannot separate the iron sulfide (Figure 2.12). Compounds cannot be separated by physical means, as can mixtures.

Example 2.10 ─────────────────────────────────

A 50.0 g sample of lead reacts with 7.8 g of sulfur to produce 57.8 g of lead sulfide. How much sulfur is required to react with 25.0 g of lead to produce lead sulfide?

Solution. The ratio of the mass of lead to the mass of sulfur to form lead sulfide must be a constant. Since one half the amount of lead is used, one half the amount of sulfur, or 3.9 g, is required.

[Additional examples may be found in **2.52–2.53** at the end of the chapter.]

Summary

The study of chemistry is based on a classification of matter according to its composition, structure, and reactions. The mass of matter per unit volume is its density. The density of matter is related to its state. The three states are gas, liquid, and solid. Gases have lower densities than liquids or solids. Solids have a definite volume and shape; liquids have a definite volume, but take the shape of their container. Gases do not have a definite shape or volume; they expand to fill the available space.

Temperature determines the direction of heat flow between samples of matter; heat flows from a hot sample to a cold sample of matter. The three temperature scales are Fahrenheit, Celsius, and Kelvin.

The energy contained in substances is called chemical energy. Chemical reactions may be exothermic or endothermic. The energy change in a chemical reaction is measured in calories. The dietetic Calorie is a kilocalorie.

Every substance can be identified based on its set of physical and chemical properties. Physical changes do not alter the composition of the substance, whereas chemical changes are associated with changes in the composition in chemical reactions. Chemical reactions are recorded in chemical equations with the reactants to the left of the arrow and the products to the right of the arrow. Chemical reactions occur with no gain or loss of mass.

Most matter exists in complex mixtures, which may be homogeneous or heterogeneous. Mixtures are of variable composition. Pure substances may be separated from mixtures by physical means. Pure substances have a definite composition and cannot be separated by physical means. The two types of pure substances are elements and compounds. Compounds are composed of elements joined by forces called bonds.

Elements are represented by chemical symbols, which may be one or two letters. The first letter is capitalized. Most elemental symbols are based on letters in the element name.

Learning Objectives

As a result of studying Chapter 2 you should be able to

- Calculate the density and specific gravity of a sample of matter.
- Convert Fahrenheit temperature to Celsius and vice versa.
- Convert Celsius temperature to Kelvin and vice versa.
- Compare the properties of the three states of matter.
- Differentiate physical properties from chemical properties.
- Distinguish between physical and chemical changes.
- Express a chemical change in terms of a chemical equation.
- Recognize the experimental consequences of the law of conservation of mass.
- Distinguish between homogeneous and heterogeneous mixtures.
- Differentiate elements from compounds.
- Demonstrate the law of definite proportions.

New Terms

A **calorie** is a unit of energy used in chemistry to describe the change in chemical energy in a chemical reaction or the change in energy of a physical process.

On the **Celsius** scale the melting point and boiling point of water are 0°C and 100°C, respectively.

A **chemical change** is a process in which the composition or structure of one or more substances is altered.

Chemical properties are characteristics of matter that involve changes in chemical composition during chemical reactions.

Chemical symbols are shorthand symbols used to represent the elements.

Compounds are pure substances composed of elements joined together by forces called bonds.

Density is the ratio of mass per unit volume.

Elements are pure substances that cannot be decomposed into any simpler substance(s) by ordinary chemical reactions.

Endothermic processes require heat energy from the surroundings.

Energy is the ability to do work.

Exothermic processes release heat energy to the surroundings.

On the **Fahrenheit** scale the melting point and boiling point of water are 32°F and 212°F, respectively.

Heterogeneous mixtures have physical and chemical properties that are not uniform throughout.

Homogeneous mixtures have physical and chemical properties that are uniform throughout.

On the **Kelvin** scale the zero point is 273.15° below the zero point on the Celsius scale.

A **mixture** can be separated into pure substances by physical means.

A **physical change** occurs without a change in the composition of matter.

Physical properties are characteristics of matter that do not involve changes in the chemical composition of substances.

Solutions are homogeneous mixtures.

Specific gravity is the ratio of the density of a substance to the density of water at the same temperature.

Sublimation is the change of a substance from a solid into a gas without passing through the liquid state.

Questions and Problems

Terminology

2.1 Explain in your own words each of the following laws.
- (a) law of conservation of energy
- (b) law of conservation of mass
- (c) law of definite proportions

2.2 Explain the difference between the two terms of each pair.
- (a) density and specific gravity
- (b) chemical and physical change
- (c) mixture and compound
- (d) element and compound
- (e) mixture and pure substance
- (f) reactant and product

Density

2.3 A patient's urine sample has a density of 1.02 g/mL. How many grams of urine are eliminated on a day in which 1250 mL is excreted?

2.4 A cube of gold that is 2.00 cm on a side has a mass of 154.4 g. What is the density of gold?

2.5 The density of hydrogen gas at 25°C and normal atmospheric pressure is 0.08 g/L. The Hindenburg dirigible contained 1.9×10^8 L. How many kilograms of hydrogen did the Hindenburg hold?

2.6 A syringe contains 5.0 mL of a solution. The mass of the solution is 5.5 g. What is the density of the solution?

2.7 The density of ether, a volatile liquid used as an anesthetic, is 0.71 g/mL. What volume would 355 g of ether occupy?

2.8 Pure silver has a density of 10.5 g/mL. A 22.8 g sample reputed to be silver displaces 2.0 mL of water. Is the sample pure silver?

Temperature

2.9 The temperature on a warm day in Columbus, Ohio, is 95°F. What is the temperature in °C?

2.10 On a cold day in Vermont, the temperature was −31°F. What was the temperature in °C?

2.11 A doctor tells an American in Europe that his oral temperature is 38.0°C. What is his temperature in °F?

2.12 A European cookbook indicates that fondant must be cooked to 115°C. To what temperature should you heat the ingredients if you have a Fahrenheit candy thermometer?

2.13 A scientist reports a study of matter at 13 K. What is the temperature in °C and °F?

2.14 Silver melts at 1234 K. What is the melting point of silver in °C and °F?

2.15 The average rectal temperature of a healthy individual is 0.9° higher than the average oral temperature on the Fahrenheit scale. What is the average rectal value in Celsius units?

2.16 Some surgical instruments are sterilized at 120°C. What is the temperature on the Fahrenheit scale?

2.17 Ethylene glycol, a compound used in antifreeze, freezes at −11.5°C. What is the temperature on the Fahrenheit scale?

Energy

2.18 How many calories are required to heat each of the following samples of water to the indicated temperature?
- (a) 1.0 g from 20°C to 100°C
- (b) 10.0 g from 20°C to 21°C
- (c) 10.0 g from 20°C to 100°C

2.19 A quantity of heat equal to 500 cal is added to 25 g of water at 20°C. What will be the final temperature?

2.20 A water bed contains 1000 L of water. How many kilocalories of heat energy are required to heat the water in the bed from 15°C to body temperature?

2.21 One ounce of cereal gives 112 Cal of energy on oxidation. How many kilograms of water can be heated from 20°C to 30°C by "burning" the cereal?

2.22 A 110 lb woman eats 10 oz of carbohydrates, 2 oz of fat, and 18 oz of protein in one day. The energy released by metabolizing 1 g of either carbohydrates or protein is 4.1 kcal. The energy released by metabolizing 1 g of fat is 9.3 kcal. Calculate the caloric content of the woman's intake.

2.23 One cup of zucchini contains about 25 Cal. Explain the meaning of this statement.

2.24 The metabolic energy requirement of a laboratory mouse is about 3.8×10^3 cal per day. How many Calories are required to feed the mouse?

States of Matter

2.25 Classify each of the following substances as solid, liquid, or gas at room temperature.

(a) cooking gas	(b) iron	(c) mercury
(d) sulfur	(e) brass	(f) salt
(g) alcohol	(h) oxygen	(i) aspirin
(j) sugar	(k) antifreeze	(l) gasoline

2.26 Why are ice, water, and steam not classified as different substances?

2.27 Nicotine has a melting point and a boiling point of −79°C and +247°C, respectively. In what state will nicotine exist at 25°C?

Law of Conservation of Mass

2.28 A 2.317 g sample of silver oxide is heated, and 2.157 g of silver results. Does this information indicate the law of conservation of mass has been violated?

2.29 A 1.06 g sample of pure aluminum reacts slowly in air to form 2.00 g of a white powder. The substance contains only aluminum and oxygen. How much oxygen reacted with the aluminum.

2.30 Silver is tarnished by sulfur. In a laboratory experiment 0.216 g of silver is heated with excess sulfur, and the excess sulfur is then burned off to form a gas, sulfur dioxide. The black solid remaining weighs 0.248 g. How much sulfur combined with the silver.

Mixtures and Pure Substances

2.31 Make a list of materials used every day by you. Identify them as mixtures, elements, or compounds.

2.32 Some people claim that they avoid all foods having "chemicals' in them. Is this really possible?

2.33 How might each of the following mixtures be separated?
(a) iron filings and sand   (b) oil and water
(c) salt and sand   (d) alcohol and water
(e) salt and water

2.34 Is tap water a pure substance or a mixture? How could you support your answer experimentally?

Chemical and Physical Properties

2.35 Indicate whether the process described is an example of a physical change or chemical change.
(a) chopping of wood
(b) boiling of water
(c) rusting of iron
(d) melting of wax
(e) digesting food
(f) distilling moonshine
(g) burning of wood
(h) melting of iron
(i) burning of gasoline
(j) photosynthesis
(k) burning of a candle
(l) grinding beef into hamburger
(m) putting sugar into tea
(n) cooking an egg
(o) burning toast
(p) lighting a match
(q) souring of milk
(r) spoiling of food
(s) breaking a glass
(t) turning on a light

2.36 What are the physical and chemical properties of this page?

2.37 Describe the physical and chemical properties of iron.

Elements and Compounds

2.38 In 1787 silica was thought to be an element because chemists could not decompose it. However, it was later discovered that silicon reacts with oxygen to form silica. Why is silica not an element?

2.39 From the following equation classify A, B, and C as elements or compounds. If a classification is impossible, indicate why.

$$A + B \longrightarrow C$$

2.40 From the following equation classify X, Y, and Z as elements or compounds. If a classification is impossible, indicate why.

$$X \longrightarrow Y + Z$$

Elemental Symbols

2.41 What elements do the following symbols represent?
(a) O   (b) S   (c) Cl   (d) I   (e) Na
(f) K   (g) F   (h) Fe   (i) Ag   (j) H
(k) N   (l) C

2.42 What are the symbols for the following elements?
(a) calcium   (b) chlorine   (c) copper
(d) carbon   (e) magnesium   (f) mercury
(g) phosphorus   (h) sulfur   (i) gold
(j) tin   (k) bromine   (l) nickel

2.43 Match the following names with their symbols.
(a) silver   (1) Ba
(b) helium   (2) As
(c) zinc   (3) Pb
(d) barium   (4) Ag
(e) uranium   (5) Li
(f) arsenic   (6) U
(g) iron   (7) Cr
(h) lead   (8) He
(i) lithium   (9) Fe
(j) chromium   (10) Zn

Element Abundance

2.44 What is the most abundant element by mass in the universe?

2.45 What are the five most abundant elements in the earth's crust?

2.46 What are the two most abundant elements in the entire earth?

2.47 What are the two most abundant elements in the atmosphere?

2.48 What are the three most abundant elements in the human body?

Chemical Equations

2.49 What does a chemical equation describe?

2.50 Look up the definition of photosynthesis and write a word equation for the reaction.

2.51 What are the reactants and products in the following reaction?

$$\text{cooking gas} + \text{oxygen} \longrightarrow \text{water} + \text{carbon dioxide}$$

Law of Definite Proportions

2.52 A 1.00 g sample of hydrogen reacts completely with 7.93 g of oxygen to produce water. What will occur when 2.00 g of hydrogen reacts with 19.00 g of oxygen?

2.53 A 50.0 g sample of lead reacts completely with oxygen to give 57.8 g of lead oxide.

(a) How much oxygen reacted with the lead?

(b) How much oxygen would be required to react with 25.0 g of lead?

(c) What will happen if 50.0 g of lead is reacted with 10.0 g of oxygen?

(d) What will happen if 60.0 g of lead is reacted with 7.8 g of oxygen?

The Structure of Matter

The structure of this abandoned railroad covered bridge is easily seen. The structure of matter must be inferred by indirect means.

3.1
What Is the Smallest Unit of Matter?

We learned in the last chapter that every pure substance has its own set of physical and chemical properties. For example, the element helium is a colorless and odorless gas of low density. Helium does not chemically react with other substances; it will not burn in air. Ethyl alcohol is a liquid that mixes with water to form solutions. Ethyl alcohol will burn in air and can be metabolized by animals. Table salt is a white solid that mixes with water to form solutions and has a characteristic "salty" taste. Table salt, chemically known as sodium chloride, has chemical properties that will not be discussed at this point. What is responsible for the properties of each substance? How small a piece of each substance will still have the described properties?

If the gas in a helium-filled dirigible and the contents of a child's helium-filled balloon are examined and compared, their physical and chemical properties will be found to be identical. If only a small part of the contents of the helium balloon is examined, its properties will also be the same. Is there a point at which the helium sample can no longer be divided and still be considered helium? The answer is yes! Eventually the smallest unit of helium is reached. This unit is called an atom.

The contents of a bottle of pure ethyl alcohol divided into smaller and smaller samples is still ethyl alcohol. However, eventually the smallest unit of ethyl alcohol is reached. This unit is called a molecule. The ethyl alcohol molecule consists of two carbon atoms, six hydrogen atoms, and one oxygen atom bonded together to form a unit that has the properties of ethyl alcohol.

A tablet of table salt can be crushed into grains. Each grain can further be subdivided. However, the smallest amount of sodium chloride that has the properties of table salt is a pair of ions, the sodium ion and the chloride ion. Ions, which are similar to atoms but are electrically charged, will be discussed further in this chapter.

In all samples of matter there comes a point at which a substance can no longer be subdivided without changing its characteristics. This point is reached when one of three types of very small particles known as atoms, molecules, or ions results. These particles are called submicroscopic, because they are very much smaller than can be seen with a microscope. We will discuss these particles in this chapter.

In this chapter you will also learn that atoms, molecules, and ions consist of even simpler particles. These particles, known as *subatomic particles, are electrons, protons,* and *neutrons.* The subatomic particles in the atom will be discussed in greater detail in Chapters 5 and 6.

3.2
The Atomic Theory

About 460 B.C., Democritus, a Greek philosopher, suggested that matter consists of a large number of individual particles. These particles were called *atomas,* meaning indivisible. Today the atomas are called atoms. Although the atomic nature of matter was accepted by many scientists by the nineteenth century, it remained for an English schoolteacher, John Dalton, to suggest simple ideas to account for the then known laws of chemistry. His ideas, published in 1803, are called the atomic theory (Sec. 1.3).

As is the case with most theories, modifications have been made to conform with additional experimental results. For example, contrary to Dalton's postulate, we now know that the atom consists of subatomic particles. Furthermore, the atom is not indestructible, since it can be split in nuclear reactions.

Atomic Mass Units

Just how small is the atom? The smallest atom is hydrogen, which has a mass of 1.67×10^{-24} g. A helium atom is four times heavier and has a mass of 6.65×10^{-24} g. The heaviest atom, hahnium, has a mass of 3.37×10^{-22} g.

The small mass of an atom is difficult to visualize. In 1.00 g of hydrogen atoms there are 6.02×10^{23} atoms!

$$1.00 \text{ g} \times \frac{1 \text{ atom}}{1.67 \times 10^{-24} \text{ g}} = 6.02 \times 10^{23} \text{ atoms}$$

To give you an idea of how large this number is, let's consider having all the people of the world count the atoms in this 1.00 g sample. If each of the 4 billion people on earth were to count one atom a second for eight hours a day it would take 1.2×10^{10} years to finish counting! Atoms are very small. Even the smallest sample of matter has many atoms.

Example 3.1

The mass of the helium atom is 6.65×10^{-24} g. How many atoms are there in a 4.00 g sample of helium gas?

Solution. The conversion factor needed has the units atom/g. When the mass of the sample is multiplied by the conversion factor, the number of atoms results.

$$4.00 \text{ g} \times \frac{1 \text{ atom}}{6.65 \times 10^{-24} \text{ g}} = 6.02 \times 10^{23} \text{ atoms}$$

Note that the number of atoms in this 4.00 g sample of helium is the same as that in the 1.00 g sample of hydrogen just discussed. This is the case because although the sample of helium has four times the mass as the sample of hydrogen the helium atoms weigh four times as much as the hydrogen atoms.

Since the masses of atoms are so small, a special method of describing mass has been selected. *The quantity 1.6603×10^{-24} g is called an* **atomic mass unit (amu).** Thus, the hydrogen atom has a mass of 1.00 amu.

$$1.67 \times 10^{-24} \text{ g} \times \frac{1 \text{ amu}}{1.6603 \times 10^{-24} \text{ g}} = 1.00 \text{ amu}$$

The advantage in using atomic mass units is that exponential notation is avoided.

Example 3.2

The mass of the helium atom is 6.65×10^{-24} g. What is its mass in amu?

Solution. In order to convert the quantity given in grams into amu, a conversion factor with the units amu/g is required.

$$6.65 \times 10^{-24} \text{ g} \times \frac{1 \text{ amu}}{1.6603 \times 10^{-24} \text{ g}} = 4.00 \text{ amu}$$

Note that the mass of helium in amu is four times that of the hydrogen atom. This must be the case because the helium atom's mass in grams is four times that of the mass of the hydrogen atom in grams.

[Additional examples may be found in **3.4–3.7** at the end of the chapter.]

Atomic Radii

Our model of an atom is a spherical particle. The radii of various atoms range from 0.9×10^{-8} to 2.4×10^{-8} cm. Because of these very small numbers, it is convenient to use Ångstrom (Å) units to express the radii of atoms. One Ångstrom unit is 10^{-8} cm. Thus, the radii of atoms range from 0.9 to 2.4 Å (Table 3.1).

Table 3.1

Atomic Radii of Some Atoms[a]

Name	Radius (Å)	Relative mass (amu)	Name	Radius (Å)	Relative mass (amu)	Name	Radius (Å)	Relative mass (amu)
helium	0.93	4.00	lithium	1.23	6.94	fluorine	0.64	19.00
neon	1.12	20.18	sodium	1.57	22.99	chlorine	0.99	35.45
argon	1.54	39.95	potassium	2.03	39.10	bromine	1.14	79.90
krypton	1.69	83.80	rubidium	2.16	85.47	iodine	1.33	126.90
xenon	1.90	131.30	cesium	2.35	132.90			

[a] For the elements in each column, the size of the atoms increases as the mass in amu increases. More details of this relationship will be given in Chapter 5.

A distance of 1 Å is difficult to comprehend. The radius of the magnesium atom is 1.60 Å. If magnesium atoms were arranged in a straight line touching each other, it would take 31 million atoms to equal 1 cm. More details about the size of atoms will be presented in Chapter 5.

3.3
Subatomic Particles

Dalton's postulate that atoms are indivisible was disproved about a century ago. We now know that atoms can be broken into more fundamental subatomic particles called electrons, protons, and neutrons. The data for these particles are summarized in Table 3.2.

Table 3.2

Properties of Subatomic Particles

Name	Mass (g)	Relative mass (amu)	Charge (coulombs)	Relative charge
proton	1.6725×10^{-24}	1.0073	1.60×10^{-19}	+1
neutron	1.6748×10^{-24}	1.0087	0	0
electron	9.109×10^{-28}	0.0005486	-1.60×10^{-19}	−1

The Electron

The electron was discovered as the result of experiments in electricity. In 1897, the English physicist J. J. Thomson was working with a cathode ray tube similar to the modern television tube. He showed that a variety of gaseous atoms in such tubes could be broken into charged particles. Each gas produced negatively charged particles called electrons. An **electron** *has a mass of* 9.109×10^{-28} g *or* 0.0005486 *amu*. This mass is 1/1837 the mass of the lightest atom, hydrogen. The mass of an electron usually can be regarded as zero when discussing the mass of the atom.

The charge of an electron is -1.60×10^{-19} *coulomb*. Such a quantity is cumbersome. However, all charges of subatomic and atomic particles are multiples of this quantity. Therefore, the electron charge is usually replaced with a relative charge of -1 without units. The hydrogen atom has only one electron, but other atoms have more electrons. The carbon atom has six electrons; the oxygen atom has eight electrons.

The Proton

Since all atoms contain electrons, they must also contain some material with a positive charge to maintain electrical neutrality. The proton is the common positive particle present in all atoms. A hydrogen atom that has one electron also has one proton. Since the mass of an electron is so small, the majority of the mass of the hydrogen atom is due to the proton. *A* **proton** *has a mass of* 1.67×10^{-24} g. This quantity is 1.00 amu.

The charge of a proton is $+1.60 \times 10^{-19}$ coulomb. This quantity is usually replaced by a relative charge of $+1$ without units.

All atoms contain protons. The number of protons in an atom is a characteristic of the element. The number of protons in an element always equals the number of electrons, since atoms are electrically neutral.

The Neutron

The mass of a helium atom is 4 amu. The helium atom contains two electrons and two protons. Based on this number of subatomic particles the helium atom should only weigh 2 amu! What accounts for the other 2 amu of the helium atom? This "missing" mass of the helium atom is due to neutrons. An English physicist, James Chadwick, showed that **neutrons** *are subatomic particles with no charge*. The mass of a neutron is 1.6748×10^{-24} g or essentially 1 amu. Therefore, the helium atom has two neutrons. All other atoms, with the exception of ordinary hydrogen, also contain neutrons.

The Nucleus and the Atom

In 1911, Ernest Rutherford, an English scientist, showed that the mass of the atom is located in a small dense region called the nucleus. Other experiments led to the conclusion that *the* **nucleus** *is composed of protons and neutrons*. The average diameter of the nucleus is 10^{-13} cm.

Since the radius of an atom is about 10^5 times larger than the nucleus, most of the rest of the atom is empty space. The electrons occupy this large volume and move in a region about the nucleus. The closest analogy to the atom is the solar system. Although the solar system is large, it is essentially empty space. The planets move about in this empty space as they revolve about the center occupied by the sun.

The structure of the atom might be easier to picture if we translate the

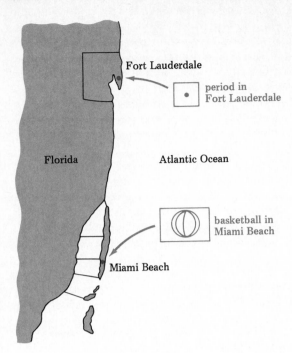

Figure 3.1
An Analogy of the Relationship Between an Electron and the Nucleus.
If the nucleus is the size of a basketball, the electron is a period 20 miles away.

distances into something we are more familiar with. If the nucleus of the hydrogen atom were scaled up to the size of a basketball, the electron would be the size of a period about 20 miles away (Figure 3.1). A more exact representation of the structure of the atom will be given in Chapter 5.

Anions and Cations

In Section 3.1 you were told that some types of matter consist of ions. *An* **ion** *results from the loss or gain of electrons from atoms.* If an atom loses one or more electrons, a positively charged ion results. *A positive ion is called a* **cation.** The sodium atom can lose an electron to form a sodium cation. Note however that the nucleus with its protons and neutrons is unchanged. Therefore, the sodium ion is not another element. If an atom gains one or more electrons, a negatively charged ion results. *A negative ion is called an* **anion.** The chlorine atom can gain an electron to form a chloride anion. Ions will be discussed in greater detail in Section 3.6.

3.4
Atomic Number, Mass Number, and Isotopes

Atoms differ from each other because of the number of subatomic particles that they contain. The number of protons actually determines the identity of the atom. For example, the hydrogen atom contains one proton, whereas the helium atom contains two protons.

The **atomic number** *of an element is equal to the number of protons in the nucleus of an atom of that element.* The atomic numbers of hydrogen and helium are 1 and 2, respectively. The atomic numbers of all the elements are listed in a Table of Elements on the inside cover of this book. Since the number of electrons in an atom equals the number of protons, the atomic number also indicates the number of electrons in an atom.

The total number of protons and neutrons in the nucleus of an atom is the **mass number.** Since helium has two neutrons and two protons in its nucleus, its mass number is 4 amu.

Elemental Symbols

Since atoms of one element differ from those of another element by the number of their subatomic particles, it is convenient to have this information included with the elemental symbol. The following symbolism is used.

A is the mass number ⎯⎯⎯⎯⎯⎯⎯⎯

Z is the atomic number ⎯⎯⎯⎯⎯⎯ $^A_Z E$ E represents the symbol of the element

The subscript Z then gives the number of protons in the atom. The mass number A is equal to the number of protons and neutrons in the atom. Thus, the number of neutrons is equal to $A - Z$.

Example 3.3

How many protons, neutrons, and electrons are contained in an atom represented by $^{14}_{6}C$?

Solution

$$^{14}_{6}C = ^A_Z E$$

The elemental symbol represents the element carbon. There are six protons in the element, as indicated by the subscript. Therefore, carbon also contains six electrons. The number of neutrons given by the quantity $A - Z$ is eight.

[Additional examples may be found in **3.16–3.22** at the end of the chapter.]

Isotopes

The second postulate of Dalton was made in the absence of any experimental evidence to the contrary. We now know that atoms of an element have different masses. These atoms, called isotopes, have the same number of protons but differ in the number of neutrons that they contain. Thus, *the* **isotopes** *of an element have identical atomic numbers but different mass numbers.* It is the number of protons in the nucleus (the atomic number) that determines the identity of the element and not the number of neutrons.

There are two naturally occurring isotopes of chlorine. Chlorine has an atomic number of 17 and contains 17 protons as well as 17 electrons. However, one type of chlorine atom or isotope has 18 neutrons in the nucleus, whereas the other isotope has 20 neutrons in the nucleus. The mass numbers of these isotopes are 35 and 37, respectively. A simplified representation of these isotopes is given by a sphere (the nucleus), which shows the tally of protons and neutrons. The electrons are listed in a partial "orbit" at a distance from the nucleus (Figure 3.2).

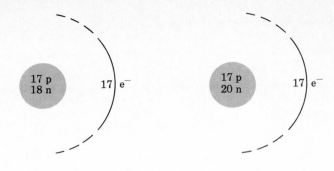

Figure 3.2
The Isotopes of Chlorine-35 and Chlorine-37. The protons and the neutrons are located in the nucleus. The electrons are in space about the nucleus.

Example 3.4

What do the symbols $^{235}_{92}U$ and $^{238}_{92}U$ mean? What relationship exists between the atoms represented by these symbols? How many neutrons are contained in each nucleus?

Solution. The subscript 92 indicates that the element represented by U, uranium, contains 92 protons. Because the symbols represent atoms that contain the same number of protons but differ in mass number, the atoms are isotopes of each other. The number of neutrons in $^{235}_{92}U$ is $235 - 92 = 143$. In $^{238}_{92}U$, the number of neutrons is $238 - 92 = 146$.

Example 3.5

The radioisotope $^{131}_{53}I$ is used to treat cancer of the thyroid gland as well as hyperthyroidism. Describe the atomic composition of this isotope.

Solution

$$^{131}_{53}I = ^A_Z E$$

There are 53 protons and also 53 electrons as $Z = 53$. The number of neutrons is $A - Z$ or $131 - 53 = 78$.

[Additional examples may be found in **3.24–3.27** at the end of the chapter.]

3.5
Molecules

There are very few pure substances that consist only of single atoms. Many elements and compounds consist of molecules.

The simplest units with the properties of the gaseous elements helium, neon, argon, krypton, xenon, and radon, are the atoms of these elements. Such elements are said to be monatomic particles.

The common gases, nitrogen, oxygen, and hydrogen consist of units that contain two atoms; they are diatomic. *A neutral unit consisting of two or more atoms that remain associated and possessing the properties of the substance is called a* **molecule.** We see that Dalton's concept of an element is not entirely correct, since many elements exist as molecules.

Among the elements that are also diatomic are fluorine, chlorine, bromine, and iodine. The molecules of some elements contain more than two atoms and are polyatomic. Phosphorus consists of molecules of four atoms, whereas sulfur exists in units of eight atoms. *A representation of the molecule of such an element uses a subscript to the right of the elemental symbol.* The diatomic molecule chlorine is represented as Cl_2. The elemental forms of phosphorus and sulfur are represented as P_4 and S_8, respectively.

$Cl_2 \longleftarrow$ subscript indicating 2 chlorine atoms
per molecule

With the exception of ionic compounds (Sec. 3.7), the molecule is the smallest possible unit of a compound. Molecules of compounds contain two or more different types of atoms, whereas molecules of elements contain only one type of atom. *A representation of a molecule of a compound that indicates the number of each kind of atom contained in the molecule is called the* **molecular formula.**

Water consists of molecules containing two hydrogen atoms and one oxygen atom. The molecular formula of water is H_2O. The subscript 2 to the right of the symbol for hydrogen indicates the number of hydrogen atoms contained in the molecule. No subscript follows the symbol for oxygen, which means that only one atom of that type is contained in the molecule. The molecule could be represented by OH_2 equally as well. It is a matter of convention that chemists have decided to use H_2O instead of OH_2.

The molecular formula of a compound gives only the composition of the substance. Further details are needed in order to determine the structure of the molecule. Consider for example the molecular formulas of the organic compounds listed in Table 3.3. The possible ways in which the many atoms in these complex molecules can be bonded is very large. Details of the structure of organic compounds will be given starting in Chapter 13.

Table 3.3

Molecular Formulas of Some Organic Compounds

Molecular formula	Common name
C_3H_8	propane
C_8H_{18}	octane
$C_6H_8O_6$	vitamin C
$C_6H_{12}O_6$	glucose
$C_8H_8O_3$	oil of wintergreen
$C_{21}H_{28}O_2$	tetrahydrocannabinol
$C_{20}H_{30}O$	vitamin A
$C_3H_5N_3O_9$	nitroglycerin
$C_{11}H_{17}NO_3$	mescaline
$C_{12}H_{12}N_2O_3$	phenobarbital
$C_{13}H_{16}N_2O_2$	novocaine
$C_{16}H_{21}NO_4$	cocaine
$C_{17}H_{17}NO_3$	morphine
$C_{16}H_{18}N_2O_4S$	penicillin G
$C_{63}H_{84}N_{14}O_{14}PCo$	vitamin B_{12}

Example 3.6 _____

One of the major components of natural gas is methane. Methane is a compound represented by the molecular formula CH$_4$. What does this molecular formula mean?

Solution. The molecular formula indicates that the compound contains carbon and hydrogen. Because no number subscript appears to the right of C, there is only one atom of carbon per molecule. There are four atoms of hydrogen per molecule. The smallest unit that has the properties of methane consists of five atoms—one of carbon and four of hydrogen.

_____

Example 3.7 _____

Aspirin, a chemical compound known to the chemist as acetylsalicylic acid, contains nine atoms of carbon, eight atoms of hydrogen, and four atoms of oxygen. Write the molecular formula.

Solution. Although we do not know the order of the symbols in the molecular formula, let us use C followed by H and finally O. The molecular formula is then given by C$_9$H$_8$O$_4$.

_____

[Additional examples may be found in **3.28–3.34** at the end of the chapter.]

3.6
Ions

As indicated in Section 3.3, atomic particles that contain more or fewer electrons than protons are called ions. A positive ion or cation results from the loss of one or more electrons and contains fewer electrons than protons. A negative ion or anion results from gaining one or more electrons and contains more electrons than protons.

Symbols of Ions
An ion is written with its relative charge as a superscript to the right of the elemental symbol. The superscript bears a positive or negative sign indicating the charge and an integer indicating how many electrons have been removed from or added to the neutral atom.

When an electron is removed from the sodium atom, the sodium ion results. This ion is symbolized as Na$^+$. Equations representing this process are shown below:

$$\left(11\,p\right)\ 11\right)e^- \longrightarrow \left(11\,p\right)\ 10\right)e^- + e^-$$

$$\text{Na}\begin{pmatrix}11\text{ protons}\\11\text{ electrons}\end{pmatrix} \longrightarrow \text{Na}^+\begin{pmatrix}11\text{ protons}\\10\text{ electrons}\end{pmatrix} + 1\,e^-\text{ (electron)}$$

$$\text{Na} \longrightarrow \text{Na}^+ + e^-$$

The symbol e^- represents the electron with its relative charge. Since the atomic number of sodium is 11, we know that the atom contains 11 protons and 11 electrons. When an electron is removed from the sodium atom, the resultant sodium ion is positively charged.

When the calcium atom (atomic number of 20) loses two electrons, the resultant calcium ion is represented by Ca^{2+}.

$$\left(20\ p\right)\ 20\Big)e^- \longrightarrow \left(20\ p\right)\ 18\Big)e^- + 2\ e^-$$

$$Ca \begin{pmatrix} 20\ protons \\ 20\ electrons \end{pmatrix} \longrightarrow Ca^{2+} \begin{pmatrix} 20\ protons \\ 18\ electrons \end{pmatrix} + 2\ e^-\ (electrons)$$

$$Ca \longrightarrow Ca^{2+} + 2\ e^-$$

As an example of an anion, consider the addition of an electron to the chlorine atom to yield the chloride ion. The atomic number of chlorine is 17.

$$1\ e^- + \left(17\ p\right)\ 17\Big)e^- \longrightarrow \left(17\ p\right)\ 18\Big)e^-$$

$$1\ e^- + Cl \begin{pmatrix} 17\ protons \\ 17\ electrons \end{pmatrix} \longrightarrow Cl^- \begin{pmatrix} 17\ protons \\ 18\ electrons \end{pmatrix}$$

$$e^- + Cl \longrightarrow Cl^-$$

In a similar manner, the oxygen atom can accept two electrons to yield the oxide ion O^{2-}.

$$2\ e^- + \left(8\ p\right)\ 8\Big)e^- \longrightarrow \left(8\ p\right)\ 10\Big)e^-$$

$$2\ e^- + O \begin{pmatrix} 8\ protons \\ 8\ electrons \end{pmatrix} \longrightarrow O^{2-} \begin{pmatrix} 8\ protons \\ 10\ electrons \end{pmatrix}$$

$$2\ e^- + O \longrightarrow O^{2-}$$

Example 3.8

The atomic number of aluminum is 13. Aluminum loses three electrons to form a cation. How many protons and electrons does the cation have? Write a representation of the ion.

Solution. The atomic number is equal to the number of protons in an atom. In addition, the atomic number of an atom indicates the number of electrons, because the number of protons and electrons are equal in a neutral atom. Aluminum has thirteen electrons. When aluminum loses three electrons, ten electrons remain. The resulting cation has three units of positive charge. This ion is represented as Al^{3+}.

[Additional examples may be found in **3.35–3.39** at the end of the chapter.]

Table 3.4

Common Simple Ions

Anion	Name	Cation	Name
F^-	fluoride	Li^+	lithium ion
Cl^-	chloride	Na^+	sodium ion
Br^-	bromide	K^+	potassium ion
I^-	iodide	Mg^{2+}	magnesium ion
O^{2-}	oxide	Ca^{2+}	calcium ion
S^{2-}	sulfide	Zn^{2+}	zinc ion
		Ba^{2+}	barium ion
		Ag^+	silver ion
		Fe^{2+}	iron (II) or ferrous ion
		Fe^{3+}	iron (III) or ferric ion
		Cu^+	copper (I) or cuprous ion
		Cu^{2+}	copper (II) or cupric ion

Table 3.5

Common Complex Ions

Formula	Name
NO_3^-	nitrate
NO_2^-	nitrite
SO_4^{2-}	sulfate
SO_3^{2-}	sulfite
HSO_4^-	bisulfate
HSO_3^-	bisulfite
PO_4^{3-}	phosphate
CO_3^{2-}	carbonate
HCO_3^-	bicarbonate
OH^-	hydroxide
CN^-	cyanide
MnO_4^-	permanganate
ClO_4^-	perchlorate
ClO_3^-	chlorate
ClO_2^-	chlorite
ClO^-	hypochlorite
NH_4^+	ammonium

P^{3-} phosptide

N^{3-} Nitride

Names of Ions

The anions derived from single atoms are named by adding -ide to the root of the element's name. A list of common anions and their names is given in Table 3.4. The common cations are also listed. Note that some elements can form more than one cation. The charge of an ion is governed by the electronic structure of the atom, which will be discussed in Chapter 5.

Ions that consist of several atoms held together by chemical bonds similar to those involved in molecules are called **polyatomic ions** *or* **complex ions.** These complex ions differ from molecules in that they bear a charge. The complex ion may be positive or negative, however negatively charged complex ions are by far the more common. A list of complex ions is given in Table 3.5.

Remember that the superscript on the right indicates the charge on the ion, whereas the subscript represents the number of atoms occurring in the ion. Thus, the nitrate ion, NO_3^-, consists of one atom of nitrogen, three atoms of oxygen, and one electron above what the individual neutral atoms contain.

You should learn the symbols for both simple and polyatomic ions. That is, you should know not only the elements present in the ion but also the charge on the ion. You will need this information in order to write correct formulas for ionic compounds.

The complex ions containing oxygen commonly have the suffixes -ate and -ite. Although the suffix does not indicate the absolute number of oxygen atoms in the complex ion, *the ion with the -ate suffix contains more oxygen atoms than that with the -ite suffix.* Thus, while nitrate and nitrite are NO_3^- and NO_2^-, respectively, sulfate and sulfite are SO_4^{2-} and SO_3^{2-}, respectively. It is for this reason that you should learn the symbols corresponding to the names of the complex ions and vice versa.

There are several combinations of two elements that give rise to more than two complex ions. Therefore, more endings than -ate and -ite are required. For example, Cl and O actually form four different complex ions. In Table 3.5 are listed perchlorate (ClO_4^-), chlorate (ClO_3^-), chlorite (ClO_2^-), and hypochlorite (ClO^-). *The prefix per- is used to mean over.* Thus perchlorate has one more oxygen atom than chlorate. *The prefix hypo- means under.* Thus hypochlorite is related to chlorite by having one less oxygen atom. These prefixes are also used with other similar ions such as hypoiodite for IO^- and perbromate for BrO_4^-.

As indicated earlier in this chapter, ions are present in some compounds such as sodium chloride. *Compounds consisting of ions are called* **ionic compounds** and are structurally different from molecular compounds. Ionic compounds are also called salts. Ionic compounds consist of oppositely charged ions held together by electrostatic forces of attraction. These forces, called ionic bonds, will be discussed further in Chapter 6.

3.7
Ionic Compounds

Structure of Ionic Compounds

Most ionic compounds are crystalline solids at room temperature. These solids consist of ions arranged in repeating three-dimensional patterns. For example, a crystal of sodium chloride (Figure 3.3) contains sodium ions surrounded by six chloride ions. Each chloride ion is also surrounded by six sodium ions. The resulting arrangement is like three-dimensional tic-tac-toe.

In a crystal of sodium chloride, the number of sodium ions equals the num-

Figure 3.3
The Structure of Sodium Chloride. (a) Schematic drawing showing the alternating pattern of the sodium and chloride ions. (b) The relative sizes of the ions. Sodium ions are smaller than chloride ions.

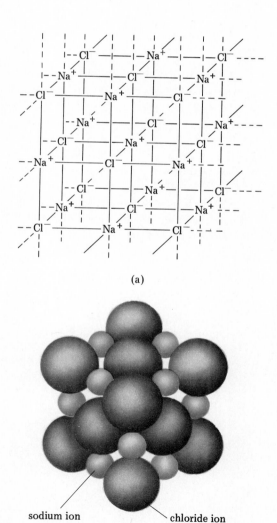

(a)

(b)

ber of chloride ions. As a whole, the crystal is electrically neutral no matter what its size. The crystal is held together by electrostatic forces between the cations and anions.

Formulas of Ionic Compounds

Ionic compounds are electrically neutral overall, since the charge of the cations is balanced by the charge of the anions. In table salt, sodium chloride, electrical neutrality is the result of an equal number of Na^+ and Cl^- ions.

$$\begin{array}{ll} Na^+ & \text{charge of } +1 \\ Cl^- & \underline{\text{charge of } -1} \\ & \text{net charge } 0 \end{array}$$

Sodium chloride is represented by NaCl even though there is no sodium chloride molecule.

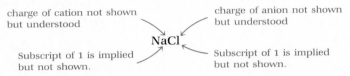

In calcium chloride there is one Ca^{2+} ion for every two Cl^- ions, and the chemists' representation is $CaCl_2$. However, there is no $CaCl_2$ molecule. A crystal of $CaCl_2$ contains Ca^{2+} and Cl^- ions in a 1:2 ratio.

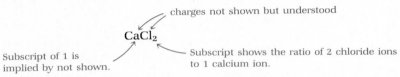

When ions are present in an ionic compound, the positive charges must balance the negative charges. Since the charge on the anion may not be equal to that on the cation, the number of anions will not always equal the number of cations.

AN ASIDE

Dental calculus, also known as tartar, is a hard deposit that eventually forms on teeth unless proper dental hygiene is practiced. Initially a thin soft film is formed; in a few days it hardens. The hard deposit retains food particles and provides an excellent environment for bacterial growth. Without proper tooth care, the tartar can cause the loss of tissue supporting the teeth and eventually the loss of teeth.

The principal components of dental calculus are complex salts of calcium. These include $Ca_3(PO_4)_2$ and $Ca_5(PO_4)_3OH$, which are not soluble in water. Thus, simply rinsing the teeth with water will not remove the calculus.

Calculus formation can be decreased somewhat by proper diet. Certain foods, such as fresh fruits and vegetables, have a detergent-type action and clean the teeth surface before calculus formation proceeds to the hardened stage. In spite of the advances of science in many areas, there is still little that can be done to prevent calculus other than the proper brushing of teeth after each meal.

positive charge + negative charge = 0

$$\left(\frac{\text{charge}}{\text{cation}} \times \begin{array}{c}\text{relative}\\\text{number}\\\text{of cations}\end{array}\right) + \left(\frac{\text{charge}}{\text{anion}} \times \begin{array}{c}\text{relative}\\\text{number}\\\text{of anions}\end{array}\right) = 0$$

For example, in $CaCl_2$, there are two chloride ions for every calcium ion.

$$\left(\frac{\text{charge}}{\text{calcium cation}} \times \begin{array}{c}\text{relative}\\\text{number}\\\text{of Ca}^{2+}\end{array}\right) + \left(\frac{\text{charge}}{\text{chloride anion}} \times \begin{array}{c}\text{relative}\\\text{number}\\\text{of Cl}^{-}\end{array}\right) = 0$$

$$(+2)(1) + (-1)(2) = 0$$

Example 3.9

Given the fact that aluminum can form a +3 cation (see Example 3.8), write the formula of aluminum oxide.

Solution. The compound that consists of Al^{3+} and O^{2-} ions must be electrically neutral. Therefore, the total charge of the cations and the anions in the formula must be balanced. The lowest common multiple of 3 and 2 is 6. Thus, three oxide ions will have a total charge of -6, and two aluminum ions will have a total charge of $+6$. By convention the formula is Al_2O_3 rather than O_3Al_2 (see discussion following on the names of ionic compounds).

Example 3.10

An oxide of chromium has the formula Cr_2O_3. What is the charge of the chromium ion?

Solution. The charge of the oxide ion is -2. For three oxide ions, the total negative charge is -6. The total charge of the chromium ions must be $+6$. Thus, the charge of each chromium ion must be $+3$.

Example 3.11

Replacement of hydroxide ion in hydroxyapatite of teeth by fluoride ion produces fluoroapatite, which is less soluble and more resistant to acids. The composition of fluoroapatite is $Ca_5(PO_4)_3F$. Show that this material is electrically neutral.

Solution. Calcium exists in ionic compounds as Ca^{2+}. The phosphate ion has a -3 charge, whereas fluoride has a -1 charge. The five Ca^{2+} ions have a total charge of $5(+2) = +10$. The three phosphate and one fluoride ion have a total charge of $3(-3) + (-1) = -10$. Thus, the compound is electrically neutral.

[Additional examples may be found in **3.45–3.51** at the end of the chapter.]

Table 3.6

Names of Some Ionic Compounds

Formula	Name
$LiClO_4$	lithium perchlorate
$LiBr$	lithium bromide
$NaNO_3$	sodium nitrate
$NaHCO_3$	sodium bicarbonate
KCl	potassium chloride
$KMnO_4$	potassium permanganate
K_3PO_4	potassium phosphate
$CaCl_2$	calcium chloride
CaS	calcium sulfide
$CaCO_3$	calcium carbonate
$CaSO_4$	calcium sulfate
$Ca_3(PO_4)_2$	calcium phosphate
MgO	magnesium oxide
MgF_2	magnesium fluoride
$Mg(CN)_2$	magnesium cyanide
$Zn(OH)_2$	zinc hydroxide
$ZnSO_4$	zinc sulfate
NH_4Cl	ammonium chloride
$(NH_4)_2SO_4$	ammonium sulfate
$(NH_4)_3PO_4$	ammonium phosphate

A number of ionic compounds are listed in Table 3.6. You should check the charges on the ions listed in Tables 3.4 and 3.5 to show that all of the compounds listed in Table 3.6 are electrically neutral. Note that whenever more than one complex ion is required to form a compound, a parenthesis is used to enclose the complex ion. The subscript to the right of the parenthesis indicates the number of such complex ions required to balance the ions of opposite charge. Thus, $Ca_3(PO_4)_2$ is an ionic compound that consists of three doubly charged calcium ions for every two triply charged phosphate ions.

Table 3.7

Medical Uses of Some Ionic Compounds

Formula	Medical use
NH_4Cl	diuretic (helps urination)
$(NH_4)_2CO_3$	smelling salts
$CaCO_3$	antacid
KNO_3	diuretic
$AgNO_3$	antiseptic (eyes of newborn)
NaF	applied to teeth by dentists
$NaCl$	saline solution
$NaBr$	sedative
NaI	treatment of iodine deficiency
$MgSO_4$	laxative
$FeSO_4$	treatment of iron deficiency
$BaSO_4$	x-ray of organs
ZnO	calamine lotion

Although many substances used in medicine are very complex compounds, there are some simple ionic compounds that are widely used. These are listed in Table 3.7.

Names of Ionic Compounds

Ionic compounds are named for the ions that they contain. The name of the positive ion is given first, then the name of the negative ion. Thus, sodium chloride is the name for NaCl. Some elements form more than one cation and thus form different compounds with the same anion. For example, $FeCl_2$ and $FeCl_3$ must be distinguished by different names. The Roman numerals following the name of the cation in iron(II) chloride and iron(III) chloride indicate the ion present in each compound.

Example 3.12

What is the name of the compound given by the formula FeS?

Solution. The sulfide ion has a charge of -2. The iron must then have a charge of $+2$. This ionic form of iron is called ferrous or iron(II). The compound name is ferrous sulfide or iron(II) sulfide.

Summary

The laws of conservation of mass and definite proportions led John Dalton to formulate an atomic theory based on an idea originally suggested by early Greeks. Atoms were hypothesized to consist of definite small particles that cannot be split apart in chemical reactions.

An atom is now known to be an electrically neutral particle that is the smallest representative unit of an element. The nucleus of an atom contains neutrons and protons; the latter subatomic particle has a positive charge. Surrounding the nucleus are a number of electrons equal to the number of protons in the nucleus. An electron is negatively charged.

Atoms are described by their atomic number and their mass number. The atomic number is the number of protons in the atom. The mass number is the sum of the number of protons and neutrons. Atoms that have the same number of protons but differ in the number of neutrons are called isotopes of the same element. Elemental symbols with a subscript to the left of the symbol representing the atomic number and a superscript to the left of the symbol to represent the mass number are used to give information about the atomic composition of the isotope.

A neutral unit consisting of atoms bonded to each other is called a molecule. Some elements exist as molecules and are represented by the elemental symbol and a subscript to the right to indicate the number of atoms in the molecule. Molecular compounds are represented by molecular formulas that indicate the number and type of each atom present in the molecule. Subscripts to the right of each symbol indicate the number of atoms of that type present.

A loss or gain of electrons by an atom yields an ion. Positive ions are cations; negative ions are anions. Ions held together by electrostatic forces of attraction result in ionic compounds. The relative number of each type of ion is balanced so that the charges of the cations equal the charges of the anions. Ionic compounds are named by using the name of the positive ion followed by the name of the negative ion.

Learning Objectives

As a result of studying Chapter 3 you should be able to
- Identify three subatomic particles by name, mass, and charge.
- Describe the structure of the atom.
- Express the number of subatomic particles by an elemental symbol.
- Represent molecules by their molecular formula.
- Distinguish between ions formed by losing or gaining electrons.

• Write formulas for ionic compounds based on the ions present.

New Terms

An **atomic mass unit** (amu) equal to 1.6603×10^{-24} g is used as the basic unit in describing atomic mass.

Anions are negatively charged atomic particles that result from the gain of electrons.

The **atomic number** of an element in amu is equal to the number of protons in the nucleus of an atom of the element.

Cations are positively charged atomic particles that result from the loss of electrons.

A **complex ion** or **polyatomic ion** is a group of bonded atoms that bear a positive or negative charge.

The **electron** is a subatomic particle with a mass of 9.109×10^{-28} g and a charge of -1.60×10^{-19} coulomb (represented by a relative charge of -1). It is present in all atoms.

Ionic compounds consist of collections of anions and cations in sufficient number to achieve electrical neutrality. The ions are attracted by electrostatic forces.

Isotopes of an element have the same number of protons but differ in the number of neutrons.

The **mass number** of an isotope is equal to the sum of the number of protons and neutrons in the nucleus of the atom.

Molecules are neutral units of atoms that remain associated to form units that possess properties of the substance.

A **molecular formula** is a representation of a molecule that indicates the number and type of each atom present in a molecule.

A **neutron** is a neutral subatomic particle that is in the nucleus of an atom. It has a mass of 1 amu.

The **nucleus** is a region in the center of an atom that contains the protons and neutrons.

A **proton** is a subatomic particle with a mass of 1 amu and a charge of 1.60×10^{-19} coulomb (represented by a relative charge of $+1$).

Questions and Problems

Terminology

3.1 Explain the difference between each of the following terms.
  (a) electron and proton
  (b) proton and neutron
  (c) mass number and atomic number
  (d) cation and anion
  (e) neutron and nucleus

3.2 Using your own words, describe the arrangement of atomic particles in an atom.

3.3 Explain how modern atomic theory differs from Dalton's atomic theory.

Properties of the Atom

3.4 Why is the mass of the hydrogen atom essentially equal to the mass of a proton?

3.5 Calculate the mass in grams of an atom of each of the following.
  (a) $^{16}_{8}O$ (b) $^{32}_{16}S$ (c) $^{65}_{30}Zn$ (d) $^{238}_{92}U$

3.6 Calculate the mass of an atom in amu from the following mass in grams.
  (a) an isotope of magnesium whose mass is 3.98×10^{-23}
  (b) an isotope of oxygen whose mass is 2.99×10^{-23}
  (c) an isotope of uranium whose mass is 3.90×10^{-22}

3.7 An atom of argon has a mass of 6.63×10^{-23} g. How many atoms are in a 40.0 g sample of argon?

3.8 The atomic radius of arsenic is $1.21\,\text{Å}$. What is its radius in cm?

3.9 Nanometers are also used in measuring the radii of atoms. What is the radius in nanometers of an atom whose atomic radius is $1.10\,\text{Å}$?

Subatomic Particles

3.10 Prepare a chart giving the relative weight and charge of the three subatomic particles.

3.11 Why are relative weights of subatomic particles used rather than the absolute masses?

3.12 Why is the mass of an electron considered to be zero on a relative mass basis?

3.13 Where are each of the subatomic particles located in the atom?

3.14 What are the sizes of the nucleus and the atom?

3.15 What relationship exists between the number of electrons and protons in an atom?

Elemental Symbols

3.16 What is the atomic number of an element? How is it represented in a symbol?

3.17 What is the mass number of an element? How is it represented in a symbol?

3.18 Indicate the number of each subatomic particle present in the following species.
  (a) $^{7}_{3}Li$ (b) $^{31}_{15}P$ (c) $^{107}_{47}Ag$ (d) $^{59}_{27}Co$
  (e) $^{69}_{31}Ga$ (f) $^{201}_{80}Hg$ (g) $^{199}_{80}Hg$ (h) $^{142}_{58}Ce$
  (i) $^{22}_{11}Na$ (j) $^{17}_{8}O$

3.19 Write the symbol for each of the following elements that contain the indicated number of protons and neutrons.
  (a) fluorine, 9 protons and 10 neutrons
  (b) silicon, 14 protons and 16 neutrons
  (c) silicon, 14 protons and 14 neutrons
  (d) iron, 26 protons and 30 neutrons
  (e) uranium, 92 protons and 145 neutrons
  (f) potassium, 19 protons and 22 neutrons

3.20 The radioisotope $^{226}_{88}Ra$ is used in cancer therapy. Indicate the number of protons, electrons, and neutrons in this isotope.

3.21 The radioisotope $^{60}_{27}Co$ is used in cobalt sources in cancer therapy. How many protons, electrons, and neutrons are in this radioisotope?

3.22 Fill in the missing quantities in each of the following.

Symbol	Mass number	Atomic number	Electrons	Protons	Neutrons
$^{14}_{7}N$					
	14	6			
		20			20
	79		35		
	35			17	

3.23 A student writes NO as the symbol for the element nobelium. Explain what is incorrect about the symbol. Explain what the symbol actually means.

Isotopes

3.24 Explain the significance of the symbols used to write the three isotopes of hydrogen: ^1_1H, ^2_1H, and ^3_1H.

3.25 The $^{14}_6C$ isotope content of ancient wooden objects is used to determine the age of the object. In what way does this isotope differ from $^{12}_6C$?

3.26 The radioisotope $^{131}_{53}I$ is used to treat cancer of the thyroid. How does this isotope differ from $^{127}_{53}I$, which is required for normal functioning of the thyroid gland?

3.27 The two naturally occurring isotopes of bromine are designated $^{79}_{35}Br$ and $^{81}_{35}Br$. Explain how these two isotopes differ.

Molecules

3.28 Indicate which of the following elements exist as molecules. How many atoms are contained in each molecule?

(a) He (b) H (c) P (d) N (e) O
(f) S (g) F (h) Cl

3.29 What is meant by each of the following formulas?

(a) H_2S (b) HCl (c) H_2SO_4
(d) C_2H_6O (e) C_2H_2 (f) $C_6H_{12}O_6$
(g) H_2O_2 (h) N_2

3.30 A molecule of TNT contains seven carbon atoms, five hydrogen atoms, three nitrogen atoms, and six oxygen atoms. Write the molecular formula of TNT.

3.31 A molecule of octane contains eight carbon atoms and eighteen hydrogen atoms. What is the molecular formula of octane?

3.32 Nicotine has the molecular formula $C_{10}H_{14}N_2$. What does this formula mean to you?

3.33 Table sugar has the molecular formula $C_{12}H_{22}O_{11}$. What does this formula mean?

3.34 The butane used in butane lighters has four carbon atoms and ten hydrogen atoms per molecule. Write the molecular formula of butane.

Ions

3.35 Sulfur can gain two electrons to form an ion. What is the symbol of the ion? How many electrons does this ion contain?

3.36 Manganese can lose two electrons to form an ion. Write the symbol for this ion. How many electrons does the ion contain?

3.37 Phosphorus can gain three electrons to form an ion. Write the symbol for this ion. How many electrons does the ion contain?

3.38 What is the difference between the ferrous ion and the ferric ion?

3.39 Cerium can form two ions. One results from the loss of three electrons and the other from the loss of four electrons. Write the symbols of these two ions. How many electrons does each ion contain?

3.40 Write the proper symbol for each of the following ions.

(a) oxide ion (b) sulfide ion
(c) iodide ion (d) chloride ion
(e) fluoride ion (f) bromide ion
(g) sodium ion (h) calcium ion
(i) potassium ion (j) magnesium ion
(k) zinc ion (l) lithium ion

3.41 Write the proper symbol for each of the following ions.

(a) sulfate ion (b) phosphate ion
(c) hydroxide ion (d) ammonium ion
(e) cyanide ion (f) carbonate ion
(g) nitrate ion (h) bicarbonate ion

3.42 What is the name of each of the following ions?

(a) O^{2-} (b) Cl^- (c) S^{2-} (d) Li^+
(e) K^+ (f) Ca^{2+}

3.43 What is the name of each of the following complex ions?

(a) OH^- (b) NH_4^+ (c) NO_3^- (d) SO_3^{2-}
(e) SO_4^{2-} (f) CN^- (g) ClO_4^- (h) ClO^-

3.44 Name the ions that are obtained from each of the following elements.

(a) iron (b) copper

Ionic Compounds

3.45 Write the correct formula for each of the following compounds.

(a) lithium fluoride (b) zinc oxide
(c) sodium cyanide (d) magnesium fluoride
(e) zinc cyanide (f) sodium nitrate
(g) sodium carbonate (h) potassium sulfide

3.46 Write the correct formulas for compounds containing the following ions.

(a) Fe^{3+} and Cl^- (b) Na^+ and OH^-
(c) Mg^{2+} and OH^- (d) Cd^{2+} and S^{2-}
(e) K^+ and Br^- (f) Li^+ and N^{3-}
(g) Ba^{2+} and NO_3^- (h) Cs^+ and ClO_4^-

3.47 Name each of the following ionic compounds.
  (a) $Ca(OH)_2$ (b) $LiClO_4$ (c) Na_3PO_4
  (d) K_2SO_4 (e) KNO_3 (f) NH_4NO_2
  (g) $MgCl_2$ (h) $LiCN$

3.48 Write the correct formula for each of the following compounds.
  (a) iron(II) chloride   (b) copper(I) oxide
  (c) lead(II) sulfide   (d) mercury(II) bromide
  (e) ferric oxide   (f) cupric sulfide
  (g) ferrous sulfide   (h) cuprous oxide

3.49 Determine how many atoms of each element are present in each of the following.

  (a) $Ca(NO_3)_2$ (b) Na_3PO_4 (c) $Ba(ClO_4)_2$
  (d) $AlPO_4$ (e) $Sc_2(SO_4)_3$ (f) $Mg(SO_3)_2$
  (g) $KMnO_4$ (h) $Fe_2(SO_4)_3$

3.50 Antacids are used to neutralize excess stomach acid. Write the formula of the indicated component of the listed antacids.
  (a) calcium carbonate found in Tums
  (b) sodium bicarbonate found in Alka-Seltzer
  (c) magnesium hydroxide found in Milk of Magnesia
  (d) aluminum hydroxide found in Maalox

3.51 The antacid in Rolaids is $NaAl(OH)_2CO_3$. Show that this substance is electrically neutral.

Chemical Arithmetic and Stoichiometry

This research chemist is weighing material required by the stoichiometry of a reaction. The balance is within a dry box, which excludes both oxygen and water.

4.1
Weights of Atoms, Molecules, and Ionic Compounds

In the preceding chapters you learned some of the concepts of chemistry. Atoms, molecules, and ions make up matter. Each material has a mass and that mass is conserved in chemical reactions. Although chemists may think in terms of this submicroscopic scale of matter, they find it necessary to work with huge numbers of particles of matter in any laboratory sample. Chemists have worked out ways of translating very small quantities of the submicroscopic realm into the everyday world of grams.

Atomic Weight

Since the majority of elements have at least two naturally occurring isotopes, a sample of an element consists of a mixture of the different isotopes. Therefore, the mass of a sample of any element is an average of the masses of its isotopes, weighted by the relative abundance of those isotopes. *The **atomic weight** of an element is the weighted average of the masses of the naturally occurring isotopes of the element expressed in atomic mass units.* The atomic weight of each element is listed on the inside cover of this text.

The atomic weight scale is based on the assignment of 12.0000 amu to the mass of the most common isotope of carbon. This isotope is known as carbon-12. Each atom of carbon-12 contains six neutrons and six protons.

The calculation of a weighted average is similar to the procedure that may have been followed in determining your grade in a course. Suppose your two examinations account for 20% and 30% of your final grade, while the final examination is 50% of your final grade. If you earned 85 on the first exam, 80 on the second exam, and 90 on the final exam, your course grade would be 86. The score is obtained by multiplying each grade by the decimal equivalent of the weighting percentage and then summing.

$$85 \times 0.20 = 17$$
$$80 \times 0.30 = 24$$
$$90 \times 0.50 = \underline{45}$$
$$86$$

Atomic weights can be calculated by multiplying the mass number of each isotope by its fractional abundance. For example, boron consists of 19.6% boron of mass 10.013 and 80.4% boron of mass 11.009 amu.

$$10.013 \times .196 = 1.96$$
$$11.009 \times .804 = \underline{8.85}$$
$$\text{atomic weight} = 10.81$$

Molecular Weight

In discussing matter and its reactions, we will be considering compounds more frequently than elements because there are many more of them. It is convenient to describe the mass of a molecule in terms of its molecular weight. *The **molecular weight** of a molecule is the sum of the masses in amu of its component atoms.* Thus, molecular weights are based on the same reference as atomic weights. The molecule methane, whose molecular formula is CH_4, has a molecular weight of 16.0 amu.

mass of 1 carbon atom	$= 1 \times 12.0$ amu	$= 12.0$ amu
mass of 4 hydrogen atoms	$= 4 \times 1.0$ amu	$= \underline{4.0 \text{ amu}}$
mass of 1 methane molecule		$= 16.0$ amu

Example 4.1 ──────────────────────────────────

What is the molecular weight of carbon dioxide, CO_2?

Solution. The molecular weight of CO_2 is the sum of the atomic weights of the constituent elements taking into account the number of each kind of atom present in the molecule.

$$\begin{array}{ll}
\text{mass of 1 carbon atom } = 1 \times 12.0 \text{ amu} = 12.0 \text{ amu} \\
\underline{\text{mass of 2 oxygen atoms } = 2 \times 16.0 \text{ amu} = 32.0 \text{ amu}} \\
\text{mass of 1 } CO_2 \text{ molecule} \qquad\qquad\quad = 44.0 \text{ amu}
\end{array}$$

Example 4.2 ──────────────────────────────────

What is the molecular weight of vitamin C, $C_6H_8O_6$?

Solution. Sum the atomic weights of the constituent atoms, taking into account the number of atoms of each kind present in the molecule.

$$\begin{array}{ll}
\text{mass of 6 carbon atoms } \quad = 6 \times 12.0 \text{ amu} = 72.0 \text{ amu} \\
\text{mass of 8 hydrogen atoms } = 8 \times 1.0 \text{ amu} = 8.0 \text{ amu} \\
\underline{\text{mass of 6 oxygen atoms } \quad = 6 \times 16.0 \text{ amu} = 96.0 \text{ amu}} \\
\text{mass of 1 } C_6H_8O_6 \text{ molecule} \qquad\qquad\quad = 176.0 \text{ amu}
\end{array}$$

The molecular weight is 176.0 amu.

[Additional examples may be found in **4.11–4.14** at the end of the chapter.]

Formula Weight

Ionic substances do not exist as molecules, but rather as collections of ions in a definite ratio described by a formula giving the simplest ratio of ions present in the compound (Sec. 3.7). *The* **formula weight** *of an ionic compound is the sum of the atomic weights of the atoms indicated by a formula unit of the substance.*

Example 4.3 ──────────────────────────────────

What is the formula weight of calcium carbonate, $CaCO_3$?

Solution. Calcium carbonate is an ionic compound consisting of calcium ions and carbonate ions in a one-to-one ratio. The formula weight is obtained by summing the atomic weights.

$$\begin{array}{ll}
\text{mass of 1 calcium atom } = 1 \times 40.1 \text{ amu} = 40.1 \text{ amu} \\
\text{mass of 1 carbon atom } \quad = 1 \times 12.0 \text{ amu} = 12.0 \text{ amu} \\
\underline{\text{mass of 3 oxygen atoms } = 3 \times 16.0 \text{ amu} = 48.0 \text{ amu}} \\
\text{mass of 1 } CaCO_3 \text{ unit} \qquad\qquad\qquad = 100.1 \text{ amu}
\end{array}$$

The formula weight is 100.1 amu.

[Additional examples may be found in **4.15–4.18** at the end of the chapter.]

4.2
Percent Composition

Percent means parts per hundred. Thus, if you scored 20 correct answers on a true–false examination consisting of 25 questions, you had 80% correct answers. This is determined by dividing the number of correct answers by the total number of questions and multiplying the quotient by 100.

$$\frac{20 \text{ correct answers}}{25 \text{ questions}} \times 100 = 80\% \text{ correct answers}$$

The percent composition of a compound is calculated and defined in a similar manner. *The* **percent composition** *of a given element in a compound is equal to the mass of that element divided by the total mass of all of the elements in the compound with the quotient multiplied by 100.* For example, an 18.00 g sample of water contains 2.00 g of hydrogen and 16.00 g of oxygen. The mass percent of hydrogen in water is 11.1 and the mass percent of oxygen is 88.9.

$$\frac{2.00 \text{ g hydrogen}}{18.00 \text{ g water}} \times 100 = 11.1\% \text{ hydrogen}$$

$$\frac{16.00 \text{ g oxygen}}{18.00 \text{ g water}} \times 100 = 88.9\% \text{ oxygen}$$

Example 4.4

A 250 mg tablet of vitamin C contains 102 mg of carbon, 12 mg of hydrogen, and 136 mg of oxygen. What is the mass percent composition of vitamin C?

Solution. To solve this problem, the mass of each individual element must be divided by the total mass and the quotient multiplied by 100.

$$\% \text{ carbon} = \frac{102 \text{ mg carbon}}{250 \text{ mg vitamin C}} \times 100 = 40.8\%$$

$$\% \text{ hydrogen} = \frac{12 \text{ mg hydrogen}}{250 \text{ mg vitamin C}} \times 100 = 4.8\%$$

$$\% \text{ oxygen} = \frac{136 \text{ mg oxygen}}{250 \text{ mg vitamin C}} \times 100 = 54.4\%$$

The percent composition can also be calculated from the molecular formula. Consider the molecule ethyl alcohol, C_2H_6O, whose molecular weight is 46.0 amu. Of this weight, 24.0 amu is carbon, 6.0 amu is hydrogen, and 16.0 amu is oxygen. The percent composition is as follows.

$$\% \text{ C} = \frac{24.0 \text{ amu carbon}}{46.0 \text{ amu ethyl alcohol}} \times 100 = 52.2\%$$

$$\% \text{ H} = \frac{6.0 \text{ amu hydrogen}}{46.0 \text{ amu ethyl alcohol}} \times 100 = 13.0\%$$

$$\% \text{ O} = \frac{16.0 \text{ amu oxygen}}{46.0 \text{ amu ethyl alcohol}} \times 100 = 34.8\%$$

Example 4.5 ————————————————————————————

What is the percent composition of glucose, $C_6H_{12}O_6$?

Solution. The molecular weight is 180.0 amu.

mass of 6 carbon atoms	$= 6 \times 12.0$ amu $=$	72.0 amu
mass of 12 hydrogen atoms	$= 12 \times 1.0$ amu $=$	12.0 amu
mass of 6 oxygen atoms	$= 6 \times 16.0$ amu $=$	96.0 amu
mass of glucose molecule		$= 180.0$ amu

The percent composition is

$$\% \text{ C} = \frac{72.0 \text{ amu carbon}}{180.0 \text{ amu glucose}} \times 100 = 40.0\%$$

$$\% \text{ H} = \frac{12.0 \text{ amu hydrogen}}{180.0 \text{ amu glucose}} \times 100 = 6.7\%$$

$$\% \text{ O} = \frac{96.0 \text{ amu oxygen}}{180.0 \text{ amu glucose}} \times 100 = 53.3\%$$

[Additional examples may be found in **4.19–4.25** at the end of the chapter.]

4.3
The Mole

In the chemical laboratory, substances are weighed in terms of grams rather than atomic mass units. Recall that 1 amu is 1.67×10^{-24} g. This value is incredibly small, and no balance can detect quantities this small. Thus, the chemist never directly weighs on a balance a single atom or even millions, billions, trillions, or quadrillions of atoms. Approximately 10^{16} atoms would be required to be detected on the most sensitive balance. Accordingly, chemists have devised a system for comparing quantities of a substance that does not depend on single atoms or molecules but on large collections of atoms or molecules.

If we weigh atoms in ratios according to their atomic weights, we will obtain samples containing the same number of atoms. The atomic weights of carbon and sulfur are 12 and 32, respectively. Thus, if we have 12 g of carbon and 32 g of sulfur, the two samples will contain the same number of atoms (Figure 4.1). The sulfur atom, whose atomic weight is 32, is 2.67 times heavier than the carbon atom, whose atomic weight is 12. Therefore, when we weigh 2.67 times as much sulfur as carbon, we obtain exactly the same number of atoms.

If we weigh molecules in ratios according to their molecular weights, the samples will contain the same number of molecules. For example, 16 g of methane and 44 g of carbon dioxide contain the same number of molecules, since their molecular weights are 16 and 44 amu, respectively. The ratio of the molecular weights is 16 to 44 for methane and carbon dioxide, and therefore any quantities of these two compounds in this ratio contain the same number of molecules. Thus, 8 g of methane and 22 g of carbon dioxide contain the same number of molecules.

The system used by chemists to deal with samples of matter is based on the number of structural units that they contain. *The* **mole** *is the quantity in grams*

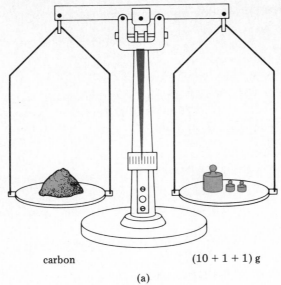

carbon   (10 + 1 + 1) g

(a)

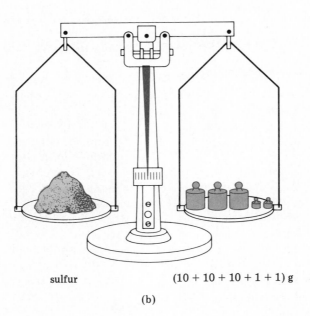

sulfur   (10 + 10 + 10 + 1 + 1) g

(b)

Figure 4.1
The Weight of Equal Numbers of Atoms.
In experiment (a) the sample of carbon weighs 12 g. In experiment (b) the sample of sulfur weighs 32 g. Both samples contain the same number of atoms because the ratio of the masses of the individual atoms is the same as the ratio of the masses of the samples.

equal to the atomic weight in the case of atoms or the molecular weight in the case of molecules. Thus, a mole of carbon has a mass of 12 g; a mole of sulfur has a mass of 32 g. A mole of methane, CH_4, has a mass of 16 g; a mole of carbon dioxide has a mass of 44 g.

Although there are no discrete units in ionic compounds that correspond to those represented in formulas such as $CaCO_3$, it is nevertheless convenient to extend the mole concept to ionic compounds. **A mole of an ionic compound has a mass in grams corresponding to the formula weight of the substance.** The formula weight of $CaCO_3$ is 100.1 amu, and the mass of one mole is 100.1 g.

Example 4.6

How many moles of gold atoms are present in a 30.000 kg ingot of gold?

Solution. The atomic weight of gold is 197.0. Therefore, one mole of gold atoms has a mass of 197.0 g. The number of moles in the ingot is calculated as follows.

$$30{,}000\ \text{g Au} \times \frac{1\ \text{mole Au atoms}}{197.3\ \text{g Au}} = 152.0\ \text{moles Au atoms}$$

Example 4.7

An average adult exhales 1000 g of carbon dioxide per day. How many moles are exhaled?

Solution. In Example 4.1, the molecular weight of carbon dioxide, CO_2, was determined to be 44.0 amu. Therefore, the number of moles in 1000 g is

$$1000\ \text{g } CO_2 \times \frac{1\ \text{mole } CO_2}{44.0\ \text{g } CO_2} = 22.7\ \text{moles } CO_2$$

Example 4.8

How many moles of calcium carbonate are contained in 25 g of the compound?

Solution. In Example 4.3 the formula weight of this ionic compound was determined to be 100.1 amu. Thus, the mass of one mole of $CaCO_3$ is 100.1 g. The number of moles in the sample is

$$25\ \text{g } CaCO_3 \times \frac{1\ \text{mole } CaCO_3}{100.1\ \text{g } CaCO_3} = 0.25\ \text{mole } CaCO_3$$

Example 4.9

Calculate the mass of 0.00100 mole of vitamin C, $C_6H_8O_6$.

Solution. In Example 4.2 the molecular weight of vitamin C was established to be 176.0 amu. Therefore, one mole has a mass of 176.0 g, and the mass of 0.00100 mole can be calculated as follows.

$$0.00100\ \text{mole } C_6H_8O_6 \times \frac{176.0\ \text{g } C_6H_8O_6}{1\ \text{mole } C_6H_8O_6} = 0.176\ \text{g } C_6H_8O_6$$

[Additional examples may be found in **4.26–4.34** at the end of the chapter.]

4.4
Avogadro's Number

The mole is an important and special measurement in chemistry. **A mole of any substance has the same number of structural units as a mole of any other substance.** Thus, a mole of carbon contains the same number of atoms as a mole of sulfur. A mole of ethyl alcohol has the same number of molecules as a mole of glucose.

The number of structural units in a mole is 6.02×10^{23} and is known as **Avogadro's number.** Amadeo Avogadro was a nineteenth century scientist who contributed to the concept of atomic weights.

A mole of helium contains 6.02×10^{23} helium atoms. A mole of argon has 6.02×10^{23} argon atoms. Since the atomic weights of helium and argon are 4 and 40 amu, respectively, a mole of helium weighs 4 g, whereas a mole of argon weighs 40 g. **The atomic weight in grams of any element, or one mole, contains Avogadro's number of atoms.**

For compounds consisting of molecules, a mole contains 6.02×10^{23} molecules. A mole of carbon dioxide contains 6.02×10^{23} carbon dioxide molecules. A mole of glucose contains 6.02×10^{23} glucose molecules. The weight of a mole, of course, depends on the substance. **The molecular weight in grams of any compound, or one mole, contains Avogadro's number of molecules.**

The subscripts in the formulas of compounds represent the number of atoms of each element present in a molecule of the substance. The subscripts also indicate the number of moles of the atoms of the elements present in a mole of molecules. Thus, in one mole of carbon dioxide there are one mole of carbon atoms and two moles of oxygen atoms. One mole of CO_2 contains 6.02×10^{23} atoms of carbon; one mole of CO_2 contains $2 \times (6.02 \times 10^{23})$ or 1.20×10^{24} atoms of oxygen.

A mole of an ionic compound contains 6.02×10^{23} formula units. Thus, one mole of $CaCO_3$ contains 6.02×10^{23} calcium ions and 6.02×10^{23} carbonate ions, which account for the 6.02×10^{23} formula units of $CaCO_3$. In one mole of sodium carbonate, Na_2CO_3, there are $2 \times (6.02 \times 10^{23})$ sodium ions and 6.02×10^{23} carbonate ions.

Example 4.10 —————————————————————————

How many molecules of vitamin C are in a 500 mg tablet of vitamin C?

Solution. In order to determine the number of molecules, it is first necessary to determine the number of moles present in the tablet, which has a mass of 0.500 g.

$$0.500 \text{ g } C_6H_8O_6 \times \frac{1 \text{ mole } C_6H_8O_6}{176.0 \text{ g } C_6H_8O_6} = 0.00284 \text{ mole } C_6H_8O_6$$

The number of molecules of $C_6H_8O_6$ in a mole is 6.02×10^{23}. In the tablet, which contains 2.84×10^{-3} mole, the number of molecules is

$$2.84 \times 10^{-3} \text{ mole } C_6H_8O_6 \times \frac{6.02 \times 10^{23} \text{ molecules } C_6H_8O_6}{1 \text{ mole } C_6H_8O_6}$$
$$= 1.71 \times 10^{21} \text{ molecules } C_6H_8O_6$$

———

[Additional examples may be found in **4.35–4.38** at the end of the chapter.]

In Section 2.7, word equations were used to express the chemical reaction of magnesium and oxygen in a flashbulb to yield magnesium oxide. Now we can use our knowledge of chemical formulas and the quantitative calculations derivable from such formulas to make these equations more precise. By substituting the correct chemical formula for each reactant and product into the equation, we can obtain a balanced chemical equation.

$$2\,Mg(s) + O_2(g) \longrightarrow 2\,MgO(s)$$

In a balanced chemical equation there are numbers placed to the left of the chemical formulas. These numbers, called **coefficients,** *indicate the number of units of the reactants and products of the reaction.* For a single unit of a reactant or product, the coefficient 1 is not shown but implied by the equation. The coefficients are chosen so that the atoms of each element, either in the elemental form or in a compound, appear in equal numbers on both sides of the equation. Thus, in the balanced equation shown there are two oxygen atoms in the one oxygen molecule of the reactant and two oxygen atoms in the two units of magnesium oxide of the product.

The physical states of the reactants and products under the reaction conditions may be shown by writing (s), (l), or (g) to the right of the symbol of each substance. The use of these symbols is optional. Many reactions occur between substances dissolved in water for which the symbol (aq), meaning aqueous solution may be used.

Now let's consider how a balanced chemical equation is obtained. Although there is no hard and fast set of steps that can be used to balance chemical equations, there are some helpful general rules. With practice and the general guidelines provided, you can balance the equations given in this text. A list of these rules is given in Table 4.1.

4.5
The Balanced Chemical Equation

Table 4.1

Guidelines to Balance Chemical Equations

1. Write the correct formulas for the reactants and products on the appropriate sides of the equation. Once the correct formula is written, do not alter it or any of its subscripts during the balancing process.
2. Disregarding hydrogen, oxygen, and polyatomic ions, find the molecule containing the largest number of atoms of a single element. Balance the number of this element by placing the proper coefficient in front of the molecule containing this element on the other side of the equation. Remember that a coefficient placed in front of a molecular formula multiplies every atom contained in the molecule.
3. Proceed to balance the atoms of other elements by the same process. Check to see if in balancing one element, others have become unbalanced. Readjust the coefficients on both sides of the equation to achieve the necessary balance.
4. Balance polyatomic ions on each side of the equation as a single unit.
5. Balance hydrogen and oxygen not previously considered in polyatomic ions.
6. Check all coefficients to insure that they are the lowest possible whole numbers. If the coefficients are fractions, multiply by a number to make the fraction a whole number. All other coefficients must be multiplied by the same quantity to maintain a balanced equation. If all the coefficients are divisible by a common factor, do so to achieve the lowest possible whole number.
7. Check the entire equation to ensure that all atoms are balanced.

Figure 4.2
Balancing an Equation.
In (a) the number of atoms of oxygen in the reactants exceeds the oxygen in the product. In (b) the number of atoms of magnesium in the product exceeds the magnesium in the reactants. In (c) the number of atoms in the product and reactants are balanced.

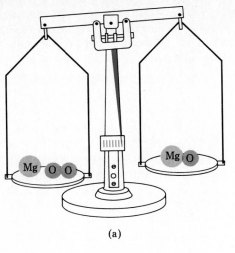

(a)

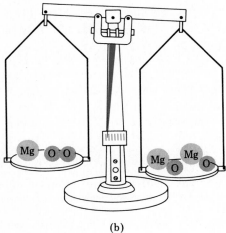

(b)

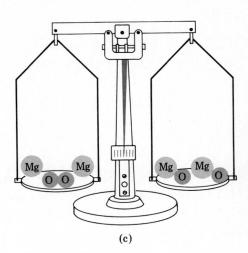

(c)

4.6
Practice in Balancing Equations

Example 4.11

Write a balanced equation for the conversion of magnesium and oxygen into magnesium oxide.

Solution. We start by properly representing the chemical constitution of the reactants and the product. Oxygen exists as a diatomic molecule and must be so represented in a chemical equation. Magnesium oxide consists of Mg^{2+} and O^{2-} ions in equal numbers and is represented as MgO (Figure 4.2). We then write

$$Mg + O_2 \longrightarrow MgO$$

To balance this equation we note that there are two oxygen atoms on the left but only one on the right. The equation cannot be balanced by changing the formula of magnesium oxide to MgO_2 as such a formula is inconsistent with the experimental facts. However, a 2 can be placed in front of MgO to indicate that two units of MgO are formed (Figure 4.2). When a coefficient 2 is placed in front of MgO, it means that there are two magnesium and two oxygen atoms. Now we have two magnesium atoms on the right but only one on the left. This situation is righted by placing a 2 as the coefficient in front of magnesium (Figure 4.2).

$$2\,Mg + O_2 \longrightarrow 2\,MgO$$

The equation is now balanced.

Note that there are other possible balanced equations that could be written.

$$Mg + \tfrac{1}{2}O_2 \longrightarrow MgO$$

$$4\,Mg + 2\,O_2 \longrightarrow 4\,MgO$$

$$6\,Mg + 3\,O_2 \longrightarrow 6\,MgO$$

These equations, however, are merely multiples of the equation in which the whole number coefficients are in the smallest ratio possible. Only the balanced equation with the smallest whole number coefficients is accepted.

Example 4.12

Aluminum hydroxide reacts with sulfuric acid to yield aluminum sulfate and water according to the following unbalanced equation.

$$Al(OH)_3 + H_2SO_4 \longrightarrow Al_2(SO_4)_3 + H_2O$$

Write the balanced equation.

Solution. Since there are two atoms of aluminum in the aluminum sulfate, it is necessary to place a coefficient 2 in front of the aluminum hydroxide. Furthermore, the three sulfate units in the aluminum sulfate are balanced by placing a coefficient 3 in front of the sulfuric acid.

$$2\,Al(OH)_3 + 3\,H_2SO_4 \longrightarrow Al_2(SO_4)_3 + H_2O$$

Only the oxygen not incorporated in the sulfate and the hydrogen remain unbalanced. There are six oxygen atoms contained in the two units of aluminum hydroxide, which may be balanced by a coefficient of 6 for H_2O.

$$2\,Al(OH)_3 + 3\,H_2SO_4 \longrightarrow Al_2(SO_4)_3 + 6\,H_2O$$

This coefficient also serves to balance the hydrogen atoms on both sides of the equation. There are six hydrogen atoms in the two units of aluminum hydroxide and six hydrogen atoms in the three units of sulfuric acid, which balance the twelve hydrogen atoms in the six water molecules.

Example 4.13

The hydrocarbon butane, C_4H_{10}, burns in oxygen to yield carbon dioxide and water.

$$C_4H_{10} + O_2 \longrightarrow CO_2 + H_2O$$

Balance the equation.

Solution. The molecular formula of butane has four carbon atoms and ten hydrogen atoms. Using guideline 2, we start our balancing process with carbon. A coefficient 4 is placed in front of carbon dioxide.

$$C_4H_{10} + O_2 \longrightarrow 4\,CO_2 + H_2O$$

Next the hydrogen is balanced as there are no other elements besides oxygen that need to be balanced. Hydrogen is chosen over oxygen as it appears in only one compound on each side of the equation. A coefficient 5 placed in front of H_2O will balance the ten hydrogen atoms of butane.

$$C_4H_{10} + O_2 \longrightarrow 4\,CO_2 + 5\,H_2O$$

Next the oxygen atoms may be balanced. The oxygen atoms in the products now total thirteen, eight in the $4\,CO_2$ and five in the $5\,H_2O$. In order to balance the diatomic oxygen molecule, a coefficient of $\frac{13}{2}$ is necessary.

$$C_4H_{10} + \tfrac{13}{2}\,O_2 \longrightarrow 4\,CO_2 + 5\,H_2O$$

Finally, the fractional coefficient is eliminated by multiplying all coefficients by 2.

$$2\,C_4H_{10} + 13\,O_2 \longrightarrow 8\,CO_2 + 10\,H_2O$$

[Additional examples may be found in **4.39–4.42** at the end of the chapter.]

4.7
Stoichiometry and the Balanced Equation

A balanced chemical equation can be used to provide a great deal more information than is directly evident from the elemental symbols and coefficients. The exact quantities of matter required in chemical reactions are provided from the equation. *The mathematical calculation of the quantities of reactants and products in a reaction given by a chemical equation is called* **stoichiometry.** The quantities of reactants and products involved in the reaction can be calculated on a mole or mass basis. The information derivable from an equation can be summarized as follows.

$$2\,H_2(g) + O_2(g) \longrightarrow 2\,H_2O(g)$$

1. The reactants are hydrogen and oxygen in the gaseous state. The product is gaseous water.
2. Two molecules of hydrogen and one molecule of oxygen react to produce two molecules of water.
3. Two moles of hydrogen molecules and one mole of oxygen molecules combine to produce two moles of water molecules.
4. In terms of grams, 4 g of hydrogen and 32 g of oxygen combine to yield 36 g of water. These quantities are the gram equivalents of the moles involved.

$$2 \text{ moles } H_2 \times \frac{2 \text{ g } H_2}{1 \text{ mole } H_2} = 4 \text{ g } H_2$$

$$1 \text{ mole } O_2 \times \frac{32 \text{ g } O_2}{1 \text{ mole } O_2} = 32 \text{ g } O_2$$

$$2 \text{ moles } H_2O \times \frac{18 \text{ g } H_2O}{1 \text{ mole } H_2O} = 36 \text{ g } H_2O$$

Figure 4.3
An Outline of the Mole
Method to Solve Stoichiome-
try Problems.

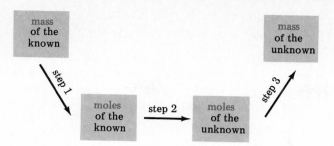

Methods of Solving Stoichiometry Problems

Although there are a variety of methods that could be used to solve stoichiometry problems, the best established one is the mole method. The mole method is rooted in the factor unit method, which was described in Chapter 1. The number of steps involved in working problems by the mole method depends on the units in which the known quantities are measured and the units required of the unknown. The first step is to convert the known quantities into their molar quantities. Then the stoichiometric relationship between the known substances and the desired unknown is established from the balanced equation. From this relationship, the moles of the unknown are determined. Finally, the moles of the unknown are converted into the desired units. The final answer must be expressed to the proper number of significant figures. The three steps of the mole method are outlined in Table 4.2 and illustrated in Figure 4.3.

Table 4.2

A Summary of the Mole Method

1. From the mass of the reactant or product given, convert to the mole equivalent.
2. Examine the balanced equation and calculate the moles of the desired unknown that can be obtained from the moles of the substance calculated in step 1.
3. From the moles of the desired substance calculated in step 2, determine the equivalent quantity in terms of mass.

The Mole Ratio

The central step, step 2, of the mole method depends on relating the number of moles of one substance to the number of moles of another substance. In order to do this, a mole ratio is used. *The **mole ratio** is a ratio between the number of moles of two substances involved in a chemical reaction.* Thus, for the reaction of hydrogen and nitrogen to give ammonia,

$$N_2 + 3\,H_2 \longrightarrow 2\,NH_3$$

there are six mole ratios that may be written. All of these ratios contain the coefficients of the corresponding substances in the equation.

$$\frac{1\ \text{mole}\ N_2}{3\ \text{moles}\ H_2} \quad \text{or} \quad \frac{3\ \text{moles}\ H_2}{1\ \text{mole}\ N_2}$$

$$\frac{1\ \text{mole}\ N_2}{2\ \text{moles}\ NH_3} \quad \text{or} \quad \frac{2\ \text{moles}\ NH_3}{1\ \text{mole}\ N_2}$$

$$\frac{3 \text{ moles } H_2}{2 \text{ moles } NH_3} \quad \text{or} \quad \frac{2 \text{ moles } NH_3}{3 \text{ moles } H_2}$$

As indicated, there are only three mole ratios that relate different pairs of substances. The remaining three are reciprocal ratios of the same quantities. Each of the mole ratios applies only to the reaction under consideration. Thus, for a reaction of nitrogen and hydrogen to produce hydrazine,

$$N_2 + 2 H_2 \longrightarrow N_2H_4$$

the mole ratios would be different.

To use the mole method for stoichiometry problems, it is necessary first to convert from the known number of moles of given starting substance to the number of moles of desired substance. These quantities can be related by the proper mole ratio.

moles of desired substance = mole ratio × moles of starting substance

The mole ratio then must be

$$\text{mole ratio} = \frac{\text{moles desired substance in balanced equation}}{\text{moles starting substance in balanced equation}}$$

Thus, by choosing the proper mole ratio, the units will be correct as will be your answer. The mole ratio, then, is the necessary factor of the factor unit method. Note that **the mole ratio is an exact number and carries any number of significant figures desired.** Thus, the number of significant figures in an answer depends only on the number of significant figures in the starting substance.

4.8
Mole–Mole Stoichiometry Problems

In this first type of stoichiometric calculation, you will find that the steps and math involved are easy. However, the principles established are essential to solution of more complex problems to be described in the next section. Therefore, make sure that you understand fully how to do the problems given in this section prior to trying the problems in the next section.

The problem is to find how to calculate the number of moles of one substance that will react with or be produced from a second substance. The necessary information is given in a balanced equation that establishes the mole relationship between all reactants and products. Recall that the number in front of each formula in an equation is called the coefficient. **The coefficient gives the number of moles of that substance in the balanced equation with respect to the number of moles of every other reactant or product in the equation.** Thus, in the reaction of nitrogen and hydrogen to produce ammonia, we may write

$$\underset{3 \text{ moles } H_2}{3 H_2} + \underset{1 \text{ mole } N_2}{N_2} \longrightarrow \underset{2 \text{ moles } NH_3}{2 NH_3}$$

Example 4.14

How many moles of ammonia, NH_3, are produced from 9 moles of hydrogen?

$$3 H_2 + N_2 \longrightarrow 2 NH_3$$

Solution. First establish the facts that the coefficients give you; 3 moles of hydrogen will yield 2 moles of ammonia. The mole ratio necessary for the solution of the problem must cancel the units of moles of hydrogen and leave moles of ammonia.

$$9 \text{ moles } H_2 \times \text{ mole ratio} = ? \text{ moles } NH_3$$

$$9 \text{ moles } H_2 \times \left(\frac{2 \text{ moles } NH_3}{3 \text{ moles } H_2} \right) = 6 \text{ moles } NH_3$$

Example 4.15

Determine the number of moles of oxygen required to burn completely 12 moles of ethane.

$$2\,C_2H_6 + 7\,O_2 \longrightarrow 4\,CO_2 + 6\,H_2O$$

Solution. The equation provides the information that 7 moles of oxygen are required to burn 2 moles of ethane. The ratio of the moles of the two substances in question is first established.

$$\text{mole ratio} = \frac{7 \text{ moles } O_2}{2 \text{ moles } C_2H_6}$$

This ratio, when multiplied by the quantity 12 moles of ethane, will give the answer in moles of oxygen.

$$12 \text{ moles } C_2H_6 \times \frac{7 \text{ moles } O_2}{2 \text{ moles } C_2H_6} = 42 \text{ moles } O_2$$

You could have quickly solved this problem by inspection of the equation and avoided the factor unit method. However, the factor unit method employing mole ratios is very helpful in solving the more difficult problems described in the next section.

[Additional examples may be found in **4.43–4.45** at the end of the chapter.]

Now you should be quite adept at solving stoichiometry problems. The technique of solving mass–mass stoichiometry problems is very similar to that described for mole–mole stoichiometry problems in the preceding section.

4.9
Mass–Mass Stoichiometry Problems

1. The mass of the known is converted to moles of the known.
2. The moles of the known and unknown are interrelated by a mole ratio.
3. The moles of the unknown are converted to mass.

Example 4.16

How many grams of oxygen are required to burn completely 1140 g of octane?

$$2\,C_8H_{18} + 25\,O_2 \longrightarrow 16\,CO_2 + 18\,H_2O$$

Solution. The number of moles of octane in the 1140 g is

$$1140 \text{ g } \cancel{C_8H_{18}} \times \frac{1 \text{ mole } C_8H_{18}}{114.0 \text{ g } \cancel{C_8H_{18}}} = 10.00 \text{ moles } C_8H_{18}$$

The number of moles of oxygen required to burn 10.00 moles C_8H_{18} is

$$10.00 \text{ moles } \cancel{C_8H_{18}} \times \frac{25 \text{ moles } O_2}{2 \text{ moles } \cancel{C_8H_{18}}} = 125.0 \text{ moles } O_2$$

Finally, the mass of oxygen required is

$$125.0 \text{ moles } \cancel{O_2} \times \frac{32.0 \text{ g } O_2}{1 \text{ mole } \cancel{O_2}} = 4000 \text{ g } O_2$$

Example 4.17

How many grams of carbon dioxide are required by a plant to produce 1.80 g of glucose by photosynthesis?

$$6 \, CO_2 + 6 \, H_2O \longrightarrow C_6H_{12}O_6 + 6 \, O_2$$

Solution. All three steps necessary to solve this problem may be combined in one mathematical expression.

$$1.80 \text{ g } \cancel{C_6H_{12}O_6} \times \underbrace{\frac{1 \text{ mole } \cancel{C_6H_{12}O_6}}{180.0 \text{ g } \cancel{C_6H_{12}O_6}}}_{\text{step 1}} \times \underbrace{\frac{6 \text{ moles } \cancel{CO_2}}{1 \text{ mole } \cancel{C_6H_{12}O_6}}}_{\text{step 2}} \times \underbrace{\frac{44.0 \text{ g } CO_2}{1 \text{ mole } \cancel{CO_2}}}_{\text{step 3}}$$

$$= 2.64 \text{ g } CO_2$$

[Additional examples may be found in **4.46–4.50** at the end of the chapter.]

Summary

The concept of a mole occupies a central position in making chemistry a quantitative science. The mole is a unit that is a count of the number of units of a substance present in matter. These units may be atoms, molecules, or ions. The number of units in a mole of a substance is called Avogadro's number.

In order to relate the count of the number of units of matter to mass as measured by balances, the individual mass of each unit of matter is needed. The atomic weight is used to measure the mass of atoms, the molecular weight to measure the mass of molecules, and the formula weight to measure the mass of ionic compounds.

The composition of a compound is given, in terms of the number of atoms present, by a formula.

The formula is used to calculate the composition by mass of each constituent atom.

A chemical equation is used to represent the reactants and products of a reaction. A balanced chemical equation provides information about the amounts of reactants and products involved. Calculations of the quantity of substances in a chemical reaction is called stoichiometry.

Learning Objectives

As a result of studying Chapter 4 you should be able to
- Calculate the atomic weight of a mixture of isotopes.
- Calculate the molecular weight of a molecule.
- Determine the formula weight of an ionic compound from its formula.

- Express the percent composition of a substance, given the chemical formula.
- Determine the number of moles and units of matter in a sample.
- Balance chemical equations.
- Calculate the number of moles and grams of substances involved in a chemical reaction.

New Terms

The **atomic weight** of an element is the weighted average of the masses of the naturally occurring isotopes of the element expressed in atomic mass units.

Avogadro's number is equal to the number of structural units of a substance in a mole.

A **coefficient** is the smallest whole number placed before a substance in a balanced chemical equation. It designates the number of units of that substance required to react with the other substance(s) involved in the reaction.

The **formula weight** is the sum of the atomic weights of the atoms indicated by the formula unit of an ionic compound.

A **mole** is a quantity in grams equal to the atomic weight in the case of atoms, molecular weight in the case of molecules, and formula weight in the case of ionic compounds. A mole contains Avogadro's number of structural units of matter.

The **mole ratio** is a ratio of the number of moles of two substances involved in a chemical reaction.

The **molecular weight** of a molecule is the sum of the atomic weights of the component atoms of the molecule.

The **percent composition** of an element in a compound is equal to the mass of that element present in the compound divided by the total mass of all elements in the compound with the quotient multiplied by 100.

Stoichiometry is the mathematical calculation of the quantities of reactants and products in a chemical reaction as indicated from a balanced chemical equation.

Questions and Problems

Terminology

4.1 Explain the differences and similarities in the terms molecular weight and formula weight.

4.2 What relationship exists between the number of moles of a substance and the number of structural units of that substance?

4.3 What information is given by a molecular formula of a compound?

Atomic Weights

4.4 Why aren't the atomic weights listed in the Table of Atomic Weights integers?

4.5 Mercury consists of seven naturally occurring isotopes. Explain how these isotopes affect the atomic weight of mercury.

4.6 Carbon exists as carbon-12 and carbon-13 in the percents 98.89 and 1.11%, respectively. Using this information, explain the listed atomic weight of carbon.

4.7 Neon exists as neon-20, neon-21, and neon-22. From the listed atomic weight of neon, determine which isotope is present in the largest amount.

4.8 The element bromine exists as the isotopes $^{79}_{35}Br$ and $^{81}_{35}Br$ in nature. The atomic weight is 79.9. Approximately how much of each isotope is present in nature?

4.9 Chlorine contains the isotopes $^{35}_{17}Cl$ and $^{37}_{17}Cl$ in 75.5 and 24.5% abundance, respectively. What is the atomic weight of chlorine?

4.10 Silicon contains the isotopes $^{28}_{14}Si$, $^{29}_{14}Si$, and $^{30}_{14}Si$ in 92.2, 4.7, and 3.1% abundance, respectively. What is the atomic weight of silicon?

Molecular Weights

4.11 Explain how the molecular weight of a compound is calculated from the molecular formula.

4.12 Why isn't the molecular weight of carbon dioxide, CO_2, equal to the sum of the atomic weights of carbon and oxygen, that is, $12 + 16 = 28$ amu?

4.13 Calculate the molecular weight of each of the following substances.

(a) SO_3 (b) P_2O_5 (c) NO_2 (d) N_2O_3
(e) CO_2 (f) N_2H_4 (g) CH_4 (h) $SOCl_2$

4.14 Calculate the molecular weight of each of the following substances.

(a) $C_6H_8O_6$ (vitamin C)
(b) $C_{20}H_{30}O$ (vitamin A)
(c) $C_{16}H_{18}N_2O_4S$ (Penicillin G)
(d) $C_{63}H_{84}N_{14}O_{14}PCo$ (vitamin B_{12})

Formula Weights

4.15 Can the formula weight of an ionic compound be calculated from its name?

4.16 Calculate the formula weight of each of the following substances.

(a) $NaCl$ (b) KBr (c) Na_2O (d) Li_2S
(e) CaO (f) $ScCl_3$ (g) $FeCl_2$ (h) $BaBr_2$
(i) MgO (j) ZnI_2

4.17 Calculate the formula weight of each of the following substances.

(a) $Ca(OH)_2$ (b) $CaCO_3$ (c) $NaOH$
(d) $Mg(OH)_2$ (e) Na_2SO_4 (f) $Al_2(SO_4)_3$
(g) Na_3PO_4 (h) KCN (i) $Na_2Cr_2O_7$
(j) $KMnO_4$

4.18 Calculate the formula weight of each of the following substances.

(a) sodium carbonate
(b) ammonium sulfate
(c) iron(III) sulfate

(d) copper(II) nitrate

(e) potassium permanganate

(f) magnesium bicarbonate

(g) calcium phosphate

(h) magnesium hydroxide

Percent Composition

4.19 Explain how the percent composition of a substance can be calculated from its molecular formula.

4.20 Calculate the percent composition of each of the following.

(a) CH_4 (b) CH_2F_2 (c) H_2O_2

(d) $C_6H_{12}O_6$ (e) P_2O_5 (f) $CaCl_2$

(g) $AgNO_3$ (h) $Al_2(SO_4)_3$

4.21 The molecular formula of aspirin is $C_9H_8O_4$. What is its percent composition?

4.22 The molecular formula of DDT is $C_{14}H_9Cl_5$. What is its percent composition?

4.23 What is the percent composition for compounds formed by combination of the elements in the indicated quantities?

(a) 2.50 g of a metal and 2.00 g of oxygen

(b) 1.40 g of an element and 1.60 g of oxygen

(c) 35.0 g of a metal and 20.0 g of sulfur

4.24 Calculate the percent composition of each of the following.

(a) sodium iodide   (b) sodium oxide

(c) calcium oxide   (d) calcium chloride

(e) aluminum oxide   (f) magnesium oxide

(g) iron(II) sulfide   (h) iron(II) chloride

4.25 Calculate the percent composition of each of the following.

(a) sodium bicarbonate   (b) ammonium sulfate

(c) magnesium carbonate   (d) calcium phosphate

(e) sodium sulfate   (f) magnesium nitrate

(g) calcium perchlorate   (h) potassium cyanide

The Mole

4.26 What relationship exists between the mole and Avogadro's number?

4.27 How do you calculate the mass of one mole of a substance?

4.28 How do you calculate the number of moles contained in a given mass of a substance?

4.29 Since the mole is so central to the study of chemistry, why hasn't a balance been made that determines the number of moles in a sample rather than the number of grams?

4.30 Calculate the number of moles of atoms in each of the following samples.

(a) 46.0 g of sodium   (b) 16.0 g of sulfur

(c) 2.0 g of mercury   (d) 24.3 g of magnesium

4.31 Calculate the number of moles of compound in each of the following samples.

(a) 180 g of H_2O (b) 0.44 g of CO_2

(c) 92 g of NO_2 (d) 14.98 g of CCl_4

4.32 Calculate the number of moles of formula units in each of the following.

(a) 4.0 g of sodium hydroxide, NaOH

(b) 22.32 g of lead(II) oxide, PbO

(c) 0.314 g of uranium hexafluoride, UF_6

(d) 0.49 g of sulfuric acid, H_2SO_4

4.33 Which of each of the following contains the larger number of moles?

(a) 2 g of helium or 10 g of argon

(b) 8 g of CH_4 or 33 g of CO_2

(c) 10 g of CaO or 10 g of CaS

4.34 Calculate the mass of each of the following.

(a) 0.10 mole of helium atoms

(b) 0.20 mole of argon atoms

(c) 2.5 moles of CO_2

(d) 5.0 moles of CH_4

(e) 0.10 mole of $C_9H_8O_4$ (aspirin)

(f) 3.0 moles of $C_{14}H_9Cl_5$ (DDT)

(g) 0.001 mole of $C_{20}H_{30}O$ (vitamin A)

(h) 0.002 mole of $C_{16}H_{18}N_2O_4S$ (Penicillin G)

Avogadro's Number

4.35 Explain why Avogadro didn't choose an easier number.

4.36 Calculate the number of atoms present in each of the following.

(a) 20.0 g of mercury   (b) 120 g of carbon

(c) 53.2 g of palladium   (d) 0.635 g of copper

4.37 Calculate the number of molecules present in each of the following.

(a) 2.8 g of carbon monoxide   (b) 90 g of water

(c) 32 g oxygen gas   (d) 0.040 g of CH_4

4.38 Calculate the number of formula units in each of the following.

(a) 28 g of calcium oxide

(b) 0.164 g of sodium phosphate

(c) 651 g of potassium cyanide

Balancing Equations

4.39 Balance each of the following equations.

(a) $Ba + O_2 \longrightarrow BaO$

(b) $Fe + Cl_2 \longrightarrow FeCl_3$

(c) $K + Br_2 \longrightarrow KBr$

(d) $H_2 + O_2 \longrightarrow H_2O$

(e) $HgO \longrightarrow Hg + O_2$

(f) $Ag_2O \longrightarrow Ag + O_2$

(g) $C + H_2 \longrightarrow CH_4$

4.40 Balance each of the following equations.

(a) $Fe_2O_3 + CO \longrightarrow Fe + CO_2$

(b) $CaO + C \longrightarrow CaC_2 + CO$

(c) $P_4O_{10} + H_2O \longrightarrow H_3PO_4$

(d) $PCl_3 + H_2O \longrightarrow H_3PO_3 + HCl$

(e) $Na_2SO_4 + C \longrightarrow Na_2S + CO_2$

(f) $Fe_2O_3 + C \longrightarrow Fe + CO$

4.41 Balance each of the following equations.
(a) $C_4H_{10} + O_2 \longrightarrow CO_2 + H_2O$
(b) $C_4H_8 + O_2 \longrightarrow CO_2 + H_2O$
(c) $C_5H_{10} + O_2 \longrightarrow CO_2 + H_2O$
(d) $C_5H_8 + O_2 \longrightarrow CO_2 + H_2O$
(e) $C_4H_{10}O + O_2 \longrightarrow CO_2 + H_2O$
(f) $C_4H_{10}O_2 + O_2 \longrightarrow CO_2 + H_2O$

4.42 Balance each of the following equations.
(a) $Ca(OH)_2 + H_2SO_4 \longrightarrow CaSO_4 + H_2O$
(b) $AgNO_3 + NaCl \longrightarrow AgCl + NaNO_3$
(c) $AgNO_3 + BaCl_2 \longrightarrow AgCl + Ba(NO_3)_2$
(d) $Ba(OH)_2 + HCl \longrightarrow BaCl_2 + H_2O$
(e) $CaO + H_3PO_4 \longrightarrow Ca_3(PO_4)_2 + H_2O$
(f) $Na_2SO_4 + BaCl_2 \longrightarrow NaCl + BaSO_4$

Mole–Mole Stoichiometry

4.43 How many moles of oxygen are required to react completely with 10 moles of sulfur dioxide to yield sulfur trioxide?

$$2\,SO_2 + O_2 \longrightarrow 2\,SO_3$$

4.44 How many moles of carbon dioxide are produced in the combustion of 0.5 mole of ethane?

$$2\,C_2H_6 + 7\,O_2 \longrightarrow 4\,CO_2 + 6\,H_2O$$

4.45 How many moles of oxygen are required to completely convert 2 moles of iron(II) sulfide to iron(III) oxide?

$$4\,FeS + 7\,O_2 \longrightarrow 2\,Fe_2O_3 + 4\,SO_2$$

Mass–Mass Stoichiometry

4.46 How many grams of copper can be produced from the reaction of 159 g of copper(II) oxide with hydrogen?

$$CuO + H_2 \longrightarrow Cu + H_2O$$

4.47 How many grams of carbon are required to produce 6.54 g of zinc by the following reaction?

$$ZnO + C \longrightarrow Zn + CO$$

4.48 How many grams of calcium carbide are required to produce 5.2 g of acetylene, C_2H_2, by the following reaction?

$$CaC_2 + 2\,H_2O \longrightarrow Ca(OH)_2 + C_2H_2$$

4.49 How many grams of carbon tetrachloride can be produced from 4.0 g of methane by the following reaction?

$$CH_4 + 4\,Cl_2 \longrightarrow CCl_4 + 4\,HCl$$

4.50 How many grams of copper can be produced by roasting 1590 g of copper(I) sulfide?

$$Cu_2S + O_2 \longrightarrow 2\,Cu + SO_2$$

The Periodic Table and Atomic Structure

A periodic table on the wall is a common feature of a chemistry classroom.

As discussed in Section 2.1, one of the important steps in developing scientific knowledge is a search for an underlying order. This natural order may become apparent if the correct classification method for the observed facts is devised. Then it may be possible to discover a reason for the underlying order that in turn may lead to a theory. The first step is to find the proper classification scheme.

Early in chemical history it was recognized that many elements have similar properties. Those properties then were used to classify the elements. The earliest classification involved two large groups of elements called metals and nonmetals.

With the exception of mercury, metals are solids at room temperature. In general, metals as a class have high densities. Metals are good conductors of heat and electricity. Most metals are ductile (can be drawn into wires) and malleable (can be rolled into sheets). Among the familiar metals are aluminum, chromium, copper, gold, iron, lead, nickel, silver, tin, and zinc. Other, less familiar, metals include barium, calcium, cobalt, magnesium, manganese, potassium, sodium, uranium, titanium, and tungsten.

About one half of nonmetals are gases such as chlorine, fluorine, nitrogen, and oxygen. Argon, helium, krypton, neon, radon, and xenon, a group of very unreactive gases known as the noble gases, are also nonmetals.

The only nonmetal that is a liquid at room temperature is bromine. Solid nonmetals, such as phosphorus and sulfur, have low melting points. In general, the nonmetals have lower densities than the metals.

There are a few elements whose properties are intermediate between metals and nonmetals. These elements, known as semimetals or metalloids, include antimony, arsenic, boron, germanium, silicon, and tellurium. Most of these elements are used in the production of semiconductors.

5.1 Classification of the Elements

The study of the chemistry of 105 elements would be difficult if it were based only on physical properties and the metal–nonmetal classification. Chemical properties provide additional criteria for classification of the elements.

Metals have little tendency to combine chemically with each other. When metals react with nonmetals, they tend to form ionic compounds in which the metal exists as a cation and the nonmetal exists as an anion. Nonmetals combine with each other to form molecular compounds such as carbon dioxide, CO_2, sulfur dioxide, SO_2, and water, H_2O.

There are additional similarities in the chemical properties of certain elements that led chemists to classify elements into smaller groups. For example, the nonmetals fluorine, chlorine, bromine, and iodine react to form similar compounds, such as NaF, NaCl, NaBr, and NaI, respectively. The metals lithium, sodium, and potassium, respectively, form LiCl, NaCl, and KCl.

Because of the similarities in both the chemical and the physical properties of elements, chemists tried to arrange or classify the elements so that their similarities could be emphasized. Eventually, in 1869, the Russian chemist Dimitri Mendeleev succeeded. He found that **when the elements are arranged in the order of increasing atomic weight, elements with similar properties occur at regular or periodic intervals.**

Just what does the term periodic mean? Consider a long fence along a field. The fence is supported by posts placed at intervals. If you walk along the fence you would periodically find a fence post. The posts resemble each other but are not exactly identical. Another analogy for the periodic concept is the order of the seasons. Every year we have the season of summer. No two summers are exactly

5.2 The Periodic Law

alike, but similar periods of the year periodically occur separated by the other seasons.

We can see the periodic pattern discovered by Mendeleev in Table 5.1. The elements are given in order of increasing atomic weight along with their properties. For example, the elements lithium, sodium, and potassium of atomic weights 6.9, 23.0, and 39.1 amu, respectively, are separated by many intervening elements, but their properties are similar. All of these elements are reactive metals. Each of these metals reacts with water to produce a metal hydroxide and hydrogen gas.

$$2\,Li + 2\,H_2O \longrightarrow 2\,LiOH + H_2$$

$$2\,Na + 2\,H_2O \longrightarrow 2\,NaOH + H_2$$

$$2\,K + 2\,H_2O \longrightarrow 2\,KOH + H_2$$

Table 5.1

Periodic Properties of Some Low-Atomic-Weight Elements

Atomic weight	Element	Property
6.9	lithium	active metal
9.0	beryllium	metal
10.8	boron	nonmetallic solid
12.0	carbon	nonmetallic solid
14.0	nitrogen	nonmetal, inactive gas
16.0	oxygen	nonmetal, active gas
19.0	fluorine	nonmetal, very active gas
23.0	sodium	very active metal, chemical properties like lithium
24.3	magnesium	active metal
27.0	aluminum	metal
28.1	silicon	nonmetallic solid
31.0	phosphorus	nonmetallic solid, chemical properties like nitrogen
32.1	sulfur	nonmetallic solid, chemical properties like oxygen
35.5	chlorine	nonmetallic gas, chemical properties like fluorine
39.1	potassium	very active metal, chemical properties like sodium
40.1	calcium	active metal, chemical properties like magnesium

There are other similarities between the elements that indicate a periodic relationship. These are best viewed by arranging the elements as shown in Figure 5.1. The elements are arranged in order of increasing atomic weight from left to right. When an element having properties similar to another element appears, it is placed in the next horizontal row. This arrangement illustrates the periodic or repeating pattern of properties.

Does the Mendeleev classification scheme represent a natural order or is it a fortuitous relationship? If a natural order has been discovered, why is mass responsible for the properties of the elements?

At the time that Mendeleev made his statement of the periodic behavior of the elements, the scientific world did not know about electrons, protons, and neutrons. Some thirty years after Mendeleev's observations, it became evident that the properties of elements are really related to their atomic number. Of course, as the atomic numbers of elements increase so do the atomic weights. However, since atomic weight depends on the number of neutrons and protons

Figure 5.1

The Periodic Relationship of Some Elements.

lithium	beryllium	boron	carbon	nitrogen	oxygen	fluorine
sodium	magnesium	aluminum	silicon	phosphorus	sulfur	chlorine
potassium	calcium					

and on the relative abundance of the isotopes, some elements are in a slightly different order when arranged by atomic weight than by atomic number. For example, argon is before potassium by atomic number, but the atomic weight of argon is larger than that of potassium. Only by arranging the elements by atomic number does a completely logical classification result. The modern **periodic law** is stated as: *the properties of elements are periodic functions of their atomic numbers.* In later sections we will see that this relationship is due to the energy of the electrons about the nucleus.

5.3
The Periodic Table

Mendeleev presented his classification of the elements in a periodic table that contained only 63 elements. He realized that some additional elements might be discovered. A blank space was left in the table if there was no known element with the proper anticipated properties and atomic weight based on the periodic concept. These blank spots helped chemists to search for the "missing" elements in nature. By 1950 all the blank spots were filled. New elements have since been made by nuclear reactions, but they all have high atomic numbers and have been added to the periodic table. Elements beyond atomic number 105 may be made by nuclear reactions in the future.

The modern periodic table is shown in Figure 5.2. Both the atomic number

Period	IA	IIA	IIIB	IVB	VB	VIB	VIIB	VIII			IB	IIB	IIIA	IVA	VA	VIA	VIIA	0
1	1 H 1.0079																	2 He 4.003
2	3 Li 6.941	4 Be 9.012											5 B 10.81	6 C 12.011	7 N 14.007	8 O 15.999	9 F 18.998	10 Ne 20.179
3	11 Na 22.990	12 Mg 24.305				transition elements							13 Al 26.982	14 Si 28.086	15 P 30.974	16 S 32.06	17 Cl 35.453	18 Ar 39.948
4	19 K 39.098	20 Ca 40.08	21 Sc 44.956	22 Ti 47.88	23 V 50.942	24 Cr 51.996	25 Mn 54.938	26 Fe 55.847	27 Co 58.933	28 Ni 58.69	29 Cu 63.546	30 Zn 65.38	31 Ga 69.72	32 Ge 72.59	33 As 74.922	34 Se 78.96	35 Br 79.904	36 Kr 83.80
5	37 Rb 85.4678	38 Sr 87.62	39 Y 88.906	40 Zr 91.22	41 Nb 92.906	42 Mo 95.94	43 Tc (98)	44 Ru 101.07	45 Rh 102.906	46 Pd 106.42	47 Ag 107.868	48 Cd 112.41	49 In 114.82	50 Sn 118.69	51 Sb 121.75	52 Te 127.60	53 I 126.904	54 Xe 131.29
6	55 Cs 132.905	56 Ba 137.3	57 * La 138.906	72 Hf 178.49	73 Ta 180.948	74 W 183.85	75 Re 186.207	76 Os 190.2	77 Ir 192.22	78 Pt 195.08	79 Au 196.966	80 Hg 200.59	81 Tl 204.383	82 Pb 207.2	83 Bi 208.980	84 Po (209)	85 At (210)	86 Rn (222)
7	87 Fr (223)	88 Ra 226.025	89 ** Ac 227.028	104 Rf (261)	105 Ha (262)	106 (263)												

	58 Ce 140.12	59 Pr 140.9077	60 Nd 144.24	61 Pm (145)	62 Sm 150.36	63 Eu 151.96	64 Gd 157.25	65 Tb 158.925	66 Dy 162.50	67 Ho 164.930	68 Er 167.26	69 Tm 168.934	70 Yb 173.04	71 Lu 174.967
lanthanides *														
actinides **	90 Th 232.0381	91 Pa 231.0359	92 U 238.02	93 Np 237.0482	94 Pu (244)	95 Am (243)	96 Cm (247)	97 Bk (247)	98 Cf (251)	99 Es (252)	100 Fm (257)	101 Md (258)	102 No (259)	103 Lr (260)

Figure 5.2

The Modern Periodic Table.

and the atomic weight are given in the rectangular box containing the symbol of the element. All of the known elements are arranged, by increasing atomic number, in horizontal rows so that elements with similar properties fall into a column.

Each vertical column in the periodic table is called a **group.** A group is sometimes called a family. For example, fluorine, chlorine, bromine, iodine, and astatine are in *Group VIIA* and *form a family of elements called the* **halogens.** Other families are also given names. For example, *elements in Groups IA and IIA are called* **alkali metals** *and* **alkaline earth metals,** respectively.

Example 5.1

The melting points of sodium and rubidium are 97.8 and 39.0°C, respectively. Predict the melting point of potassium.

Solution. Potassium is between sodium and rubidium in a column of the periodic table. The melting point should be between that of sodium and that of rubidium. The actual melting point is 63.5°C. This value is approximately the average of the other two melting points.

[Additional examples may be found in **5.43–5.46** at the end of the chapter.]

Group 0 contains the noble gases, elements that were unknown at the time Mendeleev proposed his table. The atomic numbers of the noble gases place them between the elements of the halogen and alkali metal families. This group is placed at the right side of the modern periodic table.

The horizontal rows in the periodic table are called **periods.** The seven periods are numbered sequentially from 1 through 7 from the top to the bottom of the table. Note that the periods of the table contain different numbers of elements. The first period is unusual since it contains only two elements, hydrogen and helium. The second and third periods have eight elements each, whereas the fourth and fifth periods have eighteen elements each. This means that there are ten elements in the fourth and fifth periods that have no counterparts in the two earlier periods. The table is split to accommodate these ten elements. Beyond the fifth period, there are even more elements. Elements 58–71 of the sixth period and elements 90–103 of the seventh period have no counterparts in earlier periods. They are placed outside the main table rather than splitting it again.

Elements in groups designated by Roman numerals I through VII in combination with the letter A and in Group 0 are called **representative elements.** *Elements in Groups IB through VIIB and the three columns designated VIII are called* **transition metals.** *The two rows outside the periodic table contain inner transition metals.* Elements 58 through 71 are also called lanthanides, and elements 90 through 103 are actinides.

Each element in the table belongs to a group and a period. Thus, fluorine is in Group VIIA of the second period, whereas chlorine, which is located below fluorine, is in Group VIIA of the third period. The element oxygen immediately to the left of fluorine is in Group VIA of the second period.

As knowledge of the structure of the atom and subatomic particles developed, scientists suspected that the repeating properties of elements given in the periodic table might be due to similarities in the arrangements of the subatomic particles. If the subatomic particles are regularly arranged in the atom, then a regularity of properties might result. We will return to this idea after the theory of the electronic structure of the atom is given in the next section.

5.4
The Bohr Atom

When high voltages are applied to elements in the gaseous state, colored light results. When this type of light is passed through a prism, a series of colored lines is observed. The number of lines and their color are characteristic for each element much like a fingerprint.

Niels Bohr, a Danish physicist, suggested that the light given off by the elements was due to the position and distance of the electron from the nucleus. Bohr theorized that there are a number of possible orbits for the electron at various distances from the nucleus. The electrons could be in any of these orbits but not between orbits. When an atom absorbs energy, an electron "jumps" from a low energy orbit near the nucleus to a higher energy orbit further away (Figure 5.3). When the electron "falls" to its lower orbit, energy is released in the form of light. Only certain colors of light are possible, because only certain amounts of energy can be released. This energy is equal to the energy difference between the orbits.

Bohr's theory of the atom was consistent with the data at that time. However, his work dealt with only the simpler atoms. As a result of more modern evidence the Bohr concept has been modified somewhat. It is no longer thought that electrons exist in planet-like orbits about the nucleus. Rather, groups of electrons occupy regions of space about the nucleus called shells. While electrons are in the "shell," they have a certain characteristic amount of energy. At some greater distance there is another region or shell in which there is a different group of electrons with a different amount of energy.

5.5
Shells, Subshells, and Orbitals

The modern picture of the electronic arrangement in the atom is based on a complex theoretical approach called quantum mechanics, which is beyond the learning objectives of this book. However, the conclusions of the theory are fairly simple and will be summarized here. You can use this material to understand both the periodic law and why elements form compounds (Chapter 6).

Shells

The various electrons in an atom do not all have the same energy; thus, their locations within the atom differ. In general, the higher the energy of the electron,

Figure 5.3

The Bohr Atom.
When energy is added to the atom, the electrons absorb energy and move to a higher-energy orbit. When the electrons return to a lower orbit, the energy is released as light energy. The color of light is characteristic of the change in levels.

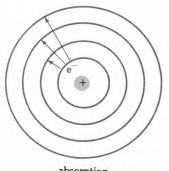

absorption
of energy

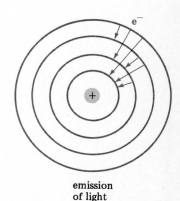

emission
of light

AN ASIDE

Many substances produce visible light when energy is added. The energy causes electronic excitation, and a higher-energy atom is produced. When the electron returns to its lower and more stable energy state, light energy results. This phenomenon is quite a common occurrence that you may observe any evening.

Modern street lights or highway lights depend on emitted light energy from electrically excited atoms. The sodium lamp and the mercury lamp are two examples. The light produced by excited sodium atoms is in the yellow region, whereas that of excited mercury atoms is in the green and blue region.

A sodium lamp is more energy efficient than a mercury lamp. For a given amount of electricity, the light intensity of the sodium lamp is greater than that of the mercury lamp. In addition, the human eye is more sensitive to yellow light. Yellow light is not as scattered by particles in the air such as water droplets in fog. Thus, yellow light penetrates fog farther than light of other colors such as the mercury lamp's.

The all too familiar neon signs used in store advertisements are additional examples of light emitted by electronically excited atoms. Neon gas produces an orange-red light. Argon gas produces a blue-purple light. Krypton gives off a near-white light. Other colors result from coating the interior of the glass tubes in the signs with special fluorescent substances.

---

the further away from the nucleus will be the electron. *The electrons in an atom exist in certain principal energy levels or* **shells.** The shells are designated by integers $n = 1, 2, 3, 4, \ldots$. The energy of the electrons in the energy levels increases in the order $1 < 2 < 3 < 4$, and so on. Therefore, the distance between the shell of electrons and the nucleus increases in the same order.

There are limitations on the number of electrons that can exist in a specific energy level. **The maximum number of electrons in a shell is given by $2n^2$, where n is the number of the energy level.** The number of electrons in each of the principal energy levels is listed in Table 5.2.

Subshells

Within the energy levels there are energy sublevels or subshells. The higher energy levels that have more electrons in them also have more subshells. Of all the known atoms, electrons exist in *four types of* **subshells,** *labeled by the lower-case letters s, p, d, and f.* (Other subshells may contain electrons when energy is added to an atom. These are labeled *g, h,* and *i* but will not be considered any further.) The shells and the subshells in an atom are represented by symbols such as $2s$, $4d$, $5p$, and so on. The number represents the energy level or shell, and the letter represents the subshell within that energy level.

There are restrictions on the number of electrons that each of the subshells can hold. **The maximum number of electrons in each of the subshells is as follows: 2 in the s subshell, 6 in the p subshell, 10 in the d subshell, and 14 in the f subshell.**

The order of the energy of the subshells is $s < p < d < f$. The relationships between shells and subshells are shown in Table 5.2. Note that the number of

Table 5.2

Location of Electrons in Shells and Subshells

Energy level, n	Sublevel	Electrons in sublevel	Total electrons possible in energy level $2n^2$
1	s	2	2
2	s	2	
	p	6	8
3	s	2	
	p	6	
	d	10	18
4	s	2	
	p	6	
	d	10	
	f	14	32
5	s	2	
	p	6	
	d	10	
	f	14	
	g	18	50
6	s	2	
	p	6	
	d	10	
	f	14	
	g	18	
	h	22	72
7	s	2	
	p	6	
	d	10	
	f	14	
	g	18	
	h	22	
	i	26	98

subshells increases with the shell number. The total number of electrons possible within the subshells is equal to the total number of electrons within the shell.

Orbitals

According to quantum mechanical theory each subshell consists of orbitals. *An* **orbital** *is a region of space in which an electron is most likely to be found.* The characteristics of orbitals are as follows.

1. All orbitals within a subshell are of the same energy.
2. Each orbital has a size and shape that are related to the energy of the electrons that it can contain.
3. The orbital volume and distance from the nucleus increase as the value of n increases.
4. Each orbital can contain a maximum of two electrons.
5. Electrons move rapidly and "occupy" the entire orbital volume.

Figure 5.4
The Shape of s Orbitals.

A useful picture of an orbital is that of an electron cloud. Clouds in the sky look quite big but do not contain much matter. The electrons within an orbital are moving rapidly and may be regarded as being everywhere. Since the electron has a small mass, the average distribution of matter within the orbital is very low or cloud-like.

The s orbitals are spherical as shown in Figure 5.4. The 2s orbital has a larger volume than the 1s orbital, and the 3s orbital is larger than the 2s orbital. However, each s orbital can still only contain two electrons.

There are three p orbitals within each p subshell, and each orbital can be occupied by two electrons. The three p orbitals are identical in shape and are located at right angles to one another (Figure 5.5). Note that each orbital consists of two teardrop shapes that meet at the nucleus. The two electrons within a p orbital can occupy any of the area shown as a p orbital. The electrons do not distribute themselves one each to the two teardrop shapes. The orbital describes a volume within which two electrons may exist. There are five d orbitals in a d subshell and seven f orbitals in an f subshell. The much more complex shapes of d and f orbitals will not be described in this text.

Electron Spin

Electrons within an orbital repel each other because their charges are identical. However, the electrons have another property, called electron spin, which causes some attraction between two electrons in the same orbital. The spin occurs about an axis much as the earth rotates about its north–south axis.

Electrons spin *or rotate in two possible directions (referred to as clockwise and counterclockwise).* These spins are often represented by arrows. An arrow pointing up (↑) means that the spin of an electron is the opposite of an electron represented by an arrow pointing down (↓).

The **Pauli exclusion principle** *states that two electrons occupying any orbital always have opposite spins.* A spinning charged particle creates a magnetic field. Thus, two electrons spinning in opposite directions in an orbital will produce opposite magnetic fields. You know that the north pole of a magnet will attract the south pole of another magnet. Thus, electrons with opposite spins will be magnetically attractive, which counters in part the electrical repulsion of the two electrons.

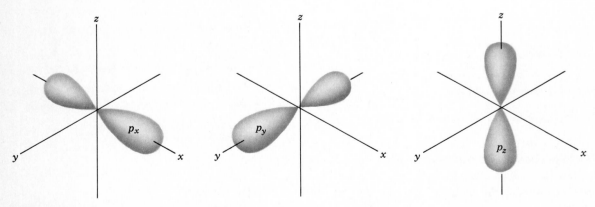

Figure 5.5
The Shape of p Orbitals.
Each orbital may contain two electrons.

The picture of the atom developed in the previous section describes where electrons may be located or "reside" in an atom. In a similar way, you could describe your residence within a city by giving the apartment building, the floor in the building, and your apartment number. In the case of the electron, the energy level is the apartment building, the subshell is the floor, and the orbital is the apartment. In the orbital "apartment" only two occupants are allowed and they must be magnetically compatible. Now let's consider the many electrons within an atom and how they are distributed.

The arrangement of the electrons within an atom is called the **electron configuration.** The rules for predicting the electron configuration are quite simple. First, electrons are located in the lowest energy orbital available. This results in the most stable atom. Second, no more than two electrons may exist in the same orbital, and then they must not have the same spin. Third, electrons must "spread out" among the orbitals within the subshell. Thus, three electrons within the p subshell will be distributed one to each of the three orbitals. Horizontal lines may be used to represent the three orbitals.

$$\underline{\uparrow}\ \underline{\uparrow}\ \underline{\uparrow}$$
preferred "spread-out" arrangement

$$\underline{\uparrow\downarrow}\ \underline{\uparrow}\ \underline{}$$
doesn't occur in the atom

The key to the "spread out" rule is the electrical repulsion between electrons. This repulsion causes them to avoid each other and is expressed as Hund's rule. **Hund's rule** *states that electrons will not enter an orbital containing another electron if an empty orbital of the same energy is available.* By having electrons in different orbitals, electrical repulsion is minimized. However, remember that for two electrons to be in the same orbital, they must have opposite spins.

The order of increasing energies of subshells is $1s$, $2s$, $2p$, $3s$, $3p$, $4s$, $3d$, $4p$, $5s$, $4d$, $5p$, $6s$, $4f$, $5d$, $6p$, $7s$, $5f$, $6d$. Examination of this order reveals that all of the subshells within a given shell will not be filled before the next shell starts to fill. For example, the $4s$ subshell intervenes between the $3p$ and the $3d$. The ordering of shells and subshells occurs in a complicated manner. Note that the d subshell of a given n energy level is always of higher energy than the s subshell of the $n + 1$ energy level and of lower energy than the p subshell of the $n + 1$ energy level.

$$4s < 3d < 4p$$
$$5s < 4d < 5p$$
$$6s < 5d < 6p$$

In addition, the f subshell of the n energy level is of higher energy than the s subshell of the $n + 2$ energy level and of lower energy than the d subshell of the $n + 1$ energy level.

$$6s < 4f < 5d$$
$$7s < 5f < 6d$$

A method of memorizing the order of energies of the subshells is given in Figure 5.6. Start from the $1s$ subshell, and follow the solid line from the start of the arrow line to the head of the arrow. Then proceed via the dotted line to the start of the next solid-line arrow. This procedure will reproduce the order of energies previously given.

We now have the information needed to predict electron configurations. A

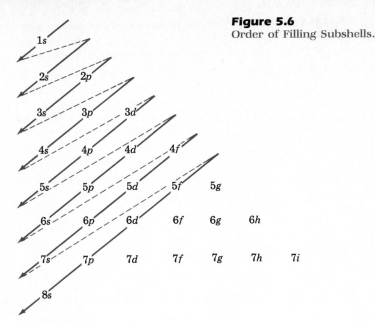

Figure 5.6
Order of Filling Subshells.

convenient way of showing energy levels, subshells, and the number of electrons that they contain is as follows.

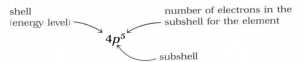

Let's predict some electron configurations. The element hydrogen contains one electron in the $1s$ subshell, which is the lowest energy subshell available. The electronic configuration of hydrogen is written $1s^1$. The superscript denotes the number of electrons in the subshell. Helium has two electrons, both occupying the $1s$ subshell. The electronic configuration of helium is $1s^2$, and the subshell is filled.

Lithium has three electrons. Two electrons fill the $1s$ subshell; the third is in the subshell with the next highest energy, which is the $2s$. The lithium electronic configuration is $1s^22s^1$. Beryllium has the electronic configuration $1s^22s^2$. The fourth electron of beryllium completes the $2s$ subshell, which has a capacity of two electrons.

In boron, the $2p$ subshell, the subshell with the next lowest energy, is occupied by a single electron. The electron configuration of boron is $1s^22s^22p^1$. A p subshell can hold a maximum of six electrons, and this subshell is filled by additional electrons added for the elements C, N, O, F, and Ne. The electron configurations for these elements are

B $1s^22s^22p^1$
C $1s^22s^22p^2$
N $1s^22s^22p^3$
O $1s^22s^22p^4$
F $1s^22s^22p^5$
Ne $1s^22s^22p^6$

Example 5.2

An isotope of strontium is radioactive and is produced in nuclear explosions. Write the electron configuration of $^{90}_{38}Sr$.

Solution. The subscript number, 38, equals the number of protons or electrons in the atom. The electrons are located in the lowest energy subshell accessible to them. The ordering of subshells is

$$1s < 2s < 2p < 3s < 3p < 4s < 3d < 4p < 5s$$

Electrons up to a total of 38 are then placed within each subshell to obtain

$$1s^2 2s^2 2p^6 3s^2 3p^6 4s^2 3d^{10} 4p^6 5s^2$$

Example 5.3

What is the arrangement of electrons in $^{60}_{27}Co$? This isotope, called cobalt-60, is used in cancer therapy.

Solution. There are 27 electrons in this isotope. Write down the order of energies of the subshells, filling each in order.

$$1s^2 2s^2 2p^6 3s^2 3p^6 4s^2$$

The above arrangement accounts for twenty electrons. The remaining seven electrons are placed in an incomplete $3d$ subshell to give

$$1s^2 2s^2 2p^6 3s^2 3p^6 4s^2 3d^7$$

Table 5.3

Illustration of Hund's Rule and Electron Configuration

Element	Atomic number	1s	2s	2p	Electron configuration
H	1	↑			$1s^1$
He	2	↑↓			$1s^2$
Li	3	↑↓	↑		$1s^2 2s^1$
Be	4	↑↓	↑↓		$1s^2 2s^2$
B	5	↑↓	↑↓	↑	$1s^2 2s^2 2p^1$
C	6	↑↓	↑↓	↑ ↑ __	$1s^2 2s^2 2p^2$
N	7	↑↓	↑↓	↑ ↑ ↑	$1s^2 2s^2 2p^3$
O	8	↑↓	↑↓	↑↓ ↑ ↑	$1s^2 2s^2 2p^4$
F	9	↑↓	↑↓	↑↓ ↑↓ ↑	$1s^2 2s^2 2p^5$
Ne	10	↑↓	↑↓	↑↓ ↑↓ ↑↓	$1s^2 2s^2 2p^6$

Now let's consider a more detailed accounting of the electrons in an atom. Consider the elements B, C, N, O, F, and Ne whose electron configurations involve electrons in the 2p subshell. In this series the 2p subshell contains from one to six electrons. According to Hund's rule, the electrons will be located as shown in Table 5.3. The lines are used to represent the orbitals, and the electron spin is represented with arrows. The order of filling the 2p subshell according to Hund's rule tells us the number of pairs of electrons and number of unpaired electrons. Nitrogen, for example, has three unpaired electrons. The number and type of bonds formed by elements are based on such considerations (Chapter 6).

Example 5.4

Consider the electron configuration of cobalt given in Example 5.3. Describe the electrons located in the 3d subshell.

Solution. There are seven electrons in a subshell that may contain up to ten electrons. There are five orbitals in the subshell. In order to "spread out" the electrons according to Hund's rule, one electron must be located in each orbital. The remaining two electrons must be paired with electrons already in orbitals. The final representation of the d subshell is

$$\uparrow\downarrow \quad \uparrow\downarrow \quad \uparrow \quad \uparrow \quad \uparrow$$

5.7
Electron-Dot Symbols

The last principal energy level of an atom containing electrons in s and p subshells is called the **valence energy level** *or* **valence shell.** The eight electrons that can occupy these subshells are known as valence electrons. Most of the chemistry of the elements to be discussed in this text is based on the fact that there is a maximum of eight valence electrons. All of the other electrons together with the nucleus comprise the core. This division is useful because the valence electrons form chemical bonds (Chapter 6). Because of the importance of valence electrons we display them in electron-dot symbols.

Electron-dot symbols are written according to the following rules.

1. The symbol of the element represents the core.
2. Dots to represent valence electrons may be placed on either side of or above or below the symbol.
3. All positions about the symbol are equivalent.
4. A maximum of two electrons per side, top, or bottom of the symbol is allowed. Thus, only eight valence electrons can be placed about the symbol.
5. Valence electrons are indicated by a pair of dots for the pairs of electrons and a single dot for an unpaired electron in the valence shell.
6. Hydrogen and helium are exceptions to the maximum of eight electrons. They have a maximum of two electrons.

These rules are derived to obtain symbols that are used as models to explain chemical observations. Several examples in Table 5.4 illustrate the use of electron dot symbols.

Table 5.4

Electron-Dot Symbols of the Second Period Elements

Element	Electron configuration	Electron-dot symbols
$^{7}_{3}Li$	$1s^2 2s^1$	Li· or Li· or ·Li or Li·
$^{9}_{4}Be$	$1s^2 2s^2$	Be: or ·Be· or ·Be· or :Be
$^{11}_{5}B$	$1s^2 2s^2 2p^1$	:B or ·B· or :B· or B·
$^{12}_{6}C$	$1s^2 2s^2 2p^2$	:C· or ·C· or :C·, etc.
$^{14}_{7}N$	$1s^2 2s^2 2p^3$	·N· or ·N: or ·N· or :N·
$^{16}_{8}O$	$1s^2 2s^2 2p^4$	·O: or ·O· or :O·, etc.
$^{19}_{9}F$	$1s^2 2s^2 2p^5$	:F· or :F: or ·F: or :F:

The periodic table is arranged in the order of increasing atomic number such that elements with similar properties occur periodically. The atomic number of an atom is equal to the number of electrons in the atom, and these electrons are systematically located in specific subshells. Thus, there is a connection between the periodic table and the electron configuration of atoms. The relationship between the energy levels of atoms and the periodic table is shown in Figure 5.7. In the first period there are only two elements, hydrogen and helium, which have one and two electrons, respectively, in the $1s$ subshell. The elements lithium (atomic number 3) and beryllium (4) have electrons in the $2s$ subshell and are separated from the other six elements in the period. Elements boron (5) through neon (10) involve the addition of six successive electrons to the $2p$ subshell. The eleventh electron of sodium and the eleventh and twelfth electrons of magnesium occupy the $3s$ subshell. Again, the six elements on the right of the periodic table involve the filling of a p subshell, which in these elements is the $3p$ subshell. Elements 19 and 20 are in the same region in which s subshells were filled in earlier periods, which in this case is the $4s$ subshell. Now, for the first time, there

5.8

Electron Configuration and the Periodic Table

Figure 5.7

Relationship Between the Periodic Table and Electron Configuration.

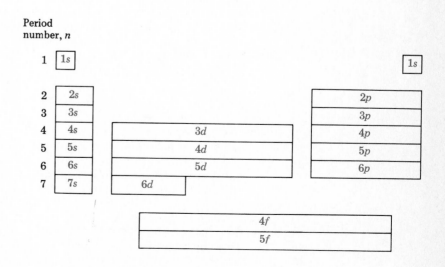

are elements in the center region of the periodic table. The electronic configurations of the ten elements scandium (21) through zinc (30) result from the addition of electrons to the $3d$ subshell. Then, elements 31–36 appear in the same region where p subshells have been filled previously. In this case it is the $4p$ subshell, since these elements are in the fourth period. The remainder of the periodic table follows in an orderly fashion in the case of s and p subshells. The d subshells being occupied are always one less than the period number. Outside the general body of the periodic table are located two groups of fourteen elements that represent the filling of the $4f$ and $5f$ subshells.

Example 5.5

What is the electron configuration of silicon?

Solution. The electron configuration of silicon (atomic number 14) can be written by going through the regions of the periodic table until silicon is reached. The configuration is $1s^2 2s^2 2p^6 3s^2 3p^2$, with the superscripts indicating the number of electrons in each subshell.

Example 5.6

What is the electron configuration of zirconium (40)?

Solution. In order to reach zirconium by reading through the periodic table, the $1s$, $2s$, $2p$, $3s$, $3p$, $4s$, $3d$, $4p$, and $5s$ regions must be traversed. Then one has to proceed by two steps into the $4d$ region. The entire electron configuration is $1s^2 2s^2 2p^6 3s^2 3p^6 4s^2 3d^{10} 4p^6 5s^2 4d^2$.

[Additional examples are given in **5.38–5.42** at the end of the chapter.]

5.9
General Information Derived from Groups

The periodic table provides information about the chemical and physical properties of the elements. In this section, some of this information is outlined.

The number of valence electrons of the representative elements is given directly by the Roman numeral of the group. Since the period number is the same as the energy level being filled, one knows the valence electron configuration at a glance. Thus, all Group IA elements, called the alkali metals, have an s^1 configuration.

$$\begin{array}{ll} \text{Li} & 1s^2 2s^1 \\ \text{Na} & 1s^2 2s^2 2p^6 3s^1 \\ \text{K} & 1s^2 2s^2 2p^6 3s^2 3p^6 4s^1 \\ \text{Rb} & 1s^2 2s^2 2p^6 3s^2 3p^6 4s^2 3d^{10} 4p^6 5s^1 \end{array}$$

Elements within groups have similar chemical properties due to the similarity of electronic configurations. Thus, all alkali metals have an s^1 configuration. All members lose one electron from the s orbital and form singly charged cations. Therefore, similar compounds such as LiCl, NaCl, KCl, RbCl, CsCl, and FrCl can be formed. For the nonmetal halogens, the electron configurations are $s^2 p^5$. These

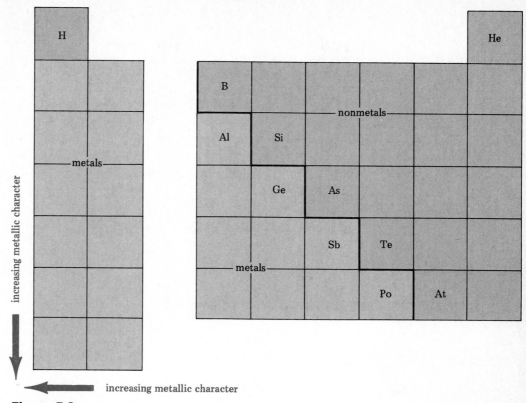

Figure 5.8
The Metals and Nonmetals of the Representative Elements.

elements tend to gain one electron to form singly charged anions. Therefore, similar compounds such as NaF, NaCl, NaBr, and NaI can be formed.

For representative elements, the metallic properties increase with increasing atomic number within a group and the nonmetallic properties decrease. Metallic properties within periods increase toward the left, and nonmetallic properties increase toward the right. Thus, within a group, the dividing line between metals and nonmetals occurs at higher atomic numbers for groups on the right than groups on the left. The stairstep line shown in Figure 5.8 separates the metals and nonmetals. Elements to the right of the line are nonmetals and elements to the left of the line are metals. The semimetals lie immediately to either side of the line. For example, in Group VA the most metallic element is bismuth; antimony and arsenic have borderline properties and are semimetals or metalloids; phosphorus and nitrogen are nonmetals.

5.10 Atomic Radii

The size of an atom reflects the number of electrons and their energies. The overall shape of the atom is pictured as spherical, and the size of an atom is expressed as the radius of a sphere. The radii of atoms are given in Ångstrom units $(1 \text{ Å} = 10^{-8} \text{ cm})$ in Figure 5.9 for the representative elements.

The atomic radii decrease from left to right in a period of the periodic table. Within a period, the nuclear charge increases in the same direction, but the electrons are located within the same energy level. This increase in nuclear

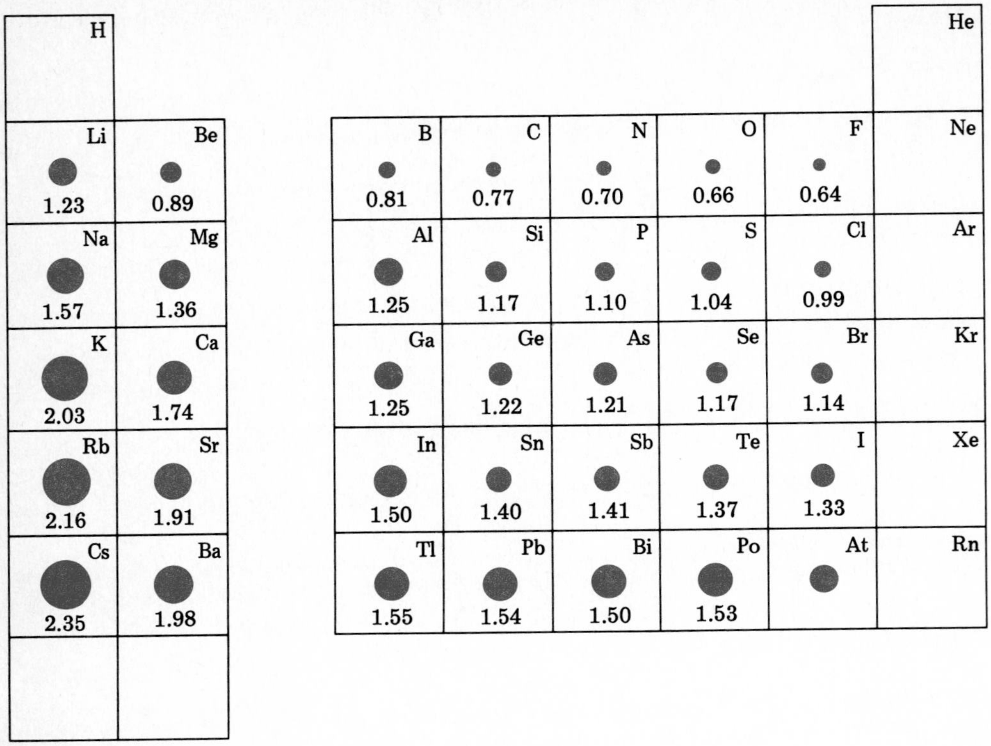

H							He
Li 1.23	Be 0.89	B 0.81	C 0.77	N 0.70	O 0.66	F 0.64	Ne
Na 1.57	Mg 1.36	Al 1.25	Si 1.17	P 1.10	S 1.04	Cl 0.99	Ar
K 2.03	Ca 1.74	Ga 1.25	Ge 1.22	As 1.21	Se 1.17	Br 1.14	Kr
Rb 2.16	Sr 1.91	In 1.50	Sn 1.40	Sb 1.41	Te 1.37	I 1.33	Xe
Cs 2.35	Ba 1.98	Tl 1.55	Pb 1.54	Bi 1.50	Po 1.53	At	Rn

Figure 5.9

Atomic Radii of the Representative Elements.

The atomic radii are given in Å, and the circles represent the relative sizes of the atoms.

charge tends to draw the electrons toward the nucleus and results in a smaller volume.

From top to bottom in a family of the periodic table, the atomic radii increase. Each successive member has one additional energy level containing electrons. Because the size of an orbital increases with the number of the energy level, the size of the atom tends to increase.

5.11
Ionization Energy

If sufficient energy is added to an atom of a gas, an electron may be removed to form a positive ion, or cation. *Electron removal from an atom is called* **ionization.** *The amount of energy required to accomplish the removal is called the* **ionization energy.** Ionization energies are expressed in electron volts per atom. The ionization energy of gaseous lithium, Li(g), is 5.4 electron volts (eV) per atom.

$$Li(g) \longrightarrow Li^+(g) + 1\,e^-$$

Because the ionization energy is a measure of the energy required to remove an electron from an element, it indicates the energy of the electron in the element. Higher energy electrons require less additional energy (lower ionization energy) to remove them from the element; conversely, low energy electrons are closely bound to the atom and require more energy for ionization. The ionization energies of the representative elements are given in Figure 5.10.

The ionization energies of the representative elements increase from left to

increasing ionization energy ⟶

increasing ionization energy ↑

H 13.6			B 8.3	C 11.3	N 14.5	O 13.6	F 17.4	He 24.6
Li 5.4	Be 9.3		B 8.3	C 11.3	N 14.5	O 13.6	F 17.4	Ne 21.6
Na 5.1	Mg 7.6		Al 6.0	Si 8.2	P 11.0	S 10.4	Cl 13.0	Ar 15.8
K 4.3	Ca 6.1		Ga 6.0	Ge 8.1	As 9.8	Se 9.8	Br 11.8	Kr 14.0
Rb 4.2	Sr 5.7		In 5.8	Sn 7.3	Sb 8.6	Te 9.0	I 10.5	Xe 12.1
Cs 3.9	Ba 5.2		Tl 6.1	Pb 7.4	Bi 7.3	Po 8.4	At	Rn 10.7
Fr	Ra 5.3							

Figure 5.10
Ionization Energies of the Representative elements.
The ionization energies of the representative elements are given in electron volts.

right within a period. This trend is due to the increase in the nuclear charge that causes an increase in the attraction for the electron. The ionization energy for the fluorine atom is 17.4 eV/atom, a value much higher than that for lithium at the start of the period. Metals have lower ionization energies than nonmetals.

Within a group there is a trend of decreasing ionization energies in agreement with the increasing metallic character of elements of higher atomic number. Thus, the ionization energy of cesium is 3.9 eV/atom, a value less than the 5.4 eV/atom of lithium.

5.12 Electronegativity

Electronegativity *is a measure of the ability of an atom to attract electrons in a bond with a second atom.* The electronegativity values of the representative elements are given in Figure 5.11. The larger the electronegativity value, the greater the tendency to attract electrons. Within a period, the electronegativities increase from left to right. Within a group, the electronegativities decrease with increasing atomic weight. Thus, nonmetals are more electronegative than metals. Metals are often referred to as being electropositive.

The electronegativities of atoms determine the types of chemical bonds that form between them. If the electronegativities of the atoms are very different, electron transfer will occur and an ionic bond results. As the electronegativities become more similar, electrons are shared. Bonding will be discussed in the next chapter.

increasing electronegativity ➡

increasing electronegativity

							He
H 2.1							
Li 1.0	Be 1.5	B 2.0	C 2.5	N 3.0	O 3.5	F 4.0	Ne
Na 0.9	Mg 1.2	Al 1.5	Si 1.8	P 2.1	S 2.5	Cl 3.0	Ar
K 0.8	Ca 1.0	Ga 1.6	Ge 1.8	As 2.0	Se 2.4	Br 2.8	Kr
Rb 0.8	Sr 1.0	In 1.7	Sn 1.8	Sb 1.9	Te 2.1	I 2.5	Xe
Cs 0.7	Ba 0.9	Tl 1.8	Pb 1.8	Bi 1.9	Po 2.0	At 2.2	Rn
Fr 0.7	Ra 0.9						

Figure 5.11
Electronegativity Values of the Representative Elements.

Summary

The periodic law and its derived periodic table provide an ordering of the elements according to their chemical and physical properties. The periodic relationships can be used to describe and compare the tendency of elements to behave as metals, metalloids, and nonmetals.

The experimental basis of the periodic table prepared by Mendeleev is now understood in terms of a description of the electron configuration of atoms. The groups of the periodic table consist of elements with similar electron configurations.

Bohr's model for the atom was devised to explain light emitted from electronically excited atoms. Electrons within an atom have certain permissible energies and are located in orbitals. The electron configuration of an atom is given by shells, subshells, orbitals, and the spin of an electron.

Learning Objectives

As a result of studying Chapter 5 you should be able to

- Illustrate the periodic law for the elements.
- Distinguish between periods and groups in the periodic table.
- Describe the shells, subshells, and orbitals within an atom.
- Write the electronic configuration of elements given the atomic number.
- Outline the periodic table in terms of the electron configuration of the elements.
- Represent the valence shell electrons of atoms by electron-dot symbols.
- Locate the metals, nonmetals, and metalloids within the periodic table.
- Identify trends in atomic radii, ionization energies, and electronegativities within the periodic table.

New Terms

The **atomic radius** of an atom is given in Angstrom units (Å) (1 Å = 10^{-8} cm).

Alkali metals are the elements of Group IA.

Alkaline earth metals are the elements of Group IIA.

The **electron configuration** of an atom is a description of the arrangement of the electrons in the atom by shells, subshells, and orbitals.

Electronegativity is a measure of the electron-attracting power of an atom.

Electron spin is a property of an electron and may be clockwise or counterclockwise.

An **electron-dot symbol** gives the number of valence electrons as dots located around the elemental symbol.

A **family** is a set of chemically related elements located in a group of the periodic table.

A **group** is a vertical column in the periodic table of elements that have similar properties.

The **halogens** are the elements of Group VIIA.

Hund's rule states that electrons in orbitals of equal energy are located singly in those orbitals before any pairing occurs.

The **ionization energy** of an atom is the energy required to remove its highest energy electron and form an ion.

An **orbital** describes the probable region where an electron may be found about an atom.

Paired electrons are electrons of opposite spin in an orbital.

The **Pauli exclusion principle** states that two electrons occupying any orbital always have opposite spins.

A **period** is a horizontal row in the periodic table.

The **periodic law** describes the periodic recurrence of properties of elements when considered in order of increasing atomic number.

The **periodic table** is an arrangement of elements with elements of similar properties grouped together.

Representative elements are elements in which the s and p subshells of the highest energy level are being filled.

The **shell** of an atom is a description of those electrons with characteristic energies. The shells are designated by an integer n.

A **subshell** is a group of orbitals of identical energy. The subshells are designated s, p, d, and f.

The **transition elements** are atoms in which the d or f subshells are being filled.

Valence shell electrons are electrons of the s and p subshells in the highest occupied energy level.

Questions and Problems

Terminology

5.1 In your own words describe the electronic organization of the atom.

5.2 Distinguish between the two terms in each of the following.
(a) period and group
(b) representative and transition element
(c) core and valence electrons
(d) shell and subshell
(e) subshell and orbital

The Periodic Table

5.3 Give an example of periodic behavior other than the two examples cited in this chapter.

5.4 How does the modern periodic law differ from Mendeleev's periodic law?

5.5 The order of the periodic table is possible because of certain facts about the elements. What facts are these?

5.6 Indicate the location of each of the following elements by period number and group designation.
(a) $_7$N (b) $_{20}$Ca (c) $_{48}$Cd (d) $_{35}$Br
(e) $_3$Li (f) $_{13}$Al (g) $_{50}$Sn (h) $_{16}$S
(i) $_{54}$Xe (j) $_{26}$Fe (k) $_{24}$Cr (l) $_{29}$Cu

5.7 What element occupies each of the indicated locations in the periodic table?
(a) period 2, Group IA (b) period 3, Group IIIA
(c) period 6, Group IIA (d) period 4, Group IVA
(e) period 5, Group VIIA (f) period 1, Group 0
(g) period 4, Group VIIB (h) period 5, Group IB
(i) period 6, Group VIB (j) period 3, Group IVB

5.8 Using the periodic table, select a metal in the sixth period that should form an ion with a single positive charge.

5.9 What halogen is in the fourth period of the periodic table?

5.10 Locate two pairs of elements that are out of order in terms of their atomic weights. Why are the elements in the order counter to their atomic weights?

5.11 Mendeleev predicted that there was a missing element in the Group IIIA between aluminum and indium. Using the atomic weights of these two elements, predict the atomic weight of the missing element.

Metals and Nonmetals

5.12 Using the periodic table, classify each of the following as a metal or a nonmetal.
(a) carbon (b) calcium (c) mercury
(d) gallium (e) iodine (f) phosphorus
(g) strontium (h) cesium (i) selenium
(j) vanadium

5.13 Using the periodic table, indicate which member of each pair is the more metallic.
(a) magnesium and silicon
(b) germanium and bromine
(c) sulfur and selenium
(d) silicon and tin
(e) phosphorus and antimony
(f) oxygen and tellurium

(g) indium and tin

(h) sodium and cesium

5.14 According to the periodic table, which element should have the most metallic character?

5.15 Explain why the metalloids are formed in a midregion of the periodic table.

5.16 Information about the properties of the transition elements is not given in this chapter. Based on their placement, what types of physical and chemical properties are expected?

Electrons and Symbols

5.17 For each of the following, indicate how many electrons are contained in one atom.

(a) $^{16}_{8}O$ (b) $^{36}_{18}Ar$ (c) $^{235}_{92}U$ (d) $^{9}_{4}Be$ (e) $^{13}_{6}C$

(f) $^{42}_{20}Ca$ (g) $^{31}_{15}P$ (h) $^{55}_{25}Mn$

5.18 How many electrons are contained in each of the following ions?

(a) $^{23}_{11}Na^{+}$ (b) $^{16}_{8}O^{2-}$ (c) $^{31}_{15}P^{3-}$ (d) $^{42}_{20}Ca^{2+}$ (e) $^{19}_{9}F^{-}$

(f) $^{35}_{17}Cl^{-}$ (g) $^{7}_{3}Li^{+}$ (h) $^{27}_{13}Al^{3+}$

Principal Energy Levels

5.19 List the number of electrons that may be located in each of the four lowest energy levels.

5.20 Why aren't more than seven energy levels discussed in this chapter or in other chemistry texts?

5.21 Explain what would be observed if an excited helium atom with an electron in the fourth energy level were converted back to ordinary helium. How would this observation differ from the conversion of an excited helium atom with an electron in the sixth energy level back to ordinary helium?

5.22 List the number of electrons contained in each of the following by principal energy levels.

(a) $^{19}_{9}F$ (b) $^{39}_{19}K$ (c) $^{42}_{20}Ca$ (d) $^{32}_{16}S$ (e) $^{14}_{7}N$

(f) $^{24}_{12}Mg$ (g) $^{35}_{17}Cl$ (h) $^{16}_{8}O$ (i) $^{10}_{4}Be$ (j) $^{28}_{14}Si$

5.23 How many electrons are contained in the valence shell of each of the following?

(a) $^{23}_{11}Na$ (b) $^{7}_{3}Li$ (c) $^{19}_{9}F$ (d) $^{35}_{17}Cl$ (e) $^{31}_{15}P$

(f) $^{14}_{7}N$ (g) $^{27}_{13}Al$ (h) $^{24}_{12}Mg$ (i) $^{12}_{6}C$ (j) $^{28}_{14}Si$

5.24 Is it necessary to list the mass number in order to ascertain the electronic arrangement of an atom? Explain.

Subshells

5.25 Can any shell contain any type of subshell?

5.26 How many subshells are there in each of the following?

(a) the second shell

(b) the fourth energy level

(c) a shell with $n = 5$

5.27 Write the entire electronic arrangement for each of the following.

(a) $^{1}_{1}H$ (b) $^{7}_{3}Li$ (c) $^{20}_{10}Ne$ (d) $^{23}_{11}Na$ (e) $^{32}_{16}S$

(f) $^{37}_{17}Cl$ (g) $^{16}_{8}O$ (h) $^{27}_{13}Al$ (i) $^{79}_{35}Br$ (j) $^{59}_{27}Co$

(k) $^{91}_{40}Zr$ (l) $^{118}_{50}Sn$

5.28 How many electrons are contained in the valence shell of each of the following? What subshells are occupied by the electrons in the valence shell?

(a) $^{1}_{1}H$ (b) $^{23}_{11}Na$ (c) $^{37}_{17}Cl$ (d) $^{7}_{3}Li$ (e) $^{19}_{9}F$

(f) $^{14}_{7}N$ (g) $^{16}_{8}O$ (h) $^{39}_{19}K$ (i) $^{32}_{16}S$ (j) $^{31}_{15}P$

(k) $^{27}_{13}Al$ (l) $^{28}_{14}Si$ (m) $^{24}_{12}Mg$ (n) $^{11}_{5}B$

Orbitals

5.29 How many orbitals are in each of the following subshells?

(a) $3s$ (b) $2p$ (c) $4d$ (d) $5f$ (e) $3p$

(f) $5d$ (g) $5s$ (h) $4f$

5.30 How many electrons may be in each of the following orbitals?

(a) the $3s$ orbital

(b) one of the orbitals in the $3p$ subshell

(c) one of the orbitals in the $4d$ subshell

(d) one of the orbitals in the $5f$ subshell

5.31 On the basis of the indicated number of electrons in each subshell, determine the number of unpaired electrons.

(a) $3d^{5}$ (b) $3p^{5}$ (c) $4s^{1}$ (d) $2p^{3}$ (e) $4d^{7}$

(f) $5f^{10}$ (g) $4p^{4}$ (h) $5d^{8}$

5.32 On the basis of the electronic configuration, predict the number of unpaired electrons in each of the following elements.

(a) $^{16}_{8}O$ (b) $^{19}_{9}F$ (c) $^{7}_{3}Li$ (d) $^{32}_{16}S$ (e) $^{27}_{13}Al$

(f) $^{28}_{14}Si$ (g) $^{31}_{15}P$ (h) $^{12}_{6}C$ (i) $^{11}_{5}B$ (j) $^{35}_{17}Cl$

Electron-Dot Symbols

5.33 What is the advantage in the use of electron-dot symbols in describing the chemical properties of the elements?

5.34 Depict each of the following by electron-dot symbols.

(a) $^{16}_{8}O$ (b) $^{23}_{11}Na$ (c) $^{19}_{9}F$ (d) $^{27}_{13}Al$ (e) $^{14}_{7}N$

(f) $^{31}_{15}P$ (g) $^{11}_{5}B$ (h) $^{35}_{17}Cl$ (i) $^{7}_{3}Li$ (j) $^{39}_{19}K$

(k) $^{12}_{6}C$ (l) $^{28}_{14}Si$ (m) $^{32}_{16}S$ (n) $^{9}_{4}Be$

5.35 What similarities are noted in the electron-dot symbols of each of the following pairs of elements?

(a) $^{19}_{9}F$ and $^{35}_{17}Cl$ (b) $^{23}_{11}Na$ and $^{39}_{19}K$

(c) $^{16}_{8}O$ and $^{32}_{16}S$ (d) $^{14}_{7}N$ and $^{31}_{15}P$

(e) $^{24}_{12}Mg$ and $^{40}_{20}Ca$ (f) $^{12}_{6}C$ and $^{28}_{14}Si$

(g) $^{11}_{5}B$ and $^{27}_{13}Al$

5.36 How many electrons are indicated in the electron-dot symbol of any halogen?

5.37 How many electrons are indicated in the electron-dot symbol of any alkali metal?

Electron Configuration and the Periodic Table

5.38 Indicate the number of valence electrons for each of the following elements.

(a) silicon   (b) selenium   (c) phosphorus

(d) krypton   (e) francium   (f) arsenic

(g) calcium   (h) aluminum

5.39 In what period and group do the elements with the following electron configurations belong?

(a) $1s^2 2s^2 2p^3$ (b) $1s^2 2s^2 2p^6 3s^1$

(c) $1s^2 2s^2 2p^6 3s^2 3p^5$ (d) $1s^2 2s^2 2p^6 3s^2 3p^6 4s^2 3d^5$

5.40 How many electrons are in the p subshell of a Group IVA element?

5.41 How many unpaired electrons are in a Group VA element?

5.42 Based on the following electron configurations, determine the group in which the element is located.

(a) $1s^2 2s^2 2p^6 3s^2 3p^2$

(b) $1s^2 2s^2 2p^5$

(c) $1s^2 2s^2 2p^6$

(d) $1s^2 2s^2 2p^6 3s^2 3p^6 4s^2 3d^{10} 4p^3$

(e) $1s^2 2s^2 2p^6 3s^2 3p^6 4s^2 3d^5$

Prediction of Physical Properties

5.43 In each of the following series, estimate the missing property of the element.

(a)
Element	Atomic radius (Å)
F	0.64
Cl	?
Br	1.14

(b)
Element	Density (g/cm³)
Ca	1.54
Sr	2.60
Ba	?

(c)
Element	Melting point (°C)
Na	?
K	63
Rb	39

5.44 In each of the following series, estimate the missing property of the compound.

(a)
Compound	Boiling point (°C)
$SiCl_4$	58
$GeCl_4$	83
$SnCl_4$	?

(b)
Compound	Melting point (°C)
H_2S	−86
H_2Se	?
H_2Te	−49

(c)
Compound	Density (g/cm³)
B_2S_3	1.6
Al_2S_3	2.4
Ga_2S_3	?

5.45 The element francium is radioactive. It is estimated that about 1 ounce of francium may exist naturally. In spite of its rarity, how can one predict its properties?

5.46 How could a chemist predict the properties of an element, atomic number 112, that might be made in the future?

Atomic Radii

5.47 Explain why the sizes of the atoms do not simply increase with increasing atomic weight.

5.48 Which atom has the smallest atomic radius?

5.49 Indicate which member of each of the following pairs of elements has the larger radius.

(a) F or Cl   (b) Si or S   (c) Li or K

(d) C or F   (e) Se or Br   (f) Ge or Pb

5.50 Predict what the radius of the oxide ion would be compared to that of the oxygen atom.

5.51 Predict what the radius of the magnesium ion would be compared to that of the magnesium atom.

Prediction of Ionization Energy

5.52 Which element in the periodic table has the highest ionization energy?

5.53 Which element in the periodic table has the lowest ionization energy?

5.54 Indicate which member of each of the following pairs of elements has the higher ionization energy.

(a) O or F   (b) Li or K   (c) Cl or Br

(d) S or Se   (e) Ge or Pb   (f) Na or Mg

5.55 How is the value of the ionization energy of metals related to their tendency to form cations?

5.56 Why don't nonmetals form cations when combined with other elements?

5.57 Which Group IIIA element has the lowest ionization energy?

5.58 Which Group VIIA element has the highest ionization energy?

Prediction of Electronegativity

5.59 Indicate which member of each of the following pairs of elements has the higher electronegativity.

(a) O or F   (b) S or Se   (c) Li or K

(d) C or N   (e) Cl or Br   (f) P or S

5.60 How is the electronegativity of nonmetals related to their tendency to form anions?

Bonding

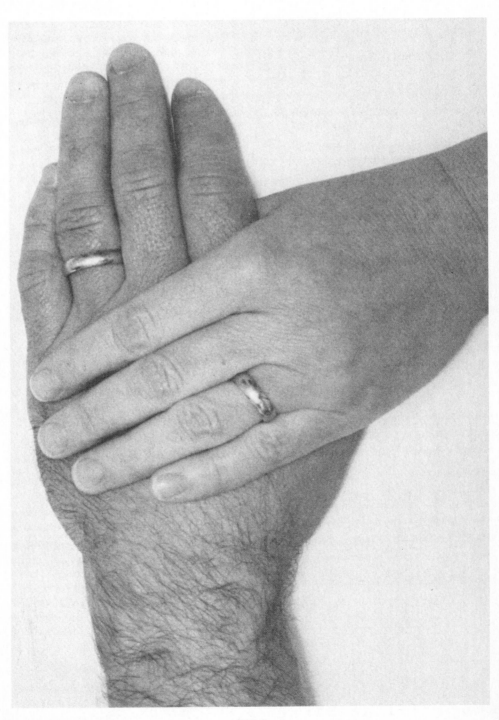

The bonds of friendship and the bond of marriage describe human relationships that hold society together. Chemical bonds hold substances together.

Most substances around us and within organisms do not consist of atoms, but of ions in ionic compounds or of atoms in combination with one another in molecules. The ionic compound most familiar to you is sodium chloride, a white crystalline solid. There are thousands of ionic compounds, but molecular compounds number in the millions. The molecules of some substances consist of very few atoms, as in the case of water, a triatomic molecule with a molecular weight of 18 amu. The molecules of most substances in living organisms are very much larger. For example, vitamin B_{12} has a molecular weight of 2×10^3 amu, and hemoglobin has a molecular weight of 6.6×10^4 amu. DNA, the substance that contains the genetic code, has a molecular weight of 2×10^{12} amu in humans.

How are the ions or atoms bonded to each other? Early chemists were puzzled by this problem, and once thought that atoms had barbs or hooks that linked them together. Thus, atoms were postulated to bond together by their own characteristic hooks in the proper number and combination. This concept is somewhat fanciful, but it was consistent with the then known facts of chemical composition and stoichiometry.

A classification of compounds based on physical properties suggests how atoms are bonded. Some substances such as sodium chloride are brittle and have a high melting point. These substances, now known as ionic compounds, have rigid crystalline structures.

The majority of substances have characteristics that are quite different from those of ionic compounds. Carbon dioxide, water, ethyl alcohol, aspirin, and a large number of medicinal compounds are examples. These compounds, which consist of molecules, tend to be gases, liquids, or low-melting solids.

The characteristics of the two broad classes of chemical compounds suggest that bonds may be of two types. Thus, the simplest approach is to describe two main bonding classes: ionic bonds and covalent bonds. In subsequent sections, the characteristics of each class will be described. Covalent bonds will be subdivided into pure covalent, polar covalent, and coordinate covalent.

The vital insight that brought the concept of bonding into the modern era came from a consideration of elements that do not form bonds! In 1916, G. N. Lewis of the University of California asked why the noble gases of Group 0 did not form compounds. All of these gases have filled valence shells. Except for helium, whose electron configuration is $1s^2$, the s and p subshells of the highest energy level contain a total of eight electrons. The stability of the filled level suggests that atoms of other elements might tend to combine to achieve an electron configuration similar to that of the noble gases.

The bonding theory of Lewis is based on the following hypotheses.

1. Only the valence electrons are involved in bonding.
2. In the ionic bond, electrons are transferred from one atom to another to form anions and cations.
3. A mutual sharing of electrons between some atoms results in molecules with covalent bonds.
4. The transfer or sharing of electrons occurs to give a noble gas electron configuration, which is an eight-electron valence shell.

These hypotheses are summarized as the **Lewis octet rule:** *Atoms other than hydrogen tend to combine and form bonds by transferring or sharing electrons until each atom is surrounded by eight valence electrons.*

In this chapter we will learn how the electrons of atoms are involved in the bonding of simple molecules. These same principles apply to very complex biochemicals. The chemical bond results from a change in the electronic structure of two or more atoms as they associate with each other. Thus, the number and type of bond formed depend on the electron configuration of the atoms involved in the bond. Therefore, a good understanding of the previous chapter is very important prior to studying bonds in compounds.

6.2
Ionic Bonds

The bond between ions is called an **ionic** *or* **electrovalent bond.** *Compounds containing ionic bonds are called* **ionic compounds** *or* **salts.** Ionic bonds are formed between two or more atoms by transfer of one or more electrons between the atoms. This electron transfer produces anions and cations. The resulting attraction between oppositely charged ions is called an ionic bond.

Let's examine the bonding in sodium chloride. The sodium atom has one valence electron ($3s^1$), whereas the chlorine atom has seven valence electrons ($3s^2 3p^5$) (Figure 6.1). In forming an ionic bond, the sodium, which has a low ionization energy, loses its valence electron to chlorine. As a result, the sodium ion has the same electron configuration as neon ($1s^2 2s^2 2p^6$). The chlorine atom, which has a high electronegativity, is converted into a chloride ion that has the same electron configuration as argon ($1s^2 2s^2 2p^6 3s^2 3p^6$). The attraction between the sodium ion and the chloride ion is an ionic bond.

A shorthand way of showing the formation of sodium chloride from the sodium and chlorine atoms involves Lewis structures. **Lewis structures** *represent only the valence electrons; unshared electron pairs are shown as a pair of dots.*

$$\text{Na} \cdot + \cdot \ddot{\underset{\cdot\cdot}{\text{Cl}}} : \longrightarrow \text{Na}^+ + : \ddot{\underset{\cdot\cdot}{\text{Cl}}} :^-$$

Sodium combines with oxygen to form the ionic compound sodium oxide, Na_2O. In this compound, each of two sodium atoms loses its valence electron and is converted into a sodium ion with a neon-like electron configuration. Thus, in total, two sodium atoms transfer their single valence electrons to oxygen. The formation of sodium oxide from the sodium and oxygen atoms is represented by electron-dot symbols as

$$2\,\text{Na} \cdot + \cdot \ddot{\underset{\cdot}{\text{O}}} : \longrightarrow 2\,\text{Na}^+ + : \ddot{\underset{\cdot\cdot}{\text{O}}} :^{2-}$$

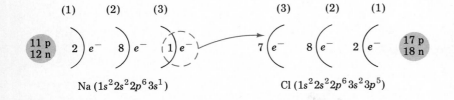

Figure 6.1
The Ionic Bond in Sodium Chloride. The transfer of an electron between sodium and chlorine atoms produces ions. (Energy levels of the atoms are in parentheses.)

In all ionic compounds, the electrons lost by one or more atoms in achieving a Lewis octet in the cation are accepted by the appropriate number of companion atoms to also achieve a Lewis octet in the anion. Ionic compounds are typically the result of combinations of metallic elements on the left side of the periodic table with nonmetals on the right side of the periodic table. The metals tend to lose electrons, because of their low ionization energy, whereas the nonmetals, because of their high electronegativity, tend to gain electrons.

Example 6.1 _____

Describe the ionic bond formed between calcium and chlorine in calcium chloride.

Solution. The electron configurations of the valence shell electrons of calcium and chlorine are $4s^2$ and $3s^2 3p^5$, respectively. Therefore, calcium can lose two electrons and achieve the stable electron configuration of argon as the Ca^{2+} cation. Chlorine can gain one electron to achieve the stable electron configuration of argon as the Cl^- anion.

$$Ca(4s^2) \longrightarrow Ca^{2+} + 2\,e^-$$

$$1\,e^- + Cl(3s^2 3p^5) \longrightarrow Cl^-(3s^2 3p^6)$$

In order to balance their mutual electronic requirements, two chlorine atoms each accept one electron from calcium and form $CaCl_2$. In terms of electron-dot symbols, the process is

$$:Ca + 2\,:\!\overset{..}{\underset{.}{Cl}}\!: \longrightarrow Ca^{2+} + 2\,:\!\overset{..}{\underset{..}{Cl}}\!:^-$$

_____

Example 6.2 _____

Describe the ionic bond formed between magnesium and oxygen in magnesium oxide.

Solution. The electron configuration of magnesium, a Group IIA element, includes the valence shell electrons given by $3s^2$. The valence shell electrons of oxygen are $2s^2 2p^4$. Magnesium can lose two electrons and form Mg^{2+}, which has a neon-like electron configuration. Oxygen can gain two electrons to form O^{2-}, which also has a neon-like electron configuration.

$$Mg(3s^2) \longrightarrow Mg^{2+}(2s^2 2p^6) + 2\,e^-$$

$$2\,e^- + O(2s^2 2p^4) \longrightarrow O^{2-}(2s^2 2p^6)$$

The +2 charge of the magnesium ion balances the −2 charge of the oxide ion, and the resulting compound is MgO. In terms of electron-dot symbols, the process is

$$Mg: + \cdot\overset{..}{\underset{.}{O}}: \longrightarrow Mg^{2+} + :\!\overset{..}{\underset{..}{O}}\!:^{2-}$$

_____

[Additional examples may be found in **6.3–6.13** at the end of the chapter.]

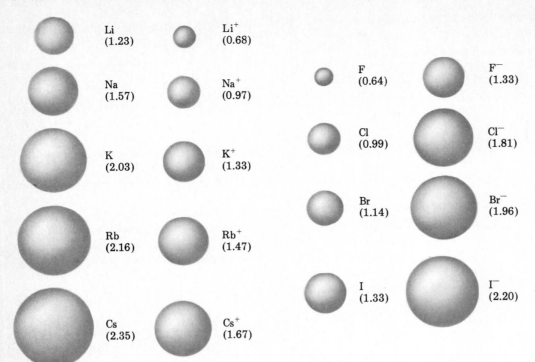

Figure 6.2

Comparison of the Sizes of Atoms and Ions.

The atoms and ions are represented as spheres within which the electrons of the atom or ion are located. The radii are given in Å.

Ionic Radii

The **ionic radius** *is the radius of a spherical ion present in ionic compounds.* The radii of ions differ from the radii of atoms. Some examples for Groups IA and VIIA are shown in Figure 6.2. For positive ions there is a decrease in radius compared to the atom. The charge on the nucleus due to protons is not changed. However, surrounding electrons are fewer in number, and there is more of a net attraction toward the nucleus. The result is a decrease in the size of the ion. For negative ions there are one or more "excess" electrons without a balancing positive charge in the nucleus. The anion then has its electrons more spread out in space.

Ion size is very important in biological systems. Several ions with the same charge do not behave the same way in biological systems. The reason for the difference is one of ionic radii. Note that the radii of Na^+ and K^+ are quite different. Many biological reactions occur with one specific ion only; no other ions are biologically active in the system.

6.3
Covalent Bonds

Covalent bonds *are electron pairs shared between atoms.* A covalent bond occurs between atoms when the difference in their electronegativities is too small for an electron transfer to occur and form ions. In a covalent bond between identical atoms, such as in H_2 or F_2, the electron pair is shared equally. In molecules containing bonds between different atoms, the sharing is unequal, and polar covalent bonds result (Section 6.4).

The hydrogen molecule is composed of two hydrogen atoms, each with one electron in a $1s$ orbital. Since both hydrogen atoms need one additional electron to complete the first energy level of the noble gas, helium, the two identical atoms join so that their two electrons are shared equally. Each atom acquires a helium-

AN ASIDE

Sodium chloride, the most common ionic compound, occurs underground in large amounts. These salt deposits, which are found in four basins in the United States, were formed when extensive primordial seas dried out millions of years ago. Sodium chloride is the most abundant dissolved compound in seawater. About 50% of the international salt production is from the solar evaporation of seawater.

Physiologically, mammals require sodium chloride to provide proper electrolyte balance. Sodium ions as well as potassium ions are widely distributed throughout the body. In fluids the amount of sodium is 30 times greater than that of potassium, but the fluid within the cells is 10 times richer in potassium than in sodium. An increase or decrease in the amount of sodium or potassium ions in the body can be very dangerous. More details on the control of ions in the body will be given in Chapter 22.

The early recognition of the necessity for sodium chloride in the diet is clear from history. Sodium chloride was used in barter, and battles have been fought over salt deposits. However, sodium chloride is obtained from many foods, and it is rarely necessary to "salt" food. In fact, our bodies cannot handle excessive amounts of salt. Sea water, which is 3% salt, will cause death if drunk because the salt will upset the proper salt balance in the body.

---

like electron configuration as long as they are bonded together. The bonding of two hydrogen atoms to form a hydrogen molecule is pictured below in a Lewis structure.

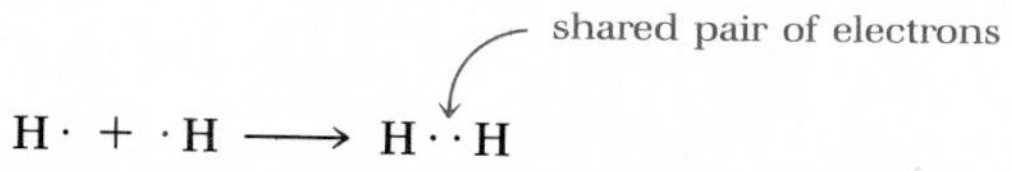

shared pair of electrons

$$H\cdot \ + \ \cdot H \longrightarrow H\overset{\cdot\cdot}{}H$$

The fluorine molecule is pictured as two fluorine atoms joined by a shared pair of electrons. Each fluorine atom requires one electron to fill the second energy level and achieve a Lewis octet or neon-like electron configuration. Because there is no difference in the ionization energy or electronegativity of the two fluorine atoms, the electron pair is shared equally. Only two of the electrons in the fluorine molecule are bonding electrons. The other pairs of electrons are called nonbonding electrons or lone pair electrons or unshared electron pairs.

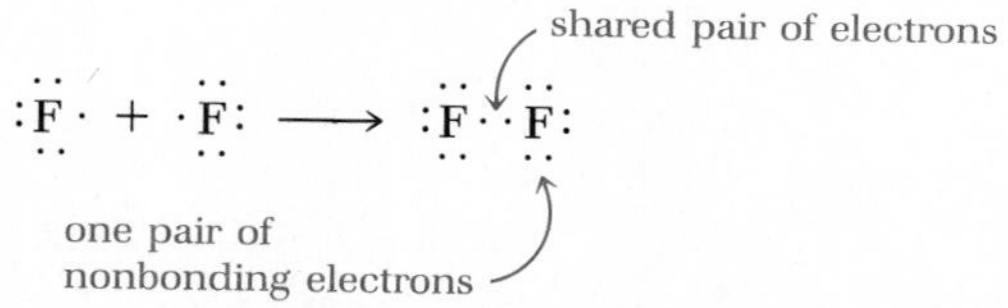

shared pair of electrons

$$:\!\overset{\cdot\cdot}{F}\!\cdot \ + \ \cdot\overset{\cdot\cdot}{F}\!: \ \longrightarrow \ :\!\overset{\cdot\cdot}{F}\overset{\cdot\cdot}{}\overset{\cdot\cdot}{F}\!:$$

one pair of
nonbonding electrons

In representing covalently bound molecules by Lewis structures, it is convenient to write a pair of electrons shared between atoms as a dash. The hydrogen and fluorine molecules are more easily written by this convention.

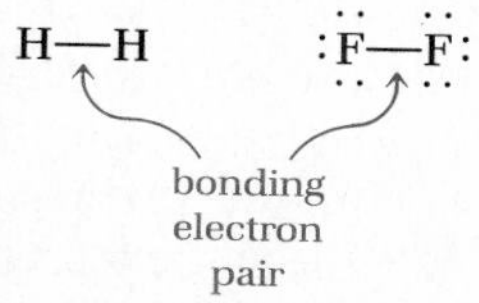

$$H\!-\!H \qquad\qquad :\!\overset{\cdot\cdot}{F}\!-\!\overset{\cdot\cdot}{F}\!:$$

bonding
electron
pair

More than one pair of electrons can be shared between pairs of atoms. *If four or six electrons are shared, the bonds are called* **double** *and* **triple bonds,** *respectively*. The nitrogen molecule consists of two nitrogen atoms bound together by six shared electrons, or a triple bond.

$$\cdot \overset{..}{N} \cdot + \cdot \overset{..}{N} \cdot \longrightarrow :N \vdots \vdots N: \quad \text{or} \quad :N \equiv N:$$

a triple bond
six shared electrons (e⁻)

unshared electron pairs

Carbon's ability to form stable bonds with other carbon atoms is responsible for the uniqueness of organic molecules. The carbon atom, which has four electrons in its valence shell, can achieve an octet of electrons by forming four covalent bonds. In ethane, ethylene, and acetylene the carbon atoms are bonded by single, double, and triple covalent bonds, respectively. The remaining bonds to carbon involve single bonds to hydrogen atoms. The structure and chemistry of these and other organic molecules will be discussed starting in Chapter 13.

single bond:
sharing of two e⁻

double bond:
sharing of four e⁻

triple bond:
sharing of six e⁻

$$\text{H}-\overset{\overset{\displaystyle\text{H}}{|}}{\underset{\underset{\displaystyle\text{H}}{|}}{\text{C}}}-\overset{\overset{\displaystyle\text{H}}{|}}{\underset{\underset{\displaystyle\text{H}}{|}}{\text{C}}}-\text{H}$$

ethane

$$\underset{\text{H}}{\overset{\text{H}}{{}}}\text{C}=\text{C}\underset{\text{H}}{\overset{\text{H}}{{}}}$$

ethylene

$$\text{H}-\text{C}\equiv\text{C}-\text{H}$$

acetylene

Example 6.3

What type of bond exists in Cl_2? Describe the electrons and orbitals involved in bond formation.

Solution. Chlorine has a $3s^2 3p^5$ electron configuration and needs only one electron to achieve a noble gas configuration. However, since both chlorine atoms are identical, the two must each share one electron and form a single covalent bond.

$$:\overset{..}{\underset{..}{Cl}}\cdot + \cdot \overset{..}{\underset{..}{Cl}}: \longrightarrow :\overset{..}{\underset{..}{Cl}}-\overset{..}{\underset{..}{Cl}}:$$

6.4
Polar Covalent Bonds

In the two previous sections we saw two extreme types of chemical bonds. The polar covalent bond is intermediate between ionic and covalent bonds. It is of considerable importance in organic chemistry and biochemistry.

A covalent bond in which the electrons are shared unequally between two nonidentical atoms is called a **polar covalent bond.** There are many molecules that contain nonidentical atoms bound together by a polar covalent bond. The hydrogen chloride molecule consists of two atoms, each of which needs one electron for a noble gas electron configuration. Chlorine has a higher electronegativity than hydrogen. However, chlorine's attraction for hydrogen's electron is not strong enough for the hydrogen to lose its electron to chlorine and form an ionic bond. Consequently, the necessary electrons are shared unequally. Hydro-

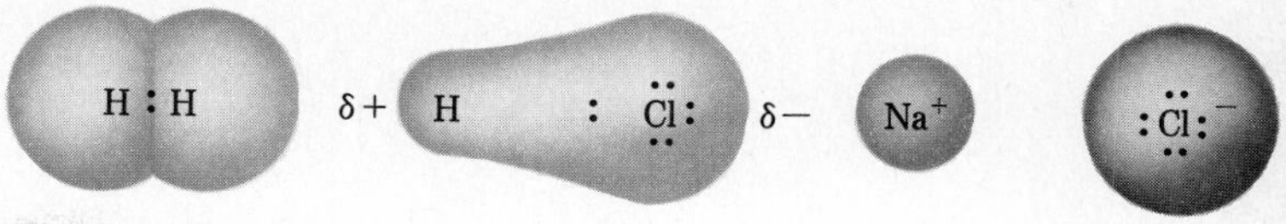

Figure 6.3

A Representation of Covalent, Polar Covalent, and Ionic Bonds. The shapes represent the volume within which electrons are located about the nuclei of the atoms. In hydrogen the volume is symmetrical about both hydrogen atoms. Note that in HCl the electron space about the hydrogen atom is diminished since the electron pair in the bond is more closely associated with the chlorine atom. In sodium chloride the electron transfer is complete and each ion is spherical.

gen chloride contains a polar covalent bond. The molecule is represented by the conventional Lewis structure, even though the shared electron pair is associated to a larger extent with chlorine.

$$^{\delta+} H—\ddot{\underset{..}{C}}l: {}^{\delta-}$$

The result of this unequal sharing is that the chlorine end of the molecule acquires a partial negative charge, whereas the hydrogen end acquires a partial positive charge. The symbol δ (the Greek lowercase letter delta) is used to denote the fractional charge located at a site within a molecule. The HCl molecule is said to possess a dipole (that is, two poles).

It should be understood that only a displacement of electrons toward chlorine occurs in HCl. There is no transfer of electrons as in NaCl. Thus, there are no ions in HCl, and the entire molecule is electrically neutral even though there are partial charges on each atom (Figure 6.3).

Example 6.4 _____

What type of bond exists in the molecule HF? Draw the Lewis structure for the molecule.

Solution. The hydrogen atom has a $1s^1$ electron configuration and requires one electron to achieve the inert gas configuration $1s^2$. The fluorine atom has a $2s^2 2p^5$ electron configuration and requires one electron to achieve an inert gas configuration $2s^2 2p^6$. If the two atoms share one electron, a covalent bond results. However, the electrons must be shared unequally as the atoms are nonidentical. The Lewis structure is

$$H—\ddot{\underset{..}{F}}:$$

_____

[Additional examples may be found in **6.28–6.30** at the end of the chapter.]

_A_ **coordinate covalent bond** _is formed when both electrons of the electron pair shared by the two atoms are derived from one atom._ However, once formed, a coordinate covalent bond is indistinguishable from any other covalent bond.

One example of the formation of a coordinate covalent bond is the reaction of ammonia with a hydrogen cation (proton) to form the ammonium ion.

6.5
Coordinate Covalent Bonds

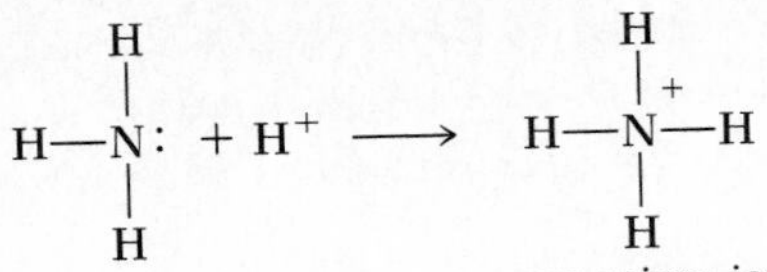

$$H-N\overset{H}{\underset{H}{\mid}}: + H^+ \longrightarrow H-\overset{H}{\underset{H}{\overset{\mid}{\underset{\mid}{N}}}}\overset{+}{-}H$$

ammonia   ammonium ion

In ammonia, the nitrogen atom shares three of its five valence electrons with hydrogen atoms. The remaining two electrons are not involved in bonding. This pair of electrons, an unshared pair of electrons, is used in forming the bond to the proton for the ammonium ion. The proton has no electrons in its valence shell and needs these two electrons to fill its shell. The proton forms a coordinate covalent bond with nitrogen. Once the coordinate covalent bond is formed, it is indistinguishable from the other three covalent bonds.

6.6
Lewis Structures for Molecules and Polyatomic Ions

In order to study chemistry, it is important that you be able to write Lewis structures of molecules and polyatomic ions easily. The following guidelines are useful.

1. The electron-dot symbols of the atoms that occur in the molecule or ion should be written.
2. The number of excess (or deficient) electrons in addition to (or less than) those present in the atoms should be indicated. For anions add one electron for each unit of negative charge. This electron may be placed on the most electronegative element. For cations remove one electron for each unit of positive charge from the least electronegative atom.
3. Each atom should be examined to determine how many electrons are required to achieve an octet.
4. Draw a single bond between each atom to be bonded by using one electron from each atom.
5. If octets are not achieved and unpaired electrons are present on bonded atoms, combine the electrons to form a second or third bond.
6. If octets are not present and only unshared pairs of electrons are available, use them to form multiple bonds and reduce the number of unshared pairs of electrons.

Example 6.5

Draw the Lewis structure for H_2O.

Solution. The electron-dot symbols of hydrogen and oxygen are as follows.

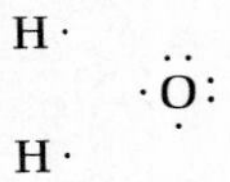

$$H\cdot \qquad \cdot\overset{..}{\underset{.}{O}}:$$
$$H\cdot$$

No extra electrons are added as the substance is electrically neutral. The oxygen atom requires two electrons to achieve an octet. Each hydrogen atom requires one electron. Thus, arranging the hydrogen atoms about the oxygen atom produces the requisite number of covalent bonds.

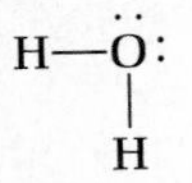

$$H-\overset{..}{\underset{\mid}{O}}:$$
$$\quad\; H$$

Example 6.6 _____

Draw the Lewis structure for CCl_4.

Solution. The electron-dot symbols of carbon and chlorine are as follows.

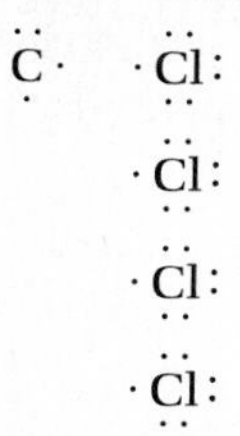

$$\dot{\underset{\cdot}{C}}\cdot \quad \cdot \ddot{\underset{\cdot\cdot}{Cl}}:$$

$$\cdot \ddot{\underset{\cdot\cdot}{Cl}}:$$

$$\cdot \ddot{\underset{\cdot\cdot}{Cl}}:$$

$$\cdot \ddot{\underset{\cdot\cdot}{Cl}}:$$

Each chlorine atom needs to gain one electron. The carbon atom requires four additional electrons, or it may donate four electrons, to achieve a noble gas electron configuration. In order to achieve a balance between carbon and chlorine, the carbon must use all four of its valence shell electrons.

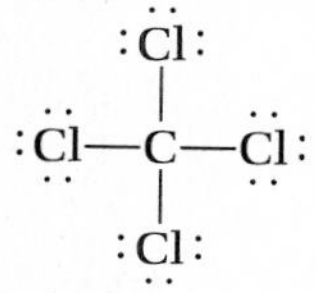

$$
\begin{array}{c}
:\ddot{Cl}: \\
| \\
:\ddot{Cl}-C-\ddot{Cl}: \\
| \\
:\ddot{Cl}:
\end{array}
$$

[Additional examples may be found in **6.19** and **6.20** at the end of the chapter.]

Example 6.7 _____

Draw the Lewis structure for the hydroxide ion, OH^-.

Solution. The electron-dot symbols of hydrogen and oxygen are as follows.

$$H\cdot \quad \cdot \ddot{\underset{\cdot}{O}}:$$

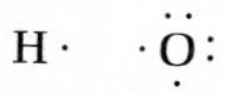

Since the ion bears a negative charge, an extra electron must be added. For clarity, a small circle, ∘, will be used for this electron, which is placed on the more electronegative element.

$$H\cdot \quad \cdot \ddot{\underset{\circ}{O}}:^-$$

Now it can be seen that oxygen with its extra electron needs only one more electron for an octet. The hydrogen and oxygen share one electron each.

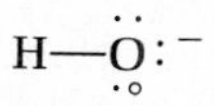

$$H-\ddot{\underset{\circ}{O}}:^-$$

Example 6.8 _____

Draw the Lewis structure for the sulfite ion, SO_3^{2-}.

Solution. The electron-dot symbols of sulfur and oxygen are as follows.

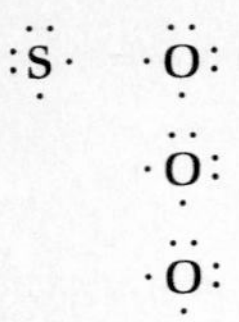

:S· ·O:

·O:

·O:

Two electrons are added to the collection of atoms to achieve the negative charge. One is added to each of two oxygen atoms, which are more electronegative than sulfur.

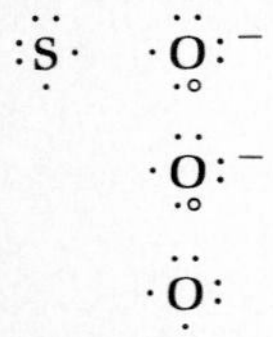

:S· ·O:⁻

·O:⁻

·O:

Sulfur requires only two electrons to achieve an octet. Two of the oxygens require one electron each, but the third oxygen requires two electrons. First the oxygens requiring only one electron are bonded to sulfur.

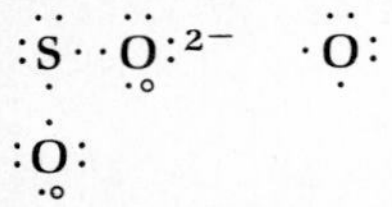

:S··O:²⁻ ·O:

:O:

The remaining oxygen atom's electrons are rearranged to produce three pairs, and it then bonds to one of sulfur's unshared pairs of electrons.

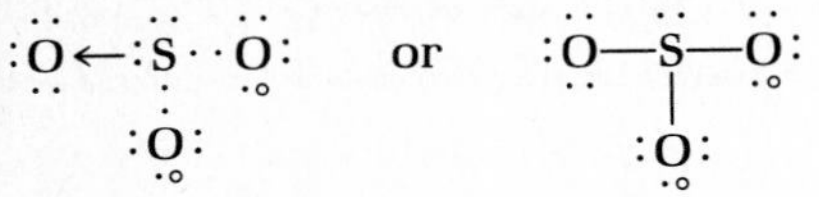

:O←:S··O: or   :O—S—O:

:O: :O:

Once the coordinate covalent bond is formed, it is no different from the other two sulfur–oxygen bonds.

[Additional examples may be found in **6.21** at the end of the chapter.]

Example 6.9

Write the Lewis structure for carbon bisulfide, CS_2.

Solution. The electron-dot symbols of the elements are

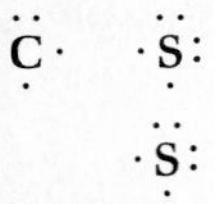

C· ·S:

·S:

Each sulfur requires two electrons and the carbon requires four electrons. If only one electron were shared between carbon and each sulfur, neither element would achieve an octet of electrons.

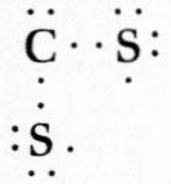

C··S:

:S·

The solution is to have each sulfur share two electrons with carbon to form double covalent bonds. The electron dots are rearranged to achieve the necessary bonds.

$$:\ddot{S}=C=\ddot{S}:$$

Example 6.10 ——————————————————————————

Write the Lewis structure for formaldehyde, H_2CO (the hydrogen atoms are bonded to carbon).

Solution. The electron-dot symbols of the elements are

H·
    ·Ċ: ·Ö:
H·

Each hydrogen atom requires one electron to achieve the helium-like config-uration. The carbon atom provides these two electrons.

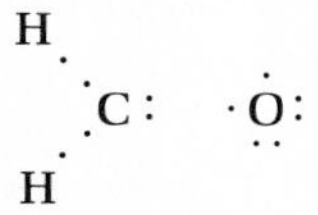

H·
   :Ċ: ·Ö:
H·

Now carbon has a total of six electrons and needs two to achieve its octet. Simi-larly, oxygen needs two electrons. The solution is to have carbon and oxygen each share two electrons for a total of four electrons.

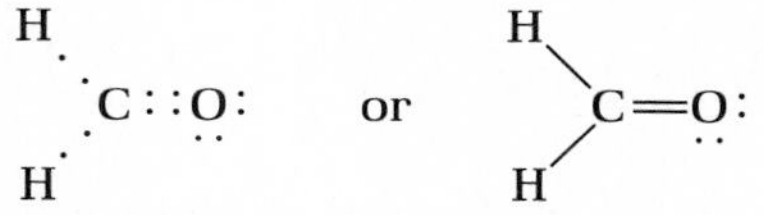

H·
   :C::Ö: or $\overset{H}{\underset{H}{{\Large{\diagdown}\atop\diagup}}}$C=Ö:
H·

[Additional examples may be found in **6.27** at the end of the chapter.]

There are cases in which a single Lewis structure for a molecule based on the Lewis octet rule is not entirely correct. By that we mean that some of the proper-ties of the molecule are different from those expected from the electron-dot structure.

Let's consider the molecule sulfur dioxide, SO_2. A structure based on the Lewis octet rule is

$$:\ddot{O}=\ddot{S}-\ddot{O}:$$

The sulfur and each oxygen have or are sharing a total of eight electrons. The structure then has both a single and a double bond between the sulfur and the oxygen atoms.

The actual SO_2 molecule has bonds to each oxygen atom that are equal in every way. The two sulfur–oxygen bonds are identical and, therefore, the Lewis structure does not describe the molecule accurately. The real structure of SO_2 can be better represented as a hybrid of two Lewis structures, neither of which is completely correct.

$$:\ddot{O}=\ddot{S}-\ddot{O}: \quad \text{and} \quad :\ddot{O}-\ddot{S}=\ddot{O}:$$

Since the actual sulfur bonds are equal, they must be intermediate between single and double bonds. The problem is that we cannot simply represent the actual molecule with a single Lewis structure. A double-headed arrow between two Lewis structures then is used to indicate that the actual structure is similar in part to the two simple structures, but lies somewhere between them. *The individual Lewis structures are called* **contributing** *or* **resonance structures.**

$$:\ddot{O}=S-\ddot{O}: \longleftrightarrow :\ddot{O}-S=\ddot{O}:$$

6.7
Oxidation Numbers

In this chapter we have seen that atoms combine by losing and gaining electrons to form ionic compounds or by sharing electrons in covalent compounds. In covalent compounds of nonidentical atoms the electrons are shared unequally. It is convenient to indicate either the loss or gain of electrons be it complete or partial in terms of a bookkeeping system that uses oxidation numbers. *The* **oxidation number** *of an atom is a positive or negative integer assigned to describe an atom as a free atom, an ion, or in a molecule.* As an ion or in an ionic compound, the charge of a cation or an anion results from the transfer of electrons away from or to the neutral atom. The positive charge on the ion indicates the number of electrons lost, whereas a negative charge indicates the number of electrons gained relative to the neutral atom.

As a reference point, **the oxidation number of an element, regardless of whether it is monatomic, diatomic, and so on, is set at zero.** When an electron is removed from an atom as Na to yield Na^+, the oxidation number of the cation is $+1$. In the case of gaining an electron, such as Cl to yield Cl^-, the oxidation number is -1. Thus, **the charge of a monatomic ion is equal to its oxidation number.** These oxidation numbers provide information about the compounds that may be formed. One sodium ion whose oxidation number is $+1$ combines with one chloride ion whose oxidation number is -1. **In any ionic compound, the sum of the oxidation numbers must be numerically equal to zero, since the compound is electrically neutral overall.**

In covalent compounds, electrons are not completely transferred, but rather are shared between atoms. The pure covalent bond that exists in elements such as H_2, F_2, P_4, and S_8 is the result of equal sharing. No single atom has gained or lost an electron, and therefore the oxidation number of each atom in the molecule is zero.

The sharing of electrons in polar covalent bonds is unequal. As a result, one atom gains a partial negative charge at the expense of another atom, which then has a partial positive charge. For the sake of convenience, the oxidation number of atoms bonded by polar covalent bonds is based on an assumed complete electron transfer to the more electronegative atom.

The most electronegative element in the polar covalent bond is assigned the negative oxidation number because it has the greatest attraction for electrons. Thus, in carbon tetrachloride, CCl_4, each electronegative chlorine atom is assigned a -1 oxidation number based on its tendency to gain an electron in its unequal sharing with carbon. Carbon is assigned a $+4$ oxidation number because it has a tendency to lose four electrons in bonding to four chlorine atoms. Since the number of electrons partially gained or lost must be equivalent, **the sum of the oxidation numbers of atoms in a molecule must equal zero.**

In order to facilitate the assignment of oxidation numbers, the rules are listed in Table 6.1.

Table 6.1

Oxidation Number Rules

1. The oxidation number of an element in its uncombined state is zero.
2. The algebraic sum of the oxidation numbers in an ionic or covalent compound is zero.
3. The oxidation number of a monatomic ion is the same as the charge of the ion.
4. The algebraic sum of the oxidation numbers in a complex ion equals the charge of the ion.
5. Metals generally have positive oxidation numbers in the combined state.
6. Negative oxidation numbers in covalent compounds of two unlike atoms are assigned to the more electronegative atom.
7. Most hydrogen compounds contain hydrogen with a +1 oxidation number.
8. In most oxygen compounds, the oxidation number is −2.
9. The oxidation number of F, Cl, Br, and I is −1 except when combined with a more electronegative element.
10. Sulfides have an oxidation number of −2.

Example 6.11

What is the oxidation number of sulfur in SO_3?

Solution. The sum of the oxidation numbers of a neutral compound must be zero (rule 2).

$$\text{oxidation number of sulfur} + \left(3 \times \text{oxidation number of oxygen}\right) = 0$$

Since the oxidation number of oxygen is −2 (rule 8), the oxidation number of sulfur may be calculated algebraically.

$$\text{oxidation number of sulfur} + 3(-2) = 0$$
$$\text{oxidation number of sulfur} - 6 = 0$$
$$\text{oxidation number of sulfur} = +6$$

Example 6.12

What is the oxidation number of nitrogen in $NaNO_2$?

Solution. The sum of the oxidation numbers of sodium, nitrogen, and oxygen must be equal to zero (rule 2).

$$\text{oxidation number of sodium} + \text{oxidation number of nitrogen} + \left(2 \times \text{oxidation number of oxygen}\right) = 0$$

The oxidation number of oxygen is −2 (rule 8). The oxidation number of sodium is positive since it is a metal (rule 5). Sodium is known to exist as only the +1 ion, and its oxidation number is equal to that charge (rule 3).

$$+1 + \genfrac{}{}{0pt}{}{\text{oxidation number}}{\text{of nitrogen}} + 2(-2) = 0$$

$$\genfrac{}{}{0pt}{}{\text{oxidation number}}{\text{of nitrogen}} + 1 - 4 = 0$$

oxidation number of nitrogen $= +3$

Example 6.13 ⎯⎯⎯⎯⎯⎯⎯⎯⎯⎯⎯⎯⎯⎯⎯⎯⎯

What is the oxidation number of manganese in the permanganate ion, MnO_4^-?

Solution. The sum of the oxidation numbers of a complex ion is equal to the charge of the ion (rule 4).

$$\genfrac{}{}{0pt}{}{\text{oxidation number}}{\text{of manganese}} + \left(4 \times \genfrac{}{}{0pt}{}{\text{oxidation number}}{\text{of oxygen}}\right) = -1$$

Since the oxidation number of oxygen is -2 (rule 8), the oxidation number of manganese can be calculated.

oxidation number of manganese $+ 4(-2) = -1$

oxidation number of manganese $- 8 = -1$

oxidation number of manganese $= +7$

[Additional examples may be found in **6.31–6.36** at the end of the chapter.]

6.8
Molecular Geometries

Up to this point, molecules have been represented in two dimensions. However, as you know, molecules are three dimensional. The three-dimensional features of molecules are important and affect both physical and chemical properties of even simple molecules such as water. In the case of the larger molecules that you will encounter in organic chemistry and biochemistry, the shapes of molecules are exceedingly important and determine their chemical reactivity.

The shape of a molecule as given by the positions of the nuclei of atoms in space is determined by the electrons that bond the atoms together as well as the unshared pairs of electrons. However, only the positions of the nuclei need to be indicated to visualize the molecular shape (Figure 6.4).

A molecule of two atoms (diatomic) is necessarily linear and has its nuclei along a line. Such a molecule is called linear. A molecule composed of three atoms, such as H_2O, must have all atoms in a plane. Such molecules could be linear or angular, but water is angular with a H—O—H bond angle of 105°. Molecules composed of more than three atoms could be linear or planar, but other shapes are also possible.

Ammonia has a nitrogen atom above a plane of three hydrogen atoms. Since the nitrogen atom is bonded to each of the hydrogen atoms, the molecule is described as trigonal pyramidal or simply pyramidal. The H—N—H bond angles are all 107°.

Molecules containing four atoms bonded to a central atom exist in a variety of shapes. In methane, CH_4 (Figure 6.4), the carbon atom is at the center of the

Figure 6.4

The Shapes of Some Simple Molecules.
The colored lines represent the geometric
shape of the molecule and are not bonds.

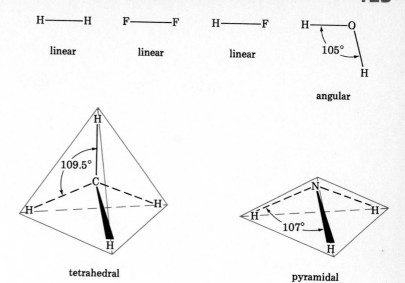

tetrahedral

pyramidal

tetrahedron with the four hydrogen atoms at the corners. The H—C—H bond
angles are all 109.5°.

While water, ammonia, and methane have shapes described as angular, pyramidal, and tetrahedral, respectively, there are some important similarities. Note that
there are four electron pairs about each central atom. Some of the electron pairs
form bonds to other atoms, while others are unshared (nonbonded). If we visualize oxygen, nitrogen, as well as carbon, as being at the center of a tetrahedron, an
interesting picture results (Figure 6.5), which has been called the **valence-shell
electron-pair repulsion theory** or **VSEPR** for short. *Electrons in chemical
bonds and unshared pairs of electrons repel one another and tend to remain as far
apart as possible.*

A good analogy for the concept of maximum separation of electron pairs
involves balloons. Consider the axis going through a balloon from the tied end to
the opposite curved surface (Figure 6.6). Now imagine four balloons tied together
closely at one point. The balloons should arrange themselves to be as uncrowded

6.9
Valence-Shell
Electron-Pair
Repulsion Theory

Figure 6.5

Location of Electron
Pairs in Some Simple
Molecules.
All electron pairs in
methane, ammonia,
and water are directed
to the corners of a tet-
rahedron. Note, how-
ever, that ammonia is
still described as py-
ramidal and water as
an angular molecules.

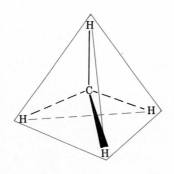
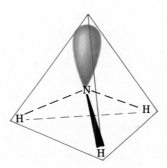
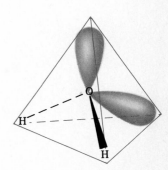

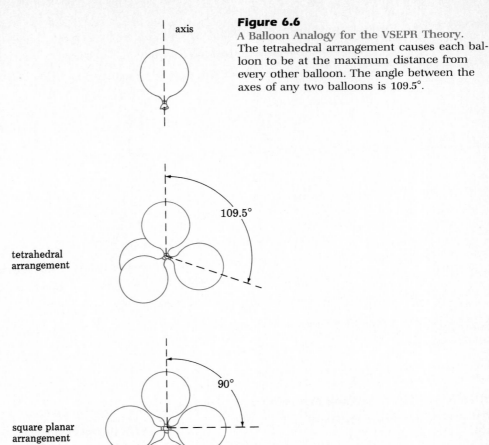

Figure 6.6

A Balloon Analogy for the VSEPR Theory. The tetrahedral arrangement causes each balloon to be at the maximum distance from every other balloon. The angle between the axes of any two balloons is 109.5°.

as possible. That arrangement is not planar because the angles between the axes are only 90°. But in a tetrahedron each angle between axes is 109.5°. This angle is equal to the H—C—H bond angle in methane. Thus, conceptually we can regard the *four* bonded pairs of electrons in methane like four equivalent balloons.

In ammonia and water, the central atom is also surrounded by four pairs of electrons. However, we do not say that these molecules have a tetrahedral shape. Shape is described by a geometric figure that results when nuclei are bonded by straight lines. The VSEPR theory describes the distribution of electron pairs including the nonbonded pairs.

Ammonia has three bonding pairs and one nonbonded pair of electrons. The bond angles are the result of the bonding pairs of electrons. However, their positions are also affected by the presence of the nonbonded electron pair. The H—N—H bond angles of 107° are somewhat less than the 109.5° tetrahedral bond angles of methane. This smaller angle can be explained by assuming that the nonbonded electron pair is more spread out (a larger balloon) and pushes the bonding electron pairs closer together.

Water has two bonded pairs and two nonbonded pairs of electrons. The two nonbonded pairs exert an even greater effect on the bonded pairs and reduce the bond angle still further to 104.5°.

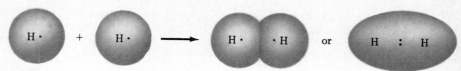

Figure 6.7

The Sharing of Electrons in a Sigma Bond of Hydrogen.
The region of space occupied by the electron pair is symmetrical about both
hydrogen atoms. Although the two electrons may be located anywhere within
the volume, it is most probable that the electrons are located between the nuclei
of the two atoms.

6.10 Orbitals and Molecular Shapes

Since bonds are formed between atoms by electrons in orbitals, the shape of
molecules depends on the location of these electrons in the orbitals. In order for
two electrons to be shared, it is necessary that they share a region of space
common to the bonding atoms. This process is pictured as an overlap or merging
of two atomic orbitals to form a molecular orbital. In the molecular orbital there
are two electrons of opposite spin.

A hydrogen molecule can be viewed as the overlap of the *s* orbitals of two
hydrogen atoms (Figure 6.7). The covalent bond is the result of the attraction of
the two nuclei for the electron pair. While the paired electrons may be anywhere
within the molecular orbital, the probability is highest that they are located between the nuclei.

The covalent bond in hydrogen is called a sigma (σ) bond. **Sigma bonds** *are
covalent bonds that are symmetrical around an axis joining the two nuclei.* If the
hydrogen molecule and its molecular orbital were rotated about the axis, their
appearance would remain unchanged. The lack of change by rotation means that
the molecule is symmetrical about the axis.

Many atoms do not use their *s* and *p* orbitals directly in bonding. Instead, *a*
hybrid *type of* **orbital** *results from the interaction of the many orbitals in the
bonded atoms.* Although the hybrid orbitals can be described by advanced mathematical equations, it is not appropriate to use such equations in this text. We
need only to know that the molecular shape is a consequence of the arrangement
of electrons about the atom. A simple nonmathematical method of understanding molecular shapes is based on the fact that electron pairs in compounds
containing only single bonds will be arranged to be as far apart as possible.

There are a number of simple but important molecules that have a central
atom surrounded by four electron pairs. Examples are the hydrogen compounds
of the second period elements carbon, nitrogen, and oxygen: CH_4, NH_3, and H_2O
(see Table 6.2).

Let's first examine CH_4 to see how its tetrahedral shape can be rationalized
by hybrid orbitals. The electronic configuration of an isolated carbon atom suggests that only two electrons are available to share with two hydrogen atoms to
form CH_2. In this way carbon would not have an octet of electrons.

$$:C—H$$
$$|$$
$$H$$

Moreover, since the unpaired electrons in carbon are in mutually perpendicular
p orbitals, the resultant bond angle should be 90°.

In order to form four bonds in CH_4 the carbon atom must contribute four
electrons. Furthermore, since CH_4 is a tetrahedral molecule, these four electrons

Table 6.2

Hybridized Atomic Orbitals in Molecules

	Electron configuration		
Isolated atom	**Hybridized atom**	**Hybridized atom and bonded atoms**	**Orbitals**
$\underset{2s}{\underline{\uparrow\downarrow}}\ \underset{2p_x}{\underline{\uparrow}}\ \underset{2p_y}{\underline{\uparrow}}\ \underset{2p_z}{\underline{}}$ one s + three p orbitals C atom	$\underset{sp^3}{\underline{\uparrow}\ \underline{\uparrow}\ \underline{\uparrow}\ \underline{\uparrow}}$ four sp^3 hybrid orbitals C atom in CH_4	$\underset{sp^3}{\underline{\uparrow\downarrow}\ \underline{\uparrow\downarrow}\ \underline{\uparrow\downarrow}\ \underline{\uparrow\downarrow}}$ CH_4: C atom and four H atoms	
$\underset{2s}{\underline{\uparrow\downarrow}}\ \underset{2p_x}{\underline{\uparrow}}\ \underset{2p_y}{\underline{\uparrow}}\ \underset{2p_z}{\underline{\uparrow}}$ N atom	$\underset{sp^3}{\underline{\uparrow\downarrow}\ \underline{\uparrow}\ \underline{\uparrow}\ \underline{\uparrow}}$ N atom in NH_3	$\underset{sp^3}{\underline{\uparrow\downarrow}\ \underline{\uparrow\downarrow}\ \underline{\uparrow\downarrow}\ \underline{\uparrow\downarrow}}$ NH_3: N atom + three H atoms	
$\underset{2s}{\underline{\uparrow\downarrow}}\ \underset{2p_x}{\underline{\uparrow\downarrow}}\ \underset{2p_y}{\underline{\uparrow}}\ \underset{2p_z}{\underline{\uparrow}}$ O atom	$\underset{sp^3}{\underline{\uparrow\downarrow}\ \underline{\uparrow\downarrow}\ \underline{\uparrow}\ \underline{\uparrow}}$ O atom in H_2O	$\underset{sp^3}{\underline{\uparrow\downarrow}\ \underline{\uparrow\downarrow}\ \underline{\uparrow\downarrow}\ \underline{\uparrow\downarrow}}$ H_2O: atom + two H atoms	

must be in orbitals at 109.5° to each other. The theory is that the four orbitals of carbon are mixed or hybridized to form four new hybrid orbitals (Table 6.2). The four new orbitals are called sp^3 hybrid orbitals. The term sp^3 means that one s and three p orbitals have been used to create the four new orbitals. The four valence electrons are located in each of the four hybrid orbitals. Each orbital then interacts with a hydrogen $1s$ orbital to form a σ bond.

In ammonia, the nitrogen atom is pictured as sp^3 hybridized. The five valence electrons are distributed in four orbitals so that one pair and three unpaired electrons result. Each of the nitrogen–hydrogen σ bonds in NH_3 is the result of an overlap of a sp^3 hybrid orbital of nitrogen with a $1s$ orbital of hydrogen.

In water, the oxygen atom is also pictured as sp^3 hybridized. The six valence electrons of oxygen are distributed in four orbitals to give two pairs of electrons and two unpaired electrons. Two σ bonds with hydrogen result.

Carbon forms double as well as triple bonds. Two examples given in Section 6.3 are ethylene, C_2H_4, and acetylene, C_2H_2. In ethylene all six nuclei lie in a plane and all the bond angles are close to $120°$ (Figure 6.8). In acetylene the four atoms are arranged to form a linear molecule. The bond angles are then $180°$ (Figure 6.9).

Each carbon atom in ethylene is pictured as having three hybridized orbitals and one remaining p orbital. The hybrid orbitals are three sp^2 orbitals, since they result from a single s orbital and two p orbitals. The four valence electrons are

6.11
Multiple Bonds and Hybrid Orbitals

Figure 6.8
The Hybridization of Carbon in a Double Bond of Ethylene.

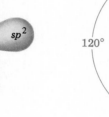

isolated carbon atom   C atom in C_2H_4

The three sp^2 orbitals lie in a plane with a $120°$ angle between them. The remaining p orbital is perpendicular to the hybrid sp^2 orbitals.

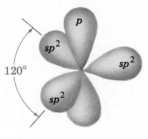

side view   top view

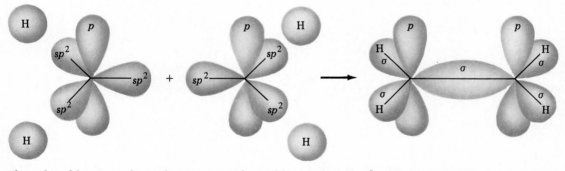

The σ bond between the carbon atoms is formed by overlap of sp^2 orbitals. The σ bonds to hydrogen are formed by overlap of the sp^2 orbitals of carbon with the s orbital of individual hydrogen atoms.

The π bond is formed by two electrons in overlapping parallel p orbitals.

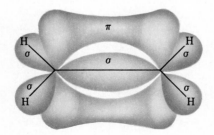

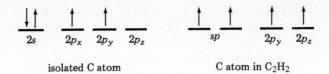

isolated C atom   C atom in C_2H_2

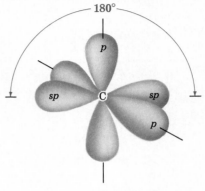

The two *sp* hybrid orbitals are at 180° to each other. The two *p* orbitals are mutually perpendicular to each other and to the axis of the *sp* orbitals.

sp hybrid orbitals

A σ bond is formed by end to end overlap of one *sp* orbital from each carbon atom. Two sets of parallel-oriented *p* orbitals form two mutually perpendicular π bonds. The remaining *sp* orbital of each carbon atom forms a σ bond to a hydrogen atom.

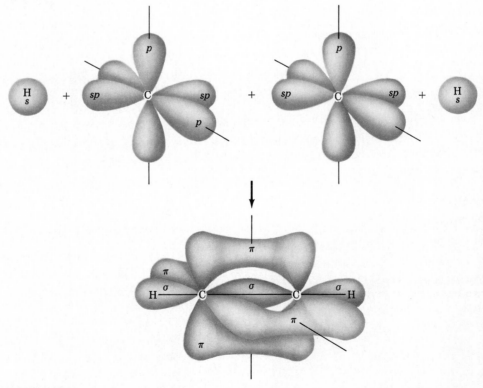

Figure 6.9
The Hybridization of Carbon in a Triple Bond of Acetylene.

distributed as indicated in Figure 6.8. Each of the three sp^2 orbitals should be separated by 120° in a plane in order to have maximum separation. Two of these orbitals containing one electron each then form bonds with hydrogen. The third orbital with one electron forms a bond with the other carbon atom in ethylene. Each of these three bonds formed by sp^2 hybrid orbitals is a σ bond.

The second bond of the double bond in ethylene results from a sidewise overlap of the p orbital of each carbon atom. The p orbital is perpendicular to the plane containing the sp^2 orbitals. One electron in the p orbital of each atom then gives an electron pair for the second bond. Note that electrons are shared in regions of space both above and below the molecule. It is nevertheless only one bond. Since the electron pair is not concentrated along an axis between the two atoms, the bond is not of the σ type. *A bond formed by sidewise overlap of p orbitals is a* **pi (π) bond.** Later, in the organic chemistry chapters, the chemical reactivity of the π bond will be discussed.

In acetylene the carbon–carbon triple bond consists of one σ bond and two π bonds. Carbon is viewed as being hybridized to form two sp hybrid orbitals with two p orbitals remaining (Figure 6.9). The two sp hybrid orbitals are at 180° angles to each other. One sp orbital and its unpaired electron form a bond with hydrogen. The other sp orbital forms a σ bond with the second carbon atom.

The second and third bonds between carbon atoms result from overlap of p orbitals. One set of p orbitals overlaps in front and back of the molecule to form one π bond. The second set of p orbitals overlaps above and below the molecule to form the second π bond. Acetylene and other organic compounds will be discussed in the chapters on organic chemistry.

Summary

The Lewis theory of bonding involves the valence electrons and the tendency of atoms to achieve a noble gas electron configuration. When the theory is applied to combinations of metals with nonmetals, it predicts the formation of the correct ions and derived formula units.

Covalently bonded molecules usually can be described structurally by sharing valence electrons in bonds according to a set of rules. At times two or more plausible structures may be drawn for a molecule. In those cases the true structure is said to be a resonance hybrid represented as a combination of the Lewis structures.

The concept of oxidation states provides a bookkeeping system for electrons. Bonded electron pairs are assigned to the more electronegative atom.

The valence-shell electron-pair repulsion (VSEPR) theory can be used to predict the shapes of molecules. The shape of a molecule depends on the distribution of bonding pairs and nonbonding pairs, which tend to achieve maximum separation from each other.

Some simple covalent molecules can be described by using the overlap of pure atomic orbitals. Many molecules have hybridized atomic orbitals that depend on the number and type of atomic orbitals required to form them. The spatial distribution of the hybrid orbitals corresponds to the distribution predicted by the VSEPR model.

There are two types of bonds that result from orbital overlap. End to end overlap along a line between the nuclei of the atoms is a sigma (σ) bond. Sidewise overlap of two p orbitals is a pi (π) bond.

Learning Objectives

As a result of studying Chapter 6 you should be able to

- Use electron configurations to explain why atoms combine.
- Classify all bonds by category.
- Formulate the electron transfers necessary between atoms to form ionic bonds.
- Arrange shared electrons to depict covalent bonds in molecules.
- Use electronegativities to predict which bonds will be polar or nonpolar.
- Calculate the oxidation number of an element in an ion or compound.

- Use the VSEPR theory to explain the shapes of molecules.
- Recognize when sp^3, sp^2, and sp hybrid orbitals are involved in bonds.
- Distinguish between σ and π bonds.

New Terms

A **coordinate covalent bond** is formed by the contribution of the electrons from just one atom.

A **covalent bond** is formed by the sharing of a pair of electrons between two atoms.

A **double bond** is formed by the sharing of two pairs of electrons between two atoms.

A **hybrid orbital** is one of a set of orbitals that replaces two or more pure atomic orbitals in forming some covalent bonds.

An **ionic bond** results from the transfer of electrons from a metal to a nonmetal.

Ionic compounds or **salts** are neutral collections of oppositely charged ions.

The **ionic radius** is the radius of a spherical ion present in ionic compounds.

The **Lewis octet** refers to the eight electrons in the valence shell of an atom or ion.

The **oxidation number** is an account of the number of electrons an atom loses, gains, or shares in bonds to other atoms.

A **polar covalent bond** is formed by sharing electrons between two atoms of unequal electronegativity.

A **pi (π) bond** results from the sidewise overlap of two p orbitals.

Resonance or **contributing structures** are structural representations of a molecule used when two or more plausible Lewis structures can be written.

A **sigma (σ) bond** results from end to end overlap of orbitals along a line joining the nuclei.

A **triple bond** is formed by the sharing of three pairs of electrons between two atoms.

The **valence-shell electron pair repulsion (VSEPR) theory** relates the shape of molecules to the distribution of electron pairs about a central atom.

Questions and Problems

Terminology

6.1 Contrast the use of the VSEPR theory with the models of hybrid orbitals used in describing geometric shapes of molecules.

6.2 Explain what experimental observations lead to the classification of bonds into two types.

Lewis Octet Theory of Ions

6.3 What group in the periodic table gave Lewis the idea that an octet of electrons is achieved when ions are formed?

6.4 Consider the electron configuration of Group IIA elements. What type of ion should be formed by these elements based on the Lewis octet theory?

6.5 Consider the electron configuration of Group VIA elements. What type of ion should be formed by these elements based on the Lewis octet theory?

6.6 Explain why carbon doesn't form C^{4+} or C^{4-} ions.

Electron Configuration of Ions

6.7 Write out the entire electron configuration of each of the following ions.

  (a) H^+ (b) H^- (c) Li^+ (d) Na^+
  (e) K^+ (f) Mg^{2+} (g) Ca^{2+} (h) F^-
  (i) Cl^- (j) Br^- (k) I^- (l) O^{2-}
  (m) S^{2-} (n) N^{3-} (o) P^{3-}

6.8 Write the electron configuration of only the valence shell electrons of the ions in **6.7.**

6.9 Write the electron-dot symbols of the ions in **6.7.**

Ionic Bonds

6.10 Write the formula for the ionic compound that results from the combination of each pair of ions.

  (a) Na^+ and Cl^- (b) Ca^{2+} and Cl^-
  (c) Mg^{2+} and O^{2-} (d) Mg^{2+} and Br^-
  (e) Ba^{2+} and F^- (f) Ba^{2+} and P^{3-}
  (g) Li^+ and N^{3-} (h) Al^{3+} and Cl^-

6.11 Write the formula for the ionic compound that results from the reaction of each pair of atoms.

  (a) Li and F   (b) Mg and Br   (c) Li and O
  (d) Mg and S   (e) Al and F   (f) Na and Se
  (g) Ca and I   (h) Na and N

6.12 Scandium combines with chlorine to form $ScCl_3$. Using the Lewis octet theory, explain this observation.

6.13 Zinc forms a +2 ion. Using the Lewis octet theory, explain this observation.

Ionic Radii

6.14 What generalization can be made about the size of an anion compared to the atom of an element?

6.15 What generalization can be made about the size of a cation compared to the atom of an element?

6.16 Which is larger in each of the following? Explain why.

  (a) Mg or Mg^{2+} (b) K or K^+ (c) Al or Al^{3+}
  (d) Br or Br^- (e) S or S^{2-} (f) N or N^{3-}

6.17 Which is larger in each of the following? Explain why.

  (a) O^{2-} or F^- (b) Na^+ or Mg^{2+}
  (c) S^{2-} or Cl^- (d) Al^{3+} or Mg^{2+}
  (e) Mg^{2+} or Ca^{2+} (f) Br^- or I^-

Covalent Bonds

6.18 Why can true covalent bonds only result from bonding of identical atoms?

6.19 Write Lewis structures for the following covalent molecules.

  (a) H_2 (b) I_2 (c) HF   (d) Br_2
  (e) Cl_2 (f) N_2 (g) HI   (h) HBr

6.20 Write Lewis structures for the following covalent molecules.

(a) H_2S (b) PH_3 (c) CF_4 (d) H_2Se
(e) CBr_4 (f) $SiCl_4$ (g) SiH_4 (h) PCl_3
(i) SbH_3

6.21 Write Lewis structures for the following ions.

(a) SH^- (b) PH_4^+ (c) H_3O^+ (d) CN^-
(e) SO_4^{2-} (f) NH_4^+

Lewis Octet Theory of Molecules

6.22 Using the Lewis octet theory, describe the Br_2 molecule.

6.23 Using the Lewis octet theory, describe the ICl molecule.

6.24 Tin forms compounds similar to those of carbon because both elements are in Group IVA. What compound should result from tin and hydrogen? Draw the Lewis structure.

6.25 Based on locations in the periodic table, predict the compound that will be formed between antimony and hydrogen. Draw the Lewis structure.

6.26 Based on locations in the periodic table, predict the compound that will be formed between silicon and chlorine.

6.27 The geometric arrangement of atoms in several molecules is given below. Using dashes to represent bonds and electron pairs, draw Lewis structures for the molecules.

(a)
```
      F
   F  C  F
      F
```

(b)
```
      H    H
   H  C  O
      H
```

(c)
```
      H
   H  C  N  H
      H  H
```

(d)
```
H
   C  O
H
```

(e)
```
Cl
   C  O
Cl
```

(f) H C N

(g)
```
      O
   H  C  O  H
```

(h)
```
H        O  H
      C  N
      H
```

Polar Covalent Bonds

6.28 How can one use electronegativity values to predict the direction of the polarity of a bond?

6.29 How might the order of the polarity of a series of bonds be predicted?

6.30 Indicate which element will be the positive end of the dipole when the following pairs of elements are bonded to each other.

(a) H and Br   (b) Br and Cl   (c) H and O
(d) O and F   (e) O and Cl   (f) Si and F
(g) N and H   (h) N and F   (i) P and Cl
(j) P and F

Oxidation Numbers

6.31 Calculate the oxidation number of the indicated element in each of the following.

(a) S in H_2S (b) Sc in ScF_3
(c) Ti in TiO_2 (d) Si in SiO_2
(e) N in Na_3N (f) Mn in $MnCl_2$
(g) Os in OsO_4 (h) Cu in CuO

6.32 Calculate the oxidation number of the indicated element in each of the following.

(a) Cu in Cu_2O (b) Fe in Fe_2O_3
(c) Cr in Cr_2O_3 (d) P in P_4O_{10}
(e) N in N_2O_3 (f) As in As_2O_5

6.33 Calculate the oxidation number of the indicated element in each of the following.

(a) S in SO_3^{2-} (b) S in SO_4^{2-}
(c) N in NO_2^- (d) N in NO_3^-
(e) Cl in ClO_4^- (f) N in NH_4^+
(g) Mn in MnO_4^- (h) Cr in $Cr_2O_7^{2-}$
(i) P in $P_2O_7^{4-}$ (j) Se in SeO_3^{2-}

6.34 Calculate the oxidation number of the indicated element in each of the following.

(a) S in HSO_4^- (b) P in $H_2PO_4^-$
(c) P in HPO_2^- (d) C in HCO_2^-
(e) C in CH_3O^- (f) P in $H_2P_2O_2^{2-}$

6.35 Calculate the oxidation number of the indicated element in each of the following.

(a) P in H_3PO_4 (b) P in H_3PO_3
(c) S in $Na_2S_2O_4$ (d) Cr in $Na_2Cr_2O_7$
(e) Mn in $KMnO_4$ (f) S in H_2SO_4
(g) Cl in $NaClO_3$ (h) S in $Al_2(SO_4)_3$

6.36 Calculate the oxidation number of carbon in each of the following.

(a)
```
     H  H
  H—C—C—H
     H  H
```

(b)
```
  H           H
     C=C
  H           H
```

(c) $H-C\equiv C-H$

Resonance

6.37 Ozone, O_3, is an angular molecule with identical bonds from each terminal oxygen atom. Draw possible Lewis structures for this substance. What are the limitations of these structures?

6.38 The nitrate ion has identical bonds between each oxygen atom and the central nitrogen atom. Draw possible Lewis structures for this ion.

Molecular Shapes

6.39 What types of bond angles are present in each of the compounds listed below? What is the geometry of each molecule?

(a) SiH_4 (b) CCl_4 (c) $SiCl_4$ (d) NF_3
(e) $HOCl$ (f) OF_2 (g) GeH_4

6.40 The molecule BF_3 does not have a nonbonded electron pair on boron. Draw a Lewis structure of this molecule. Based on this structure, explain why all four atoms are in a single plane. What is the expected F—B—F bond angle?

6.41 Draw a Lewis structure for $BeCl_2$. Would this molecule be expected to be linear or angular?

6.42 Multiple bonds can be considered as one region of space in the VSEPR theory. On this basis explain why HCN is a linear molecule.

6.43 Formaldehyde, H_2CO, has two hydrogen atoms and one oxygen atom bonded to a central carbon atom. Using the fact that a multiple bond can be considered as one region of space in the VSEPR theory, describe the expected geometry of formaldehyde.

6.44 Predict the shapes of the following ions according to the VSEPR theory.
(a) SO_4^{2-} (b) CO_3^{2-} (c) NH_4^+ (d) H_3O^+
(e) ClO_4^- (f) NH_2^-

6.45 Predict the shape of ethylene and acetylene in terms of the VSEPR theory (see Section 6.3).

6.46 Predict the shape of each of the following by the VSEPR theory.
(a) CF_4 (b) NF_3 (c) OF_2 (d) HOCl
(e) GeH_4 (f) SbH_3

Hybrid Orbitals

6.47 Indicate the hybridization of the orbitals in the bonds of each of the ions given in **6.44.**

6.48 Indicate the hybridization of the orbitals in the bonds of each of the molecules given in **6.46.**

6.49 Describe the types of bonds in the molecules (a)–(c) of **6.27.**

6.50 Describe the types of bonds in the molecules (d)–(h) of **6.27.**

Gases

This scuba diver may remain under water for an hour with the air supply in the gas tank under a pressure of 3000 psi.

Although we cannot see most gases, we can make observations that tell us that matter exists as gases. The odors of spring flowers or decaying matter both are the result of gases reaching our nose. We can also see at least the result of gases moving about as the wind blows against a tree. Gases can exert a force against an object that is every bit as strong as that exerted by a solid object such as a car hitting a tree.

We can pump gases into a container; for example, oxygen into a tank or air into a tire. When this is done, the matter within exerts a pressure on the walls of the container. If too much gas is put into the container, the resulting pressure can rupture the walls, which will result in an explosion and the escape of the gas.

Let's consider what we mean by the term pressure. **Pressure** *is a force per unit area on an object*. The fact that pressure is a ratio makes it a somewhat more difficult concept for some students. Let's consider some examples of pressure.

The pressure that we exert on our shoes depends on our mass and the area of the soles of our feet. For two individuals of the same mass who wear different size shoes, the individual with the larger shoe size exerts less force per unit area on the soles of his shoes, although the total force involved is the same in both cases. The effectiveness of snowshoes on snow or water skis on water is due to the distribution of force over a wide area to give a low pressure. The pressure exerted by a ballerina also illustrates the phenomenon of pressure. While she is on toe, the pressure on the points is substantially greater than that exerted in a rest position where the force is distributed over the entire slipper (Figure 7.1).

7.2
Measurement of Pressure

A gas in a container exerts a pressure against a wall by colliding with it. This "pushing" against the wall can be measured by a pressure gauge that also experiences the pressure caused by the collisions.

Most people do not realize that they live under gas pressure. We live at the bottom of an ocean of air and are under pressure much like the pressure on a diver in the ocean. However, since we have lived under air pressure all of our lives we don't feel this pressure.

The pressure of our atmosphere is the result of the weight of air from the outer edge of the atmosphere down to the surface of the earth. The weight of a column of air above an area of 1 square inch is 14.7 pounds at sea level. We say the

Figure 7.1
A Comparison of Force and Pressure. The pressure exerted by a 100 lb ballerina depends on the area of contact with the floor. On the toe having an area of 1 square inch the pressure is 100 lb/in.^{2}. In a rest position the area of both feet in contact with the floor is 50 square inches and the pressure is 2 lb/in.^{2}.

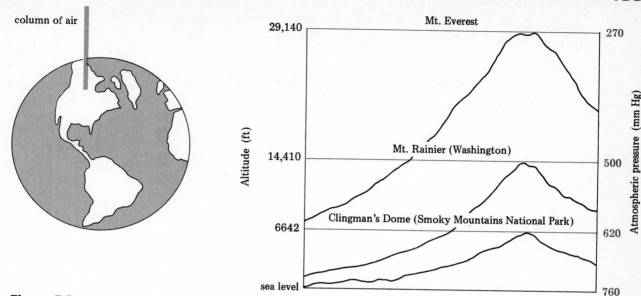

Figure 7.2

Atmospheric Pressure.

A column of air from the outer atmosphere to a given point on earth contains gases that exert a force equal to 14.7 lb on each square inch. This pressure is 1 atm or 760 mm Hg. At higher altitudes the pressure is less since the mass of the column of air at that point to the outer atmosphere is less. Some examples of pressure on some mountains are given on the right.

pressure is 14.7 lb/in.^{2} or 14.7 psi. We have a tremendous weight pushing down on our bodies! At sea level, your hand is under approximately 300 lb of force from the atmosphere. At higher altitudes the pressure decreases because the column of air from that point on the earth's surface to the edge of the atmosphere weighs less (Figure 7.2).

Some organisms live under extremely high pressures. A fish at one mile down in the ocean lives under a pressure of 2500 lb/in.^{2}. The body fluids of such a fish must exert an outward pressure equal to the ocean pressure if it is to survive. When divers descend to 200 feet, the pressure is about 100 lb/in.^{2}, and they must have air at that pressure in order to breathe.

The **barometer,** which *is the standard device for measuring the pressure of the atmosphere,* was invented by Evangelista Torricelli in the seventeenth century. It consists of a long tube that is closed at one end. The tube is filled with mercury and inverted in a vessel of mercury. If the tube is more than 76 cm long, part of the mercury will run out of the tube when it is inverted. A column of mercury approximately 76 cm high will remain in the tube (Figure 7.3).

A column of mercury remains in the tube of the barometer because the atmosphere exerts a pressure on the surface of the mercury. This pressure, transmitted through the mercury to the base of the column, supports the mercury in the tube. Therefore, the pressure at the base of the mercury column is due to the mass of the mercury in the column. The mercury falls until the pressure exerted by its weight equals the atmospheric pressure. A balance or equilibrium results.

Pressure should be expressed in units of force per unit area. In the barometer, the downward force exerted by the mercury column is proportional to the mass of the liquid supported in the column, which, in turn, is proportional to the

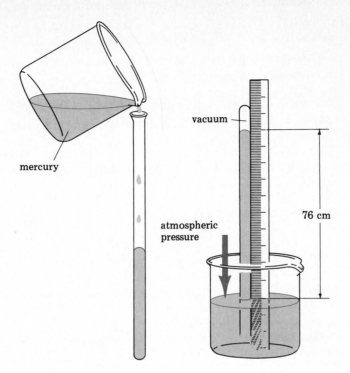

Figure 7.3
The Barometer.
The construction of a barometer: a glass tube, filled with mercury, is closed and inserted in a beaker of mercury. When the end of the column is opened, the column drops to about 76 cm.

vacuum

atmospheric
pressure

mercury

76 cm

height of the column. Therefore, the pressure can be expressed in centimeters or millimeters of mercury. It is understood that pressure is not the same thing as length, but by convention the term centimeter, or millimeter, of mercury (Hg) indicates the pressure or force per unit area exerted by the column of mercury.

Atmospheric pressure depends on altitude and the local weather, and it fluctuates from day to day. *At 0°C at sea level, the average atmospheric pressure supports a column of mercury 760 mm high.* This pressure is called a **standard atmosphere** and is referred to as 1 atm. The unit of pressure, 1 mm Hg, is also called 1 torr. Therefore, the standard pressure (1 atm) is 760 torr.

Example 7.1 ────────────────────────────────

The pressure on a mountain is 0.700 atm. What is the pressure in torr?

Solution. First determine two possible conversion factors relating atmosphere and torr.

$$\frac{1 \text{ atm}}{760 \text{ torr}} \quad \text{and} \quad \frac{760 \text{ torr}}{1 \text{ atm}}$$

The second factor will cancel the given unit of atm and give the required torr unit.

$$0.700 \text{ atm} \times \frac{760 \text{ torr}}{1 \text{ atm}} = 532 \text{ torr}$$

[Additional examples may be found in **7.3–7.10** at the end of the chapter.]

AN ASIDE

The pressure that blood exerts within blood vessels is an important indicator of our health. Blood pressure is measured by using an air cuff wrapped around the arm above the elbow. The air cuff is pumped up, and the pressure within the cuff can stop blood flow through the brachial artery (Figure 7.4). When the pressure within the cuff exceeds the pressure within the artery, the artery is pressed closed. Air is then released from the cuff and air pressure is measured by a pressure device called a sphygmomanometer.

A stethoscope is placed on the artery below the elbow. When the first faint sound of a blood pulse or flow is heard, the pressure on the sphygmomanometer is noted. This pressure is equal to the systolic pressure. *Systolic pressure is the maximum pressure exerted within the artery when the heart contracts.*

As the air is released and the pressure in the cuff is decreased, the blood will be able to flow even when the pressure is at its lowest. When the sound of the blood pushing past the cuff levels out, the lowest pressure or diastolic pressure is recorded. *Diastolic pressure is the pressure within arteries when the heart relaxes between beats.*

A normal systolic pressure range for a young person is 110–140 mm Hg, while the diastolic range is 70–90 mm Hg. Blood pressure is usually written as 120/80 or stated as "120 over 80." The ratio is systolic over diastolic pressure. Blood pressure varies with age, physical fitness, health, stress, and diet.

Blood pressure readings are taken on the upper arm because it is close to the heart where the pressure is the highest. The pressure at the artery end of a small capillary may be as little as 30 mm Hg because of the distance from the heart. High blood pressure, *hypertension*, can be caused by increased resistance in blood vessels to the passage of blood. One cause of this type of hypertension is the buildup of fat deposits, which cause a narrowing of the arteries (atherosclerosis).

Figure 7.4
Measurement of the Blood Pressure.

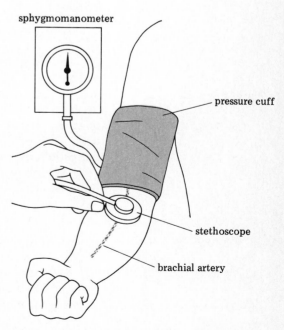

7.3
Kinetic Theory of Matter

The phenomenon of pressure as well as other properties of gases can be explained by the kinetic molecular theory. The concept that atoms are moving is known as the kinetic theory of matter. By extension, it can also apply to the liquid and solid states.

The assumptions made in the kinetic theory of matter can be summarized as follows.

1. Gases are composed of atoms or molecules that are widely separated from one another. The space occupied by the atoms or molecules is extremely small compared with the space accessible to them.
2. The atoms or molecules are moving rapidly and randomly in straight lines. Their direction is maintained until they collide with a second atom or molecule or with the walls of the container.
3. There are no attractive forces between molecules or atoms of a gas.
4. Collisions of molecules or atoms are elastic; that is, there is no net energy loss upon collision, although transfer of energy between molecules or atoms may occur in the collision.
5. In a specific gas, individual atoms or molecules move at different speeds. However, the average velocity and average kinetic energy remain constant. The kinetic energy of a particle is given by the expression $KE = \frac{1}{2}mu^2$, where m is the mass of the particle and u is its average velocity. As the temperature increases, the average kinetic energy increases and therefore the average velocity also increases. The average kinetic energy is directly proportional to the temperature on the Kelvin scale.

A hypothetical gas that conforms to all of the assumptions of the kinetic theory is an **ideal gas.** Thus, the ideal gas particles have very small volume compared to the total volume of the gas sample. Furthermore, the gas particles are not attracted to each other, and any collisions that occur are elastic. An ideal gas, then, could never be condensed to form a liquid. However, we know that all real gases, including components of our atmosphere such as oxygen and nitrogen, can be converted to liquids. **Real gases are any gaseous substances that actually exist.**

Under conditions of high temperature and low pressure, most real gases will conform to the behavior predicted for the ideal gas. At low pressures, such as exist at high altitudes, the gas particles are far apart, as given by assumption 1 of the kinetic theory. At high temperatures the gas particles have high kinetic energies and are moving so rapidly that the attractive forces existing in real gases can exert very little influence. In this chapter we will assume an ideal gas behavior for all problems.

At high pressures and low temperatures, all real gases can be condensed to liquids. Thus, they do not behave ideally under these conditions.

7.4
Boyle's Law

In experiments with vacuums and vacuum pumps in 1662, Robert Boyle made the first systematic study of the relationship between volume and pressure in gases. He measured what he termed "the spring of air," the pressure with which a gas sample pushes back when it is compressed. Boyle found a relationship between gas volume and pressure. This relationship has been verified many times with a variety of gases and is known as **Boyle's law.** *At constant temperature, the volume of a given quantity of gas varies inversely with the pressure exerted on it.* Thus, if the pressure of a given volume of gas is doubled, the volume will be decreased by one half. Conversely, if the pressure is decreased by one half, the volume will double.

A mathematical expression of Boyle's law is

$$V = k \times \frac{1}{P}$$

where V stands for volume, P for pressure, and k is a proportionality constant. The value of k depends on the temperature and the quantity of gas being studied. An alternative expression, obtained by multiplying both sides of the equation by P, is

$$PV = k$$

Therefore, **the product of the volume and pressure of a constant amount of gas at a constant temperature is constant.** If the quantity of the gas sample and the temperature remain the same, the product of the pressure and volume will remain unchanged under different experimental conditions. The mathematical statement given below is another way of expressing Boyle's law.

$$P_2 \times V_2 = k = P_1 \times V_1$$

An example of some data illustrating Boyle's law is given in Figure 7.5.

For a change in pressure, the new volume may be calculated from the rearranged equation.

$$V_2 = V_1 \times \frac{P}{P_2} \qquad \text{or} \qquad V_2 = V_1 \times P_{\text{factor}}$$

Figure 7.5
An Illustration of Boyle's Law Data.

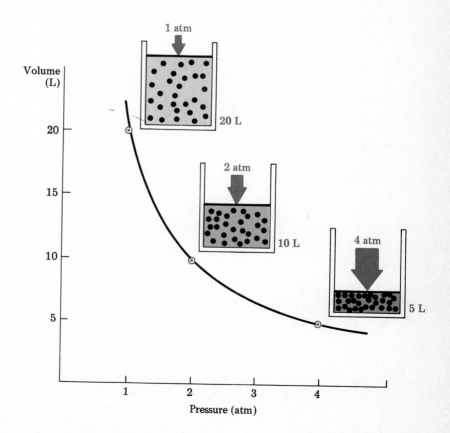

Similarly, if a new pressure must be calculated to effect a desired volume change, then

$$P_2 = P_1 \times \frac{V_1}{V_2} \quad \text{or} \quad P_2 = P_1 \times V_{\text{factor}}$$

In either a pressure or a volume calculation, the necessary volume factor or pressure factor may be determined by recalling the inverse proportionality of Boyle's law. Thus, if a pressure is increased, the volume must decrease. Therefore, only a pressure factor less than unity will give the correct answer. Memorization of the formulas given is not necessary.

Example 7.2

A gas sample occupies 15 L at 25°C and 3.0 atm. If the pressure is then decreased to 1.0 atm, what will be the volume of the gas?

Solution. Since the temperature and the sample of the gas are unchanged in the described process, we may use Boyle's law. First arrange the information in a clear form so that you can analyze how to solve the problem.

$P_1 = 3.0$ atm $\qquad\qquad V_1 = 15$ L

$P_2 = 1.0$ atm $\qquad\qquad V_2 = ?$ L
a pressure decrease $\qquad$ Volume must be increased.

The pressure has decreased from 3.00 to 1.00 atm, and the volume must increase by a pressure factor composed of these two numbers.

$V_2 = V_1 \times P_{\text{factor}}$

$V_2 = 15 \text{ L} \times \dfrac{3.00 \text{ atm}}{1.00 \text{ atm}}$

$\quad\;\; = 15 \text{ L} \times 3.00$

$\quad\;\; = 45 \text{ L}$

[Additional examples may be found in **7.14–7.18** at the end of the chapter.]

AN ASIDE

Our thoracic cavity is an area sealed from outside air. In this cavity are our two lungs, which are inflated with outside air rushing in when we inhale and deflated when we exhale. The intrapleural space within the thoracic cavity is equal to the volume of the cavity minus the volume occupied by the lungs (Figure 7.6).

At the bottom of the thoracic cavity is the diaphragm. When the diaphragm is moved down and the ribcage expands, there is a temporary expansion in the volume of the thoracic cavity. As a result, there is an increase in the intrapleural space, and the pressure will decrease, as predicted by Boyle's law. However, as soon as the pressure starts to drop, air rushes into the lungs. The lungs expand,

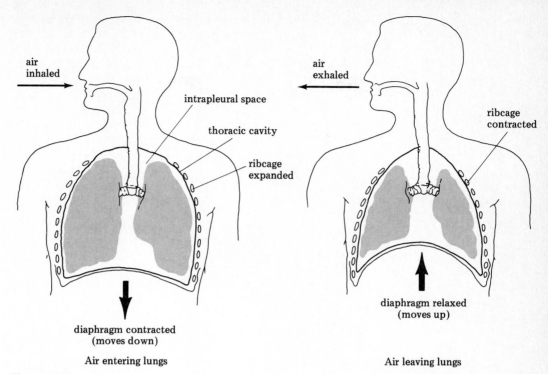

air
inhaled

intrapleural space

thoracic cavity

ribcage
expanded

diaphragm contracted
(moves down)

Air entering lungs

air
exhaled

ribcage
contracted

diaphragm relaxed
(moves up)

Air leaving lungs

Figure 7.6
The Mechanism of Breathing.
The volume of the thoracic cavity remains constant as diaphragm and ribcage positions are changed. Air may enter the lungs to allow them to expand or leave to allow them to contract.

the intrapleural space is decreased back to its original volume, and the intrapleural pressure is restored.

When the diaphragm relaxes and the ribcage contracts, the thoracic cavity decreases in volume. As a result, the intrapleural pressure should increase as predicted by Boyle's law. However, as soon as the pressure starts to increase, the same pressure is exerted on the lungs, which then decrease in size. This decrease in volume would increase the pressure in the lungs, but air is forced out or exhaled.

Medical respirators such as the iron lung change the volume of the respiratory cavity by a mechanically operated diaphragm. With volume changes the pressure on the individual is alternately increased and decreased. An increase in pressure forces the individual to exhale, whereas a decrease in pressure causes the individual to inhale.

Artificial respiration is designed to simulate the sequence of pressure experienced by the lungs in normal breathing. By expanding and contracting the thoracic cavity, air can be made to enter and leave the lungs.

The Heimlich maneuver, an emergency procedure used to remove food stuck in the trachea, also works on the principle of Boyle's law. The procedure involves wrapping one's arms below the ribcage of the choking individual. A sharp upward bearhug results in an increase in the pressure of the lungs, which in turn can expel the lodged material.

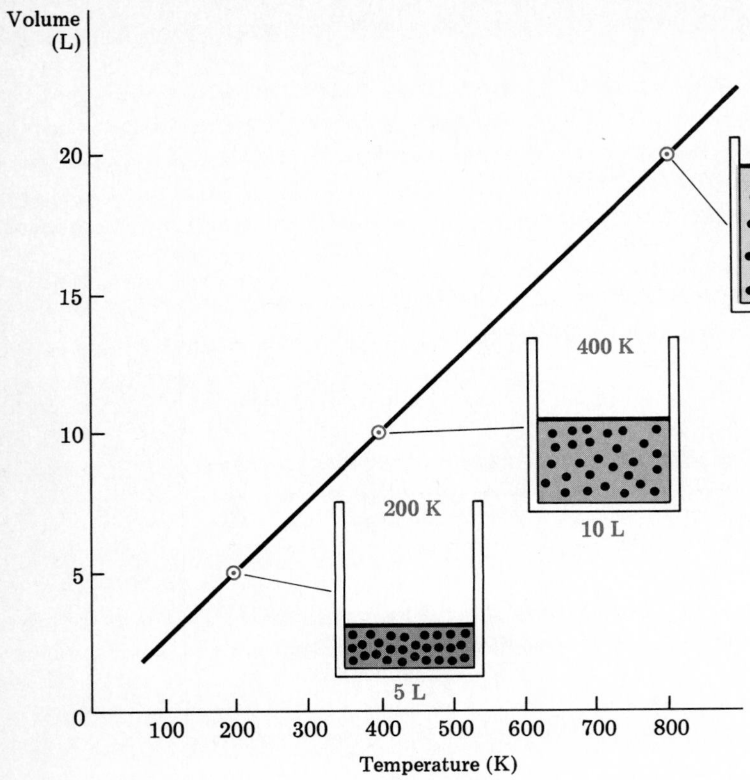

Volume (L)

Figure 7.7
An Illustration of Charles' Law Data.

Temperature (K)

7.5

Charles' Law

Gases expand when heated under constant pressure and contract when cooled. In 1787, the French physicist J. A. C. Charles observed this relationship between volume and temperature at constant pressure. From the data of Charles and other scientists, **Charles' law** states that *at constant pressure the volume of a fixed mass of a gas is directly proportional to the absolute (Kelvin) temperature.* An example of such data is given in Figure 7.7.

The mathematical expression with a proportionality constant, k, is

$$V = kT$$

The proportionality constant is dependent only on the pressure and the sample of gas considered. Dividing both sides of the equation by T gives

$$\frac{V}{T} = k$$

Since the constant remains unchanged, the relationship given below must be obeyed under any experimental conditions. It is an expression giving Charles' law.

$$\frac{V_2}{T_2} = k = \frac{V_1}{T_1}$$

For a change in temperature, the new volume may be calculated by the use of a temperature factor.

$$V_2 = V_1 \times \frac{T_2}{T_1} \quad \text{or} \quad V_2 = V_1 \times T_{factor}$$

Alternatively, if it is necessary to calculate the temperature required to alter a volume, we use a volume factor.

$$T_2 = T_1 \times \frac{V_2}{V_1} \quad \text{or} \quad T_2 = T_1 \times V_{\text{factor}}$$

Remember, in the use of either formula, **the temperature must be in Kelvin units.** The formulas need not be memorized if one recalls that the proportion is a direct one.

Hot air balloons operate because of the increase in volume caused by heating air. In fact, Charles rose to a height of about 5 miles in his hot air balloon in 1783. The increase of temperature results in a decrease in density of the air in the balloon, which then "floats" in the surrounding more dense cooler air.

Example 7.3

A sample of a gas occupies 600 mL at 27°C and 1 atm. What will be its volume at 127°C if the pressure is kept constant?

Solution. Since the pressure and quantity of the gas remain constant, Charles' law applies to the solution of the problem. Arrange the data in tabular form, and calculate the temperature in Kelvin units.

$$T_1 = 27 + 273 = 300 \text{ K} \qquad V_1 = 600 \text{ mL}$$

$$T_2 = 127 + 273 = 400 \text{ K} \qquad V_2 = ? \text{ mL}$$
$$\text{a temperature increase} \qquad \text{Volume must be increased.}$$

The volume must increase as a consequence of the increase in the temperature. Therefore, the volume is multiplied by a temperature factor to increase the volume.

$$V_2 = V_1 \times T_{\text{factor}}$$

$$V_2 = 600 \text{ mL} \times \frac{400 \cancel{K}}{300 \cancel{K}}$$

$$= 600 \text{ mL} \times \frac{400}{300}$$

$$= 800 \text{ mL}$$

[Additional examples may be found in **7.19–7.23** at the end of the chapter.]

Gay-Lussac's law states that *at constant volume the pressure of a fixed mass of a gas is directly proportional to the absolute (Kelvin) temperature.* Mathematically stated, the direct proportionality between pressure and temperature is given by

7.6
Gay-Lussac's Law

$$P = kT$$

The value of k depends on the volume of the sample and the quantity of gas.

An alternative expression is obtained by dividing both sides of the equation by T.

$$\frac{P}{T} = k$$

Therefore, the quotient of the pressure divided by the temperature will be a constant for a specific volume containing a given mass of gas, regardless of the pressure or temperature.

$$\frac{P_2}{T_2} = k = \frac{P_1}{T_1}$$

For a change in temperature, the effect on the pressure may be calculated from the following equivalent equation.

$$P_2 = P_1 \times \frac{T_2}{T_1} = P_1 \times T_{factor}$$

Similarly, for a change in pressure, the related necessary temperature change may be calculated from an alternative rearranged equation.

$$T_2 = T_1 \times \frac{P_2}{P_1} = T_1 \times P_{factor}$$

In either case, you only need remember that there is a direct proportion between the Kelvin temperature and the pressure.

Heating a gas in a closed container may result in an explosion if the container cannot stand the pressure. For this reason, spray cans of deodorants, paint, and so on, have a warning label stating that the can should not be heated and should be stored at temperatures below a stated limit.

Example 7.4

What will happen to the pressure of a 5 L sample of a gas at 5 atm if it is heated from 250 to 300 K and the volume is held constant?

Solution. Gay-Lussac's law conditions are stated and the data may be arranged as follows.

$$T_1 = 250 \text{ K} \qquad P_1 = 5 \text{ atm}$$

$$T_2 = 300 \text{ K} \qquad P_2 = ? \text{ atm}$$

a temperature increase   The pressure must increase.

The pressure increases by a temperature factor. Therefore, we write

$$P_2 = P_1 \times T_{factor}$$

$$P_2 = 5 \text{ atm} \times \frac{300 \text{ K}}{250 \text{ K}}$$

$$= 5 \text{ atm} \times \frac{300}{250}$$

$$= 6 \text{ atm}$$

[Additional examples may be found in **7.24–7.28** at the end of the chapter.]

Both Boyle's and Charles' laws may be combined into one mathematical expression, which then also gives the Gay-Lussac law as well.

$$\frac{PV}{T} = k$$

Therefore, for a fixed sample of a gas, it follows that

$$\frac{P_2 V_2}{T_2} = k = \frac{P_1 V_1}{T_1}$$

This equation need not be memorized because any variable can be changed by factors of the other two variables.

$$P_2 = P_1 \times V_{factor} \times T_{factor}$$

$$V_2 = V_1 \times P_{factor} \times T_{factor}$$

$$T_2 = T_1 \times P_{factor} \times V_{factor}$$

In order to solve problems in which two variables change, it is only necessary to consider separately what effect each variable will have. Thus, for a volume to change as a consequence of pressure and temperature changes, you determine first what effect the pressure change will have on the volume. If the pressure increases, the volume will decrease, and vice versa. After the proper pressure factor has been determined, the temperature factor can be considered. If the temperature decreases, the volume will decrease, and vice versa. Application of the pressure factor and the temperature factor gives the final correct answer.

Example 7.5

A 1.0 L sample of a gas at $-73°C$ and 2 atm is heated to $123°C$ and the pressure is decreased to 0.5 atm. What will be the final volume?

Solution

$$V_1 = 1.0 \text{ L} \qquad V_2 = ? \text{ L}$$

$$T_1 = 200 \text{ K} \qquad T_2 = 400 \text{ K}$$

$$P_1 = 2 \text{ atm} \qquad P_2 = 0.5 \text{ atm}$$

Because of the temperature increase, the volume must increase by a factor of 400/200. Because of the pressure decrease, the volume must increase by a factor of 2/0.5.

$$V_2 = V_1 \times T_{factor} \times P_{factor}$$

$$V_2 = 1.0 \text{ L} \times \frac{400 \text{ K}}{200 \text{ K}} \times \frac{2 \text{ atm}}{0.5 \text{ atm}}$$

$$= 1.0 \text{ L} \times \frac{400}{200} \times \frac{2}{0.5}$$

$$= 8 \text{ L}$$

[Additional examples may be found in **7.29–7.33** at the end of the chapter.]

7.8
Avogadro's Hypothesis

Avogadro's hypothesis (1811) states that *equal volumes of gases under the same conditions of temperature and pressure contain the same number of submicroscopic particles.* For example, equal volumes of helium and argon at 1 *atm pressure and 0°C* (*called* **standard temperature and pressure** *or* **STP**) contain the same number of atoms.

Avogadro's hypothesis can be used to determine the relative weight of atoms and molecules. Because equal volumes of gases under the same conditions of temperature and pressure contain the same number of submicroscopic particles, the weight of the gas particles must be in the same ratio as the gas densities. The densities of the monatomic gases helium and argon at standard temperature and pressure are 0.179 and 1.79 g/L, respectively, and indicate that one argon atom is ten times as massive as one helium atom.

The concept of the mole and Avogadro's hypothesis allow the determination of the molecular weight of an unknown compound. The volume occupied by 1 mole (32 g) of oxygen molecules at standard temperature and pressure has been determined to be 22.4 L. **Since equal volumes of gases contain the same number of particles, 22.4 L must be the volume occupied by 1 mole of any gaseous substance at standard temperature and pressure.**

Example 7.6

A 1.12 L sample of a gas at 2.00 atm and 0°C has a mass of 3.00 g. What is the molecular weight of the gas?

Solution. It is first necessary to change the stated experimental conditions to STP in order to make use of the molar volume quantity of 22.4 L/mole.

$$V_2 = V_1 \times P_{factor}$$

$$V_2 = 1.12 \text{ L} \times \frac{2.00 \text{ atm}}{1.00 \text{ atm}} = 2.24 \text{ L}$$

Now it is possible to calculate the number of moles present in the sample.

$$\text{number of moles} = 2.24 \text{ L} \times \frac{1 \text{ mole}}{22.4 \text{ L}} = 0.100 \text{ mole}$$

Once the number of moles that is present in the sample is known, the mass of the sample can be used to calculate the mass of a mole.

$$\text{mass of one mole} = \frac{3.00 \text{ g}}{0.100 \text{ mole}} = 30.0 \text{ g/mole}$$

The molecular weight is then 30.0 amu.

Example 7.7

Calculate the density of CO_2 at STP.

Solution. The molecular weight of carbon dioxide is

$$\begin{aligned}
\text{molecular weight of CO}_2 &= \text{atomic weight of C} + (2 \times \text{atomic weight of O}) \\
&= 12.0 \text{ amu} + (2 \times 16.0) \\
&= 44.0 \text{ amu}
\end{aligned}$$

Therefore, 22.4 L of CO_2 at STP must have a mass of 44.0 g. The density of CO_2 can be calculated as follows.

$$\text{density of } CO_2 = \frac{44.0 \text{ g}}{22.4 \text{ L}} = 1.97 \text{ g/L}$$

[Additional examples may be found in **7.36–7.42** at the end of the chapter.]

Now that Avogadro's hypothesis has been discussed, the significance of a constant value of k for the same initial volumes of various gases under the same conditions becomes evident. The same initial volumes, according to Avogadro's hypothesis, contain the same number of atoms or molecules.

**7.9
Ideal Gas Law**

The gas laws can be combined as a more general expression known as the **ideal gas law equation.** The term n represents the number of moles of a gas, and R is a proportionality constant.

$$\frac{PV}{T} = k = nR \qquad \text{or} \qquad PV = nRT$$

The proportionality constant R is called the universal gas constant. It can be evaluated numerically from the knowledge that 1 mole of a gas at standard temperature and pressure occupies 22.4 L.

$$R = \frac{PV}{nT} = \frac{1 \text{ atm} \times 22.4 \text{ L}}{1 \text{ mole} \times 273 \text{ K}} = 0.082 \frac{\text{L atm}}{\text{mole deg}}$$

The value for R is 0.082, and the units are (liter $\times$ atmosphere) $\div$ (mole $\times$ degree) (L atm/mole deg), read as "liter atmosphere per mole per degree." The term R could be evaluated for any units of P and V desired, but atmospheres and liters are most commonly used.

Example 7.8 ————————————————————

A mass of 1.34 g of a gas occupies 2.0 L at 91°C and 0.50 atm. What is the molecular weight of the gas?

Solution. The number of moles in this sample is

$$n = \frac{PV}{RT}$$

$$n = \frac{0.50 \text{ atm} \times 2.0 \text{ L}}{0.082 \text{ L atm/mole deg} \times 364 \text{ K}} = 0.033 \text{ mole}$$

Because 0.033 mole of the gas has a mass of 1.34 g, the mass of 1 mole is

$$\frac{1.34 \text{ g}}{0.033 \text{ mole}} = 40 \text{ g/mole}$$

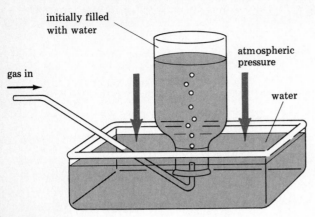
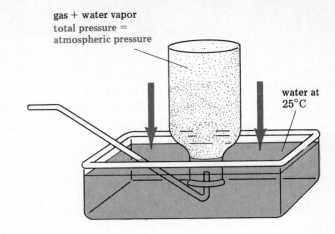

Figure 7.8
Collection of a Gas over Water and Dalton's Law of Partial Pressures.
The total pressure of the gaseous phase over the water is equal to the pressure exerted by the collected gas plus the
pressure exerted by the water vapor. The pressure of the collected gas may be determined by using Dalton's law.

7.10
Dalton's Law of Partial Pressures

John Dalton, of atomic theory fame, also studied some of the properties of gases.
His **law of partial pressures** states that *each gas in a gas mixture exerts a partial
pressure that is equal to the pressure it would exert if it were the only gas present
under the experimental conditions; the sum of the partial pressures of all of the
gases is equal to the total pressure.* Dalton's law expressed mathematically is

$$P_{total} = P_1 + P_2 + P_3 + \cdots$$

where the subscripted P values are the partial pressures of the gases in the mixture.

Gases are commonly collected by water displacement in college chemistry
laboratories (Figure 7.8). One such gas is oxygen. Potassium chlorate, when
heated, decomposes to produce potassium chloride and oxygen.

$$2\,KClO_3 \longrightarrow 2\,KCl + 3\,O_2$$

The oxygen, which is collected by downward displacement of water from an
inverted jar, is not pure but contains some water vapor mixed with it. The pressure of the collected gas is equal to the external atmospheric pressure when the
level of the water inside the jar is equal to the level in the large water bath. It is
a sum of partial pressures of the two gases contained in the jar, which are water
vapor and oxygen.

$$P_{total} = P_{atm} = P_{O_2} + P_{H_2O}$$

In order to determine the pressure of the gas, it is necessary to subtract the
partial pressure of the water vapor at that temperature.

Example 7.9

A 200 mL sample of oxygen is collected at 26°C over water. The vapor pressure of water is 25 torr at 26°C, and the atmospheric pressure is 750 torr at the
time of the experiment. Calculate the partial pressure of oxygen, and determine
the volume that the dry oxygen would occupy at 26°C and 750 torr.

Solution. The partial pressure of oxygen is determined by use of the law of partial pressures.

$$P_{total} = P_{atm} = P_{O_2} + P_{H_2O}$$
$$750 \text{ torr} = P_{O_2} + 25 \text{ torr}$$
$$P_{O_2} = 750 \text{ torr} - 25 \text{ torr} = 725 \text{ torr}$$

Now the problem may be treated as a Boyle's law type. If 200 mL of dry oxygen at 725 torr was subjected to a new pressure of 750 torr, the volume would decrease.

$$V_2 = 200 \text{ mL} \times \frac{725}{750} = 193 \text{ mL}$$

This means that only 193 mL of the 200 mL is actually O_2.

[Additional examples may be found in **7.43–7.47** at the end of the chapter.]

Another illustration of the law of partial pressures is the air of our atmosphere. Air consists of nitrogen and oxygen and, in lesser amounts, water vapor and carbon dioxide. The total pressure exerted by air is a sum of the partial pressures of the individual gases. At 1 atm, the partial pressures of nitrogen, oxygen, and water vapor are approximately 595, 159, and 6 torr, respectively; that of CO_2 is very low. At higher altitudes, the total pressure of air decreases as do the partial pressures of all the components of air. For example, at an altitude of one mile, the total pressure is 630 torr, and the partial pressures of nitrogen, oxygen, and water vapor are 494, 132, and 4 torr, respectively. An individual coming from sea level cannot operate as efficiently and becomes tired easily at higher altitudes. This change occurs because at this partial pressure of oxygen, the red blood cells absorb a smaller amount of oxygen. However, if one stays at the high altitude for a time, the body adapts by producing more red blood cells.

Respiration involves an exchange of two blood gases, oxygen and carbon dioxide. This exchange is the result of different partial pressures at various points in the body (Table 7.1). Gases naturally flow from regions of high pressure to those of low pressure. The individual partial pressures of blood gases have similar effects.

Table 7.1

Partial Pressure of Blood Gases in the Atmosphere and in the Body (mm Hg)

Gas	Atmosphere	Inspired air	Expired air	Alveolar air
O_2	159	149	116	106
CO_2	0.3	0.3	28	40
H_2O	6	47	47	47
N_2	595	564	569	573

Atmospheric oxygen at a partial pressure of 159 mm Hg will diffuse into the alveoli of the lungs where the partial pressure is about 106 mm Hg. The alveoli are small air sacs at the very ends of the air passages in the lungs. Diffusion of oxygen into the red blood cells of arterial blood results in a partial pressure of 100 mm

If too little oxygen is available to body tissues, even for brief periods, coma and/or irreversible tissue damage may occur. Brain tissue is especially sensitive to oxygen levels. Inhalation therapists are trained to administer gases to treat a number of diseases. For example, if a patient has difficulty breathing, the respiration therapist may use a respirator to increase the rate of breathing and thus increase the volume of oxygen reaching the patient's lungs.

Increasing the partial pressure of oxygen above that normally contained in air may also be a prescribed treatment. The extra oxygen may be provided by an oxygen tent or by a face mask.

A patient may be placed in a hyperbaric chamber that contains air or additional oxygen at higher than atmospheric pressure. The resulting larger partial pressure of oxygen is useful in treating gangrene or in enhancing the effect of radiation in cancer therapy. Gangrene is caused by bacteria that are very sensitive to oxygen and their growth is inhibited by oxygen. In the case of cancer therapy, the cancerous cells, which grow more rapidly than normal cells, are destroyed more effectively by x-rays under high partial pressures of oxygen. The effect of the x-rays on normal cells is proportionately less.

An inhalation therapist must also maintain the proper partial pressure of water vapor in the lungs. We remove water vapor from membranes when breathing. If our inhaled air is too dry, the membranes will tend to dry out. Therefore gases administered in inhalation therapy must contain sufficient water vapor.

Hg. When the oxygenated blood passes through tissues where the partial pressure of O_2 is about 40 mm Hg, oxygen is released. It is this oxygen that allows metabolic reactions to occur. The deoxygenated blood or venous blood has a partial pressure of 40 mm Hg and is then returned to the lungs to be oxygenated again.

Removal of carbon dioxide from the tissues via venous blood to the lungs and then to the atmosphere is also the result of partial pressures. Carbon dioxide is also removed by some chemical reactions to be discussed in Chapter 11.

Carbon dioxide produced in tissues by metabolic reactions is at 50 mm Hg. Diffusion of CO_2 from tissues into blood occurs because the partial pressure of CO_2 in arterial blood is 40 mm Hg. Venous blood with a partial pressure of CO_2 at 46 mm Hg returns to the lungs where the partial pressure is about 30 mm Hg. Diffusion of CO_2 from venous blood to the lungs followed by expiration removes the CO_2 from our bodies.

Summary

Gases may be described by pressure, volume, temperature, and amount of the gas. The gas pressure is measured by comparison to the pressure exerted by a column of mercury. The units of pressure are atmospheres, cm Hg, and mm Hg (or torr).

The theoretical basis for understanding the behavior of gases is the kinetic molecular theory. Different gases at the same temperature move with the same average kinetic energy although not at the same average velocity. Real gases behave according to the theory only at high temperatures and low pressures. Under these conditions, gas molecules are widely separated and experience only low attractive forces.

Relationships between pairs of the four gas variables, the other two remaining constant, are given by names of individuals. These are Boyle's law, which relates pressure and volume; Charles' law, which relates volume and temperature; Gay-Lussac's law, which relates pressure and temperature; and Avogadro's hypothesis, which relates volume and the amount of the gas.

Combination of the individual gas laws gives a combined gas law. By using a constant called the universal gas law constant, a general expression known as the ideal gas law is obtained.

Dalton's law of partial pressure relates the properties of gas mixtures. The properties of blood gases as well as the science of respiration therapy depend on the law of partial pressure.

Learning Objectives

As a result of studying Chapter 7 you should be able to
- Convert units of pressure.
- Calculate changes in gas volume resulting from pressure changes and vice versa.
- Calculate changes in gas volume resulting from temperature changes and vice versa.
- Calculate changes in gas pressure resulting from temperature changes and vice versa.
- Illustrate how the gas laws are involved in breathing.

New Terms

An **atmosphere** is the pressure exerted by the atmosphere under normal conditions, which supports a column of mercury 760 mm high.

Avogadro's hypothesis states that equal volumes of gases under the same conditions have the same number of molecules.

A **barometer** is a device to measure atmospheric pressure.

Boyle's law states that the volume of a fixed amount of a gas at constant temperature varies inversely with pressure.

Charles' law states that the volume of a fixed amount of a gas at constant pressure is directly proportional to the absolute temperature.

Dalton's law of partial pressures states that the sum of the partial pressures of the gases in a mixture equals the total pressure of the mixture.

Gay-Lussac's law states that at constant volume, the pressure of a gas is directly proportional to the absolute temperature.

An **ideal gas** is one whose behavior is predicted by the ideal gas law and that conforms to the kinetic molecular theory.

The **ideal gas law equation** relates the pressure, volume, temperature, and number of moles of a gas.

The **kinetic molecular theory** is a model that describes gas behavior based on the concept of an average kinetic energy that is directly proportional to the absolute temperature.

A **partial pressure** is the pressure exerted by a single gas in a gaseous mixture.

Pressure is the force per unit area. In gases the pressure is measured in terms of the height of a column of mercury that can be maintained by the gas.

STP or standard temperature and pressure is 1 atm and 0°C.

Questions and Problems

Terminology
7.1   How is force different from pressure?
7.2   How is a real gas different from the ideal gas?

Pressure
7.3   What is one atmosphere?
7.4   How many mm Hg equal 1 torr?
7.5   How many cm Hg are equal to 1 atm?
7.6   Convert each of the following pressures into the indicated unit.
   (a) 380 mm Hg into atm
   (b) 0.500 atm into cm Hg
   (c) 0.100 atm into torr
7.7   Explain why the mercury does not run out of a barometer.
7.8   What is the equivalent of 1 atm in pounds per square inch?
7.9   How does the location beneath the ocean affect the pressure on an organism?
7.10 Explain why the pressure on Mt. McKinley is only 350 torr.

Kinetic Molecular Theory
7.11 What relationship exists between the average kinetic energy of a gas and its temperature?
7.12 Is the average velocity of H_2 equal to the average velocity of He at the same temperature?
7.13 A sample of a gas is cooled from 127°C to −73°C. How is the average kinetic energy affected?

Boyle's Law
7.14 A sample of a gas at 600 torr and 25°C occupies 300 mL. What volume will the gas occupy at 800 torr and 25°C?
7.15 A sample of a gas at 0.5 atm and 100°C occupies 2 L. What pressure is necessary to cause the gas to occupy 0.5 L?
7.16 Calculate the volume of a gas in liters at 1.0 atm if it occupies 2.0 L at 158 cm Hg.

7.17 Calculate the pressure required to convert 500 mL of a gas at 0.200 atm into a volume of 10 mL.

7.18 Gases in cylinders used in chemistry labs may be at 2000 psi. What volume would 10 L of such a compressed gas occupy at 1 atm?

Charles' Law

7.19 A sample of a gas at −91°C and 1 atm occupies 2.0 L. What volume will the gas occupy at 0°C at the same pressure?

7.20 A sample of a gas at 300 K and 700 torr occupies 60 mL. What temperature is necessary to increase the volume to 75 mL at the same pressure?

7.21 A 2.0 L sample of air at −23°C is warmed to 100°C. What is the new volume if the pressure remains constant?

7.22 In order to change the volume of a 100 mL sample of a gas at 250 K to 300 mL, what temperature must be reached, assuming the pressure is kept constant?

7.23 A 300 mL sample of a gas at 127°C is cooled to 150 K. What is the volume if the pressure is kept constant?

Gay-Lussac's Law

7.24 A sample of a gas in a rigid container is heated from 273 K to 273°C. If the gas was initially at 1 atm, what is the final pressure?

7.25 A sample of a gas in a rigid container at −23°C is at a pressure of 500 torr. What temperature will be necessary to change the pressure to 800 torr?

7.26 The pressure in an aerosol container is 2.7 atm at 27°C. What will be the pressure if the temperature is lowered to −73°C?

7.27 The pressure in an automobile tire is 30 lb/in.² at 27°C at the start of a trip. At the end of a trip the pressure is 33 lb/in.². What is the temperature of the tire?

7.28 An aerosol container has a pressure of 3.0 atm at 27°C. The can is left in the sun at 100°F. What is the pressure in the can?

Combined Gas Law

7.29 A gas occupies 250 mL at 700 torr and 300 K. What volume will the gas occupy at 350 torr and 450 K?

7.30 A gas occupies 800 mL at 1 atm and 250 K. At what pressure will the gas occupy 400 mL at 500 K?

7.31 A gas occupies 2 L at 127°C and 2 atm. At what temperature will the gas occupy 6 L at 1 atm?

7.32 A 2.0 L sample of air at −50°C has a pressure of 700 torr. What will be the new pressure if the temperature is raised to 50°C and the volume is increased to 4.0 L?

7.33 At STP a gas has a volume of 5.0 L. What volume will the gas occupy at 80°C and 800 torr?

Avogadro's Hypothesis

7.34 What statement can be made about the number of molecules contained in a 2 L sample of H_2 molecules at STP and the number of atoms in a 2 L sample of helium at STP?

7.35 Why are the densities of gases under the same pressure and temperature different?

Molar Volumes

7.36 How many moles of a gas are present under each of the following conditions?
(a) 1.12 L at 2.00 atm and 0°C
(b) 560 mL at 5.00 atm and 182°C
(c) 22.4 L at 273°C and 2 atm

7.37 How many molecules of sulfur dioxide (SO_2) are contained in a 2 mole sample of the gas?

7.38 A 1.12 L sample of a gaseous compound has a mass of 2.2 g at STP. What is the mass of a molecule of this compound in amu?

7.39 A 5.6 L sample of a gas at 182°C and 38 cm of mercury has a mass of 4.8 g. What is the mass of a mole of this gas?

7.40 Calculate the density of carbon monoxide, CO, at STP.

7.41 Calculate the density of C_2H_6 at 0°C and 2 atm.

7.42 Calculate the density of CH_4 at 273°C and 2 atm.

Dalton's Law of Partial Pressures

7.43 A mixture of three gases has the following partial pressures: oxygen, 100 torr; nitrogen, 300 torr; hydrogen, 150 torr. What is the total pressure of the mixture? If carbon dioxide were added until the pressure reached 750 torr, what would be the partial pressure of the carbon dioxide?

7.44 A nitrogen sample collected over water at 26°C and 775 torr has a volume of 300 mL. The vapor pressure of water at 26°C is 25 torr. What volume of nitrogen at 775 torr has been collected?

7.45 If 300 mL of nitrogen gas at 760 torr was bubbled through water at 26°C and collected, what would be the volume of the wet gas sample?

7.46 The partial pressure of nitrogen on Mt. McKinley is 288 torr on a day when the atmospheric pressure is 0.460 atm. What is the partial pressure of oxygen?

7.47 The concentration of carbon monoxide in a city one day is 100 ppm. Calculate the partial pressure of the carbon monoxide.

The Air We Breathe

7.48 You may inhale and exhale 6 L of air a minute. What is the volume of oxygen available to body tissues in each minute?

7.49 Why are oxygen masks worn by pilots at high altitudes?

7.50 How does your blood differ from that of individuals who live all their lives at high altitudes?

7.51 Why is the partial pressure of oxygen lower in venous blood than in arterial blood?

7.52 Which gas law governs the effect of the diaphragm on the action of the lungs in breathing?

Liquids and Solids

There are a variety of solids about us, but water is the most common liquid. This waterfall in Yosemite National Park has cut a path through solid rock.

8.1
Kinetic Theory of Liquids and Solids

Since the temperature of a gas is proportional to the average kinetic energy of its molecules, the molecules will move more slowly as the gas is cooled. **Condensation,** *the conversion of a gas into a liquid,* occurs at some temperature. During condensation the molecules tend to stick together when they collide. This point occurs when the kinetic energy becomes similar to the average forces of attraction between molecules. These forces are always present, but they cannot cause the formation of a liquid if the kinetic energy is too high. The reverse of condensation, **vaporization,** *is the conversion of a liquid into a gas.*

Since liquids consist of molecules that are close together, their densities are larger than those of gases. In addition, the cluster of moving liquid molecules has a definite volume. The balance between attractive forces and the kinetic energy of the molecules results in the liquid state, a state that has cohesion but no rigid structure. This model explains why a liquid can be poured and will assume the shape of its container.

Cooling a liquid decreases the kinetic energy of the atoms or molecules and allows the attractive forces between neighboring particles to exert a greater influence. Thus, cooling results in the formation of crystals in which the motion of the particles has been largely overcome so that they can vibrate only about fixed positions. The decreased motion of the molecules results in a rigid structure with a definite volume and shape. *The conversion of a liquid into a solid is called* **freezing.**

In this chapter and Chapter 9, dealing with physical changes, and Chapters 10 and 11, dealing with chemical reactions, the concept of equilibrium is involved. *An* **equilibrium** *is a dynamic state in which two or more opposing processes or forces are taking place at the same time and are in balance.* As a result, a system at equilibrium does appear to be static.

Consider your finances as an example of equilibrium. If the deposits to and withdrawals from your checking account are controlled to give a constant balance, you are in financial equilibrium. Money is still changing hands. Your balance can remain unchanged and appear to be static, but the same dollars are not always there and your finances are dynamic. We can depict this equilibrium with arrows that show the two directions of money flow.

$$\text{student} \underset{\text{withdrawals}}{\overset{\text{deposits}}{\rightleftharpoons}} \text{checkbook balance}$$

An understanding of equilibrium is important in developing a model for the liquid and solid states. In these states the balance between attractive forces and average kinetic energy is affected by temperature and pressure. First the liquid state and the equilibrium with the gaseous state will be considered. We will use equilibrium concepts for both physical and chemical processes in this and subsequent chapters.

8.2
Vapor Pressure of Liquids

You are probably familiar with the phenomenon of evaporation, which involves the transfer of matter from the liquid phase to the gas phase. In a liquid, individual particles are traveling at different speeds. Those particles of high velocity possess sufficient kinetic energy to break away from the attractive forces of their neighbors and enter the gas or vapor phase. Their departure causes a decrease in the average kinetic energy of the remaining particles and results in a lowered temperature of the liquid.

Any individual who has exercised and perspired or who has stood in a

Figure 8.1

Evaporation of a Liquid.
The molecules that enter the
gas phase tend to wander
away from the beaker and
not return to the liquid
phase. As a consequence,
eventually all of the liquid
will evaporate.

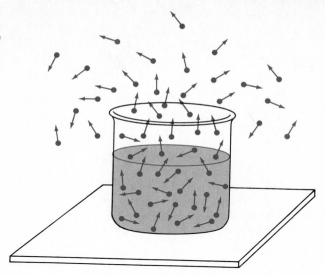

breeze immediately after stepping out of the swimming pool is well acquainted
with the cooling effect of evaporation. The water molecules that leave your skin
surface most readily are the most energetic particles, and the remaining liquid is
cooler. If a heat source is available, the temperature of the liquid will be maintained during evaporation. A glass of water in a room will evaporate without any
noticeable cooling (Figure 8.1). As the most energetic particles leave the liquid
phase, heat is transferred from the surroundings to the liquid, and the temperature is maintained.

Evaporation of matter from the liquid phase is less likely as the temperature
decreases because the average kinetic energy of the particles, and hence their
ability to escape into the gas phase, decreases. This is easily verified by experience. For example, it is more difficult to dry clothes on a cool day than on a
hot day.

Even at the same temperature, different liquids evaporate at different rates.
For example, gasoline evaporates faster than lubricating oil. The ease of evaporation in the liquid phase varies with the liquid. Because the average kinetic energies of particles in two different liquids at the same temperature are identical, the
escaping tendency depends on the attractive forces between neighbors. If attractive forces between neighboring particles are large, the escaping ability of a given
particle is retarded by its neighbors.

Whenever a liquid is placed in a closed vessel, particles leave the liquid
phase and enter the gas phase. As the vapor particles become more numerous,
they are more likely to collide with the liquid surface and return to it. Eventually a
balance is achieved: the rates at which the particles leave and return to the liquid
phase become equal (Figure 8.2). When such a balance occurs, the system is in
equilibrium. Evaporation and condensation occur continuously, but in such a
manner that they balance each other.

$$\text{liquid} \underset{\text{condensation}}{\overset{\text{evaporation}}{\rightleftarrows}} \text{vapor}$$

The particles in the gas phase exert a pressure like those of any gas. *The
pressure of a gas in equilibrium with its liquid phase is called the* **vapor pressure**

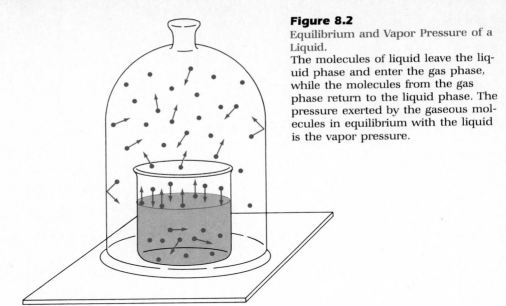

Figure 8.2
Equilibrium and Vapor Pressure of a Liquid.
The molecules of liquid leave the liquid phase and enter the gas phase, while the molecules from the gas phase return to the liquid phase. The pressure exerted by the gaseous molecules in equilibrium with the liquid is the vapor pressure.

of the liquid. The vapor pressure is a measurable quantity, just as is the pressure of a gas. It indicates the escaping tendency of the liquid, which in turn is characteristic of the individual liquid and its temperature.

Figure 8.3 shows the vapor pressures of diethyl ether (an anesthetic), ethyl alcohol, and water at various temperatures. The molecular weights of diethyl ether, ethyl alcohol, and water are 74, 46, and 18 amu, respectively. Clearly, molecular weight is not the determining feature of vapor pressures. If it were, the order of increasing vapor pressure would be diethyl ether, ethyl alcohol, and water. Experimentally, the order of vapor pressures at a given temperature is water < ethyl alcohol < diethyl ether. Therefore, attractive forces between particles are an important factor in determining the ability of a molecule to escape from the liquid phase. Water molecules attract one another very strongly, and their escaping tendency is low. We will return to discuss this very common but unusual liquid.

8.3
Boiling Point of Liquids

On heating to a specific temperature, any liquid eventually will undergo a very pronounced transformation. Bubbles are formed throughout the liquid, and they rise rapidly to the surface, bursting and releasing vapor in large quantities. This process is called boiling, and the temperature at which it occurs is called the boiling point of the liquid. *At the* **boiling point** *the vapor pressure of the liquid equals the atmospheric pressure.*

The vapor pressure of water at 80°C is 355 mm Hg. If we lived on a planet with a normal atmospheric pressure of 355 mm Hg, water would boil at 80°C. At this temperature the vapor pressure of the liquid would equal the atmospheric pressure. Water can boil at any temperature if the external pressure is increased or decreased appropriately. To avoid ambiguity, the *standard or* **normal boiling point** *is the boiling point at one standard atmosphere.*

The fact that liquids boil at lower temperatures under reduced pressures can be confirmed by anyone who has attempted to cook food in boiling water at high altitude. At higher altitudes the atmospheric pressure is lower. In Figure 8.4, the

Figure 8.3
Vapor Pressure of Liquids and Temperature.

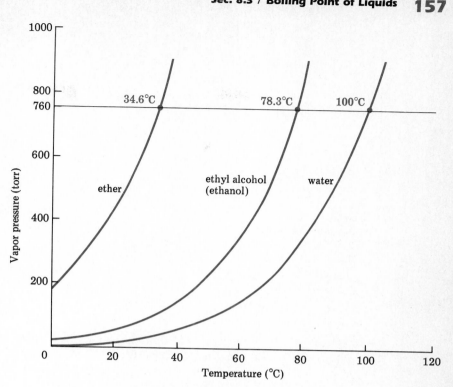

Temperature (°C)	Vapor pressure (torr)		
	Ether (74 amu)	Ethyl alcohol (46 amu)	Water (18 amu)
0	185	12	5
20	442	43	18
40	920	132	55
60	1730	347	149
80	3000	814	355
100	4865	1780	760
120	7495	3535	1489

effect of altitude on the boiling point of water is illustrated. Since water, or any liquid, at its boiling point, stays at the temperature of boiling until there is no liquid left, no amount of fuel can raise the temperature above the low boiling point. Food requires longer to cook at the lower temperature, and it is not uncommon to have to use cooking times twice as long as normal at altitudes of 7000 ft above sea level.

Conversely, food can be cooked faster at the high pressures obtained in a pressure cooker. When the pressure gauge is set for 5 psi above atmospheric pressure, the boiling point of the water in the pressure cooker is about 108°C. Under these conditions, food cooks twice as fast as at 100°C.

The average kinetic energies of particles in gaseous and liquid phases are equal at the boiling point. However, energy is required in order to maintain boiling and transfer matter from the liquid to the gaseous phase. The heat added does not increase the temperature of the liquid at the boiling point, but provides

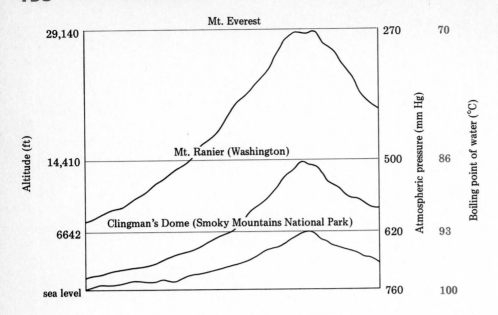

Figure 8.4
The Effect of Pressure on the Boiling Point of Water.
At higher altitudes the atmospheric pressure decreases. When the vapor pressure of water is equal to the atmospheric pressure, the water boils.

the energy necessary for the most energetic particles to continue to escape. *The quantity of heat energy required to transform 1 g of a substance at its boiling point from a liquid into a gas is called its* **heat of vaporization.** The heat of vaporization of water is 540 cal/g, a value that is rather large and reflects the strong attractive forces between neighboring water molecules in the liquid phase.

8.4
Solids

When heat energy is added to a solid, the temperature increases until the solid starts to melt. *That temperature at which the added heat energy is used only to melt the solid without raising the temperature of the solid or liquid is called the* **melting point.** At the melting point, the solid and liquid states can exist in equilibrium. Particles from the solid state, which consist of ordered arrangements, escape and enter the more random liquid state, while particles from the liquid may be deposited on the surface of the solid.

$$\text{solid} \underset{\text{freezing}}{\overset{\text{melting}}{\rightleftharpoons}} \text{liquid}$$

The effect of pressure on the melting point is not as dramatic as its effect on the boiling point. For most solids, the melting point of a substance increases with pressure. Water is atypical; its melting point decreases with increasing pressure. The melting point decrease is approximately 0.01°C/atm. Skaters take advantage of this phenomenon when they skate on the surface of the ice. Under the pressure exerted by the narrow edge of a hollow-ground skate blade, the ice melts to provide water as a lubricant.

The amount of heat energy required to transform 1 g of a solid into a liquid at the melting point is called the **heat of fusion.** The heat of fusion of water is 80 cal/g. Like the heat of vaporization for water, this value is higher than that for many solids, yet another indication that water has strong attractive forces between neighboring molecules. The melting point of a solid can be considered an approximate indication of its intermolecular attractive forces. For substances of similar molecular weight, those with the higher melting point have the stronger intermolecular forces. However, there are many more things that affect the melting point of a solid. Foremost among these is the packing or the geometrical arrangement of one particle with respect to its neighbors.

The French chemist Henri Le Châtelier suggested in 1888 a simple generalization about systems at equilibrium. *If an external force is applied to a system at equilibrium, the system will readjust to reduce the stress or change on it.* Le Châtelier's principle is one that you should find intuitively acceptable based on your own observations.

Consider how you react to changes in temperature. If the air temperature drops, you experience a change that your body then attempts to counteract. Your body may respond by shivering in the cold in order to produce heat. It would not respond by perspiring when it is cold. If the air temperature increases, your body cools itself by perspiring. It isn't reasonable to expect the body to produce heat when it is hot.

Le Châtelier's principle explains the effect of pressure on the boiling point of a liquid and the melting point of a solid. Placing pressure on matter should make it tend to occupy the smallest volume possible. In the case of evaporation of a liquid, there is a tremendous increase in volume. Application of pressure will cause the equilibrium system of a liquid and its vapor at its normal boiling point to shift toward the liquid state, because the resultant decrease in volume tends to diminish the pressure on the equilibrium system. Therefore, in order to boil a liquid at higher pressures, higher temperatures are necessary.

In the case of the melting point of a solid, most substances decrease in volume in going from the liquid to the solid state, and application of pressure will cause an equilibrium system of liquid and solid to shift toward the solid state. This shift decreases the volume of the system and counteracts the applied pressure. Therefore, in order to melt most solids under pressure, higher temperatures are necessary. Only small increases in melting points are usually observed due to the small difference in volume between solids and liquids. The abnormal behavior of water is the result of a lower density of ice compared to water. Thus, there is an increase in volume in going from water to ice. An increase in pressure shifts the ice–water equilibrium toward water.

The strong forces or bonds that hold atoms or ions together in compounds are **intramolecular forces.** The prefix *intra* means within, as in intramural sports within a college. In order to break a chemical bond, large amounts of energy must be added. To a break a hydrogen molecule into hydrogen atoms requires 104 kcal/mole.

There are weaker forces that exist between molecules. *The forces between molecules are* **intermolecular forces,** and they are responsible for holding molecules close together in the liquid and solid states. These forces are less than 10 kcal/mole. Intermolecular forces are of three types: London forces, dipole–dipole forces, and hydrogen-bonding forces.

London Forces

Nonpolar compounds and elements can be liquefied by decreasing the temperature. There then must be intermolecular forces even between nonpolar molecules and atoms such as neon and helium. What is responsible for these forces?

On the average, the electrons in a nonpolar molecule or atom are distributed uniformly. However, at some instant the electrons may be distributed closer to

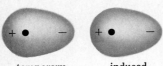

Figure 8.5
Van der Waals Attractive Forces.
The electron distribution in an atom becomes distorted and a temporary dipole results. An adjacent atom then has a dipole induced by the movement of electrons to result in a net attraction between atoms.

temporary dipole   induced dipole

one nucleus in a molecule or toward one side of an atom. At that instant a temporary dipole is present (Figure 8.5). A temporary dipole will exert an influence on nearby molecules or atoms. The effect is to polarize neighboring molecules and induce a dipole in them. *The resulting attractive forces between a temporary dipole and an induced dipole are called* **London forces.**

The strength of London forces depends on the number of electrons in a molecule or atom. The size and shape of a molecule are also important. The more electrons there are and the farther away they are from the nucleus, the more polarizable is the molecule.

Consider the boiling points of the nonpolar molecules Cl_2, Br_2, and I_2 given in Figure 8.6. As you can see, the boiling points increase with increasing molecular weight. A plot of boiling point versus molecular weight gives essentially a straight line. The relationship shown is the result of increased London forces with increasing molecular weight. These forces tend to hold the molecules together, and therefore the temperature required to boil the substance is increased.

Example 8.1

The boiling points of liquid krypton and liquid argon are -152.9 and $-185.8°C$, respectively. Suggest a reason for these facts and predict the boiling point of xenon.

Solution. The atomic numbers of krypton and argon are 36 and 18, respectively. The electrons in krypton are in higher energy levels and further from the nucleus than the electrons in argon. As a result, the electrons in krypton are more polarizable, and larger London forces result. One would predict the boiling point of xenon to be still higher. It is $-108.1°C$.

[Additional examples may be found in **8.26–8.36** at the end of the chapter.]

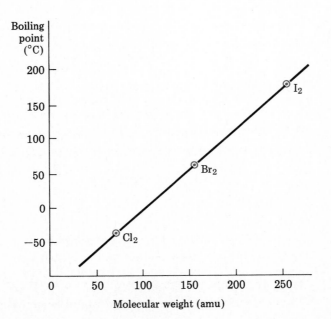

Figure 8.6
Correlation Between Boiling Point and Molecular Weight.

Figure 8.7
Dipole–Dipole Intermolecular Forces.
The positive iodine atom is attracted to the negative chlorine atom of a neighboring ICl molecule. The dipole is the result of a difference in the electronegativities of the two atoms.

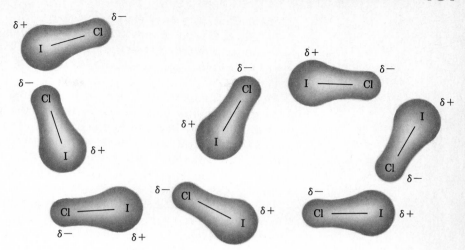

Dipole–Dipole Forces

If a molecule has a negative end and a positive end as a result of the unequal sharing of electrons, it is said to be polar and to have a dipole. Polar molecules attract one another. The positive end of one molecule is attracted to the negative end of another and vice versa. Polar molecules then tend to associate closely, and their physical properties reflect this association (Figure 8.7).

The molecular weight of ICl (162 amu) is close to the molecular weight of Br_2 (160 amu). Yet the boiling point of ICl is 97°C, 38°C higher than that of Br_2. If the boiling point and molecular weight of ICl were plotted on the graph given in Figure 8.6, the point would be above the line. Why does ICl boil at a higher temperature than Br_2? In Br_2 the pair of bonding electrons is shared equally between the two atoms and a nonpolar bond results. In ICl the bonding pair of electrons is attracted toward chlorine, the more electronegative element. The bond in ICl is then polar covalent, and the molecules are attracted to one another. In order to boil ICl, it is necessary to heat the molecules to a higher temperature than that needed for a nonpolar molecule of comparable molecular weight.

$$^{\delta+}I{-}Cl^{\delta-} \qquad ^{\delta+}I{-}Cl^{\delta-}$$

An attraction occurs between the positive I and negative Cl.

Example 8.2

The boiling points of nitrogen and carbon monoxide, CO, are 77 and 83 K, respectively. Suggest a reason for this order of boiling points.

Solution. The molecular weights of the two substances are the same. Nitrogen is a nonpolar molecule, and the only intermolecular forces are of the London type. Carbon monoxide is a slightly polar compound as a result of the difference in electronegativity between carbon and oxygen. The dipole–dipole forces cause an attraction between the positive carbon atom and the negative oxygen atom of a neighboring molecule.

$$^{\delta+}{:}C{\equiv}O{:}^{\delta-} \qquad {:}N{\equiv}N{:}$$

[Additional examples may be found in **8.26–8.36** at the end of the chapter.]

Hydrogen-Bonding Forces

Compounds containing hydrogen bonded to F, O, and N have very much higher intermolecular forces than expected. This is best seen in Figure 8.8, which gives the boiling points of the hydrogen compounds of the Group IV, V, VI, and VII elements.

The boiling point of SnH_4 is higher than other compounds of Group IV because of the large London forces. As the molecules become smaller, the electrons of the atoms are less polarizable, and the attractive forces are decreased. The CH_4 molecule then has the lowest boiling point.

In the compounds of the Group V, VI, and VII elements there is an obvious anomaly. The compounds NH_3, H_2O, and HF all have very much higher boiling points than expected. The three molecules NH_3, H_2O, and HF have two structural features in common. Each has at least one hydrogen joined by a polar covalent bond to a very electronegative atom. Each has at least one unshared pair of electrons (Figure 8.9). The very small hydrogen atom has a partial positive charge as a result of the bonding electrons being attracted toward an electronegative element. Consequently, an intermolecular attraction exists between hydrogen and an electron pair on a neighboring molecule. *The "bridging" of a hydrogen atom between two electronegative atoms is called a* **hydrogen bond.** Hydrogen bonds are quite strong and in the order of 3–10 kcal/mole.

The order of decreasing hydrogen bond strength with the electronegative

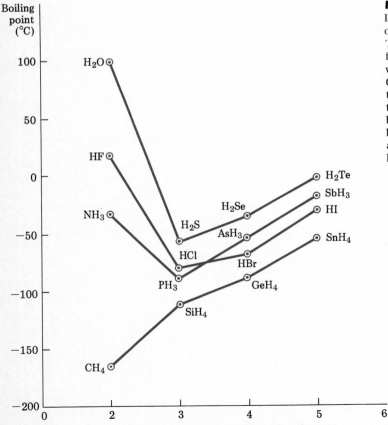

Figure 8.8

Illustration of the Effect of Hydrogen Bonding on Boiling Points.

The boiling points of compounds within a family increase with increasing molecular weight for elements in periods 3, 4, and 5. Only CH_4 of Group IV also is a part of this trend. The elements fluorine, oxygen, and nitrogen form compounds with hydrogen that boil considerably higher than anticipated based on London forces. The special force of attraction in these compounds is called a hydrogen bond.

Figure 8.9
Molecules That Form
Hydrogen Bonds.
Hydrogen fluoride is lim-
ited in forming hydrogen
bonds by its single hydrogen atom in spite of its
three electron pairs. Am-
monia is limited in forming
hydrogen bonds by its single electron pair in spite of
its three hydrogen atoms.
Only water has an equal
number of hydrogen atoms
and electron pairs.

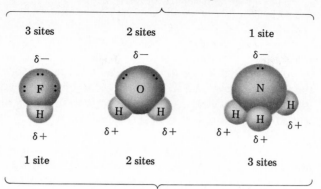

atoms might be expected to be $F > O > N$. However, in Figure 8.8 it appears that
the anticipated order is incorrect. HF boils at a lower temperature than H_2O.
Actually the hydrogen bonds in HF are stronger than those in H_2O. The reason for
the apparent anomaly is that although the hydrogen bond to fluorine is strong,
the molecule HF can only form one hydrogen bond per molecule (Figure 8.9). The
fluorine atom has three unshared electron pairs, but only one hydrogen atom. In
the case of water there are two hydrogen atoms and two unshared electron pairs
per molecule and the extent of aggregation can be greater than in HF. Ammonia,
NH_3, has three hydrogen atoms, but only one electron pair to share. Water has the
proper balance of hydrogen atoms and electron pairs to form the maximum
number of hydrogen bonds per mole.

Many molecules of biological importance contain O—H or N—H bonds that
can form hydrogen bonds to water or with each other within the molecule. Both
proteins and DNA form hydrogen bonds that affect their molecular shapes. Hydrogen bonding will be discussed at several other points in this text.

Example 8.3 _____

The boiling points of ethyl alcohol and dimethyl ether are 78.5 and $-24°C$,
respectively. The compounds have the same molecular formula, C_2H_6O, but have
different structures. Explain the large difference in the boiling points of the two
compounds.

$$H—\overset{\overset{\displaystyle H}{|}}{\underset{\underset{\displaystyle H}{|}}{C}}—\overset{..}{\underset{..}{O}}—\overset{\overset{\displaystyle H}{|}}{\underset{\underset{\displaystyle H}{|}}{C}}—H \qquad H—\overset{\overset{\displaystyle H}{|}}{\underset{\underset{\displaystyle H}{|}}{C}}—\overset{\overset{\displaystyle H}{|}}{\underset{\underset{\displaystyle H}{|}}{C}}—\overset{..}{\underset{..}{O}}:$$

dimethyl ether   ethyl alcohol

Solution. Ethyl alcohol has a —OH group as part of its structure. This group can
form hydrogen bonds via its hydrogen atom or the unshared pairs of electrons on
the oxygen atom. Dimethyl ether may not form hydrogen bonds to itself because
it does not have a positively charged hydrogen atom attached to oxygen. The
hydrogen bonding in ethyl alcohol is responsible for the higher boiling point.

[Additional examples may be found in **8.26–8.36** at the end of the chapter.]

8.7
Water—An Unusual Compound

The hydrogen bonds in water result in some interesting physical properties that are important in life processes. We will consider two of these properties, density and surface tension.

Density

At 100°C the density of water is 0.958 g/mL. The density increases as the temperature decreases until at 4°C the density of water is 1.0000 g/mL, which is its maximum density. The density then decreases below 4°C until the density of 0.99987 g/mL is reached at 0°C (Figure 8.10).

Ice at 0°C has a density of 0.917 g/mL and is less dense than liquid water. For that reason ice displaces its own weight and floats. The volume of ice floating above the surface of the water is approximately 8%. Since 92% of a mass of ice is below the surface, the visible part of an iceberg is indeed only "the tip of the iceberg." There is much more ice hidden beneath the sea than the part of an iceberg which we can see.

The facts that ice is less dense than water and that the maximum density of liquid water occurs at 4°C rather than 0°C have profound ecological significance. As the temperature of the water near the surface of a lake is lowered in cold weather, the cooled and more dense water tends to sink toward the bottom of the lake. The warmer water in the lake rises to the top, since it is less dense. This circulation of water continues until the overall water temperature is 4°C. Further cooling results in a lower density and the cooler water no longer sinks to the bottom of the lake. Thus, the warmest water is 4°C at the lake bottom.

On top of the lake, the surface water freezes. Since ice is less dense than water, the ice remains on the surface. If ice were more dense than water, the ice

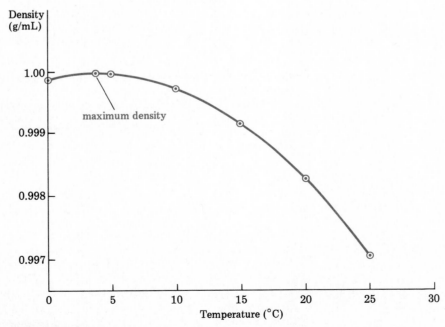

Figure 8.10
Density of Water and Temperature.
The density of water is at a maximum at 4°C. The density decreases from 4°C to 0°C.

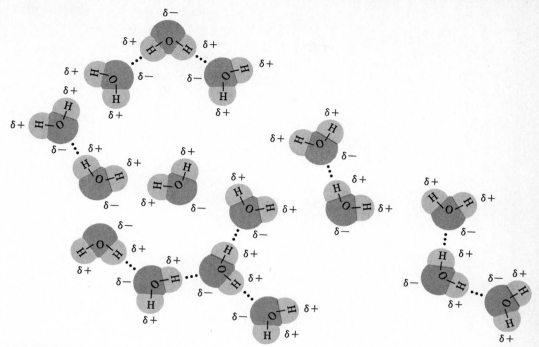

Figure 8.11
Hydrogen Bonding in Liquid Water.
The $\delta+$ and $\delta-$ charges represent the polarity of the hydrogen–oxygen bonds. In liquid water the molecules can move about in a somewhat restricted manner. The hydrogen bonds of clumps of associated water molecules are formed and broken easily. As the temperature is lowered, the kinetic energy decreases and the liquid volume contracts. The density then increases.

would sink to the bottom. A continued cold spell would cause the lake to freeze from the bottom to the top and living organisms in the lake would not survive.

The lower density of ice results in a sheet of ice that protects the water beneath its surface from further decreases in temperature. The ice thickness will slowly increase, but the water temperature will be maintained comfortable enough for aquatic life to survive.

The change in density of water with the temperature below 4°C and the fact that ice is less dense than liquid water are due to hydrogen bonding. Most liquid materials become more dense as the temperature decreases. Then they become still more dense when the solid state is formed.

For most liquids the decrease in temperature lowers the kinetic energy and allows more of matter to become aggregated or tightly packed. In the solid state the molecules then get as close as possible and leave a minimum of free space.

In water the aggregation as a result of hydrogen bonding has some unique structural requirements. As long as the "clumps" are small (Figure 8.11), at higher temperatures, water behaves like other liquids and becomes more dense as the temperature decreases. However, at 4°C a pattern of molecules emerges as the number of hydrogen bonds is increased. Then, as freezing occurs, the pattern of hydrogen-bonded molecules crystallizes. The pattern is an open cage-like network. "Holes" in the cages then mean that the density of ice is less than water (Figure 8.12).

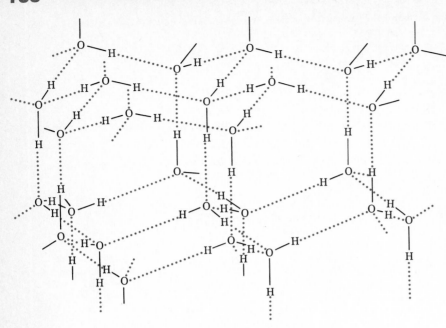

Figure 8.12
Structure of Ice and Hydrogen Bonds.
The three-dimensional structure shows each covalent oxygen–hydrogen bond as a solid line. The hydrogen bonds are shown by color dotted lines. Each oxygen atom in the structure has two covalent bonds and two hydrogen bonds. Each hydrogen atom is covalently bonded to one oxygen atom and hydrogen bonded to another oxygen atom. The void or space created by the structure accounts for the lower density of ice than of water.

Surface Tension

Surface tension is a phenomenon where a liquid seems to have a thin membrane on its surface. As a result, objects that are denser than the liquid may be made to "float" on the membrane. However, if the object is pushed beneath the surface, it will sink. Water has a high surface tension, and even steel needles and razor blades can be carefully set afloat on the surface of water.

The high surface tension of water is also a result of hydrogen bonding. The water molecules within the liquid are attracted in all directions by other water molecules. However, at the surface, a water molecule is attracted only to other surface molecules and those beneath it (Figure 8.13). There is no attraction in an upward direction. Surface tension is the result of a net downward attraction making the surface molecules act as a membrane coating the liquid.

High surface tension decreases the ability of water to wet a surface. You have observed this phenomenon when a drop of water is placed on a glass or on the

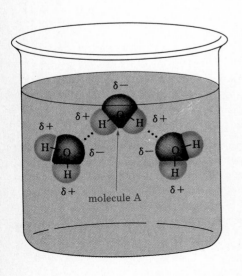

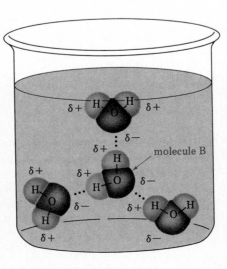

Figure 8.13
Surface Tension Is the Result of Attractive Forces Pulling Surface Molecules Inward.
Molecule A is attracted to molecules within the liquid and to other molecules on the surface of the liquid. The result is a net downward pull. Molecule B within the liquid has balanced attractions to its neighboring molecules.

windshield of a car. Water resists increasing its surface area and tends to cohere and form drops. Since medicines and foods must be dispersed in water, the surface tension of water must be decreased so that the water molecules can "wet" the material. *Surface-active substances or surfactants reduce the surface tension of water.*

Surfactants are molecules that may displace water molecules at the surface. As a result, the membrane is broken, and the water molecules may act individually to wet another object. The best-known examples of surfactants are detergents and soaps. Water containing such a surfactant will spread over a glass, while pure water will simply form drops.

Bile stored in the gallbladder behaves as a surfactant. When it is released into the duodenum during digestion, bile aids in dispersing fats into the water of the digestive juices. Without bile, the fats, which are nonpolar, would exist as globs containing many molecules. The surrounding digestive juices could not react efficiently with the fats. However, bile decreases the surface tension and allows the fats to become "wet" and disperses the molecules. For this reason persons whose gallbladders are removed may have to restrict their fat intake.

Summary

When the temperature of a gas is lowered, the average kinetic energy of the molecules decreases, and a point is reached at which intermolecular forces cause condensation to form a liquid. As the temperature is further lowered, the liquid solidifies and a rigid structure of arranged molecules results.

A liquid has a characteristic vapor pressure that depends on the temperature. When the vapor pressure equals the atmospheric pressure, the liquid boils. The normal boiling point of a liquid is the temperature at which the vapor pressure equals one atmosphere. The energy required to transform one gram of a liquid to a gas at its boiling point is the heat of vaporization.

The temperature at which a solid is converted to a liquid in equilibrium is called the melting point. Pressure exerts little effect on the melting point.

Le Châtelier's principle provides an explanation for the physical changes of the states of matter.

Intermolecular forces are of three types; London, dipole–dipole, and hydrogen bonding. Compounds of nitrogen, oxygen, and fluorine with hydrogen form hydrogen bonds. The unusual properties of water are the result of hydrogen bonds.

Learning Objectives

As a result of studying Chapter 8 you should be able to

- Describe the liquid and solid states using the kinetic molecular theory.
- List several equilibria between the states of matter.
- Use Le Châtelier's principle to explain the physical properties of liquids and solids.
- Relate vapor pressure and temperature to the boiling point of a liquid.
- Give examples of the effect of three types of intermolecular forces on physical properties of matter.
- Give a summary of the unique properties of water.
- Describe the phenomenon of surface tension.

New Terms

Boiling is a process in which a liquid is vaporized to a gas when the vapor pressure is equal to atmospheric pressure.

Condensation is a process in which a gas is converted to a liquid.

An **equilibrium** is a dynamic state in which opposing processes are in balance.

Freezing is a process in which a liquid is converted to a solid. This process occurs at the same temperature at which melting of a solid occurs.

Heat of fusion is the amount of energy released as one gram of a liquid is converted into a solid at its melting point.

Heat of vaporization is the amount of energy required to convert one gram of liquid to a gas at its boiling point.

A **hydrogen bond** is an intermolecular attraction between an electropositive hydrogen atom and a nonbonded electron pair of an electronegative atom of a neighboring molecule.

An **induced dipole** is a separation of charge within a molecule caused by a temporary dipole in the vicinity.

An **temporary dipole** is a separation of charge produced

momentarily in an otherwise nonpolar region of a substance.

Intermolecular forces are forces of attraction between separate molecules.

Intramolecular forces hold atoms together in compounds.

Le Châtelier's principle states that a system will readjust to reduce the stress placed on it.

London forces are a type of intermolecular force involving instantaneous and induced dipoles.

The **melting point** is the temperature at which a solid is converted into a liquid.

Normal boiling point is the temperature at which the vapor pressure of a liquid equals one atmosphere.

Vaporization is the process of converting molecules from the liquid to the solid state.

Vapor pressure is the pressure of a vapor in equilibrium with its liquid form.

Questions and Problems

Terminology

8.1 Using the kinetic molecular theory, distinguish between liquids and solids.

8.2 Distinguish between boiling point and normal boiling point.

8.3 How are vapor pressure and boiling point related?

Evaporation of Liquids

8.4 What factors control the drying of clothes at temperatures above the freezing point of water?

8.5 Explain the difference felt by a swimmer emerging from the water on a windy day versus a calm day, assuming the temperature is the same.

8.6 The rate of evaporation of water from a beaker inside a larger sealed container at constant temperature decreases with time. Explain this phenomenon and contrast it with the rate of evaporation from the same beaker in a room at constant temperature.

8.7 A beaker of water in a closed room at constant temperature does not decrease in volume on a given day. Why does the water not evaporate?

Vapor Pressure of Liquids

8.8 The space above the column of mercury in a barometer is considered to be a vacuum. Is this idea strictly correct?

8.9 Why is mercury used in a barometer rather than another liquid such as water?

8.10 The vapor pressure of liquid A at 25°C is equal to that of liquid B at 50°C. If the molecular masses of A and B are identical, which of the two liquids has the stronger intermolecular attractive forces?

8.11 Is the vapor pressure of water at 25°C at an altitude of 10,000 ft in the mountains less than, equal to, or more than the vapor pressure of water at sea level at 25°C?

Boiling Points of Liquids

8.12 What is the highest temperature at which water vapor will condense to yield water under 1 atm pressure? At what temperature will water boil under pressure of 148.9 cm of mercury?

8.13 Explain the operation of a pressure cooker.

8.14 Explain why the time required to fry meat at a high altitude is no different than at sea level, whereas the time required to boil potatoes varies significantly as a function of altitude.

Heat of Vaporization

8.15 Steam at 100°C causes more severe burns than water at 100°C. Why?

8.16 Calculate the amount of heat required to convert 10 g of water at its normal boiling point into steam.

8.17 How many calories are released by 10 g of steam when it condenses to water at 100°C?

Melting Points of Solids

8.18 Why does solid A melt at 200°C, whereas solid B melts at 400°C?

8.19 A highly trained figure skater skates on only a small portion of the edge of a blade and may exert 300 atm pressure. What is the melting point of the ice under this pressure?

8.20 A speed skating blade is long but very thin. Explain why this design is chosen.

8.21 Mercury thermometers cannot be used below −39°C. Explain why.

Heat of Fusion

8.22 How many calories are required to melt 1000 g of ice?

8.23 How many calories must be removed from water at 0°C to freeze 100 g of water?

Le Châtelier's Principle

8.24 Explain the prediction of the effect of pressure on boiling point using Le Châtelier's principle.

8.25 Explain the effect of pressure on the melting point of solids using Le Châtelier's principle.

Intermolecular Forces

8.26 List and describe the three types of intermolecular forces.

8.27 If water were a "normal" compound, what would be its expected boiling point based on London and dipole–dipole forces?

8.28 The compounds CF_4 and CCl_4 are tetrahedral molecules. Neither is polar. Explain why the boiling points of CF_4 and CCl_4 are −129 and 76.8°C, respectively.

8.29 The boiling points of ethyl alcohol and ethylene glycol (used in antifreeze) are 78.5 and 190°C, respectively. Explain the high boiling point of ethylene glycol.

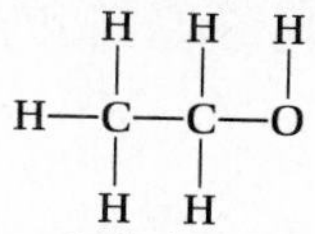

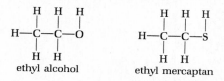

ethyl alcohol 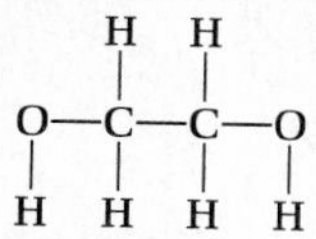   ethylene glycol

8.30 Predict which of the following compounds would have the higher boiling point. On what type of intermolecular forces did you base your prediction?

ethyl alcohol   ethyl mercaptan

8.31 Why doesn't methane form hydrogen bonds?

8.32 Should methane or ethane have the higher boiling point?

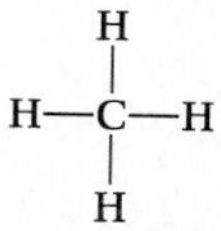

methane 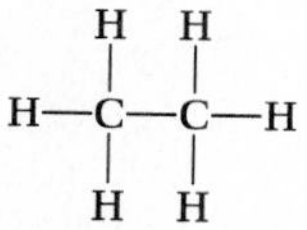   ethane

8.33 Predict the order of boiling points of O_2 and N_2.

8.34 Predict the order of boiling points of helium and neon.

8.35 The boiling points of propane and butane are −42 and −0.5°C, respectively. Explain this order.

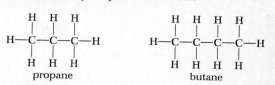

propane   butane

8.36 Describe the special intermolecular forces that exist in water.

8.37 Explain the difference in the principles involved when a wooden matchstick floats on water as compared with a steel needle.

8.38 Explain how hydrogen bonding is responsible for the high surface tension of water.

8.39 Why does water have a relatively low vapor pressure and high boiling point?

8.40 Why does ice float on water?

8.41 Explain why the maximum density of water allows water at 4°C to exist at the bottom of a lake.

8.42 Why is the lower density of ice over water important to aquatic life?

8.43 Use the structure of ice to explain its being lower in density than water.

8.44 What are surfactants? Explain their action.

8.45 What surfactant is involved in digestion, and what is its source?

Water and Solutions

STOPPER UNDER
HIGH PRESSURE
AND WILL SELF-
EJECT WITH GREAT
FORCE WHEN WIRE
CLOSURE IS LOOSENED

The solubility of gases in water is increased by pressure. This label on a bottle of champagne warns of the pressure of the dissolved carbon dioxide.

The water molecule is one of the simplest yet one of the most extraordinary structures in terms of its importance to the maintenance of life on this planet. In addition, there is more water present on earth than any other substance. Water surrounds us as water vapor in the air, as a liquid in rivers, lakes, and oceans, and as a solid, ice, both on land and in the oceans. About 70% of the surface of the earth is covered with liquid water. The oceans, containing 97% of the earth's water, are the major reservoirs (Table 9.1).

Water is a heat regulator. Large quantities of solar energy are absorbed by the oceans and then released. As a result, the temperature of the land is moderated by nearby water. Without water, dramatic temperature changes would occur, and life forms would not have as comfortable an environment. Water also controls the temperature of organisms and allows the temperature of cells to remain reasonably constant in spite of fluctuations in the temperature of the environment. Without careful temperature regulation, the many chemical reactions in the human body go out of control and death results.

Water balance is critical to all living organisms—even more important than food. Humans can live for weeks without food, but only a few days without water. The major component of our body, as it is for other organisms, is water. Bacterial cells may contain as little as 50% water. A human contains about 70% water, whereas marine invertebrates are as much as 95% water.

Body water remains fairly constant because water loss is usually balanced by water intake unless a severe biological disturbance occurs. We lose water by perspiring and by excreting body wastes. Sources of water include water itself, other beverages, and foods such as fruit, vegetable, meat, bread, and dairy products (Table 9.2).

Pure water is colorless, odorless, and tasteless. Small amounts of dissolved minerals cause changes in these properties. In fact, the taste of spring water, which people regard as pure, is really due to the impurities present in it. Pure water tastes bland and is not very appealing to drink.

All of the biological chemistry of life requires water. Water transports food and oxygen from site to site so that reactions may occur, and then water removes the waste products. Furthermore, water is a reactant in many of these reactions. It plays a major role in acid–base reactions (Chapter 11).

Water has an extraordinary ability to form mixtures with a large number of substances. These mixtures are classified as solutions, colloidal dispersions, and suspensions (Section 9.2). All of these types of mixtures are found in parts of your body. In fact, within a cell, all three types of mixtures are found. Certain simple cations, anions, carbohydrates, and amino acids are present as solutions. Globules of fat and protein molecules are present as colloids, whereas some parts of the cell called organelles are present as suspensions in water. In order to understand the chemistry of the life processes that occur in water, it is first necessary to study the principles and concepts of simpler solutions.

9.1
Water—An Important Compound

Table 9.1
Distribution of Water

Location	Percent
oceans	97.2
glaciers and polar caps	2.16
subsurface water	0.62
lakes and rivers	0.019
atmospheric water	0.001

Table 9.2
Water Content of Food

Food	Percent water
apples	85
broccoli	90
cheddar cheese	35
eggs	75
grapes	80
milk	87
potatoes	78
steak	73
tuna fish	60

Water forms three types of mixtures. These mixtures are called suspensions, solutions, and colloidal dispersions. The distinction among these mixtures is one of the size of the particles suspended, dispersed, or dissolved in the water.

9.2
Water Mixtures

Suspensions

If small chunks of matter larger than 10^{-5} cm in diameter, such as clay, are shaken with water, they become suspended to form nontransparent muddy

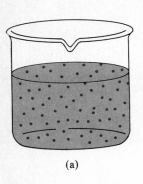

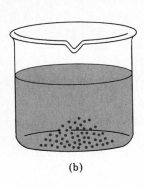

(a)   (b)

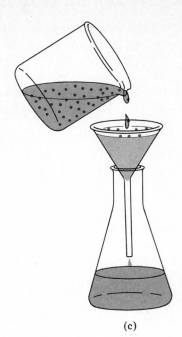

(c)

Figure 9.1
Separation of a Suspension.
Suspended particles (a) are distributed in an agitated liquid.
However, with time the particles will settle (b). The particles
may be separated (c) by filtering the suspension through filter
paper. The liquid that passes through the filter is called the fil-
trate.

water. The clusters of matter contain many millions of molecular units and can
be seen under an optical microscope. Eventually the particles settle as a result of
gravitational forces. Passing the suspension through a filter will separate the clay
from the water (Figure 9.1). Suspensions are nontransparent; light is not transmitted through them.

Solutions

Molecules and ions have diameters less than 10^{-7} cm. When they are uniformly distributed throughout water, a solution results. Solutions differ from suspensions in being uniform down to the molecular level. There are no aggregates
of molecules present. The particles in a solution cannot be filtered and do not
settle out. Although they may be colored, solutions are transparent and transmit
light. Solutions will be discussed in subsequent sections of this chapter.

Colloidal Dispersions

Colloidal dispersions *are mixtures intermediate between suspensions and
solutions.* Particles that are 10^{-7} to 10^{-5} cm in diameter may contain up to a
thousand molecular units. Such clumps of matter cannot be seen with an optical
microscope. Mixtures with particle sizes in this range are called colloidal dispersions.

When a beam of light is passed through a true solution, no light can be seen
at right angles to the path of the light (Figure 9.2). However, a beam of light
passing through a colloid can be observed at right angles to the beam. A colloid
contains particles that are large enough to scatter light. This light scattering is
called the **Tyndall effect.** The same scattering occurs in air containing dust,
moisture (fog), or smoke; these substances are gaseous colloids.

The Tyndall effect is controlled by the size and shape of the particles in the
colloid. This scattering phenomenon produces the vivid sunsets seen in the
desert or on the ocean. In each location, the long distance to the horizon enables

Figure 9.2
The Tyndall Effect.
Light is transmitted through a solution and is not scattered. Light is scattered by a colloid and can be seen at angles to the path of the light.

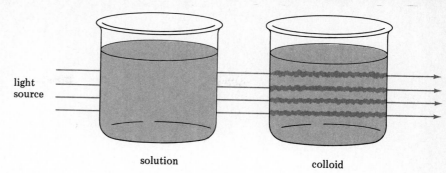

light source

solution   colloid

the observer to see light transmitted through miles of air containing colloidal matter.

The particles dispersed in the colloid are called the **dispersed phase;** the material in which they are dispersed is the **dispersing phase** or medium. Some examples of colloids are given in Table 9.3.

Table 9.3

Types of Colloids

Dispersing medium	Dispersed phase	Name	Example
gas	liquid	liquid aerosol	fog
gas	solid	solid aerosol	smoke
liquid	gas	foam	whipped cream
liquid	liquid	emulsion	milk
liquid	solid	sol	latex paints
solid	gas	solid foam	foam rubber
solid	liquid	gel	jello
solid	solid	solid sol	ruby

A gas can serve as the dispersing phase for either solids or liquids. Small dust particles in the air in major metropolitan areas are a well-known example of a solid dispersed in a gas. Fog, a colloid of liquid water dispersed in air, forms when water condenses from saturated air. Solids and liquids dispersed in a gas are called *solid aerosols* and *liquid aerosols*, respectively.

Liquids, too, serve as the dispersing phase for many colloids. Starch in a pudding is a solid dispersed in a liquid. Colloids of solids dispersed in liquids are called *sols*. Colloids of a liquid in a liquid are called *emulsions*. Mayonnaise, cream, and milk are all emulsions. In the homogenization of milk, the dispersed fat is reduced to such a small size that the emulsion becomes virtually permanent. Examples of gases dispersed in liquids include whipped cream and meringue. Gases dispersed in liquids are called *foams*.

Solids can serve as the dispersing phase of colloids. Foam rubber and styrofoam are examples of a gas dispersed in a solid. Such colloids are called *solid foams*. The dispersal of a liquid in a solid is called a *gel*. Common examples are jellies and cheeses. Colored glass and gems such as ruby and turquoise are examples of a solid dispersed in a solid. These colloids are called *solid sols*.

A **solution** is a uniform mixture of substances that can vary in composition. The substance present in the largest quantity in a solution is the **solvent;** the substance dissolved in the solvent is called the **solute.** The solvent may be a gas, liquid, or

9.3
Solutions

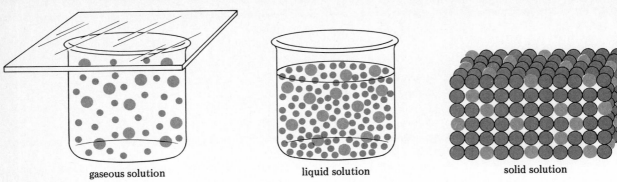

gaseous solution   liquid solution   solid solution

Figure 9.3
Solutions and the States of Matter.
Solutes may be dissolved in solvents in any of the three states.

solid (Figure 9.3). Each classification of solvent will be discussed briefly before we concentrate on the more common liquid solution in which water is the solvent.

Gaseous Solutions

All gases that do not react with each other will mix with each other in all proportions. In such solutions, the individual molecules are relatively far apart (Figure 9.3) and move independently. Air is the most common example of a solution consisting only of gases. Dry air contains approximately 78% nitrogen molecules, 21% oxygen molecules, and 1% argon atoms.

There are no true solutions of a liquid or a solid in a gas. Such systems are colloids.

Liquid Solutions

Gases, liquids, and solids may dissolve in liquid solvents. Because the molecules of liquids are close together, they are less independent of one another (Figure 9.3). Oxygen dissolved in water, which maintains aquatic life, is an example of a solution of a gas in a liquid. Another example is carbon dioxide in carbonated beverages. Two common examples of liquids that dissolve in water are ethyl alcohol and acetic acid. Ethyl alcohol is a component of all alcoholic beverages; acetic acid is present in vinegar. Both salt and sugar are well-known examples of solids that will dissolve in water. Although all gases dissolve in each other, not all gases, liquids, and solids dissolve in all liquids. This limitation is called the solubility of the solute and will be discussed in Section 9.6.

Solid Solutions

Although solid solutions might appear to be less common, there are, in fact, numerous examples of this type of solution. Many metals dissolve in one another to form solid solutions called alloys (Figure 9.3). Examples of such alloys are brass (zinc and copper) and sterling silver (silver and copper). Alloys of importance to us include mercury alloys used for dental restoration and vinertia alloy (67% Co, 27% Cr, 6% Mo) used to replace body joints such as hips and knees.

Concentrations

Solutions also may be classified in terms of concentration. The concentration is a measure of the amount of solute in a given amount of solution. Concentration units will be treated quantitatively in the next section.

A **concentrated solution** *contains a large amount of solute per given amount of solvent or solution. A* **dilute solution** *contains a small amount of solute per given amount of solvent or solution.* There is no sharp line between these two qualitative descriptions of solutions. However, one would judge a solution of 1 g of sugar in 1 L of water to be dilute, whereas a solution of 100 g of sugar in 1 L of water would be considered concentrated.

When a few milligrams of NaCl is added to a test tube containing 10 mL of water, only a small amount of stirring is required to dissolve the solid. If more and more NaCl is added, a point at which no additional salt will dissolve is reached. *At this point, when added solute remains undissolved, the solution is* **saturated.** *As long as more solute will dissolve, the solution is* **unsaturated.** Thus, the term saturated means "full," whereas unsaturated means that the "solution could hold more."

Sodium chloride is quite soluble, and 36 g is required to saturate 100 mL of water. On the other hand, only 9.0×10^{-5} g of AgCl is required to saturate 100 mL of water. The term saturated should not be confused with concentrated nor should unsaturated be confused with dilute. Solutions of 9.0×10^{-5} g of NaCl and 9.0×10^{-5} g of AgCl in 100 mL of water are both dilute solutions. Only the AgCl solution is saturated!

9.4 Concentration of Solutions

The qualitative comparative terms dilute versus concentrated or unsaturated versus saturated are descriptively useful. However, more quantitative indications of concentrations are needed in many fields. For example, the effective administration of a medicine usually requires a prescribed amount of the therapeutic agent. Quantitative expressions of the amount of solute in a solvent are required in chemical laboratories, industries, hospitals, and pharmacies.

Our discussion of concentration will involve only water as the solvent. *A solution of a solute in water is called an* **aqueous solution.** A variety of methods are used to express concentration. Each has been chosen for convenience under some particular set of circumstances.

Percent Concentration

The concentration of a solution is easily expressed using the percent of the solute in the solution. Since percent means parts per hundred, the method can be used by anyone. You don't have to be a chemist to understand percent concentrations.

Three percent means 3 parts out of a total of 100 parts. There are, of course, 97 other parts present to make the 100 parts. It is not necessary that the percent be an integer. If something has a 0.9% concentration, there is 0.9 part of that item and 99.1 parts of something else present. Notice that the percent refers to the parts of the solute. A 5% solution means 5 parts of the solute.

Percent concentrations may be expressed with units of mass or volume. Three derived concentrations commonly used are weight/weight, weight/volume, and volume/volume.

The weight/weight percent (w/w %) is given by the following equation.

$$\text{w/w \%} = \frac{\text{weight of solute}}{\text{weight of solute + weight of solvent}} \times 100$$

The weights are usually in grams.

Figure 9.4
Preparation of 0.9% NaCl Solution.
(a) Weigh 9 g of sodium chloride.
(b) Add to a 1.00 L flask. (c) Add
water to the mark while shaking
slightly until the NaCl dissolves completely.

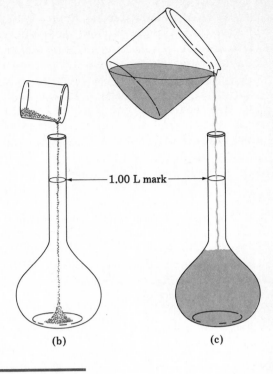

1.00 L mark

9 g NaCl

(a)   (b)   (c)

Example 9.1

What is the w/w % of a solution of 5.0 g of sugar in 45.0 g of water?

Solution. Both quantities are in common units and may be substituted into the
percentage equation for w/w %.

$$\frac{5.0 \text{ g sugar}}{5.0 \text{ g sugar} + 45.0 \text{ g water}} \times 100 = 10\% \text{ sugar solution}$$

A common way of expressing percents in clinical situations is weight/
volume percent (w/v %), which compares the weight of the solute to the total
volume of the solution.

$$w/v \% = \frac{\text{grams of solute}}{100 \text{ mL of solution}} \times 100$$

For example, the normal saline solution used to dissolve drugs for intravenous
therapy is 0.9% aqueous sodium chloride. The solution is prepared by dissolving
0.9 g of NaCl in water to give 100 mL of solution. To prepare a liter (1000 mL) of
normal saline, 9 g of NaCl is weighed out (Figure 9.4) and then enough water is
added to give 1000 mL of solution.

Example 9.2

A 2.00 mL sample of blood plasma contains 6.80 mg of sodium ions. Calculate
the concentration in mg/100 mL.

Solution. First calculate the weight/volume ratio in the units given.

$$\frac{6.80 \text{ mg sodium}}{2.00 \text{ mL plasma}} = 3.40 \text{ mg/mL}$$

In a 100 mL volume there would be 100 times the mass of sodium ions.

$$3.40 \text{ mg/mL} \times 100 \text{ mL} = 340 \text{ mg}$$

Therefore, there are 340 mg of sodium ions in 100 mL of blood plasma and the concentration is 340 mg/100 mL. The unit mg/100 mL is used to report sodium ion concentration in the blood.

Example 9.3

How many grams of sodium chloride are present in 250 mL of a salt solution that is 0.90% sodium chloride?

Solution. A 0.90% solution contains 0.90 g of sodium chloride for every 100 mL of solution. Therefore, the amount of sodium chloride in any quantity of sodium chloride solution may be calculated by using a factor.

$$250 \text{ mL solution} \times \frac{0.90 \text{ g NaCl}}{100 \text{ mL solution}} = 2.2 \text{ g NaCl}$$

[Additional examples may be found in **9.14–9.17** at the end of the chapter.]

Volume/volume percent (v/v %) is useful when dealing with liquid solutes. The equation is as follows.

$$\text{v/v \%} = \frac{\text{volume of solute}}{\text{total volume of solution}} \times 100$$

The volumes are usually given in milliliters. Note that the v/v % concentration expresses directly the milliliters of solute per 100 mL of solution. This method is used to express the concentration of alcohol in beverages. A wine that is 12% alcohol contains 12 mL of alcohol in 100 mL of solution. Proof, a term that is also used to express the alcohol content in alcoholic beverages, is equal to twice the v/v % of the alcohol. An alcoholic beverage that is 40% alcohol is 80 proof. Pure alcohol is 100% or 200 proof.

Example 9.4

A solution is prepared by dissolving 200 mL of ethyl alcohol in sufficient water to produce 1 L of solution. What are the v/v % concentration and proof of the solution?

Solution. Both volumes must be expressed in common units of mL.

$$\frac{200 \text{ mL of ethyl alcohol}}{1000 \text{ mL of solution}} \times 100 = 20\% \text{ alcohol}$$

The proof is twice that of the concentration or 40 proof.

Parts per Million

Some solutes, such as pollutants in air or water, are in such low concentrations that decimal percentages would have to be used. A unit that avoids the use of decimals is parts per million (ppm). Like percent concentration, the ppm unit may refer to w/w, w/v, or v/v. A **ppm** *is one part of solute per million parts of solution.* For example, a pollutant in a water source might be present at 0.0005 g per 100 mL of solution. The percent would then be 0.0005%. However, based on parts per million the concentration would be

$$\frac{0.0005 \text{ g}}{100 \text{ mL}} = \frac{5 \text{ g}}{1,000,000 \text{ mL}} = 5 \text{ ppm}$$

Example 9.5

The federal Food and Drug Administration (FDA) has set the human tolerance for mercury in fish at 0.5 ppm. A 5.0 g sample of fish contains 20 μg of mercury. Calculate the concentration of mercury in ppm, and determine if the FDA tolerance is exceeded.

Solution. First convert 20 μg into grams so that the mass of mercury and that of the fish are in the same units. The ppm will be on a w/w basis.

$$20 \, \mu g \times \frac{1 \text{ g}}{10^6 \, \mu g} = 20 \times 10^{-6} \text{ g} = 2.0 \times 10^{-5} \text{ g}$$

Now determine the ratio and multiply by 10^6 to obtain ppm.

$$\frac{2.0 \times 10^{-5} \text{ g Hg}}{5.0 \text{ g fish}} \times 10^6 = 4 \text{ ppm}$$

The concentration exceeds the tolerance limit of the FDA.

[Additional examples may be found in **9.18–9.20** at the end of the chapter.]

Molarity

Although percent composition solutions are easy to prepare, they have limitations in chemistry. Solutions that have the same percent composition contain the same weight of solute. However, they do not necessarily contain the same number of moles because the molecular weights of the solutes may not be the same.

In chemistry the number of moles in a sample is more important than the weight of the sample. The number of moles is related to the number of molecules, and it is this quantity that is used in discussing chemical reactions (Chapter 10). Therefore, a concentration known as molarity is convenient for chemistry.

Molarity *is defined as the number of moles of solute per liter of solution.* The abbreviation for molarity is M.

$$\text{molarity} = \frac{\text{moles of solute}}{\text{L of solution}} = M$$

In order to prepare a 1 M solution, it is necessary to add sufficient solvent to 1 mole of the solute to make a total volume of solution of exactly 1 L. Molarity is a very convenient concentration unit because it gives information about the volume of the solution that must be used to obtain the desired number of moles of solute. Volumes are easily measured with volumetric laboratory glassware such as a graduated cylinder or pipet.

Example 9.6 _____

What is the molarity of a solution prepared from 9.8 g of sulfuric acid, H_2SO_4, and sufficient water to yield 200 mL of solution?

Solution. The number of moles of sulfuric acid is 0.10.

$$9.8 \text{ g } H_2SO_4 \times \frac{1.0 \text{ mole } H_2SO_4}{98 \text{ g } H_2SO_4} = 0.10 \text{ mole } H_2SO_4$$

The volume of solution in liters is

$$200 \text{ mL} \times \frac{1 \text{ L}}{1000 \text{ mL}} = 0.200 \text{ L}$$

The molarity is calculated from the number of moles of solute and the volume of solution.

$$\frac{0.10 \text{ mole } H_2SO_4}{0.200 \text{ L solution}} = 0.50 \, M \; H_2SO_4$$

Example 9.7 _____

How many grams of sodium hydroxide are required to produce 250 mL of a $0.20 \, M$ solution?

Solution. First the number of moles of NaOH required for this volume and molarity must be determined, and then the number of grams may be calculated. The number of moles is

$$250 \text{ mL} \times \frac{1 \text{ L}}{1000 \text{ mL}} \times \frac{0.20 \text{ mole}}{1 \text{ L}} = 0.050 \text{ mole}$$

The number of grams is

$$0.050 \text{ mole} \times \frac{40.0 \text{ g}}{\text{mole}} = 2.0 \text{ g}$$

Example 9.8 _____

Calculate the number of liters of a $0.10 \, M$ glucose solution required to provide 90 g of glucose, $C_6H_{12}O_6$.

Solution. First the number of moles contained in 90 g must be calculated.

$$90 \text{ g} \times \frac{1 \text{ mole}}{180 \text{ g}} = 0.50 \text{ mole}$$

Now the volume that contains 0.50 mole of glucose may be calculated by using the inverse factor of the molarity.

$$0.50 \text{ mole} \times \frac{1 \text{ L}}{0.10 \text{ mole}} = 5.0 \text{ L}$$

[Additional examples may be found in **9.24–9.26** at the end of the chapter.]

9.5
Electrolytes and Nonelectrolytes

Very pure water is essentially a nonconductor. This can be illustrated by the use of the apparatus shown in Figure 9.5. Two electrodes are immersed in water and are not in direct contact with each other. When the switch is closed, electricity would flow and the light bulb would glow if water could conduct the electricity between the two electrodes. However, the light does not glow with pure water in the beaker because pure water is not a conductor.

Any substance that dissolves in water to form a solution that conducts electricity is called an **electrolyte.** The solution of NaCl in water does conduct electricity. When this solution is used in the apparatus in Figure 9.5, the light glows

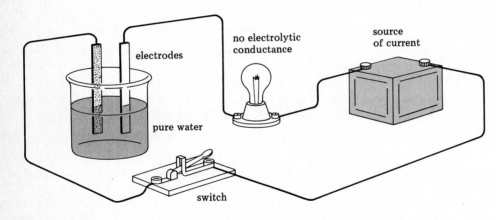

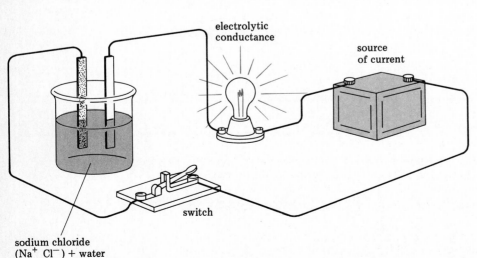

Figure 9.5
Conductivity of a Solution. The current flows only if ions exist in solution. Substances that provide ions in solution are called electrolytes.

brightly. The explanation for this phenomenon was provided by the Swedish scientist Svante Arrhenius in 1884. He suggested that substances whose aqueous solutions conduct electricity form ions in the solution. **The cations in the solution migrate to the negative electrode, called the cathode, while the anions migrate to the positive electrode, called the anode.** Such migrations provide the necessary contact between the electrodes and allow for passage of electricity.

When ionic substances are dissolved in water, their ions are separated from the crystal structure by the water molecules. *This separation of ions already present in another phase is called* **dissociation.** There are covalent substances that produce ions when dissolved in water. *The formation of ions upon dissolution in water is called* **ionization.** For example, the gaseous covalent molecule hydrogen chloride yields ions when dissolved in water. The ions formed are given by the equation

$$HCl + H_2O \longrightarrow H_3O^+ + Cl^-$$

The details of the ionization process will be given in Chapter 11.

Any substance that when dissolved in water does not allow the passage of a current is called a **nonelectrolyte.** Ordinary cane sugar, known as sucrose, $C_{12}H_{22}O_{11}$, is a nonelectrolyte. Sucrose does not contain ions in the solid state, and therefore dissociation does not occur. Furthermore, water cannot produce ions from sucrose. Another common substance that is a nonelectrolyte is ethyl alcohol, C_2H_6O.

Electrolytes are divided into two classes called *strong* and *weak electrolytes.* Strong electrolytes enable the ready passage of electricity in the device shown in Figure 9.5, causing the light bulb to glow brightly. The degree of brightness is related to the number of ions present in solution. All ionic substances that are soluble in water are strong electrolytes. Some covalent compounds ionize completely in water. Because these compounds produce ions that cause the bulb to glow brightly, they are also called strong electrolytes. Other covalent substances do not ionize completely in water. These substances cause only a dull glow and are called weak electrolytes. Weak electrolytes will be discussed in Chapter 11.

9.6 Solubility

The process that occurs when a solute dissolves in a solvent can be described by the kinetic molecular theory. In an ionic solid, such as sodium chloride, the sodium and chloride ions are attracted to each other. In the process of dissolving in water, these ions are separated from one another. This separation occurs when the polar water molecules approach and eventually surround the ions (Figure 9.6). The ions are removed from the site of the sodium chloride crystal by the water molecules. The electronegative oxygen of water is oriented toward the positive sodium ion, while the electropositive hydrogens of water are oriented toward the negative chloride ion.

The amount of sodium chloride that dissolves in water depends on the temperature. If 30 g of sodium chloride is placed in 100 g of water at 25°C, the solid dissolves completely. If another 6 g of sodium chloride is added, it too dissolves, although at a somewhat slower rate. Now, if more sodium chloride is added, the solid remains at the bottom of the container. At this point, the solution is saturated. Thus, 36 g of sodium chloride in 100 mL of water at 25°C is a saturated solution.

Although no further change is apparent in a saturated solution in contact with undissolved solid, two processes are continuing. Some of the solid is still

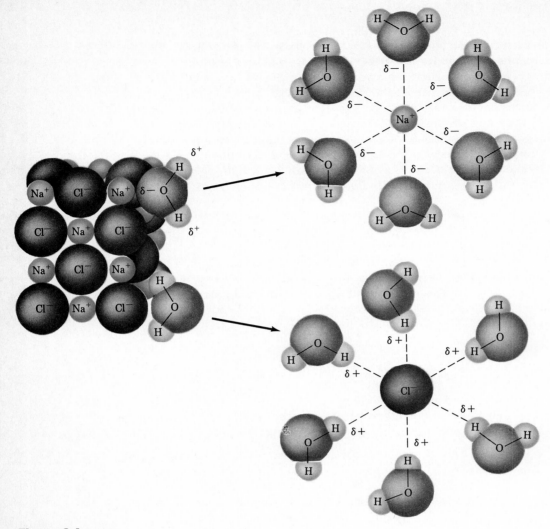

Figure 9.6
Dissolving an Ionic Compound in Water.
The attractive forces between ions and the polar water molecules are sufficient in soluble compounds to overcome the attractive forces within the crystal. The cations are approached by the negative part of the water molecule and enter solution to be surrounded by water molecules. The anions are approached by the positive part of the water molecule and enter solution to be surrounded by water molecules.

dissolving in the solvent, while some of the dissolved solute is crystallizing. There is a state of equilibrium. Thus, a saturated solution is one in which the rate of dissolution of the solute equals the rate of crystallization of the solute. The conditions that determine this equilibrium are temperature, pressure, and the identities of the solute and solvent.

Effect of Pressure on Solubility

As was pointed out in Chapter 7, there are large changes in volume that occur with pressure changes in the gaseous state. Similarly, the solubility of gaseous solutes in liquids show a large dependence on pressure.

AN ASIDE

The condition called the bends that deep-sea divers may experience if they do not readjust to the lower surface pressure on returning from the ocean's depth, is due to the effect of pressure on the solubility of a gas. The air that divers breathe is compressed. As a consequence, large quantities of all the gases in air are dissolved in the blood and other body fluids. Thus, for example, nitrogen, which is not used by the body, is dissolved in the blood. If the pressure on the diver is suddenly decreased by a rapid return to the surface, the nitrogen in his blood will come out of solution rapidly and form small bubbles that may block capillaries and cut off blood flow. The result of this condition is great pain and sometimes death.

If the diver returns to the surface slowly, the gases are released slowly enough to be expelled by the respiratory system as they escape from the blood. For each atmosphere of pressure, which is equal to about 30 ft of water, the time period known as the decompression time is about 20 min.

Artificial mixtures for breathing consisting of helium and oxygen are used by divers for lengthy stays in the ocean at great depths. Helium is less soluble than nitrogen in water. When the pressure on the diver is decreased, there is less dissolved gas to form bubbles and the bends can be prevented more easily.

The solubility of a gas is directly proportional to the pressure of that gas above the surface of the solution. This relationship is known as Henry's law.

The solubility of CO_2 in water demonstrates Henry's law. All carbonated beverages are bottled under a pressure of carbon dioxide. When the bottle is opened, there is a low partial pressure of CO_2 above the liquid, since there is little CO_2 in the atmosphere. As a result, the solubility of CO_2 decreases and the solution effervesces as the CO_2 bubbles off. Table 9.4 lists the solubilities for some common gases in water. The values are for a pressure of 1 atm of the specific gas. Thus, the solubility of O_2 is 4.0×10^{-2} g/L at 25°C under a pressure of 1 atm of oxygen. Since the partial pressure of oxygen in air is 0.2 atm, the solubility of oxygen is only 8×10^{-3} g/L. This solubility would be insufficient to support human life if we had to depend on this amount of oxygen in our blood. However, hemoglobin molecules in our blood cells chemically bind oxygen molecules, and this chemical process greatly increases the amount of oxygen available to the tissues.

Table 9.4

Solubility of Gases in 1 L of Water at 0°C and 1 atm

Gas	Mass (g)	Moles
He	0.0016	0.00038
N_2	0.018	0.00066
O_2	0.040	0.0013
CO_2	0.14	0.032
Ar	0.06	0.0015

Effect of Temperature on Solubility

The solubility of many gases in water decreases with increasing temperature. As water is warmed, the dissolved air forms bubbles and escapes from the liquid. Many industrial plants and power plants use water for cooling, and the hot water is passed into lakes and streams. As these waterways warm up, oxygen is less soluble, and the water is said to be thermally polluted.

Fish, like other nonmammals, cannot adapt to rapid temperature changes. An increase of 10°C in the temperature doubles the metabolic rate. This increase requires more oxygen at a time when the supply of oxygen has been decreased by the higher temperature.

There is no general rule for the solubility changes of liquids and solids with temperature. Quite often solubility increases with increasing temperature. For example, the solubility of potassium chloride, KCl, increases from 28 g/100 g of

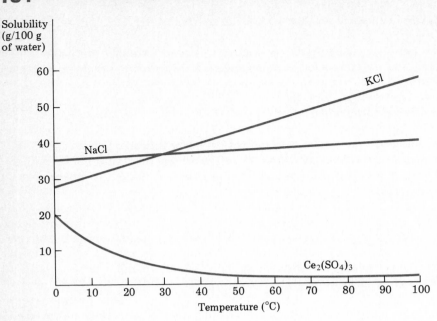

Figure 9.7
Solubility of Solids in Water as a
Function of Temperature.

water at 0°C to 57 g/100 g of water at 100°C. Similarly, the solubility of sodium
chloride, NaCl, increases from 35 g/100 g to 40 g/100 g of water over the same
temperature range. The solubilities of other salts, such as cerium sulfate,
$Ce_2(SO_4)_3$, decrease with increasing temperature (Figure 9.7).

AN ASIDE

Vitamin C is one of the water-soluble vitamins, whereas vitamin A is a fat-
soluble vitamin.

The vitamin C molecule is small and has many —O—H groups that can form
hydrogen bonds to water. The vitamin A molecule, with the exception of one
—O—H group, is nonpolar. The vitamin A molecule is not "like" water and has a
very low solubility in water. On the other hand, vitamin A, which structurally
resembles the carbon compounds in fats, will dissolve in fatty tissue.

Because of its water solubility, vitamin C is not stored in the body and should
be taken in as part of your daily diet. Unneeded vitamin C is eliminated from the
body.

Fat-soluble vitamins are stored to be used by the body. Thus, if large quantities of fat-soluble vitamins are consumed in vitamin supplements, illness can
result. The condition is known as hypervitaminosis.

Effect of Solvent Polarity on Solubility

A maxim of the chemistry laboratory is that "likes dissolve likes." This generalization is reasonable, since molecules of solute that are similar to molecules of solvent are expected to be better able to coexist. Water, which is a polar solvent, is a good solvent for polar solutes, ionic compounds, and substances that can produce ions in water. Carbon tetrachloride, CCl_4, a nonpolar substance, will not dissolve sodium chloride. However, fats and waxes readily dissolve in this nonpolar solvent because they are relatively nonpolar substances.

Liquids that dissolve in each other are **miscible.** Liquids that do not dissolve in each other are immiscible. Ethyl alcohol is miscible with water, but carbon tetrachloride is immiscible with water. The miscibility of ethyl alcohol with water can be rationalized based on its structure.

$$\begin{array}{ccc} H & H & \\ | & | & \\ H-C-C-O \\ | & | & | \\ H & H & H \end{array}$$

The structural feature —O—H resembles water. Ethyl alcohol is polar, as is water. The nonbonded electron pairs on the oxygen atom in ethyl alcohol will form hydrogen bonds with water, thus making it soluble.

The properties of a solvent containing a solute are different from those of the pure solvent. *The properties that depend on the number of particles dissolved and not on their chemical identity are known as* **colligative properties.** Some of these colligative properties include vapor pressure, boiling point, freezing point, and osmotic pressure.

A solution containing a nonvolatile solute has a lower vapor pressure than the pure solvent. This lowering of vapor pressure is a consequence of the solute molecules or ions occupying positions on the surface of the solution. For example, dissolved sodium chloride in water decreases the escaping tendency of water to the gaseous phase. The decrease in vapor pressure is directly proportional to the concentration of the solute particles present in the solution.

The decrease in vapor pressure of solvent means that the solution must be raised to a higher temperature to reach the boiling point. For a $1\,M$ solution of sucrose in water, the boiling point is $100.52°C$.

9.7
Colligative Properties

Example 9.9 ───────────────────────────

The boiling point of a $0.5\,M$ solution of NaCl in water is the same as the boiling point of a $1.0\,M$ solution of sucrose. Explain why.

Solution. A $0.5\,M$ solution of NaCl in water contains 0.5 mole of Na^+ per liter as well as 0.5 mole of Cl^- per liter. Thus, the solution contains the same total number of particles as are in a $1.0\,M$ solution of sucrose. The boiling point depends on the vapor pressure of the solvent, which in turn depends only on the number of solute particles present in solution.

─────────────────────────────

The solute in a solvent decreases the rate at which solvent molecules can crystallize as a solid. However, the solvent molecules in the pure solid can escape to the liquid phase at a rate that is not affected by the solute particles in the liquid. Thus, addition of a solute will cause the solid to melt. In order to freeze a solution, a lower temperature is required. The freezing point is depressed. This phenomenon can be observed by spreading salt on ice in the winter.

Osmosis

We will focus our attention on osmosis and osmotic pressure as an example of a colligative property. This phenomenon is extremely important in cells, the simplest unit of life. Furthermore, osmosis is related conceptually to dialysis, a medical approach to saving the lives of people with kidney failure.

If a solution such as sugar in water is separated from pure water by a **semipermeable membrane,** *a membrane through which water can pass but not other large molecules such as sugar* (Figure 9.8), the volume of the solution increases while the volume of the pure water decreases. The water remains pure because sugar molecules cannot pass through the membrane. The water molecules can cross the membrane and do so in both directions. *The process of*

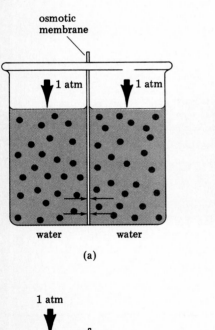

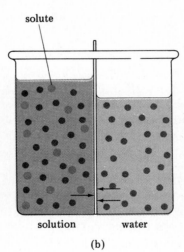

(a)

(b)

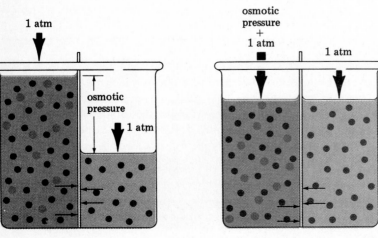

(c)

(d)

Figure 9.8
Osmotic Pressure of a Solution.
(a) Equal rates of transfer of water result in an equilibrium between pure water on both sides of the membrane. (b) The rate of transfer of water from the solution side is decreased because of the presence of solute. The rate of transfer from the pure water is unchanged. A net transfer of water results, and the level of water increases on the solution side of the membrane.
(c) When the level of the liquid on the solution side is sufficient to reestablish the balance of rates of transfer from one side to the other, a new equilibrium results. The difference in the levels of the two liquids is the osmotic pressure. (d) By applying pressure on the solution side equal to the osmotic pressure the liquid levels are made equal again.

transfer of molecules through a semipermeable membrane is called **osmosis.**

The tendency of water molecules to pass out of the solution side to the pure water side of the membrane is diminished by the presence of solute molecules. Since there is no such restriction on passage toward and through the membrane from the pure water side of the membrane, a net transfer of water molecules occurs. The result of the transfer is to dilute the sugar solution.

Eventually no further net transfer occurs, and the level of the solution remains constant at some height above that of the water. The difference in the levels of the solution and the pure solvent is a measure of the net tendency of water to go through the membrane to dilute the solution. There is a "back pressure" due to the height of the column of liquid on the solution side of the membrane. In effect the column of liquid provides a counter-push to the tendency of water molecules to push across the membrane.

If pressure is mechanically applied on the side of the tube containing the solution, the net flow of water molecules can be reversed, and the height of the two columns of liquid made equal. *The pressure required to maintain equal levels of the water and the solution is called the* **osmotic pressure.**

The greater the concentration of the solution, the higher is the osmotic pressure. A $0.2 M$ sugar solution has twice the osmotic pressure of a $0.1 M$ sugar solution. However, the osmotic pressure depends not only on the molarity of the solution but also on the number of moles of solute particles present. For example, a $0.1 M$ NaCl solution has the same osmotic pressure as a $0.2 M$ sugar solution. The reason for the phenomenon is that a $0.1 M$ NaCl solution is $0.1 M$ in sodium ions and $0.1 M$ in chloride ions. Thus, in 1 L of solution there is 0.2 mole of ions. In a sugar solution the sugar is present as covalent molecules. A $0.2 M$ solution then has 0.2 mole of molecules per liter.

The relationship between osmotic pressure, π, and the molarity of the solute particles is similar to the ideal gas law.

$$\pi V = nRT$$

In this equation n is the number of moles of all solute particles. V is the volume of the solution. The equation can be rearranged to obtain the quotient n/V.

$$\pi = \frac{n}{V} \times RT$$

The quotient n/V corresponds to the molarity of all solute particles. For a molecular substance the quotient is equal to the molarity of the solution. For an ionic compound it is equal to the sum of the molarities of all ions.

Physiologists use the osmole and a related term, osmolarity, to calculate osmotic pressure. *An* **osmole** *is the molecular or formula weight of the substance in grams divided by the number of particles provided by the material in solution.* Thus, 58.5 g of NaCl is equal to one mole but two osmoles.

$$\frac{58.5 \text{ g/mole}}{\dfrac{2 \text{ moles ions}}{\text{mole NaCl}}} = 29.2 \text{ g/mole ion} = 29.2 \text{ g/osmole}$$

A $1.0 M$ solution of NaCl contains 58.5 g NaCl/L. A 1.0 osmolar solution of NaCl contains 29.2 g NaCl/L and is $0.5 M$ in Na^+ and $0.5 M$ in Cl^-. The concentration as osmolarity can be used directly in place of n/V in the osmotic pressure equation.

Example 9.10

What is the osmotic pressure of a physiological saline solution that is 0.15 M NaCl at 37°C?

Solution. The osmolarity of the solution is twice the molarity, since 1 mole of NaCl yields 2 moles of ions in solution.

$$\frac{n}{V} = 2 \times (0.15 \text{ mole/L}) = 0.30 \text{ osmole/L}$$

$$\pi = 0.30 \, \frac{\text{osmole}}{\text{L}} \times 0.082 \, \frac{\text{L atm}}{\text{osmole deg}} \times 310 \text{ K}$$

$$= 7.6 \text{ atm}$$

Example 9.11

The osmotic pressure of a solution containing 5.0 g of insulin per liter of solution is 16.3 mm Hg at 25°C. Calculate the approximate molecular weight of insulin. (Insulin is a covalent molecule.)

Solution. The osmotic pressure in atm is first calculated.

$$16.3 \text{ mm Hg} \times \frac{1 \text{ atm}}{760 \text{ mm Hg}} = 0.0214 \text{ atm}$$

The osmolarity of the solution is calculated as follows.

$$n/V = \frac{\pi}{RT} = \frac{0.0214 \text{ atm}}{0.082 \, \dfrac{\text{L atm}}{\text{osmole deg}} \times 298 \text{ K}} = 1.10 \times 10^{-4}$$

The 5.0 g sample is equal to 1.10×10^{-4} mole, and the molecular weight is 4.5×10^4 amu.

$$\frac{5.0 \text{ g}}{1.10 \times 10^{-4} \text{ mole}} = 4.5 \times 10^4 \text{ g/mole}$$

Solutions of different osmolarity separated by a semipermeable membrane will undergo a net flow of water. When the osmotic pressures of solutions are unequal, the tendencies of water to cross the membrane are not equal. Water will flow from less concentrated to more concentrated solutions.

Osmotic pressure differences account for the wilting of plant forms. For example, if the salad dressing on a salad has an osmotic pressure higher than the fluid in the lettuce cells, these cells will lose the water that gives the lettuce its crispness. Similarly, a flower placed in a solution will wilt as it loses water from its cells. However, if it is then placed in pure water, it will regain its shape.

Figure 9.9
Effect of Osmotic Pressure
of Solutions on Red Blood
Cells.
(a) In an isotonic solution there is no
change in the shape of a cell since
water enters and leaves at the same
rate.

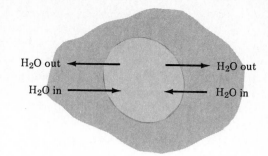

(b) In a hypotonic solution the rate of
water passing into the cell exceeds the
rate of loss of water. As a result the cell
distends and may burst.

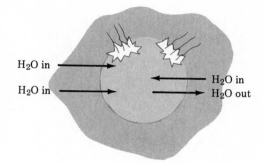

(c) In a hypertonic solution the rate at
which water leaves the cell exceeds the
rate of water entering the cell. This
causes the cell to shrink.

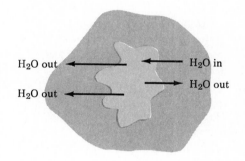

Osmosis through a semipermeable membrane is a process of great importance in maintaining life processes. Both animals and plants contain membranes through which water passes. If the concentrations of the solutes in the water solution are not properly balanced, water transport will occur and may damage cells. *Rupture of the red blood cell is* **hemolysis;** *shriveling is* **crenation.** Water will transfer from solutions of low osmotic pressure to solutions of high osmotic pressure. If the osmotic pressure of a solution surrounding blood cells is less than that within the cells, water will enter the cells and the cell walls may hemolyze. If the osmotic pressure of a solution is greater than that within the blood cells, the water will leave the cells and cause them to crenate. Both hemolysis and crenation are damaging biological events (Figure 9.9).

Cell membranes are very complex. They allow some substances but not others to diffuse across them in the direction from high to low concentration. This process is governed by differences in concentration and requires no cellular energy. However, some substances also move from low to high concentration in

9.8
Isotonic Solutions

cells. This process requires the expenditure of energy by the cell and is called active transport. Active transport will be discussed in Chapter 22.

The intravenous administration of dextrose or the replacement of ionic constituents of body fluids must be carefully controlled. *A solution whose concentration of solute has an osmotic pressure equal to that within cells is* **isotonic.** *A solution with an osmotic pressure less than that within a cell is* **hypotonic;** *if the solution's osmotic pressure is greater than that within the cell, it is* **hypertonic.** The osmotic pressure of the red blood cell is 7.7 atm at the body temperature of 98.6°F. This high pressure can be balanced by relatively low concentrations of solute in surrounding fluid. A 0.9% sodium chloride solution has an osmotic pressure equal to that of a red blood cell.

It is sometimes medically necessary to administer intravenously controlled concentrations of hypertonic or hypotonic solutions to achieve the desired water balance in a patient. A hypertonic solution will cause the net transfer of water from tissues to the blood, after which the kidneys will remove the water. A hypotonic solution may be used to cause water to transfer out of the blood into the surrounding tissues and thus decrease the blood pressure.

9.9
Dialysis

Dialysis *is the physical separation of small molecules and ions from a colloid.* The procedure uses a dialysis membrane. The membrane allows solvent molecules as well as ions and small molecules to cross it, while retaining the colloid on one side of the membrane. The materials that do cross tend to move toward the more dilute side of the membrane. As in the case of osmosis, the direction of flow tends to balance the concentration of the solute. The use of a dialysis membrane in a colloid is shown in Figure 9.10. By constantly circulating water in the bath, the undesired ions and small molecules will eventually all be removed.

Our kidneys clean our blood of the waste products of metabolism. When the kidneys fail, the buildup of waste causes uremia and eventually death. The artificial kidney (Figure 9.11) uses dialysis to purify the blood of patients with kidney failure. Tubes, which are permeable to the harmful products of metabolism, are used to circulate the patient's blood through a bath into which the products may diffuse. Some vital substances would also tend to diffuse out if the bath were pure water. In order to maintain the balance of the normal constituents of blood, the

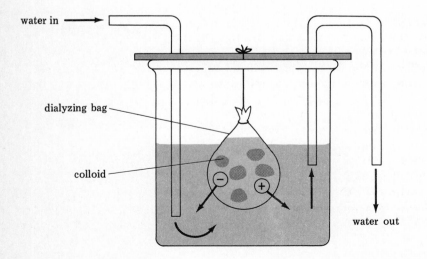

Figure 9.10
The Dialysis of a Colloid.
Only the small ions and water may pass through the dialysis membrane. The colloid remains inside the membrane.

water in →

dialyzing bag

colloid

water out

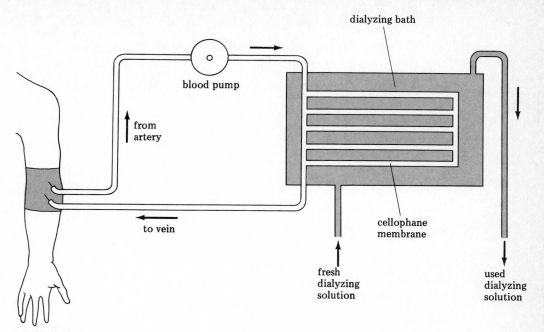

Figure 9.11
Schematic of an Artificial Kidney Machine.
The long coiled cellophane membrane has a pore size such that toxic waste substances
enter the dialyzing solution and are removed. The dialyzing solution is isotonic in the materials required to maintain blood balance of electrolytes.

bath contains materials that are in the same concentration as in blood. The bath
contains 0.6% NaCl, 0.04% KCl, 0.2% NaHCO$_3$, and 1.5% dextrose. These concentrations give a bath that is isotonic in all the solutes that must remain in the
blood. No net transfer occurs for any of the vital materials of the blood. For
example, for every chloride ion that leaves the blood to enter the bath, a chloride
leaves the bath to enter the blood.

Summary

In describing a solution, we indicate the relative
amounts of solute and solvent. This is done by a variety of expressions for concentration. Percent concentration units include weight/weight, weight/volume,
and volume/volume. Molarity is a concentration unit
that provides direct information about the number of
moles of solute.

Solutes in aqueous solution may be electrolytes
or nonelectrolytes. Electrolytes are substances that
are ionic or are converted into ions when dissolved in
water. Nonelectrolytes are substances that exist as
molecules in solution.

Colligative properties depend on the number of
solute particles in solution and not on their identity.
Solutes lower the vapor pressure, raise the boiling
point, decrease the freezing point, and increase the
osmotic pressure of a solvent.

Osmosis is the net flow of water across a semipermeable membrane. Dialysis is the flow of solvent and
selected solute molecules across a membrane. Transfer of material across membranes is an important biological process at the cellular level.

Learning Objectives

**As a result of studying Chapter 9 you should be
able to**
- Differentiate between solutions and colloids.
- Identify colloids involving all three states of matter.
- Identify solutions involving all three states of matter.
- Calculate the concentration of solutions using percent,
  parts per million, and molarity.

- Explain the effects of pressure, temperature, and the solvent on solubility.
- Distinguish between electrolytes and nonelectrolytes.
- Interpret the effect of solute on the osmotic pressure of a solution.
- Demonstrate the effect of osmotic pressure of a solution on cells.
- Describe dialysis and compare it with osmosis.

New Terms

An **aqueous solution** is a mixture of a solute in water.

A **colloid** is a mixture whose particles are intermediate between those of a suspension and a solution.

A **colligative property** is a property of a solvent that depends on the number of dissolved particles.

A **concentrated** solution has a high concentration of solute.

Crenation is the shrinking of red blood cells caused by net water loss by the cell to the surrounding medium.

Dialysis is the selective passage of small molecules and ions across a membrane.

A **dilute** solution has a low concentration of solute.

Dissociation is the separation of ions in water.

The **dispersed phase** is the dispersed substance in a colloid.

The **dispersing phase** is the substance that disperses the dispersed phase in a colloid.

Electrolytes are ionic compounds or compounds that form ions when dissolved in water.

An **emulsion** is a dispersion of a liquid in a liquid.

A **foam** is a dispersion of a gas in a solid.

Hemolysis is a rupture of red blood cells resulting from net water transfer into the cell.

A **hypertonic solution** has a higher osmotic pressure than a second solution such as may exist within a cell.

A **hypotonic solution** has a lower osmotic pressure than a second solution such as may exist within a cell.

Ionization is the formation of ions upon dissolution in water.

An **isotonic solution** has the same osmotic pressure as another solution such as may exist within a cell.

Molarity is equal to the ratio of moles of solute per liter of solution.

A **nonelectrolyte** is a neutral covalent compound that does not form ions in solution and does not conduct electricity.

An **osmole** is the molecular weight in grams divided by the number of particles formed in solution.

Osmosis is the net flow of solvent molecules through a semipermeable membrane.

Osmotic pressure is the pressure required to stop the net transfer of solvent across a semipermeable membrane.

A **ppm** is a part per million, a unit of concentration.

Proof, a concentration unit used for alcohol solution, is twice the v/v% concentration.

A **saturated solution** is a solution in which additional solute will not dissolve.

A **semipermeable membrane** is a material that allows only certain material to pass through it.

A **sol** is a dispersion of a solid in a liquid.

Solubility is a measure of the amount of solute that can dissolve in a solvent.

The **solute** is the minor component of a solution.

The **solvent** is the major component of a solution.

A **solution** is a homogeneous mixture of two or more substances.

An **unsaturated solution** has a lower concentration of a solute than its solubility.

Questions and Problems

Terminology

9.1 Explain the difference between each pair of terms.
   (a) solute and solvent
   (b) solution and colloid
   (c) saturated and unsaturated
   (d) hypertonic and hypotonic
   (e) osmosis and dialysis

9.2 Explain how a saturated solution may be considered to be a dilute solution.

9.3 Explain why some covalent substances are electrolytes, whereas others are nonelectrolytes.

Water Mixtures

9.4 Why is a mixture of fine sand and water not considered a solution?

9.5 Describe the difference in the passage of light through a solution, a colloid, and a suspension.

9.6 Our blood is both a solution and a colloidal suspension. How is this possible?

9.7 What is an emulsion? Give some examples.

9.8 What is a sol? Give some examples.

Solutions

9.9 Give one example of each of the following solutes in a solvent.
   (a) a gas in a gas   (b) a gas in a liquid
   (c) a liquid in a liquid   (d) a solid in a liquid
   (e) a solid in a solid

9.10 What is the solvent in a vinertia alloy?

9.11 What is the solute in vinegar?

9.12 The solubility of sodium chloride is 36 g/100 mL at 25°C. A solution contains 6 g of NaCl in 20 mL at 25°C. Is the solution saturated?

9.13 What qualitative terms are used to describe the amount of solute dissolved in a solvent?

Concentrations

9.14 Calculate the weight/volume percent concentration of each of the following.
   (a) 10.0 g of NaH_2PO_4 in 1.00 L of solution

(b) 16.0 g of $NaNO_3$ in 400 mL of solution

(c) 1.50 g of NaCl in 50 mL of solution

(d) 4.0 g of $NaHCO_3$ in 5.00 L of solution

9.15 Calculate the volume/volume percent concentration of each of the following.

(a) 100 mL of ethyl alcohol in 500 mL of solution

(b) 3 mL of acetic acid in 60 mL of vinegar solution

(c) 0.003 mL of alcohol in 1 mL of blood

9.16 Calculate the number of grams of solute needed to make each of the following w/w % solutions.

(a) 1.00 kg of 5.00% NH_4Cl

(b) 500 g of 3.00% H_2SO_4

(c) 100 g of 1% $C_6H_{12}O_6$ (glucose)

9.17 Calculate the volume of solution required to provide the indicated mass of solute from the given w/v % solutions.

(a) 18.0 g of NaCl from a 0.900% solution

(b) 30 g of $C_6H_{12}O_6$ from a 5.0% glucose solution

(c) 60.0 g of Na_2CO_3 from a 1.0% solution

9.18 You breathe 1×10^4 L of air a day. The concentration of sulfur dioxide, a pollutant in the air, is 0.1 ppm. What volume of SO_2 molecules will you breathe?

9.19 When the carbon monoxide concentration in air reaches 100 ppm, dizziness and headaches occur. Express this concentration in volume/volume percent.

9.20 Some solutions are so dilute that parts per million concentrations are in the 0.001–0.009 ppm range. What type of concentration unit might be used to eliminate the decimal and the zeros?

9.21 A person has 0.25% alcohol in his blood. Using v/v % concentration, calculate the volume of alcohol in the blood, assuming a blood volume of 6 L.

9.22 A sherry is 20% alcohol. What concentration unit is being used?

9.23 A vodka is 100 proof. What is its alcohol concentration?

9.24 Calculate the molarity of each of the following solutions.

(a) 4.0 g of NaOH in 500 mL of solution

(b) 2.0 g of NaOH in 2000 mL of solution

(c) 49.0 g of H_2SO_4 in 1 L of solution

(d) 4.9 g of H_3PO_4 in 50 mL of solution

9.25 How many grams of solute are there in each of the following solutions?

(a) 250 mL of 1.00 M glucose, $C_6H_{12}O_6$

(b) 250 mL of 1.00 M sucrose, $C_{12}H_{22}O_{11}$

(c) 1.50 L of 0.250 M NaOH

(d) 0.50 L of 0.14 M $NaHCO_3$

(e) 25.0 mL of 0.100 M NaCl

9.26 What volume will provide the required number of moles of solute with the given concentration of solution?

(a) 0.20 mole of NaOH from 0.50 M NaOH

(b) 1.00 mole of H_2SO_4 from 0.100 M H_2SO_4

(c) 0.010 mole of NaCl from 0.100 M NaCl

(d) 0.100 mole of glucose from 0.500 M glucose.

9.27 How would each of the following solutions be prepared?

(a) 5 L of 0.5 M NaCl

(b) 2 kg of 1.5% NaCl

(c) 250 mL of 0.1 M H_2SO_4

(d) 4000 mL of 1.0% Na_2SO_4 4000 kg of 1.0% w/wt

9.28 Which solution of each of the following pairs is the more concentrated?

(a) 10% or 1 M KCl

(b) 0.1% or 0.1 M NaCl

(c) 2% or 2 M NH_3

9.29 A 2.00 mL sample of blood contains 0.490 mg of calcium ions. Calculate the molarity of calcium ions.

Electrolytes

9.30 Hydrogen bromide, HBr, is a covalent molecule, but an aqueous solution of HBr conducts electricity. Write an equation that explains this observation.

9.31 Ethyl alcohol is a nonelectrolyte. What does this tell us about ethyl alcohol?

9.32 Silver chloride, AgCl, is an ionic compound. A saturated solution of AgCl does not show appreciable conductivity. Explain why.

9.33 What criterion is used to distinguish between strong and weak electrolytes.

Solubility

9.34 The solubility of sodium chloride is 36 g/100 mL at 25°C. Is a solution that contains 150 g of NaCl in a liter saturated or unsaturated?

9.35 The solubility of barium sulfate is 0.00024 g/100 mL at 20°C. A solution contains 0.00120 g of $BaSO_4$ in 500 mL. Is the solution saturated or unsaturated?

9.36 Is anything observed when a solid is added to a saturated solution of the solid? Is anything actually happening?

9.37 Why do bubbles escape from a bottle of beer after the cap is removed?

9.38 Why does an open bottle of soft drink go "flat" at room temperature faster than one stored in a refrigerator?

9.39 How is the oxygen content of river water affected by industrial plants that use the water for cooling?

9.40 Why is the concentration of oxygen in water increased by an increase in the partial pressure of oxygen above the surface of water?

9.41 On hot days, fish tend to swim to lower depths in a lake. Recalling the fact that colder water has a higher density, explain why the fish prefer the depths of the lake.

9.42 Explain what would happen to a saturated solution in contact with undissolved solute for each of the following as the temperature is increased from 25 to 100°C.

(a) NaCl   (b) KCl   (c) $Ce_2(SO_4)_3$

9.43 Ammonia, NH_3, is very soluble in water. Explain why.

9.44 Some ionic solids are insoluble in water. What does this fact indicate?

9.45 Iodine is more soluble in CCl_4 than in water. Suggest a reason.

9.46 Bromine is less soluble than iodine in CCl_4. Suggest a reason.

9.47 DDT has the following structure. Although its solubility is only about 10^{-5} g/100 mL, DDT concentrates in the body. Explain why.

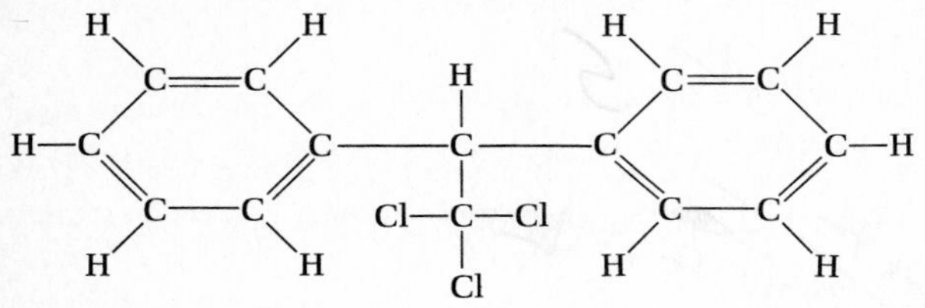

9.48 Glycerol has the following structure. Predict its solubility in water and CCl_4.

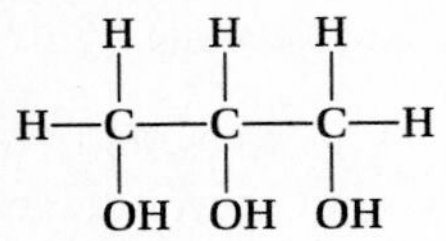

Colligative Properties

9.49 Why does fresh water evaporate more rapidly than sea water at the same temperature?

9.50 How are the boiling point and freezing point of water affected by the concentration of solute?

9.51 The freezing points of a $0.10\,M$ KCl solution and a $0.10\,M$ glucose solution are not equal. Explain why.

9.52 Why would $CaCl_2$ be more effective than NaCl in melting ice in the winter?

Osmosis

9.53 Celery that is kept in a refrigerator for a long time can become limp. The crispness can be restored by putting the celery in water. Explain why.

9.54 Explain what would happen if red blood cells were placed in each of the following solutions.
(a) 0.9% NaCl (b) 0.15% NaCl (c) $0.15\,M$ NaCl
(d) 1.5% NaCl (e) $1.5\,M$ NaCl

9.55 Salt spread on driveways to melt ice may be injurious to nearby plants and shrubs. Explain why.

9.56 A 5% glucose, $C_6H_{12}O_6$, solution is isotonic. Compare the number of particles present in this solution with those present in 0.9% NaCl.

9.57 Explain what would happen if a 0.4% NaCl solution were separated by a semipermeable membrane from a 0.8% NaCl solution.

9.58 Explain what would happen if an isotonic glucose solution were separated by a semipermeable membrane from an isotonic NaCl solution.

9.59 Amylose, a soluble starch, has a molecular weight of approximately 30,000 amu. What would be the osmotic pressure of a solution of 2.0 g of amylose dissolved in 1 L of water?

9.60 The osmotic pressure of a solution of 10 g of a milk protein in 1 L of water is 5 mm Hg at 25°C. What is the molecular weight of the protein?

9.61 The osmotic pressure of 0.100 g of hemoglobin in 10.0 mL of solution is 2.87 torr at 25°C. Calculate the molecular weight of hemoglobin.

9.62 Why must the dialysis bath be changed after a time interval for effective treatment of a patient?

9.63 What would happen if a dialysis bath were improperly prepared and were hypertonic?

Chemical Reactions

Combustion is a chemical reaction used to provide energy for transportation. Even these hot-air balloons depend on combustion to provide the hot air.

195

10.1
Chemical Change

Chemical changes or chemical reactions are what chemistry and indeed even life is all about. From the burning of natural gas to photosynthesis by plants and the metabolism of food by animals, we are immersed in chemical change. Reading this page involves many complex chemical reactions in our eyes and brain. The transmission of electrical impulses by nerves and muscle contractions also are based on chemical reactions.

Each chemical reaction is a unique event in which certain bonds in specific reactants are broken and certain bonds in specific products are formed. The number of known and potential reactions among the millions of compounds is astronomically large. In order to understand these reactions, it is useful to classify them according to defined similar processes. However, unlike bonding, for which a few classes suffice, chemical reactions involve numerous classes. Even the methods of classification differ within the various specialized fields of chemistry such as organic chemistry and biochemistry. A few of the classes of reactions of organic chemistry and biochemistry will be presented briefly in this chapter.

In the previous chapter you learned that the physical properties of water and its solutions are important in maintaining life. Water also plays a central role in biochemical processes that occur in aqueous solutions. In this chapter we will deal with three types of reactions—precipitation, acid–base neutralizations, and oxidation–reduction reactions—that occur in aqueous solutions. Acid–base neutralizations will be discussed in greater detail in Chapter 11. Oxidation–reduction reactions, also known as redox reactions, will be discussed at many points in the chapters about organic chemistry and biological chemistry.

There is more to understanding chemical reactions than just classifying them. Chemical reactions occur at a variety of rates. Slow reactions are involved in the formation of oil over centuries, whereas the burning of gasoline in a car engine occurs with explosive speed. Chemists are interested in how the speed of chemical reactions may be changed. *The study of the speed or rate of chemical reactions is called* **kinetics.**

The kinetics of chemical reactions is of concern in industry. It is important to understand the factors that allow the rapid formation of a desired chemical product in preference to other products. Rapidly produced pure products are the goal dictated by economics.

An understanding of kinetics is also important in biochemistry. The reactions that occur in cells are quite rapid and very efficient. However, for good health a large number of related reactions must continue to be efficient. Conditions that slow or stop one reaction cause problems in all other reactions that are dependent on that one reaction. Enzymes are responsible for the speed and efficiency of biochemical processes. These substances will be considered in more detail in Chapter 24.

It is important to determine under what conditions reactants may be converted quantitatively to products. We will find that chemical reactions are reversible and can result in an equilibrium mixture. An understanding of these equilibrium processes is important in both industrial and biological chemistry. In industry, it is necessary to find the experimental conditions that ensure that as much reactant is converted to product as possible. The inefficient conversion of chemicals is costly and also requires that the unwanted material be removed so that the product is pure. In biological chemistry, our concern for equilibrium also has practical consequences. Our equilibrium processes may be disturbed as a result of sickness, improper diet, or the introduction of foreign chemicals. It is then necessary to understand how to help restore the normal equilibrium state.

Soluble ionic compounds exist in aqueous solutions as cations and anions. When two solutions containing dissolved ionic compounds are mixed, the ions of both substances are intermingled. If any of these ions can react to form an insoluble substance, that substance will separate from the solution. *The combination of ions in solution to form a solid is called* **precipitation.** *The insoluble solid that deposits from solution is called a* **precipitate.**

When a solution of sodium sulfate is added to a solution of barium chloride, a precipitate of barium sulfate is formed. The formation of a precipitate is shown by a down arrow ($\downarrow$) in an equation. The other product, sodium chloride, remains in solution.

$$Na_2SO_4 + BaCl_2 \longrightarrow 2\,NaCl + BaSO_4\downarrow$$

The equations for this and all other precipitation reactions can be written in simpler form by using only the ions actually involved in the reaction. *Equations that use the formula of the predominant species present in solutions and include only those species that undergo chemical change are called* **net ionic equations.**

For the reaction of $BaCl_2$ with Na_2SO_4, we write the formula of each species as it actually exists in solution. Since barium sulfate is not in solution, it is written as $BaSO_4$.

$$2\,Na^+ + SO_4{}^{2-} + Ba^{2+} + 2\,Cl^- \longrightarrow 2\,Na^+ + 2\,Cl^- + BaSO_4\downarrow$$

Notice that the sodium ions and chloride ions appear on both sides of the equation and neither reacted. This often is the case in reactions of ionic compounds in solution. *Ions that are present in solution but not involved in a reaction are called* **spectator ions.**

The net ionic equation for a reaction in solution does not include the spectator ions. Only the ions that react or are produced are used in the equation. The net ionic equation for the reaction under consideration is

$$SO_4{}^{2-} + Ba^{2+} \longrightarrow BaSO_4\downarrow$$

It should be remembered that other ions are present in solution. One cannot obtain a solution containing only sulfate ion or only barium ion. Other ions must

---

Table 10.1 ──────────────────────────────

General Rules for the Water Solubilities of Common Ionic Compounds

1. All common salts of the alkali (IA) metals and of the ammonium ($NH_4{}^+$) ion are **soluble.**
2. All nitrates ($NO_3{}^-$), chlorates ($ClO_3{}^-$), perchlorates ($ClO_4{}^-$), and acetates ($C_2H_3O_2{}^-$) are **soluble.**
3. The chlorides (Cl^-), bromides (Br^-), and iodides (I^-) of most metals are **soluble.** The principal **exceptions** are those of Pb^{2+}, Ag^+, and $Hg_2{}^{2+}$.
4. All sulfates ($SO_4{}^{2-}$) are **soluble** except for those of Sr^{2+}, Ba^{2+}, Pb^{2+}, and $Hg_2{}^{2+}$.
5. All carbonates ($CO_3{}^{2-}$), chromates ($CrO_4{}^{2-}$), and phosphates ($PO_4{}^{3-}$) are **insoluble** except for those of the alkali metals and ammonium salts.
6. The group IA metal hydroxides (OH^-) are **soluble.** The hydroxides of Ca^{2+}, Sr^{2+}, and Ba^{2+} are **moderately soluble.** The rest of the hydroxides are **insoluble.**
7. The sulfides of all metals are **insoluble** except for those of $NH_4{}^+$ and the IA and IIA metals.

be present to maintain electrical neutrality. For every sulfate ion in a solution of sodium sulfate there are two sodium ions present. Similarly, every barium ion in a solution of barium chloride has two chloride ions present. The objective of a net ionic equation is to focus only on the species that participate in the reaction.

In order to predict when precipitation reactions will occur and to write net ionic equations, some general solubility rules are necessary (Table 10.1).

Example 10.1 _____

Write an ionic equation for the reaction of aqueous solutions of silver nitrate and sodium chloride.

Solution. The reaction produces insoluble AgCl. The molecular equation is

$$AgNO_3(aq) + NaCl(aq) \longrightarrow AgCl\!\downarrow + NaNO_3(aq)$$

According to the rules for expressing ionic equations, the chemicals are written as

$$Ag^+ + NO_3^- + Na^+ + Cl^- \longrightarrow AgCl\!\downarrow + Na^+ + NO_3^-$$

The Na^+ and NO_3^- ions are not involved in the reaction and are eliminated.

$$Ag^+ + Cl^- \longrightarrow AgCl$$

In this case the equation is balanced as it stands. Both the numbers of atoms and the net charges on each side of the equation are the same.

[Additional examples may be found in **10.4–10.9** at the end of the chapter.]

10.3
Acid–Base Reactions

An **acid** is a substance that may donate or transfer a proton to another substance. A **base** is a substance that may accept a proton from another substance. For example, the acid hydrogen chloride can donate a proton to the base sodium hydroxide.

$$HCl + NaOH \longrightarrow NaCl + H_2O$$

The combination of an acid and a base to form water is called **neutralization.**
When hydrogen chloride, a polar covalent substance, dissolves in water, a solution of ions results. The reaction involves a donation of a proton from HCl to water. The products are the hydronium ion and the chloride ion; the solution is known as hydrochloric acid.

$$HCl + H_2O \longrightarrow H_3O^+ + Cl^-$$

Further details of this reaction will be presented in Chapter 11.
Sodium hydroxide is an ionic compound whose ions are separated from the solid crystal when it dissolves in water. Thus, the reaction with HCl and NaOH in water involves ions. The hydroxide ion accepts a proton from the hydronium ion.

$$H_3O^+ + OH^- \longrightarrow 2\,H_2O$$

This net ionic equation is the central feature of acid–base reactions in aqueous solution.

Materials other than OH^- are bases. For example, ammonia acts as a base by accepting a proton from the acid HCl.

$$HCl + NH_3 \longrightarrow NH_4^+ + Cl^-$$

In aqueous solution the net ionic equation is

$$H_3O^+ + NH_3 \longrightarrow H_2O + NH_4^+$$

One of the important acid–base reactions in the bloodstream is the reaction of bicarbonate ion, HCO_3^-, with acid. When acid is produced in excessive amounts in metabolic processes, the hydronium ion reacts with HCO_3^- in blood.

$$H_3O^+ + HCO_3^- \longrightarrow H_2CO_3 + H_2O$$

The reaction neutralizes the acid and produces carbonic acid, which is unstable and decomposes into water and carbon dioxide. Carbon dioxide is then expelled through the lungs.

$$H_2CO_3 \longrightarrow H_2O + CO_2$$

Example 10.2 _____

Milk of Magnesia is an antacid containing magnesium hydroxide. Stomach acid is dilute hydrochloric acid. Write an ionic equation for the reaction of Milk of Magnesia with stomach acid.

Solution. Since magnesium hydroxide is insoluble in water, the compound is written as a formula unit. Stomach acid is written as hydronium and chloride ions. Since there are 2 moles of hydroxide ions per mole of $Mg(OH)_2$, 2 moles of hydronium ions are required for neutralization.

$$Mg(OH)_2 + 2\,H_3O^+ + 2\,Cl^- \longrightarrow Mg^{2+} + 4\,H_2O + 2\,Cl^-$$

The chloride ions are not directly involved and are not included in the net ionic equation.

$$Mg(OH)_2 + 2\,H_3O^+ \longrightarrow Mg^{2+} + 4\,H_2O$$

[Additional examples may be found in **10.11** and **10.12** at the end of the chapter.]

10.4
Oxidation–Reduction Reactions

At one time oxidation meant exclusively the reaction of an element, such as a metal, with oxygen. Reactions such as the one that occurs with magnesium in a flashbulb and the rusting of iron are examples.

$$2\,Mg + O_2 \longrightarrow 2\,MgO$$

$$4\,Fe + 3\,O_2 \longrightarrow 2\,Fe_2O_3$$

In both cases the metal is oxidized.

The burning of natural gas (methane) involves the combination of carbon with oxygen and is also an oxidation reaction.

$$CH_4 + 2\,O_2 \longrightarrow CO_2 + 2\,H_2O$$

Note that in this case the carbon atom also loses its hydrogen atoms. Thus, oxidation has also been viewed as the loss of hydrogen.

The opposite of oxidation is known as reduction. Reduction at one time meant the removal of oxygen from a compound. For example, the removal of oxygen from rust by carbon involves the reduction of iron.

$$2\,Fe_2O_3 + 3\,C \longrightarrow 4\,Fe + 3\,CO_2$$

Note, however, that while iron is reduced, carbon is oxidized. Oxidation and reduction are always involved together. Oxidation cannot occur without reduction and vice versa.

Reduction has also been viewed as the combination of hydrogen with the oxygen of a compound or the direct combination of hydrogen with a substance. Two such examples are

$$CuO + H_2 \longrightarrow Cu + H_2O$$

$$N_2 + 3\,H_2 \longrightarrow 2\,NH_3$$

Today chemists have consolidated all oxidation and reduction reactions (also called redox reactions) by a new definition that has wider applications than the older definitions. This definition removes the necessity for dealing only with hydrogen or oxygen.

Oxidation *is the complete or partial loss of electrons.* **Reduction** *is the complete or partial gain of electrons.* The complete loss or gain of electrons occurs in ions. Partial gain or loss refers to changes in electron distribution about an atom as a result of the formation of polar covalent bonds.

In the reaction of CuO with H_2 to form Cu and H_2O, the copper in CuO exists as a Cu^{2+} ion. The ion is reduced by gaining two electrons to form copper metal. Hydrogen gas has the two electrons of the molecule shared equally between the two hydrogen atoms. In water, each hydrogen atom has partially lost its electron to the oxygen atom in the polar covalent bond. Hydrogen is then oxidized.

Oxidation, or the loss of electrons, results in an increase in the oxidation number (Section 6.7) of an ion or an atom in a molecule. The change in the oxidation number of magnesium in its reaction with oxygen is from 0 to +2.

$$\underset{(0)}{2\,Mg} + O_2 \longrightarrow \underset{(+2)}{2\,MgO}$$

Reduction, or the gain of electrons, results in a decrease in the oxidation number of an ion or an atom in a molecule. In the reaction of nitrogen with hydrogen, the nitrogen in N_2 is in the 0 oxidation state. In ammonia, the oxidation number of nitrogen is −3.

$$\underset{(0)}{N_2} + 3\,H_2 \longrightarrow \underset{(-3)}{2\,NH_3}$$

Oxidizing agent and reducing agent are two additional terms of use in describing oxidation–reduction reactions. These terms recognize the fact that whenever a substance is oxidized by losing electrons, another substance is reduced as it gains those electrons. *The material that gains electrons and becomes reduced is also called the* **oxidizing agent.** Similarly, *the material that loses electrons and becomes oxidized is the* **reducing agent.** In the reaction of magnesium with oxygen, the oxygen atoms in O_2 gain electrons and act as the oxidizing agent for magnesium. The magnesium atom loses electrons and is the reducing agent for oxygen. In the case of covalent substances such as ammonia, electrons are not completely lost or gained. However, nitrogen is more electronegative than hydrogen and gains the larger share of the pairs of electrons in the polar covalent bonds. Nitrogen is then the oxidizing agent in the reaction, while hydrogen is the reducing agent.

Example 10.3 _____

Consider the reaction of sodium and chlorine to form sodium chloride. Which substance is oxidized, and which is reduced? Which substance is the oxidizing agent, and which is the reducing agent?

$$2\,Na + Cl_2 \longrightarrow 2\,NaCl$$

Solution. The oxidation number of elemental sodium is 0, whereas that of sodium in sodium chloride is +1. The increase in oxidation number corresponds to an oxidation.

The oxidation number of elemental chlorine is 0, whereas that of chlorine in sodium chloride is −1. The decrease in oxidation number corresponds to a reduction.

Since sodium is oxidized and loses electrons, it serves as a reducing agent. Chlorine, which is reduced and gains electrons, serves as the oxidizing agent.

[Additional examples may be found in **10.13–10.20** at the end of the chapter.]

10.5 Organic and Biochemical Redox Reactions

Oxidation is a common reaction of organic compounds. These substances will be discussed starting in Chapter 13. Many such reactions involve the loss of hydrogen to an oxidizing agent as in the case of the conversion of ethyl alcohol to acetaldehyde.

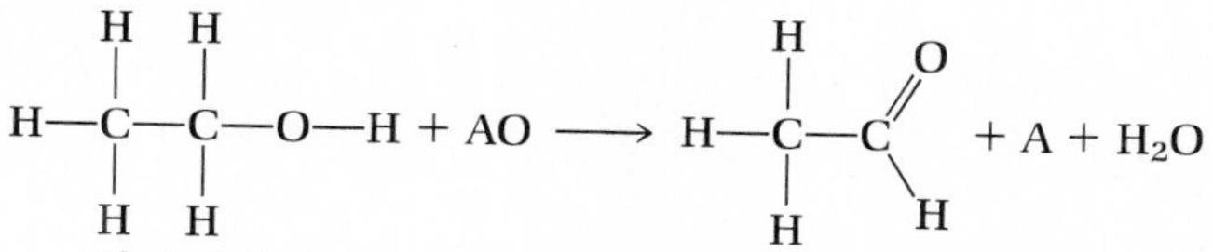

ethyl alcohol   acetaldehyde

The oxidizing agent is not specified and is represented by AO. Note that the oxidizing agent becomes reduced in the reaction, since the oxidation number of A is +2 in AO but 0 in A. The oxidation number of the carbon atom changes from −1 to +1.

Oxidation reactions in organic compounds may also involve a gain of oxygen. The conversion of acetaldehyde to acetic acid is an example.

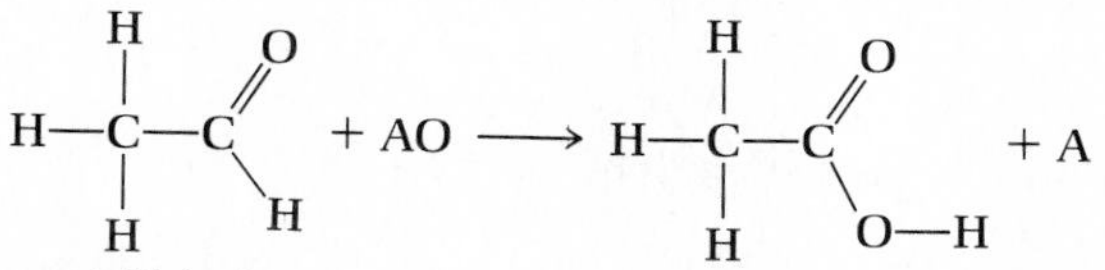

acetaldehyde   acetic acid

Although the oxidizing agent is unspecified, it is reduced from AO to A as the organic material is oxidized. The oxidation number of the affected carbon atom changes from +1 in acetaldehyde to +3 in acetic acid.

Oxidation–reduction reactions are important in biochemistry. In photosynthesis, the carbon in carbon dioxide is reduced. The oxidation number of carbon

Since there are millions of organic compounds, the classification of their reactions has been vital to making sense out of a very large number of observations. Starting with Chapter 13, you will learn just how a complex and varied subject can be simplified by classification. At this point a few types of organic reactions will be given.

A **substitution reaction** *is the substitution or replacement of one atom or group of atoms in a molecule by some other atom or group of atoms.* For example, a hydrogen atom of methane can be substituted by a chlorine atom when heated with chlorine gas at a high temperature.

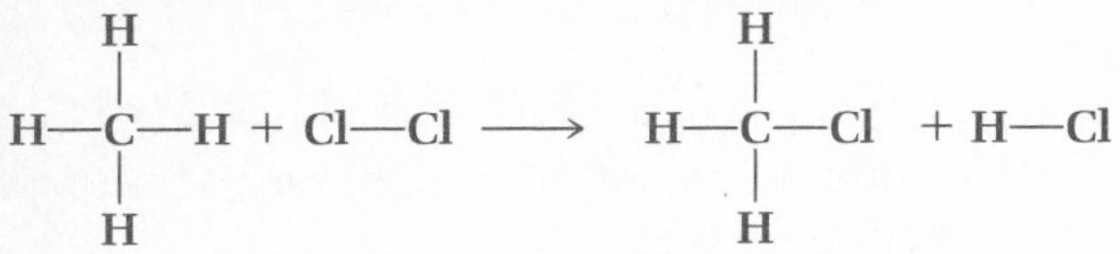

Groups of atoms, such as in the hydroxide ion, may also substitute for another group, as in the case of the reaction of iodomethane with aqueous base.

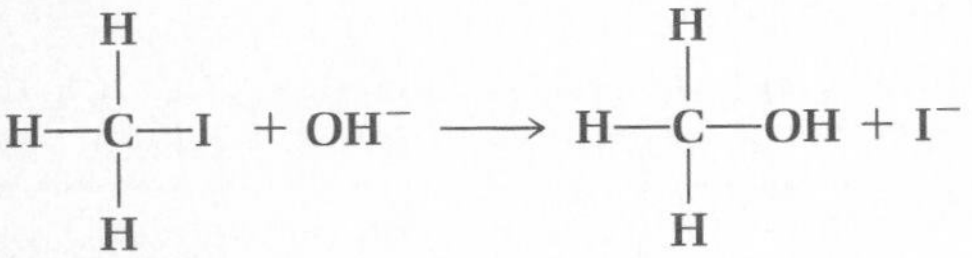

Addition reactions *involve the attachment or addition of an atom or group of atoms to a molecule.* In addition reactions, no atoms or groups of atoms are released by the molecule.

For a molecule to undergo an addition reaction, multiple bonds must be present. Such compounds are called unsaturated because the multiple bonds are the result of not having enough other atoms with which to bond. Ethylene is an unsaturated compound. Ethane, with only single bonds, is saturated because carbon atoms cannot form any additional single bonds. Hydrogen can add to ethylene to form ethane.

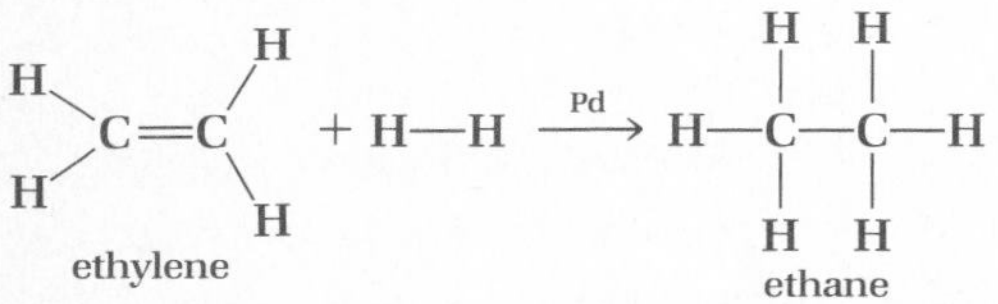

Condensation reactions *involve the joining together of two or more relatively small molecules to form a larger molecule and the elimination of a very small molecule like water or ammonia.* The reaction of glucose to give glycogen is a condensation reaction. A large number of glucose molecules, designated by the coefficient n, combine to form water and glycogen.

$$n \text{ glucose} \longrightarrow n \text{ H}_2\text{O} + \text{glycogen}$$

Glycogen is stored predominately in the liver and is used as a source of quick biochemical energy (Chapter 26).

Hydrolysis reactions *involve the separation of a molecule into two or more smaller molecules when reacted with water.* The water is incorporated in the product molecules. Protein digestion involves an enzyme-catalyzed hydrolysis to give mixtures of amino acids (Chapter 28).

$$\text{protein} + \text{H}_2\text{O} \longrightarrow \text{amino acids}$$

in carbon dioxide is +4, but the average oxidation number of carbon in glucose is decreased to 0.

$$6\,CO_2 + 6\,H_2O \longrightarrow C_6H_{12}O_6 + 6\,O_2$$

Thus, carbon accepts electrons from the oxygen in water, which, in turn, is oxidized to molecular oxygen. Chemical energy is stored in glucose, an energy-rich compound, in photosynthesis.

When cells in living organisms metabolize glucose, the carbon atoms are reoxidized to carbon dioxide, an energy-poor compound, and molecular oxygen is reduced to form water. In this process, the chemical energy from the sun, stored by photosynthesis, is released.

Oxidation in metabolic reactions does not directly involve reaction of oxygen with the energy-rich compounds. One of the typical steps in oxidation involves NAD^+ (nicotinamide adenine dinucleotide), which is reduced to NADH. A general reaction in which H_2M represents a biochemical molecule containing two covalently bonded hydrogen atoms is given below.

$$H_2M + NAD^+ \longrightarrow M + NADH + H^+$$

The reduced NADH reacts in a complex series of steps known as the electron transport chain (Chapter 25). The net result is the oxidation of NADH and the reduction of O_2.

$$O_2 + 2\,NADH + 2\,H^+ \longrightarrow 2\,NAD^+ + 2\,H_2O$$

The relationship of the electron transport chain to metabolic reactions will be presented in Chapter 25.

Although many chemical equations can be written, they do not necessarily represent real chemical reactions that proceed to convert reactants to products. *Reactions that occur without an outside source of energy are called* **spontaneous reactions.** *A reaction that does not give the indicated product unless energy is added is* **nonspontaneous.** What determines what makes some reactions go, whereas others do not? An analysis of all spontaneous reactions reveals that two features control them: enthalpy and entropy. Enthalpy is a chemical term used to represent heat energy or some equivalent energy form. Entropy is a new term for you, but its meaning can be explained qualitatively.

Enthalpy

All substances contain stored potential energy in their bonds. When reactants are converted to products, the stored chemical energies are not the same because the number and types of bonds are altered. If the products of a reaction contain less chemical energy than the reactants, the net difference can be released as heat. *A reaction that gives off heat energy is* **exothermic** (Figure 10.1). The combustion of methane (natural gas) is exothermic and releases 211 kcal/mole of methane.

$$CH_4 + 2\,O_2 \longrightarrow CO_2 + 2\,H_2O \qquad (211\ \text{kcal released})$$

When the products of a reaction contain more stored chemical energy than the reactants, the net difference must be supplied during the reaction. *A reaction that requires or absorbs heat is* **endothermic.**

The energy difference between the products and the reactants is called a

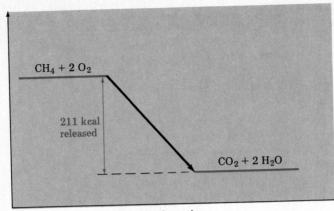

Chemical
potential
energy

CH₄ + 2 O₂

211 kcal
released

CO₂ + 2 H₂O

Progress of reaction
reactants ⟶ products

Figure 10.1
Potential Energy and the Energy of an Exothermic Reaction.
The chemical potential energies of the reactants are higher than those of the products. When the reaction of the combustion of methane occurs, the energy difference is released as heat energy.

change in enthalpy, $\Delta H°$. By convention, the release of energy in an exothermic process is given a negative sign. Thus, the combustion of methane has a $\Delta H°$ of -211 kcal/mole.

The energy released or absorbed in a chemical reaction need not be in the form of heat. Electrical or light energy may be involved. Chemical reactions in a battery produce an electric current. Other reactions, such as the electrolysis of water to produce hydrogen and oxygen, require electrical energy.

Light can be released by some reactions as occurs in the firefly. In the reverse sense, some reactions occur when light energy is added. The most important example of the latter case is photosynthesis, in which plants produce carbohydrates from carbon dioxide and water. Light energy equivalent to 686 kcal/mole is required for each mole of glucose formed (Figure 10.2).

$$6 CO_2 + 6 H_2O \longrightarrow C_6H_{12}O_6 + 6 O_2 \qquad (686 \text{ kcal required})$$

Endothermic processes are given a positive sign when expressed as an enthalpy change. For the photosynthesis process, $\Delta H° = +686$ kcal/mole of glucose formed.

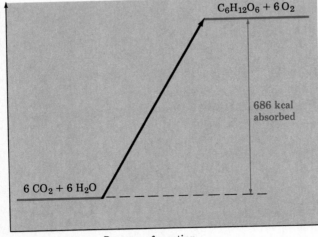

Chemical
potential
energy

C₆H₁₂O₆ + 6 O₂

686 kcal
absorbed

6 CO₂ + 6 H₂O

Progress of reaction
reactants ⟶ products

Figure 10.2
Potential Energy and the Energy of an Endothermic Reaction.
The chemical potential energies of the reactants are lower than those of the products. When the reaction of photosynthesis occurs, it is necessary to have energy added equal to the net energy difference. This energy is supplied as light energy from the sun.

Entropy

The second controlling feature of all physical processes and chemical reactions is the tendency to achieve the most random or disordered arrangement possible. This feature is important in the expansion of a gas or the vaporization of a liquid. *The degree of randomness or disorder is called the entropy of a system.* By definition, *the **entropy change**, $\Delta S°$, is positive for increasing disorder.*

A positive entropy change counterbalances unfavorable enthalpy changes in some reactions. If the increase in the degree of disorder is great, an endothermic process can occur. In addition, a reaction leading to increasing order (a negative entropy change) can occur only by the expenditure of energy (a negative enthalpy change). The formation of ordered life forms is such an example.

Temperature is significant in determining the importance of enthalpy and entropy contributions to a system that can undergo change. At absolute zero, all substances are ordered and entropy differences between two or more substances are zero. Therefore, only their relative energies determine their relative stabilities. With an increase in temperature, molecular motions become possible, and the tendency toward disorder varies from substance to substance. The differences in entropy can play a variable role as a function of temperature. At extremely high temperatures, the entropy differences between substances may play a dominant role in the course of a reaction.

Free Energy

The relationship between enthalpy changes and entropy changes is given by the following expression, in which $\Delta G°$ symbolizes the change in free energy of a system at constant pressure.

$$\Delta H° - (T \times \Delta S°) = \Delta G°$$

The **free energy change** *is a measure of the tendency of a reaction to proceed spontaneously.* When $\Delta G°$ is negative, a chemical or physical process occurs. The negative enthalpy change previously described can be seen to contribute toward making $\Delta G°$ more negative. Similarly, a positive entropy change contributes toward making $\Delta G°$ negative. Note from the expression that $\Delta H°$ is more important at low temperatures and $\Delta S°$ becomes more important at high temperatures.

A summary of the contributions of $\Delta H°$ and $\Delta S°$ to $\Delta G°$ and the resultant prediction of the spontaneity of a chemical reaction is given in Table 10.2.

Table 10.2

Contributions of $\Delta H°$ and $\Delta S°$ to $\Delta G°$

$\Delta H°$	$\Delta S°$	$\Delta G°$	Result
negative	positive	negative at all temperatures	reaction will proceed
positive	negative	positive at all temperatures	reaction will not proceed
negative	negative	negative if the temperature is sufficiently low	reaction might proceed at sufficiently low temperature
positive	positive	negative if the temperature is sufficiently high	reaction might proceed at sufficiently high temperature

10.7
Kinetics

Kinetics is a study of the velocity of chemical reactions. The reaction velocity is a measure of the rate of conversion of reactants into products. The factors that influence the reaction velocity are the structures of the reactants, the temperature, the concentrations of the reactants, and the presence of substances called catalysts.

Reactants and Reaction Rates

The transformation of reactants into products involves the rupture of some bonds and the formation of others. Therefore, the nature of the chemical substances involved is the most important feature controlling the reaction.

Reaction velocity varies considerably depending on structure. Some biological reactions occur incredibly fast. Changes in the light-absorbing compound in the eye occur in picoseconds (10^{-12} s). Enzyme-catalyzed reactions may occur in milliseconds.

Even simple molecules can react at dramatically different rates. The colorless compound nitric oxide, NO, which may be produced in an internal combustion engine, reacts very quickly with oxygen at room temperature to form a reddish brown compound, nitrogen dioxide, NO_2.

$$2\,NO + O_2 \longrightarrow 2\,NO_2$$

Oxides of nitrogen, such as nitrogen dioxide, are responsible in part for the smog in some metropolitan areas.

When carbon monoxide, CO, is produced from incomplete combustion of gasoline in a car engine, it does not react very fast with oxygen at room temperature.

$$2\,CO + O_2 \longrightarrow 2\,CO_2$$

As a consequence, the carbon monoxide can build up in the atmosphere in urban areas. If the reaction of CO were as rapid as that of NO, carbon monoxide poisoning would not occur.

Whereas the reactions of NO and CO with oxygen appear similar, there is a large difference in the rates of the two reactions. Nitric oxide is reactive, whereas carbon monoxide is not. The difference in reactivity must be due to differences in the bonding in these two molecules. The carbon atom in CO has an octet of electrons. Nitric oxide does not have an octet of electrons about nitrogen, and its reactivity may be attributed to this deficiency of electrons.

$$\cdot \ddot{N}\!=\!\ddot{O}\!: \qquad\qquad :C\!\equiv\!O\!:$$

Concentration and Reaction Rates

Two common reactions are the burning of wood and the rusting of iron. Both reactions involve the reaction of oxygen gas with a solid. The reactants must come in contact with each other to react. The reaction velocity increases with an increase in surface area. If wood is chopped into fine kindling or if iron is ground into powder, more of the solid comes in contact with air and the reaction velocity increases.

In any reaction there is a similar necessity for physical contact of the reactants. As the concentration of reactants in either a gas or liquid is increased, the reaction velocity increases. In a gas, the reaction velocity can be increased either by increasing the amount of reactants in a constant volume or by decreasing the volume (increasing the pressure) containing the reactants. In the liquid phase,

AN ASIDE _____

Chemical reactions in living organisms are dramatically affected by both increases and decreases in temperature. Increased temperature results in an increased metabolic rate. Lowering of the temperature slows down the reaction rates of the body.

The metabolic rates of aquatic life in streams near power plants are increased by the release of warm water effluents. At high metabolic rates, the fish require more oxygen. The higher temperature also causes a decreased concentration of oxygen dissolved in the water. The result of faster metabolism and the decreased oxygen supply may cause the death of the fish.

In humans, a fever causes an increased basal metabolism rate. For each degree Fahrenheit above the normal temperature, the metabolism rate increases by 5%. As a consequence, either food intake must be increased or a weight loss results. Animals that hibernate are able to survive because their body temperatures decrease to some temperature above the freezing point of water, thus decreasing their need for food. A woodchuck's heart rate decreases from about 75 beats a minute in the nonhibernating state to about 5 beats a minute while hibernating. The rates of all chemical reactions dependent on stored body fat and oxygen are decreased. The woodchuck can then survive the hibernation period.

The technique of lowering the body temperature (hypothermia) of surgical patients is used by doctors to enable them to perform certain operations and avoid deterioration of vital tissue. Without oxygen from blood circulation, the brain is irreparably damaged in a few minutes at normal body temperature. However, if the body temperature is lowered from 37 to 20°C, the brain can be deprived of oxygen for about 1 hour and not be affected. Thus, surgeons can perform operations on a still heart, which can then be restarted after repairs have been made.

reactant concentrations may be increased by adding reactants. The result is to bring reacting particles closer together.

Temperature and Reaction Rates

All reaction velocities increase with a rise in temperature. This occurs because the kinetic energies of the reactant particles increase with increasing temperature. The faster the particles move, the greater the chance they will collide and react with other particles. Some reaction rates are very sensitive to temperature changes, whereas others are only slightly affected. However, **a general rule that can be used is that a 10°C rise in temperature usually doubles the reaction rate.**

Example 10.4 _____

Milk is known to spoil at room temperature, but to remain suitable for drinking if stored in a refrigerator. Explain why.

Solution. The spoiling of milk involves chemical reactions. These reactions occur faster at room temperature than at the temperature of the refrigerator. However, given sufficient time, even milk stored in a refrigerator will spoil.

Catalysts and Reaction Rates

A **catalyst** *is a substance that increases a reaction rate when it is present in the reaction mixture.* A catalyst is said to catalyze the reaction, and its effect is known as catalysis. Catalysts are usually required only in trace amounts. The catalyst is present in the same amount before and after the reaction takes place, even though it may interact with the reactant at a given step in the reaction.

In college chemistry laboratories, oxygen is produced from potassium chlorate. When potassium chlorate, $KClO_3$, is heated, it slowly decomposes into oxygen and potassium chloride, KCl.

$$2\ KClO_3 \longrightarrow 2\ KCl + 3\ O_2$$

If a small amount of manganese dioxide, MnO_2, is added, the rate of decomposition of potassium chlorate is increased. The manganese dioxide serves as a catalyst in the reaction.

AN ASIDE

Catalysts in plants and animals allow these organisms to carry out reactions at rates sufficient for the organism to survive. *Catalysts in organisms are called* **enzymes.** There are many enzymes in any species. Each enzyme is highly specific.

In the body, glucose is metabolized efficiently at $37°C$.

$$C_6H_{12}O_6 + 6\ O_2 \longrightarrow 6\ H_2O + 6\ CO_2$$

The actual conversion occurs in several reactions, each catalyzed by a specific enzyme. Outside the body, the same combustion can occur rapidly only above $600°C$. At body temperature, the chemical conversion without enzymes would require months. This rate is too slow to provide the energy necessary to support life.

The enhancement of the rate of a reaction can be seen in the decomposition of hydrogen peroxide, H_2O_2.

$$2\ H_2O_2 \longrightarrow 2\ H_2O + O_2$$

At room temperature, hydrogen peroxide decomposes so slowly that there is no evidence of oxygen being produced. When platinum is added, the reaction rate speeds up by a factor of 3×10^4! Platinum is then a catalyst. When the enzyme catalase is present, the reaction is increased by a factor of about 10^{10}! Catalase is present in the human blood. Its catalytic action on hydrogen peroxide is seen when a hydrogen peroxide solution is used to clean and disinfect wounds. The solution on the skin bubbles as oxygen is given off.

Enzymes are as common in the plant kingdom as in the animal kingdom. One of the very important processes catalyzed by enzymes is nitrogen fixation. The nitrogen gas in the atmosphere ultimately is the principal natural source of ammonia and other nitrogen compounds essential for protein formation. The gas is quite unreactive, but small quantities of nitrogen are converted into nitrogen compounds by lightning discharges. However, under substantially less dramatic conditions, the bacteria in the roots of leguminous plants such as alfalfa, beans, clover, and peas readily convert nitrogen gas into ammonia. These bacteria contain enzymes that catalyze the conversion under mild conditions and provide the largest source of nitrogen compounds to organisms.

In the chemical industry, catalysts are used for economical conversion of reactants into a desired product. The catalysts chosen are usually specific. A **specific catalyst** *accelerates one chemical reaction, while not facilitating other possible competitive reactions.* Since catalysts are only needed in small amounts and are not used up, they are economically desirable in industrial processes.

Depending on the catalyst employed and the experimental conditions chosen, a set of reactants can be made to produce different products. For example, the reaction of carbon monoxide and hydrogen can produce either methane, CH_4, or methanol, CH_3OH, if the appropriate catalyst is chosen. Note that the catalysts are written over the arrow in chemical equations.

$$CO + 3\,H_2 \xrightarrow{\text{Ni}} CH_4 + H_2O$$

$$CO + 2\,H_2 \xrightarrow{\text{ZnO + Cr}_2\text{O}_3} CH_3OH$$

10.8
Activation Energy

In a chemical reaction, reactant molecules are converted into product molecules by breaking some bonds and forming other bonds. For example, in the combustion of methane, carbon–hydrogen bonds in methane and the oxygen–oxygen bond in oxygen are broken. Additionally carbon–oxygen bonds in carbon dioxide and hydrogen–oxygen bonds in water are formed.

For bonding changes to occur, reactant molecules must be brought together with considerable energy. Only if the energy is high, can the molecules be forced close enough together to overcome the repulsion between the electrons around the nuclei in the reactants. *The minimum energy required for a successful collision leading to a chemical reaction is called the* **activation energy.** Molecules with less than the activation energy rebound without reaction. Each reaction has its own characteristic activation energy based on the number and type of bonds broken and formed in the reaction.

Because there is a specific activation energy for each reaction, a certain temperature is required for the reaction to occur at an appreciable rate. The temperature controls the average kinetic energy of the molecules. As the average kinetic energy increases, the chances of molecular collisions with energy equal to the activation energy increase. A chemical reaction then occurs more frequently, and the observed rate of reaction increases.

In the combustion of methane, a match increases the temperature so that more methane molecules have the activation energy. This activation energy is shown in the reaction-progress diagram of Figure 10.3. The potential energy of the molecules must be increased by the activation energy to boost them to the top of the energy hump shown. Once the molecules reach this point, conversion to products can occur and energy is released. The energy released is equal to the activation energy originally added plus an amount equal to that characteristic for the exothermic reaction. In total, the net release of energy for the reaction is 211 kcal/mole. The energy released in this exothermic reaction is sufficient to continue to increase the kinetic energy of the remaining reactant molecules. The reaction is then self-sustaining and the methane continues to burn.

Endothermic reactions also require an input of energy to allow the molecules to undergo a chemical reaction. Consider the photosynthesis of glucose shown in the reaction-progress diagram in Figure 10.4. Light energy is used to activate the molecules. Some energy is released as products are formed, but it is less than the activation energy. The total energy difference is equal to the amount

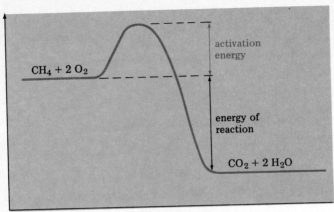

Progress of reaction ⟶

Figure 10.3
Activation Energy of an Exothermic Reaction. After the highest energy point is reached by adding energy, the activation energy and energy of reaction are released. Since the activation energy was needed to cause the reactants to reach the high point of chemical energy, the net energy release is equal to the heat of the reaction.

by which the reaction is endothermic. Thus, if the light energy is not continually provided, the photosynthetic reaction ceases.

When the temperature is increased for either an exothermic or an endothermic reaction, the reaction rate increases. This rate increase reflects both the increased frequency of collision of faster-moving molecules and the fact that a higher fraction of molecules has the necessary activation energy for reaction when they do collide.

In the preceding section the importance of catalysts in increasing the rate of reaction was noted. The effect of the catalyst is to provide a different path for the progress of the reaction. The path starts at the reactants and concludes at the products. However, the path for the catalyzed reaction has a different activation energy (Figure 10.5). An analogy for a catalyst is that of a pathfinder who locates a different route from one altitude to some other altitude via a lower mountain range.

To illustrate the effect of a catalyst on the path of a reaction, consider the hypothetical reaction of A and B.

$$A + B \longrightarrow X$$

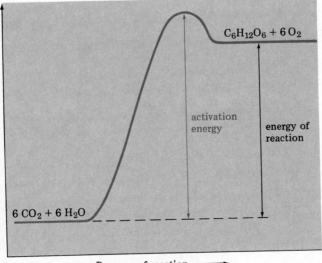

Progress of reaction ⟶

Figure 10.4
Activation Energy of an Endothermic Reaction. After the highest energy point is reached by adding the activation energy, some energy is released. The energy released is less than the activation energy. Thus a net energy input is required for the reaction.

Figure 10.5

Activation Energy of a Reaction and the Effect of Catalysis.

The catalyst provides an alternate pathway for a reaction. The activation energy for the process is lower than for the uncatalyzed reaction. This lower energy requirement results in a faster reaction.

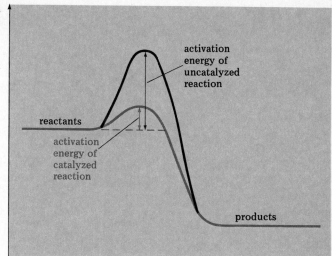

In order for the reaction to occur, the activation energy must be provided. In the presence of a catalyst, such as an enzyme represented by E, the following reactions may occur.

step 1: $A + E \longrightarrow A\text{---}E$

step 2: $A\text{---}E + B \longrightarrow X + E$

The enzyme may combine with A in a reaction with a low activation energy. Similarly, the reaction of A—E with B may require little energy. If the activation energy of each step is low, more molecules will be able to react via this enzyme-catalyzed pathway than could without the enzyme (Figure 10.6).

Figure 10.6

Activation Energy of an Enzyme-Catalyzed Reaction.

The alternate pathway for a catalyzed reaction with an enzyme may involve several steps. Each step has a low activation energy.

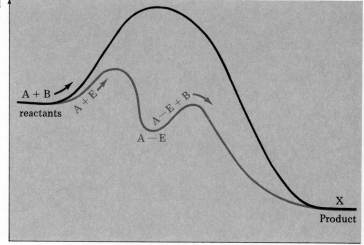

10.9
Chemical Equilibrium

In previous chapters the concept of an equilibrium was presented in terms of physical processes. Thus, water vapor can exist in equilibrium with liquid water. At equilibrium, the rates at which molecules leave and return to the liquid are equal. Similarly, in osmosis, water molecules cross the membrane in both directions, until an equilibrium is achieved based on concentration and pressure. All equilibria are dynamic situations in which processes are still occurring, although there is no net change.

Chemical reactions do not always proceed in one direction. As a reaction occurs, product molecules are formed and they collide. Some collisions lead to the formation of the original reactants. Thus, two opposing reactions occur. **When the rate of formation of products is equal to the rate of formation of reactants, an equilibrium is established.**

The question that can now be asked is, How far does the reaction go in the forward direction to yield products? In general, *the reaction is favored in the exothermic direction. The more exothermic the reaction, the more complete is the reaction.*

Many reactions in living organisms are reversible. That is, they may proceed in both forward and reverse directions. For example, in the lungs, oxygen reacts with hemoglobin, Hb, to form oxyhemoglobin, HbO_2. The solubility of oxygen is only 5 mL per liter of water. However, the solubility of oxygen in blood is 250 mL of gas per liter of blood.

$$Hb + O_2 \longrightarrow HbO_2$$

The oxyhemoglobin is transported by the blood to cells that require oxygen where the reverse reaction occurs. The reaction that liberates oxygen is

$$HbO_2 \longrightarrow Hb + O_2$$

Since one reaction is the reverse of the other, we may combine both equations.

$$Hb + O_2 \rightleftharpoons HbO_2$$

The two half-headed arrows indicate that the reaction may proceed in either direction.

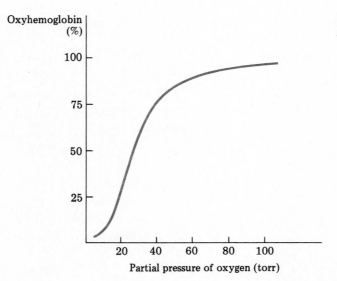

Figure 10.7
Association of Oxygen with Hemoglobin.

A plot of the percentage of hemoglobin converted into oxyhemoglobin versus the partial pressure of oxygen is shown in Figure 10.7. Note that at a partial pressure of 100 mm Hg, hemoglobin is very substantially converted into oxyhemoglobin. This is the partial pressure of oxygen in the lungs. As the partial pressure of oxygen decreases, the percentage of oxyhemoglobin decreases. However, the change in the percentage is gradual initially and only begins to decrease substantially at about 60 mm Hg. Thus, a considerable change in altitude is required before the decreased percentage of oxyhemoglobin is great enough to affect the body. In Chapter 23, the details of the formation of oxyhemoglobin will be given. Hemoglobin actually combines with four molecules of oxygen.

For a balanced general equation at equilibrium

$$m\,A + n\,B \rightleftharpoons p\,X + q\,Y$$

the following expression is a constant at a specific temperature. The brackets indicate concentrations.

$$\frac{[X]^p[Y]^q}{[A]^m[B]^n} = K$$

The numerical value of the expression for the relationship between the concentrations of reactants and products is called the **equilibrium constant** *and is given by the symbol K.*

The equilibrium constant expression for any reaction can be written by inspection of the balanced equation. For example, carbon dioxide and ammonia can combine to form urea and water. The following simple equilibrium expression can be written.

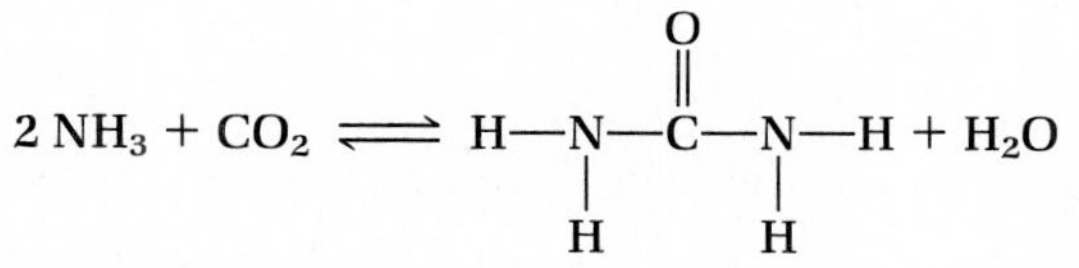

$$2\,NH_3 + CO_2 \rightleftharpoons \underset{\underset{H}{|}}{H-N}-\overset{\overset{O}{\|}}{C}-\underset{\underset{H}{|}}{N}-H + H_2O$$

According to the generalized from, the equilibrium constant expression is

$$K = \frac{[NH_2CONH_2][H_2O]}{[NH_3]^2[CO_2]}$$

The concentration of ammonia is squared because the coefficient of ammonia in the balanced equation is 2.

10.10
Equilibrium Constant

Example 10.5

Ammonia is produced commercially by a reaction of nitrogen and hydrogen gases at high pressure and temperature in the presence of a catalyst. Write an equilibrium constant expression for the reaction

$$N_2 + 3\,H_2 \rightleftharpoons 2\,NH_3$$

Solution. First write the concentration of ammonia, the product, in the numerator and square it because the coefficient of ammonia in the equation is 2.

$$[NH_3]^2$$

Next, place the concentration of nitrogen, a reactant, in the denominator. The power of 1 is assumed.

$$\frac{[NH_3]^2}{[N_2]}$$

Finally, place the concentration of hydrogen, another reactant, in the denominator and cube it because the coefficient of hydrogen in the equation is 3.

$$\frac{[NH_3]^2}{[N_2][H_2]^3} = K$$

The equilibrium constant gives the relationship between the concentrations of reactants and products. For an equilibrium in which the concentrations of the materials raised to the appropriate powers in the numerator are larger than the concentration of the materials raised to the appropriate powers in the denominator, the equilibrium constant is large. Conversely, for a small equilibrium constant, the numerical quantities in the numerator are less than the numerical quantities in the denominator. A large equilibrium constant such as 10^5 indicates that there is little reactant left at equilibrium. For an equilibrium constant of 10^{-5}, there is little product present at equilibrium.

One of the many steps in the metabolism of carbohydrates (Chapter 26) is the conversion of glucose 1-phosphate into glucose 6-phosphate.

$$\text{glucose 1-phosphate} \rightleftharpoons \text{glucose 6-phosphate}$$

At equilibrium the ratio of the concentration of glucose 6-phosphate to glucose 1-phosphate is 20:1. We then write

$$\frac{[\text{glucose 6-phosphate}]}{[\text{glucose 1-phosphate}]} = 20$$

Catalysts do not change the position of a chemical equilibrium or the value of the equilibrium constant. The catalyst facilitates both the forward and reverse reactions equally and no positional difference results.

Example 10.6

Consider the following equilibrium reaction, which started with $1.0 \times 10^{-2}\,M$ carbon monoxide and $1.0 \times 10^{-2}\,M$ water at 1000°C. The equilibrium concentrations are given. What is the equilibrium constant?

$$\underset{5.6\,\times\,10^{-3}}{\text{CO}} + \underset{5.6\,\times\,10^{-3}}{\text{H}_2\text{O}} \rightleftharpoons \underset{4.4\,\times\,10^{-3}}{\text{CO}_2} + \underset{4.4\,\times\,10^{-3}}{\text{H}_2}$$

Solution. First write the equilibrium constant expression.

$$\frac{[CO_2][H_2]}{[CO][H_2O]} = K$$

Next, substitute the equilibrium concentrations into the expression and solve the equation.

$$\frac{(4.4 \times 10^{-3}\,M)(4.4 \times 10^{-3}\,M)}{(5.6 \times 10^{-3}\,M)(5.6 \times 10^{-3}\,M)} = K = 0.62$$

Table 10.3

Effect of Temperature on the Equilibrium Constant

Exothermic reaction $CO + H_2O \rightleftharpoons CO_2 + H_2$ $\Delta H° = -9900$ cal		Endothermic reaction $N_2 + O_2 \rightleftharpoons 2\ NO$ $\Delta H° = 43{,}200$ cal	
T (°C)	*K*	*T* (°C)	*K*
600	3.5	1600	2.5×10^{-4}
800	1.23	1800	7.3×10^{-4}
1000	0.60	2000	1.9×10^{-3}
1200	0.35	2200	4.0×10^{-3}

(increase *T*, decrease *K* for exothermic; increase *T*, increase *K* for endothermic)

Equilibrium constants depend on temperature. **For reactions that are exothermic in the forward direction, the equilibrium constant decreases with increasing temperature. For reactions that are endothermic in the forward direction, the equilibrium constant increases with increasing temperature** (Table 10.3).

10.11 Chemical Equilibrium and Le Châtelier's Principle

Physical or chemical systems at equilibrium consist of two or more processes that occur at rates to balance each other. However, such systems can be disrupted in many ways by outside forces. The point of equilibrium then is altered, but there is a natural tendency to reestablish the equilibrium at some new position.

Le Châtelier's principle allows us to predict how equilibrium systems will respond to changes imposed on them. **If an external force is applied on an equilibrium system, the system, if possible, will readjust to reduce the stress imposed on it.** In chemical reactions, the "stress" is a change in conditions that can cause a change in an equilibrium. These changes are alterations in concentration, pressure, and temperature.

Concentration Changes

Concentration changes do not change the equilibrium constant but rather only the individual concentrations of the substances in equilibrium with each other. If additional reactant is added to a chemical reaction at equilibrium, the concentrations of reactants and products change so as to maintain the equilibrium constant. This is done by decreasing the concentration of the reactants and increasing the concentration of the products. In short, the force imposed on the system—the addition of material—is countered by causing it to react. If a substance is removed from a chemical reaction at equilibrium, reactions will occur to produce more of that substance. Regardless of the condition imposed on the chemical equilibrium, the concentrations ultimately are changed so that the same value of the equilibrium constant is obtained.

Consider the equilibrium that exists between glucose in the blood and glycogen stored in the liver. While glycogen is also stored in muscles, the liver is an exporter of glucose to other organs.

$$n \text{ glucose} \rightleftharpoons n\ H_2O + \text{glycogen}$$

Glycogen is a very large molecule consisting of many covalently bonded glucose molecules. These bonds can be broken with water to give free glucose molecules.

At equilibrium we have 65–100 mg of glucose per 100 mL of blood. When we do work, our blood glucose level decreases. One of the ways our system responds is to produce more glucose from stored glycogen in the liver. By removing a substance on the left side of the equation, the reverse or right-to-left reaction is favored. When an excess of glucose is ingested, the forward reaction is favored as the liver converts the excess blood glucose into glycogen. The actual glucose-to-glycogen and glycogen-to-glucose conversions are complex multistep reactions. The structure of glycogen will be given in Chapter 21.

Pressure Changes

Pressure changes in chemical systems also are countered in a way predicted by Le Châtelier's principle. If the pressure of a gas is increased, the system will respond by processes that tend to decrease the pressure. In the hemoglobin–oxyhemoglobin reaction, the amounts of the two compounds are controlled by the pressure of oxygen.

$$Hb + O_2 \rightleftharpoons HbO_2$$

High pressures of oxygen such as in a hyperbaric chamber cause conversion of hemoglobin into oxyhemoglobin as the system seeks to decrease the pressure of oxygen. The benefit to an individual is that the increased amount of oxyhemoglobin allows more oxygen to be transported to the cells. Conversely, at lower oxygen concentrations at high altitudes, the amount of oxyhemoglobin is lowered and the body cannot function as efficiently. However, the body does eventually respond by forming more hemoglobin.

Temperature Changes

Changes in temperature change the value of the equilibrium constant. We can understand the change by considering heat as a reactant or product. For a reaction that is exothermic in the forward direction, heat is a product.

$$A \rightleftharpoons B + heat$$

If the system is then heated, it will respond by using up the added heat energy; the reverse reaction is favored, and B is consumed. Thus, the value of the equilibrium constant is decreased. If the system is cooled, it will react to produce heat much as our body does; the amount of B increases as does the equilibrium constant.

Summary

Chemical reactions with similar characteristics are classified into reaction types in order to facilitate the study of chemical change. Reactions in water that play an important role in life processes include precipitation, acid–base neutralization, and oxidation–reduction reactions.

Precipitation reactions are the result of the limited solubility of certain combinations of anions and cations. These reactions are written as net ionic equations. Acid–base reactions are the result of proton transfer from an acid to a base. Oxidation–reduction reactions involve electron transfer. A substance that loses electrons is oxidized and acts as a reducing agent. A substance that gains electrons is reduced and acts as an oxidizing agent.

A spontaneous reaction has a natural tendency to occur. All spontaneous reactions release free energy by a combination of an enthalpy change and an entropy change.

Reaction rates are determined by an activation energy: the higher the activation energy, the slower the reaction. Increasing the temperature of a reaction increases the reaction rate. Catalysts cause an increase in reaction rates by providing a reaction pathway with a lower activation energy.

When the rates of conversion of reactants to products and vice versa are equal, a chemical equilibrium results. The equilibrium constant gives a measure of the concentrations of reactants and products at equilibrium. Altering the concentrations of one or more substances in the reaction causes a shift in the concentrations of all reactants and products so as to maintain the equilibrium constant. The equilibrium constant is affected by temperature changes, but not by the presence of a catalyst.

Learning Objectives

As a result of studying Chapter 10 you should be able to

- List and discuss the characteristics of precipitation, acid–base, and oxidation–reduction reactions.
- Write net ionic equations.
- Relate the spontaneity of a reaction to enthalpy changes and entropy changes involved in the process.
- Relate the rate of reactions to reactants, concentrations of reactants, temperature, and catalysts.
- Relate the rate of chemical reactions to activation energies.
- Illustrate the utility of catalysts in industrial and biological processes.
- Express an equilibrium reaction in terms of an equilibrium constant.
- Demonstrate the use of Le Châtelier's principle in chemical equilibria.

New Terms

An **acid** is a substance that may donate a proton.

The **activation energy** is the energy that molecules must possess so that collisions between them can cause a chemical reaction.

An **addition reaction** occurs when atoms bond to points of unsaturation in a molecule.

A **base** is a substance that may accept a proton.

Catalysis is the speeding up of a reaction in the presence of a substance called a **catalyst** that changes the path of a reaction to one of lower energy.

A **condensation reaction** involves the joining together of two molecules to form a larger molecule and a small molecule such as water or ammonia.

An **enthalpy change,** symbolized by $\Delta H°$, is the energy difference between two states or substances. In chemical reactions the $\Delta H°$ is the heat of the reaction and may be positive or negative.

An **entropy change,** symbolized by $\Delta S°$, is a measure of the degree of disorder in a system. A positive $\Delta S°$ means an increase in disorder.

An **equilibrium constant** describes numerically the relationship between the concentrations of reactants and products at equilibrium.

The **free energy change,** symbolized by $\Delta G°$, measures the spontaneity of a chemical reaction. A spontaneous reaction has a negative $\Delta G°$.

A **hydrolysis reaction** involves the separation of a molecule into two or more smaller molecules when reacted with water.

Kinetics is the study of the rates of chemical reactions.

A **net ionic equation** gives only the species actually participating in a chemical reaction.

Neutralization is the combination of an acid and a base to form water.

In **oxidation** some atoms lose one or more electrons and undergo an increase in oxidation number.

An **oxidizing agent** is a substance that is reduced and allows a second substance to become oxidized.

A **precipitation reaction** is a process in which a **precipitate** forms by the combination of ions in solution.

In **reduction** some atoms gain one or more electrons and undergo a decrease in oxidation number.

A **reducing agent** is a substance that is oxidized and allows a second substance to become reduced.

Spectator ions are present in solution but are not involved in a reaction.

Spontaneous reactions occur without an outside source of energy.

A **substitution reaction** involves the replacement of one atom or group of atoms in a molecule by some other atom or group of atoms.

Questions and Problems

Terminology

10.1 Explain how an acid and a base are chemically related to each other.

10.2 Explain why it is not possible to have an oxidation reaction without a reduction reaction.

10.3 What feature determines the degree of spontaneity of a reaction?

Precipitation Reactions

10.4 Identify which of the following compounds are insoluble in water.
  (a) Na_2SO_4 (b) $AgBr$ (c) $(NH_4)_2SO_4$
  (d) Na_2CO_3 (e) $CaCl_2$ (f) $NaNO_3$
  (g) $BaCO_3$ (h) CdS

10.5 Identify which of the following compounds are soluble in water.
  (a) $Ca_3(PO_4)_2$ (b) NH_4I (c) $BaSO_4$

(d) $CaCO_3$ (e) $PbCl_2$ (f) Na_2CrO_4

(g) $LiClO_4$ (h) $PbCrO_4$

10.6 Predict whether the following pairs of solutions will give a precipitate when mixed.

(a) $NH_4I + AgNO_3$ (b) $K_2SO_4 + Ba(NO_3)_2$

(c) $NH_4Br + NaNO_3$ (d) $Pb(NO_3)_2 + NaCl$

(e) $Hg(NO_3)_2 + Na_2S$ (f) $Pb(NO_3)_2 + Na_2CrO_4$

(g) $Ba(NO_3)_2 + NaCl$ (h) $RbCl + MgBr_2$

Net Ionic Equations

10.7 Silver bromide is insoluble in water. Based on this fact, write net ionic equations for the expected reaction when solutions of each of the following substances are mixed.

(a) $AgNO_3 + NaBr$

(b) $AgNO_3 + HBr$

(c) $AgNO_3 + NH_4Br$

10.8 Lead sulfate is insoluble in water. Based on this fact, write net ionic equations for the expected reaction when solutions of each of the following substances are mixed.

(a) $Pb(NO_3)_2 + Na_2SO_4$

(b) $Pb(NO_3)_2 + H_2SO_4$

(c) $Pb(NO_3)_2 + (NH_4)_2SO_4$

10.9 Write a net ionic equation for the reaction between each of the following sets of reactants.

(a) $NH_4I + AgNO_3$

(b) $K_2SO_4 + Ba(NO_3)_2$

(c) $Pb(NO_3)_2 + NaCl$

(d) $Hg(NO_3)_2 + Na_2S$

(e) $Pb(NO_3)_2 + Na_2CrO_4$

(f) $Ba(NO_3)_2 + Na_2CO_3$

Acid–Base Reactions

10.10 What is a neutralization reaction?

10.11 Each of the following is an acid–base reaction. Balance the equation if needed.

(a) $HNO_3 + NaOH \longrightarrow NaNO_3 + H_2O$

(b) $HNO_3 + Mg(OH)_2 \longrightarrow Mg(NO_3)_2 + H_2O$

(c) $H_2SO_4 + NaOH \longrightarrow Na_2SO_4 + H_2O$

(d) $H_3PO_4 + NaOH \longrightarrow Na_3PO_4 + H_2O$

(e) $H_2SO_4 + Mg(OH)_2 \longrightarrow MgSO_4 + H_2O$

10.12 Sodium bicarbonate, $NaHCO_3$, is used as an antacid to reduce stomach acid, which is $0.1\,M$ HCl. Describe how the antacid works.

Oxidation–Reduction Reactions

10.13 What is the oxidation number of the indicated element in each of the following substances?

(a) S in SO_2 (b) S in SO_3 (c) Cu in $CuBr_2$

(d) N in NH_3 (e) N in N_2H_4 (f) W in WO_3

10.14 Indicate what substances are oxidized and reduced in each of the following reactions.

(a) $2\,SO_2 + O_2 \longrightarrow 2\,SO_3$

(b) $Cu + Br_2 \longrightarrow CuBr_2$

(c) $2\,Cl^- + I_2 \longrightarrow Cl_2 + 2\,I^-$

(d) $WO_3 + 3\,H_2 \longrightarrow W + 3\,H_2O$

10.15 Indicate what substances are the oxidizing agent and the reducing agent in **10.14.**

10.16 What type of substance is required to convert acetaldehyde into ethanol?

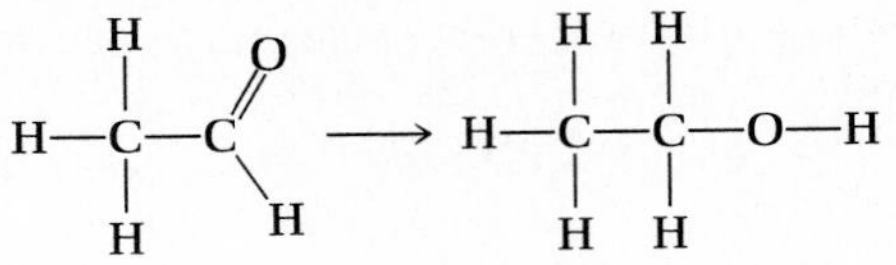

10.17 The oxidation number of a substance increases in a chemical reaction. Has the substance been oxidized or reduced?

10.18 The oxidation number of a substance decreases in a chemical reaction. Has the substance served as an oxidizing agent or a reducing agent?

10.19 Fill in each of the following blanks.

(a) A substance that loses electrons becomes _____.

(b) A substance whose oxidation number increases in a chemical reaction becomes _____.

(c) The substance that gains electrons in a chemical reaction is the _____ agent.

(d) The substance that has a lower oxidation number after a chemical reaction is the _____ agent.

10.20 During the metabolism of carbohydrates and fats, succinic acid is converted into fumaric acid. Is this reaction an oxidation or a reduction?

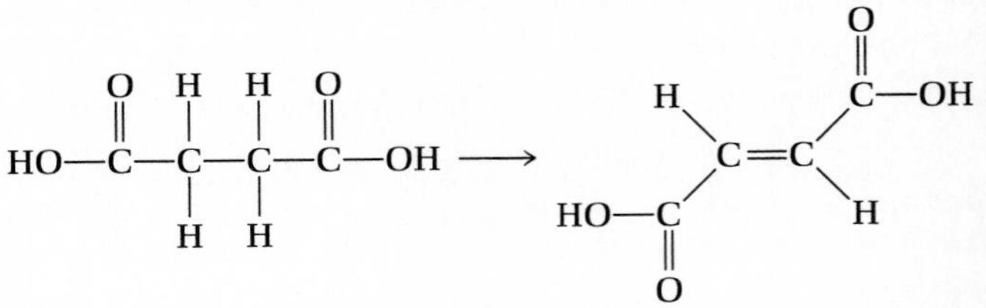

Substitution Reactions

10.21 What products are expected from each of the following substitution reactions?

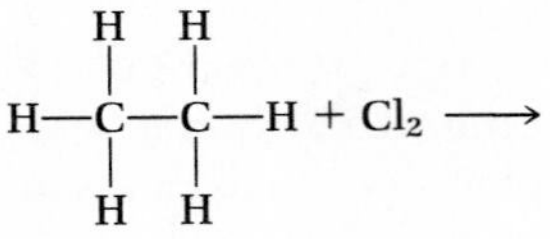

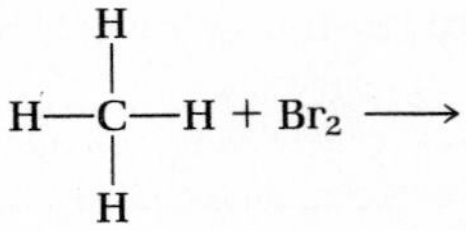

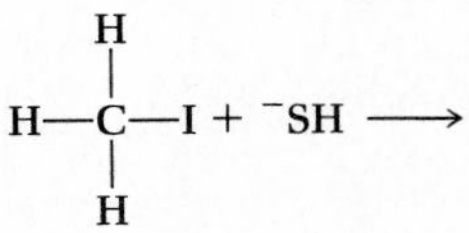

(d)
$$
\begin{array}{cc}
\text{H} & \text{H} \\
| & | \\
\text{H}-\text{C}-\text{C}-\text{I} + {}^-\text{OH} \longrightarrow \\
| & | \\
\text{H} & \text{H}
\end{array}
$$

Addition Reactions

10.22 Bromine undergoes an addition reaction with ethylene. Write the structure of the product.

10.23 Acetylene reacts with hydrogen with palladium as a catalyst. What product is formed?

$$\text{H}-\text{C}{\equiv}\text{C}-\text{H} + \text{H}_2 \xrightarrow{\text{Pd}}$$

10.24 Hydrogen bromide adds to ethylene. What product is formed?

$$
\begin{array}{c}
\text{H} \qquad\qquad \text{H} \\
\diagdown \qquad \diagup \\
\text{C}{=}\text{C} \quad + \text{HBr} \longrightarrow \\
\diagup \qquad \diagdown \\
\text{H} \qquad\qquad \text{H}
\end{array}
$$

Condensation and Hydrolysis Reactions

10.25 Classify the following reactions.

(a)
$$
\begin{array}{ccc}
\text{H} & & \text{O} \\
| & & \parallel \\
\text{H}-\text{C}-\text{C}-\text{O}-\text{H} + \text{H}-\text{O}-\text{C}-\text{C}-\text{H} \longrightarrow \\
| & & \\
\text{H} & &
\end{array}
$$

$$
\begin{array}{c}
\text{H} \quad\;\; \text{O} \quad\; \text{H} \;\; \text{H} \\
| \quad\;\; \parallel \quad\; | \;\;\; | \\
\text{H}-\text{C}-\text{C}-\text{O}-\text{C}-\text{C}-\text{H} + \text{H}_2\text{O} \\
| \qquad\qquad\;\; | \;\;\; | \\
\text{H} \qquad\qquad\; \text{H} \;\; \text{H}
\end{array}
$$

(b)
$$
\begin{array}{ccc}
\text{H} & & \text{O} \\
| & & \parallel \\
\text{H}-\text{C}-\text{C}-\text{OH} + \text{H}-\text{N}-\text{C}-\text{H} \longrightarrow \\
| & & \\
\text{H} & &
\end{array}
$$

$$
\begin{array}{c}
\text{H} \quad\;\; \text{O} \quad\; \text{H} \;\; \text{H} \\
| \quad\;\; \parallel \quad\; | \;\;\; | \\
\text{H}-\text{C}-\text{C}-\text{N}-\text{C}-\text{H} + \text{H}_2\text{O} \\
| \qquad\qquad\;\; | \;\;\; | \\
\text{H} \qquad\qquad\; \text{H} \;\; \text{H}
\end{array}
$$

(c)
$$
\begin{array}{c}
\quad\;\; \text{O} \\
\quad\;\; \parallel \\
\text{H}-\text{C}-\text{N}-\text{H} + \text{H}_2\text{O} \longrightarrow \\
\quad\;\; | \\
\quad\;\; \text{H}
\end{array}
$$

$$
\begin{array}{c}
\quad\;\; \text{O} \\
\quad\;\; \parallel \\
\text{H}-\text{C}-\text{OH} + \text{NH}_3
\end{array}
$$

(d)
$$
\begin{array}{c}
\quad\;\; \text{O} \;\;\; \text{H} \\
\quad\;\; \parallel \;\;\; | \\
\text{H}-\text{C}-\text{O}-\text{C}-\text{H} + \text{H}_2\text{O} \longrightarrow \\
\quad\qquad\;\;\; | \\
\quad\qquad\;\;\; \text{H}
\end{array}
$$

$$
\begin{array}{c}
\quad\;\; \text{O} \qquad\qquad \text{H} \\
\quad\;\; \parallel \qquad\qquad | \\
\text{H}-\text{C}-\text{OH} + \text{H}-\text{O}-\text{C}-\text{H} \\
\quad\qquad\qquad\qquad\;\;\; | \\
\quad\qquad\qquad\qquad\;\;\; \text{H}
\end{array}
$$

Enthalpy, Entropy, and Free Energy

10.26 What is meant by a spontaneous reaction?

10.27 Is a reaction with a negative $\Delta H°$ exothermic or endothermic?

10.28 Under what conditions may an endothermic reaction be spontaneous?

10.29 A reaction has a negative $\Delta S°$. What is meant by this fact?

10.30 What relationships exist among $\Delta G°$, $\Delta H°$, and $\Delta S°$?

10.31 What predictions may be made about reactions with the following combinations of $\Delta H°$ and $\Delta S°$?
  (a) positive $\Delta H°$ and negative $\Delta S°$
  (b) negative $\Delta H°$ and negative $\Delta S°$
  (c) positive $\Delta H°$ and positive $\Delta S°$
  (d) negative $\Delta H°$ and positive $\Delta S°$

10.32 Explain why the energy required in the photosynthesis of glucose is equal to the energy released in the metabolism of glucose.

10.33 Are the products in an exothermic reaction at a higher or lower potential energy than the reactants?

10.34 How is the potential energy stored in chemicals? Why does this potential energy differ in two compounds such as H_2S and H_2O?

10.35 Why is the temperature inside a compost pile higher than that of the surrounding air?

Kinetics

10.36 Why might an industry choose to carry out a reaction at a high temperature in spite of the higher costs of heat?

10.37 Substances burn more rapidly in pure oxygen than air. Explain why. This phenomenon was responsible for the tragedy during a ground test of an Apollo spacecraft in which three American astronauts died.

10.38 Meat will spoil at room temperature, but remain unspoiled for longer periods in a refrigerator. Explain why.

10.39 Explain why persons with elevated temperature for a prolonged period must increase their food intake to avoid losing weight.

10.40 Ripened tomatoes are stored in the refrigerator, whereas unripened tomatoes are left at room temperature. Explain why.

10.41 Certain antibiotic drugs must be stored under refrigeration. Suggest a reason for this requirement.

10.42 List three experimental factors that can be used to speed up a reaction.

Activation Energies

10.43 Which reaction will occur at a faster rate at 20°C: a reaction with an activation energy of 15 kcal/mole or one with an activation energy of 25 kcal/mole?

10.44 Explain why chemical reactions increase in rate as the temperature is increased.

10.45 Draw a progress-of-reaction diagram for the metabolism of glucose to yield carbon dioxide and water.

10.46 The activation energy for the decomposition of hydrogen peroxide, H_2O_2, is 18 kcal/mole. In the presence of platinum the activation energy is 12 kcal/mole. Draw the two progress-of-reaction diagrams. What is the function of platinum?

Equilibrium Constants

10.47 Write the equilibrium constant expression for each of the following reactions.
   (a) $3 O_2 \rightleftharpoons 2 O_3$
   (b) $N_2 + 3 H_2 \rightleftharpoons 2 NH_3$
   (c) $CH_4 + Cl_2 \rightleftharpoons CH_3Cl + HCl$
   (d) $2 CO + O_2 \rightleftharpoons 2 CO_2$

10.48 How do each of the following experimental changes affect the equilibrium constant, if at all?
   (a) increase the concentration of a reactant
   (b) increase the concentration of a product
   (c) increase the temperature of an exothermic reaction
   (d) add a catalyst

10.49 How does the magnitude of the equilibrium constant indicate the amounts of product and reactant present at equilibrium?

10.50 How is the concentration of oxyhemoglobin available for the operation of cells controlled by oxygen pressure?

10.51 What occurs when the blood glucose level decreases because of extreme exercise?

Le Châtelier's Principle

10.52 Carbonic acid reacts with hydroxide ion according to the following equation.

$$H_2CO_3 + {}^-OH \longrightarrow HCO_3^- + H_2O$$

Carbonic acid is in equilibrium with carbon dioxide and water. Will an increase of CO_2 aid or hinder the above reaction?

10.53 Consider the following reaction. What effect would each of the changes in condition listed have on the reaction?

$$O_3 + NO \rightleftharpoons O_2 + NO_2$$

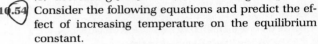

   (a) increasing $[O_3]$ (b) increasing $[O_2]$
   (c) decreasing $[NO]$ (d) decreasing $[NO_2]$

10.54 Consider the following equations and predict the effect of increasing temperature on the equilibrium constant.
   (a) $N_2 + 2 O_2 + 16\,kcal \rightleftharpoons 2 NO_2$
   (b) $2 CO + O_2 \rightleftharpoons 2 CO_2 + 135\,kcal$

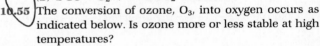

10.55 The conversion of ozone, O_3, into oxygen occurs as indicated below. Is ozone more or less stable at high temperatures?

$$2 O_3 \longrightarrow 3 O_2 + 64.8\,kcal$$

Acids and Bases

11

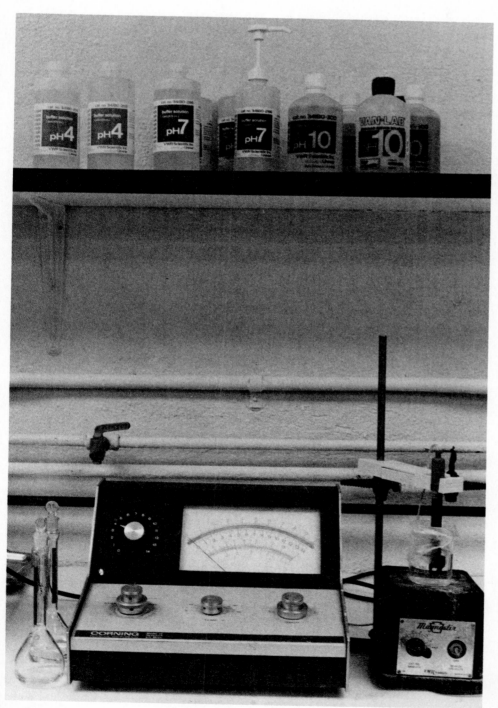

A pH meter is used to measure the pH of solutions.
Buffer solutions have characteristic pH values.

In Chapter 10, acids were defined as proton donors and bases were defined as proton acceptors. Only a limited description of the reactions of acids and bases was given at that time. However, acids and bases play important roles in many diverse areas. Thus, the chemistry of these substances will be presented in some detail in this chapter.

Many common household items are either acids or bases. Citric acid in lemon juice, acetic acid in vinegar, and acetylsalicylic acid of aspirin tablets are acids. Baking soda, household ammonia, antacids, and "lye" are bases. The list of products used in the home that are acids or bases increases each year. Among the potentially dangerous products that are bases are drain cleaners and oven cleaners.

In agriculture, the acidity of the soil is important in determining what crops may be grown. Fertilizers containing acids or bases can be applied to make the soil more suitable for particular crops.

Acid–base reactions play significant roles in the chemistry of your body. Most of the food we eat contains compounds that are acids or can be converted into acids. Metabolism of food produces organic acids that are eventually converted into CO_2 and H_2O. The types and amounts of acids metabolically produced in diabetics differ from those produced in other individuals.

Most body fluids are acidic; however, blood is slightly basic. As the chemistry of life processes occurs, acids are produced or used and their concentration must be precisely controlled. Very slight changes in chemical equilibria involving acids and bases can disrupt physiological reactions. Hemoglobin–oxygen binding in respiration is affected by acid concentration. Increased or decreased respiration affects the acid–base balance of the blood.

Gastric acid, a solution of hydrochloric acid, is required for the digestion that

Table 11.1

Composition of Some Antacids

Compound	Comments
aluminum hydroxide, $Al(OH)_3$	An effective and nonhazardous antacid with no dosage restriction. The compound is combined with other substances in Di-Gel, Gelusil and Maalox.
dihydroxyaluminum sodium carbonate, $NaAl(OH)_2CO_3$	An effective antacid found in Rolaids. However, usage must be restricted by individuals on a low sodium diet.
calcium carbonate, $CaCO_3$	This antacid, used in Tums and Pepto Bismol, may cause constipation. Excessive usage can lead to a high blood calcium level and cause kidney stones.
magnesium hydroxide, $Mg(OH)_2$	Small dosages are used as an antacid, but larger amounts act as a laxative. As a suspension the compound is known as Milk of Magnesia.
sodium bicarbonate, $NaHCO_3$	Also known as ordinary baking soda and found in Alka-Seltzer. It gives fast and effective antacid action. Usage should be restricted by individuals on a low sodium diet. Two Alka-Seltzer tablets contain a gram of sodium.

occurs in the stomach. However, excessive production of acid can cause an ulcer. Antacids, used to treat conditions of excess stomach acidity, are bases. The compositions of some common antacids are listed in Table 11.1.

Acids turn the color of a dye called litmus from blue to red. The dye is impregnated in a strip of paper called blue litmus paper. The presence of an acid may be shown by placing a drop of the solution on the paper. A blue to red color change indicates that the solution contains an acid.

Many acids have a sour taste. Examples include the dilute solutions of citric acid in lemons and acetic acid in vinegar. Most acids that you will encounter in the laboratory are aqueous solutions. Concentrated solutions of acids are very corrosive and should not be tasted. Acids cause serious chemical burns if spilled on your skin because they react with proteins. Any affected area must be immediately flooded with water to decrease the damage. In the eye area response time is critical. For this reason, safety glasses should be worn in all laboratories. An eyewash should also be available.

Bases turn the color of red litmus dye to blue. Red litmus is available in impregnated strips of paper that are used to test for bases.

Most common bases are solid ionic compounds containing hydroxide ions. Examples include sodium hydroxide and potassium hydroxide. Solutions of bases have a slippery feeling and a bitter taste. Bases can react with fats and oils and change them into soluble compounds. Thus, solutions of bases are useful as cleaning agents. However, they will also react with materials that make up cell membranes. Therefore, contact with living tissue should be avoided. In fact, although not generally known, bases are more destructive to the eyes than acids, so immediate irrigation of your eyes is necessary if a base should be splashed on your face.

11.2 Properties of Acids and Bases

11.3 Common Acids and Bases

Although there are many acids and bases, only a few of the more important ones that you might encounter are discussed here. The description of their uses is selective because the applications of acids and bases in industry, commercial products, agriculture, and many other areas are too numerous to describe fully. The concentrations of some common acids are given in Table 11.2.

Table 11.2

Properties of Concentrated Solutions of Common Acids

Name	Formula	Concentration		Density (g/mL)
		w/v %	M	
hydrochloric acid	HCl	37	12	1.19
nitric acid	HNO_3	70	16	1.42
sulfuric acid	H_2SO_4	96	18	1.82
phosphoric acid	H_3PO_4	85	15	1.70
acetic acid	CH_3CO_2H	100	17	1.05

Sulfuric Acid

Sulfuric acid, H_2SO_4, is the acid produced in the largest commercial quantity in the world. It is used directly or indirectly in a large number of industrial processes. In fact, one measure of the development of a modern industrial nation is its sulfuric acid production and consumption. Sulfuric acid is the "battery acid" found in automobile batteries. Sulfuric acid is a powerful dehydrating agent (ability to remove water). If sulfuric acid is in contact with tissues, this dehydration process is very harmful.

Hydrochloric Acid

A 12 M hydrochloric acid, HCl, solution is made by dissolving gaseous HCl in water. (As little as 0.1% hydrogen chloride in air can cause death.) The gastric juice in your stomach is 0.1 M hydrochloric acid, but the lower concentration of the acid and the special mucosal tissue present in your stomach prevent any damage. However, excessive secretion of HCl, hyperchlorhydria, can cause ulcers over an extended time. Mild and temporary increases in acid may be treated with antacids, but the prolonged use of antacids can also upset the acid–base balance of the stomach.

Nitric Acid

Nitric acid, HNO_3, is a very corrosive acid. If you spill it on your skin, a yellow stain results. The nitric acid reacts with parts of your protein molecules to produce a colored compound that is then part of your skin. The color is lost only as newly formed skin replaces it. Nitric acid is used to make products as diverse as fertilizers and explosives.

Phosphoric Acid

Phosphoric acid, H_3PO_4, is a thick, viscous liquid. It is used in dilute form in soft drinks and in the production of detergents and fertilizers. Phosphoric acid combined with organic molecules plays an important role in the chemistry of cells.

Acetic Acid

Acetic acid, CH_3CO_2H, is an organic acid and will be discussed in later chapters. Only one of the four hydrogen atoms is acidic. Those bonded to carbon are not acidic.

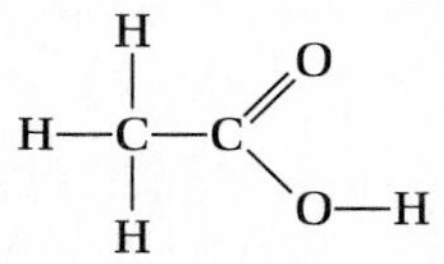

$$H-\underset{\underset{H}{|}}{\overset{\overset{H}{|}}{C}}-C\overset{O}{\underset{O-H}{\diagup}}$$

Vinegar is a dilute solution (5%) of acetic acid in water.

Carbonic Acid

Carbonic acid is formed in an equilibrium reaction with carbon dioxide and water.

$$H_2O + CO_2 \rightleftharpoons H_2CO_3$$

Only about 1% of the carbon dioxide dissolved in water exists as carbonic acid. Carbonic acid is present in carbonated beverages.

Sodium Hydroxide

Sodium hydroxide, NaOH, commonly called lye or caustic soda, is a solid. Both the solid and aqueous solutions of sodium hydroxide are to be handled with care.

Sodium hydroxide is used in soap manufacture, paper production, textile manufacturing, and many other industrial processes. It is also the active ingredient in oven and drain cleaners.

Ammonia

Ammonia, NH_3, is a gas produced in industrial quantities second only to sulfuric acid. It is used in the production of fertilizer. Ammonia dissolves in water to form small amounts of ammonium and hydroxide ions. For this reason ammonia solutions are sometimes called ammonium hydroxide. However, the major component of the solution is ammonia.

$$NH_3 + H_2O \rightleftharpoons NH_4^+ + OH^-$$

The 27% solution of ammonia sold commercially will cause skin burns. A 2% solution of ammonia is used as an inhalant to revive people who have fainted. Although less common now than a few years ago, aqueous ammonia may be used to clean items such as windows.

Over the years, acids and bases have been classified in a variety of ways. However, the most convenient definition for use in this text is that described by Brønsted and Lowry. *An* **acid** *is a substance that can donate a proton,* H^+. *A* **base** *is a substance that can accept a proton.*

11.4
Brønsted–Lowry Concept of Acids and Bases

Acids

In aqueous solutions of acids, the proton is donated to water and the hydronium ion, H_3O^+, is formed. For example, when gaseous hydrogen chloride is dissolved in water, a solution of hydrochloric acid is formed.

$$HCl + H_2O \longrightarrow H_3O^+ + Cl^-$$

The transfer of a proton from an acid to water is called **ionization.** An illustration of this process is shown in Figure 11.1. In the hydronium ion, the proton can be viewed as attached to oxygen by means of the nonbonded electrons on oxygen.

$$H^+ + \overset{..}{:}\overset{..}{O}-H \longrightarrow H-\overset{..}{O}-H^+$$
$$\qquad\quad | \qquad\qquad\qquad |$$
$$\qquad\quad H \qquad\qquad\qquad H$$

Chemists may interchangeably use the terms proton, hydrogen ion, or hydro-

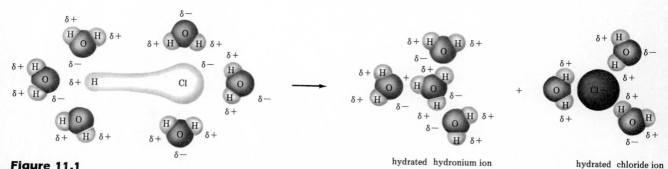

Figure 11.1

Ionization of HCl.

The polar covalent bond in HCl becomes stretched and is broken by the action of surrounding water molecules. The proton becomes attached to a water molecule to produce a hydronium ion.

nium ion in describing aqueous acid solutions. However, the preferred term is hydronium ion, since that is what exists in solution. Like all ions in solution, the hydronium ion is surrounded by water molecules.

Hydrochloric acid, nitric acid, and perchloric acid, $HClO_4$, are monoprotic acids. **Monoprotic acids** *yield a 1:1 ratio of hydronium ion to the negative ion.*

$$H_2O + HCl \longrightarrow H_3O^+ + Cl^-$$

$$H_2O + HNO_3 \longrightarrow H_3O^+ + NO_3^-$$

$$H_2O + HClO_4 \longrightarrow H_3O^+ + ClO_4^-$$

Sulfuric acid is a diprotic acid. It can react to donate two protons to water.

$$H_2SO_4 + H_2O \longrightarrow H_3O^+ + HSO_4^-$$

$$HSO_4^- + H_2O \longrightarrow H_3O^+ + SO_4^{2-}$$

Transfer of one proton yields the hydrogen sulfate ion (or bisulfate ion), HSO_4^-. The second transfer results in the formation of the sulfate ion, SO_4^{2-}.

Carbonic acid is a diprotic acid that plays an important role in respiration and in the acid–base balance of the blood.

$$H_2CO_3 + H_2O \rightleftharpoons HCO_3^- + H_3O^+$$

$$HCO_3^- + H_2O \rightleftharpoons CO_3^{2-} + H_3O^+$$

Phosphoric acid is a triprotic acid. It may donate three protons to water.

$$H_3PO_4 + H_2O \rightleftharpoons H_3O^+ + H_2PO_4^-$$

$$H_2PO_4^- + H_2O \rightleftharpoons H_3O^+ + HPO_4^{2-}$$

$$HPO_4^{2-} + H_2O \rightleftharpoons H_3O^+ + PO_4^{3-}$$

All of the acids described produce hydronium ions in solution. It is the hydronium ion that is responsible for the acidic properties of acids. In fact, the hydronium ion is an acid because it has a proton that it may donate to a base.

Bases

The most common base is the hydroxide ion. The hydroxide ion exists as an ion in the compounds such as NaOH, KOH, and $Mg(OH)_2$. When dissolved in water, the ions of these compounds are separated from the crystal and distributed in the solution. The hydroxide ion is a base because it can accept a proton from an acid such as the hydronium ion. *The reaction of hydronium ions with hydroxide ions is called a* **neutralization reaction.**

$$H_3O^+ + OH^- \longrightarrow H_2O + H_2O$$

Ions other than the hydroxide ion can be bases, as can some neutral molecules. Ammonia is a base because it can accept a proton from an acid such as HCl.

$$NH_3 + HCl \longrightarrow NH_4^+ + Cl^-$$

Conjugate Acids and Bases

The forward and reverse reactions of acid–base reactions are related in an interesting way. When an acid and a base react through the transfer of a proton, another base and acid are produced. This relationship is emphasized by considering a substance and the related substance formed after proton transfer as a conjugate pair. *When an acid loses a proton, the material formed is called the*

conjugate base *of the acid.* The conjugate base of HCl is the chloride ion. *When a base accepts a proton, the new substance formed is called the* **conjugate acid** *of the base.* The conjugate acid of ammonia is the ammonium ion.

$$NH_3 + HCl \longrightarrow NH_4^+ + Cl^-$$

base   acid   conjugate   conjugate
                  acid   base

Amphoterism

Substances that can either lose or gain a proton are **amphoteric substances** *and can act as either an acid or a base.* Consider the diprotic acid sulfuric acid.

$$H_2SO_4 + H_2O \rightleftharpoons H_3O^+ + HSO_4^-$$

$$HSO_4^- + H_2O \rightleftharpoons H_3O^+ + SO_4^{2-}$$

The bisulfate ion or hydrogen sulfate ion, HSO_4^-, is the conjugate base of sulfuric acid, but it is also the conjugate acid of sulfate ion. Bisulfate ion can accept a proton to form H_2SO_4 or lose a proton to form the sulfate ion. Bisulfate ion is amphoteric.

Water itself is amphoteric. It may function as an acid as, for example, when it donates a proton to ammonia. Water can also accept protons as, for example, in its reaction with HCl.

$$NH_3 + H_2O \rightleftharpoons NH_4^+ + OH^-$$

$$HCl + H_2O \rightleftharpoons H_3O^+ + Cl^-$$

The ions $H_2PO_4^-$ and HCO_3^- are amphoteric substances that are important to acid–base balance in the chemistry of life. These ions will be discussed later in this chapter.

Example 11.1

What are the conjugate acid and the conjugate base of the amphoteric $H_2PO_4^-$ ion?

Solution. When a substance acts as an acid, it donates a proton to a base such as water. The material related to the acid is called the conjugate base. The conjugate base of $H_2PO_4^-$ is HPO_4^{2-}.

$$H_2PO_4^- + H_2O \longrightarrow HPO_4^{2-} + H_3O^+$$

When a substance acts as a base, it accepts a proton from an acid such as water. The material related to the base is called the conjugate acid. The conjugate acid of $H_2PO_4^-$ is H_3PO_4.

$$H_2PO_4^- + H_2O \longrightarrow H_3PO_4 + OH^-$$

[Additional examples may be found in **11.18–11.22** at the end of the chapter.]

The words strong and weak are used to refer to the degree of ionization of acids and bases. These terms do not refer to the concentrations of acids and bases. An acid such as HCl, which is strong, is strong regardless of whether it is in a dilute or concentrated solution. Similarly, a weak acid, such as acetic acid, is weak regardless of its concentration in water.

11.5
Strengths of Acids and Bases

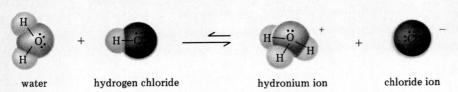

water + hydrogen chloride ⇌ hydronium ion + chloride ion

Figure 11.2
Ionization of a Strong Acid in Water. Transfer of a proton from the acid hydrogen chloride to water produces the hydronium ion. Ionization is essentially complete.

Strong Acids

When hydrogen chloride is dissolved in water, virtually no covalent HCl remains. The transfer of a proton from HCl to water occurs essentially completely. This fact is indicated by arrows of unequal length in the equilibrium equation (Figure 11.2).

$$HCl + H_2O \rightleftharpoons H_3O^+ + Cl^-$$

Acids that essentially completely transfer their protons to water are called **strong acids.** From the position of the equilibrium, we conclude that HCl has a stronger tendency to donate protons than does H_3O^+ and is therefore a stronger acid than H_3O^+. In addition, the position of the equilibrium reflects the relative abilities of H_2O and Cl^- to accept protons. The water molecule is a stronger base than Cl^-.

Table 11.3
Brønsted–Lowry Conjugate Acid–Base Pairs

Acid	Base
Strongest	Weakest
$HClO_4$	ClO_4^-
H_2SO_4	HSO_4^-
HCl	Cl^-
HNO_3	NO_3^-
H_3O^+	H_2O
HSO_4^-	SO_4^{2-}
H_3PO_4	$H_2PO_4^-$
HF	F^-
CH_3CO_2H	$CH_3CO_2^-$
H_2CO_3	HCO_3^-
H_2S	HS^-
$H_2PO_4^-$	HPO_4^{2-}
NH_4^+	NH_3
HCO_3^-	CO_3^{2-}
HPO_4^{2-}	PO_4^{3-}
H_2O	OH^-
HS^-	S^{2-}
OH^-	O^{2-}
Weakest	Strongest

From a consideration of the equilibria between acids and bases and their conjugate bases and acids, it can be concluded that a strong acid, with its great tendency to lose protons, is conjugate to a weak base that has a low affinity for protons. **The stronger the acid, the weaker is its conjugate base.** Strong bases attract protons strongly and are conjugate to weak acids that do not readily lose protons. The stronger a base, the weaker is its conjugate acid. A list illustrating the relationship between common acid–base pairs is given in Table 11.3.

Weak Acids

Most acids, especially organic acids, do not transfer their protons completely to water and few ions are produced. *Acids that are only partially ionized in water are called* **weak acids.**

Acetic acid, an organic acid, is a weak acid. At 25°C, a 1 M solution of acetic acid is approximately 0.4% ionized and the concentration of ions is very low.

$$CH_3CO_2H + H_2O \rightleftharpoons H_3O^+ + CH_3CO_2^-$$

Acetic acid is a weaker acid than H_3O^+, and $CH_3CO_2^-$ is a stronger base than H_2O. In this reaction, the equilibrium lies to the side containing the weaker acid and weaker base. This statement is general and quite logical because the proton must reside with the weaker acid or the substance that has the smallest tendency to lose it. Furthermore, the proton remains with the acid because the base on that side of the equation does not have much tendency to remove it.

In order to compare the strengths of acids, it is necessary to measure their tendencies to transfer protons to a base, usually water. The order of acid strengths can be established by measuring the acid ionization constant. For an acid with the general formula HA, the equilibrium constant for ionization is obtained from the equation for ionization.

$$HA + H_2O \rightleftharpoons H_3O^+ + A^-$$

$$K = \frac{[H_3O^+][A^-]}{[HA][H_2O]}$$

The concentration of water is about 55 M and is so large compared to that of the other components of the equilibrium that its value changes very little when the

Table 11.4

Acidity of Acids

Acid	K_a (mole/L)	Percent ionization
HSO_4^-	1.3×10^{-2}	11
H_3PO_4	7.5×10^{-3}	8.3
HF	6.6×10^{-4}	2.5
CH_3CO_2H	1.8×10^{-5}	.42
H_2CO_3	4.3×10^{-7}	.065
H_2S	1.3×10^{-7}	.036
$H_2PO_4^-$	6.2×10^{-8}	.025
HCN	4.0×10^{-10}	.0020
HCO_3^-	4.8×10^{-11}	.00069
HPO_4^{2-}	2.2×10^{-13}	.000047
HS^-	7.1×10^{-15}	.0000084

acid HA is added. Therefore, it is included in a constant called the acid ionization constant.

$$K_a = K[H_2O] = \frac{[H_3O^+][A^-]}{HA}$$

The acid ionization constants of some acids and the percent ionization of a 1 M solution are given in Table 11.4. The larger the value of K_a, the larger is the percent ionization at the same concentration.

Example 11.2

Lactic acid is a monoprotic organic acid produced in metabolic reactions. A 1.0 M solution of lactic acid is about 1% ionized. Calculate the K_a for lactic acid.

Solution. The concentration of hydronium ion will be 0.01 times that of the lactic acid. Thus, the concentration of hydronium ion is

$$[H_3O^+] = (0.01)(1.0\,M) = 0.01\,M$$

The process of ionization of a monoprotic acid produces hydronium ions and the conjugate base of the acid in a 1:1 ratio. If the acid is represented as HL, then the concentration of L^- is 0.01 M.

The concentration of lactic acid is slightly diminished by ionization. The concentration of lactic acid is 0.99 M. The acid constant is then calculated.

$$\frac{[H_3O^+][L^-]}{[HL]} = \frac{[0.01][0.01]}{[0.99]} = 1 \times 10^{-4}$$

[Additional examples may be found in **11.29–11.34** at the end of the chapter.]

Strengths of Bases

A **strong base** *completely removes the proton of an acid.* The most common strong base is the hydroxide ion, which will remove and accept protons from even weak acids such as acetic acid.

$$OH^- + CH_3CO_2H \rightleftharpoons H_2O + CH_3CO_2^-$$

Weak bases do not have a very large attraction for protons of an acid. Only a small fraction of a weak base in a sample will accept protons at equilibrium. Ammonia is the most common example of a weak base. When ammonia dissolves in water, hydroxide ions are formed as a result of the abstraction of a proton from water by ammonia.

$$NH_3 + H_2O \rightleftharpoons NH_4^+ + OH^-$$

11.6
Self-ionization of Water

In Chapter 8, water was discussed as a molecule that is extensively hydrogen-bonded. The hydrogen of one molecule, which is slightly positive because of the polar covalent bond, is attracted to an unshared electron pair of a neighboring water molecule. At that time you might have wondered why the proton was not

Figure 11.3
Self-ionization of Water.
A proton is transferred from one
water molecule to another to produce equal quantities of hydroxide
and hydronium ions. The reaction
proceeds to only a limited extent.

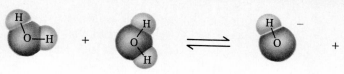

hydroxide ion   hydronium ion

transferred to give ions. In fact, this self-ionization reaction does occur to a small extent. This reaction is shown with the shorter arrow toward the ionic products.

$$H_2O + H_2O \rightleftharpoons H_3O^+ + OH^-$$

In water at 25°C there are both hydronium ions and hydroxide ions (Figure 11.3). Note that for every hydronium ion formed, there must be one hydroxide ion formed. Water is then neither acidic nor basic. At equilibrium the concentration of H_3O^+ is $1 \times 10^{-7}\,M$, the same as the concentration of OH^-.

We can write the equilibrium for the self-ionization of water as follows.

$$K = \frac{[H_3O^+][OH^-]}{[H_2O]^2}$$

Since the concentration of water is hardly affected by the slight extent of the reaction, it is essentially constant and is included in an ion product constant of water, K_w.

$$K[H_2O]^2 = K_w = [H_3O^+][OH^-]$$

The value of K_w is 1×10^{-14}, since the $[H_3O^+]$ and $[OH^-]$ are both $1 \times 10^{-7}\,M$.

$$K_w = [H_3O^+][OH^-] = [1 \times 10^{-7}][1 \times 10^{-7}] = 1 \times 10^{-14}$$

When a strong acid such as HCl is dissolved in water, the hydronium ion concentration becomes very large compared to what it is in pure water. For example, 0.01 M HCl, which contains $1 \times 10^{-2}\,M$ H_3O^+, is 100,000 times as concentrated as H_3O^+ in pure water. *We call the solution acidic since it has a higher concentration of hydronium ions than pure water.*

How is the concentration of hydroxide ions affected by the increased concentration of hydronium ions? Le Châtelier's principle gives the answer.

$$2\,H_2O \rightleftharpoons H_3O^+ + OH^-$$

As hydronium ions are added, the hydroxide ions will tend to resist this change by reacting with them. The OH^- concentration then must decrease. For $1 \times 10^{-2}\,M$ H_3O^+ solution, the hydroxide ion concentration may be calculated as follows.

$$K_w = [H_3O^+][OH^-] = 1 \times 10^{-14}$$

$$[1 \times 10^{-2}][OH^-] = 1 \times 10^{-14}$$

$$[OH^-] = \frac{1 \times 10^{-14}}{1 \times 10^{-2}}$$

$$= 1 \times 10^{-12}$$

Thus, the hydroxide ion concentration is reduced from $1 \times 10^{-7}\,M$ in pure water to $1 \times 10^{-12}\,M$ in a 0.01 M HCl solution.

Now let's consider what occurs in a basic solution. If 0.001 mole of sodium hydroxide is added to sufficient water to produce 1 L of solution, the OH^- concentration derived from this completely ionized base is $10^{-3} M$. The H_3O^+ concentrations of the solution can be calculated from the K_w value of water.

$$[H_3O^+][OH^-] = 1 \times 10^{-14}$$

$$[H_3O^+][1 \times 10^{-3}] = 1 \times 10^{-14}$$

$$[H_3O^+] = \frac{1 \times 10^{-14}}{1 \times 10^{-3}}$$

$$= 1 \times 10^{-11} M$$

A basic solution then has a hydronium ion concentration that is lower than that of pure water.

Example 11.3 _____

The hydronium ion concentration of a blood sample is $4.5 \times 10^{-8} M$. What is the value of the hydroxide ion concentration?

Solution. The blood is slightly basic, since the hydronium ion concentration is less than $1 \times 10^{-7} M$. The hydroxide ion concentration must be greater than $1 \times 10^{-7} M$. Substitution into the K_w expression gives the expected answer.

$$[H_3O^+][OH^-] = 1.0 \times 10^{-14}$$

$$(4.5 \times 10^{-8})[OH^-] = 1.0 \times 10^{-14}$$

$$[OH^-] = 2.2 \times 10^{-7} M$$

11.7
The pH Scale

Expressing hydronium ion concentrations in exponential notation is somewhat complex for many common applications involving acids and bases. The pH scale was developed as a more convenient method of expressing hydronium ion concentration. The definition is

$$pH = -\log[H_3O^+]$$

In pure water, the hydronium ion concentration is $1 \times 10^{-7} M$, and the pH is 7.

$$pH = -\log[H_3O^+]$$
$$= -\log(10^{-7})$$
$$= -(-7) = 7$$

At the higher hydronium ion concentrations, that is, in acidic solutions, the pH is smaller. In a 0.01 M HCl solution, the hydronium ion concentration is $1 \times 10^{-2} M$, and the pH is 2.

$$pH = -\log[10^{-2}]$$
$$= -(-2) = 2$$

In basic solutions, the hydronium ion concentrtion is lower than in pure water. For a 0.001 M NaOH solution, the hydroxide ion concentration is $1 \times 10^{-3} M$, and the hydronium ion concentration is $1 \times 10^{-11} M$. The pH is then 11.

$$pH = -\log[10^{-11}]$$
$$= -(-11) = 11$$

Table 11.5

The pH Scale

$[H_3O^+]$	pH	$[OH^-]$	
10^0	0	10^{-14}	
10^{-1}	1	10^{-13}	
10^{-2}	2	10^{-12}	
10^{-3}	3	10^{-11}	Acidic solution
10^{-4}	4	10^{-10}	
10^{-5}	5	10^{-9}	
10^{-6}	6	10^{-8}	
10^{-7}	7	10^{-7}	Neutral solution
10^{-8}	8	10^{-6}	
10^{-9}	9	10^{-5}	
10^{-10}	10	10^{-4}	
10^{-11}	11	10^{-3}	Basic solution
10^{-12}	12	10^{-2}	
10^{-13}	13	10^{-1}	
10^{-14}	14	10^0	

In general, for any hydronium ion concentration of 1×10^{-x} the pH is x. The pH scale is shown in Table 11.5. Note that **acidic solutions have pH values smaller than 7, whereas basic solutions have pH values higher than 7.**

The pH scale in Table 11.5 involves only changes of integer units. Note that such changes involve changes by a factor of 10 in the hydronium ion concentration. Thus, fluctuations of a few pH units actually involve large changes in hydronium ion concentration. A pH change of three units corresponds to a factor of 10^3 or 1000 in altering the hydronium ion concentration.

Dealing with pH for solutions with hydronium ion concentrations that can be written as 1×10^{-x} is fairly simple. Now let's consider a blood sample with a hydronium ion concentration equal to $4.5 \times 10^{-8} M$. What is the pH? In order to calculate the pH, you must know the following mathematical identity of logarithms.

$$\log(n \times p) = \log n + \log p$$

For a solution with $[H_3O^+] = 4.5 \times 10^{-8} M$ we need to know the logarithms of 4.5 and 10^{-8}.

$$\begin{aligned} pH &= -\log [H_3O^+] \\ &= -\log(4.5 \times 10^{-8}) \\ &= -\log 4.5 - \log 10^{-8} \end{aligned}$$

Hand calculators that have a key for base 10 logarithms can be used to calculate pH. Consult your instruction booklet for directions on how to calculate a logarithm on your calculator. The logarithm of 4.5 is 0.65321251 as displayed on a calculator. However, only 0.65 need be used in your calculation.

$$\begin{aligned} pH &= -\log 4.5 - \log 10^{-8} \\ &= -0.65 - (-8) \\ &= 7.35 \end{aligned}$$

The pH of blood is slightly on the basic side, as expected from the hydronium ion concentration.

Example 11.4

The hydronium ion concentration of a sample of urine is $2 \times 10^{-6}\,M$. What is its pH?

Solution. Note that the sample is more acidic than a solution with a hydronium ion concentration of $1 \times 10^{-6}\,M$. Therefore, the pH must be less than 6. Now substitute into the pH formula, and determine the logarithm of 2 on your calculator.

$$\begin{aligned}
\text{pH} &= -\log[\text{H}_2\text{O}^+] \\
&= -\log(2 \times 10^{-6}) \\
&= -\log 2 - \log 10^{-6} \\
&= -0.30 - (-6) \\
&= 5.70
\end{aligned}$$

The value of the pH is somewhat less than 6, as expected.

[Additional examples may be found in **11.39–11.40** at the end of the chapter.]

Measuring pH

A pH measurement is an important procedure in biochemical experiments. Pepsin, a protein-digesting enzyme in the stomach, is most effective in acidic solutions, whereas another protein-digesting enzyme, trypsin, found in the small intestine, is most effective in slightly basic solution. Therefore, studies of kinetics must be done within the pH range in which the enzyme functions.

The pH values of several of our body fluids are given in Table 11.6 Except for gastric juices in which the main acid is HCl, the majority of body fluids have pH values near 7. No body fluid is very basic. This fact is discussed further in connection with buffers (Section 11.8).

Table 11.6
pH Values of Body Fluids

blood	7.35–7.45
gastric juices	1.6–1.8
bile	7.8–8.6
urine	5.5–7.0
saliva	6.2–7.4
interstitial fluid	7.4
muscle intracellular fluid	6.1
liver intracellular fluid	6.9
pancreatic juice	7.8–8.0

The pH meter is an instrument that presents the pH either by a needle reading on a scale or a digital display. The instrument measures the voltage when an electric current passes between electrodes immersed in the solution. Such meters can measure pH to the nearest 0.01 pH unit.

A somewhat older and less accurate method of determining pH uses chemical indicators. Chemical indicators are certain weak acids or bases. Each indica-

Table 11.7

Colors of Indicators

Indicator	Approximate pH at which color changes	Color at lower pH	Color at higher pH
methyl green	1	yellow	blue
thymol blue	2	red	yellow
erythrosine	3	orange	red
congo red	4	blue	red
methyl red	5	red	yellow
alizarin	6	yellow	red
bromothymol blue	7	yellow	blue
α-naphtholphthalein	8	orange-yellow	green-blue
phenolphthalein	9	colorless	red
thymolphthalein	10	colorless	blue
alizarin yellow	11	yellow	red

tor can exist as an acid and its conjugate base. Each member of the conjugate pair is a different color, so that an indicator changes color as the concentrations of the two substances forming the conjugate pair change. For example, phenolphthalein is colorless in acidic solution, where it exists in its acid form, and is pink in basic solution, where it exists in its basic form. The colors of a number of other indicators are listed in Table 11.7 along with the approximate pH at which the color change occurs.

Use of the pH Scale

Although the mathematics of pH is not understood by the general population, there are nevertheless many practical uses of this scale. Because some crops flourish under acidic conditions, while others require more basic conditions, agriculture depends on the proper soil pH. Soil is often tested to determine whether acidic or basic fertilizers are required for a particular crop. Soils in humid areas are acidic, whereas soils in arid areas tend to be basic or neutral. If the soil is too acidic, it can be "limed" by adding calcium carbonate.

$$2\,H^+ + CaCO_3 \longrightarrow Ca^{2+} + H_2O + CO_2$$

When the soil is too basic, ferric sulfate, $Fe_2(SO_4)_3$, is applied to release protons.

$$2\,Fe^{3+} + 3\,SO_4^{2-} + 6\,H_2O \longrightarrow 2\,Fe(OH)_3 + 6\,H^+ + 3\,SO_4^{2-}$$

In Table 11.8 are listed the pH values of a number of common foods. All of those listed are acidic.

Table 11.8

pH of Foods

apples	2.9–3.3
cabbage	5.2–5.4
corn	6.0–6.5
grapes	3.5–4.5
lemons	2.2–2.4
milk	6.3–6.6
oranges	3.0–4.0
peas	5.8–6.4
potatoes	5.6–6.0
tomatoes	4.0–4.4

The term to buffer means to prevent changes or to lessen the shock of changes. In chemistry, *a **buffer** is a solution that will prevent a rapid pH change. A chemical buffer is prepared by dissolving both a weak acid and a salt of its conjugate base in water,* for example, acetic acid and sodium acetate. The buffer will react with any strong acid or base that may be added. Hydronium ions will react with the acetate ions. Hydroxide ions will react with the acetic acid.

11.8
Buffers

$$H_3O^+ + CH_3CO_2^- \longrightarrow H_2O + CH_3CO_2H$$

$$OH^- + CH_3CO_2H \longrightarrow H_2O + CH_3CO_2^-$$

Thus, either H_3O^+ or OH^-, which could drastically change the pH, is neutralized by reaction with a component of the buffer.

Buffers may be prepared from any ratio of concentrations of weak acids and the salt of the weak acid. A 1:1 ratio of acid to salt is the most efficient for handling the addition of either a base or an acid. If the buffer contains much more acid than the conjugate base, it will be less efficient in handling an acid. Alternatively, a buffer with much more of the conjugate base than of acid cannot efficiently counteract the addition of a base. For example, the H_2CO_3/HCO_3^- buffer in blood has a 1:20 ratio of acid to salt. Blood is then more effective in counteracting acids, which are the products of metabolism.

A variety of pH values may be achieved for various ratios of a weak acid and the salt of the weak acid. A relationship that gives the pH in terms of the acid ionization constant and the two components of the buffer can be derived as follows.

$$K_a = \frac{[H_3O^+][A^-]}{[HA]}$$

$$[H_3O^+] = K_a \frac{[HA]}{[A^-]}$$

$$\log[H_3O^+] = \log K_a + \log \frac{[HA]}{[A^-]}$$

$$pH = -\log K_a + \log \frac{[A^-]}{[HA]}$$

This equation will be further developed in Chapter 18 on carboxylic acids and in Chapter 23 on amino acids.

11.9
Buffers in the Body

Normal metabolic reactions in muscles as a result of exercise continuously produce a variety of acids. One such acid is lactic acid.

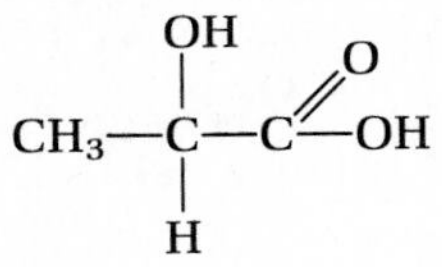

$$CH_3-\overset{\overset{\displaystyle OH}{|}}{\underset{\underset{\displaystyle H}{|}}{C}}-\overset{\overset{\displaystyle O}{\diagup\!\!\!\diagup}}{C}-OH$$

On the average, 10 moles of a variety of acids are produced in your body each day.

There are two mechanisms for decreasing the concentration of an acid resulting from metabolic reactions. One is by breathing and expelling carbon dioxide formed from carbonic acid. The second is by excretion of acid in the urine (Table 11.6) Both mechanisms depend on buffers. The carbonate buffer, H_2CO_3/HCO_3^-, is present in the blood, whereas the phosphate buffer, $H_2PO_4^-/HPO_4^{2-}$, is involved in kidney functions.

Blood has a normal pH range of 7.35–7.45, which corresponds to hydronium ion concentrations of 4.5×10^{-8} to 3.6×10^{-8} mole/L. Any acids formed in the blood react with bicarbonate ion to give carbonic acid, which then decomposes to give carbon dioxide.

$$HCO_3^- + H_3O^+ \rightleftharpoons H_2CO_3 + H_2O$$

$$H_2CO_3 \rightleftharpoons H_2O + CO_2$$

The carbon dioxide is removed from the blood by the lungs and exhaled.

Although it is a less common occurrence, the bicarbonate/carbonic acid buffer can prevent an increase in base concentration. The carbonic acid can react to neutralize a base.

$$H_2CO_3 + {}^-OH \rightleftharpoons HCO_3^- + H_2O$$

However, the capacity of the buffer to handle a base is much more limited than its capacity to neutralize acid because the $[H_2CO_3]/[HCO_3^-]$ ratio is about 0.05.

The $H_2PO_4^-/HPO_4^{2-}$ buffer is important within cells. Many reactions in cells involve complex compounds with covalently bonded phosphate groups. The proper pH in cellular fluids is maintained by the reaction of a base with $H_2PO_4^-$ and the reaction of an acid with HPO_4^{2-}.

$$H_2PO_4^- + {}^-OH \rightleftharpoons HPO_4^{2-} + H_2O$$

$$HPO_4^{2-} + H_3O^+ \rightleftharpoons H_2PO_4^- + H_2O$$

However, the formation of an acid is the more common result of metabolism, and HPO_4^{2-} is the more important component of the buffer. The normal $[H_2PO_4^-]/[HPO_4^{2-}]$ ratio in the cell is about $1:4$, which allows the buffer to neutralize acid more efficiently than base. The $H_2PO_4^-$ formed from the reaction of HPO_4^{2-} with acid is then eliminated by excretion in the urine.

11.10 Acidosis and Alkalosis

When the blood pH is lower than 7.35 ($[H_3O^+] > 4.5 \times 10^{-8} M$) for a significant period of time, the condition is known as **acidosis.** The condition results from (a) excessive production of acid, (b) decreased concentration of HCO_3^-, or (c) increased concentration of H_2CO_3. Note that according to Le Châtelier's principle, either increasing the $[H_2CO_3]$ or decreasing the $[HCO_3^-]$ results in increased $[H_3O^+]$.

$$H_2CO_3 + H_2O \rightleftharpoons HCO_3^- + H_3O^+$$

Increased acid production from metabolic reactions, called **metabolic acidosis,** occurs both in diabetes mellitus and in some low carbohydrate–high protein diets. Normal metabolism of glucose (a carbohydrate) produces acids that the normal buffer action of the body can handle. Lack of insulin prevents glucose metabolism, and glucose is excreted in the urine. The body metabolizes more of its stored fats, which yields larger amounts of substances such as acetoacetic acid and β-hydroxybutyric acid, known as ketone bodies.

acetoacetic acid β-hydroxybutyric acid

The continued formation of these acids and the excretion of them as salts increases the hydronium ion concentration and affects the acid–base balance. Coma and death can result if the diabetic is not treated.

Of course, increased ingestion of acids will also tend to increase the hydronium ion concentration. Similarly, kidney failure will cause an increase in $[H_3O^+]$ as the acid is not excreted in the required amount to maintain the acid–base balance.

The converse, decreased concentration of bicarbonate, is unusual, but does occur in some cases of severe diarrhea.

The specific condition of decreased blood pH as a result of increased carbonic acid is called **respiratory acidosis.** Several conditions that cause a reduced respiration rate lead to the increased concentration of H_2CO_3, for it is not then eliminated as CO_2. Emphysema, pneumonia, poliomyelitis, anesthesia, and heart failure all cause an increase in carbonic acid.

Temporary acidosis results from holding your breath for a long period of time. As your body continues to produce acid, the result is a decreased blood pH. Only by exhalation of carbon dioxide can the pH be returned to normal. The nervous system forces you to breathe, and you cannot voluntarily hold your breath for too long.

When the blood pH is higher than 7.45 ($[H_3O^+] < 3.5 \times 10^{-8} M$) for a significant period of time, the condition is known as **alkalosis.** Although less common than acidosis, alkalosis can result from the ingestion of bases such as antacids or from the loss of stomach acid through severe vomiting. In such cases, the kidneys excrete alkaline urine containing Na_2HPO_4. Remember that the normal excretion process involves removal of excess acid in the form of NaH_2PO_4.

Excessive loss of carbonic acid results in a condition known as **respiratory alkalosis.** This condition results from hyperventilation, which may occur in extreme fevers or severe hysteria. Hyperventilation can increase the pH to 7.6 in a few minutes. The breakdown of H_2CO_3 and expulsion as CO_2 result in a shift of the H_2CO_3/HCO_3^- equilibrium to decrease the $[H_3O^+]$ or increase the pH. In order to compensate, the body tends to lower the respiration rate. In some cases of hysteria, fainting results and the respiration rate is thus lowered.

11.11
Normality

In Chapter 9, it was pointed out that there are numerous units in which concentration can be expressed. For acid–base concentrations, normality is appropriate. **Normality (N)** *is the number of equivalents of acid or base per liter of solution.*

$$N = \frac{\text{equivalents}}{\text{liter of solution}}$$

An **equivalent** *is equal to the weight of an acid containing one mole of protons. An equivalent of a base is that weight which will react with one mole of protons sup-*

Table 11.9

Equivalent Weights of Acids

Compound	Molecular weight	Equivalent weight
HCl	36.5	$\frac{36.5}{1} = 36.5$
H_2SO_4	98.0	$\frac{98.0}{2} = 49.0$
H_3PO_4	98.0	$\frac{98.0}{3} = 32.7$
NaOH	40.0	$\frac{40.0}{1} = 40.0$
$Ca(OH)_2$	74.0	$\frac{74.0}{2} = 37.0$
$Al(OH)_3$	78.0	$\frac{78.0}{3} = 26.0$

plied by an acid. The equivalent weight of an acid is calculated by dividing the molecular weight of the acid by the number of moles of hydrogen atoms contained in one mole of the acid. Similarly, the equivalent weight of a base is calculated by dividing the molecular weight of the base by the number of moles of hydrogen atoms that will react with one mole of the base. A listing of equivalent weights of several acids and bases is given in Table 11.9.

The advantage of the normality unit is that one equivalent of an acid will react with one equivalent of a base. If molarity is used, a factor must always be applied before determining whether equal numbers of equivalents of acid or base are present for neutralization. For example, while 1 L of a 1 M solution of HCl will neutralize 1 L of a 1 M solution of NaOH, a 1 M solution of H_2SO_4 contains twice as much hydrogen and will require 2 L of a 1 M NaOH solution. However, 1 L of a 1 N solution of any acid will neutralize 1 L of a 1 N solution of any base.

Example 11.5

What is the normality of a solution of H_3PO_4 obtained by dissolving 49.0 g of the acid in sufficient water to produce 3.00 L of solution?

Solution. First the equivalent weight of H_3PO_4 must be used to determine how many equivalents are present in 49.0 g.

molecular weight of H_3PO_4 = 98.0 g

equivalent weight of H_3PO_4 = 98.0 g ÷ 3 = 32.7 g

$$49.0 \, \cancel{g} \times \frac{1 \text{ equivalent}}{32.7 \, \cancel{g}} = 1.50 \text{ equivalents}$$

Now the normality may be calculated.

$$\frac{1.50 \text{ equivalents}}{3.00 \text{ L}} = 0.500 \, N$$

Example 11.6

Calculate the number of grams of H_2SO_4 present in 100 mL of a 0.20 N H_2SO_4 solution.

Solution. First the number of equivalents of H_2SO_4 in the solution must be calculated.

$$\frac{0.20 \text{ equivalent}}{1 \, \cancel{L}} \times 0.100 \, \cancel{L} = 0.020 \text{ equivalent}$$

The equivalent weight of H_2SO_4 is one-half the molecular weight.

$$\frac{98.0 \text{ g}}{\cancel{\text{mole}}} \times \frac{1 \, \cancel{\text{mole}}}{2 \text{ equivalents}} = \frac{49.0 \text{ g}}{\text{equivalent}}$$

Now the number of grams in the sample may be calculated.

$$0.020 \, \cancel{\text{equivalent}} \times \frac{49.0 \text{ g}}{\cancel{\text{equivalent}}} = 0.98 \text{ g}$$

Example 11.7

What is the molarity of a 0.020 N solution of $Ca(OH)_2$?

Solution. In a mole of $Ca(OH)_2$ there are two equivalents.

$$\frac{1 \text{ mole } Ca(OH)_2}{2 \text{ equivalents } Ca(OH)_2}$$

Therefore, the definition of normality can be related to molarity by the use of this factor.

$$\frac{0.02 \cancel{\text{ equivalent}}}{1 \text{ L}} \times \frac{1 \text{ mole}}{2 \cancel{\text{ equivalents}}} = \frac{0.01 \text{ mole}}{1 \text{ L}} = 0.01 \text{ M}$$

[Additional examples may be found in **11.50–11.52** at the end of the chapter.]

11.12
Titration

Titration *is a procedure in which the concentration or amount of an acid or base is determined by reacting a sample of it with a known amount of a base or acid to achieve neutralization.* The point of neutralization is known as the endpoint or neutralization point. The endpoint is detected by using an indicator.

In a titration, a measured volume or mass of an acid or base, whose quantity is the unknown, is placed in a flask and a small amount of an indicator is added. The indicator chosen must undergo a change in color at the pH of the solution of the products of the titration. A solution of a base or acid of known concentration is then added dropwise from a buret. When the indicator changes color, the addition is terminated and the volume of solution added from the buret is recorded (Figure 11.4).

The number of equivalents of base or acid added from the buret is calculated by multiplying the volume used by its normality. This number of equivalents must be equal to the number of equivalents of the unknown because one equivalent of an acid is required to neutralize one equivalent of a base.

Frequently, in a laboratory, the volumes of solutions used are not large. Therefore, the units of equivalents and liters are not convenient. For this reason, normality can also be regarded as the number of milliequivalents per milliliter.

$$N = \frac{\text{milliequivalents}}{\text{milliliter}} = \frac{\text{meq}}{\text{mL}}$$

A milliequivalent is 1/1000 of an equivalent and is abbreviated meq.

Example 11.8

A 5.00 mL sample of household ammonia is titrated with 48.0 mL of a 0.200 N solution of hydrochloric acid to achieve a methyl red endpoint. What is the molarity of the household ammonia?

Solution. The acid solution contains 0.200 meq of hydrochloric acid per milliliter. Thus the number of milliequivalents of acid used is

$$\frac{0.200 \text{ meq}}{\cancel{\text{mL}}} \times 48.0 \cancel{\text{ mL}} = 9.60 \text{ meq}$$

Figure 11.4

Titration.

(a) A solution of an unknown concentration of an acid (or a base) is placed in a flask. (b) A titrating solution of a known concentration of a base (or an acid) is added dropwise from a buret. (c) The titration is stopped when the amount of acid and base are equal, as shown by a change in the color of the indicator.

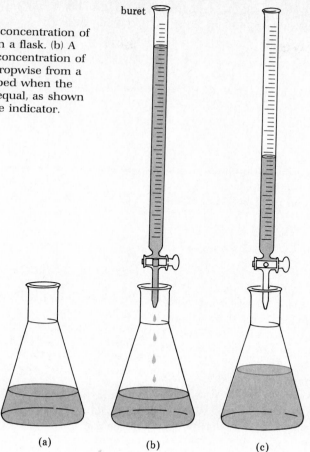

buret

(a)   (b)   (c)

The number of milliequivalents of ammonia must also be equal to 9.60. This quantity is contained in 5.00 mL of the ammonia solution, and the normality must be

$$\frac{9.60 \text{ meq}}{5.00 \text{ mL}} = 1.92 \, N$$

For ammonia, which only reacts with one equivalent of acid per mole, the equivalent weight is the same as the molecular weight. Therefore, the molarity is the same as the normality, or $1.92 \, M$.

Example 11.9

Baking soda, which is sodium hydrogen carbonate, may be used to establish the normality of an acid solution such as HCl by the following reaction.

$$NaHCO_3 + HCl \longrightarrow NaCl + CO_2 + H_2O$$

A 0.0420 g sample of sodium bicarbonate is dissolved in water in a flask, and 20.0 mL of a hydrochloric acid solution is used to neutralize it. What is the normality of the HCl solution?

Solution. The equivalent weight of $NaHCO_3$ is 84.0. A milliequivalent will have a mass of 0.0840 g. Thus, the number of milliequivalents in the sample is

$$\frac{0.0420 \text{ g}}{0.0840 \text{ g/meq}} = 0.500 \text{ meq}$$

The number of milliequivalents in the 20.0 mL of HCl must also be 0.500 meq. Thus, the normality is

$$\frac{0.500 \text{ meq}}{20.0 \text{ mL}} = 0.025 \ N$$

[Additional examples may be found in **11.54–11.57** at the end of the chapter.]

Summary

The Brønsted–Lowry concept of acid–base reactions focuses on proton transfer from an acid to a base. The tendency of acids and bases to lose and gain protons, respectively, determines the direction of an acid–base reaction.

Water undergoes self-ionization to produce small concentrations of hydronium and hydroxide ions. The addition of acids or bases to water affects the concentrations of both hydronium ions and hydroxide ions in a manner predicted by Le Châtelier's principle. The concentration of hydronium ions may be expressed in pH units.

Acid strength refers to the degree of ionization of an acid. Strong acids are completely ionized. An equilibrium exists between weak acids and their conjugate bases. The acid ionization constant describes the degree of ionization.

Buffers have components, a weak acid and a salt of its conjugate base, that will react with added amounts of either a base or an acid. The effectiveness of a buffer against an added acid or base depends on the ratio of the concentrations of the two components of the buffer. Buffers play a critical role in controlling the pH of fluids in living organisms.

Concentrations of acids or bases may be expressed in normality units. This concentration indicates the number of equivalents of an acid or base in a liter of solution.

Titration is a procedure in which an acid and a base reach a point of neutralization. An acid–base indicator is used in a titration.

Learning Objectives

As a result of studying Chapter 11 you should be able to

- Identify strong and weak acids, identify strong and weak bases.
- Identify conjugate acid–base pairs according to the Brønsted–Lowry concept.
- Calculate the pH of a solution given the hydronium or hydroxide ion concentration.
- Explain the action of a chemical buffer.
- Calculate the amount of an acid or base present in a sample from titration data.
- Calculate the normality of an acid or base solution.
- Explain the action of buffers in the body.

New Terms

An **acid** is a proton donor.

The **acid ionization constant** is the equilibrium constant for the ionization of a weak acid.

Acidosis is a condition when the pH of blood is lower than 7.35 for a significant period of time.

Alkalosis is a condition when the pH of blood is higher than 7.45 for a significant period of time.

An **amphoteric** substance can act as either an acid or a base.

A **base** is a proton acceptor.

A **buffer** is a solution that resists a change in pH and consists of a weak acid and a salt of its conjugate base.

A **conjugate acid** is formed when a base gains a proton.

A **conjugate base** is formed when an acid donates a proton.

An **equivalent** of an acid contains one mole of protons.

The **hydronium ion** is the principal form in which protons are found in aqueous solution.

Ionization of an acid involves transfer of a proton to water.

A **neutralization reaction** is the reaction of hydronium ions with hydroxide ions.

The **normality** of a solution indicates the number of equivalents of acid or base in a liter of solution.

The **pH** of a solution is a logarithmic expression of the hydronium ion concentration.

A **strong acid** essentially completely transfers its proton to water.

A **strong base** completely removes the proton of an acid.

Titration is a procedure to determine the amount of an acid or base by neutralization.

A **weak acid** is only partially ionized in water.

Questions and Problems

Terminology

11.1 Distinguish between each pair of terms.
(a) acid and base
(b) hydroxide ion and hydronium ion
(c) strong acid and weak acid
(d) molarity and normality
(e) acidosis and alkalosis

11.2 Describe the relationship between each of the following pairs.
(a) an acid and its conjugate base
(b) a base and its conjugate acid

Properties of Acids and Bases

11.3 Explain why acids such as HCl are electrolytes in aqueous solution.

11.4 Explain why bases such as NaOH are electrolytes in aqueous solution.

11.5 What is the difference between hydrogen chloride and hydrochloric acid?

11.6 How can you distinguish between acids and bases by using litmus paper?

11.7 What precautions should you take when working with acids and bases in the laboratory?

11.8 Why should you wear eye protection when working with drain cleaner or oven cleaner?

Common Acids and Bases

11.9 What is the molecular formula for each of the following acids and bases?
(a) phosphoric acid
(b) acetic acid
(c) sulfuric acid
(d) calcium hydroxide
(e) potassium hydroxide
(f) aluminum hydroxide

11.10 Glucose, $C_6H_{12}O_6$, contains hydrogen, but it is not an acid. What test could you use to confirm this fact?

11.11 Sodium bicarbonate, $NaHCO_3$, is often kept in laboratories to place on acid or base spills. What reaction occurs when an acid reacts with $NaHCO_3$? What reaction occurs when a base reacts with $NaHCO_3$?

11.12 Name one acid and one base that have wide industrial uses.

Brønsted–Lowry Theory

11.13 Classify each of the following as a monoprotic, diprotic, or triprotic acid.
(a) $HClO_4$
(b) H_3PO_4
(c) HNO_3
(d) H_2SO_4
(e) H_2CO_3
(f) CH_3CO_2H

11.14 Can a substance act as either an acid or a base? Explain, giving examples.

11.15 What is meant by the term amphoterism?

11.16 Write equations for the stepwise ionization of H_2CO_3.

11.17 Write equations for the stepwise ionization of H_3PO_4.

Conjugate Acids and Bases

11.18 What is the conjugate acid of each of the following?
(a) ClO_4^-
(b) Cl^-
(c) NO_3^-
(d) CO_3^{2-}
(e) $CH_3CO_2^-$
(f) HSO_4^-
(g) NH_3
(h) H_2O
(i) HCO_3^-

11.19 What is the conjugate base of each of the following?
(a) H_2O
(b) HCl
(c) HCO_3^-
(d) H_2CO_3
(e) HNO_3
(f) H_2SO_4
(g) HSO_4^-
(h) H_3PO_4
(i) $H_2PO_4^-$
(j) HPO_4^{2-}
(k) CH_3CO_2H

11.20 Pyruvic acid, an acid produced in metabolic reactions, has the following structure. What is the molecular formula of the conjugate base? What is the charge of this ion?

11.21 Methylamine is an organic base. From its structure write the conjugate acid. What is the charge of the ion?

11.22 Aspartic acid is a diprotic amino acid with the molecular formula $C_4H_7NO_4$. What is the molecular formula and charge of the ion that results from the loss of two protons?

Acid–Base Strength

11.23 Ammonia is very soluble in water. Why is a 27% solution of ammonia still considered to be a solution of a weak base?

11.24 A 0.1 M solution of HCl is a solution of a strong acid. Why is the term strong used for a solution of such low concentration?

11.25 Fill in the blanks in each of the following statements:
  (a) A strong acid has a _____ conjugate base.
  (b) A strong base has a _____ conjugate acid.
  (c) A weak acid has a _____ conjugate base.
  (d) A weak base has a _____ conjugate acid.

11.26 Classify each of the following as a weak or strong acid.
  (a) CH_3CO_2H (b) HCl   (c) HNO_3
  (d) HCN   (e) H_2SO_4 (f) H_2O

11.27 Classify each of the following as a weak or strong base.
  (a) NH_3 (b) NaOH   (c) H_2O (d) $Mg(OH)_2$

11.28 Vinegar is about 0.8 M acetic acid. Could 0.8 M HCl be used as a substitute?

Acid Ionization Constant

11.29 Write the expression for the acid ionization constant of the weak acid HCN.

11.30 Oxalic acid has the following structure. Write equilibrium reactions and acid ionization constant expressions for this diprotic acid.

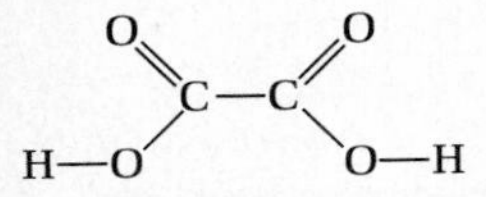

11.31 Write an acid ionization constant expression for each ionization step of the triprotic acid H_3PO_4.

11.32 A 0.040 M solution of HF is 12% ionized. Calculate K_a.

11.33 A 0.10 M ammonia solution has an equilibrium NH_4^+ concentration equal to 0.0013 M. Calculate K for the following equilibrium.

$$NH_3 + H_2O \rightleftharpoons NH_4^+ + OH^-$$

11.34 Urea, a metabolic product found in urine, is a weak base. The concentrations of urea and its conjugate base are $4.0 \times 10^{-1} M$ and $4.0 \times 10^{-8} M$, respectively, in a solution whose hydronium ion concentration is $7.6 \times 10^{-8} M$. Calculate the equilibrium constant for the base.

Ionization of Water

11.35 Explain why water is not acidic or basic.

11.36 What are the hydronium ion and hydroxide ion concentrations in each of the following solutions?
  (a) 0.1 M HCl   (b) $10^{-2} M$ HNO_3
  (c) $10^{-3} M$ KOH   (d) 0.05 M H_2SO_4
  (e) $10^{-4} M$ NaOH   (f) 0.005 M $Mg(OH)_2$

11.37 Explain what occurs to the concentration of hydronium ions in water when solid sodium hydroxide is added.

11.38 Explain what occurs to the concentration of hydroxide ions when gaseous hydrogen chloride is bubbled into water.

pH Values

11.39 Calculate the pH of the following solutions.
  (a) $10^{-3} M$ HCl   (b) $2 \times 10^{-3} M$ HNO_3
  (c) $10^{-2} M$ NaOH   (d) $3 \times 10^{-3} M$ NaOH

11.40 Calculate the pH of the following items.
  (a) a soft drink with $[H_3O^+] = 2 \times 10^{-4} M$
  (b) milk with $[H_3O^+] = 2 \times 10^{-7} M$
  (c) ammonia with $[H_3O^+] = 2 \times 10^{-12} M$
  (d) gastric juice with $[H_3O^+] = 0.1 M$

11.41 Some natural fruit juices change color when a base is added. Explain why.

11.42 How does an indicator work?

Buffers

11.43 Explain how a buffer works.

11.44 Could a solution of HCl and NaCl be a buffer?

11.45 Consider the structure for the amino acid glycine. How could it be a buffer?

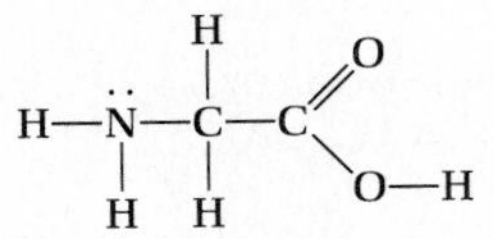

11.46 How does the acid/conjugate base ratio affect each of the following?
  (a) the pH of the buffer
  (b) the effectiveness of the buffer toward acid or base

Acidosis and Alkalosis

11.47 Describe the effects that hyperventilation and hypoventilation have on blood pH.

11.48 What two processes are used by the body to eliminate the acids that form in normal metabolic reactions?

11.49 How do antacids work?

Normality

11.50 Why is normality rather than molarity more commonly used for expressing the concentrations of acids and bases?

11.51 How many milliequivalents of acid are present in each of the following solutions?
  (a) 50 mL of 0.01 N HCl
  (b) 5 mL of 0.1 N HNO_3
  (c) 1 L of 1 N HCl

11.52 What is the normality of each solution?
  (a) 4.0 g NaOH in 500 mL of solution
  (b) 3.65 g HCl in 1 L of solution
  (c) 4.9 g H_2SO_4 in 100 mL of solution
  (d) 0.327 g H_3PO_4 in 10 mL of solution
  (e) 0.0037 g $Ca(OH)_2$ in 500 mL of solution

Titration

11.53 What is titration? How is an indicator used in a titration?

11.54 What quantity of $0.10\,N$ NaOH is required to titrate 25.0 mL of a $0.060\,N$ HCl solution?

11.55 A 50.0 mL solution of $Ca(OH)_2$ is titrated with 36.0 mL of a $0.05\,N$ HCl solution to an endpoint. What is the molarity of the $Ca(OH)_2$ solution?

11.56 A $0.1060\,g$ sample of Na_2CO_3 is neutralized with 48.0 mL of a $HClO_4$ solution. What is the normality of the $HClO_4$ solution?

11.57 A $0.100\,g$ sample of NaOH is titrated to an endpoint with 12.5 mL of HCl. What is the normality of the HCl?

Nuclear Chemistry

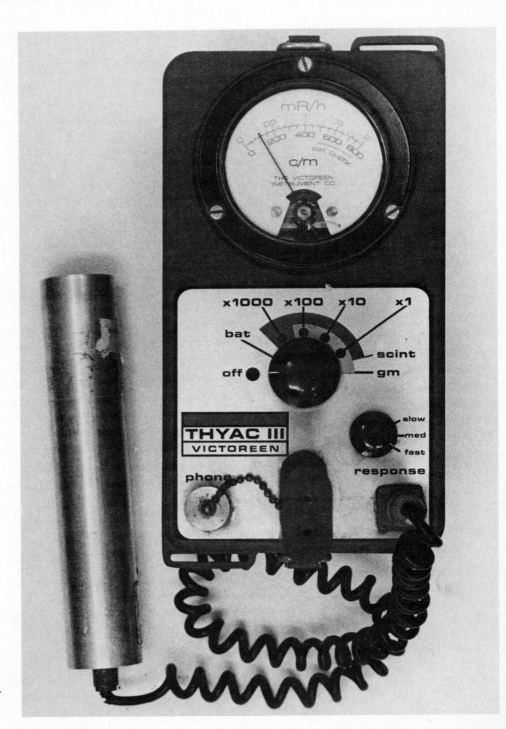

A Geiger counter may be used to monitor radiation. The cylinder at the left contains a gas that can be ionized by radiation.

Up to this point, the elements have been discussed in terms of the number and arrangement of electrons about the nucleus of an atom. In ordinary chemical reactions, changes in the arrangement of the electrons occur in the reactants to yield products. The energy changes in these chemical reactions are in the order of kilocalories.

Now we will examine reactions that involve the protons and neutrons in the nucleus of the atom. The energy required to make a stable nucleus undergo a change is extremely high and cannot be achieved by ordinary chemical reactions. Furthermore, when unstable nuclei of radioactive isotopes react in high concentrations, a rapid chain reaction or atomic explosion occurs. The energy released in an atomic explosion is beyond the comprehension of anyone who has not viewed one.

Today's generation of college students has grown up in a nuclear age in which the chemistry of the nucleus is part of modern life. On the one hand, we live in fear of a nuclear holocaust, and on the other, we enjoy many benefits of radioisotopes.

The atomic bombs that were used to devastate two Japanese cities and end World War II were small compared to the hydrogen bombs of today. This aspect of nuclear chemistry continues as the major political powers try to control their respective arms and to prevent the further spread of nuclear devices to smaller countries.

Nuclear energy for peaceful uses, such as the generation of power, is a continuing and controversial story. The world is limited in its supply of fossil fuels, and nuclear energy is currently being used as an alternative for the generation of electricity. However, there is much concern about the radioactivity hazards and environmental side effects as the number of nuclear power plants increases.

One major benefit from nuclear chemistry is the development of radioisotopes for purposes other than war or energy production, particularly for uses in medicine. In this chapter the principles of nuclear chemistry are presented and the practical aspects of nuclear processes are outlined.

12.1
Nuclear Chemistry—A Perspective

Recall from Section 3.4 that isotopes of an element have the same number of protons but different numbers of neutrons. In the general symbolism, $^A_Z E$, Z is the atomic number or number of protons, and A is the mass number or the sum of the numbers of protons and neutrons. In this chapter we will also use the symbol Na-24 or sodium-24. The number following the elemental symbol or element name is the mass number of the isotope.

Some elements exist as only one or two isotopes, whereas others have many more isotopes. Fluorine exists only as fluorine-19, whereas tin exists in 10 isotopic forms. As a result of multiple isotopes of the elements, there are over 300 naturally occurring isotopes. Table 12.1 gives a few of the elements and their isotopes.

Some naturally occurring isotopes contain nuclei that are stable for indefinite periods of time, whereas other isotopes undergo nuclear decay reactions. *The reaction nucleus or parent nucleus decays to form another nucleus known as the* **daughter nucleus.** In this process some high-energy particles collectively known as radiation are formed. The daughter nucleus formed is more stable than the original nucleus. However, the daughter nucleus itself may undergo further decay reactions. Thus, a whole series of decay processes may occur before a very stable nucleus is ultimately formed.

12.2
Radioactivity

Table 12.1

Naturally Occurring Isotopes of Some Selected Elements

Element	Isotope	Percent abundance	Element	Isotope	Percent abundance
neon	$^{20}_{10}$Ne	90.92	selenium	$^{74}_{34}$Se	0.87
	$^{21}_{10}$Ne	0.26		$^{76}_{34}$Se	9.02
	$^{22}_{10}$Ne	8.82		$^{77}_{34}$Se	7.58
silicon	$^{28}_{14}$Si	92.21		$^{78}_{34}$Se	25.52
	$^{29}_{14}$Si	4.70		$^{80}_{34}$Se	49.82
	$^{30}_{14}$Si	3.09		$^{82}_{34}$Se	9.19
potassium	$^{39}_{19}$K	93.10	tin	$^{112}_{50}$Sn	0.96
	$^{40}_{19}$K	0.01		$^{114}_{50}$Sn	0.66
	$^{41}_{19}$K	6.89		$^{115}_{50}$Sn	0.35
zinc	$^{64}_{30}$Zn	48.89		$^{116}_{50}$Sn	14.30
	$^{66}_{30}$Zn	27.81		$^{117}_{50}$Sn	7.61
	$^{67}_{30}$Zn	4.11		$^{118}_{50}$Sn	24.03
	$^{68}_{30}$Zn	18.57		$^{119}_{50}$Sn	8.58
	$^{70}_{30}$Zn	0.62		$^{120}_{50}$Sn	32.85
				$^{122}_{50}$Sn	4.92
				$^{124}_{50}$Sn	5.94

Isotopes that undergo nuclear change and produce radiation are called **radioisotopes.** Some radioisotopes occur in nature. In addition, many radioisotopes have been produced by nuclear reactions in nuclear laboratories. Regardless of the source of radioactive substances, radiation can cause extensive biological damage. However, under controlled conditions radioisotopes are used for medical diagnosis and therapy.

The three common types of nuclear radiation emitted from radioactive isotopes are alpha (α), beta (β), and gamma (γ). A radioactive isotope does not emit all three types of radiation in a single process. However, gamma radiation may be produced simultaneously with either alpha or beta radiation. A single sample of radioactive material may emit all three types of radiation, if a succession of several radioactive processes is occurring.

Alpha Particles

An **alpha particle** *is a helium nucleus, which consists of two protons and two neutrons*, and may be represented as $^4_2$He^{2+}. However, the $+2$ charge is ordinarily omitted in writing balanced nuclear equations. The nuclear equation deals only with the content of the nucleus.

When alpha particles are emitted from a nucleus of an atom, they travel at approximately $\frac{1}{10}$ the speed of light, or 18,600 miles per second. These particles have little penetrating power and can be stopped by this sheet of paper (Figure 12.1). A 0.05 mm layer of dead cells on the surface of skin will stop alpha particles. However, an intense dose of alpha radiation can cause skin burns. If radioactive dust particles get inside the body, the alpha particles emitted can affect normal cells and biological damage will occur.

Beta Particles

Beta particles *are high-energy electrons produced within a radioactive nucleus.* Although the electrons do not exist as such in the nucleus, they are pro-

Figure 12.1
An Illustration of the
Penetration Power of
Radiation.

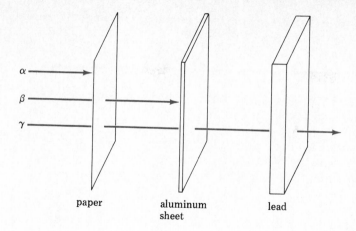

paper   aluminum   lead
               sheet

duced during a process in which a neutron is transformed into a proton and a beta particle.

$$\text{neutron} \longrightarrow \text{proton} + \text{electron}$$

$$^1_0 n \longrightarrow {}^1_1 H + {}^0_{-1} e$$

A beta particle is written $^0_{-1}e$ in nuclear equations. The superscript corresponds to the position of the atomic mass in elements. Since the mass of an electron is very much smaller than that of any element, it is assigned a value of zero. The -1 subscript occupies the position of the atomic number of an element. Since the atomic number is equal to the charge on the nucleus, the -1 value for the electron emphasizes its charge.

Beta particles travel at $\frac{9}{10}$ the speed of light, or about 165,000 miles per second. While beta particles have greater penetrating power than alpha particles, they are stopped by thin sheets of metal (Figure 12.1). Beta particles can penetrate about 4 mm into living tissue. While radiation burns result, the vital internal organs are not affected unless a radioactive source is ingested.

Gamma Rays

Gamma rays do not consist of particles, but *are high-energy radiation similar to x-rays*. When gamma rays are emitted by a nucleus, they travel at the speed of light and have high penetrating power. Even 50 cm of tissue will only reduce the intensity of gamma radiation by 10%. A 5 cm sheet of lead will not stop all gamma rays (Figure 12.1).

Unlike ordinary chemical reactions, nuclear reactions lead to changes in the composition of the nucleus. Thus, both atomic numbers and mass numbers must be included with the elemental symbol in a balanced nuclear equation. In a balanced nuclear equation, the sum of the mass numbers of the reactants must equal the sum of the mass numbers of the products. Furthermore, the sum of the atomic numbers of reactants and products must also be equal.

The proper symbols for alpha and beta particles are also included in balanced nuclear equations. Gamma rays do not have any mass or charge and do not have to be used to balance the equation.

Several examples of approaches to balancing nuclear equations follow. The

12.3
Balancing Nuclear Equations

only tools necessary for balancing are simple arithmetic and a list of elements and their atomic numbers.

Example 12.1

The isotope polonium-212 undergoes alpha decay to yield one alpha particle per atom and a single element. What is the element?

Solution. First, write the symbol for polonium-212 to the left of an arrow using the information from a list of elements that its atomic number is 84. Place ^4_2He on the right of the arrow along with ^A_ZE to represent the unknown element.

$$^{212}_{84}\text{Po} \longrightarrow {}^4_2\text{He} + {}^A_Z\text{E}$$

The sum of the atomic numbers of the products must equal that of the single reactant, polonium.

$$84 = 2 + Z$$
$$82 = Z$$

Similarly, the sum of the mass numbers of the products must equal the mass of polonium.

$$212 = 4 + A$$
$$208 = A$$

Now it is established that the element has an atomic number of 82 and a mass number of 208. In the list of elements one finds that lead has an atomic number of 82; therefore, the element is lead-208.

$$^A_Z\text{E} = {}^{208}_{82}\text{Pb}$$

The balanced equation is

$$^{212}_{84}\text{Po} \longrightarrow {}^4_2\text{He} + {}^{208}_{82}\text{Pb}$$

Example 12.2

The isotope carbon-14 is unstable and emits a beta particle per atom. What single element is produced?

Solution. First write the unbalanced equation.

$$^{14}_6\text{C} \longrightarrow {}^0_{-1}\text{e} + {}^A_Z\text{E}$$

The atomic number must be given by the equation

$$6 = -1 + Z$$
$$7 = Z$$

The mass number of the element is equal to the mass number of carbon, as the beta particle has a negligible mass.

$$14 = 0 + A$$
$$14 = A$$

The element is nitrogen, as its atomic number is 7.

$$^{14}_{7}E = {}^{14}_{7}N$$

The balanced equation is

$$^{14}_{6}C \longrightarrow {}^{14}_{7}N + {}^{0}_{-1}e$$

[Additional examples may be found in **12.7–12.13** at the end of the chapter.]

All radioactive decay reactions occur at rates characteristic for the specific parent nucleus. The rate is described in terms of the time necessary for the parent to decay to give a daughter nucleus. *The time required for one-half of a given amount of radioisotope to decay is called its* **half-life.**

The half-life of uranium-238 is 4.5×10^9 yr! This means that a 100 g sample of uranium-238 obtained now will require 4.5×10^9 yr to be converted to 50 g of the isotope. The mass that is "missing" is actually in the decay products and the daughter nucleus. In another 4.5×10^9 yr, the 50 g of the isotope will have decreased to one-half of that value or 25 g. A list of half-life values is given in Table 12.2.

12.4
Half-lives

Table 12.2

Half-life Periods of Selected Naturally Occurring Isotopes

Element	Isotope	Half-life (yr)
potassium	$^{40}_{19}K$	1.3×10^9
vanadium	$^{50}_{23}V$	6.0×10^{14}
samarium	$^{147}_{62}Sm$	1.1×10^{11}
samarium	$^{148}_{62}Sm$	1.2×10^{13}
samarium	$^{149}_{62}Sm$	4.0×10^{14}
lead	$^{204}_{82}Pb$	1.4×10^{19}
radium	$^{226}_{88}Ra$	1.6×10^{19}

A graph that illustrates the decay pattern of radioactive materials is given in Figure 12.2. The isotope strontium-90 has a half-life of 28.1 yr. This isotope is one of many produced in an atomic bomb explosion. Tests in the atmosphere in the 1950s produced quantities of strontium-90. As shown in Figure 12.2, for 80 mg of the isotope produced, the decay process leaves 40 mg after 28.1 yr. In another 28.1 yr 20 mg remains. Thus, in the year 2006, there will still be 25% of the isotope from a 1950 test.

Example 12.3

Sodium-24 is used in medicine to determine the effectiveness of blood circulation. Its half-life is 15 h. How much of the radioisotope will remain after 60 h?

Figure 12.2
Illustration of the Half-life
of Strontium-90.

Time (yr)	Amount (mg)
0	80
28.1	40
56.2	20
84.3	10
112.4	5

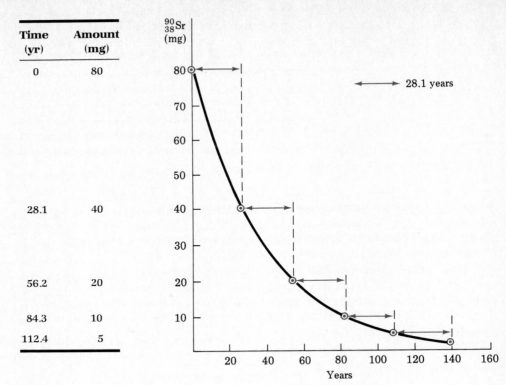

Solution. After each half-life, the amount remaining will be one-half of the amount remaining after the previous half-life. A table can be prepared to show this behavior.

Time (h)	Amount (%)
0	100
15	50
30	25
45	12.5
60	6.25

[Additional examples may be found in **12.14–12.20** at the end of the chapter.]

One of the best-known applications of the radioisotope half-life is the carbon-14 dating method. The age of ancient materials made of plant or animal matter can be established on the basis of their $^{14}_{6}C$ content. Carbon dioxide in the atmosphere consists of carbon-12 with trace amounts of carbon-14, which is radioactive and decays. However, the concentration of carbon-14 does not decrease because it is constantly being formed in the atmosphere from the action of cosmic rays on nitrogen, $^{14}_{7}N$. All plants absorb carbon dioxide from the atmosphere; as long as the plant is living, the amount of carbon-14 incorporated into the molecules it produces and uses will be a constant fraction of the amount of carbon present. When the plant dies, the carbon compounds in its cells are no longer interacting with the CO_2 of the atmosphere. The amount of carbon-14 in these plants therefore diminishes with time as the carbon-14 decays. Since the half-life of carbon-14 is 5570 yr, it is possible to measure the age of an object by determining the amount of radioactive carbon remaining in it. A wooden dish that contains only 25% of the carbon-14 in trees living today is therefore approximately 11,000 yr old. The carbon-14 dating technique cannot be used accurately for objects that are older than 50,000 yr. After many half-lives have elapsed, such a small amount of carbon-14 remains that accurate measurement is not possible.

12.5
Ionizing Radiation and Its Biological Effects

As is generally known by the public, radiation is dangerous to humans and all other life forms. What is the reason for the effect of radiation on cells?

All radiation travels with high speed when ejected from a nucleus. As a consequence of the high energy associated with radiation, chemical reactions can occur when the radiation interacts with matter. Radiation causes atoms and, more importantly, molecules to lose electrons and form ions. The electron and the cation produced are then reactive.

Consider the effect of radiation on the covalent bond of a hypothetical molecule, A—B.

$$A\text{—}B \longrightarrow A\cdot B^+ + e^-$$

$$A\cdot B^+ \longrightarrow A\cdot + B^+$$

$$A\cdot B^+ \longrightarrow A^+ + B\cdot$$

After an electron is "knocked out" of the covalent bond, the resultant cation does not have the necessary bonding electrons to remain together. Decomposition of the cation results in another cation and a species called a radical, which has an unpaired electron.

The disruption of the otherwise stable A—B molecule by radiation produces reactive species. The subsequent reaction of cations and radicals with other molecules results in abnormal processes that can seriously disrupt the normal functions of cells.

All of us are exposed to radiation in many forms. The most common source of radiation is from the sun. Excessive exposure to the ultraviolet light in sunlight can cause skin cancer. The ionizing radiation of x-rays is a potentially more hazardous form of radiation. Low dosages of x-rays are necessary for certain medical and dental diagnoses. However, in recent years the diagnostic use of x-rays has been cut back. Prolonged exposure can cause cell damage and, over a period of many years, cancer. For this reason x-ray technicians must take precautions for their health safety.

An understanding of the effects of radiation on living material has developed in the last three decades. Radiation can affect any living cells that it reaches. Gamma radiation, with its great penetrating power, can literally plow through your entire body, and for this reason it is the most damaging type of radiation. Regardless of the type of radiation, the chemical effect on a cell is one of breaking molecules apart. The disruption of a stable organic molecule results in the formation of reactive materials that may recombine with themselves or react with neighboring molecules. Thus, chemical reactions can occur that are foreign to the cell and cause damage that may be serious. Cells are able to repair themselves up to a point, and low radiation exposure might not cause permanent biological damage. However, no exposure to radiation is ever 100% risk free.

The most serious effect of radiation is the mutation of a species. If radiation affects the nuclei of germ cells that produce eggs or sperm, then the offspring may be mutants. The seriousness of the mutation depends on the extent of the radiation.

The nucleus of any cell contains DNA that is responsible for producing RNA and passing on genetic information for the formation of new cells. Excessive radiation could terminate DNA's ability to reproduce, and the death of the cell would result. If an alteration of the DNA occurred, then the RNA molecules that direct synthesis of enzymes and other proteins could be altered. Faulty enzymes could lead to serious health consequences and the death of the organism.

Tissues that reproduce at a rapid rate, such as bone marrow, lymph nodes, and embryonic tissue, are the most sensitive to radiation damage. Since bone marrow is the site of red cell formation, one of the early signs of a radiation overexposure is a drop in the red cell count. A pregnant woman must keep her radiation exposure low to protect the genetic health of her unborn child.

12.6
Radiation Detection

The method of detecting radiation depends on both the amount and the source of radiation. Methods in use include the Geiger counter, the film badge, and the scintillation counter.

A Geiger counter consists of a metal tube containing a gas at low pressure (Figure 12.3). The gas chamber is connected to a battery. In the absence of radiation, there is no current flow through the gas chamber. When the gas is exposed to radiation, ions and electrons are produced. As a consequence, a current flows through the chamber. The higher the intensity of radiation, the larger is the current flow. Either an audible clicking sound or swing of a needle in the Geiger counter gives a measure of the current flow that is proportional to the intensity of the radiation.

A film badge is used by workers to measure the amount of radiation that the body receives. The film badge consists of photographic film in a holder sealed from light. Radiation can pass through the holder, and the film becomes exposed in a manner similar to taking pictures. As the amount of radiation increases, the developed film will be darker. Film badges are periodically processed to check the amount of exposure of the wearer to radiation.

Scintillation counters are important in detecting radiation in medical tests. A scintillation counter contains crystals of some chemicals that can produce scintillations or flashes of light when hit by radiation. An electronic device called a photomultiplier records and amplifies the scintillations. Scintillation counters

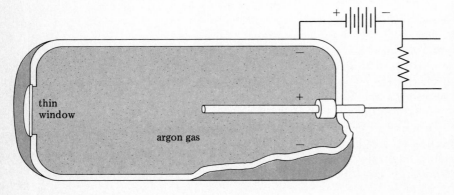

Figure 12.3
The Geiger Counter.
Radiation enters the Geiger tube through the thin window at the end of the tube. A voltage difference is maintained between the metal rod in the center of the tube and the lining of the tube. When ionizing radiation enters the tube, the argon atoms lose electrons and cations are formed. An electrical discharge results, and a current flows. The current is amplified and recorded either as a clicking sound or on a meter.

thin window

argon gas

are used to scan patients to measure the location and amount of radioisotopes present in the body.

Radiation has been measured quantitatively by two different methods. One method measures the intensity of the radiation emitted from the radioactive source. The other method is chemical or biochemical and measures the radiation absorbed by matter, such as air or tissue.

The Curie

The Curie is a unit of radiation intensity. *One* **Curie** *(Ci) is equal to* 3.7×10^{10} *nuclear disintegrations per second.* This number was chosen because it is the number of disintegrations that 1 g of radium undergoes in 1 s. The mass of other substances that produce 1 Curie may be more or less than 1 g.

The Roentgen

The roentgen is a chemical measure of exposure to gamma radiation or x-rays. *One* **roentgen** *(R) is the dose of radiation that produces* 2.1×10^9 *ions in* 1 cm^3 *of dry air at 0°C and* 1 atm *pressure.* Geiger counters, which operate by detecting ionized air, are calibrated to read in milliroentgens per hour (mR/h).

The Rad

The rad, short for radiation absorbed dose, was selected to relate energy absorbed by tissue. *The* **rad** *is defined as the absorption of* 100 ergs $(2.4 \times 10^{-6}$ cal) *of energy per gram of absorbing tissue.* Although the roentgen and the rad are defined differently, they are numerically similar. One roentgen of gamma radiation causes 0.97 rad in muscle tissue. In bone, 1 roentgen delivers 0.92 rad. A dose of about 500 rads of gamma radiation would be lethal for a human.

RBE and the Rem

Even the dosage actually absorbed does not necessarily indicate the biological damage to an organ. In other words, a 10 rad dose of one type of radiation may result in different biological consequences than a 10 rad dose of another type of radiation. For this reason, factors known as dose equivalents are defined. These are multiplicative factors that are multiplied by the rad dose.

The **relative biological effectiveness (RBE)** *accounts for differences in biological damage caused by radiation.* The standard for RBE is gamma radiation from cobalt-60. The RBE of a radiation source is the ratio of the dose of gamma radiation from cobalt-60 to the dose of the radiation in question required to cause the same biological effect. For example, one rad of alpha particles may have the same effect as 10 rads of gamma radiation from cobalt-60. The RBE of alpha particles is then 10.

Highly charged and heavy particles such as alpha particles cause more ionization in matter than lighter and singly charged beta particles. The RBE of beta particles is about 1.

The **rem,** short for roentgen equivalent for man, *is defined as the product of rads and RBE.* For alpha particles, 1 rad equals 10 rem, whereas for beta particles 1 rad equals 1 rem.

rem = rad × RBE

Monitoring the radiation exposure of workers in radiation laboratories is done in rem, since it accounts for the effect of radiation regardless of the source. Radiation accumulated then can be added in rem units, and medical records are more easily maintained. Monitoring of even millirem (mrem) is important.

12.8
Radiation Sickness and Radiation Safety

Radiation sickness is the result of overexposure to ionizing radiation that creates ions and radicals. These particles cause cellular chemical changes that may be minor or cause death.

Overexposure to radiation can be the result of a single large dose or many cumulative small doses. It is estimated that in one year the average American receives 200 mrem of radiation, half of which is from natural environmental background and cannot be decreased. The second major radiation source is x-rays used in medical diagnosis.

The exact effects of radiation can only be stated in terms of probability of biological consequences. A listing is provided in Table 12.3.

Table 12.3
Effect of Short-Term Radiation Doses

Dose (rem)	Probable effect
<25	no detectable short-term effect
25–100	decrease in white blood cell count, which lowers resistance to infection
100–200	reduction in number of blood cells, fatigue, nausea
200–300	nausea and vomiting on first day of exposure; fever, diarrhea, loss of hair by the third week
300–500	nausea, vomiting, and diarrhea in a few hours; probability of death about 50%
>500	vomiting, severe changes in blood and gastrointestinal system; death within 2 months in essentially 100% of the cases

Exposure of the entire body to less than 25 rem results in no detectable clinical short-term effect. The consequences of increased exposure, 25–100 rem, are evident in the blood-forming tissue of bone marrow. These cells divide very rapidly and are easily affected by chemical changes caused by radiation. The medical result is a marked decrease in white blood cells, which lowers resistance to infection.

A dose of 100–200 rem results in longer-term reduction of some blood cells and will cause fatigue and nausea. A dose of 200–300 rem results in nausea and vomiting on the first day of exposure. By the third week, fever, diarrhea, loss of hair, and a 50% chance of death occur. Exposures greater than 300 rem result in the same symptoms. In a few hours, nausea, vomiting, and diarrhea result. The probability of death is about 50%. The central nervous system, blood formation system, and gastrointestinal system are all disrupted by radiation doses above 500 rem. Death results in essentially all persons receiving such doses.

Radiation exposure limits have been proposed by the National Council on Radiation Protection. Although the average American is not exposed to more than 200 mrem (0.2 rem) in a year, the recommended maximum is 0.5 rem. Some individuals are exposed to radiation in their occupation, and a 5 rem limit per year is suggested for whole-body exposure. Those in regular contact with radiation wear protective clothing, gloves, and masks. In addition, a film badge is used to monitor dosage, since radiation cannot be seen and you cannot tell if you have been exposed.

Shielding material can provide substantial protection from radiation. Such precautions are required for technicians who routinely use x-ray machines. Both the thickness and composition of the shielding material are important in decreasing the chances of radiation damage. As indicated earlier, alpha particles can be stopped by a sheet of paper. Beta particles can be stopped by a thin sheet of metal or about 1 in. of wood. Gamma rays require concrete or lead shielding. The thickness required depends on the sources of the γ- radiation. For example, one-half of the radiation of technetium-99 will be stopped by 0.2 mm of lead, whereas iodine-131 requires 6 mm for the same protection, and 12 mm to stop 75% of its radiation. Thus, a thick shielding material is required for good protection.

Exposure to radiation is related to the distance between the individual and the radiation source. The dosage decreases as the square of the distance. Thus, at 2 ft the dosage is $(\frac{1}{2})^2$ or $\frac{1}{4}$ the dosage at 1 ft. Every increase of distance provides increased protection. At 4 ft the dosage is $(\frac{1}{4})^2 = \frac{1}{16}$ the dosage at 1 ft. This relationship can be stated as an inverse square law with I for intensity and d for distance.

$$\frac{I_a}{I_b} = \frac{d_b^2}{d_a^2}$$

Example 12.4 ───────────────────────────────

The dosage at 5 ft from a radiation source is 50 mrem. What is the dosage at 25 ft from the same source?

Solution. The inverse square law for radiation can be used and the values substituted into the equation.

$$\frac{I_a}{I_b} = \frac{d_b^2}{d_a^2}$$

$$\frac{50 \text{ mrem}}{x \text{ mrem}} = \frac{(25 \text{ ft})^2}{(5 \text{ ft})^2}$$

$$x \text{ mrem} = 50 \text{ mrem} \times \frac{25 \text{ ft}^2}{625 \text{ ft}^2} = 2 \text{ mrem}$$

[Additional examples may be found in **12.29–12.35** at the end of the chapter.]

12.9
Use of Radioisotopes in Diagnosis

There are five criteria a radiologist uses in selecting a radioisotope for diagnosis. The first is that the radioisotope must be effective in diagnosis at as low a concentration as possible and yet still be reliably detected. Second, in order to minimize any radiation damage to the organism, the radioisotope must have a short half-life. This will result in a large number of detectable nuclear disintegrations in a short time. The third criterion is the ready elimination of the radioisotope from the body after the diagnosis is complete. The nature of the chemical species chosen and its function in the body form the fourth criterion. Ideally, the chemical will be selectively transmitted to the part of the body where diagnosis is desired. Finally, the fifth criterion is dependent on the type of radiation emitted by the radioisotope. In order to detect the radioisotope the radiation must have sufficient penetrating power to reach the instruments placed outside the body. Both alpha and beta particles have low penetrating power. Thus, only radioisotopes that emit gamma rays are appropriate for diagnosis. Some of the radioisotopes used in diagnosis are given in Table 12.4.

Table 12.4

Radioisotope Half-lives and Uses in Diagnosis

Isotope	Half-life	Part of body	Use in diagnosis
barium-131	11.6 d	bone	detection of bone tumors
chromium-51	27.8 d	blood	determination of blood volume and red blood cell lifetime
		kidney	assessment of kidney activity
copper-64	12.8 h	liver	diagnosis of Wilson's disease
gold-198	64.8 h	kidney	assessment of kidney activity
iodine-125	60 d	blood	determination of hormone level in blood
iodine-131	8.05 d	brain	detection of fluid buildup in the brain
		kidney	location of cysts
		lung	location of blood clots
		thyroid	assessment of iodine uptake by thyroid
iron-59	45 d	blood	evaluation of iron metabolism in blood
krypton-79	34.5 h	blood	evaluation of cardiovascular system
mercury-197	65 h	spleen	evaluation of spleen function
		brain	brain scans
selenium-75	120 d	pancreas	determination of size and shape of pancreas
technetium-99	6.0 h	brain	detection of brain tumors, hemorrhages, or blood clots
		spleen	measurement of size and shape of spleen
		thyroid	measurement of size and shape of thyroid
		lung	location of blood clots
xenon-133	5.3 d	lung	determination of lung volume

Iodine-131 is useful in diagnosing thyroid activity. A patient drinks water containing the radioisotope in the form of sodium iodide. If the thyroid is functioning normally, about one-sixth of the radioisotope accumulates in the thyroid gland within 24 hours. A radiation detector placed at the neck is used to measure the concentration of iodine in the thyroid. If a hypothyroid condition exists, the amount accumulated is less than normal. If a greater than average amount accumulates, then a hyperthyroid condition exists. A thyroid scan can show the distribution of the iodine-131 in the thyroid as a radiation "picture." If any part of the thyroid is missing from the "picture," isotopic iodine was not absorbed. The absence of radiation in an area is called a "cold spot."

Example 12.5 _____

Iodine-131 and a beta particle are produced by bombarding an element with a neutron. What element is used?

Solution. Write the nuclear equation using $_Z^A E$ as the symbol for the element.

$$_Z^A E + _0^1 n \longrightarrow _{53}^{131} I + _{-1}^0 e$$

The mass number (A) of the element plus 1 for the neutron must equal 131.

$$A + 1 = 131 + 0$$
$$A = 130$$

The atomic number (Z) of the element must be equal to

$$Z + 0 = 53 + (-1)$$
$$Z = 52$$

The atomic number of tellurium is 52. The element is tellurium-130.

_____

[Additional examples may be found in **12.36–12.39** at the end of the chapter.]

Technetium-99 is a useful radioisotope to scan the brain for tumors. The ionic compound sodium pertechnetate, $NaTcO_4$, is administered intravenously in an isotonic sodium chloride solution. Normal brain chemistry prevents the ionic pertechnetate salts from entering the brain. However, in the case of a tumor, the pertechnetate enters and concentrates in the area of the tumor. As a result, a scan of the brain will reveal the presence, size, and location of the tumor as a "hot spot."

Body fluids can also be studied with radioisotopes. The radioisotope Na-24 is used to determine the effectiveness of blood circulation. A dose of sodium-24 is injected as NaCl directly into the blood, and its progress is monitored. If it takes a longer time than average to reach a certain part of the body, then impaired circulation is indicated. In a similar manner, the cerebrospinal fluid that surrounds the spinal cord and brain can be labeled by using iodine-131 in a spinal tap. Any blockage causing buildup of fluid can then be detected.

12.10
Use of Radioisotopes in Therapy

The objectives in using radioisotopes for therapy are somewhat different than those for diagnosis. The radioisotope must be placed in the proper part of the body and must selectively destroy cells or tissue. There is no need to monitor the radioisotope by an external detector. Thus, gamma ray emitters are not needed. Indeed, such radiation would be destructive to healthy tissue and organs in the high doses used for therapy. This means that only alpha or beta emitters are used because they will destroy tissue or cells in a localized area.

Cells that are in the process of dividing are the most susceptible to radiation damage. Thus, cancer that involves rapid uncontrolled division of abnormal cells can be affected easily by radiation. It is this principle that is applied in cancer therapy. Choice of a radioisotope that will do maximum damage to the cancerous cells while producing minimum damage to healthy cells is the goal of the radiologist.

Salts of the radioisotope $_{88}^{226} Ra$ were among the earliest used in cancer therapy. In addition to alpha particles, the radioactive decay results in highly penetrating gamma rays as well.

$$^{226}_{88}\text{Ra} \longrightarrow {}^{222}_{86}\text{Rn} + {}^{4}_{2}\text{He}$$

Very high intensity treatment of cancers is now done with cobalt-60. This radioisotope emits beta particles and gamma rays. Its half-life is 5.2 yr. Cobalt-60 in therapy involves using a beam of radiation that is focused on the small area of the body where the tumor is located (Figure 12.4). The radiation is carefully directed at the site of the cancer cells, but unfortunately some normal cells are always destroyed as well. The patient will suffer the effects of radiation sickness while undergoing treatment for cancer by this method.

Radiation therapy can also be accomplished by using sources of radiation located inside the body. For example, radium may be implanted in the body. Very small hollow gold needles containing the radium salt are inserted into the tumor. However, this therapeutic method has several disadvantages. Radium is very expensive and has a half-life of 1620 yr. Therefore, the implant has to be removed at some later date to prevent continued radiation.

Yttrium-90 has considerable advantages over radium-226. It is a beta emitter with a half-life of 64 h. An implant of an yttrium salt results in localized treatment of the tumor. Since no gamma rays are produced, healthy tissues a few centimeters away are unaffected. The short half-life is also useful because the radioisotope decays rapidly and is rendered harmless.

$$^{90}_{39}\text{Y} \longrightarrow {}^{90}_{40}\text{Zr} + {}^{0}_{-1}\text{e}$$

Some radioisotopes are injected into the body or ingested. The treatment is based on the fact that certain substances tend to be concentrated at specific locations in the body. For example, iodine tends to be concentrated in the thyroid gland. An overactive thyroid gland may be treated by administering orally a solution of NaI containing some iodine-131. When this radioisotope is concentrated in the thyroid, it causes the destruction of some cells. Iodine-131 has a half-life of 8.1 d, which allows large doses to be used with safety. The radioactivity decreases rapidly after the treatment.

Phosphorus-32 has been used in the form of phosphate ions to treat some forms of leukemia. The phosphate ion can be transported to many parts of the body and can eventually be eliminated in the urine. However, phosphate ions are incorporated in bones. The radioisotope in the bone will affect the production of white blood cells in the bone marrow.

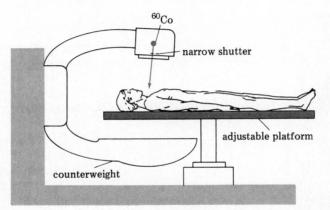

Figure 12.4
Use of Cobalt-60 Source in Radiation Therapy.
The cobalt source is heavily shielded with a metal alloy to prevent stray radiation from damaging healthy tissue.

In order to have radioisotopes available for medical and other uses, nuclear scientists must make them in the laboratory. High-energy nuclear particles are forced to collide with a target of some stable isotope. *Collision of a high-energy particle with a nucleus to produce a different nucleus is called nuclear* **transmutation.**

In nuclear transmutation *the high-energy particle is called the projectile.* Electrons ($_{-1}^{0}e$), neutrons ($_{0}^{1}n$), protons $_{1}^{1}H$), deuterons ($_{1}^{2}H$), and alpha particles ($_{2}^{4}He$) have been used as projectiles.

The stable isotope used as a reactant in nuclear transmutation is called the **target nucleus.** Most stable isotopes available in nature have been used as targets. As a result, another 1000 artificial isotopes have been made in the laboratory from about 300 isotopes.

Reaction of nitrogen-14 with high-energy neutrons produces carbon-14.

$$^{14}_{7}N + {}^{1}_{0}n \longrightarrow {}^{14}_{6}C + {}^{1}_{1}H$$

Carbon-14 is a beta emitter. One important use of carbon-14 is serving as a radioactive "tag" in organic molecules. The carbon-14 atoms are incorporated in an organic molecule, which then is allowed to proceed through either an organic or a biochemical reaction. As the compound reacts to form products, the location of the carbon-14 "tag" is followed. The process of photosynthesis was studied by such a method.

Radioactive isotopes needed for both medical diagnosis and therapy are also made by nuclear transmutation. Cobalt-60 is made from cobalt-59 by bombardment with neutrons.

$$^{59}_{27}Co + {}^{1}_{0}n \longrightarrow {}^{60}_{27}Co$$

By nuclear transmutation reactions, scientists have been able to produce elements of higher atomic number than exist in nature. Before the nuclear age, uranium was the element with the highest atomic number. *Those elements with higher atomic number than uranium are called* **transuranium elements.** A list of some of the isotopes of transuranium elements, with the transmutation reactions used to prepare them, is given in Table 12.5.

12.11
Nuclear Transmutations

Table 12.5

Transmutation Reactions to Form Transuranium Elements

Element	Atomic number	Reaction
neptunium, Np	93	$^{238}_{92}U + {}^{1}_{0}n \longrightarrow {}^{239}_{93}Np + {}^{0}_{-1}e$
plutonium, Pu	94	$^{238}_{92}U + {}^{2}_{1}H \longrightarrow {}^{238}_{93}Np + 2{}^{1}_{0}n$
		$^{238}_{93}Np \longrightarrow {}^{238}_{94}Pu + {}^{0}_{-1}e$
americium, Am	95	$^{239}_{94}Pu + {}^{1}_{0}n \longrightarrow {}^{240}_{95}Am + {}^{0}_{-1}e$
curium, Cm	96	$^{239}_{94}Pu + {}^{4}_{2}He \longrightarrow {}^{242}_{96}Cm + {}^{1}_{0}n$
berkelium, Bk	97	$^{241}_{95}Am + {}^{4}_{2}He \longrightarrow {}^{243}_{97}Bk + 2{}^{1}_{0}n$
californium, Cf	98	$^{242}_{96}Cm + {}^{4}_{2}He \longrightarrow {}^{245}_{98}Cf + {}^{1}_{0}n$
einsteinium, Es	99	$^{238}_{92}U + 15{}^{1}_{0}n \longrightarrow {}^{253}_{99}Es + 7{}^{0}_{-1}e$
fermium, Fm	100	$^{238}_{92}U + 17{}^{1}_{0}n \longrightarrow {}^{255}_{100}Fm + 8{}^{0}_{-1}e$
mendelevium, Md	101	$^{253}_{99}Es + {}^{4}_{2}He \longrightarrow {}^{256}_{101}Md + {}^{1}_{0}n$
nobelium, No	102	$^{246}_{96}Cm + {}^{12}_{6}C \longrightarrow {}^{254}_{102}No + 4{}^{1}_{0}n$
lawrencium, Lr	103	$^{252}_{98}Cf + {}^{10}_{5}B \longrightarrow {}^{257}_{103}Lr + 5{}^{1}_{0}n$

12.12
Nuclear Fission

The conversion of a nucleus of a heavy element into two or more lighter elements is called **nuclear fission.** One example of nuclear fission is the bombardment of uranium-235 with neutrons. The unstable uranium-236 nucleus, which forms first, rapidly splits into nuclei of lower atomic mass. Although many pairs of elements are actually produced, $^{89}_{37}Rb$ and $^{144}_{55}Cs$ are typical products.

$$^{235}_{92}U + {}^{1}_{0}n \longrightarrow {}^{236}_{92}U \longrightarrow {}^{89}_{37}Rb + {}^{144}_{55}Cs + 3\,{}^{1}_{0}n$$

The fission process can continue because the neutrons produced by the fission reaction can collide with other uranium atoms causing them to split apart. Because more neutrons are produced than are used in initiating the reaction, the fission process rapidly becomes a self-sustaining chain reaction that releases large amounts of energy in a short time. This principle is used in the construction of atomic bombs. The chain process is illustrated in Figure 12.5.

Although atomic energy in the arsenals of numerous countries is potentially destructive, controlled nuclear energy can be of service in a world of limited resources. The limited fossil-derived fuels (oil, gas, and coal) are being used at an ever-increasing rate, and this source will be exhausted in the future.

There is a limit to the availability of uranium-235. This isotope is only 0.71% of naturally occurring uranium. With the estimated world reserves and anticipated growth of nuclear power, it is possible that the supply of uranium-235 may be exhausted in two to three decades.

The uranium in nuclear reactors is of lower purity than weapons-grade ura-

Figure 12.5
Nuclear Chain Reaction.
Each nuclear event produces more neutrons than are used. The rate of the reaction then accelerates as more and more nuclei undergo fission.

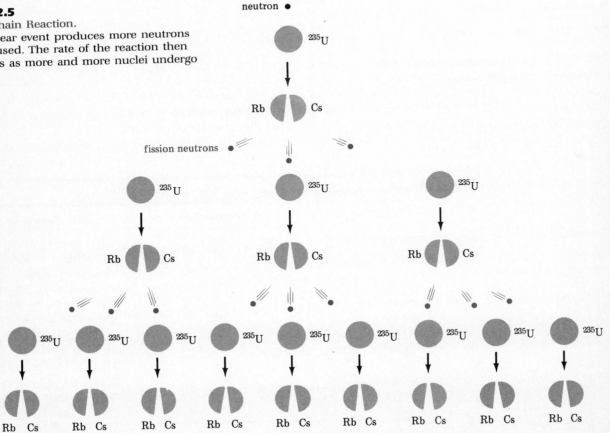

Figure 12.6

Generation of Electricity from a Nuclear Power Plant.
The hot radioactive water is pumped to a heat exchanger that converts water
in a second system into steam. The steam drives a turbine to produce electricity.

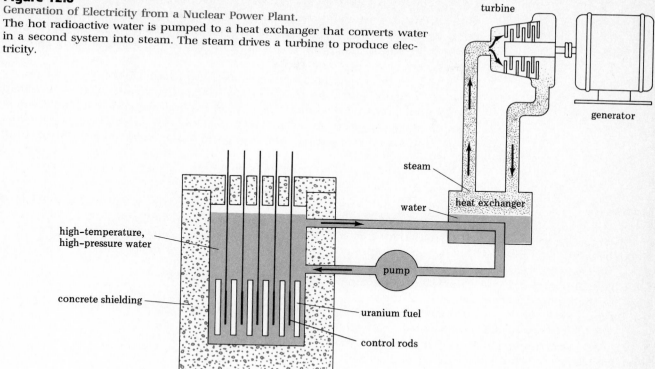

nium. Furthermore, the rate of the chain reaction is controlled by a moderator
and control rods. Water is a moderator because it can slow down the neutrons.
For this reason, the uranium is suspended in water in nuclear reactors. Control
rods are usually made of cadmium metal, which can absorb some of the neutrons
generated in the nuclear reaction. The control rods are positioned between the
uranium fuel samples (Figure 12.6). When the reaction needs to be slowed, the
rods are lowered. To speed up the reaction, the rods are raised.

One of the important considerations in using nuclear fission reactions for
energy is the problem of the disposal of the spent material. The waste products
are themselves radioactive and have quite long half-lives. The technological problem of separation and disposal of these materials is complicated by the considerable political problem of where to store the material.

*The formation of a heavy element from two lower-mass elements by nuclear
changes is called* **nuclear fusion.** Nuclear fusion is a source of even greater energy than nuclear fission. However, unlike nuclear fission, fusion requires extremely high temperature (40,000,000°C) for initiation. Fusion occurs in the sun,
where the temperature is high enough to convert hydrogen into helium. The
steps proposed to account for the reaction are

12.13
Nuclear Fusion

$$2\,{}^{1}_{1}\text{H} \longrightarrow {}^{2}_{1}\text{H} + {}^{0}_{1}\text{e}$$

$$ {}^{2}_{1}\text{H} + {}^{1}_{1}\text{H} \longrightarrow {}^{3}_{2}\text{He}$$

$$2\,{}^{3}_{2}\text{He} \longrightarrow {}^{4}_{2}\text{He} + 2\,{}^{1}_{1}\text{H}$$

On earth, nuclear fusion has been achieved by using atomic bomb "triggers" to provide sufficient energy to initiate fusion. This is the principle upon which the hydrogen bomb is based. Since hydrogen is so readily available from water, controlled nuclear fusion for practical purposes would eliminate any concern about future sources of energy.

Obviously, atomic bombs cannot be used to produce fusion power in power plants. The fusion reaction will have to be initiated and controlled by technical means not yet devised. The United States has expended about a billion dollars in research to develop power from fusion reactors. Although progress is being made, some scientists estimate that fusion power may not become available until after the year 2000.

Summary

Many isotopes that occur naturally have existed since the earth was formed. Other isotopes are undergoing radioactive transformations that generate radiation. Radiation from natural sources includes alpha particles, beta particles, and gamma radiation.

Radioactive isotopes may be made in the nuclear chemistry laboratory by the bombardment of stable nuclei with subatomic particles. This method has been used to produce both isotopes for medical applications and new elements.

All radioisotopes decay at a rate that is given by its half-life. One of the applications of half-life data is the use of carbon-14 in dating ancient wooden materials.

Radioactive substances cause surrounding material to ionize and undergo abnormal reactions. The effect of radiation on living tissue can be minor or so severe as to damage DNA to the extent that mutation of the species or death results.

Methods to detect radiation include the Geiger counter, film badges, and scintillation counters. The unit measuring the number of radioactive disintegrations is the Curie. Roentgen, rad, and rem are used to measure dosage. The use of radioisotopes requires shielding, but the effect of radiation also depends on the distance between the worker and the radiation source.

Radioisotopes are used in medicine for both diagnosis and therapy. The choice of radioisotope is dictated by the purpose as well as the target organ. The radioisotopes are produced by nuclear transmutation reactions.

Two types of nuclear reaction, fusion and fission, are alternative fuel sources for the future. Fission processes are currently being used, but are limited by the availability of uranium-235 and the problem of waste disposal. Fusion has not been achieved in a self-sustaining reaction, but research continues to develop this promising source of energy.

Learning Objectives

As a result of studying Chapter 12 you should be able to
- Distinguish among the types of radiation.
- Describe the units of the measurement of radiation.
- Balance nuclear equations.
- Calculate the fraction of a radioisotope remaining in a reaction given its half-life.
- Describe medical uses of radioisotopes.
- Compare nuclear fusion and nuclear fission.

New Terms

An **alpha particle** consists of two protons and two neutrons and is identical to the nucleus of the helium atom.

A **beta particle** is an electron formed in the nucleus as the result of conversion of a neutron into a proton.

The **Curie** is a quantity of radioactive material that produces 3.7×10^{10} nuclear disintegrations per second.

A **daughter nucleus** is the product of a nuclear disintegration reaction.

A **gamma ray** is a form of radiation, similar to x-rays, that is emitted from certain radioactive nuclei.

A **Geiger counter** is a device to detect and measure ionizing radiation.

The **half-life** of a radioisotope is the time required for it to decay to one-half of its original amount.

Nuclear fission is a radioactive decay process in which a heavy nucleus breaks apart into smaller nuclei and one or more neutrons.

Nuclear fusion is a nuclear process in which two small nuclei combine into a larger nucleus.

The **rad** or radiation absorbed dose is the absorption of 100 ergs of energy per gram of absorbing tissue.

A **radioisotope** is an isotope that undergoes nuclear change and produces radiation.

The **RBE** or relative biological effectiveness is a multiplicative factor to account for the differing consequences of the source of radiation.

The **rem** or roentgen equivalent for man is a product of RBE and rad. The unit accounts for the varying effects of types of radiation on tissue.

The **roentgen** is a dose of radiation that produces 2.1×10^9 ions in 1 cm^3 of dry air at 0°C and 1 atm pressure.

A **target nucleus** is an isotope used as a reactant in nuclear transmutation.

Nuclear transmutation is the conversion of one element into another by the collision of a high-energy particle.

A **transuranium element** is an element with an atomic number larger than that of uranium.

Questions and Problems

Terminology

12.1 Distinguish between the terms in each of the following.
(a) alpha particle and beta particle
(b) rem and rad
(c) fusion and fission

Nuclear Symbols

12.2 Write the symbols used in balancing nuclear equations for each of the following particles.
(a) neutron (b) proton (c) alpha (d) beta
(e) electron

12.3 Write the nuclear symbol for each of the following isotopes.
(a) carbon-14 (b) boron-10 (c) uranium-238
(d) chlorine-37 (e) zinc-65 (f) sodium-24
(g) curium-243 (h) cobalt-60

12.4 Indicate the number of protons and neutrons contained in each of the following.
(a) $^{206}_{82}Pb$ (b) $^{235}_{92}U$ (c) $^{127}_{53}I$ (d) $^{13}_{7}N$
(e) $^{25}_{11}Na$ (f) $^{3}_{1}H$

Types of Radiation

12.5 Give the symbol, charge, and mass in amu of each of the following types of radiation.
(a) alpha (b) beta (c) gamma

12.6 Describe the penetrating power of alpha, beta, and gamma radiation.

Balancing Nuclear Equations

12.7 Write a balanced nuclear equation for beta emission from each of the following.
(a) fluorine-21 (b) silicon-32
(c) magnesium-28

12.8 Write a balanced nuclear equation for alpha emission from each of the following.

(a) polonium-212 (b) curium-240
(c) einsteinium-252

12.9 Supply the correct symbol for the product of each of the following reactions.
(a) $^{20}_{8}O \longrightarrow {}_{-1}^{0}e + ?$ (b) $^{22}_{9}F \longrightarrow {}_{-1}^{0}e + ?$
(c) $^{140}_{56}Ba \longrightarrow {}_{-1}^{0}e + ?$

12.10 Supply the correct symbol for the product of each of the following reactions.
(a) $^{243}_{96}Cm \longrightarrow {}_{2}^{4}He + ?$ (b) $^{222}_{86}Rn \longrightarrow {}_{2}^{4}He + ?$
(c) $^{234}_{94}Pu \longrightarrow {}_{2}^{4}He + ?$

12.11 Supply the correct symbol required to balance each of the following equations.
(a) $^{140}_{56}Ba \longrightarrow ? + {}^{140}_{57}La$ (b) $^{245}_{96}Cm \longrightarrow ? + {}^{241}_{94}Pu$
(c) $^{30}_{13}Al \longrightarrow ? + {}^{30}_{14}Si$

12.12 In 1919 Ernest Rutherford proved the existence of a nuclear particle by the following experiment. What is the nuclear particle?

$$^{14}_{7}N + {}^{4}_{2}He \longrightarrow {}^{17}_{8}O + ?$$

12.13 In 1932 James Chadwick proved the existence of a nuclear particle by the following experiment. What is the nuclear particle?

$$^{9}_{4}Be + {}^{4}_{2}He \longrightarrow {}^{12}_{6}C + ?$$

Half-life

12.14 What percentage of an element will remain after each of the following number of half-lives?
(a) one (b) two (c) three (d) four
(e) five (f) six

12.15 A wood object has 12.5% of the usual abundance of carbon-14 remaining. How old is the object?

12.16 The half-life of mercury-203 is 47 d. Estimate how much will remain after 1 yr.

12.17 The half-life of $^{32}_{15}P$ is 14 d. How many grams of the isotope in a 2.00 g sample will remain after 56 d?

12.18 The half-life of an isotope is 30 min. How many hours are required for a 10.00 g sample to decay to 0.62 g of this isotope?

12.19 The half-life of iodine-123 is 13 h. If 10 ng is given to a patient, how much will remain after 3 d and 6 h?

12.20 A wooden object found in a cave in Greece has 6.2% of the normal abundance of carbon-14. How old is the object?

Biological Effects of Radiation

12.21 Explain the term "ionizing radiation."

12.22 Why is ionizing radiation harmful to living cells?

12.23 Write equations for the products that can be formed from water by ionizing radiation.

Detection of Radiation

12.24 Explain the operation of the Geiger counter.

12.25 How are film badges processed to detect radiation received by workers in nuclear laboratories?

Units of Radiation

12.26 The Becquerel (Bq) is a new unit of radiation intensity. The equivalence is 1 Bq = 1 disintegration/s. How many Bq are in one Curie (Ci)?

12.27 The Gray (Gy) is a new unit of radiation absorbed dose. It is the absorption of 1 joule of energy per kilogram of tissue. A joule is equal to 10^7 ergs. Show that 1 Gy = 100 rad.

12.28 Explain why rem is used instead of rad in describing the radiation received by a human.

Radiation and Safety

12.29 Is there a minimum radiation that will have no biological effect? Explain how individuals working with radiation can minimize their exposure.

12.30 Which is more dangerous, a radioisotope that emits alpha particles or one that emits beta particles, providing they have the same half-life?

12.31 Which is more dangerous, a radioisotope with a short half-life or one with a long half-life, if they emit the same type of radiation?

12.32 A person receives 20 rem of radiation per year for 15 yr and does not show any biological effect. However, a person who receives 300 rem in one dose may die. Explain the difference.

12.33 Explain why it is not advisable to eat in a room where radioisotopes are located.

12.34 A radiologist measures 128 mrem at a distance of 2 m from a source. At what distance will the radiation measure 2 mrem?

12.35 A radiologist measures 25 mrem radiation at 10 m. What will the radiation be at 1 m?

Radiation in Diagnosis

12.36 Explain why radioisotopes used in diagnosis usually emit gamma radiation.

12.37 Explain why radioisotopes used in diagnosis should have a short half-life.

12.38 Phosphorus-32 for treatment of leukemia is made by bombarding an isotope with neutrons. What is the element?

$$? + {}_0^1n \longrightarrow {}_{15}^{32}P + {}_1^1H$$

12.39 A radioisotope required for a diagnostic procedure is produced by the following reaction. What is the radioisotope?

$$ {}_{47}^{109}Ag + {}_2^4He \longrightarrow ? + 2\,{}_0^1n$$

Radiation and Therapy

12.40 How do isotopes used in therapy differ from isotopes used in diagnosis?

12.41 Radioisotopes can cause cancer and yet are used in therapy for cancer. Explain this apparent contradiction.

12.42 Explain how phosphorus-32 is used to treat leukemia.

12.43 Explain how iodine-131 is used to treat an overactive thyroid.

12.44 Cobalt-60, used in radiation therapy, is made from cobalt-59. What nuclear particle is used in the reaction? Write the balanced nuclear equation.

Transmutation of Elements

12.45 Nickel-58 produces an element and an alpha particle when bombarded with a proton. What is the elemental symbol of the product?

12.46 The bombardment of ${}_3^7Li$ by an alpha particle produces a neutron and an element. What is the element?

12.47 Bombardment of ${}_{92}^{238}U$ by ${}_7^{14}N$ yields five neutrons and an element. What is the element?

12.48 Gold-197 can yield mercury-197 and a neutron when bombarded with the proper particle. What is the particle?

12.49 Complete the following equations.
 (a) ${}_{11}^{23}Na + ? \longrightarrow {}_{11}^{24}Na + {}_1^1H$
 (b) ${}_5^{10}B + ? \longrightarrow {}_7^{13}N + {}_0^1n$
 (c) ${}_{27}^{59}Co + ? \longrightarrow {}_{25}^{56}Mn + {}_2^4He$

12.50 Complete the following equations.
 (a) ${}_{92}^{235}U + {}_1^2H \longrightarrow ? + {}_{-1}^0e$
 (b) ${}_{13}^{27}Al + {}_2^4He \longrightarrow ? + {}_0^1n$
 (c) ${}_{52}^{130}Te + {}_0^1n \longrightarrow ? + {}_{-1}^0e$

12.51 Complete the following equations.
 (a) ${}_{92}^{238}U + {}_7^{14}N \longrightarrow ? + 6\,{}_0^1n$
 (b) ${}_{98}^{252}Cf + {}_5^{10}B \longrightarrow ? + 5\,{}_0^1n$
 (c) ${}_{92}^{238}U + 15\,{}_0^1n \longrightarrow ? + 7\,{}_{-1}^0e$
 (d) ${}_{96}^{246}Cm + {}_6^{13}C \longrightarrow ? + 5\,{}_0^1n$

Nuclear Fusion and Fission

12.52 What is the difference between the materials used for nuclear reactors and for atomic bombs?

12.53 How are control rods used in nuclear reactors?

12.54 How many neutrons are produced by each of the following nuclear disintegrations?
 (a) ${}_{92}^{235}U + {}_0^1n \longrightarrow {}_{56}^{137}Ba + {}_{36}^{94}Kr + ?$
 (b) ${}_{92}^{235}U + {}_0^1n \longrightarrow {}_{42}^{103}Mo + {}_{50}^{131}Sn + ?$
 (c) ${}_{92}^{235}U + {}_0^1n \longrightarrow {}_{53}^{135}I + {}_{39}^{97}Y + ?$

12.55 How is a hydrogen bomb detonated?

12.56 Why haven't nuclear power plants employing nuclear fusion been built?

Introduction to Organic Chemistry

The structure of molecules such as glucose can be shown with models. The dark, medium, and light spheres represent carbon, oxygen, and hydrogen atoms, respectively.

13.1
Inorganic and Organic Compounds

As early as 1800, chemistry was a distinct field of science that dealt with the composition, structure, physical properties, and chemical reactions of substances. It was evident at that time that substances could be divided into two broad classes called inorganic and organic compounds. The composition of inorganic compounds was quite varied and included most of the elements. Organic compounds always contained carbon and only a few other elements such as hydrogen, oxygen, nitrogen, or sulfur. Although the concept of structure was not strongly developed in the early period of chemistry, it was clear that the structures of organic compounds had to be immensely more complex than those of inorganic compounds. Compounds such as sulfuric acid, H_2SO_4, ammonia, NH_3, and sodium chloride, NaCl, contain so few atoms that their structures can easily be postulated and represented by drawings. In contrast, how does the chemist picture the structure of cholesterol, $C_{27}H_{46}O$, the fat glycerol tripalmitate, $C_{51}H_{98}O_6$, the sugar sucrose, $C_{12}H_{22}O_{11}$, or vitamin B_2, $C_{17}H_{20}N_4O_6$? Clearly the problem of determining the order in which atoms are bonded in organic compounds has to be quite difficult.

The physical properties of inorganic and organic compounds differ substantially. Inorganic compounds generally have high melting points; sodium chloride melts at 801°C and magnesium oxide melts at 2800°C! Many organic compounds are gases or liquids. Those that are solids usually have low melting points. Glycerol tripalmitate melts at 65, cholesterol at 148, sucrose at 185, and vitamin B_2 at 280°C. The solubilities of inorganic and organic compounds also differ. Most inorganic compounds are soluble in water, whereas most organic compounds are not. On the other hand, inorganic compounds are seldom soluble in organic liquids such as ether, alcohol, or carbon tetrachloride, whereas organic compounds easily dissolve in one or more of these solvents.

In the light of more modern methods, we can also see a substantial difference between inorganic and organic compounds based on electrical conductivity. Most inorganic compounds, such as NaCl, exist in water as ions that conduct an electric current. The organic compounds that do dissolve in water, such as sucrose, seldom conduct electricity. The reason for this difference is one of structure—most organic compounds consist of covalently bonded molecules.

13.2
Vital Force and Chemical Synthesis

Early chemists felt that organic compounds could only be produced by the living cells of plants or animals. Sucrose can be made by plants and glycerol tripalmitate by animals. Prior to 1830, no organic compound had been produced in the laboratory from the elements or inorganic compounds.

In 1810, Berzelius suggested that there was a "vital force" in living matter that was necessary to form organic compounds. Although the nature of this vital force was not understood, the concept was not seriously questioned.

In 1829, the German chemist Friedrich Wohler accidentally synthesized the first organic compound in the laboratory. He attempted to produce ammonium isocyanate from solutions of inorganic compounds, ammonia and cyanic acid.

$$NH_3 + HOCN \longrightarrow NH_4^+ + OCN^-$$

In the process of evaporating the water, he overheated the material and produced urea.

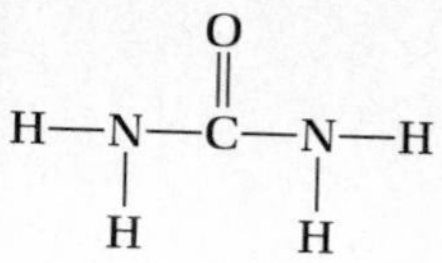

```
        O
        ‖
H—N—C—N—H
    |       |
    H       H
      urea
```

Urea had been isolated decades earlier as one of the end products of metabolism. It is formed in the liver, removed by the kidneys, and excreted in urine. The synthesis of urea in the laboratory then was a direct contradiction of the vital force theory. The synthesis of other organic compounds followed in other laboratories, and it soon became clear that the uniqueness of organic compounds was not a vital force but the presence of carbon in these compounds. Therefore, today we define organic chemistry as the chemistry of carbon compounds. The term organic is used even for those compounds of carbon produced in the laboratory that have never been formed by a living organism.

An immense number of substances contain carbon. Approximately 5 million organic compounds are known, and approximately 5000 new compounds are being produced each week. Why does carbon form so many compounds? Why doesn't any other element combine to form a series of compounds that is large enough to be considered a separate field of chemistry?

13.3 The Unique Carbon Atom

Carbon is unique among the elements for three reasons.

1. Carbon forms quite stable single, double, and triple bonds.
2. Carbon forms strong covalent bonds with many other atoms such as H, O, N, and S.
3. Carbon atoms can concatenate.

Concatenation *is the formation of networks of identical atoms bonded to each other.* With the exception of carbon no element concatenates to form stable molecules of more than a few atoms. Silicon, which is also a member of Group IVA, has been concatenated to form molecules with fewer than 10 silicon atoms, but they burn spontaneously in air.

With its four valence-shell electrons, carbon may form four covalent bonds to itself or to other atoms and achieve an octet. The other atoms then must contribute a total of four electrons. A single carbon atom can form a variety of bond types, including combinations of types.

1. Four single bonds.
2. Two single bonds and a double bond.
3. One single bond and a triple bond.
4. Two double bonds.

The first three classes will be examined extensively in this text. The last type is less common, although it is present in carbon dioxide.

Structures of carbon that include four single bonds are many and varied. Compounds with carbon bonded to hydrogen as well as to chlorine, oxygen, sulfur, and nitrogen are shown in Figure 13.1. For convenience, the bonds for compounds of carbon containing four single bonds are drawn at right angles to each other. However, you should recall from Chapter 6 that the carbon bond angles in such compounds are 109.5°. The carbon orbitals are sp^3 hybridized, and the bonds to carbon are directed to form a tetrahedral shape.

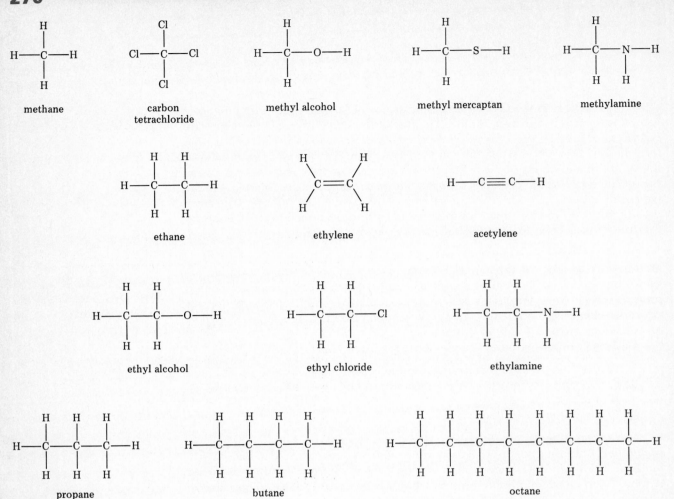

Figure 13.1
Structural Formulas of Some Carbon Compounds.

Carbon may bond to itself by sharing one, two, or three pairs of electrons.

$$\cdot \overset{\cdot}{\underset{\cdot}{C}}\!-\!\overset{\cdot}{\underset{\cdot}{C}} \cdot \qquad\qquad :C\!=\!C: \qquad\qquad \cdot C\!\equiv\!C \cdot$$

single bond   double bond   triple bond

In the single-bond case, carbon is sp^3 hybridized, and there is one σ bond. In the double-bond case, carbon is sp^2 hybridized (Section 6.9), and there are one σ bond and one π bond. In the triple-bond case, carbon is sp hybridized, and there are one σ bond and two π bonds (Section 6.9). The remaining unpaired electrons in each case can then be shared with hydrogen to give ethane, ethylene, or acetylene (Figure 13.1).

Since carbon can bond to itself, it is easy to see how the units pictured with unpaired electrons could be joined to form longer chains of carbon atoms. Propane, used as a fuel, and butane, used in some cigarette lighters, contain chains of three and four carbon atoms, respectively. The carbon skeletons of propane and butane are as follows.

$$C\!-\!C\!-\!C \qquad\qquad C\!-\!C\!-\!C\!-\!C$$

Extension of the chain of carbon atoms yields many more compounds. Octane is such an example (Figure 13.1).

The molecular formula of a compound gives its atomic composition. For example, the molecular formula of butane is C_4H_{10} and that of ethyl alcohol is C_2H_6O. However, in order to understand the chemistry of organic compounds, it is necessary to know their structure. *The structure of a molecule is represented by a* **structural formula** *that shows the arrangement of atoms and bonds.* The structural formulas of both butane and ethyl alcohol are given in Figure 13.1.

Chemists find it convenient to draw shorthand or condensed versions of structural formulas. **Condensed structural formulas** *show only specific bonds; other bonds are implied.* The degree of condensation depends on which bonds are shown and which are implied. For example, since hydrogen forms only a single bond to carbon, such bonds need not be shown.

$$CH_3{-}CH_2{-}CH_2{-}CH_3$$

The C—H bonds are implied. Since carbon always forms four bonds, the carbon atoms at each end of butane are understood to have three single bonds to hydrogen. Those carbon atoms in the interior of the molecule have two carbon–carbon bonds shown, whereas the two carbon–hydrogen bonds are understood. Note that the hydrogen atoms are written after the carbon atom. The following condensed formula gives correct information, but by convention such structures are generally not used.

$$H_3C{-}H_2C{-}H_2C{-}H_3C$$

In a further condensation of a structural formula, both the C—H and C—C bonds can be "left out" and understood to be present.

$$CH_3CH_2CH_2CH_3$$

In this representation of butane, the carbon atom on the left is understood to be bonded to the three hydrogen atoms and to the carbon atom to the right. The second carbon atom from the left is bonded to the two hydrogen atoms to the right. In addition, that carbon atom is bonded to the two carbon atoms to the immediate right and left, since carbon must have four bonds.

In the most condensed formula of butane, similar structural units are grouped together within parentheses. The number of times the unit is repeated is shown by a subscript following the closing parenthesis.

$$CH_3(CH_2)_2CH_3$$

The —CH_2— unit is called a methylene group. It occurs twice in butane. Since the methylene units are linked together in a repeating chain, they are placed within the parentheses. Thus, $(CH_2)_2$ in a condensed formula means

$$-CH_2{-}CH_2{-} \quad \text{or} \quad \begin{array}{c} H \quad\; H \\ | \quad\; | \\ -C{-}C{-} \\ | \quad\; | \\ H \quad\; H \end{array}$$

Example 13.1 _____

The structural formula of octane is given in Figure 13.1. Write three condensed structural formulas for octane.

Solution. With the C—H bonds understood, we may write

CH_3—CH_2—CH_2—CH_2—CH_2—CH_2—CH_2—CH_3

With both the C—H and C—C bonds understood, we may write

$CH_3CH_2CH_2CH_2CH_2CH_2CH_2CH_3$

In the most condensed version, the six methylene units may be represented within parentheses.

$CH_3(CH_2)_6CH_3$

[Additional examples may be found in **13.13–13.15** at the end of the chapter.]

Ethyl alcohol can be written with two differing degrees of condensation.

CH_3—CH_2—OH CH_3CH_2OH

In the first, both the C—H and O—H bonds are understood, while both the C—C and C—O bonds are shown. In the second, the C—C and C—O bonds are implied. The carbon atom on the right is bonded to two hydrogen atoms, to the carbon atom on the left, and to the oxygen atom on the right.

Example 13.2 _____

Write the completely expanded structural formula for the structure given by $CH_3(CH_2)_2OH$.

Solution. The units within parentheses can be expanded to give

$CH_3CH_2CH_2OH$

which subsequently can be written to show C—C and C—O bonds to give

CH_3—CH_2—CH_2—OH

and finally to show all bonds.

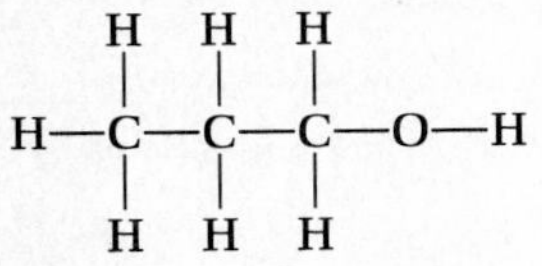

```
    H   H   H
    |   |   |
H—C—C—C—O—H
    |   |   |
    H   H   H
```

[Additional examples may be found in **13.16–13.19** at the end of the chapter.]

13.5
Structural Models

When you look at a structural formula of a molecule shown in two dimensions, you should understand what is represented. However, perhaps more important, you should know what is not conveyed by the two-dimensional representation of a three-dimensional molecule. For example, it is incorrect to view methane as a planar molecule with carbon at the center of a square.

On paper, the three-dimensional shape of methane can be shown by using a wedge and a dashed line (Figure 13.2). The wedge is viewed as a bond extending

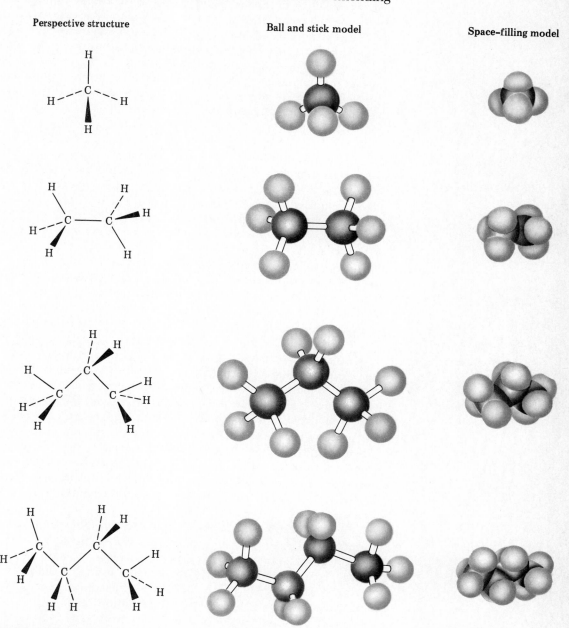

Figure 13.2
Perspective Structures and Molecular Models.

out of the plane of the page toward the reader. The dashed line represents a bond directed behind the plane of the page. The other two lines are bonds in the plane of the page. *Representations of the wedge–dashed line type are* **perspective structural formulas.** Perspective structures of more complex organic molecules are given in Figure 13.2. In each case, the three-dimensional shape of the molecule is shown.

Chemists use models of molecules that can be constructed manually and then viewed from a variety of angles. You may find it useful of purchase a molecular model kit to help you understand the structures of organic molecules. Two types of molecular models are ball and stick and space filling. Each has certain advantages and disadvantages. The ball-and-stick models show the molecular framework and bond angles clearly: the balls represent the nuclei of the atom, and the sticks represent the bonds (Figure 13.2). The actual molecular volume is shown unrealistically. Space-filling models show the entire space occupied by the electrons surrounding each atom. As a consequence, the carbon skeleton and its bond angles are obscured (Figure 13.2).

13.6
Conformations

When you are building a molecular model, you may sometimes arrive at representations that appear to be different. Consider ethane as shown in Figure 13.3. In both examples, the two carbon atoms are bonded to each other, and each carbon atom is bonded to three hydrogen atoms. The two representations differ in how the hydrogen atoms of one carbon are positioned relative to the hydrogen atoms of the other carbon atom. Which of the two forms represents ethane? The answer is that both do to some degree.

When an sp^3 hybrid orbital of one carbon atom overlaps with an sp^3 hybrid orbital of another carbon atom, it does so along the axis between the two nuclei. Each carbon atom and its three bonded hydrogen atoms can rotate about the resulting C—C σ bond. The two carbon atoms are still bonded (Figure 13.3). Only the positions of the hydrogen atoms on the adjacent carbon atoms are altered.

Molecules like ethane are in constant rotational and vibrational motion. The motion could be compared to the twisting and turning of your body while dancing. You may look different, but all parts of your body are still attached in a specific sequence. Only the orientation of your limbs is changing. *A molecule can exist in different orientations or conformations by rotation about single bonds. The individual molecules in a given conformation are called* **conformers.**

Ethane can exist in an infinite number of conformations as the two CH_3 portions of the molecule rotate with respect to each other. However, the staggered conformation is of lower energy. In this conformation, the hydrogen atoms are as far apart as possible, and there is less repulsion between them. There is a larger fraction of ethane in the staggered conformation than in any other conformation.

The higher-energy conformation has the hydrogen atoms closest to each other. Each C—H bond on one carbon atom eclipses a C—H bond on another carbon atom, much as the earth may eclipse the moon. The energy of the eclipsed conformation is 3 kcal/mole higher than that of the staggered conformation. The difference in energy between eclipsed and staggered conformations is quite small. Thus, we can regard the rotation about the C—C bond as free or unrestricted. Conformers of ethane cannot be separated from the sample. The properties of ethane represent those of a mixture of conformers.

When writing structural formulas of molecules you may write two or more

Figure 13.3

Conformations of Ethane.
Rotation of the carbon on the right by 60°
converts the staggered conformation into
the eclipsed conformation. The rotation
occurs about the σ bond between the carbon atoms.

staggered

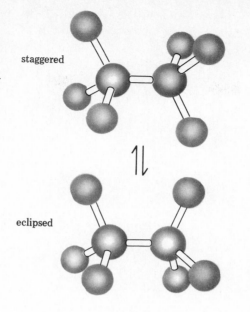

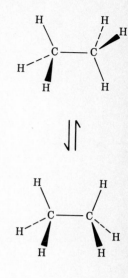

eclipsed

ball and stick models

perspective structures

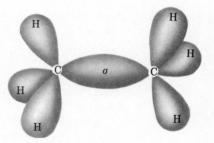

representations of the same molecule. Consider butane as shown in Figure 13.4.
Each structure consists of four carbon atoms connected in a chain. Only the
orientation of the chain differs in these structures, which represent different
conformations. Butane is a three-dimensional molecule in which rotation can
occur about each C—C bond. Just as there are numerous conformers possible,
we can write many structural formulas for butane. By convention, chemists
choose to write two-dimensional representations of carbon chains in a straight
line, rather than as a bent chain. Each "link" of the chain is straightened out by
pulling at the end "links" (carbon atoms) to produce a straight line.

You may have noticed that ammonium isocyanate, NH_4OCN, and urea,
NH_2CONH_2, consist of the same number of identical atoms. Each can be represented by CH_4N_2O. There are many other cases in which two or more compounds have the same molecular formulas. *Compounds that have the same molecular formula but differ in physical properties are called* **isomers.**

The phenomenon of isomeric substances is one that contributes to the large
number of organic compounds. Although there is only one substance repre-

**13.7
Isomers**

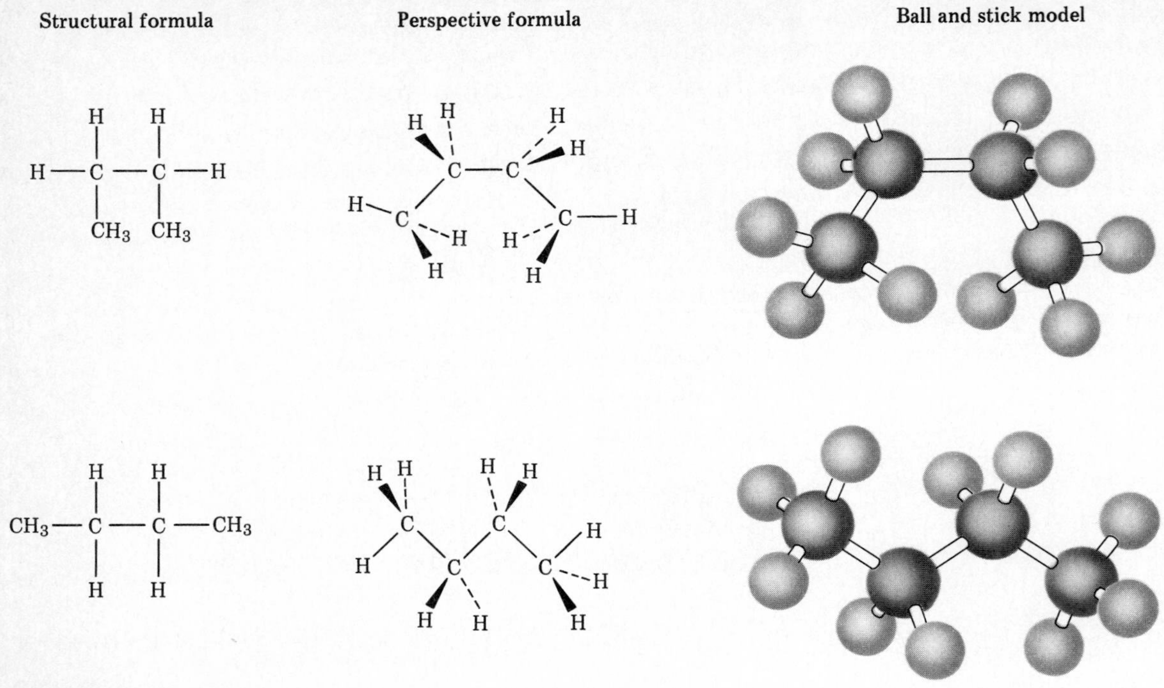

Structural formula   Perspective formula   Ball and stick model

Figure 13.4
Structural Formulas and Conformations of Butane.

sented by CH_4, C_2H_6, or C_3H_8, there are two substances with the molecular formula C_4H_{10}. There are 75 compounds or isomers with a molecular formula $C_{10}H_{22}$. Although very few of them have been isolated or synthesized, it is calculated that there are 62,491,178,805,831 isomers of $C_{40}H_{82}$!

Isomers have the same composition, but they differ in structure, that is, in the sequence in which the atoms are bonded. In order to understand this phenomenon, let's consider the two isomers of C_4H_{10}. One isomer is butane, which consists of an uninterrupted chain of carbon atoms (Figure 13.5). In the second isomer, called isobutane, only three carbon atoms are connected in sequence, while the fourth forms a branch. Note that butane and isobutane are not conformers. One isomer cannot be converted into another isomer by rotation about C—C bonds. Interconversion or isomerization could occur only by breaking and forming a different sequence of bonds.

Butane and isobutane have different physical properties and can be separated if mixed together. Butane melts at $-138°C$ and boils at $-1°C$, whereas isobutane melts at $-160°C$ and boils at $-12°C$.

In Section 13.4 you were given the structural formula CH_3CH_2OH for ethyl alcohol. The molecular formula is C_2H_6O. There is a second substance called dimethyl ether, which has the molecular formula C_2H_6O; therefore, dimethyl ether and ethyl alcohol are isomers. How do they differ? The answer is that while the compositions are identical, the structures are different. The structures are given by structural formulas shown in Table 13.1. If we neglect the hydrogen atoms, we find that the atomic sequence is C—C—O in ethyl alcohol and C—O—C in dimethyl ether. Note that the physical properties are dramatically

Figure 13.5
Isomers of C_4H_{10}.

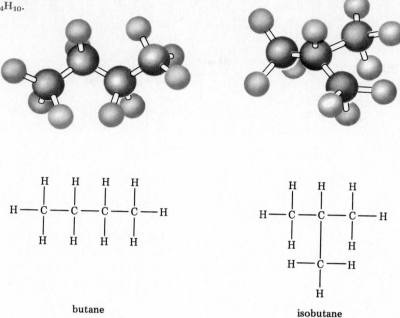

butane   isobutane

different. In addition there is an important difference in chemical properties. For example, ethyl alcohol reacts with sodium to produce hydrogen gas, whereas dimethyl ether does not react with sodium.

It is important that you be able to distinguish between conformers of a compound and isomers of it. For example, 1,2-dichloroethane can exist in an infinite number of conformers resulting from rotation about the C—C bond. A few of the two-dimensional structural formulas are shown on the next page.

Table 13.1

Structure and Properties of C_2H_6O Isomers

	Ethyl alcohol	**Dimethyl ether**
Molecular formula	C_2H_6O	C_2H_6O
Structural formula		
Condensed structural formula	CH_3—CH_2—O—H	CH_3—O—CH_3
Boiling point	78.5°C	−24°C
Melting point	−117°C	−138.5°C
Solubility in H_2O	completely soluble in all proportions	slightly soluble
Reactivity with Na	vigorous reaction; H_2 evolved	no reaction

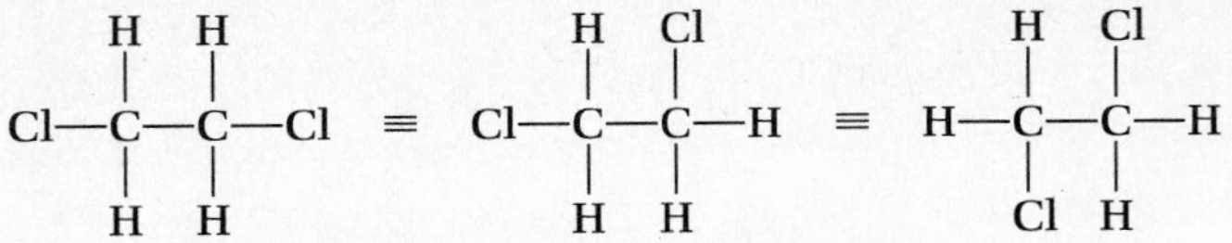

$$
\begin{array}{ccc}
\underset{\substack{|\\H}}{\overset{\substack{H\;\;\;H}}{Cl-C-C-Cl}} & \equiv & \underset{\substack{|\\H\;\;\;H}}{\overset{\substack{H\;\;\;Cl}}{Cl-C-C-H}} \equiv \underset{\substack{|\\Cl\;\;\;H}}{\overset{\substack{H\;\;\;Cl}}{H-C-C-H}}
\end{array}
$$

1,2-dichloroethane, CH_2ClCH_2Cl

In each formula, by neglecting the C—H bonds, we see that the bonding sequence is Cl—C—C—Cl.

The isomer of 1,2-dichloroethane is 1,1-dichloroethane. That the two compounds are isomers can be seen by noting the location of the two chlorine atoms.

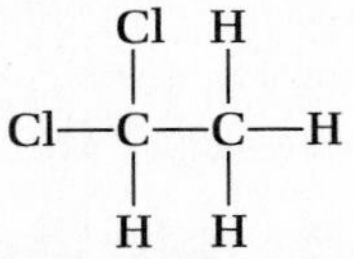

$$
\underset{\substack{|\\H\;\;\;H}}{\overset{\substack{Cl\;\;\;H}}{Cl-C-C-H}}
$$

1,1-dichloroethane, $CHCl_2CH_3$

In 1,1-dichloroethane, the two chlorine atoms are bonded to the same carbon atom. In 1,2-dichloroethane, the two chlorine atoms are bonded to different carbon atoms. This fact is emphasized in the different condensed structural formulas, $CHCl_2CH_3$ and CH_2ClCH_2Cl. The boiling point of 1,1-dichloroethane is 57°C, whereas that of 1,2-dichloroethane is 84°C.

Example 13.3

Consider the following structural formulas. Do they represent isomers?

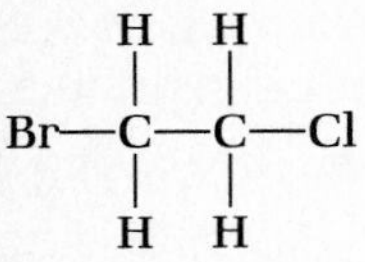

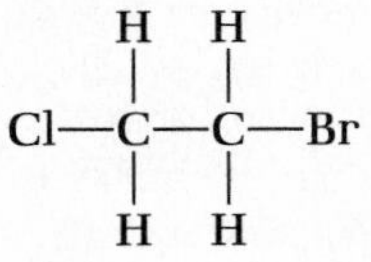

$$
\underset{\substack{|\\H\;\;\;H}}{\overset{\substack{H\;\;\;H}}{Br-C-C-Cl}} \qquad \underset{\substack{|\\H\;\;\;H}}{\overset{\substack{H\;\;\;H}}{Cl-C-C-Br}}
$$

Solution. The atomic compositions given in these structural formulas are identical. The molecular formula is C_2H_4ClBr. The condensed structural formula of the first structure is $BrCH_2CH_2Cl$ and the second is $ClCH_2CH_2Br$. However, the two structural formulas do not represent isomers. The order of bonds between atoms is Br—C—C—Cl in both cases. The second structural formula is merely the first structural formula written "in reverse".

Example 13.4

Consider the following structural formulas. Do they represent isomers?

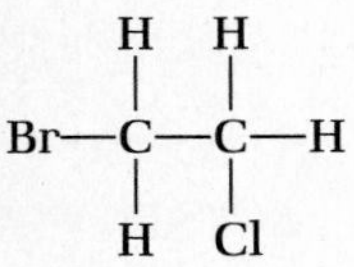

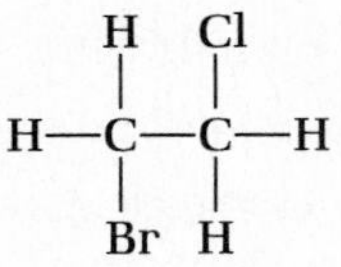

$$
\underset{\substack{|\\H\;\;\;Cl}}{\overset{\substack{H\;\;\;H}}{Br-C-C-H}} \qquad \underset{\substack{|\\Br\;\;\;H}}{\overset{\substack{H\;\;\;Cl}}{H-C-C-H}}
$$

Solution. The atomic compositions given in these structural formulas are identical, and therefore the molecular formula is C_2H_4ClBr. The condensed structural formula of the first structure is $BrCH_2CH_2Cl$, and the second is also $BrCH_2CH_2Cl$.

The two structural formulas do not represent isomers. The order of the bonds between bonded atoms is Br—C—C—Cl in both cases. Since the atoms bonded to a carbon can be variously located in space because of rotation about the carbon–carbon bond, numerous conformations are possible.

[Additional examples may be found in **13.22–13.25** at the end of the chapter.]

13.8
Functional Groups

Classification, a technique used in many subject areas of chemistry, is used in organic chemistry as well. It is virtually impossible to study the millions of organic compounds without some sort of grouping or classification. The classification is based on the functional groups contained in molecules. *Atoms or groups of atoms and their bonds that confer on a molecule similar physical and chemical properties are called* **functional groups.** The chapters in the organic section of this text are arranged by functional groups.

A carbon–carbon double bond is an example of a functional group. In ethylene, the double bond reacts in an addition reaction with hydrogen in the presence of a platinum catalyst.

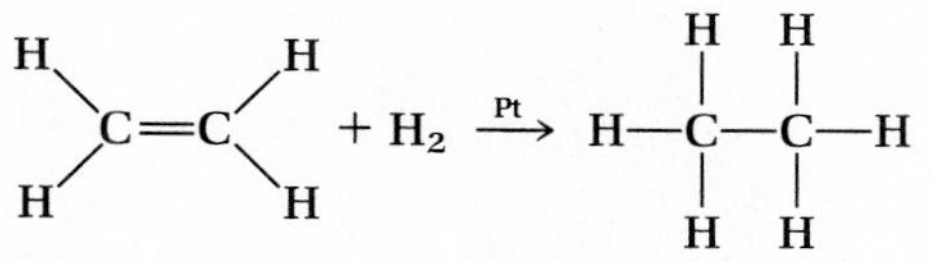

More complex compounds that contain a carbon–carbon double bond also react to add hydrogen. The chemistry of the carbon–carbon double bond will be presented in Chapter 15.

Functional groups often contain elements other than carbon; the most common elements are oxygen and nitrogen, although sulfur or the halogens may also be present. A dozen of the most common functional groups are listed in

Example 13.5 ───────────────────────────────

Classify the following compound by its functional group. Suggest a reaction of this compound.

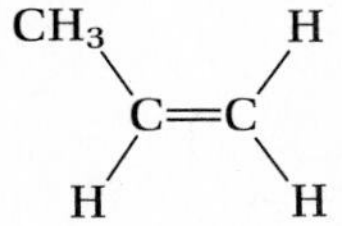

Solution. The compound has a carbon–carbon double bond and is classified as an alkene. This compound should react similarly to ethylene. The double bond can react with hydrogen in a process known as an addition reaction.

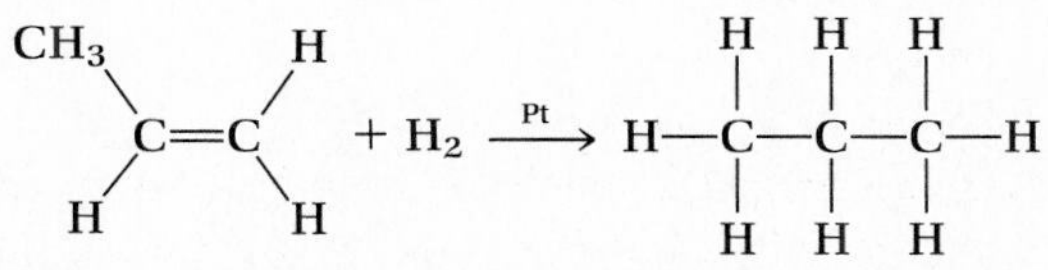

Table 13.2

Classes and Functional Groups of Organic Compounds

Class	Functional group	Example of expanded structural formula	Example of condensed structural formula
alkene	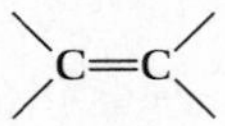 $>C=C<$	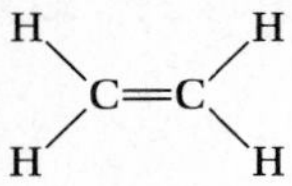 H₂C=CH₂ (expanded)	$CH_2{=}CH_2$
alkyne	$-C{\equiv}C-$	$H-C{\equiv}C-H$	$CH{\equiv}CH$
alcohol	$-O-H$	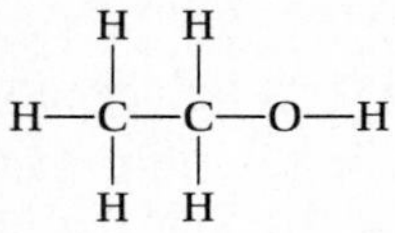 H–C–C–O–H	$CH_3CH_2{-}OH$
ether	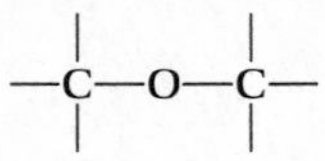 $-C-O-C-$	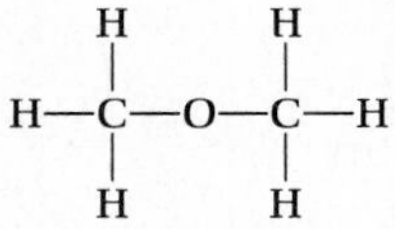 H–C–O–C–H	$CH_3{-}O{-}CH_3$
aldehyde	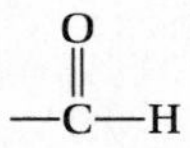 $-\overset{O}{\overset{\|}{C}}-H$	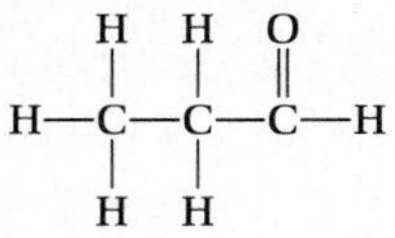 H–C–C–$\overset{O}{\overset{\|}{C}}$–H	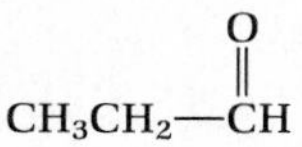 $CH_3CH_2{-}\overset{O}{\overset{\|}{C}}H$
ketone	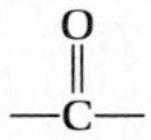 $-\overset{O}{\overset{\|}{C}}-$	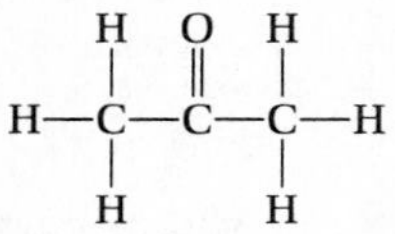 H–C–$\overset{O}{\overset{\|}{C}}$–C–H	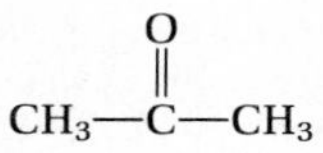 $CH_3{-}\overset{O}{\overset{\|}{C}}{-}CH_3$
carboxylic acid	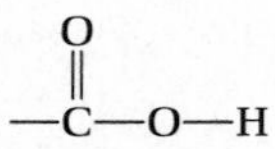 $-\overset{O}{\overset{\|}{C}}-O-H$	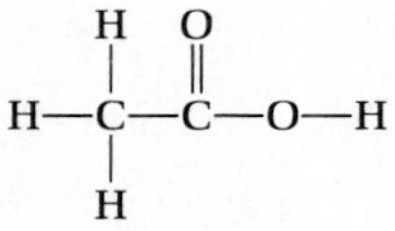 H–C–$\overset{O}{\overset{\|}{C}}$–O–H	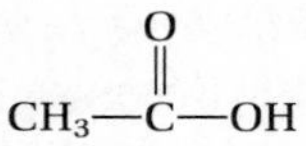 $CH_3{-}\overset{O}{\overset{\|}{C}}{-}OH$
ester	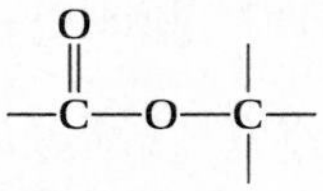 $-\overset{O}{\overset{\|}{C}}-O-C-$	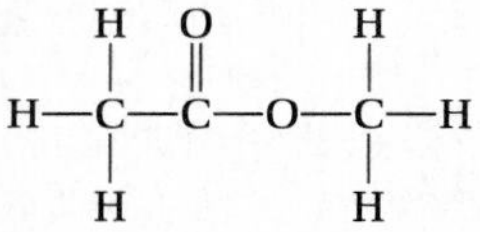 H–C–$\overset{O}{\overset{\|}{C}}$–O–C–H	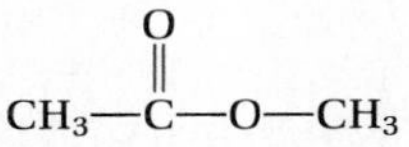 $CH_3{-}\overset{O}{\overset{\|}{C}}{-}O{-}CH_3$
amine	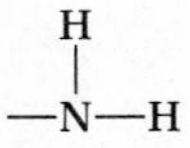 $-N-H$ (with H on top)	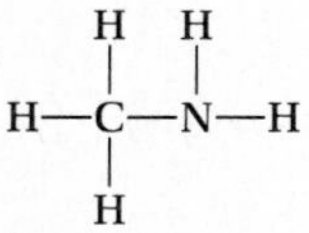 H–C–N–H	$CH_3{-}NH_2$
amide	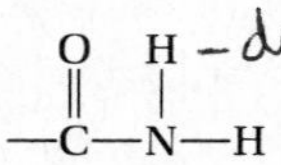 $-\overset{O}{\overset{\|}{C}}-N-H$	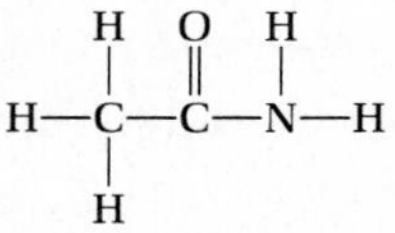 H–C–$\overset{O}{\overset{\|}{C}}$–N–H	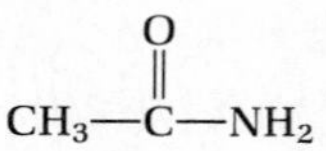 $CH_3{-}\overset{O}{\overset{\|}{C}}{-}NH_2$
halide	$-X$ (X = F, Cl, Br, I)	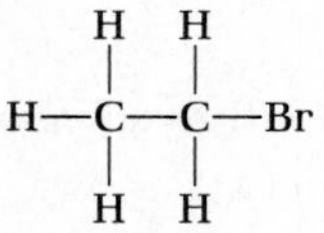 H–C–C–Br	$CH_3CH_2{-}Br$
mercaptan or thiol	$-S-H$	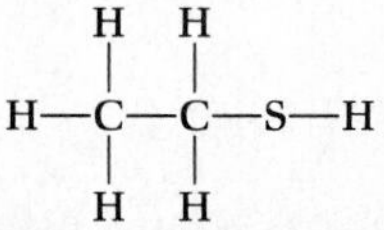 H–C–C–S–H	$CH_3CH_2{-}SH$

H — doesn't always have to have H

✶ amine also can be $-\overset{|}{N}-C$ or $N\overset{-H}{\underset{-C}{}}$

Table 13.2. Remember that the reason for using functional groups is to classify molecules that have similar properties. Once you learn the properties and reactions of one functional group, you will know the properties and reactions of thousands of compounds in that class.

Example 13.6 _____

Consider the following two compounds. Suggest a single reagent that will react with one compound but not the other.

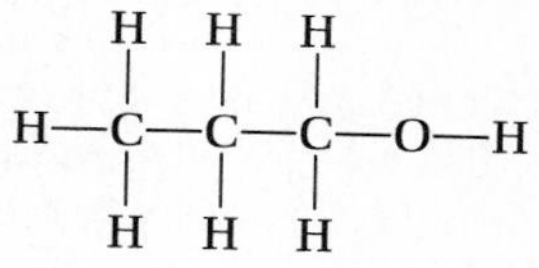

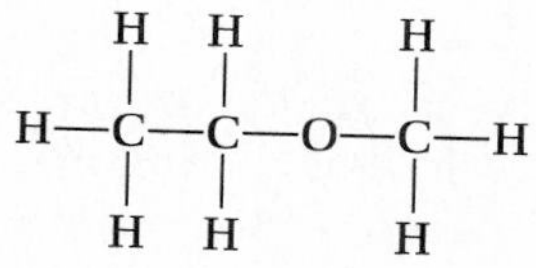

Solution. The first compound has an O—H grouping like ethyl alcohol and is classified as an alcohol. The second compound has a C—O—C grouping and is an ether. Ethyl alcohol reacts with sodium, and it would be expected that the first compound would as well. Dimethyl ether does not react with sodium, and thus it would be expected that the second compound also would not react with sodium.

_____

[Additional examples may be found in **13.26–13.30** at the end of the chapter.]

13.9 Nomenclature

Nomenclature *is a system of naming materials.* In chemistry, the nomenclature of compounds is exceedingly important. Without a system of names, the science of organic chemistry with its millions of compounds would be difficult to comprehend. The phenomenon of isomerism easily illustrates this important point. There are two isomeric C_4H_{10} compounds called butane and isobutane. There are two isomeric C_2H_6O compounds called ethyl alcohol and dimethyl ether.

When few organic compounds were known, chemists gave them common names such as butane and isobutane. However, how could common names be devised and memorized for the 75 isomers of $C_{10}H_{22}$ or the 62,491,178,805,831 isomers of $C_{40}H_{82}$? A systematic method of naming compounds needed to be developed.

There was yet another reason for developing the nomenclature of organic compounds. Often different names had been given to the same compound. For example, CH_3CH_2OH had been called simply alcohol, but also spirit, grain alcohol, ethyl alcohol, methyl carbinol, and ethanol. Furthermore, a variety of names were developed in each language.

In 1892, at a meeting in Geneva, Switzerland, chemists systematized the nomenclature of all compounds including organic compounds. The system continues as part of the activities of the International Union of Pure and Applied Chemistry (**IUPAC**). The rules result in a clear and definitive name for each compound. Once the rules are applied, only one name results for each given structure and one structure for a given name. This system will be discussed in subsequent chapters. IUPAC names in the figures will be in color.

In spite of the IUPAC system, many common names are so well established that both common and IUPAC names must be recognized. The IUPAC name for CH_3CH_2OH is ethanol, but ethyl alcohol is also still used as a common name.

Example 13.7

Although you have not been given any of the IUPAC rules for nomenclature, suggest the IUPAC name for the following compound, whose common name is methyl alcohol.

$$\begin{array}{c} \text{H} \\ | \\ \text{H}-\text{C}-\text{O}-\text{H} \\ | \\ \text{H} \end{array}$$

Solution. The compound is an alcohol containing one carbon atom. The name for the alcohol with two carbon atoms is ethanol. Ethanol can be compared structurally to ethane, except it has an O—H functional group in place of a hydrogen atom. The name ethanol may be considered to be derived from ethane minus the terminal -e plus -ol. Thus, the alcohol with one carbon atom might be named after methane. Removal of the terminal -e and adding -ol gives methanol. This is the IUPAC name for the compound.

Summary

Organic compounds have chemical and physical properties that are consequences of the carbon–carbon covalent bond. The great variety of organic compounds is the result of stable single and multiple bonds between carbon atoms as well as the incorporation of other atoms into bonds with carbon. Certain groupings of atoms called functional groups are used to classify organic compounds. These functional groups confer common chemical reactivities on a class of compounds.

Isomerism is common among organic compounds. Isomers have identical molecular formulas but different structures. Two or more models or written structural representations for the same organic compound that represent different orientations of atoms in space are called conformers. Conformers can be interconverted by rotation about carbon–carbon single bonds.

Organic compounds have specific names assigned by the International Union of Pure and Applied Chemistry. There is a one-to-one correspondence between a structure and its name.

Learning Objectives

As a result of studying Chapter 13 you should be able to

- List some of the differences between inorganic and organic compounds.

- Describe the unique bonding characteristics of the carbon atom.
- Draw structures of simple organic molecules.
- Distinguish between molecular and structural formulas.
- Write condensed structural formulas from a complete structural formula.
- Write a complete structural formula from a condensed structural formula.
- Represent a molecule in a three-dimensional perspective structure.
- Define and illustrate the phenomenon of conformations.
- Recognize isomers from structural formulas.
- Distinguish between conformers and isomers.
- Recognize major functional groups.

New Terms

Concatenation is the formation of networks of identical atoms joined to each other.

A **condensed formula** is a simplified structural formula in which some of the bonds are not shown but implied.

Conformers refer to different spatial arrangements of atoms in space as a result of rotation about single bonds.

A **functional group** is an atom or group of atoms that confers specific properties to an organic molecule.

Isomers have the same molecular formula but different structural formulas.

The **IUPAC** is an acronym for the International Union of Pure and Applied Chemistry, a scientific organization that devises rules for selecting unique names for chemical compounds.

Nomenclature is a system of naming materials.

Perspective formulas are structural formulas written in

two dimensions that impart some three-dimensional aspects to the representation.

Structural formulas represent the spatial arrangement of atoms and bonds.

Questions and Problems

Terminology

13.1 Explain the difference between a molecular formula and a structural formula.

13.2 Explain the difference between conformers and isomers.

Bonding in Carbon

13.3 What kinds of bonds are common in organic compounds?

13.4 Describe the four ways in which carbon can form covalent bonds to other carbon atoms.

13.5 List a few of the atoms that may form bonds to carbon.

Molecular Formulas

13.6 What is the molecular formula of each of the following simple organic compounds?
(a) methane   (b) ethane
(c) ethylene   (d) acetylene
(e) propane   (f) butane
(g) ethyl alcohol   (h) dimethyl ether

13.7 Write the molecular formula for each of the following structures.
(a) $CH_3-CH_2-CH_2-CH_2-CH_3$
(b) $CH_3-CH_2-CH_2-CH_3$
(c) $CH_3CH_2CH_2CH_2CH_2CH_3$
(d) $CH_3CH_2CH_2CH_2CH_2CH_2CH_2CH_3$

13.8 Write the molecular formula for each of the following structures.
(a) $CH_3-CH_2-C{\equiv}C-H$
(b) $CH_3-CH_2-C{\equiv}C-CH_3$
(c) $CH_3-CH_2-CH_2-CH{=}CH_2$
(d) $CH_2{=}CH-CH_2-CH_3$

13.9 Write the molecular formula for each of the following structures.
(a) $CH_3-CH_2-CH_2-OH$
(b) $CH_3-CH_2-O-CH_2-CH_3$
(c) $CH_3-CH_2-O-CH_3$
(d) $HO-CH_2-CH_2-CH_2-CH_3$

13.10 Write the molecular formula for each of the following structures.
(a) $CH_3-CH_2-CHCl_2$
(b) $CH_3-CCl_2-CH_3$
(c) $Br-CH_2-CH_2-Br$
(d) $Br-CH_2-CH_2-CH_2-Cl$

13.11 Write the molecular formula for each of the following structures.
(a) CH_3-CH_2-S-H
(b) $CH_3-CH_2-S-CH_3$
(c) $CH_3-CH_2-CH_2-NH_2$
(d) $CH_3-CH_2-NH-CH_3$

13.12 Write the molecular formula for each of the following structures.

(a) 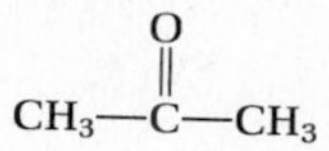
$$CH_3-\overset{\overset{\displaystyle O}{\|}}{C}-CH_3$$

(b)
$$CH_3-CH_2-CH_2-\overset{\overset{\displaystyle O}{\|}}{C}-O-CH_3$$

(c)
$$CH_3-CH_2-CH_2-CH_2-\overset{\overset{\displaystyle O}{\|}}{C}-OH$$

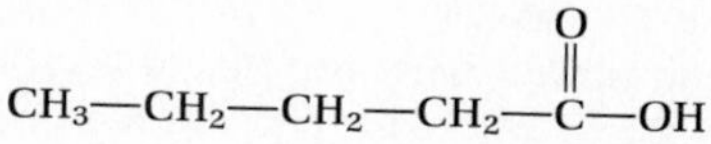

(d) 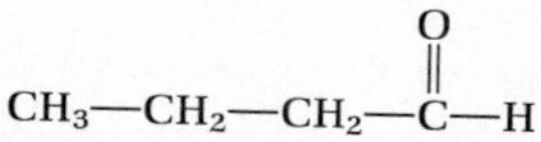
$$CH_3-CH_2-CH_2-\overset{\overset{\displaystyle O}{\|}}{C}-H$$

Condensed Structural Formulas

13.13 Write condensed structural formulas in which only the bonds to hydrogen are not shown.

(a) 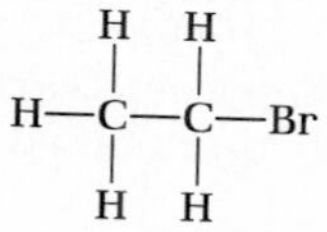
$$\begin{array}{ccc} H & H \\ | & | \\ H-C-C-Br \\ | & | \\ H & H \end{array}$$

(b)
$$\begin{array}{ccc} H & Br \\ | & | \\ H-C-C-H \\ | & | \\ H & H \end{array}$$

(c)
$$\begin{array}{ccc} H & H \\ | & | \\ Br-C-C-Br \\ | & | \\ H & H \end{array}$$

(d)
$$\begin{array}{ccccc} H & H & H & H & H \\ | & | & | & | & | \\ H-C-C-C-C-C-H \\ | & | & | & | & | \\ H & H & H & H & H \end{array}$$

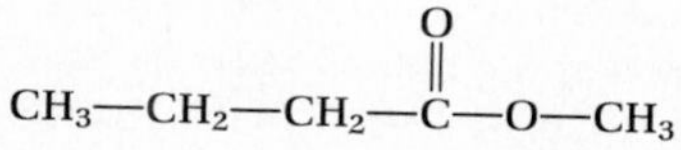

(e)
$$\begin{array}{cccc} H & H & H \\ | & | & | \\ H-C-C-C-S-H \\ | & | & | \\ H & H & H \end{array}$$

(f)
$$\begin{array}{cccc} H & H & H \\ | & | & | \\ H-C-C-C-N-H \\ | & | & | & | \\ H & H & H & H \end{array}$$

(g)
$$\begin{array}{cccc} H & H & H & H \\ | & | & | & | \\ H-C-C-C-C-OH \\ | & | & | & | \\ H & H & H & H \end{array}$$

(h)
$$\begin{array}{ccccc} H & H & H & H & H \\ | & | & | & | & | \\ H-C-C-C-C-C-Cl \\ | & | & | & | & | \\ H & H & H & H & H \end{array}$$

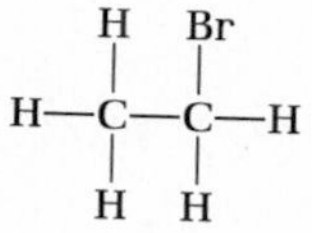

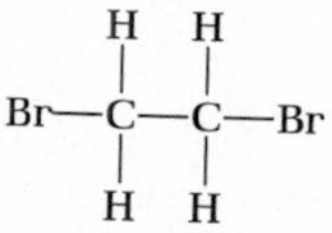

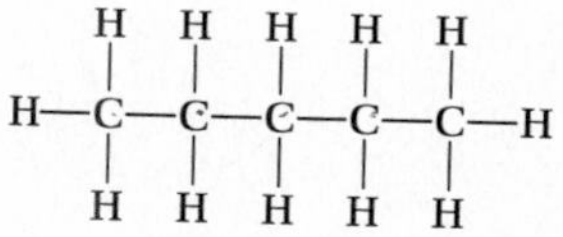

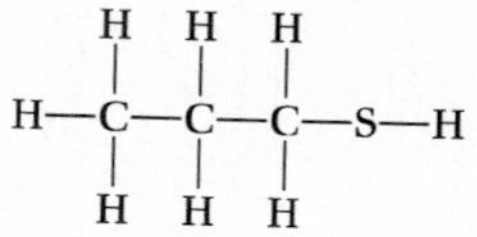

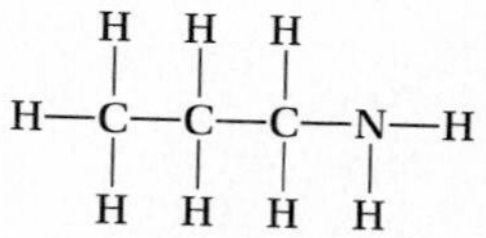

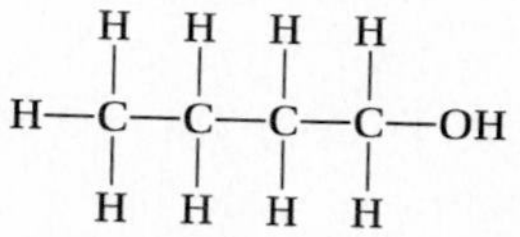

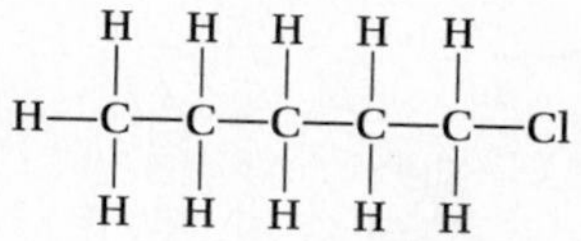

13.14 Write a condensed structural formula in which no bonds are shown for each substance in **13.13.**

13.15 Write completely condensed formulas for (d), (g), and (h) of **13.13.**

Structural Formulas

13.16 Write a complete structural formula showing all bonds for each of the following condensed formulas.
(a) $CH_3CH_2CH_2CH_3$ (b) $CH_3CH_2CH(CH_3)_2$
(c) $CH_3CH(CH_3)CH_2CH_3$ (d) $(CH_3)_3CCH_2CH_3$

13.17 Write a complete structural formula showing all bonds for each of the following condensed formulas.
(a) $CH_3CH_2CHBr_2$
(b) $CH_3CHBrCH_2CH_2CH_3$
(c) $CH_3CH_2CCl_2CH_2CH_3$
(d) $CH_2ClCH_2CHClCH_3$

13.18 Write a complete structural formula showing all bonds for each of the following condensed formulas.
(a) CH_3CHCH_2 (b) $CH_3CH_2CHCCl_2$
(c) $CH_3CH_2CCCH_3$ (d) $CH_3CH_2CH_2CCH$

13.19 Write a complete structural formula showing all bonds for each of the following condensed formulas.
(a) $CH_3CH_2CH_2CH_2SH$
(b) $CH_3CH_2OCH_2CH_2CH_3$
(c) $CH_3CH_2CH_2CH_2OH$
(d) $CH_3SCH_2CH_2CH_3$
(e) $CH_3CH_2NHCH_3$
(f) $CH_3CH_2CH_2CH_2NH_2$

Perspective Structures

13.20 Draw a three-dimensional structure for each of the following.
(a) CH_3Cl (b) CH_2Br_2 (c) CH_3F
(d) CHF_3 (e) CH_2ClBr (f) CCl_4

13.21 Draw a three-dimensional structure for each of the following.
(a) CH_3CH_3 (b) CH_3NH_2 (c) CH_3OH
(d) CH_3SH (e) CH_3CH_2Cl (f) CH_3CHCl_2

Isomers

13.22 Which of the following pairs consist of isomers?
(a) CH_3OH and CH_3CH_2OH
(b) $CH_3C{\equiv}CH$ and $CH_3CH{=}CH_2$
(c) $CH_3CH_2CH_2OH$ and $CH_3CH_2OCH_3$
(d) $CH_3{-}\overset{\overset{O}{\|}}{C}{-}CH_3$ and $CH_3{-}CH_2{-}\overset{\overset{O}{\|}}{C}{-}H$
(e) $CH_3{-}CH_2{-}CH_3$ and $CH_3{-}\underset{\underset{CH_3}{|}}{CH_2}$
(f) $CH_3CHClCH_3$ and $CH_3CH_2CH_2Cl$
(g) $CH_2ClCH_2CH_2Cl$ and $CH_3CH_2CHCl_2$
(h) $CH_3CHClCH_2Cl$ and $CHCl_2CH_2CH_3$

13.23 There are two isomers for each of the following molecular formulas. Draw their structural formulas.
(a) $C_2H_4Br_2$ (b) C_2H_6O (c) C_2H_4BrCl
(d) C_2H_6S (e) C_3H_7Cl (f) C_2H_7N
(g) $C_2H_3Br_3$ (h) C_3H_6O

13.24 There are three isomers for each of the following molecular formulas. Draw their structural formulas.
(a) $C_2H_3Br_2Cl$ (b) C_3H_8O (c) C_3H_8S
(d) C_5H_{12}

Isomers and Conformers

13.25 Indicate whether the following pairs of structures are isomers or conformers.

(a) $H{-}\overset{\overset{Br}{|}}{\underset{\underset{H}{|}}{C}}{-}\overset{\overset{H}{|}}{\underset{\underset{H}{|}}{C}}{-}Br$ and $Br{-}\overset{\overset{H}{|}}{\underset{\underset{H}{|}}{C}}{-}\overset{\overset{H}{|}}{\underset{\underset{H}{|}}{C}}{-}Br$

(b) $CH_3{-}\underset{\underset{CH_2{-}Cl}{|}}{CH_2}$ and $CH_3{-}CH_2{-}CH_2{-}Cl$

(c) $CH_3{-}\underset{\underset{CH_3}{|}}{CHCl}$ and $CH_3{-}CH_2{-}CH_2{-}Cl$

(d) $CH_3{-}\underset{\underset{CH_3}{|}}{CH}{-}CH_2{-}CH_3$ and $CH_3{-}CH_2{-}\underset{\underset{CH_2{-}CH_3}{|}}{CH_2}$

Functional Groups

13.26 Name the class of compounds to which each of the following belongs.
(a) $CH_3{-}\overset{\overset{O}{\|}}{C}{-}O{-}CH_2{-}CH_3$
(b) $CH_3{-}CH_2{-}O{-}CH_2{-}CH_3$
(c) $CH_3{-}CH_2{-}\overset{\overset{O}{\|}}{C}{-}H$
(d) $CH_3CH_2CH{=}CH_2$
(e) $CH_3{-}C{\equiv}C{-}CH_3$
(f) $CH_3{-}CH_2{-}CH_2{-}OH$
(g) $CH_3{-}CH_2{-}S{-}H$
(h) $CH_3{-}CH_2{-}CH_2{-}CH_2{-}NH_2$

13.27 Identify all of the functional groups present in each of the following compounds.
(a) $CH_3{-}\overset{\overset{O}{\|}}{C}{-}\overset{\overset{O}{\|}}{C}{-}OH$ (b) $\underset{\underset{OH}{|}}{CH_2}{-}\underset{\underset{OH}{|}}{CH}{-}\overset{\overset{O}{\|}}{C}{-}H$
(c) $CH_3{-}\underset{\underset{NH_2}{|}}{CH}{-}\overset{\overset{O}{\|}}{C}{-}OH$ (d) $CH_3{-}\overset{\overset{O}{\|}}{C}{-}NH{-}CH_3$

13.28 Classify the following compounds and suggest a substance that will react with one but not the other.

$$CH_3CH_2CH_2CH_2CH_2OH \qquad CH_3CH_2OCH_2CH_2CH_3$$

13.29 Recalling the acid–base properties of ammonia from Chapter 11, predict the product of the following reaction.

$$CH_3NH_2 + HCl \longrightarrow$$

13.30 Skunks emit the following substance to ward off their enemies.

$$CH_3CH_2CH_2CH_2SH$$

On this basis, what generalization can be made about the properties of mercaptans?

Saturated Hydrocarbons

Kerosene is obtained from petroleum. The liquid is a mixture of saturated hydrocarbons.

Recall that when the term saturated is used to describe a solution, it means that the solution contains the maximum amount of a solute. When the term saturated is used in organic chemistry, it refers to the type of bonds to carbon. *A hydrocarbon is* **saturated** *if it has the maximum possible number of hydrogen atoms bonded to its carbon atoms.* This occurs only if there are no multiple bonds present. A saturated hydrocarbon cannot undergo addition reactions, since it has no multiple bonds. Ethane, a saturated hydrocarbon, does not react with hydrogen, whereas both ethylene and acetylene, which are unsaturated hydrocarbons, do react.

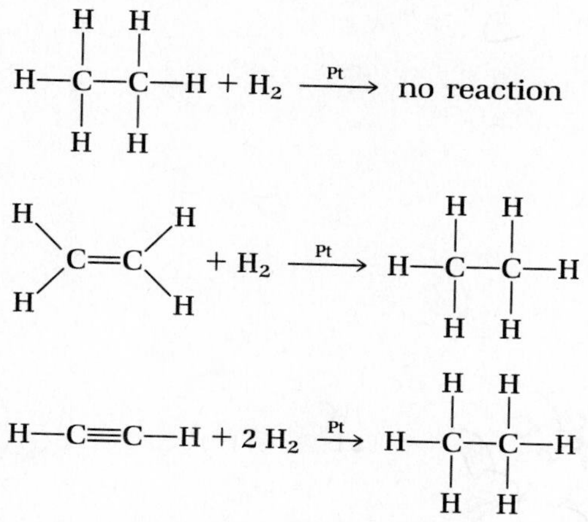

Most organic molecules of interest contain functional groups. These functional groups are responsible for the characteristic chemical reactivity of organic molecules. In subsequent chapters, the chemistry of several functional groups will be discussed. However, it is first necessary to examine the structures, names, physical properties, and chemical properties of the hydrocarbons that form the backbone of organic molecules.

Saturated hydrocarbons are of two types: alkanes and cycloalkanes. Alkanes contain carbon atoms bonded in chains of atoms. In cycloalkanes, some of the carbon atoms are bonded to form a ring of atoms.

Alkanes *contain carbon atoms bonded either to other carbon atoms or to hydrogen atoms by single covalent bonds.* Each carbon atom has four covalent bonds in a tetrahedral arrangement about it. Each carbon atom is sp^3 hybridized, and the four sp^3 orbitals form sigma bonds to other atoms.

Normal Alkanes

The names and condensed structural formulas of 10 alkanes are given in Table 14.1. These compounds, called normal alkanes, consist of carbon atoms bonded one to another like links in a chain. They are also referred to as "straight-chain" alkanes, even though their actual structure is "zig-zag" with bond angles of 109.5°. More precisely these substances have a continuous chain of carbon atoms, but the term straight chain is still widely used in textbooks.

The general molecular formula for "straight-chain" alkanes is C_nH_{2n+2}. The reason why this general representation can be used can be seen by looking at the structural formula of hexane, C_6H_{14}.

Table 14.1

Structure and Nomenclature of Normal Alkanes

Molecular formula	IUPAC prefix	IUPAC name	Structural formula
CH_4	meth-	methane	CH_4
C_2H_6	eth-	ethane	CH_3CH_3
C_3H_8	prop-	propane	$CH_3CH_2CH_3$
C_4H_{10}	but-	butane	$CH_3CH_2CH_2CH_3$
C_5H_{12}	pent-	pentane	$CH_3CH_2CH_2CH_2CH_3$
C_6H_{14}	hex-	hexane	$CH_3CH_2CH_2CH_2CH_2CH_3$
C_7H_{16}	hept-	heptane	$CH_3CH_2CH_2CH_2CH_2CH_2CH_3$
C_8H_{18}	oct-	octane	$CH_3CH_2CH_2CH_2CH_2CH_2CH_2CH_3$
C_9H_{20}	non-	nonane	$CH_3CH_2CH_2CH_2CH_2CH_2CH_2CH_2CH_3$
$C_{10}H_{22}$	dec-	decane	$CH_3CH_2CH_2CH_2CH_2CH_2CH_2CH_2CH_2CH_3$

$$
\begin{array}{c}
\;H\;\;\;H\;\;\;H\;\;\;H\;\;\;H\;\;\;H \\
\;|\;\;\;\;|\;\;\;\;|\;\;\;\;|\;\;\;\;|\;\;\;\;| \\
H-C-C-C-C-C-C-H \\
\;|\;\;\;\;|\;\;\;\;|\;\;\;\;|\;\;\;\;|\;\;\;\;| \\
\;H\;\;\;H\;\;\;H\;\;\;H\;\;\;H\;\;\;H
\end{array}
$$

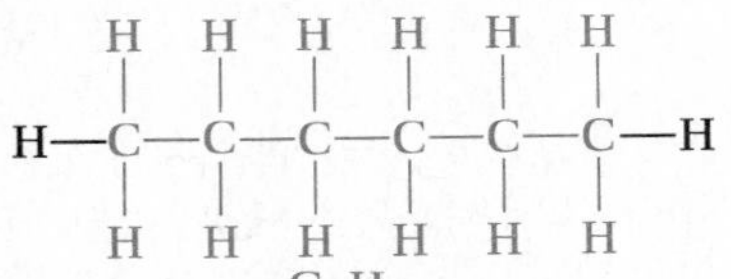

$$C_nH_{2n}$$

$$H_1 \quad + \quad C_6H_{12} \quad + \quad H_1 = C_6H_{14}$$

With the exception of the two carbon atoms at the ends of the chain, each carbon atom has two hydrogen atoms bonded to it, accounting for the $2n$ in the general formula. Each of the two end carbon atoms has an additional hydrogen atom bonded to it. This accounts for the $+2$ in the subscript on hydrogen. Each compound in Table 14.1 differs from the one above or below it by a —CH_2— unit. *A series of compounds that differs from adjacent members by a repeating unit is called a* **homologous series.**

Branched Alkanes

Alkanes also exist as branched-chain alkanes. *In* **branched-chain alkanes,** *one or more carbon atoms is bonded to more than two other carbon atoms.* The carbon atom that is bonded to three or four other carbon atoms is the branching point. *The group of carbon atoms attached to the main chain of carbon atoms at the branching point is called an* **alkyl group.**

Isobutane is an example of a branched-chain alkane. There are three atoms in the main chain, and the alkyl group consists of a single carbon atom with three bonded hydrogen atoms.

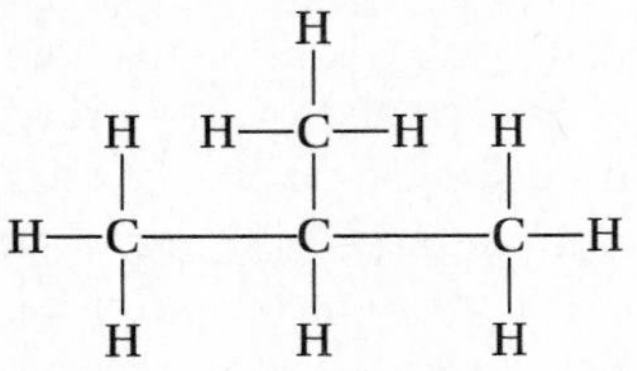

$$
\begin{array}{c}
H \\
| \\
H\;\;\;\;H-C-H\;\;\;H \\
|\;\;\;\;\;\;\;\;|\;\;\;\;\;\;\;\;| \\
H-C-\!-\!-\!-C-\!-\!-\!-C-H \\
|\;\;\;\;\;\;\;\;|\;\;\;\;\;\;\;\;| \\
H\;\;\;\;\;\;\;H\;\;\;\;\;\;H
\end{array}
$$

You may find it useful to regard the main carbon chain as a main street in a town. The branch then is a side street.

The general formula for a branched-chain alkane is also C_nH_{2n+2}. Each time a

Figure 14.1
Structural Represen-
tations of Propane.
The two structural
formulas are equiva-
lent. There is only
one propane com-
pound. The structure
on the right is pre-
ferred by convention.

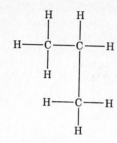

branch is present, the number of hydrogen atoms at the branching point, which
is 2*n* in straight-chain hydrocarbons, is decreased by one. However, the branch
now provides another "end" to the molecule, which means there is an additional
hydrogen in the branch above the 2*n* value.

Example 14.1

What is the molecular formula of an alkane containing 25 carbon atoms?

Solution. The value of *n* is 25. There must be $2(25) + 2$ hydrogen atoms. The
molecular formula is $C_{25}H_{52}$. Any alkane, normal or branched, with 25 carbon
atoms would have 52 hydrogen atoms.

[Additional examples may be found in **14.6–14.8** at the end of the chapter.]

Representing Alkanes

A model of propane is shown in Figure 14.1. There are many conformations
that can result from rotation about either of the two carbon–carbon bonds. Recall, however, that conformers are not isomers. There is only one substance of
molecular formula C_3H_8.

Propane and other alkanes of higher molecular weight can be written in a
variety of two-dimensional representations. Two such drawings are given for propane. While they are equally correct, it is less confusing to write linear arrangements of carbon atoms. Remember, however, that all bond angles in propane are
109.5°, and not 90° and 180° as they appear in the two-dimensional structures.
The structure for propane with a 90° angle between carbon atoms is really the
same as the linear arrangement. Both structures show that the three carbons
atoms are connected in a chain.

The two isomeric alkanes with the formula C_4H_{10} were illustrated in Chapter 13.
The compounds that constitute the butane family were shown in Figure 13.5. The
common name for the isomer with a continuous chain of carbon atoms is normal
butane, *n*-butane, or just butane. The other isomer has only three carbons in a
continuous chain and has a "branch" of one carbon atom. The branched isomer
is called isobutane.

As the number of carbon atoms in an alkane increases, the number of isomers also increases. A list of the number of isomers is given in Table 14.2. Many of
these possible isomers have never been found in petroleum or produced in a
chemistry laboratory. However, any of them could be made if it were needed.

14.3
Isomers of Alkanes

Table 14.2

Number of Isomers of the Alkanes

Molecular formula	Number of isomers
CH_4	1
C_2H_6	1
C_3H_8	1
C_4H_{10}	2
C_5H_{12}	3
C_6H_{14}	5
C_7H_{16}	9
C_8H_{18}	18
C_9H_{20}	35
$C_{10}H_{22}$	75
$C_{20}H_{42}$	336,319
$C_{30}H_{62}$	4,111,846,763
$C_{40}H_{82}$	62,491,178,805,831

Models and two-dimensional structures of the three isomers of the pentane family are given in Figure 14.2. Note that the preferred method of illustration is to draw the longest chain of carbon atoms in a horizontal line.

Example 14.2

Are the following representations of the same substance or isomers?

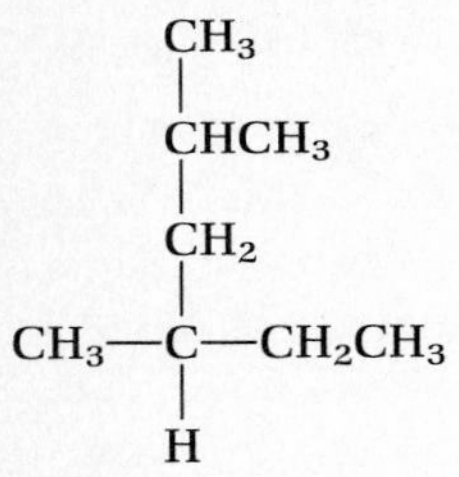

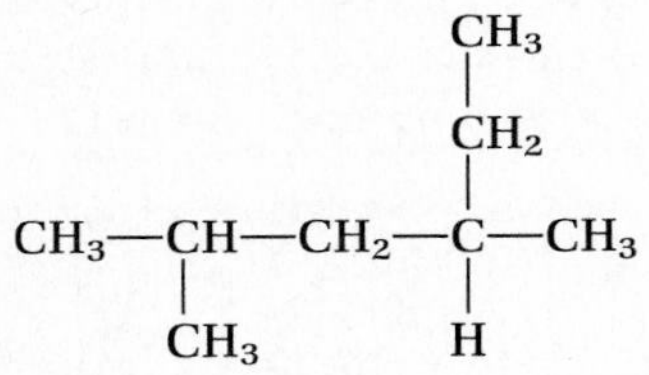

$$
\begin{array}{ccc}
CH_3 & & CH_3 \\
| & & | \\
CHCH_3 & & CH_2 \\
| & & | \\
CH_2 & & CH_3{-}CH{-}CH_2{-}C{-}CH_3 \\
| & & | \quad\quad\quad | \\
CH_3{-}C{-}CH_2CH_3 & & CH_3 \quad\quad\ H \\
| & & \\
H & &
\end{array}
$$

Solution. The longest chain in each representation contains six carbon atoms. Both have two branches located at the same points. The structures are both equivalent to

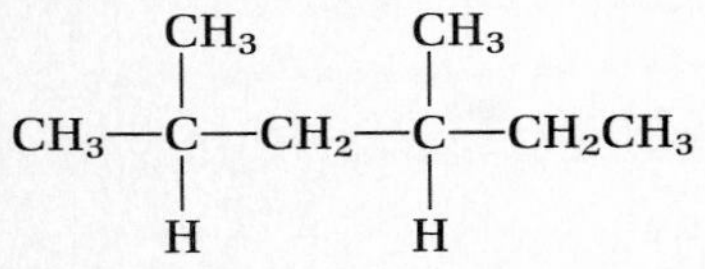

$$
\begin{array}{cc}
CH_3 & CH_3 \\
| & | \\
CH_3{-}C{-}CH_2{-}C{-}CH_2CH_3 \\
| & | \\
H & H
\end{array}
$$

[Additional examples may be found in **14.9** and **14.10** at the end of the chapter.]

14.4

Nomenclature of Alkanes

The International Union of Pure and Applied Chemistry (IUPAC) rules are used to describe and name alkanes. If these rules are followed, a unique name is obtained for every compound. The name will then tell you the exact structure of a compound, although you may never have heard of it before.

Figure 14.2
The Isomeric C_5H_{10}
Compounds.

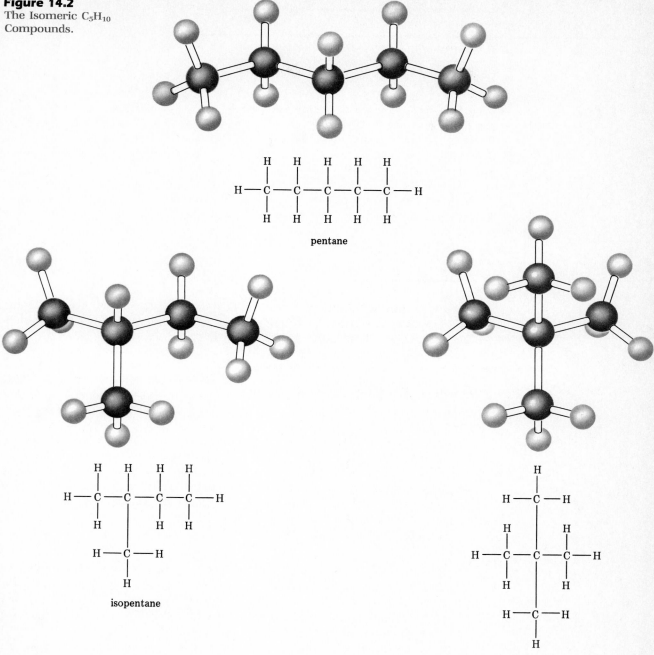

pentane

isopentane

neopentane

The IUPAC method for naming alkanes depends on identifying and number-ing the longest carbon chain in a molecule. Groups such as halogens or alkyl groups, which are attached to this chain, are then designated by name and loca-tion. The system is then analogous to naming a street, numbering the houses on the street, and listing who lives at each address.

The names of the simpler alkyl groups are given in Table 14.3. A general shorthand representation of an alkyl group is R—. The alkyl group resembles an

Table 14.3 ——

The Structure of Alkyl Groups

Parent alkane	Alkyl group	Name
CH_4	CH_3-	methyl
CH_3CH_3	CH_3CH_2-	ethyl
$CH_3CH_2CH_3$	$CH_3CH_2CH_2-$	propyl
	$\begin{array}{c} H_3C \\ \\ H_3C \end{array}\!\!\!\!>\!CH-$	isopropyl
$CH_3CH_2CH_2CH_3$	$CH_3CH_2CH_2CH_2-$	butyl
	$\overset{\displaystyle CH_3}{\underset{\displaystyle \vert}{CH_3CH_2CH-}}$	*sec*-butyl (secondary butyl)
$\overset{\displaystyle CH_3}{\underset{\displaystyle \vert}{CH_3CHCH_3}}$	$\overset{\displaystyle CH_3}{\underset{\displaystyle \vert}{CH_3CHCH_2-}}$	isobutyl
	$CH_3-\overset{\displaystyle CH_3}{\underset{\displaystyle \vert}{\overset{\displaystyle \vert}{C}}}-$ $\underset{\displaystyle CH_3}{}$	*t*-butyl (tertiary butyl)

alkane that is missing one hydrogen atom. *Alkyl groups are named by replacing the -ane ending of an alkane by -yl.* Thus, CH_3- is the methyl group. Note that there are two different alkyl groups with three carbon atoms, depending on which hydrogen atom of propane is missing.

Since there are two isomeric compounds with the formula C_4H_{10}, the alkyl groups derived from each structure must be different. There are two alkyl groups derived from butane and two from isobutane. There are four C_4H_9 alkyl groups.

Other groups such as halogens may be substituted for hydrogen on a chain of carbon atoms. In the case of halogens, the groups are called halo groups. Thus, $Cl-$ and $Br-$ are chloro and bromo, respectively.

The IUPAC rules for naming alkanes are as follows.

1. The longest continuous chain of carbon atoms is selected as the base. This chain is named according to the stem name listed in Table 14.1 plus the suffix -ane.
2. The carbon atoms in the chain are numbered consecutively from that end of the chain nearest a branch.
3. Each branch is located by the number of the atom to which it is attached on the chain.

$$\overset{}{\underset{4321}{CH_3CH_2-\overset{\displaystyle CH_3}{\underset{\displaystyle H}{\overset{\displaystyle \vert}{\underset{\displaystyle \vert}{C}}}}-CH_3}} \quad \text{is} \quad \text{2-methylbutane}$$

4. If two or more of the same types of branches occur, the number of them is

indicated by the prefixes di-, tri-, etc., and the location of each on the main chain is indicated by a number.

$$
\overset{1}{C}H_3 \overset{2}{-} \overset{3}{\underset{\underset{CH_3}{|}}{C}H} \overset{}{-} \overset{}{\underset{\underset{CH_3}{|}}{\overset{\overset{CH_3}{|}}{C}H}} \overset{4}{-} CH_3 \quad \text{is} \quad \text{2,3-dimethylbutane}
$$

5. The numbers for the positions of the alkyl groups are placed immediately before the names of the groups, and hyphens are placed before and after the numbers. If two or more numbers occur together, commas are placed between them.

6. When several different alkyl groups are present, they are placed in alphabetical order and prefixed onto the stem name of the longest continuous chain. The whole name is written as a single word.

Although the systematic names are more cumbersome than the common names, the structure corresponding to the IUPAC name can be written readily even if that particular name is unfamiliar. The name contains all the information needed to specify the complete structure. The rules can be applied to complex alkanes as well, for example,

$$
\overset{1}{C}H_3\overset{2}{C}H_2 \overset{3}{-} \overset{}{\underset{\underset{\underset{CH_3}{|}}{CH_2}}{C}H} \overset{4}{-} \overset{\overset{CH_3}{|}}{C}H \overset{5}{-} \overset{6}{C}H_2\overset{}{C}H_2 \overset{7}{-} \overset{}{\underset{\underset{\underset{CH_3}{|}}{CH_2}}{C}H} \overset{8}{-} \overset{9}{C}H_2\overset{10}{C}H_2CH_3
$$

1. Because the longest chain has ten carbons atoms, this is a decane.
2. The chain is numbered from left to right because the branch nearest the end is located on the left. This numbering will result in the branches being assigned low position numbers.
3. There are ethyl groups on carbon atoms 3 and 7. These are named and numbered -3,7-diethyl.
4. A methyl group is located at position 4 and is named -4-methyl.
5. Assembling all components, we have 3,7-diethyl-4-methyldecane. The hyphen preceding the 3 is dropped.

Be careful to locate the longest possible chain and also to number the chain correctly. The correct name and an incorrect one are given for two examples. Only if the rules are carefully applied can each structure correspond to a unique name and each name correspond to a unique structure.

$$
\overset{5}{C}H_3\overset{4}{C}H_2\overset{3}{C}H_2 \overset{2}{-} \overset{\overset{CH_3}{|}}{\underset{\underset{CH_3}{|}}{C}} \overset{1}{-} CH_3
$$

2,2-dimethylpentane
(*not* 4,4-dimethylpentane)

$$
CH_3 \overset{3}{-} \overset{}{\underset{\underset{\underset{\underset{^1CH_3}{|}}{^2CH_2}}{|}}{C}H} \overset{4}{-} \overset{\overset{CH_3}{|}}{C}H \overset{5}{-} \overset{6}{C}H_2CH_3
$$

3,4-dimethylhexane
(*not* 2-ethyl-3-methylpentane)

Example 14.3

Name the following compound.

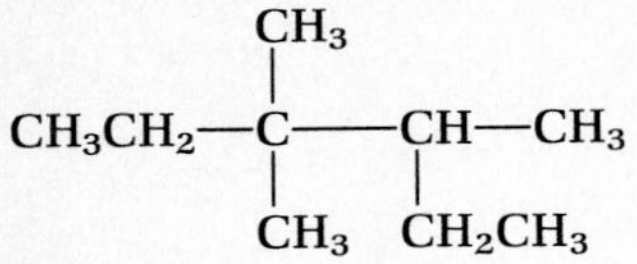

$$CH_3CH_2-\overset{\overset{\displaystyle CH_3}{|}}{\underset{\underset{\displaystyle CH_3}{|}}{C}}-\overset{}{\underset{\underset{\displaystyle CH_2CH_3}{|}}{CH}}-CH_3$$

Solution. The longest continuous chain has six carbon atoms. The chain is numbered so that the three methyl groups are located at positions 3, 3, and 4. The compound is 3,3,4-trimethylhexane. The name 2-ethyl-3,3-dimethylpentane is incorrect.

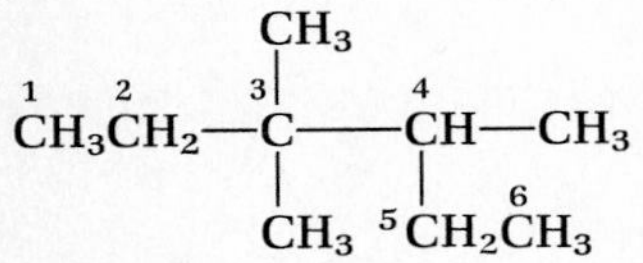

$$\overset{1}{C}H_3\overset{2}{C}H_2-\overset{\overset{\displaystyle CH_3}{|}{\scriptstyle 3}}{\underset{\underset{\displaystyle CH_3}{|}}{C}}-\overset{4}{\underset{\underset{\underset{\displaystyle CH_2CH_3}{5\;\;\;\;\;\;6}}{|}}{CH}}-CH_3$$

[Additional examples may be found in **14.11–14.19** at the end of the chapter.]

14.5 Classification of Carbon and Hydrogen Atoms

Because there are many different arrangements of carbon and hydrogen atoms possible in hydrocarbons, it is convenient to designate parts of the structures according to the number of carbon and hydrogen atoms that are attached to a specific carbon atom. *A* **primary carbon atom** *is one that is directly bonded to only one other carbon atom.*

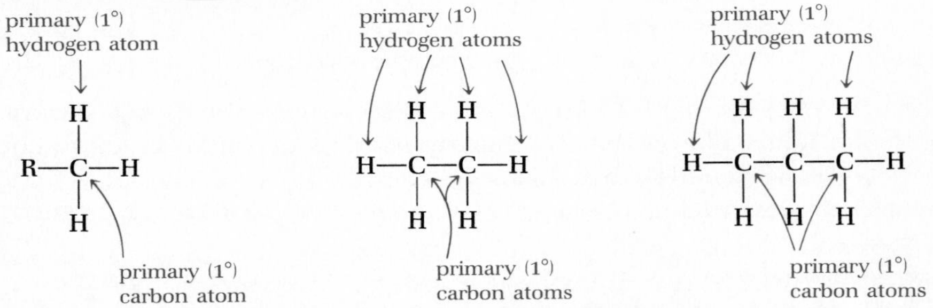

Ethane contains two primary carbon atoms. Propane contains two primary carbon atoms; the internal carbon atom in propane is not primary, for it is bonded to two other carbon atoms. *The hydrogen atoms bonded to primary carbon atoms are called primary hydrogens.* Ethane contains six primary hydrogen atoms, as does propane.

A **secondary carbon atom** *is bonded to two other carbon atoms*, as is the case for the internal carbon atom of propane. *The hydrogen atoms attached to a secondary carbon atom are called secondary hydrogen atoms.* Propane contains two secondary hydrogen atoms.

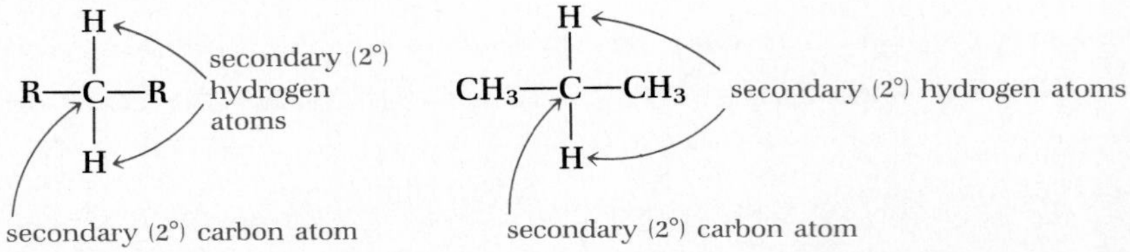

A **tertiary carbon atom** *is bonded to three other carbon atoms.* One of the four carbon atoms in 2-methylpropane is a tertiary carbon atom. The other three carbon atoms are primary. *A hydrogen atom directly bonded to a tertiary carbon atom is a tertiary hydrogen atom.* There is one tertiary hydrogen atom in 2-methylpropane.

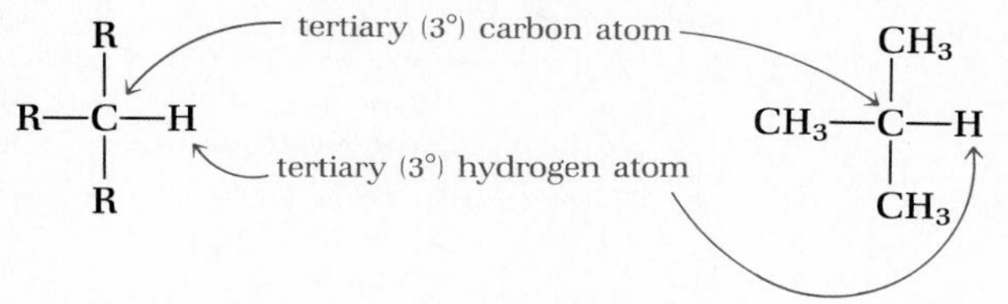

A **quaternary carbon atom** *is bonded to four other carbon atoms.*

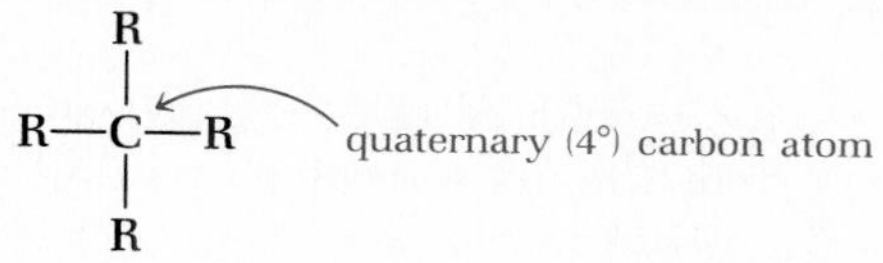

Example 14.4

What types of carbon atoms are present in 2-methylbutane?

Solution. The structure when properly drawn has three primary carbon atoms, one secondary carbon atom, and one tertiary carbon atom.

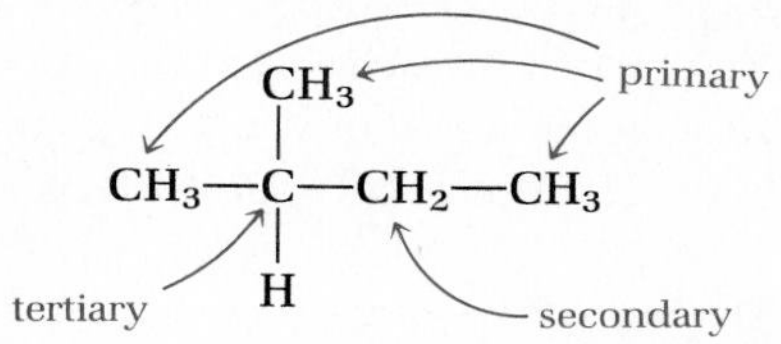

[Additional examples may be found in **14.24–14.29** at the end of the chapter.]

14.6
Physical Properties of Alkanes

The boiling points of the normal alkanes increase with increasing molecular weight (Table 14.4). Branching results in a decrease in boiling point (Table 14.5). Both of these trends are the result of London forces. As the molecular weight and chain length increase, there are more polarizable bonds and the attractive forces increase. With increased branching, the molecules are more compact and the possibility of polarizable sites attracting each other decreases.

Pure alkanes are colorless, tasteless, and nearly odorless. You may have noted, however, that gasoline does have an odor and some color. This is because dyes are added to gasoline by refiners to indicate composition. Gasoline also contains aromatic compounds (Section 15.9), which have characteristic odors.

Alkanes are not soluble in water because they are nonpolar substances. The alkanes will dissolve other nonpolar organic materials such as fats and oils. Inhalation of alkane vapors causes severe damage to lung tissue because the fatty material in cell membranes is dissolved. Similarly, long-term contact between low-molecular-weight alkanes and skin will remove skin oils and can cause soreness and blisters.

Table 14.4
Physical Properties of Some Normal Alkanes

Name	Formula	Melting point (°C)	Boiling point (°C)
methane	CH_4	−183	−162
ethane	C_2H_6	−173	−89
propane	C_3H_8	−187	−42
butane	C_4H_{10}	−138	0
pentane	C_5H_{12}	−130	36
hexane	C_6H_{14}	−95	69
heptane	C_7H_{16}	−91	98
octane	C_8H_{18}	−57	126
nonane	C_9H_{20}	−54	151
decane	$C_{10}H_{22}$	−30	174

Table 14.5
Boiling Points of C_6H_{14} Isomers

Compound	Boiling point (°C)
hexane	69
3-methylpentane	63
2-methylpentane	60
2,3-dimethylbutane	58
2,2-dimethylbutane	50

High-molecular-weight alkanes have some medicinal value. Mixtures of C_{20} to C_{30} alkanes are used in some skin and hair lotions to replace natural oils. Mineral oil, a purified mixture of C_{20} to C_{30} alkanes, can be used as a laxative. Petroleum jelly, a mixture of C_{20} to C_{40} alkanes, is used to protect skin from irritants and is a base for some medicated salves.

14.7
Reactions of Alkanes

The carbon–carbon and carbon–hydrogen bonds of alkanes are not very reactive. For this reason, *alkanes are also called paraffins, a name derived from Latin* (parum affinis, *little activity*). This low reactivity can be attributed to the type of bonds present in alkanes. The carbon–carbon bonds are σ bonds, and the bonding electrons are tightly held in a small region of space between the carbon atoms. Thus, the electrons are not readily available to other substances. Furthermore, the carbon–carbon bonds are nonpolar and do not attract other molecules.

The carbon–hydrogen bonds are only weakly polar, but they are located about the carbon skeleton and are more susceptible to reaction. Although the bonds are also sigma, they can react, but usually under extreme conditions.

Two important reactions of alkanes are combustion, an oxidation process, and halogenation, a substitution reaction.

Oxidation

The world currently depends on the combustion of petroleum as a major source of energy. Oil companies spend billions of dollars to drill wells up to 6 miles deep in such diverse places as deserts, jungles, frozen Arctic regions, and the oceans.

Methane, the major component of natural gas, yields 192 kcal/mole when burned.

$$CH_4 + 2\,O_2 \longrightarrow CO_2 + 2\,H_2O + 192\,kcal$$

Any alkane can be oxidized, but the amount of heat liberated will be different. As the molecular weight increases, so does the heat released in the oxidation reaction.

If alkanes are burned with insufficient oxygen, carbon monoxide is produced.

$$2\,CH_4 + 3\,O_2 \longrightarrow 2\,CO + 4\,H_2O$$

Carbon monoxide is produced in internal combustion engines where low oxygen-to-hydrocarbon ratios sometimes occur. Cigarette smoke also contains high levels of carbon monoxide.

Inhalation of carbon monoxide reduces the oxygen-carrying capacity of blood, since carbon monoxide binds to hemoglobin more strongly than oxygen does. The equilibrium constant for binding of carbon monoxide is 200 times that for oxygen.

$$Hb + O_2 \xrightleftharpoons{K_{O_2}} HbO_2$$

$$Hb + CO \xrightleftharpoons{K_{CO}} HbCO \qquad K_{CO} = 200 K_{O_2}$$

Thus, even at low partial pressures, carbon monoxide will bind preferentially to hemoglobin in competition with a higher partial pressure of oxygen. At low carbon monoxide concentrations, there is an effect on the central nervous system. Higher concentrations can cause a coma and death. A hyperbaric chamber (Section 7.10) can be used to increase the concentration of oxygen available for hemoglobin.

Our bodies also convert organic compounds to carbon dioxide and water and release stored chemical energy. However, our bodies do not metabolize simple alkanes. Materials such as fats and carbohydrates are used, and the body oxidizes these compounds in a controlled series of enzyme-catalyzed reactions as energy is needed.

Halogenation

Under proper reaction conditions, the carbon–hydrogen bond can undergo a substitution reaction. One such example is the halogenation of alkanes in which a halogen atom replaces or substitutes a hydrogen atom. Such reactions are exothermic, but require large activation energies.

When methane and chlorine are heated to a high temperature or are exposed to ultraviolet light, a chlorination reaction occurs.

$$CH_4 + Cl_2 \longrightarrow \underset{\substack{\text{chloromethane} \\ \text{(methyl chloride)}}}{CH_3Cl} + HCl$$

The reaction can continue to produce more extensively chlorinated materials.

$$CH_3Cl + Cl_2 \longrightarrow \underset{\substack{\text{dichloromethane} \\ \text{(methylene chloride)}}}{CH_2Cl_2} + HCl$$

$$CH_2Cl_2 + Cl_2 \longrightarrow \underset{\substack{\text{trichloromethane} \\ \text{(chloroform)}}}{CHCl_3} + HCl$$

$$CHCl_3 + Cl_2 \longrightarrow \underset{\substack{\text{tetrachloromethane} \\ \text{(carbon tetrachloride)}}}{CCl_4} + HCl$$

Halogenated hydrocarbons are used for many industrial purposes. Unfortunately such compounds can cause severe biological disturbances such as liver damage and cancer. Chloroform was used as an anesthetic, and carbon tetrachloride was used as a drycleaning solvent. Other compounds have replaced these compounds today.

Chlorination of higher-molecular-weight alkanes results in mixtures of chlorinated alkanes. For example, propane will yield two isomers.

$$
\text{CH}_3\text{CH}_2\text{CH}_3 + \text{Cl}_2 \longrightarrow \underset{\text{2-chloropropane}}{\text{CH}_3\overset{\displaystyle \overset{\text{Cl}}{|}}{\text{C}}\text{HCH}_3} + \text{HCl}
$$

$$
\text{CH}_3\text{CH}_2\text{CH}_3 + \text{Cl}_2 \longrightarrow \underset{\text{1-chloropropane}}{\text{CH}_3\text{CH}_2\text{CH}_2\text{Cl}} + \text{HCl}
$$

Therefore, halogenation reactions are generally not used if a single product is desired. Only if the compounds can be separated conveniently and both are useful is halogenation used as a preparative reaction.

Example 14.5

How many mono-, di-, and trichlorinated compounds can result from the chlorination of ethane?

Solution. Both carbon atoms in ethane are equivalent, and only one monochlorinated compound can result.

CH$_3$CH$_2$Cl
chloroethane

After the first chlorine is substituted, the two carbon atoms are not equivalent. The second chlorine may substitute on the same or different carbon atom.

CH$_3$CHCl$_2$ ClCH$_2$CH$_2$Cl
1,1-dichlorethane   1,2-dichloroethane

Three chlorine atoms may be located on the same carbon atom as in 1,1,1-trichloroethane. However, two chlorine atoms may be on one carbon atom and one on the other.

CH$_3$CCl$_3$ Cl$_2$CHCH$_2$Cl
1,1,1-trichloroethane   1,1,2-trichloroethane

[Additional examples may be found in **14.30–14.33** at the end of the chapter.]

14.8
Cycloalkanes

Carbon atoms may be bonded to form a ring or cyclic structure. Such cyclic hydrocarbons, called **cycloalkanes,** *have the general formula* C$_n$H$_{2n}$. The reason for the smaller number of hydrogen atoms in cycloalkanes than in alkanes is that an additional carbon–carbon bond is needed to form the ring. The cycloalkanes are considered saturated hydrocarbons, since they contain only carbon–carbon single bonds.

The structural formulas of four cycloalkanes are shown in Table 14.6. The fully condensed formulas are simple polygons. It is understood that each corner is a carbon atom that contains the correct number of carbon–hydrogen bonds.

Cyclopropane is a sweet-smelling, colorless gas used as an inhalation anesthetic. It produces unconsciousness in a few seconds. Cyclohexane is a widely used industrial solvent.

Cyclopropane and to some extent cyclobutane are more reactive than the other cycloalkanes. This reactivity stems from the small bond angles, which are

Table 14.6 _____
Cycloalkanes

Molecular formula	Name	Structural formula	Abbreviated notation
C_3H_6	cyclopropane	CH_2 / CH_2 CH_2	△
C_4H_8	cyclobutane	CH_2-CH_2 / CH_2-CH_2	▢
C_5H_{10}	cyclopentane	CH_2 / CH_2 CH_2 / CH_2-CH_2	⬠
C_6H_{12}	cyclohexane	CH_2 / CH_2 CH_2 / CH_2 CH_2 / CH_2	⬡

less than the 109.5° tetrahedral bond angle. The small C—C—C bond angles in these molecules cause strain, which then causes the bonds to be reactive. Both cyclopropane and cyclobutane will react with hydrogen at high temperatures.

△ $+ H_2 \xrightarrow{120°C} CH_3CH_2CH_3$

▢ $+ H_2 \xrightarrow{200°C} CH_3CH_2CH_2CH_3$

The C—C—C bond angles in cyclopentane are about 108°, which is close to the tetrahedral value. As a result, bonds in the cyclopentane ring do not react with hydrogen except at extremely high temperature.

The atoms in the cyclopentane ring can twist or partially rotate about the bonds, and some bending or puckering of the ring is possible. The cyclohexane ring has normal tetrahedral bond angles and exists in a puckered shape (Figure 14.3). *The shape resembles a reclining chair and is called the chair conformation.*

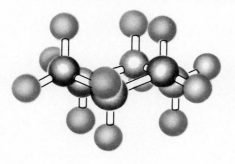

ball and stick model

space-filling model

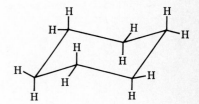

perspective structure

Figure 14.3
The Chair Conformation of Cyclohexane.

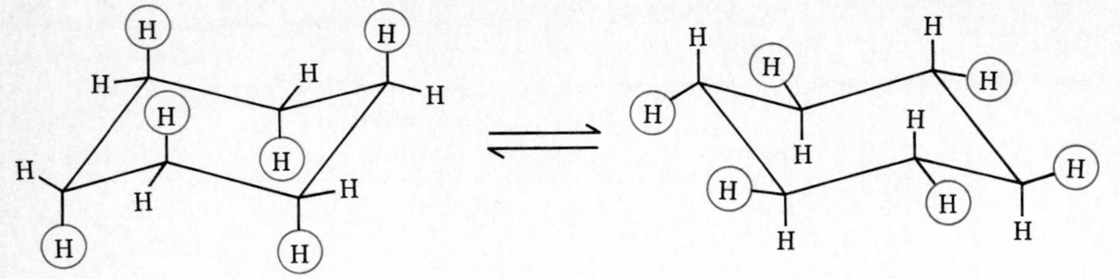

Figure 14.4
Interconversion of Chair Conformations.
The axial hydrogen atoms in structure A are circled in color. When this chair conformation is converted to an alternate chair conformation, given by structure B, the circled hydrogen atoms become equatorial.

structure A   structure B

At room temperature, about 99.99% of cyclohexane molecules are in the chair conformation.

If we consider the rough "plane" of the six carbon atoms in a ring, we see that there are two types of carbon–hydrogen bonds. *Six of the hydrogen atoms are approximately in the plane and are called equatorial hydrogen atoms.* The other six are *perpendicular to the plane and are called axial hydrogen atoms.* However, when one chair conformation is converted to another by rotation about the carbon–carbon bond, the carbon–hydrogen bonds that were originally axial become equatorial and vice versa (Figure 14.4).

Geometric Isomerism

In Chapter 13 you learned that isomerism is possible for compounds with different carbon skeletons, different functional groups, and different functional group locations. In each case, isomerism is the result of the different sequential arrangement of atoms.

Now let's consider a different type of isomerism. **Geometric isomerism** *occurs in compounds that have the same sequential arrangement of atoms but a different spatial arrangement.* Geometric isomers can occur when there is restricted rotation in a molecule, which then results in a specific spatial placement of atoms or groups of atoms attached at two sites within the molecule. Such a situation occurs in cycloalkanes and alkenes (Chapter 15). In either class of substances, there must be two groups of nonidentical atoms attached to each of two sites.

Because of the lack of free rotation in cycloalkanes, groups attached to the

Table 14.7

Geometric Isomerism in Cycloalkanes

	Three-dimensional representation	Abbreviated notation
cis isomer	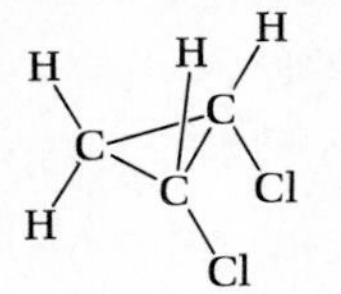	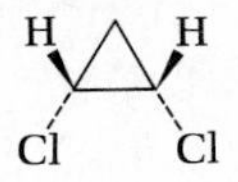
trans isomer	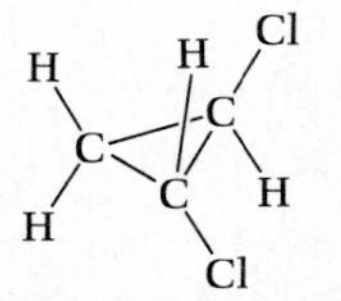	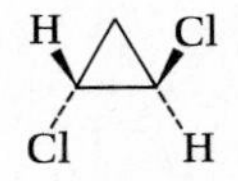

ring may be held "above" or "below" the plane of the ring. When two groups are present, they may be on the same side (*cis*) or opposite sides (*trans*) of the ring. Thus, 1,2-dichlorocyclopropane exists as two isomers, *cis* and *trans* (Table 14.7). Note that these are not two conformations of the same molecule. It is impossible to convert one isomer into the other without breaking a bond. Even in puckered molecules we can still think of "sides" to the molecule. *If groups of atoms are on the same side of the approximate plane of the ring, the substance is a **cis** isomer. The **trans** isomer has groups of atoms on the opposite sides of the approximate plane of the ring.*

Example 14.6

Can the following substances exist as *cis* and *trans* isomers?

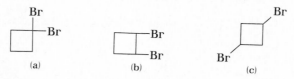

Solution. Structure (a) has only one possible arrangement for the two bromine atoms attached to the same carbon atom.

Structure (b) has the two bromine atoms on different carbon atoms. These atoms may be arranged to be on the same or opposite sides of the ring.

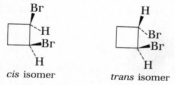

Structure (c) has the two bromine atoms on different carbon atoms. These bromine atoms may be arranged on the same or opposite sides of the ring.

Nomenclature of Cycloalkanes

Cycloalkanes are named by the IUPAC system according to the number of carbon atoms in the ring, with the prefix cyclo. When there is only one position containing a functional group, alkyl group, or single atom, there is only one possible compound and therefore no number is necessary. Thus, bromocyclopropane (Figure 14.5) defines the molecule completely.

When there is more than one group attached to the ring, the ring is numbered. One substituent is always given the number 1 and the others are given the next lowest possible number. The numbering must be done in a clockwise or counterclockwise direction so as to give the lowest combination of numbers for the groups attached to the ring.

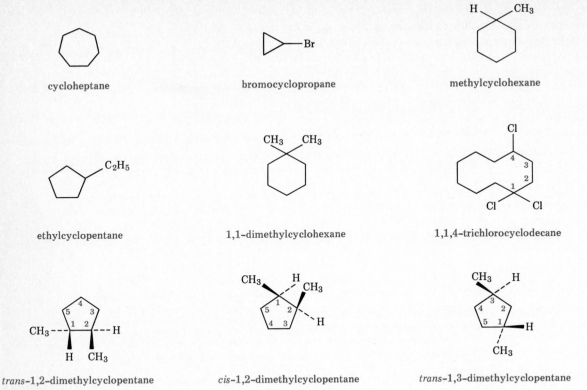

cycloheptane

bromocyclopropane

methylcyclohexane

ethylcyclopentane

1,1–dimethylcyclohexane

1,1,4–trichlorocyclodecane

trans–1,2–dimethylcyclopentane

cis–1,2–dimethylcyclopentane

trans–1,3–dimethylcyclopentane

Figure 14.5
IUPAC Nomenclature of Cycloalkanes.

Geometric isomers have the prefix *cis*- or *trans*- before the name. Care must be used to show the spatial arrangement of substituents so that the proper name may be assigned. Examples of cycloalkane nomenclature are given in Figure 14.5.

Example 14.7 _____

What is the name of the following compound?

Solution. The ring must be numbered by starting from a carbon atom with a chlorine atom. Then the direction of numbering must proceed to result in the lowest possible number for the other carbon atom bonded to a chlorine atom. Starting from the carbon atom at the "4 o'clock" position and numbering clockwise gives the number 3 to the other carbon atom with a chlorine atom. The two chlorine atoms are *cis* and the name is *cis*-1,3-dichlorocyclohexane.

Alternatively starting from the "8 o'clock" position and numbering counterclockwise gives the number 3 to the other carbon atom with a chlorine atom. The name *cis*-1,3-dichlorocyclohexane again results.

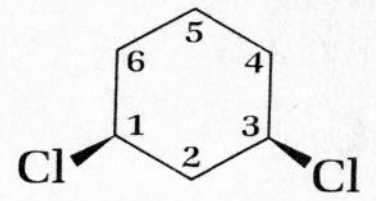

[Additional examples may be found in **14.34–14.36** at the end of the chapter.]

Cycloalkanes in Nature

Cycloalkanes containing five- and six-membered rings are quite common in nature. One such substance that also illustrates the phenomenon of geometric isomers is menthol (Figure 14.6). Although they are less common, there are compounds containing rings of a large number of atoms. Civetone, the sex attractant of the male civet cat, contains 17 carbon atoms. Civetone is used in the formulation of some expensive perfumes.

Multiple rings are common in steroids. One such example is cholesterol.

Example 14.8

What is the molecular formula for the following structure?

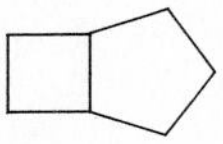

Solution. The structure indicates that there are seven carbon atoms. Each carbon atom, except those at positions 1 and 5, has two carbon–carbon bonds and two carbon–hydrogen bonds. Carbon atoms 1 and 5 have three carbon–carbon bonds and one carbon–hydrogen bond. Therefore the molecular formula is C_7H_{12}.

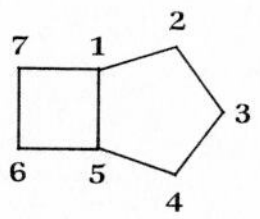

[Additional examples are given in **14.37** and **14.38** at the end of the chapter.]

Figure 14.6
Naturally Occurring Cycloalkanes.

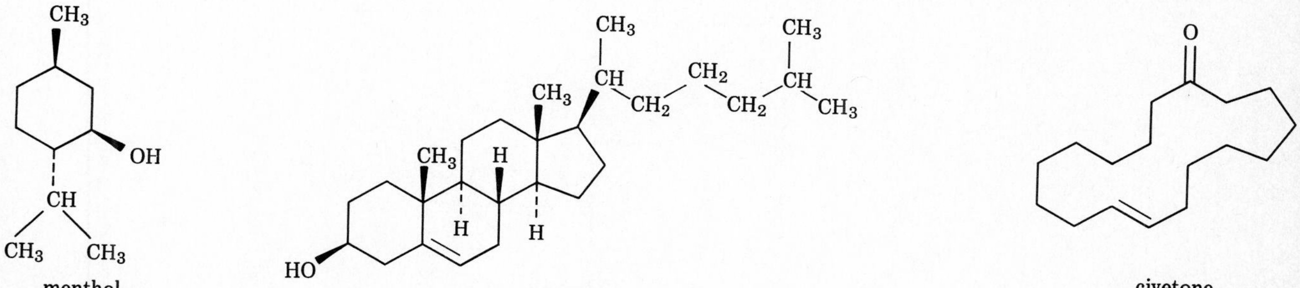

menthol
(mint oil)

cholesterol
(steroid in human body)

civetone
(sex attractant of
the male civet cat)

Note that there are several sites where the geometry of groups is indicated. The structural formulas of multiple ring compounds such as cholesterol are usually shown without indicating the ring atoms. Each juncture of two or more lines represents a carbon atom. Since each carbon atom must have four covalent bonds, any "missing" bonds are accepted as bonds to hydrogen atoms.

AN ASIDE

Petroleum is a complex mixture consisting predominately of alkanes, cycloalkanes, alkenes, and aromatic compounds. The latter two classes of compounds will be discussed in the next chapter. Small quantities of oxygen-, nitrogen-, and sulfur-containing compounds are also found in petroleum. The actual composition of petroleum varies with its location. Pennsylvania crude oils are largely alkanes, whereas western crude oils have high concentrations of aromatic compounds.

The first step in refining petroleum is the separation of the crude oil into simpler mixtures or fractions based on boiling points. The higher-molecular-weight components of the crude oil have the higher boiling points. Upon heating to 400°C, the majority of the components of the mixture are vaporized. The unvaporized residue material is used as the asphalt in surfacing roads. In a fractionating tower, the vapors rise and condense at various collecting points. The vapors of the higher-boiling materials are condensed at lower points in the column. Uncondensed low-molecular-weight material is saved as a gas from the top of the column (Figure 14.7). The fractions of petroleum and their boiling ranges are listed in Table 14.8.

Figure 14.7
Fractionating Column for Separation of Components of Petroleum.

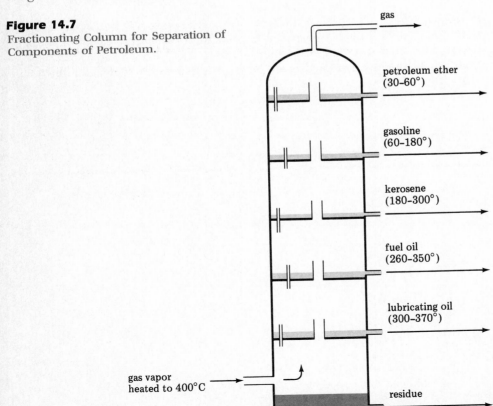

Table 14.8

Fractions of Petroleum

Fraction	Number of carbon atoms	Boiling point range (°C)	Uses
gas	C_1–C_4	−160–20	cooking gas
petroleum ether	C_5–C_6	30–60	solvent for organic chemicals
gasoline	C_6–C_{12}	60–180	automobile fuel
kerosene	C_{11}–C_{16}	180–300	rocket and jet fuel
fuel oil	C_{14}–C_{18}	260–350	domestic heating
lubricating oil	C_{15}–C_{24}	300–370	lubricants for automobiles and machines

Gasoline is the best known component of petroleum and is itself a mixture of hydrocarbons containing mostly alkanes and cycloalkanes. However, some chemical modifications are made of the structure of some of the alkanes in order to make it a better fuel in automobile engines. Similarly, other fractions such as fuel oil may be processed to remove sulfur before being used in a power station or large industrial plant.

Unbranched alkanes are not suitable as fuel for an automobile engine. These compounds tend to burn too rapidly and cause a knocking or pinging sound. This knocking phenomenon reduces the power derived from gasoline. Branched hydrocarbons, because they burn more smoothly and slowly, are the most efficient fuel.

The burning efficiency of gasolines is rated by an *octane number* scale. An octane number of 100 is assigned to 2,2,4-trimethylpentane, an excellent fuel. Heptane, a poor fuel, has an octane number of 0. Gasoline with the same burning characteristics as a 90% mixture of 2,2,4-trimethylpentane and 10% heptane is rated at 90 octane.

Octane numbers decrease with increasing molecular weight. For isomeric compounds, increased branching increases the octane number (Table 14.9).

Table 14.9

Octane Numbers of Some Alkanes

Formula	Compound	Octane number
C_4H_{10}	butane	94
C_5H_{12}	pentane	62
	2-methylbutane	94
C_6H_{14}	hexane	25
	2-methylpentane	73
	2,2-dimethylpentane	92
C_7H_{16}	heptane	0
	2-methylhexane	42
	2,3-dimethylpentane	90
C_8H_{18}	octane	−19
	2-methylheptane	22
	2,3-dimethylhexane	71
	2,2,4-trimethylpentane	100

Summary

Saturated hydrocarbons contain only carbon–carbon single bonds and carbon–hydrogen bonds. Each carbon atom of a saturated hydrocarbon uses four sp^3 hybrid orbitals to form four single covalent bonds with other atoms. The atoms bonded to each carbon atom form a tetrahedral shape.

There are two classes of saturated hydrocarbons; alkanes and cycloalkanes. The general molecular formula for an alkane is C_nH_{2n+2}. Each ring of a cycloalkane results in two fewer hydrogen atoms. The general formula for a cycloalkane is C_nH_{2n}.

The International Union of Pure and Applied Chemistry (IUPAC) has a set of rules for naming organic compounds. An alkane is named by selecting the longest continuous chain of carbon atoms as the parent compound. A group of carbon atoms attached to the chain is an alkyl group. The name of the alkyl group is related to an alkane containing the same number of carbon atoms. Both the position and the identity of the alkyl group are prefixed to the parent name.

There is free rotation about each carbon–carbon bond in an alkane. Arrangements of atoms in space as a result of rotation of single bonds are called conformations and do not represent different substances.

Geometric or *cis–trans* isomers have the same order of atoms bonded to each other, but have different orientations of the atoms in space. The *cis–trans* isomers cannot be interconverted by rotation about the carbon–carbon bonds. A substituted cycloalkane that has two substitutents on the same side of a ring is called the *cis* isomer, whereas a cycloalkane with the two substituents on the opposite sides of the ring is the *trans* isomer.

Saturated hydrocarbons have only weak London forces between molecules. The compounds have low boiling points and melting points. Boiling points of alkanes increase with increasing molecular weight and decrease with increasing branching of isomeric substances.

Saturated hydrocarbons are quite unreactive. Two important reactions of saturated hydrocarbons are halogenation and combustion.

Learning Objectives

As a result of studying Chapter 14 you should be able to

- Distinguish between saturated and unsaturated hydrocarbons.
- Give the IUPAC names for alkanes.
- Draw the structures of alkanes from IUPAC names.
- Recognize isomers of an alkane as compared to different representations.
- Classify carbon and hydrogen atoms in alkanes by type.
- Write the products of halogenation of an alkane.
- Give the IUPAC names for cycloalkanes.
- Draw representations of geometric isomers of cycloalkanes.

New Terms

An **alkane** is a compound containing only carbon and hydrogen atoms, which are bonded only by single bonds.

An **alkyl group** is a group of carbon and hydrogen atoms that resembles an alkane but has one less hydrogen atom. An alkyl group may be bonded to the parent chain of an alkane.

A **branched alkane** has an alkyl group bonded to a parent alkane.

A *cis* isomer has two groups of atoms oriented on the same side of a structural feature such as a cycloalkane ring.

A **cycloalkane** is a hydrocarbon that contains a ring of carbon atoms bonded by single covalent bonds.

Geometric isomers are isomers that have the same sequence of atoms, but whose orientation in space differs.

A **homologous series** is a series of compounds that differ from adjacent members by a repeating unit.

The **octane number** is a rating scale of the burning efficiency of hydrocarbons.

A **primary carbon atom** is directly bonded to only one other carbon atom.

A **saturated hydrocarbon** has only carbon–carbon single bonds.

A **secondary carbon atom** is bonded to two other carbon atoms.

A **substituent** is an atom or group of atoms attached to a skeleton of carbon atoms.

A **tertiary carbon atom** is bonded to three other carbon atoms.

A *trans* isomer has two groups of atoms oriented on opposite sides of a structural feature such as a cycloalkane ring.

Questions and Problems

Terminology

14.1 Explain what is meant by the term saturated in organic chemistry.

14.2 How do branched and normal alkanes differ?

14.3 What do the terms primary, secondary, and tertiary mean?

Saturation

14.4 Classify each of the following as saturated or unsaturated.

(a) CH$_3$CHCH$_3$
         |
         CH$_3$

                           CH$_3$
                            |
(b) CH$_3$—C=CH$_2$

(c) CH$_3$C≡CCH$_3$

(d) C$_2$H$_4$

(e) C$_2$H$_2$

(f) cyclohexane

(g)
        CH$_2$
       /   \
      CH   CH$_2$
      || |
      CH   CH$_2$
       \   /
        CH$_2$

(h) ⬠

(i)
   ⬡—CH=CH$_2$

14.5 Write a balanced chemical reaction for the reaction of hydrogen with each of the following compounds.

(a) CH$_3$CH=CH$_2$

(b) CH$_3$C≡CCH$_3$

(c) CH$_2$=CH—CH=CH$_2$

(d) CH$_2$=CH—C≡CH

(e) ⬠

14.6 Which of the following molecular formulas for noncyclic hydrocarbons represent saturated compounds?

(a) C$_6$H$_{12}$ (b) C$_5$H$_{12}$ (c) C$_8$H$_{18}$ (d) C$_{10}$H$_{22}$

(e) C$_{18}$H$_{34}$ (f) C$_{20}$H$_{40}$

14.7 How many hydrogen atoms are contained in an alkane with the following number of carbon atoms?

(a) 10   (b) 8   (c) 5   (d) 15   (e) 6

(f) 4   (g) 7   (h) 21

14.8 Which of the following formulas cannot correspond to an actual molecule?

(a) C$_6$H$_{14}$ (b) C$_{10}$H$_{23}$ (c) C$_7$H$_{18}$ (d) C$_5$H$_{14}$

(e) C$_{10}$H$_{20}$ (f) C$_5$H$_{10}$

Structural Formulas

14.9 Draw each of the following so that the longest continuous chain is written horizontally.

(a) CH$_3$—CH$_2$
            |
            CH$_2$—CH$_3$

(b) CH$_2$—CH$_2$—CH—CH$_2$CH$_3$
     | |
     CH$_3$ CH$_2$
                       |
                      CH$_3$

(c) CH$_3$—CH—CH$_2$—CH$_3$
          |
          CH$_2$
          |
          CH$_3$

(d)   CH$_3$
            |
    CH$_3$—CH—CH—CH$_3$
              |
              CH$_2$
              |
              CH$_3$

(e) CH$_3$—CH—CH$_2$
          | |
          CH$_3$ CH$_3$

                CH$_3$ CH$_3$
                 | |
(f) CH$_3$—CH—CH$_2$—CH—CH
                         | |
                         CH$_2$ CH$_2$
                         | |
                         CH$_3$ CH$_3$

(g) CH$_3$—CH—CH$_2$—CH$_2$
          | |
          CH$_2$ CH$_3$
          |
          CH$_3$

        CH$_3$
         |
(h) CH$_3$—CH—CH$_2$
              |
              CH$_2$
              |
      CH$_3$—CH$_2$

14.10 Which of the following represent the same compound drawn in different perspectives?

             CH$_3$CHCH$_3$
(a) CH$_3$CH$_2$CH$_2$CHCH$_3$

                                  CH$_3$
                                   |
(b) CH$_2$CHCH$_2$CHCH$_3$
     | |
     CH$_3$ CH$_3$

                      CH$_3$
                       |
(c) CH$_3$CHCH$_2$CH$_2$CHCH$_3$
     |
     CH$_3$

              CH$_3$
               |
       CH$_3$CHCH$_2$
(d) CH$_3$CH$_2$CHCH$_3$

        CH$_3$
         |
  CH$_3$CH
(e) CH$_2$CHCH$_3$
     |
     CH$_2$CH$_3$

                    CH$_3$
                     |
(f) CH$_3$CHCHCH$_2$CH$_2$
           | |
           CH$_3$ CH$_3$

(g) CH$_3$CHCH$_2$CHCH$_3$
     | |
     CH$_3$ CH$_3$

Naming of Alkyl Groups

14.11 Name each of the following alkyl groups.

(a) CH$_3$—

(b) CH$_3$—CH$_2$—

          CH$_3$
           |
(c) CH$_3$—C—
           |
          CH$_3$

(d) CH$_3$—CH—CH$_2$—CH$_3$
          |

(e) CH$_3$—CH—CH$_3$
          |

(f)
$$CH_3$$
$$|$$
$$CH_3—CH—CH_2—$$

(g) $CH_3—CH_2—CH_2—$

(h) $CH_3—CH_2—CH_2—CH_2—$

14.12 Name each of the following alkyl groups.

(a)
$$—CH_2$$
$$|$$
$$CH_2$$
$$|$$
$$CH_3—CH_2$$

(b)
$$CH_3—CH—$$
$$|$$
$$CH_2$$
$$|$$
$$CH_3$$

(c)
$$CH_3$$
$$|$$
$$—C—CH_3$$
$$|$$
$$CH_3$$

(d)
$$CH_3—CH_2$$
$$|$$
$$CH_2—$$

(e)
$$CH_3—CH—$$
$$|$$
$$CH_3$$

(f)
$$CH_3$$
$$|$$
$$CH_3—CH_2—CH—$$

Nomenclature

14.13 Give the IUPAC name for each of the following compounds.

(a)
$$\begin{array}{ccccc} H & H & H & H & H \\ | & | & | & | & | \\ H—C & —C & —C & —C & —C—H \\ | & | & | & | & | \\ H & H & H & H & H \end{array}$$

(b)
$$CH_3$$
$$|$$
$$CH_3—C—CH_3$$
$$|$$
$$CH_3$$

(c)
$$CH_3$$
$$|$$
$$CH_3—C—CH_2—CH_3$$
$$|$$
$$CH_3$$

(d)
$$CH_3—CH—CH_2—CH—CH—CH_3$$
$$\qquad |\qquad\qquad |\quad\ |$$
$$\qquad CH_3\qquad\ CH_2\ CH_3$$
$$\qquad\qquad\qquad\quad |$$
$$\qquad\qquad\qquad\ CH_3$$

14.14 Give the IUPAC name for each of the following compounds.

(a)
$$CH_3$$
$$|$$
$$CH_3—CH$$
$$|$$
$$CH_3$$

(b)
$$CH_3—CH_2$$
$$|$$
$$CH_3$$

(c)
$$CH_3—CH—CH_3$$
$$|$$
$$CH_3—CH_2$$

(d)
$$CH_3—CH—CH_2$$
$$\qquad\quad |\quad\ |$$
$$\qquad CH_3\ CH_3$$

(e)
$$CH_3—CH—CH_2$$
$$\qquad\quad |$$
$$\qquad\quad CH_2$$
$$\qquad\quad |$$
$$\qquad CH_3—CH_2$$

(f)
$$CH_3$$
$$|$$
$$CH_3—CH—CH—CH_3$$
$$\qquad\quad |$$
$$\qquad\quad CH_2$$
$$\qquad\quad |$$
$$\qquad\quad CH_3$$

(g)
$$CH_2CH_3$$
$$|$$
$$CH_3—CH—CH_2$$
$$\qquad\quad |$$
$$\qquad\quad CH_2$$
$$\qquad\quad |$$
$$\qquad CH_3—CH_2$$

(h)
$$\begin{array}{cc} CH_3 & CH_3 \\ | & | \\ CH_2 & CH_2 \\ | & | \\ CH_2—CH—CH_2 \\ & | \\ & CH_2 \\ & | \\ & CH_3 \end{array}$$

14.15 Name each of the following compounds.

(a)
$$CH_3CH_2CH_2CHCH_3$$
$$\qquad\qquad\quad |$$
$$\qquad\qquad\ CH_3$$

(b)
$$CH_3CH_2CH_2CHCH_2CH_3$$
$$\qquad\qquad\quad |$$
$$\qquad\qquad\ CH_2CH_2CH_3$$

(c)
$$CH_3CH_2CHCH_2CH_2CHCH_3$$
$$\qquad\quad |\qquad\qquad\quad |$$
$$\qquad\ CH_3\qquad\qquad CH_3$$

(d)
$$CH_3CH_2CH_2CHCH_2CH_2CH_3$$
$$\qquad\qquad\qquad |$$
$$\qquad\qquad\ CHCH_3$$
$$\qquad\qquad\qquad |$$
$$\qquad\qquad\ CH_3$$

14.16 Write a structural formula for each of the following compounds.
(a) 3-methylpentane
(b) 3,4-dimethylhexane
(c) 2,2,3-trimethylpentane
(d) 4-ethylheptane
(e) 2,3,4,5-tetramethylhexane

14.17 Indicate why each of the following is an incorrect name.
(a) 4-methylpentane
(b) 2-ethylbutane
(c) 2-dimethylhexane
(d) 3,4-dimethylbutane
(e) 2,2,2-trimethylhexane

14.18 Name each of the following compounds.
(a) CH_3CHCl_2 (b) $ClCH_2CH_2CH_2Cl$
(c) $CH_3CH_2CHCl_2$ (d) $CCl_3CH_2CH_2CH_3$
(e) $CH_2FCH_2CH_2CHF_2$ (f) $CH_3CBr_2CH_3$

14.19 Write the structural formula for each of the following compounds.
(a) 1,2,3-trichloropropane
(b) 2,2-dichloropentane
(c) 1,1,1-trichloropropane
(d) 1,1,2,3-tetrachlorobutane

Isomers

14.20 Draw structural formulas for the five isomers of C_6H_{14}. Name each compound.

14.21 Draw the structural formulas for the four isomers of $C_3H_6Cl_2$. Name each compound.

14.22 Draw the structural formulas for the four isomers of C_4H_9Cl. (*Hint:* Start with the two isomeric C_4H_{10} alkanes and consider the classification of the carbon atoms in each molecule.) Name each compound.

14.23 Draw the eight isomers of $C_5H_{11}Cl$.

Classification of Carbon Atoms

14.24 Classify each carbon atom in the following compounds.

(a) $CH_3CH_2CH_2CH_2CH_3$　(b) $CH_3\overset{\displaystyle CH_3}{\underset{\displaystyle CH_3}{C}}{-}CH_2CH_3$

(c) $CH_3\underset{\displaystyle CH_3}{CH}CH_2CH_3$

14.25 Draw the structure of each compound, with the indicated types of carbon atoms.
(a) a C_5H_{12} compound with one quaternary and four primary carbon atoms.
(b) a C_8H_{18} compound with two quaternary and six primary carbon atoms.
(c) a C_6H_{14} compound with two tertiary and four primary carbon atoms.

14.26 Determine the number of primary carbon atoms in each of the following.
(a) $CH_3{-}CH_2{-}CH_2{-}CH_3$
(b) $CH_3{-}CH_2{-}CH_2{-}CH_2{-}CH_3$

(c) $CH_3\overset{\displaystyle CH_3}{\underset{\displaystyle CH_3}{-}\!\!{C}}{-}CH_3$

(d) $CH_3\overset{\displaystyle CH_3}{-}CH{-}CH_3$

(e) $CH_3{-}CH_2\overset{\displaystyle CH_3}{-}CH{-}CH_3$

(f) $CH_3{-}CH_2\overset{\displaystyle CH_3}{\underset{\displaystyle CH_3}{-}\!\!{C}}{-}CH_2{-}CH_3$

(g) $CH_3\overset{\displaystyle CH_3}{\underset{\displaystyle CH_3}{-}\!\!{C}}{-}CH_2\overset{\displaystyle CH_3}{\underset{\displaystyle CH_3}{-}\!\!{C}}{-}CH_3$

(h) $CH_3\overset{\displaystyle CH_3}{\underset{\displaystyle CH_3}{-}\!\!{C}}{-}CH_2\overset{\displaystyle CH_3}{-}CH{-}CH_3$

(i) $CH_3{-}CH_2\overset{\displaystyle CH_3}{-}CH{-}CH_2\overset{\displaystyle CH_3}{-}CH{-}CH_3$

(j) $CH_3\overset{\displaystyle CH_3}{-}CH{-}CH_2\overset{\displaystyle CH_3}{-}CH\underset{\underset{\displaystyle CH_3}{\displaystyle CH_2}}{-}CH{-}CH_3$

14.27 Determine the number of secondary carbon atoms in each structure of **14.26**.

14.28 Determine the number of tertiary carbon atoms in each structure of **14.26**.

14.29 Determine the number of quaternary carbon atoms in each structure of **14.26**.

Halogenation of Alkanes

14.30 Write all possible products resulting from the halogenation of methane.

14.31 Write all possible products resulting from the halogenation of ethane. Indicate which substances are isomers.

14.32 How many products can result from the substitution of one hydrogen atom by a chlorine atom in each of the following compounds.
(a) propane　　　　　(b) butane
(c) methylpropane　　(d) pentane
(e) 2-methylbutane　　(f) 2,2-dimethylbutane
(g) 2,3-dimethylbutane

14.33 How many products can result from the substitution of one hydrogen atom by a chlorine atom in each of the following compounds?

(a) $CH_3\overset{\displaystyle CH_3}{\underset{\displaystyle CH_3}{-}\!\!{C}}{-}CH_3$

(b) $CH_3\overset{\displaystyle CH_3}{-}CH{-}CH_3$

(c) $CH_3\overset{\displaystyle CH_3}{\underset{\displaystyle CH_3}{-}\!\!{C}}{-}CH_2\overset{\displaystyle CH_3}{-}CH{-}CH_3$

(d) $CH_3\overset{\displaystyle CH_3}{\underset{\displaystyle CH_3}{-}\!\!{C}}{-}CH_2{-}CH_3$

(e) $CH_3{-}CH_2\overset{\displaystyle CH_3}{\underset{\displaystyle CH_3}{-}\!\!{C}}{-}CH_2{-}CH_3$

(f) $CH_3\underset{\displaystyle CH_3}{CH}{-}\underset{\displaystyle CH_3}{CH}{-}CH_3$

Cycloalkanes

14.34 Write fully condensed formulas for each of the following compounds.
(a) chlorocyclopropane
(b) 1,1-dimethylcyclobutane
(c) cycloheptane

14.35 There are three isomeric dichlorocyclopropanes. Draw the structural formula of each compound.

14.36 Draw the structural formulas for *cis-* and *trans-*1,3-dibromocyclobutane.

14.37 Name each of the following compounds.

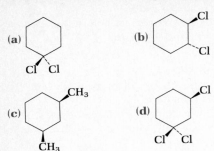

(a)

(b)

(c)

(d)

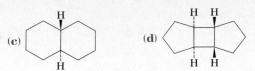

(c)

(d)

14.40 Consider the following three representations of 1,2-dibromocyclohexane. Is (b) an isomer or conformer of (a)? Is (c) an isomer or conformer of (a)?

(a)

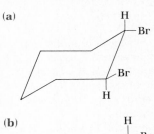

14.38 What is the molecular formula of each of the following compounds?

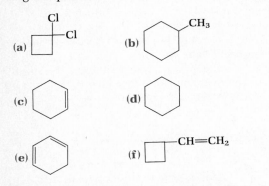

(a)

(b)

(c)

(d)

(e)

(f)

(b)

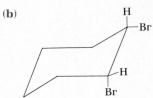

(c)

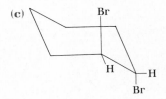

14.39 What is the molecular formula of each of the following compounds?

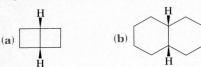

(a)

(b)

Unsaturated Hydrocarbons

Computers and word processors operate very fast. However, the secretary's visual process is based on very fast reactions of unsaturated compounds.

15.1
Types of Unsaturated Hydrocarbons

One of the reasons given in Chapter 13 for the structural diversity and the large number of organic compounds is the ability of carbon to form single, double, and triple bonds. In the preceding chapter, the structures of alkanes and cycloalkanes that contain carbon–carbon single bonds were presented. Now, in the next logical step, you will see how double and triple bonds may be located in an organic molecule.

Organic compounds with carbon–carbon multiple bonds contain fewer than the largest number of hydrogen atoms possible based on the number of carbon atoms. For this reason we say that compounds that contain carbon–carbon multiple bonds are **unsaturated.** You may have heard the term unsaturated on television commercials as part of the word polyunsaturated. Polyunsaturated refers to the presence of several multiple bonds in a compound. Oils that contain several double bonds are polyunsaturated and will be discussed in Chapter 22.

Recall that in Chapter 6 the bonding of the simple molecules ethylene and acetylene was presented. The multiple bonds in these molecules were described in terms of sigma (σ) and pi (π) bonds. The π bond is the result of a side-by-side overlap of p orbitals on adjacent carbon atoms. Both the structure and the chemical reactivity of unsaturated molecules are a consequence of the π bond. Therefore, it is important to review the bonding principles and the concept of hybridization presented earlier.

There are three main classes of unsaturated compounds. These are based on the type of multiple bonds present in the molecules.

1. **Alkenes** *are compounds that contain a carbon–carbon double bond.*
2. **Alkynes** *have a carbon–carbon triple bond.*
3. **Aromatic hydrocarbons** *contain a benzene ring or structural units like a benzene ring.*

15.2
Alkenes

The presence of a double bond in an alkene decreases the number of hydrogen atoms in a molecule by two compared to the number in alkanes. As a result, the general formula for an alkene is C_nH_{2n}. Note that this general formula is the same for cycloalkanes, which are thus isomeric with alkenes. Alkenes were formerly called olefins, meaning oil forming, because alkenes undergo addition reactions with chlorine to produce oily liquids.

The homologous series of alkenes (C_nH_{2n}) is similar in its physical properties to the homologous series of alkanes (C_nH_{2n+2}). Alkenes are nonpolar molecules and have low boiling points. The members of the series containing fewer than five carbon atoms are gases at room temperature. As in the case of alkanes, the boiling points increase with an increase in the number of carbon atoms in the molecule.

The structure of the simplest alkene, ethylene, is shown in Figure 15.1. All six atoms, two carbon atoms and four hydrogen atoms, are located in the same plane. The plane may be chosen to be that of the printed page or one perpendicular to it. If the plane is perpendicular to the printed page, the carbon–hydrogen bonds must project in front of and in back of the page. Wedge-shaped lines are used for bonds in front of the page and dotted lines for those behind the page. The bond angles H—C—H and H—C—C of ethylene are 118° and 121°, respectively. These values are close to the ideal value of 120° for a sp^2-hybridized carbon atom.

Molecules containing more than one double bond are quite common in nature. Several of these substances are shown in Figure 15.2. The presence of the double bonds in these compounds is, of course, responsible for their chemical

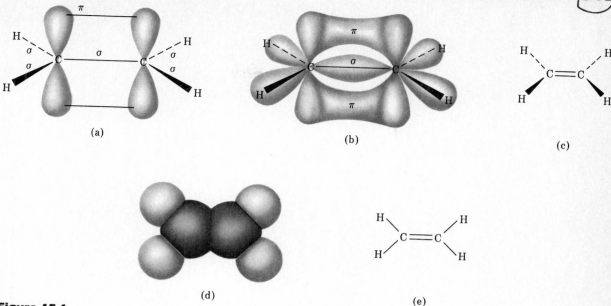

(a)

(b)

(c)

(d)

(e)

Figure 15.1

Structure of Ethylene.

(a) p orbitals that overlap to form a π bond. (b) π bond with electrons located above and below the plane of the molecule. (c) Perspective formula with the plane of the molecule perpendicular to the page. (d) Space-filling model of ethylene in the plane of the page. (e) Perspective formula with the plane of the molecule in the plane of the page.

reactivity. Note the resemblance of vitamin A to carotene. A biochemical reaction splits and oxidizes carotene into two molecules of vitamin A.

Example 15.1

What is the molecular formula for a hydrocarbon with six carbon atoms that contains a ring and a double bond?

Solution. The presence of a ring decreases the number of hydrogen atoms by 2 below that of an alkane to give C_nH_{2n}. If a double bond is present, another two hydrogen atoms must be subtracted to give C_nH_{2n-2}. For six carbon atoms the molecular formula must be C_6H_{10}.

[Additional examples may be found in **15.7–15.9** at the end of the chapter.]

In an alkane there is free rotation about the carbon–carbon bonds, and the molecule can then twist and adopt many conformations. Thus, any of the following structures can represent 1,2-dichloroethane.

15.3
Geometric Isomerism

314

Figure 15.2
Polyunsaturated Substances
in Nature.

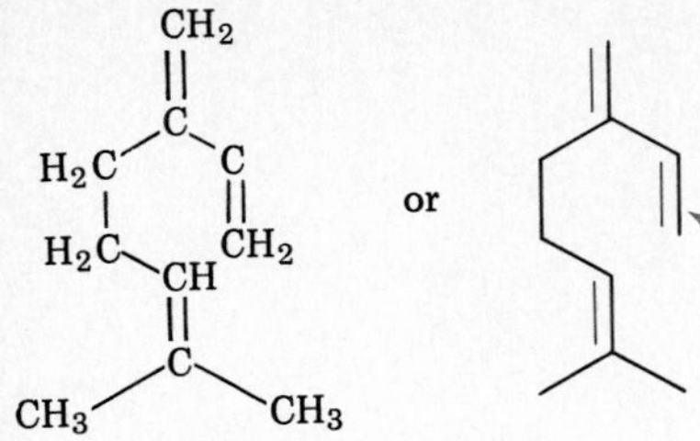

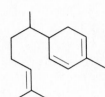

myrcene
(from bay oil)

This structure illustrates the method
of drawing condensed structures. Each
single line has a carbon at each end.

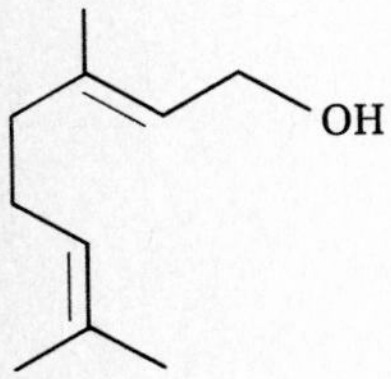

geraniol
(from roses and
other flowers)

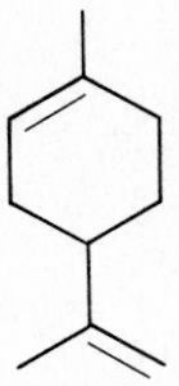

limonene
(lemons)

zingiberene
(from oil of ginger)

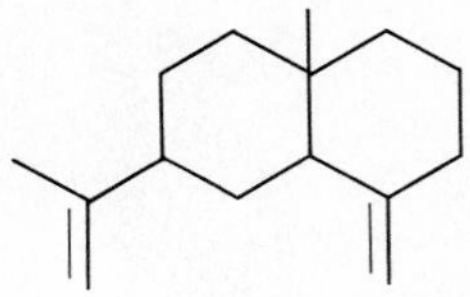

β-selinene
(from oil of celery)

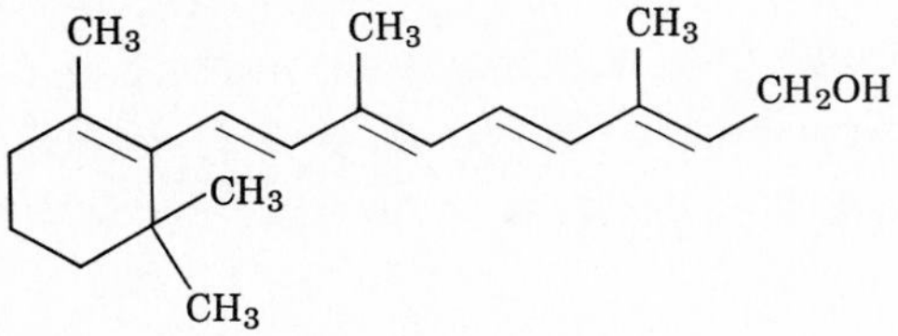

vitamin A

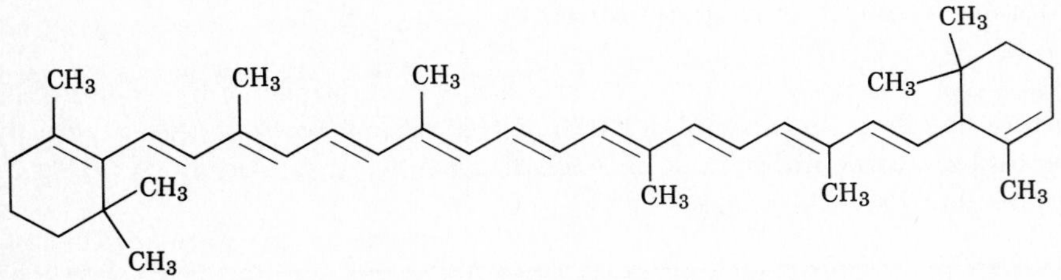

β-carotene
(vegetables such as carrots)

The positions of the chlorine atoms in space vary constantly, and it doesn't matter which way the molecule is drawn.

The double bond does not permit free rotation. In order to maintain the π
bond of an alkene, the two p orbitals must remain side by side. If the two carbon
atoms are rotated about the bond axis, the σ bond would remain but the π bond
would be broken. This process can occur only if 60 kcal/mole is supplied (Figure
15.3) to break the π bond. Under ordinary conditions, the two carbon atoms do
not rotate in relation to each other.

As a result of the restriction of rotation, alkenes have a rigid geometry for the

Figure 15.3
Restricted Rotation of a Double Bond.
(a) The two electrons can be shared between the two carbon atoms when the *p* orbitals are parallel. (b) If the carbon on the right is rotated, the two electrons can no longer be shared to form a bond.

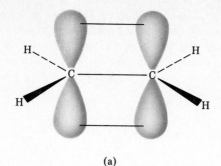
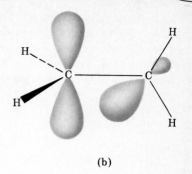

(a)   (b)

groups attached to the carbon atoms of the double bond. The four atoms directly attached to the two double-bonded carbon atoms lie in the same plane. These atoms can exist in different spatial or geometric arrangements, and isomerism about the double bond occurs. *Isomers that differ from each other in the geometry of the molecules and not in the bonding order of the atoms are geometric isomers.* They are also known as *cis–trans* isomers.

There are two geometric isomers of 1,2-dichloroethylene. The structure with the chlorine atoms on the same "side" of the molecule is called the *cis* isomer. The structure with the chlorine atoms on opposite "sides" is called the *trans* isomer.

$$\begin{array}{cc} \text{H} \qquad \text{H} & \text{H} \qquad \text{Cl} \\ \text{C}=\text{C} & \text{C}=\text{C} \\ \text{Cl} \qquad \text{Cl} & \text{Cl} \qquad \text{H} \end{array}$$

cis-1,2-dichloroethylene   *trans*-1,2-dichloroethylene

The two isomers cannot be converted into each other without breaking the π bond and rotating about the σ bond. Such reactions occur only at high temperatures. The physical properties of the geometric isomers differ. For example, the boiling points of the above *cis* and *trans* isomers are 60 and 47°C, respectively.

Geometric isomers do not occur for all alkenes. Only alkenes that have two different groups attached to each of the doubly bonded carbon atoms can exist as *cis–trans* isomers. 1,2-Dichloroethylene contains a chlorine atom and a hydrogen atom on one unsaturated carbon atom. These groups are different. Similarly the groups on the other unsaturated carbon atom are different—a hydrogen atom and a chlorine atom.

different groups $\Big<$
$$\begin{array}{cc} \text{H} & \text{H} \\ \text{C}=\text{C} \\ \text{Cl} & \text{Cl} \end{array}$$
$\Big>$ different groups

Neither chloroethylene nor 1,1-dichloroethylene can exist as a *cis–trans* pair of geometric isomers. There is only one chloroethylene and one 1,1-dichloroethylene.

different groups $\Big<$
$$\begin{array}{cc} \text{H} & \text{H} \\ \text{C}=\text{C} \\ \text{Cl} & \text{H} \end{array}$$
$\Big>$ identical groups

identical groups $\Big<$
$$\begin{array}{cc} \text{Cl} & \text{H} \\ \text{C}=\text{C} \\ \text{Cl} & \text{H} \end{array}$$
$\Big>$ identical groups

Mammals have an enzyme that splits β-carotene (Figure 15.2) in half to give two molecules of an aldehyde named retinal. The retinal consists of a series of alternating single and double bonds. Geometric isomers can exist about each of the double bonds. However, the all-*trans* compound and the isomer with a *cis* orientation about the indicated double bond play an important role in vision.

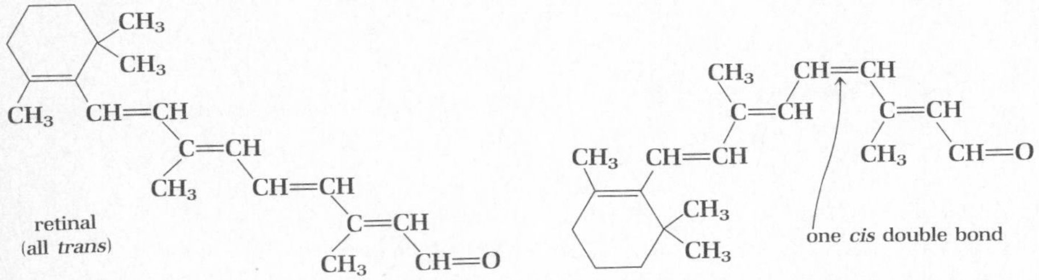

retinal (all *trans*)

one *cis* double bond

The *cis*-9 retinal reacts with a protein in the retina called opsin to form a substance called rhodopsin. The shape of *cis*-9-retinal is such that it "fits" into the protein.

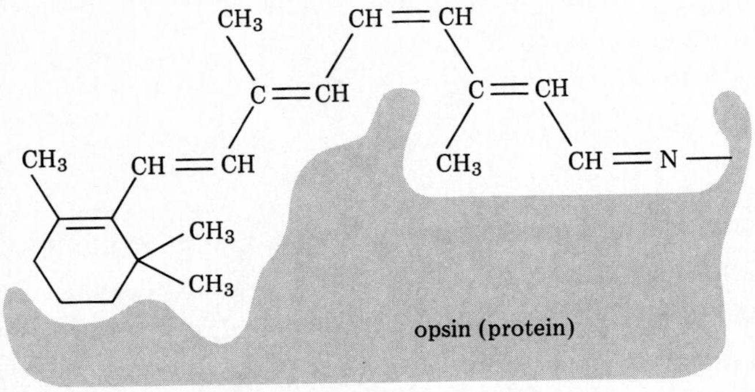

opsin (protein)

When light strikes the rhodopsin, the *cis* double bond is isomerized to a *trans* double bond. The resulting retinal no longer fits and is released from the protein in a complex series of reactions. As part of these reactions, an electrical impulse travels from the retina to the brain, which allows us to see.

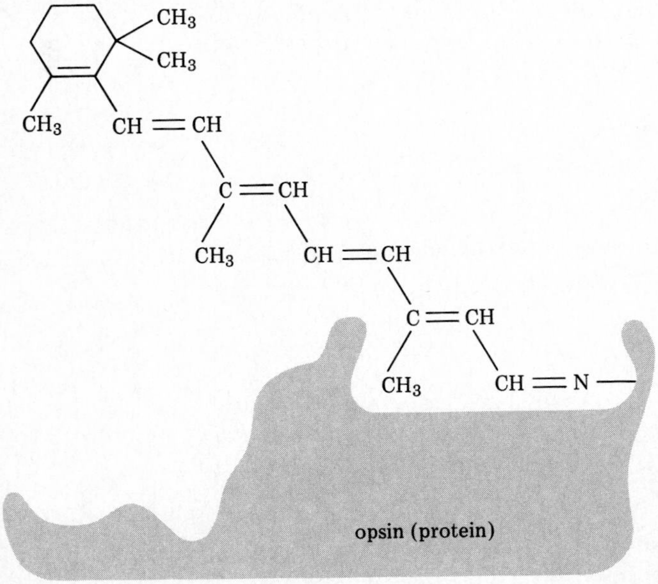

opsin (protein)

You can check to see if geometric isomers are possible by mentally inter-changing the two groups bonded to one carbon atom. If the H and Cl are inter-changed in *cis*-1,2-dichloroethylene, a different structure results. However, if the two H atoms on the same carbon atom of chloroethylene are interchanged, an identical structure results.

The IUPAC rules for naming alkenes are similar to those for alkanes, but include an indication of the position of the double bond in the chain and the geometric arrangement about the double bond.

15.4
Nomenclature of Alkenes

1. The longest continuous chain containing the double bond is used as the parent.

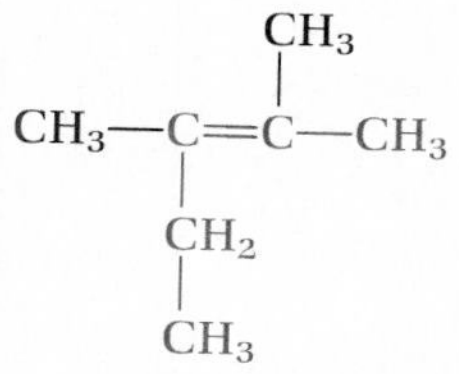

$$CH_3-\underset{\underset{\underset{CH_3}{|}}{\overset{\overset{CH_3}{|}}{C}}}{C}=\underset{}{C}-CH_3$$

2. The longest chain is given the same stem name as an alkane but with -ene replacing -ane. The longest chain is in color in the above structure and the parent name is pentene.

3. The carbon atoms are numbered consecutively from the end nearest the double bond. The number of the first carbon atom of the double bond is used as a prefix to the parent name separated by a hyphen. The compound is, therefore, a substituted 2-pentene.

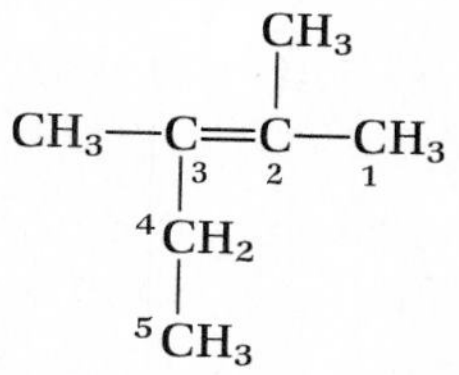

$$CH_3-\overset{\overset{CH_3}{|}}{\underset{\underset{4CH_2}{|}}{C}}\underset{3}{=}\overset{}{\underset{2}{C}}-\overset{}{\underset{1}{CH_3}}$$
$$^4CH_2$$
$$^5CH_3$$

4. Alkyl groups and other substituents are named and their positions on the chain are determined by the numbering established by the above rule. Names and numbers are prefixed to the parent name.

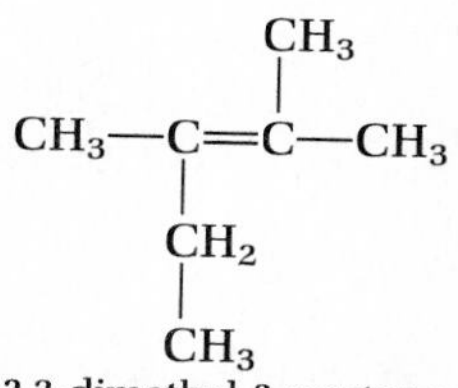

$$CH_3-\overset{\overset{CH_3}{|}}{\underset{\underset{CH_2}{|}}{C}}=\overset{}{C}-CH_3$$
$$CH_3$$

2,3-dimethyl-2-pentene

5. If the compound can exist as a *cis* or *trans* isomer, the appropriate prefix followed by a hyphen is placed in front of the name. There are two identi-cal CH_3 groups attached to the carbon at position 2 in 2,3-dimethyl-2-pentene so there is no possibility of *cis* or *trans* isomers.

6. If more than one double bond is present, the location of each double

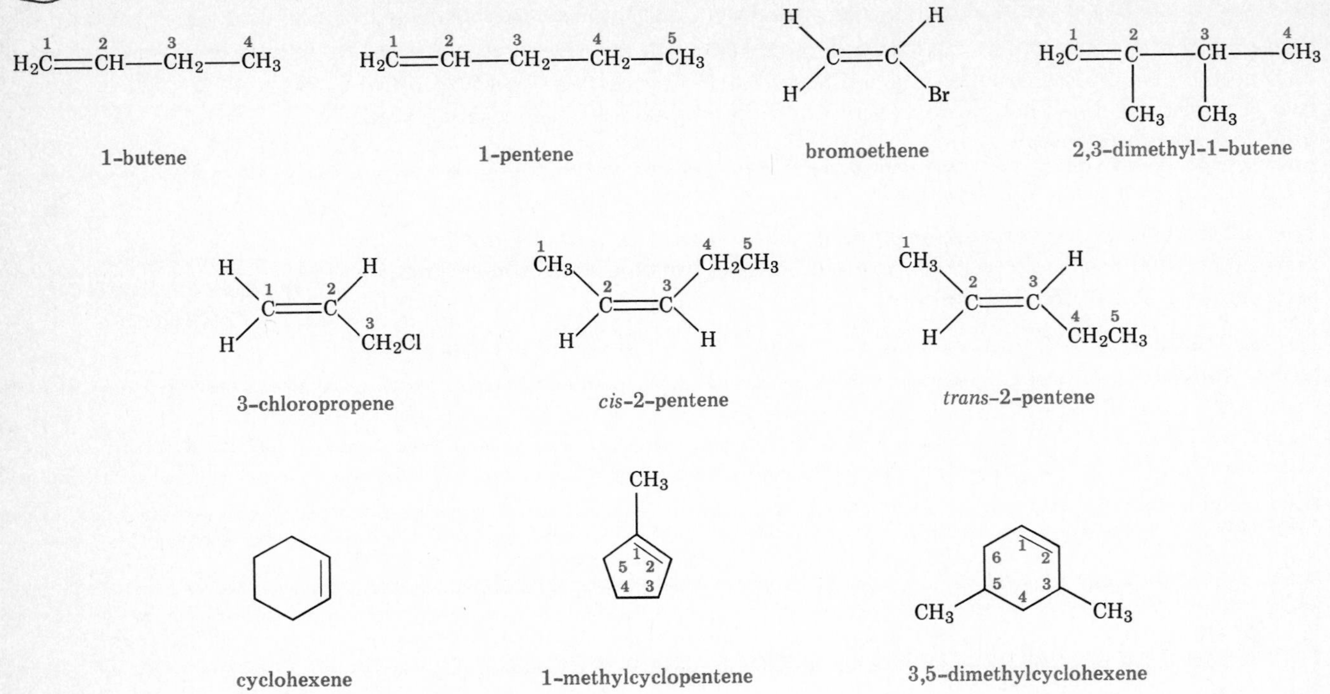

$$\overset{1}{H_2C}{=}\overset{2}{CH}{-}\overset{3}{CH_2}{-}\overset{4}{CH_3}$$

1-butene

$$\overset{1}{H_2C}{=}\overset{2}{CH}{-}\overset{3}{CH_2}{-}\overset{4}{CH_2}{-}\overset{5}{CH_3}$$

1-pentene

$$\underset{H}{\overset{H}{>}}C{=}C\underset{Br}{\overset{H}{<}}$$

bromoethene

$$\overset{1}{H_2C}{=}\underset{CH_3}{\overset{2}{C}}{-}\underset{CH_3}{\overset{3}{CH}}{-}\overset{4}{CH_3}$$

2,3-dimethyl-1-butene

$$\underset{H}{\overset{H}{>}}\overset{1}{C}{=}\overset{2}{C}\underset{CH_2Cl}{\overset{H}{<}}$$

3-chloropropene

$$\underset{H}{\overset{\overset{1}{CH_3}}{>}}\overset{2}{C}{=}\overset{3}{C}\underset{H}{\overset{\overset{4}{}\overset{5}{CH_2CH_3}}{<}}$$

cis-2-pentene

$$\underset{H}{\overset{\overset{1}{CH_3}}{>}}\overset{2}{C}{=}\overset{3}{C}\underset{\overset{4}{}\overset{5}{CH_2CH_3}}{\overset{H}{<}}$$

trans-2-pentene

cyclohexene

1-methylcyclopentene

3,5-dimethylcyclohexene

Figure 15.4
Nomenclature of Alkenes.

bond is indicated by a number. The suffix must indicate the number of double bonds.

$$\overset{1}{CH_2}{=}\overset{2}{CH}{-}\overset{3}{CH}{=}\overset{4}{CH_2}$$
1,3-butadiene

$$\overset{1}{CH_2}{=}\overset{2}{CH}{-}\overset{3}{CH}{=}\overset{4}{CH}{-}\overset{5}{CH}{=}\overset{6}{CH}{-}\overset{7}{CH_3}$$
1,3,5-heptatriene

7. In naming cycloalkenes, number the ring to give the double-bonded carbon atoms the numbers 1 and 2. Choose the direction of numbering so that the substituents receive the lowest numbers. The position of the double bond is not given because it is known to be between the number 1 and 2 carbon atoms.

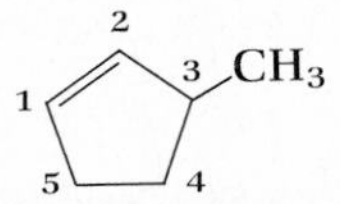

3-methylcyclopentene

Some examples of alkene nomenclature are given in Figure 15.4.

Example 15.2

Name the following compound.

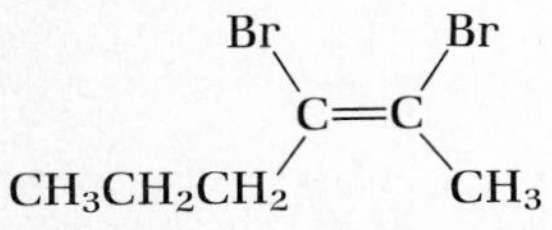

$$\underset{CH_3CH_2CH_2}{\overset{Br}{>}}C{=}C\underset{CH_3}{\overset{Br}{<}}$$

Solution. There are six carbon atoms in the longest chain. It is numbered from right to left so that the double bond is at the carbon atom in position 2. The parent is then 2-hexene. The bromine atoms are at the 2 and 3 positions.

$$\underset{6}{CH_3}\underset{5}{CH_2}\underset{4}{CH_2}\quad\underset{3}{\overset{Br}{C}}=\underset{2}{\overset{Br}{C}}\quad\underset{1}{CH_3}$$

2,3-dibromo-2-hexene

Since there are two different groups of atoms on each unsaturated carbon atom, there are possible *cis* and *trans* isomers. Inspection of the molecule indicates that this is *cis*-2,3-dibromo-2-hexene.

[Additional examples may be found in **15.22–15.27** at the end of the chapter.]

Whereas the oxidation of alkanes by oxygen is destructive and occurs to give carbon dioxide and water, alkenes can be oxidized more selectively. The π electrons can easily be attacked by oxidizing agents, while the single bonds of the molecule remain unchanged. Potassium permanganate oxidizes an alkene to a dialcohol or diol (two OH groups).

15.5
Oxidation and Reduction of Alkenes

$$3\,CH_3\!-\!CH\!=\!CH\!-\!CH_3 + 2\,KMnO_4 + 4\,H_2O \longrightarrow 3\,CH_3\!-\!\underset{OH}{CH}\!-\!\underset{OH}{CH}\!-\!CH_3 + 2\,MnO_2 + 2\,KOH$$

Potassium permanganate is purple in aqueous solution. Manganese dioxide (MnO_2) is a brown solid that precipitates from solution. Since there is a color change, the oxidation of alkenes by potassium permanganate is a reaction that can be used to test visually for the presence of a double bond. Alkanes and cycloalkanes are not oxidized by potassium permanganate.

The reaction of hydrogen with an unsaturated molecule results in a saturated molecule. While the process is one of reduction, the reaction is also called hydrogenation. The hydrogenation of 1-octene yields octane.

$$\underset{\text{1-octene}}{CH_3(CH_2)_5\!-\!CH\!=\!CH_2} + H\!-\!H \xrightarrow{\;Pt\;} \underset{\text{octane}}{CH_3(CH_2)_5\!-\!CH_2CH_3}$$

In order for the reaction to occur, finely divided platinum is used as a catalyst. Other catalysts that can be used include nickel and palladium, which are also members of Group VIIIB of the periodic table.

Hydrogenation is used commercially to convert liquid oils into solid fats.

$$3\,H_2 + \begin{array}{l} H_2C\!-\!O\!-\!\underset{O}{\overset{\|}{C}}\!-\!(CH_2)_7\!-\!CH\!=\!CH\!-\!(CH_2)_7\!-\!CH_3 \\[6pt] HC\!-\!O\!-\!\underset{O}{\overset{\|}{C}}\!-\!(CH_2)_7\!-\!CH\!=\!CH\!-\!(CH_2)_7\!-\!CH_3 \\[6pt] H_2C\!-\!O\!-\!\underset{O}{\overset{\|}{C}}\!-\!(CH_2)_7\!-\!CH\!=\!CH\!-\!(CH_2)_7\!-\!CH_3 \end{array} \xrightarrow[200°]{Ni} \begin{array}{l} H_2C\!-\!O\!-\!\underset{O}{\overset{\|}{C}}\!-\!(CH_2)_7\!-\!CH_2\!-\!CH_2\!-\!(CH_2)_7\!-\!CH_3 \\[6pt] HC\!-\!O\!-\!\underset{O}{\overset{\|}{C}}\!-\!(CH_2)_7\!-\!CH_2\!-\!CH_2\!-\!(CH_2)_7\!-\!CH_3 \\[6pt] H_2C\!-\!O\!-\!\underset{O}{\overset{\|}{C}}\!-\!(CH_2)_7\!-\!CH_2\!-\!CH_2\!-\!(CH_2)_7\!-\!CH_3 \end{array}$$

a liquid oil   a solid fat

Oils tend to be unsaturated, whereas fats are more saturated. As the degree of saturation increases, the melting point increases.

Example 15.3 ————————————————————————————————

The molecular formula of both cyclohexane and 1-hexene is C_6H_{12}. How can a chemist distinguish one compound from the other?

Solution. Cyclohexane is saturated and will not react with $KMnO_4$. When $KMnO_4$ is added dropwise to 1-hexene, the purple color is lost and a brown precipitate is formed.

Example 15.4 ————————————————————————————————

How many moles of hydrogen gas will react with the following compound? What is the molecular formula of the product?

Solution. There are two double bonds in the compound. One mole of the compound will react with two moles of hydrogen gas.

The product is a cycloalkane containing ten carbon atoms. Since the general molecular formula for a cycloalkane is C_nH_{2n}, the molecular formula of the compound is $C_{10}H_{20}$.

[Additional examples may be found in **15.36** at the end of the chapter.]

15.6
Addition Reactions of Alkenes

An **addition reaction** *occurs when two substances join together to form a compound containing all atoms present in the original substances.* The carbon–carbon double bond provides the site for addition reactions of alkenes with other substances. The reactions of ethylene with some common reagents are

$$CH_2{=}CH_2 + Br_2 \longrightarrow BrCH_2CH_2Br$$

$$CH_2{=}CH_2 + HCl \longrightarrow CH_3CH_2Cl$$

$$CH_2{=}CH_2 + H_2O \xrightarrow{\text{H}^+} CH_3CH_2OH$$

Note that one part of the reagent becomes bonded to one carbon atom, while the second part is bonded to the second carbon atom. As a result, the double bond is converted to a single bond.

In the case of the addition of bromine, the reaction is easily seen. Bromine is reddish; reaction leads to a colorless organic compound. This disappearance of the reddish color of bromine is then a useful test in determining if a compound is unsaturated. Drops of Br_2 dissolved in CCl_4 can be added to a compound. If the color disappears, the compound is unsaturated.

Reagents that add to alkenes can be classified into two types. **Symmetrical reagents** *consist of two identical groups, each of which can become attached to the carbon atoms of the double bond.* Bromine is an example of a symmetrical reagent. *An* **unsymmetrical reagent** *consists of different groups, each of which can become attached to the carbon atoms of the double bond.* Both HCl and H_2O are examples of unsymmetrical reagents.

There is only one possible product of the addition of a symmetrical reagent to an alkene. For example, bromine will react with propene to yield one specific compound.

$$CH_3CH{=}CH_2 + Br_2 \longrightarrow CH_3\overset{\displaystyle Br}{\overset{\displaystyle |}{C}}HCH_2Br$$

propene   1,2-dibromopropane

It doesn't make any difference which bromine atom becomes attached to which carbon atom. Bromine is a symmetrical molecule, and there is no way to distinguish the bromine atoms from each other.

When an unsymmetrical reagent, such as HCl, is added to an alkene, there are two possible products that can result if the alkene is not symmetrical. For a symmetrical alkene, such as ethylene, only one product is possible because the two carbon atoms are identical.

$$CH_2{=}CH_2 + HCl \longrightarrow \underline{H{-}CH_2{-}CH_2{-}Cl} \text{ or } \underline{Cl{-}CH_2{-}CH_2{-}H}$$

identical
carbon
atoms

identical structures

While two possible products could result from the addition of HCl to an unsymmetrical alkene, only one is actually formed. Addition of HCl to propene could yield either 1-chloropropane or 2-chloropropane. However, only the latter is formed. The X written through one reaction arrow indicates that the reaction does not occur.

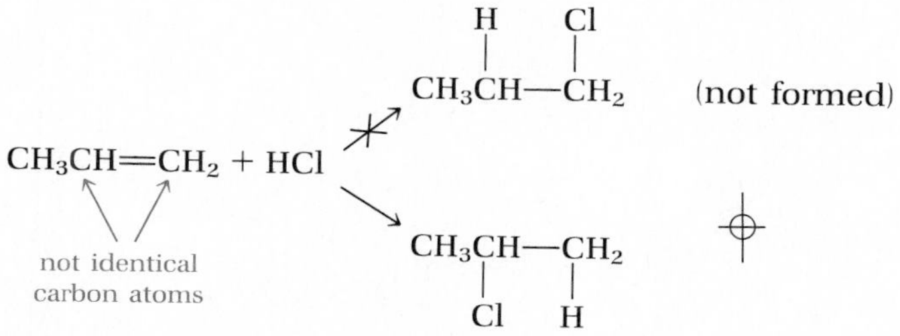

$$CH_3\overset{\displaystyle H}{\overset{\displaystyle |}{C}}H{-}\overset{\displaystyle Cl}{\overset{\displaystyle |}{C}}H_2 \quad \text{(not formed)}$$

$$CH_3CH{=}CH_2 + HCl$$

not identical
carbon atoms

$$CH_3\overset{\displaystyle |}{\underset{\displaystyle Cl}{C}}H{-}\overset{\displaystyle |}{\underset{\displaystyle H}{C}}H_2$$

The Russian chemist Markovnikov observed that unsymmetrical reagents add to unsymmetrical double bonds in a specific way. He stated that

A molecule of the general formula HX adds to a double bond so that the hydrogen atom bonds to the unsaturated carbon atom that has the greater number of directly bonded hydrogen atoms.

In the case of propene, one carbon atom has two hydrogen atoms attached, whereas the other has only one hydrogen atom. Thus HCl is added as follows.

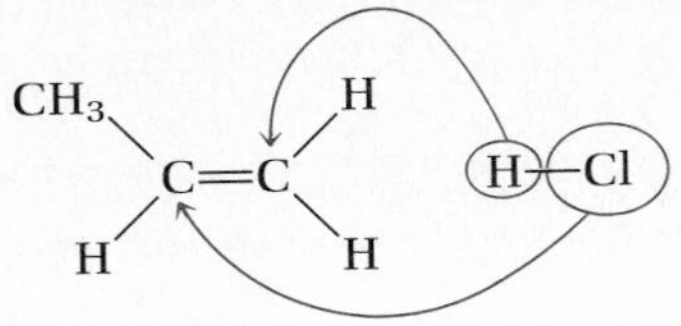

As another example, H_2O will add to methylpropene to produce only one of the two possible isomers. The addition reaction, which is catalyzed by acid, occurs as follows.

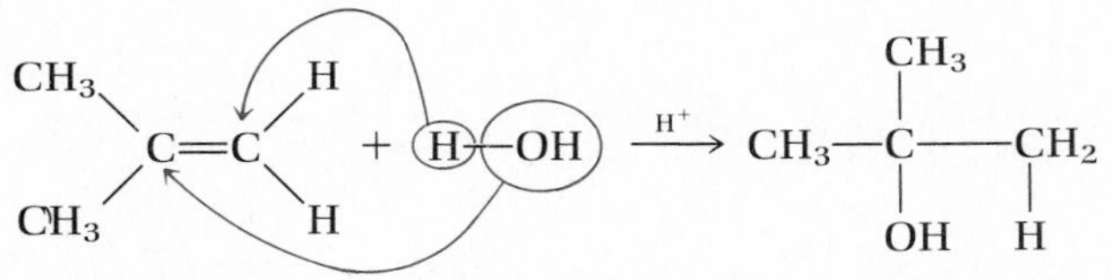

Addition reactions occur in biological systems that are controlled by enzymes. In the metabolism of fats, the following addition reaction with water occurs.

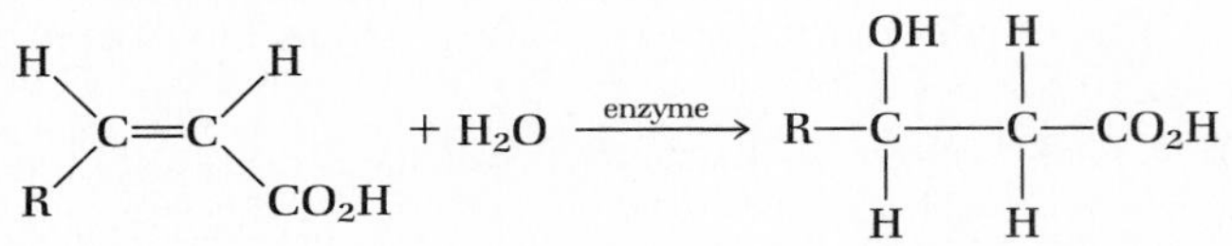

Note that while two possible products could result, the enzyme catalyzes the reaction to form only one isomer.

Example 15.5

Predict the product formed when HCl adds to 1-methylcyclopentene.

Solution. One unsaturated carbon atom has one attached hydrogen atom, while the other carbon atom has no attached hydrogen atom. The product is then 1-chloro-1-methylcyclopentene.

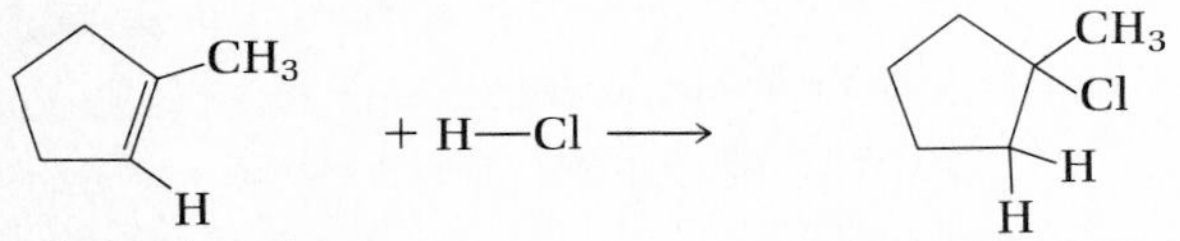

1-methylcyclopentene $+ H—Cl \longrightarrow$ 1-chloro-1-methylcyclopentane

[Additional examples may be found in **15.32–15.34** at the end of the chapter.]

15.7
Polymerization of Alkenes

Polymers *are high-molecular-weight molecules made from thousands of repeating units, which are low-molecular-weight molecules.* The individual molecules that are used to produce the polymer are called **monomers** (meaning one part). *The process of joining monomers to produce polymers is called* **polymerization.**
Alkenes can be polymerized by a multiple addition reaction. For example,

with the appropriate catalyst, ethylene can be made to add to itself. Each ethylene molecule successively adds to a neighbor forming a carbon–carbon bond. Polyethylene is the product.

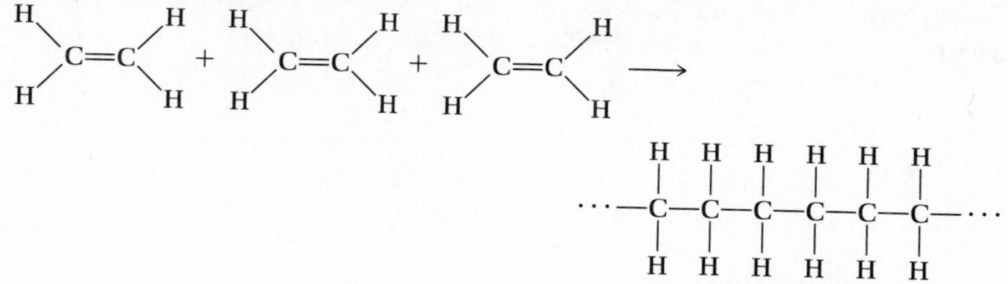

polyethylene

When $CH_2=CHCl$, vinyl chloride, is polymerized, the resultant polymer is polyvinyl chloride (PVC).

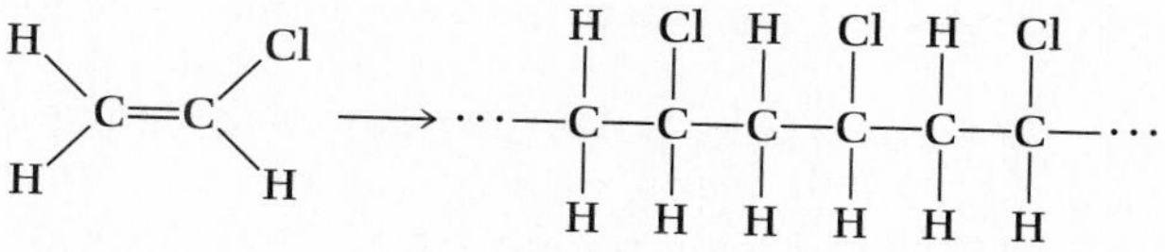

An exact formula for a polymer cannot be written, since the molecules are so large. Furthermore, the actual size of the molecule may vary depending on how it is formed. There is no single "polyethylene" or "polyvinyl chloride" molecule,

Table 15.1

Uses and Structures of Polymers

Monomer	Polymer	Use
propylene $H_2C=CHCH_3$	polypropylene $-CH_2CHCH_2CHCH_2CH-$ with CH_3 CH_3 CH_3	carpet fibers, heart valves, bottles
vinyl chloride $H_2C=CHCl$	polyvinyl chloride (PVC) $-CH_2CHCH_2CHCH_2CH-$ with Cl Cl Cl	floor covering, records, garden hoses
tetrafluoroethylene $F_2C=CF_2$	Polytetrafluoroethylene $-CF_2CF_2CF_2CF_2CF_2CF_2-$	utensil coverings, bearings
acrylonitrile $H_2C=CHCN$	polyacrylonitrile $-CH_2CHCH_2CHCH_2CH-$ with CN CN CN	Orlon, Acrilan
styrene $H_2C=CHC_6H_5$	polystyrene $-CH_2CH-CH_2CH-CH_2CH-$ with C_6H_5 C_6H_5 C_6H_5	toys, styrofoam
methyl methacrylate H_3C O $H_2C=C-COCH_3$	polymethyl methacrylate CH_3 CH_3 CH_3 $-CH_2C-CH_2-C-CH_2-C-$ $COCH_3$ $COCH_3$ $COCH_3$ O O O	Plexiglas, Lucite

since a mixture of compounds with a range of molecular weights results. For these reasons polymers are represented by placing the repeating units derived from the monomer within a set of parentheses. For ethylene, the unit is ($—CH_2CH_2—$). To show that a large number of units are present, the subscript n is used. For the polymerization of ethylene, we write

$$n\ CH_2{=}CH_2 \longrightarrow [CH_2{-}CH_2]_n$$

The properties of polymers depend on the monomer used and the molecular weight of the product. A list of some useful polymers is given in Table 15.1. Vinyl chloride has been linked to a form of liver cancer. Chemical workers must take special precautions when producing vinyl chloride. Polyvinyl chloride is not carcinogenic and is widely used in plastic products.

15.8
Alkynes

The triple bond in an alkyne decreases the number of hydrogen atoms in the molecule by four compared to alkanes. As a result, the general molecular formula for alkynes is C_nH_{2n-2}.

The simplest alkyne, C_2H_2, is commonly called acetylene. Unfortunately, the common name ends in -ene, which then might suggest that the compound contains a double bond. Such confusion is one reason why IUPAC names are so important for clear communication in chemistry. The IUPAC name for C_2H_2 is eth*yne*.

The structure of ethyne is shown in Figure 15.5. All four atoms lie in a straight line. Each H—C—C bond angle is 180°. In other alkynes, the two triple-bonded carbon atoms and the two atoms directly attached to them all lie in a straight line.

Nomenclature of Alkynes

The rules for naming alkynes by the IUPAC system are analogous to those for alkanes.

1. The longest continuous chain containing the triple bond is used as the parent.

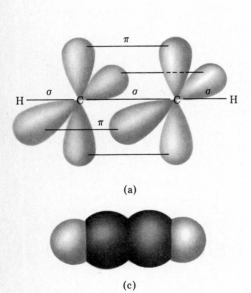

(a)

(c)

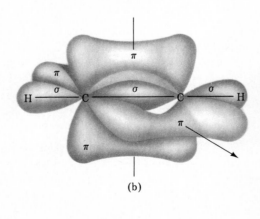

(b)

$$H{-}C{\equiv}C{-}H$$

(d)

Figure 15.5
The Structure of Ethyne.
(a) *p* orbitals that overlap to form two π bonds. (b) One π bond is above and below the molecule; the second π bond is in front and in back of the molecule. (c) Space-filling model. (d) Perspective formula with all atoms in the plane of the page.

2. The stem name of the alkane is changed by dropping the *-ane* suffix and adding *-yne.*

3. The carbon atoms are numbered consecutively from the end nearest the triple bond. The number of the first carbon atom with the triple bond is used as a prefix separated by a hyphen from the parent name.

4. Alkyl groups are named and their positions on the chain are determined by the numbering established by the above rule.

Examples of the correct nomenclature follow.

$$\overset{5}{CH_3}-\overset{4}{CH_2}-\overset{3}{C}\equiv\overset{2}{C}-\overset{1}{CH_3}$$

2-pentyne
(*not* 3-pentyne)

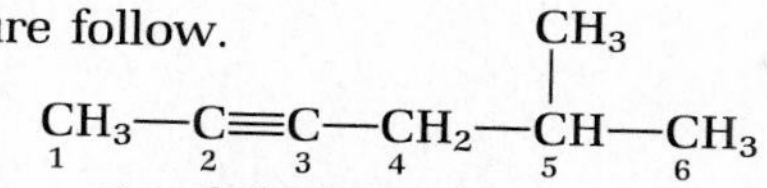

$$\overset{1}{CH_3}-\overset{2}{C}\equiv\overset{3}{C}-\overset{4}{CH_2}-\overset{5}{CH}-\overset{6}{CH_3} \;\; \overset{CH_3}{|}$$

5-methyl-2-hexyne
(*not* 2-methyl-4-hexyne)

Reactions of Alkynes

The triple bond is made up of one σ bond and two π bonds, and its reactivity is similar to a double bond in an alkene. Thus, triple bonds undergo addition reactions just like double bonds. However, they can add two molecules of a specific reagent. For example, propyne reacts with hydrogen to give propane.

$$CH_3C\equiv C-H \xrightarrow[Pt]{H_2} CH_3CH=CH_2 \xrightarrow[Pt]{H_2} CH_3CH_2CH_3$$

While the reaction can be controlled with special catalysts to stop at the alkene, complete hydrogenation of an alkyne produces an alkane.

Addition of two molecules of halogens also occurs to produce a compound containing four halogen atoms.

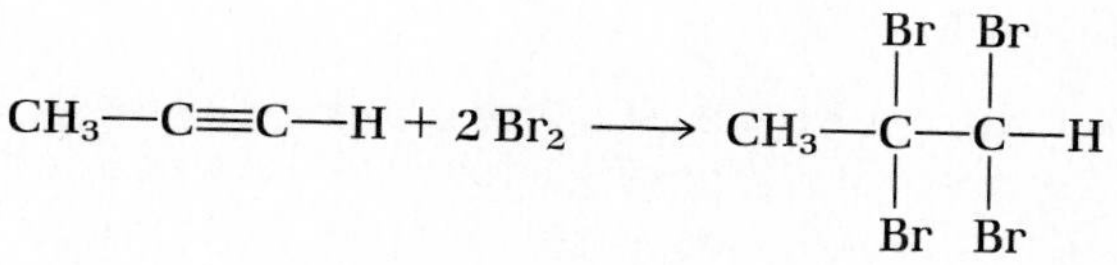

$$CH_3-C\equiv C-H + 2\,Br_2 \longrightarrow CH_3-\underset{\underset{Br}{|}}{\overset{\overset{Br}{|}}{C}}-\underset{\underset{Br}{|}}{\overset{\overset{Br}{|}}{C}}-H$$

Example 15.6

What is the molecular formula for the following compound? How many moles of hydrogen will react with the compound?

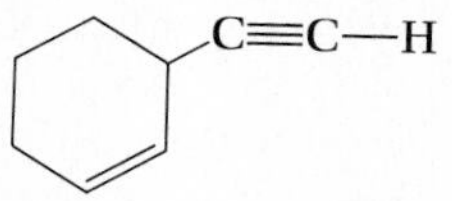

Solution. The compound contains eight carbon atoms. An eight-carbon cyclo-alkane without double or triple bonds would have the molecular formula C_8H_{16} based on the general molecular formula C_nH_{2n}. A double bond creates a deficiency of two hydrogen atoms, and a triple bond creates a deficiency of four hydrogen atoms. Thus, the molecular formula is C_8H_{10}.

One mole of hydrogen per mole of compound is required for the double bond. Two moles of hydrogen per mole of compound are required for the triple bond. A total of three moles of hydrogen per mole of compound will react.

[Additional examples may be found in **15.37–15.39** at the end of the chapter.]

The product of the addition of hydrogen halides is that predicted by Markovnikov's rule.

$$CH_3-C{\equiv}C-H + \overset{}{H}{-}Cl \longrightarrow CH_3-\underset{\underset{Cl}{|}}{C}{=}\overset{\overset{H}{|}}{C}-H$$

2-chloropropene

$$\underset{\underset{Cl}{\diagup}}{\overset{CH_3\diagdown}{}}C{=}\overset{\overset{H}{\diagup}}{\underset{H}{C}} + \overset{}{H}{-}Cl \longrightarrow CH_3-\underset{\underset{Cl}{|}}{\overset{\overset{Cl}{|}}{C}}{-}\underset{\underset{H}{|}}{\overset{\overset{H}{|}}{C}}-H$$

2,2-dichloropropane

15.9
Aromatic Hydrocarbons

Aromatic hydrocarbons are a class of unsaturated compounds that do not easily undergo addition reactions. While this definition may seem to be a contradiction based on what we know about unsaturated compounds, let's proceed to look at an aromatic compound.

Benzene, C_6H_6, has a low ratio of hydrogen to carbon. Compared to hexane, benzene is deficient by eight hydrogen atoms. However, benzene does not add bromine like alkenes and alkynes do. In fact, benzene will react with bromine only in the presence of ferric bromide (or iron, which reacts with bromine to form ferric bromide) and then substitution rather than addition occurs.

$$C_6H_6 + Br_2 \xrightarrow{\text{FeBr}_3} C_6H_5Br + HBr$$

In 1865, a German chemist, August Kekulé, suggested that benzene contains a ring of six carbon atoms and a series of alternating single and double bonds. He proposed that the double bonds and single bonds are in rapid oscillation around the ring. Kekulé wrote two representations of benzene in which the positions of the double bonds were alternated.

The rapid oscillation was thought somehow to make the molecule resistant to addition reactions.

Modern physical measurements have shown that benzene is a planar molecule in which all carbon–carbon bonds are of equal length. The bond angles of the ring are all 120°. Based on this data, benzene is pictured as consisting of sp^2-hybridized carbon atoms. Each atom shares one electron in each of its three σ bonds. Two of the σ bonds are with adjacent carbon atoms. The third σ bond is with a hydrogen atom. The fourth electron is in a p orbital perpendicular to the planar benzene ring (Figure 15.6). A set of six p orbitals, one from each carbon atom, and their six electrons overlap to share electrons in a π bond over the

Figure 15.6
Bonding in the Benzene Ring.
(a) There are six *p* orbitals in benzene, and they are all mutually parallel. (b) Overlap and delocalization over the molecule of the six electrons.

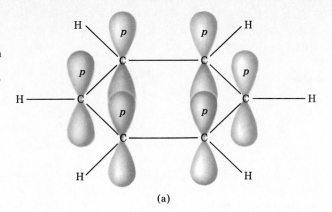

(a)

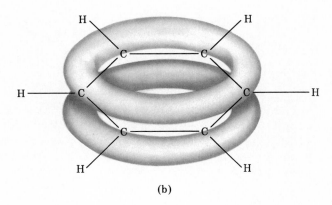

(b)

entire ring. *The sharing of electrons over many atoms is called* **delocalization.** It is this delocalization of electrons that accounts for the unique chemical stability of benzene. In order to write the structure for benzene, two resonance structures are sometimes used. These two structures are the Kekulé structures and are said to contribute to the structure of benzene. These Kekulé structures do not exist, but the benzene molecule can be viewed as a resonance hybrid of these two structures. Some chemists prefer to represent benzene by a circle drawn within a hexagon, using the circle to indicate the six electrons in the π system that is spread out above and below the plane of the ring. Each corner of the hexagon represents a carbon atom and one attached hydrogen atom.

All twelve atoms of benzene are in a plane. When a substituent group replaces a hydrogen atom, the atom of the group bonded to the ring is also in the plane of the ring.

15.10
Nomenclature of Aromatic Hydrocarbons

The IUPAC system of naming substituted aromatic hydrocarbons uses the name of the substituents as a prefix to benzene. Examples include the halogen-substituted benzenes.

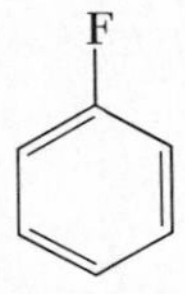

fluorobenzene 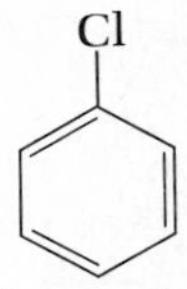   chlorobenzene 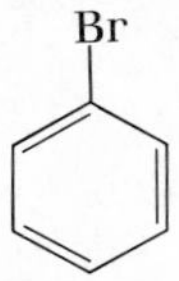   bromobenzene

All of the compounds shown have the substituent at a "12 o'clock" position. However, all six positions on the benzene ring are equivalent, and you should recognize a compound such as bromobenzene no matter where the bromine atom is written.

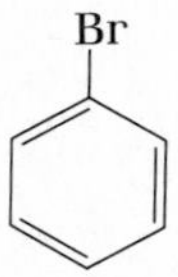

is the same as 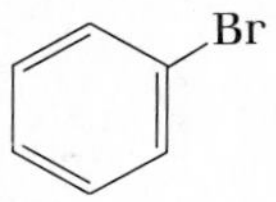   is the same as

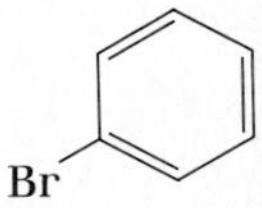

Disubstituted compounds result when two hydrogen atoms are replaced by groups. Three disubstituted isomers are possible. The three isomers of dichlorobenzene are designated ortho, meta, and para. These terms are abbreviated as *o*, *m*, and *p*. Alternatively a numbering system also may be used to give the lowest possible numbers to the carbon atoms bearing chlorine atoms.

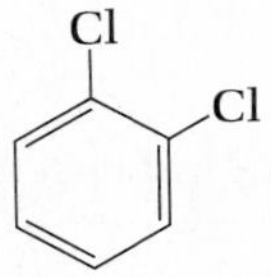

o-dichlorobenzene
1,2-dichlorobenzene

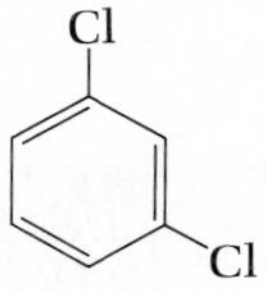

m-dichlorobenzene
1,3-dichlorobenzene

p-dichlorobenzene
1,4-dichlorobenzene

When two groups are adjacent on the ring, they are in ortho positions. When the groups are separated by one carbon atom, they are in meta positions. When the groups are separated by two carbon atoms, they are in para positions. It is important to realize that any of these isomers could be written in a different orientation on the page without changing their identity or name.

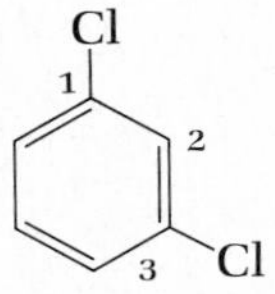

is the same as 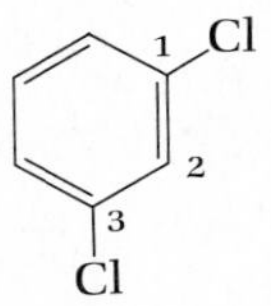   is the same as

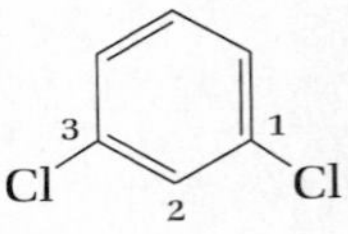

When three or more substituents are attached on the benzene ring, the positions are numbered. Note in the following examples that the numbers are chosen to assign the lowest set of numbers for the substituents.

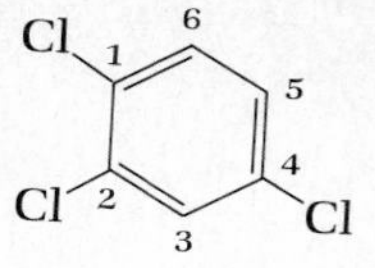

1,2,4-trichlorobenzene

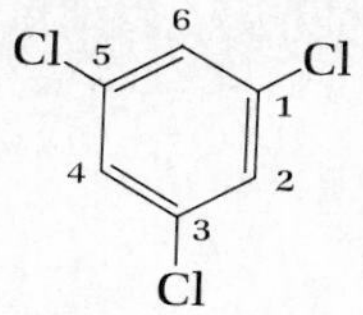

1,3,5-trichlorobenzene

Many compounds are better known by their common names than by their systematic names. These include toluene, phenol, and aniline.

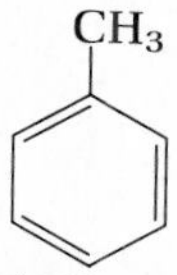

methylbenzene
(toluene)

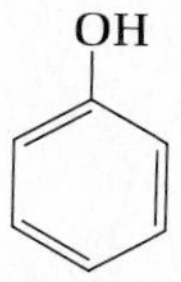

hydroxybenzene
(phenol)

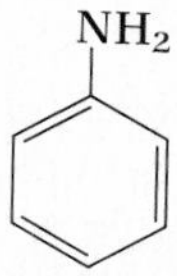

aminobenzene
(aniline)

Sometimes compounds are named with the common name of the monosubstituted aromatic compound as the parent. Three such cases are given below.

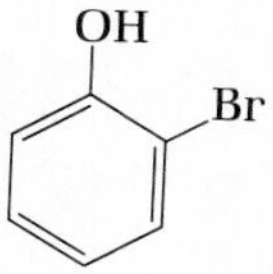

o-bromophenol

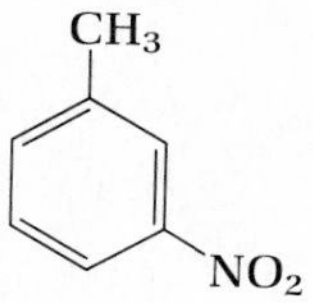

m-nitrotoluene

p-chloraniline

Other compounds with special names include the xylenes, cresols, and toluidines.

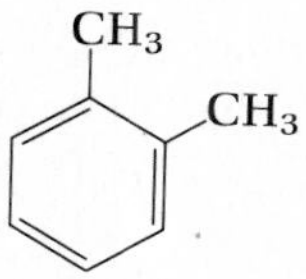

o-xylene

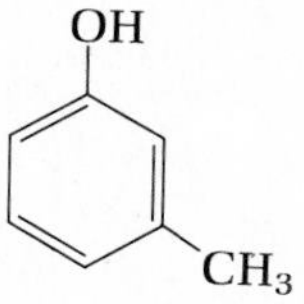

m-cresol

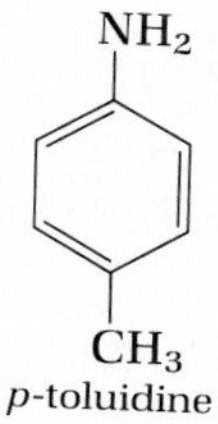

p-toluidine

In complex molecules, the benzene ring is often named as a substituent on a parent chain of carbon atoms. The aromatic group C_6H_5— derived from benzene is called a phenyl group. The general name for aromatic groups that are to be treated as substituents is aryl, symbolized Ar.

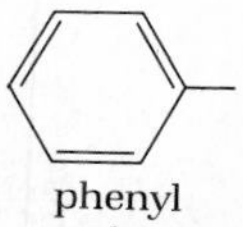

$$CH_3CH_2CH_2CH_2CHCH_2CH_3$$

phenyl
(an aryl group)

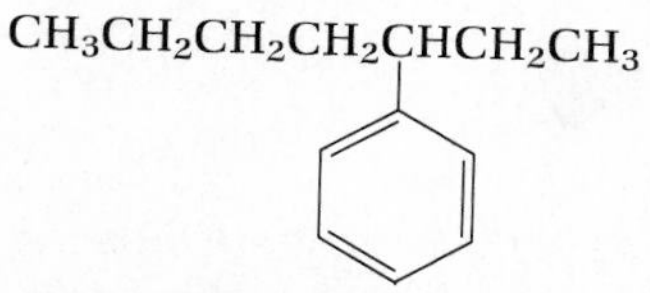

3-phenylheptane

Example 15.7 _____

What is the name of the following compound?

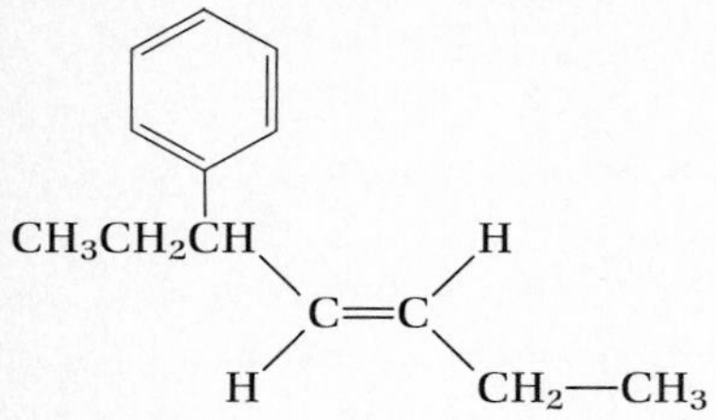

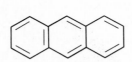

Solution. The compound is an alkene with an aromatic ring as a substituent. First, determine that there are seven carbon atoms in the chain. Next, number the chain from right to left so that the double bond is assigned the number 3. The phenyl group is then located on the number 5 carbon atom. Finally, note that the compound is the *trans* isomer. The complete name is *trans*-5-phenyl-3-heptene.

_____

[Additional examples may be found in **15.46** and **15.47** at the end of the chapter.]

15.11
Polycyclic and Heterocyclic Aromatic Hydrocarbons

There are aromatic compounds that contain two or more rings "fused" together. **Fusion** *of carbon rings means that two carbon atoms are common to two rings.* Compounds of this type are called polycyclic aromatic hydrocarbons. Several examples of these compounds are shown in Figure 15.7. Note that all carbon atoms contain a bond to a hydrogen atom, except those that are at the points of fusion. The molecules are planar; all atoms in the ring and those directly attached to the rings are in a plane.

Some of the fused polycyclic aromatic hydrocarbons, with an "angle" in the series of rings, are carcinogenic. Two of the most potent carcinogens are 1,2,5,6-

Figure 15.7
Structure of Polycyclic Aromatic Compounds.

naphthalene, $C_{10}H_8$

anthracene, $C_{14}H_{10}$

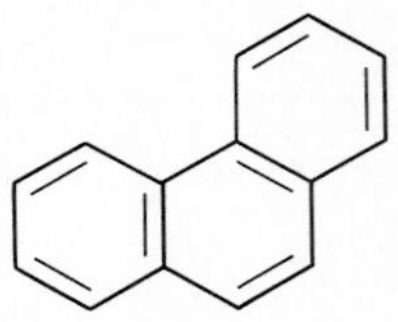

phenanthrene, $C_{14}H_{10}$

1,2–benzanthracene

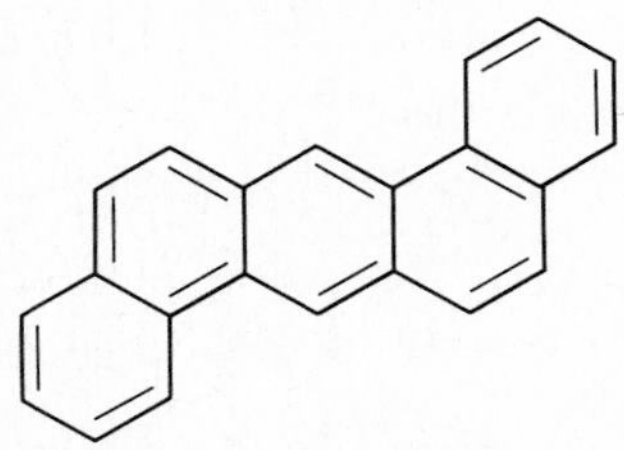

1,2,5,6–dibenzanthracene

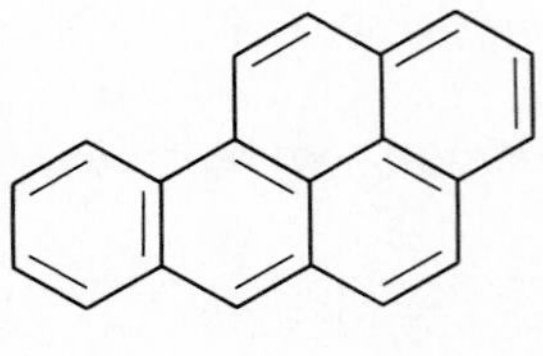

3,4–benzpyrene

dibenzanthracene and 3,4-benzpyrene. Applied to the skin of mice, small amounts of these angular fused-ring aromatic hydrocarbons cause cancer in about a month. These compounds are present in effluent from coal-burning power plants and in automobile exhaust. In addition, they are found in tobacco smoke and in meat cooked over charcoal. The incidence of lung cancer among smokers and inhabitants of large urban areas may be partially the result of breathing in these airborne compounds in minute amounts.

Many biologically important compounds have aromatic rings of five or six atoms in which one or more nitrogen atoms replace carbon atoms. Examples that will be discussed in Chapter 29 include pyrimidine and purine. Derivatives of these compounds are parts of the structures of DNA and RNA.

pyrimidine   imidazole   purine

Summary

The alkenes are one class of unsaturated hydrocarbons. They contain a carbon–carbon double bond that consists of a σ and a π bond. The double-bonded carbon atoms are sp^2 hybridized.

Arrangement of atoms bonded to the double-bonded carbon atoms is spatially restricted. Two geometric isomers are possible for suitably substituted alkenes. These are called *cis* and *trans* isomers.

The IUPAC system for naming alkenes uses the -ene suffix and designates the position of the double bond by a number. The number of substituents is dictated by the numbers assigned to double-bonded carbon atoms.

The characteristic reaction of alkenes is addition to the double bond. Examples of reagents that react are bromine, water, and the hydrogen halides, such as HCl and HBr.

Alkynes are a second class of unsaturated hydrocarbons. They contain a carbon–carbon triple bond that consists of a σ bond and two π bonds. The triple-bonded carbon atoms are sp hybridized.

The IUPAC system of naming alkynes uses the -yne suffix. The atoms in a compound are numbered so as to assign the lowest numbers to the triple-bonded carbon atoms.

The reactions of alkynes are additions across the triple bond. The reactions resemble those of the alkenes, but two moles of reagent may add to one mole of the alkyne.

Aromatic hydrocarbons are the third type of unsaturated hydrocarbon. These compounds have a special stability as a result of delocalization of π electrons about a ring of carbon atoms. Each carbon atom in the ring is sp^2 hybridized.

Learning Objectives

As a result of studying Chapter 15 you should be able to

- Distinguish between alkenes, alkynes, and aromatic compounds from molecular formulas.
- Determine from a structural formula if an alkene can exist as geometric isomers.
- Name alkenes and alkynes by the IUPAC method.
- Distinguish by chemical tests between saturated and unsaturated hydrocarbons.
- Calculate the molecular formula of a substance from the number and type of carbon–carbon bonds present.
- Write equations and the structures of the products for the oxidation or hydrogenation of alkenes.
- Predict the product of addition of an unsymmetrical reagent to an alkene or alkyne.
- Represent the structures of polymers from the monomers.
- Explain why benzene is chemically less reactive than alkenes.
- Name monosubstituted, disubstituted, and polysubstituted benzene compounds.
- Recognize the common names of monosubstituted benzene compounds.
- Describe the structures of some common fused aromatic ring compounds.

New Terms

An **addition reaction** occurs when two substances join together to form a compound containing all atoms present in the original substances.

Alkenes are hydrocarbons that contain a carbon–carbon double bond.

Alkynes are hydrocarbons that contain a carbon–carbon triple bond.

Aromatic compounds contain a benzene ring or structural units resembling a benzene ring.

Delocalization is the sharing of electrons over many atoms.

Fused carbon rings means that a compound has two carbon atoms in common to two rings.

Markovnikov's Rule states that when a compound HX adds to a double bond, the hydrogen atom bonds to the unsaturated carbon atom that has the greater number of directly bonded hydrogen atoms.

A **monomer** is a low-molecular-weight molecule used to produce a polymer.

Polymers are high-molecular-weight molecules made from thousands of repeating units, which are low-molecular-weight molecules.

A **symmetrical reagent** consists of two identical groups, each of which can become attached to the carbon atoms of a double bond in an addition reaction.

Questions and Problems

Terminology
15.1 What is an unsaturated hydrocarbon?
15.2 Distinguish between structural and geometric isomerism.
15.3 Explain how a π bond differs from a σ bond.
15.4 Describe the structural similarities and differences between alkenes and alkynes

Classification of Hydrocarbons
15.5 Classify the following compounds as saturated or unsaturated.
  (a) $CH_3CH_2CH_3$ (b) $CH_3CH_2CH=CH_2$
  (c) $CH_3CH_2CH(CH_3)_2$ (d) $HC\equiv CCH_2CH_3$

  (e) $CH_2=CHCH_2CH=CH_2$ (f)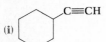

15.6 Classify each of the compounds in **15.5** according to a structural class.

Molecular Formulas
15.7 What is the molecular formula for each of the compounds with the following structural features?
  (a) six carbon atoms and one double bond
  (b) five carbon atoms and two double bonds
  (c) seven carbon atoms, a ring, and a double bond
  (d) four carbon atoms and a triple bond
  (e) four carbon atoms and two triple bonds
  (f) four carbon atoms, a double bond, and a triple bond
  (g) ten carbon atoms and two rings
  (h) ten carbon atoms, two rings, and five double bonds

15.8 Write the molecular formula for each of the following compounds.

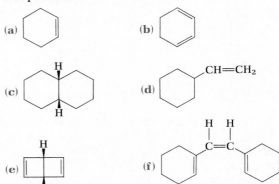

15.9 Write the molecular formula for each of the following compounds.

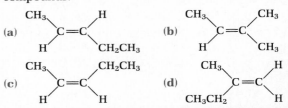

15.10 The sex attractant released by the female codling moth has the following structural formula. What is the molecular formula of the compound?

Isomers
15.11 There are four isomeric alkenes with the molecular formula C_4H_8. Draw structural formulas for the compounds, and name them.
15.12 There are two isomeric cycloalkanes with the molecular formula C_4H_8. Write the structures of these compounds.

15.13 There are six isomeric alkenes with the molecular formula C_5H_{10}. Write the structures, and name the compounds.

15.14 There are two isomeric alkynes with the molecular formula C_4H_6. Write the structures, and name the compounds.

15.15 Can an alkadiene and an alkyne be isomeric? Explain.

Geometric Isomers

15.16 Why are there two geometric isomers of 2-butene but not for 1-butene?

15.17 Which of the following molecules can exist as *cis* and *trans* isomers?
(a) $CH_3CH=CHBr$ (b) $CH_2=CHCH_2Br$
(c) $CH_3CH=CHCH_2Cl$ (d) $(CH_3)_2C=CHCH_3$
(e) $CHBr=CHCl$ (f) $CBr_2=CCl_2$

15.18 Draw the structural formula for the *cis* isomer of each of the following alkenes.
(a) $CH_3CH=CHCH_2CH_3$ (b) $CHCl=CHBr$
(c) $CH_2ClCH=CHCH_2Br$ (d) $CH_3CH=CHBr$

15.19 Which of the following compounds can exist as *cis* and *trans* isomers?
(a) 1-hexene   (b) 3-heptene
(c) 4-methyl-2-pentene   (d) 2-methyl-2-butene
(e) 3-ethyl-3-hexene

15.20 The structural formula for the sex attractant of the silkworm moth is given below. What is the geometry about each double bond?

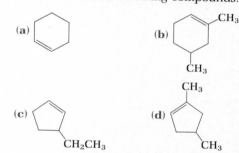

15.21 How many geometric isomers are there for the compound in **15.20?**

Nomenclature of Alkenes

15.22 Name each of the following compounds.

(a) $CH_3-\overset{\overset{\displaystyle CH_3}{|}}{C}=CH_2$

(b) $\underset{CH_3}{\overset{CH_3}{}}C=C\underset{CH_3}{\overset{CH_3}{}}$

(c) $CH_3-\overset{\overset{\displaystyle CH_3}{|}}{C}=CHCH_3$

(d) $CH_3\overset{\overset{\displaystyle CH_3}{|}}{C}=CHCH_2\overset{\overset{\displaystyle CH_3}{|}}{C}HCH_3$

(e) $\underset{H}{\overset{H}{}}C=C\underset{Cl}{\overset{H}{}}$

(f) $\underset{H}{\overset{H}{}}C=C\underset{CH_2-Br}{\overset{H}{}}$

(g) $(CH_3)_2C=CH-CH=C(CH_3)_2$

15.23 Name each of the following compounds.

(a) $\underset{H}{\overset{CH_3}{}}C=C\underset{CH_2CH_3}{\overset{H}{}}$

(b) $\underset{CH_3}{\overset{H}{}}C=C\underset{\underset{CH_3}{\overset{|}{CHCH_2CH_3}}}{\overset{H}{}}$

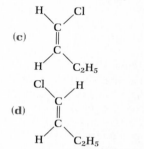

15.24 Name each of the following compounds.

(a) [cyclohexene structure]

(b) [3-methyl-5-methylcyclohexene structure with CH_3 groups]

(c) [cyclopentene with CH_2CH_3 structure]

(d) [methylcyclopentene with CH_3 groups structure]

15.25 Write the structural formula for each of the following compounds.
(a) 2-methyl-2-pentene
(b) 1-hexene
(c) *cis*-2-methyl-3-hexene
(d) *trans*-5-methyl-2-hexene

15.26 Write the structural formula for each of the following compounds.
(a) *trans*-1-chloropropene
(b) *cis*-2,3-dichloro-2-butene
(c) 3-chloropropene
(d) 4-chloro-2,4-dimethyl-2-hexene

15.27 Write the structural formula for each of the following compounds.
(a) cyclohexene
(b) 1-methylcyclopentene
(c) 1,2-dibromocyclohexene
(d) 4,4-dibromocyclohexene

Nomenclature of Alkynes

15.28 Name each of the following compounds.
    (a) $CH_3CH_2CH_2C\equiv CH$
    (b) $(CH_3)_3CC\equiv CCH_2CH_3$
    (c) $Cl(CH_2)_2C\equiv C(CH_2)_3CH_3$
    (d) $CH_3CHBrCHBrC\equiv CCH_3$

    (e) 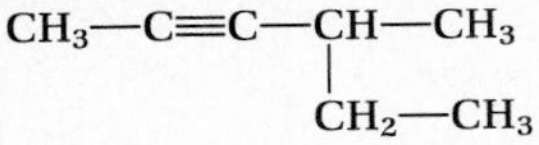 $CH_3\!-\!C\equiv C\!-\!\underset{\underset{\textstyle CH_2-CH_3}{|}}{CH}\!-\!CH_3$

    (f) 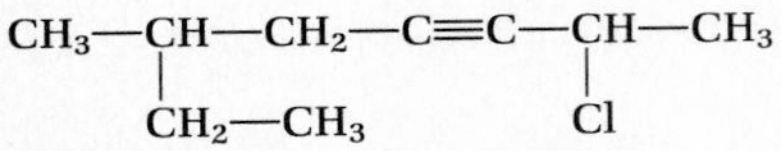 $CH_3\!-\!\underset{\underset{\textstyle CH_2-CH_3}{|}}{CH}\!-\!CH_2\!-\!C\equiv C\!-\!\underset{\underset{\textstyle Cl}{|}}{CH}\!-\!CH_3$

15.29 Write the structural formula for each of the following compounds.
    (a) 2-hexyne
    (b) 3-methyl-1-pentyne
    (c) 5-ethyl-3-octyne
    (d) 4,4-dimethyl-2-pentyne

Chemical Reactions of Alkenes

15.30 Describe the observation that should be made when 3-hexene reacts with Br_2. How can bromine be used to distinguish between 3-hexene and cyclohexane?

15.31 Describe the observation that should be made when 3-hexene reacts with potassium permanganate.

15.32 Complete the following equations.
    (a) $CH_3CH_2CH\!=\!CH_2 + Br_2 \longrightarrow$
    (b) $CH_3CH_2CH\!=\!CH_2 + HBr \longrightarrow$
    (c) $CH_3CH_2CH\!=\!CH_2 + H_2O \xrightarrow{\text{H}^+}$

    (d) 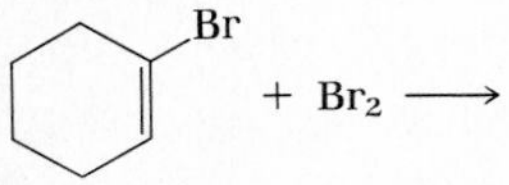 [cyclohexene with Br substituent] $+ Br_2 \longrightarrow$

15.33 Classify the following addition reagents as symmetrical or unsymmetrical.
    (a) Cl_2 (b) H_2O (c) HBr   (d) Br_2 (e) HCl

15.34 Write the product of the reaction of each of the reagents in **15.33** with propene. Do the same with methylpropene.

15.35 How many moles of bromine will react with each compound in **15.8?**

15.36 How many moles of hydrogen gas will react with the compound in **15.10?**

Chemical Reactions of Alkynes

15.37 Complete the following reactions
    (a) $CH_3C\equiv CH + 2 H_2 \longrightarrow$
    (b) $CH_3C\equiv CH + 2 Br_2 \longrightarrow$
    (c) $CH_3C\equiv CH + HBr \longrightarrow$
    (d) $CH_3C\equiv CH + 2 HBr \longrightarrow$

15.38 How many moles of bromine will react with each compound in **15.29?**

15.39 How many moles of hydrogen will react with each of the following compounds?
    (a) $CH_3CH\!=\!CHC\equiv CH$

    (b) $HC\equiv C\!-\!C\equiv CH$

    (c) 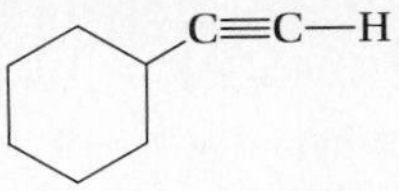 [cyclohexane with $C\equiv C\!-\!H$ substituent]

    (d) $HC\equiv C\!-\!CH\!=\!CH\!-\!C\equiv CH$

Aromatic Compounds

15.40 Indicate which dibromobenzene is ortho, meta, and para.

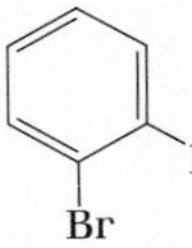

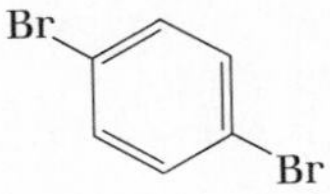

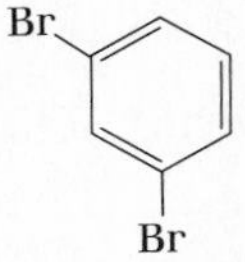

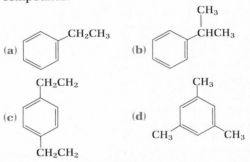

15.41 Give the common names of each of the following compounds.

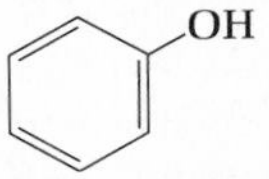

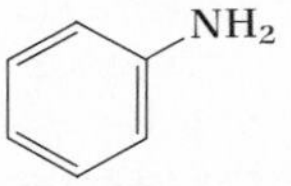

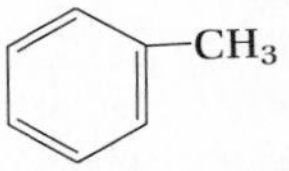

15.42 Name each of the following alkyl-substituted benzene compounds.

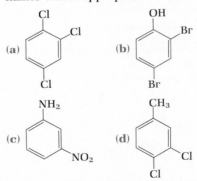

15.43 Name each of the following using accepted common names where appropriate.

15.44 Give the common name for each of the following compounds.

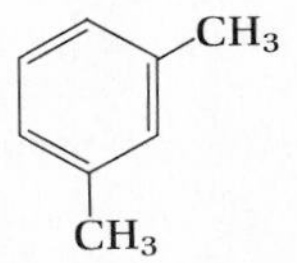

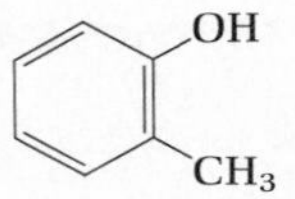

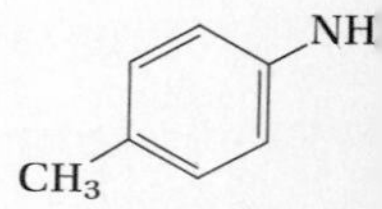

15.45 The determination of the amount of glucose in urine is made by reacting glucose with *o*-toluidine to give a colored substance whose concentration can be measured optically. What is the structure of *o*-toluidine?

15.46 Name the following compounds.

(a) 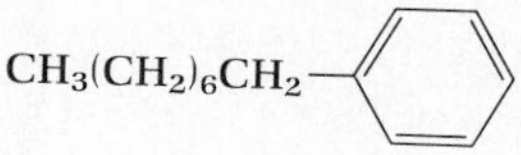 $CH_3(CH_2)_6CH_2$— (benzene ring)

(b) 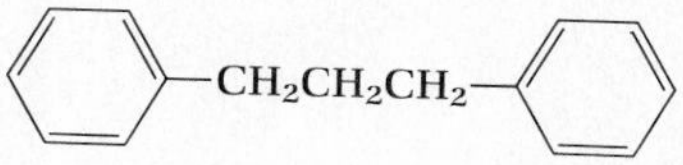 (benzene ring)—$CH_2CH_2CH_2$—(benzene ring)

(c) 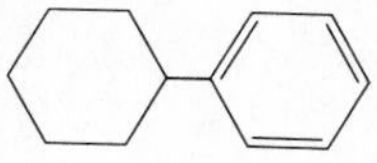 (cyclohexane ring)—(benzene ring)

(d) 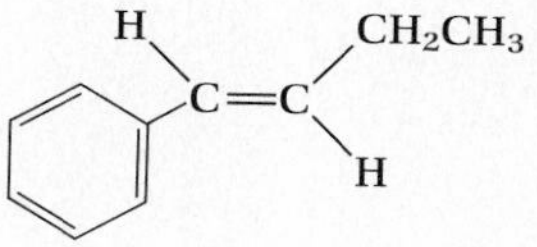 (benzene ring)—$\underset{H}{\overset{H}{C}}=\underset{H}{\overset{CH_2CH_3}{C}}$

15.47 Write the structure of each of the following compounds.
(a) 3-phenylheptane
(b) 1,1-diphenylethane
(c) 1-chloro-2-phenylpentane
(d) 1,2-diphenylethane

15.48 What is the molecular formula for the three compounds lacking molecular formulas in Figure 15.7?

Polymers

15.49 If the average molecular weight of a polyethylene sample is 40,000 amu, how many ethylene monomers are contained in the average molecule?

15.50 Draw a representation of the polymer produced from each of the following monomers.
(a) $CH_2=CHCH_3$
(b) $CH_2=CHCN$
(c) $CH_2=CCl_2$
(d) $CH_2=C(CH_3)_2$

16 Alcohols, Phenols, and Ethers

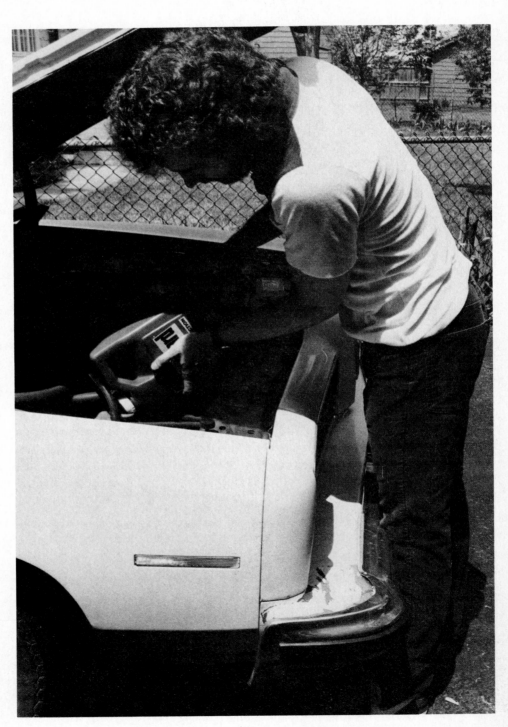

Antifreeze is an alcohol called ethylene glycol. It is soluble in water, and its solutions have low freezing points and high boiling points.

Oxygen is most familiar to you as a component of air, a gaseous mixture. The oxygen that we breathe is, of course, necessary to sustain life. However, in living systems, the role of oxygen is the result of how oxygen bonds to other atoms. Water, a compound of oxygen and hydrogen, is exceedingly important in life processes and was discussed in Chapter 8. In water, the oxygen atom shares its two unpaired electrons with two hydrogen atoms. Two covalent single bonds result. The oxygen atom is sp^3 hybridized and the H—O—H bond angle is 105°.

In combination with carbon, oxygen forms several important families of organic compounds. In addition, very complex biochemical molecules, which include carbohydrates, fats, proteins, and nucleic acids, contain oxygen. In order to understand the function of biochemical molecules containing oxygen, it is first necessary to study simpler organic compounds in which oxygen is part of a functional group. Families of organic compounds that contain oxygen include alcohols, phenols, ethers, aldehydes, ketones, acids, esters, and amides. Alcohols will be the main focus of this chapter. Two other structurally related classes of compounds, phenols and ethers, will be discussed briefly. A few examples of alcohols, phenols, and ethers are given in Table 16.1. Aldehydes and ketones are the subject of Chapter 17; acids and esters are presented in Chapter 18. Amides are presented in Chapter 19.

Alcohols *are compounds that contain the hydroxyl* (—OH) *functional group bonded to a saturated carbon atom.* In methanol, the simplest alcohol, the hydroxyl group is bonded to a one-carbon alkyl group, the methyl group. In methanol, both the carbon atom and the oxygen atom are sp^3 hybridized. The H—C—O bond angle is 110°, whereas the C—O—H bond angle is 106°. The O—H bond in methanol is quite polar, owing to the difference in electronegativity between oxygen and hydrogen, and it resembles the O—H bonds in water. The C—H bonds are nonpolar and resemble those in hydrocarbons. Alcohols can be viewed as the organic analogs of water, in which one hydrogen atom is replaced by an alkyl group (Figure 16.1). Alcohols have some chemical and physical properties resembling water, but also some properties resembling hydrocarbons.

Compounds in which the hydroxyl group is bonded to an aromatic ring are called **phenols.** The removal of one hydrogen atom from an aromatic ring carbon results in an aryl group, symbolized by Ar. A phenol then can be viewed as an organic analog of water in which one hydrogen atom is replaced by an aryl group (Figure 16.1). Phenols will be contrasted to alcohols in this chapter. You should anticipate that there will be some similarity in chemical properties because of the

Table 16.1 —————————————

Structure of Alcohols, Phenols, and Ethers

Alcohols	Phenols	Ethers
CH_3—O—H	(phenol with OH)	CH_3—O—CH_3
(benzyl)—CH_2—O—H	(naphthol with OH)	CH_3—O—(phenyl)

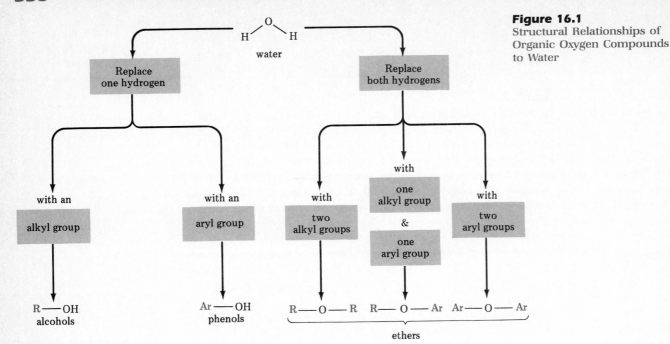

Figure 16.1
Structural Relationships of Organic Oxygen Compounds to Water

hydroxyl group, but differences will exist because of the difference between an alkyl and an aryl group.

In dimethyl ether, both hydrogen atoms of water are replaced with methyl groups. Other more complex groups of carbon atoms may bond to oxygen. **Ethers** *contain two groups, which may be alkyl or aryl groups, bonded to oxygen* (Figure 16.1). Ethers are different from alcohols because they do not have an —OH group. Whereas alcohols and phenols resemble water and react in much the same way, ethers have quite different chemical properties.

16.2
Nomenclature and Classification of Alcohols

The lower-molecular-weight alcohols have common names that frequently are used. The names consist of the alkyl group name plus the term alcohol. Several examples of this nomenclature are given in Figure 16.2.

The IUPAC system of naming alcohols is as follows.

1. The longest continuous chain of carbon atoms that includes the hydroxyl group is chosen as the parent chain.
2. The parent name is obtained by substituting the ending -ol for the -e of the corresponding alkane.
3. The position of the hydroxyl group is indicated by the number of the carbon atom to which it is attached, with the numbering arranged such that the hydroxyl group receives the lowest possible number.

$$\underset{4}{CH_3}-\underset{3}{\overset{\overset{\displaystyle CH_3}{|}}{\underset{|}{C}}}-\underset{2}{\overset{\overset{\displaystyle H}{|}}{\underset{|}{C}}}-\underset{1}{CH_3}$$

3-methyl-2-butanol (correct)
2-methyl-3-butanol (incorrect)

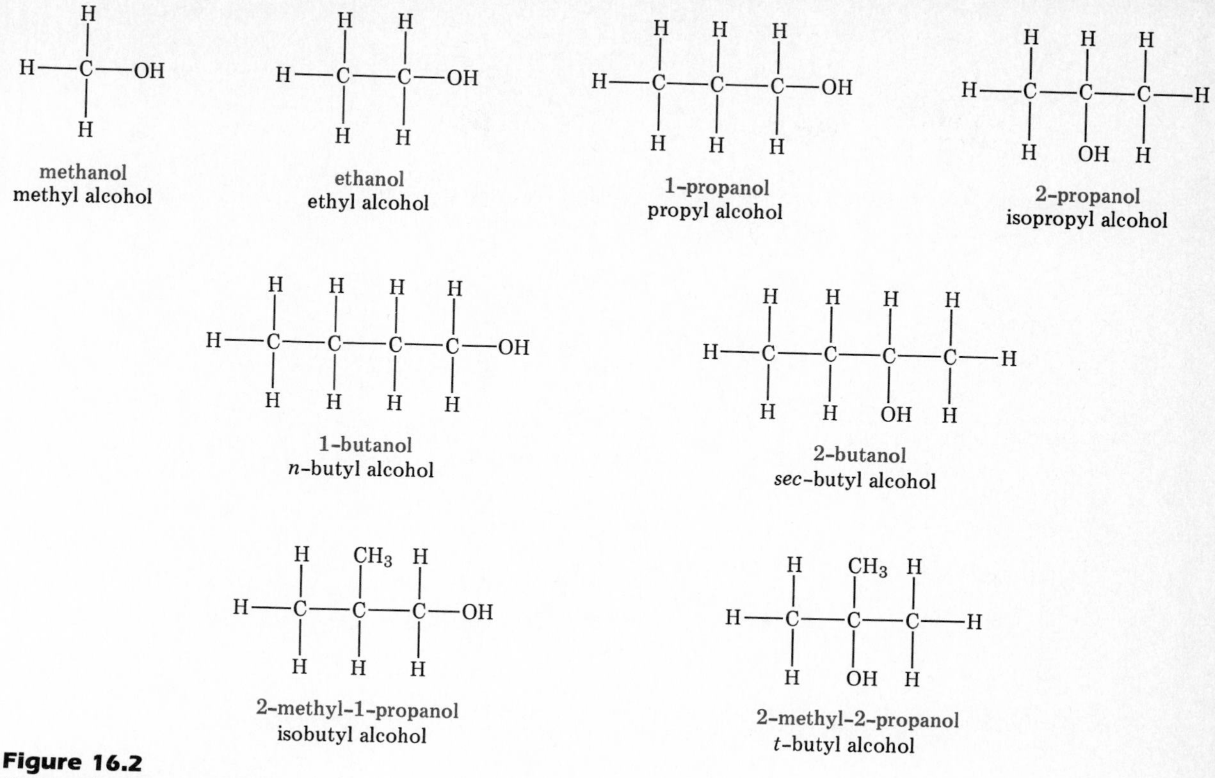

methanol
methyl alcohol

ethanol
ethyl alcohol

1–propanol
propyl alcohol

2–propanol
isopropyl alcohol

1–butanol
n–butyl alcohol

2–butanol
sec–butyl alcohol

2–methyl–1–propanol
isobutyl alcohol

2–methyl–2–propanol
t–butyl alcohol

Figure 16.2
Nomenclature of Alcohols.

4. The ring of cyclic alcohols is numbered starting with the carbon atom bearing the hydroxyl group. Numbering continues to give the lowest numbers to carbon atoms bearing alkyl groups. The number 1 is not used in the name (Figure 16.3).

5. Compounds containing two or more hydroxyl groups are called diols, triols, and so on. The terminal -e of the alkane is retained, and the suffix -diol or -triol is added. Numbers are used to designate the positions of the hydroxyl groups. Several examples are given in Figure 16.3. Compounds containing two hydroxyl groups bonded to the same carbon atom are usually quite unstable.

Example 16.1

Name the following compound.

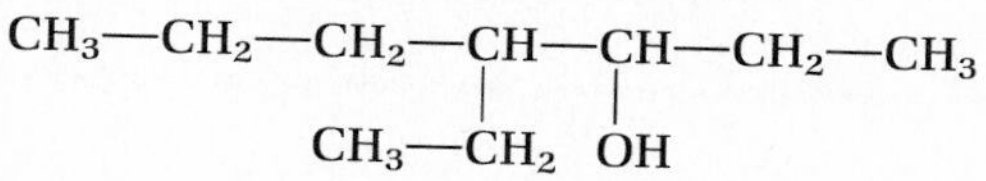

$$CH_3-CH_2-CH_2-CH-CH-CH_2-CH_3$$
$$\qquad\qquad\qquad\quad | \qquad |$$
$$\qquad\qquad\quad CH_3-CH_2 \quad OH$$

Solution. The longest chain contains seven carbon atoms. The hydroxyl group is on the third carbon atom when the chain is numbered from right to left. The ethyl group is at position 4. The name is 4-ethyl-3-heptanol.

[Additional examples may be found **16.4–16.15** at the end of the chapter.]

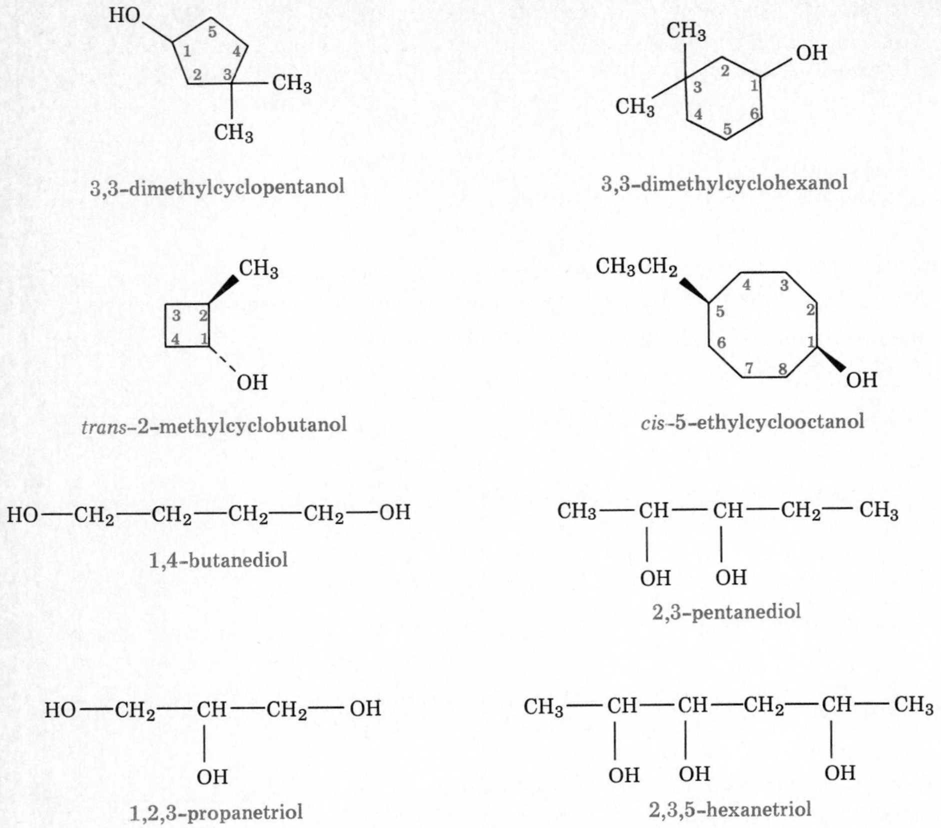

3,3-dimethylcyclopentanol

3,3-dimethylcyclohexanol

Figure 16.3
Nomenclature of Cyclic Alcohols, Diols, and Triols.

trans-2-methylcyclobutanol

cis-5-ethylcyclooctanol

$$HO-CH_2-CH_2-CH_2-CH_2-OH$$

1,4-butanediol

2,3-pentanediol

1,2,3-propanetriol

2,3,5-hexanetriol

Alcohols are classified according to the structure of the alkyl group to which the hydroxyl group is attached. A **primary (1°) alcohol** *is a compound in which the hydroxyl group is bonded to a primary carbon atom. In a* **secondary (2°) alcohol,** *the hydroxyl group is bonded to a secondary carbon atom. In a* **tertiary (3°) alcohol,** *the hydroxyl group is bonded to a tertiary carbon atom.*

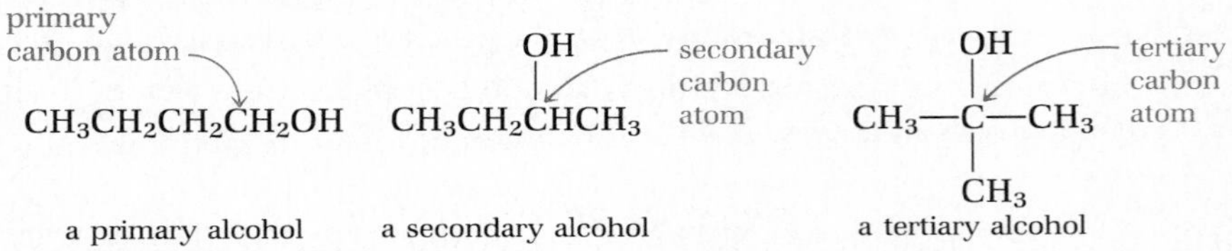

a primary alcohol   a secondary alcohol   a tertiary alcohol

This classification is used in discussing the chemical properties of alcohols.

Example 16.2

Classify the following alcohol.

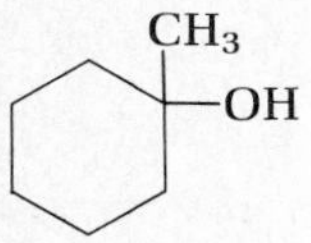

Solution. Although the compound is a cyclic alcohol, it is still classified on the basis of the carbon atom bearing the hydroxyl group. This carbon atom is bonded to a methyl group and to carbon atoms 2 and 6 of the ring. Thus, the number 1 carbon atom is tertiary and the compound is a tertiary alcohol.

[Additional examples may be found in **16.16–16.22** at the end of the chapter.]

As a class, alcohols boil at higher temperatures than alkanes of comparable molecular weight (Table 16.2). For example, ethanol, CH_3CH_2-OH, of molecular weight 46 boils at 78°C, whereas propane of molecular weight 44 boils at −42°C. This dramatic difference in boiling points is due to hydrogen bonding of alcohols in the liquid state. The attraction between a hydrogen atom in one molecule and the oxygen atom of a neighboring molecule is called a hydrogen bond. A similar effect on the boiling point of water was discussed in Chapter 8.

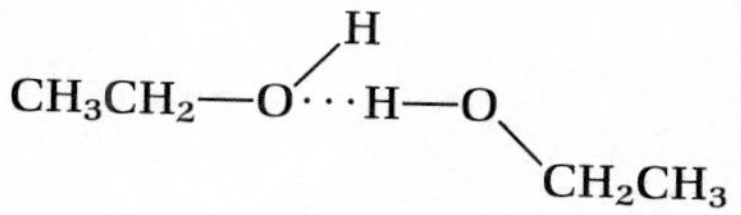

16.3
Hydrogen Bonding in Alcohols

$$CH_3CH_2-O\overset{H}{\cdots}H-O$$
$$CH_2CH_3$$

The solubility of alcohols in water is very different from the solubility of hydrocarbons. Table 16.3 lists the solubilities of some alcohols containing normal alkyl groups. The lower-molecular-weight alcohols are completely soluble in water, as they closely resemble it. However, with increasing size of the alkyl group, the alcohol more closely resembles an alkane, and the hydroxyl group becomes a smaller contributor of the overall physical properties of the alcohols. As a result the solubility of alcohols decreases with the increasing size of the alkyl group.

$$H-O$$
$$CH_3-O\overset{H}{\underset{H}{}}$$
$$CH_3-H$$

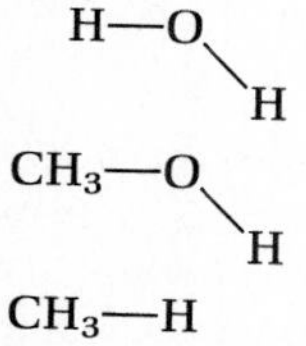

Methanol is much like water.

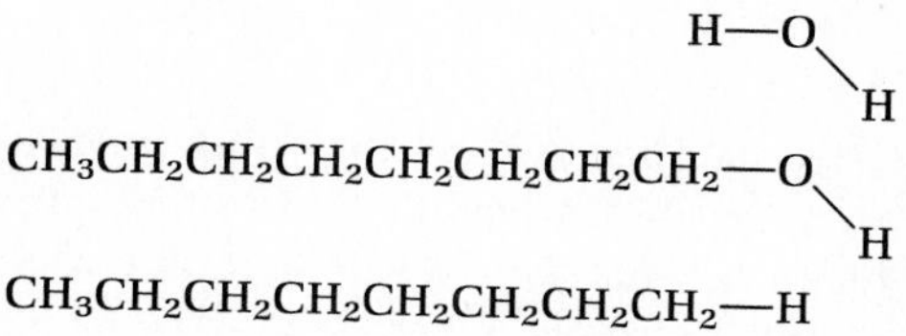

$$CH_3CH_2CH_2CH_2CH_2CH_2CH_2CH_2-O\overset{H-O}{\underset{H}{}}$$
$$CH_3CH_2CH_2CH_2CH_2CH_2CH_2CH_2-H$$

1-Octanol resembles octane more closely than water.

Table 16.2

Boiling Points of Some Alkanes and Alcohols

Name	Structure	Molecular weight	Boiling point (°C)
methyl alcohol	CH_3-OH	32	65
ethane	CH_3CH_3	30	−88
ethyl alcohol	CH_3CH_2-OH	46	78
propane	$CH_3CH_2CH_3$	44	−42
n-propyl alcohol	$CH_3CH_2CH_2-OH$	60	97
n-butane	$CH_3CH_2CH_2CH_3$	58	−0.5

Table 16.3

Boiling Points and Solubilities of Alcohols

Name	Formula	Boiling point (°C)	Solubility (g/100 mL in water)
methanol	CH_3OH	65	miscible
ethanol	CH_3CH_2OH	78	miscible
1-propanol	$CH_3CH_2CH_2OH$	97	miscible
1-butanol	$CH_3CH_2CH_2CH_2OH$	118	7.9
1-pentanol	$CH_3CH_2CH_2CH_2CH_2OH$	138	2.7
1-hexanol	$CH_3(CH_2)_4CH_2OH$	157	0.59
1-heptanol	$CH_3(CH_2)_5CH_2OH$	176	0.09
1-octanol	$CH_3(CH_2)_6CH_2OH$	195	insoluble
1-nonanol	$CH_3(CH_2)_7CH_2OH$	213	insoluble
1-decanol	$CH_3(CH_2)_8CH_2OH$	231	insoluble

16.4
Common Alcohols

Methanol

Methanol is sometimes called wood alcohol because it was originally obtained by heating wood to a high temperature in the absence of air. Under these conditions, decomposition occurs, and methanol is one of the substances produced. Methanol is now made from carbon monoxide and hydrogen.

$$CO + 2\,H_2 \xrightarrow[\text{200 atm, 350°C}]{\text{ZnO–Cr}_2\text{O}_3} CH_3OH$$

Methanol is a toxic substance. Temporary blindness, permanent blindness, or death can result from the consumption of methanol present as an impurity in illegal sources of ethanol. As little as 30 mL (1 fluid ounce) of pure methanol can cause death. The prolonged breathing of methanol vapors also is a serious health hazard. Methanol was extensively used as a radiator antifreeze prior to its replacement by ethylene glycol. Methanol is used in car windshield washer fluids.

Ethanol

Ethanol is the substance popularly known as alcohol. Fermentation of almost any substance containing sugar will produce ethanol but only in concentrations up to about 14%. The rate of the reaction is inhibited as the alcohol concentration increases. For this reason most red, white, or rose table wines contain 12% alcohol. The color of the wine is controlled by the variety of grape used and the manner in which the grape juice and skins are treated in the crushing process. Wines such as Madeira, port, and sherry, which contain 20% alcohol, cannot be produced from direct fermentation. A distilled grape brandy must be added in order to increase the alcohol concentration.

Distillation of alcohol–water mixtures increases the concentration up to 95%. To increase the concentration to 100%, it is necessary to distill with a small amount of benzene. The resultant 100% alcohol is called *absolute alcohol*, but it contains a small amount of benzene. The benzene is not detrimental for most reactions in the laboratory. However, the body cannot metabolize benzene, and it can affect the kidneys.

Alcohol concentrations are also given in proof units (Section 9.5). The proof value is twice the alcohol concentration. Thus, a wine containing 12% alcohol is 24 proof; an 80 proof vodka is 40% alcohol.

Physiologically, ethanol is a depressant and not a stimulant as some people believe. In moderate amounts ethanol causes drowsiness and depresses brain functions. As a consequence, activities requiring judgment or coordination are impaired. In larger quantities ethanol causes nausea and vomiting.

Ethanol is widely used as an industrial solvent. Special tax-free permits are issued for this use. However, substances are added in small amounts to render such alcohol unfit for drinking. Alcohol containing adulterants is called *denatured alcohol*. Methanol is one substance used to denature ethanol. All denaturants are difficult to remove.

Isopropyl Alcohol

Isopropyl alcohol is commonly known as rubbing alcohol. It is soluble in all proportions in water and is an excellent industrial solvent. Its low freezing point and solubility in water allow this compound to be used as a deicer for airplane wings and to prevent fuel-line freezeup in cars. In addition, it is used in inks, paints, and cosmetic preparations.

Polyhydroxy Alcohols

Ethylene glycol is a compound containing two hydroxyl groups, one each on the adjacent carbon atoms of ethane.

$$\begin{array}{cc} CH_2 & \!\!-\!\! & CH_2 \\ | & & | \\ OH & & OH \end{array}$$

It has now replaced methanol as an automobile antifreeze. One of the largest commercial uses of ethylene glycol is in the production of Dacron and Mylar film, the latter of which is used in tapes for recorders and computers.

Glycerol, also known as glycerin, contains three hydroxyl groups, one each on the three carbon atoms of propane.

$$\begin{array}{ccccc} CH_2 & \!\!-\!\! & CH & \!\!-\!\! & CH_2 \\ | & & | & & | \\ OH & & OH & & OH \end{array}$$

Obtained originally as a by-product in the formation of soap, glycerol is now in such great demand that a synthetic process using propene has been developed. The moisture-retaining properties of glycerol make it useful in skin lotions, inks, and pharmaceuticals.

Glycerol is present as the backbone of several important biological compounds. These include phospholipids, which are present in cell membranes, and fats and oils, which are used as a source of energy in metabolic reactions. These substances will be discussed in Chapter 22.

Reactions of alcohols can be separated into four classes based on the bond broken.

1. Breaking the oxygen–hydrogen bond.
2. Breaking the carbon–oxygen bond.
3. Breaking both the oxygen–hydrogen bond and the carbon–hydrogen bond at the carbon atom bearing the hydroxyl group.
4. Breaking both the carbon–oxygen bond and the carbon–hydrogen bond at a carbon atom adjacent to the carbon atom bearing the hydroxyl group.

16.5
Reactions of Alcohols

Reactions of types 1 and 2 will be discussed in this section. Types 3 and 4 will be discussed in subsequent sections.

Reaction with Active Metals

Alcohols react with alkali metals the same way water does. In this reaction an oxygen–hydrogen bond is broken.

$$2\,Na + 2\,CH_3OH \longrightarrow H_2 + 2\,CH_3O^- + 2\,Na^+$$

$$2\,Na + 2\,H_2O \longrightarrow H_2 + 2\,HO^- + 2\,Na^+$$

Hydrogen gas is evolved, and an alkoxide of the metal is formed. In the above reaction, sodium methoxide is produced. The reaction is somewhat less vigorous than with water. Nevertheless, the evolution of hydrogen gas is a good qualitative test to indicate the presence of an —OH group in an organic molecule.

Alkoxides are somewhat stronger bases than the hydroxide ion. Alkoxides are used as bases in organic solvents, since they are more soluble than hydroxide salts. Alkoxides react with water to produce an alcohol and the hydroxide ion. The net ionic reaction is

$$CH_3CH_2O^- + H_2O \rightleftharpoons CH_3CH_2OH + OH^-$$

This is an acid–base reaction in which the proton is transferred to give the less acidic substance, the alcohol. Recall the relationship between acids and conjugate bases (Chapter 11). Since alkoxides are stronger bases than hydroxide ion, alcohols are weaker acids than water.

Substitution Reactions

Recall that in a substitution reaction one atom or group of atoms is substituted for another atom or group of atoms. The hydroxyl group of alcohols can be substituted or replaced by halogens. Consider the following two reactions.

$$CH_3CH_2CH_2OH + HCl \xrightarrow{ZnCl_2} CH_3CH_2CH_2Cl + H_2O$$

$$3\,CH_3CH_2CH_2OH + PBr_3 \longrightarrow 3\,CH_3CH_2CH_2Br + H_3PO_3$$

Example 16.3

How could 1-pentanol and 2-pentanol be distinguished?

Solution. The reaction of 2-pentanol with HCl and $ZnCl_2$ should occur readily at room temperature although more slowly than that of a tertiary alcohol. The reaction with 1-pentanol should be extremely slow or not occur at room temperature.

$$CH_3CH_2CH_2CH_2CH_2OH$$

1-pentanol
(a primary alcohol)

$$CH_3CHCH_2CH_2CH_3$$
$$|$$
$$OH$$

2-pentanol
(a secondary alcohol)

[Additional examples may be found in **16.34–16.39** at the end of the chapter.]

In these substitution reactions the carbon–oxygen bond is broken, and the halogen replaces the hydroxyl group. These reactions are a more efficient way of obtaining haloalkanes than by halogenating alkanes, which produces a mixture of products. The rate for the reaction with HCl, in the presence of a $ZnCl_2$ catalyst, decreases in the order of tertiary > secondary > primary alcohols. The reaction is used for classification purposes and is called the *Lucas test*.

Haloalkanes are insoluble in water, so they are easily detected in the Lucas test. If an alcohol is added to Lucas reagent and turbidity occurs immediately, the alcohol is probably tertiary. The turbidity results from the insoluble halide formed in the reaction. If the turbidity develops within 5–10 min, a secondary structure is indicated. Primary alcohols may never cause turbidity at room temperature.

16.6
Dehydration of Alcohols

The removal of water from an alcohol is a **dehydration reaction.** An acid catalyst such as sulfuric acid or phosphoric acid is used. The reaction is illustrated by the formation of ethylene from ethyl alcohol.

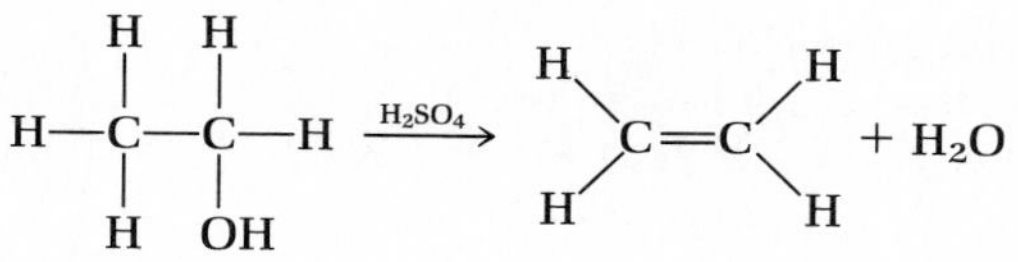

The reaction involves breaking a carbon–oxygen bond and the carbon–hydrogen bond of an adjacent carbon atom. For more complex alcohols, such as 2-butanol, the dehydration produces a mixture of products because there are two neighboring carbon atoms adjacent to the carbon atom bearing the hydroxyl group. In such cases, the isomers that contain the greatest number of alkyl groups attached to the double bond predominate in the mixture.

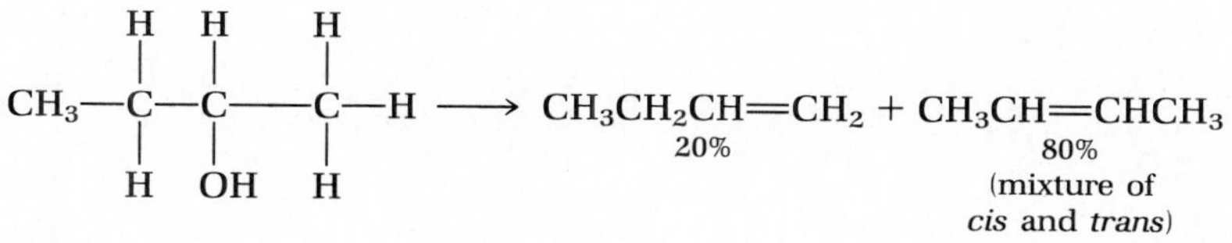

Dehydration reactions occur in biological systems catalyzed by enzymes. One such example is the dehydration of citric acid, catalyzed by the enzyme aconitase, to give *cis*-aconitic acid.

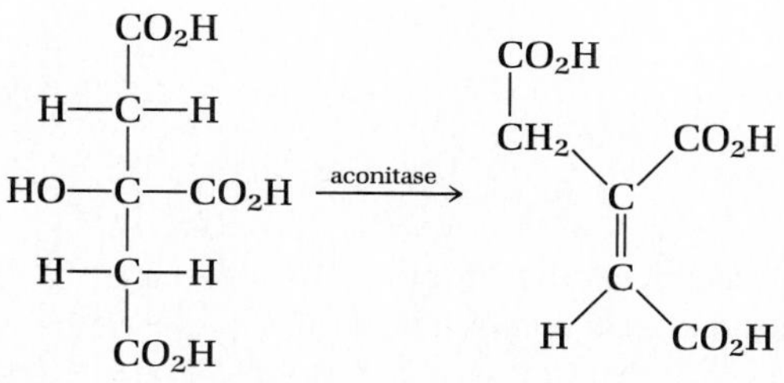

citric acid   *cis*-aconitic acid

Example 16.4 ——————————————————————————————

What is the product of the dehydration of 1-methylcyclohexanol?

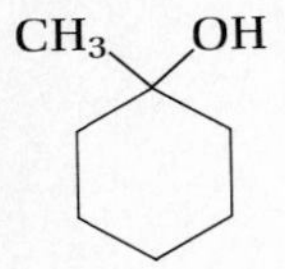

CH₃ OH

Solution. This tertiary alcohol has three adjacent carbon atoms, which may lose a hydrogen atom in the dehydration. However, the loss of a hydrogen atom from either carbon 2 or carbon 6 will result in the same product. Thus, only two isomers result.

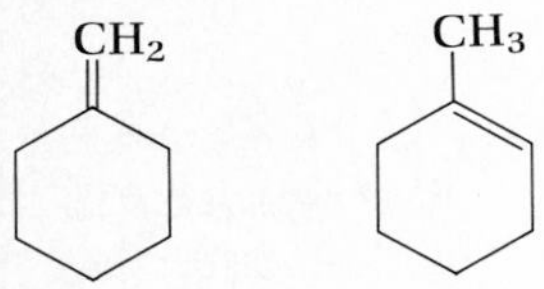

CH₂   CH₃

The second structure with the double bond within the six-membered ring predominates. This compound has three carbon atoms attached to the two double-bonded carbon atoms.

———————————————————————————————

[Additional examples may be found in **16.40–16.42** at the end of the chapter.]

16.7
Oxidation of Alcohols

Like the hydrocarbons, alcohols will burn when heated. The lower-molecular-weight alcohols, which are already partially oxidized, burn with quite clean flames, as evidenced by the ignition of warm brandy on cherries jubilee and other such gourmet specialities.

$$CH_3CH_2OH + 3\,O_2 \xrightarrow{\Delta} 2\,CO_2 + 3\,H_2O$$

The alcohols can be oxidized in a less destructive manner in the laboratory to produce oxidized compounds. The various classes of alcohols react in individually characteristic ways with an oxidizing agent such as potassium dichromate.

Primary alcohols, given by the general formula RCH_2OH, are oxidized to compounds called aldehydes.

$$3\,RCH_2OH + K_2Cr_2O_7 + 4\,H_2SO_4 \longrightarrow 3\,RC\overset{O}{\underset{\text{(aldehyde)}}{\diagup}}H + Cr_2(SO_4)_3 + K_2SO_4 + 7\,H_2O$$

While the above balanced equation gives the stoichiometry of the reaction, we will seldom balance these and other organic reactions. The primary interest is usually in the changes of the organic compounds. Thus, the oxidation of a primary alcohol is simply written with the inorganic reagent displayed above the reaction arrow. Note that both the hydrogen atom of the —OH group and a hydrogen from the carbon bearing the —OH group are removed. As pointed out in Chapter 10, oxidation often involves the removal of hydrogen atoms as in the case of conversion of primary alcohols to aldehydes.

primary alcohol — aldehyde

1-butanol — butanal

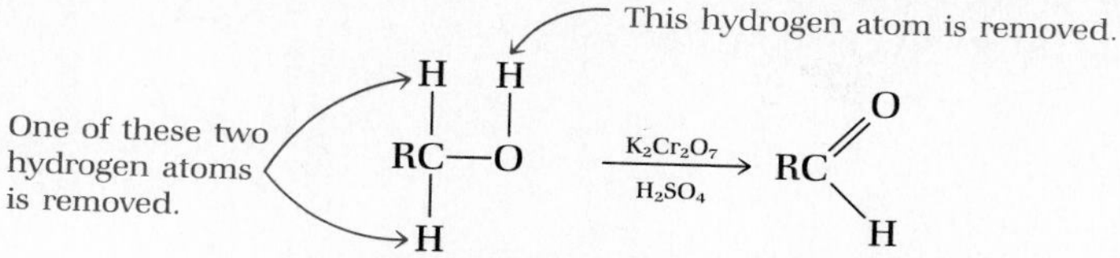

Aldehydes are easily oxidized further to produce acids. In this step, the oxidation is the result of a gain of an atom of oxygen.

butanal — butanoic acid

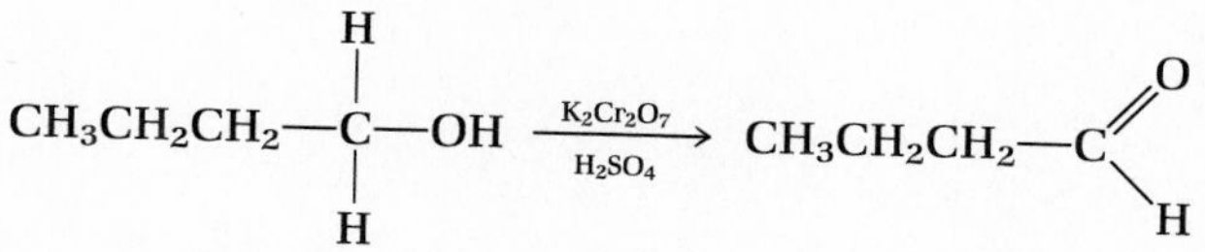

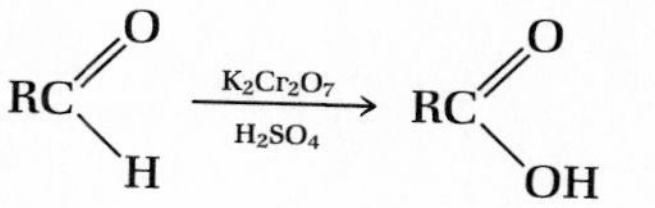

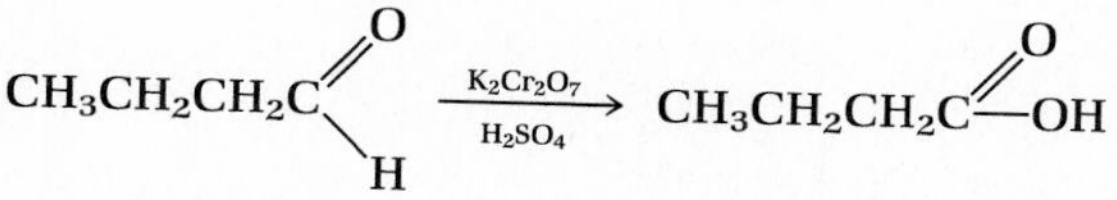

Secondary alcohols are oxidized to form ketones, which are not further oxidized.

3-pentanol — 3-pentanone

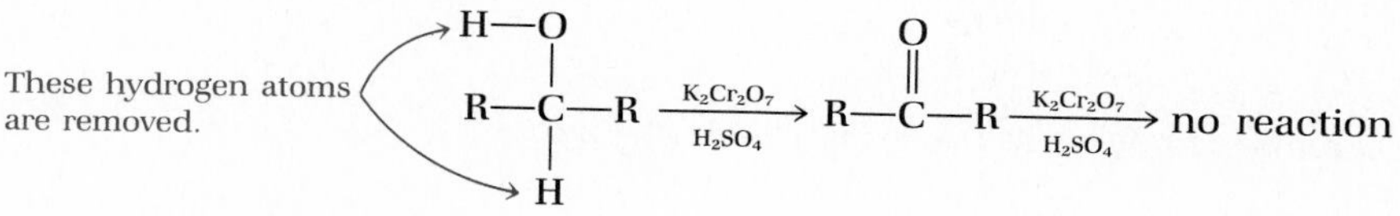

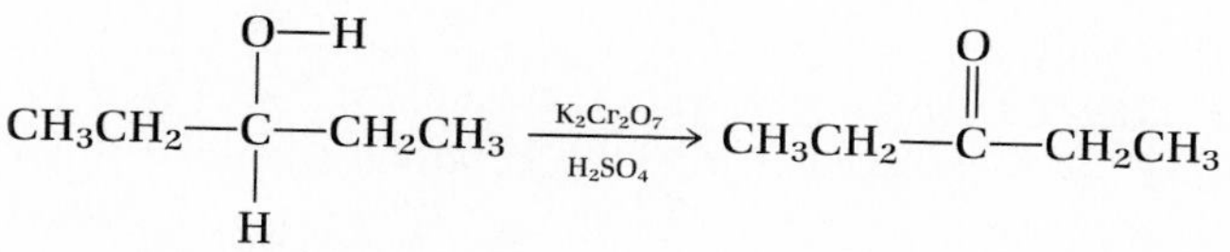

The reason for the stability of ketones is the absence of a hydrogen atom on the oxygen-bearing carbon atom as compared to the aldehydes.

Tertiary alcohols are not oxidized by potassium dichromate. In tertiary alcohols, the carbon atom bearing the —OH group has no hydrogen atom that can be removed by the oxidizing agent.

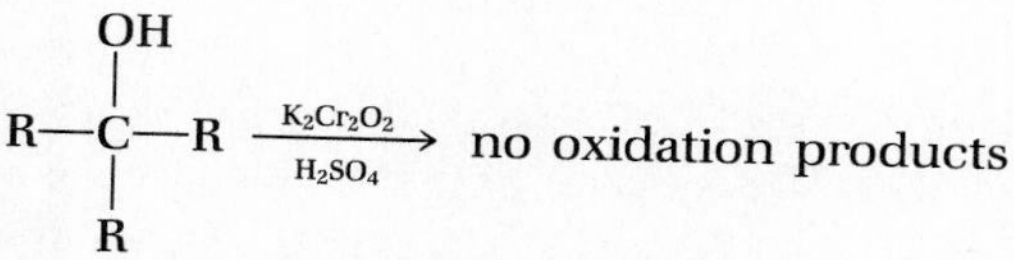

AN ASIDE

Oxidation of alcohols occurs as one step in many sequences of biological reactions. For example, in the oxidation of fats, a β-hydroxy acid is oxidized to a β-keto acid.

$$\underset{\substack{\beta\text{-carbon} \qquad \alpha\text{-carbon}}}{R-\overset{\displaystyle OH}{\underset{\displaystyle H}{C}}-\overset{\displaystyle H}{\underset{\displaystyle H}{C}}-\overset{\displaystyle O}{C}\diagdown_{OH}} + NAD^+ \longrightarrow R-\overset{\displaystyle O}{C}-\overset{\displaystyle H}{\underset{\displaystyle H}{C}}-\overset{\displaystyle O}{C}\diagdown_{OH} + NADH + H^+$$

The β position of the acid is the second one down the carbon chain after the carbon atom of the acid. The oxidizing agent is a coenzyme, nicotinamide adenine dinucleotide (NAD^+). One of the steps in the citric acid cycle is the oxidation of the secondary alcohol of malic acid. The oxidizing agent is again NAD^+.

$$\underset{\text{malic acid}}{\overset{\displaystyle CO_2H}{\underset{\displaystyle CO_2H}{\overset{\displaystyle |}{\underset{\displaystyle |}{H-\overset{|}{C}-OH}} \atop H-\overset{|}{C}-H}}} + NAD^+ \longrightarrow \underset{\text{oxaloacetic acid}}{\overset{\displaystyle CO_2H}{\underset{\displaystyle CO_2H}{\overset{\displaystyle |}{\underset{\displaystyle |}{C=O} \atop H-\overset{|}{C}-H}}}} + NADH + H^+$$

In the body, ethanol is oxidized in the liver by an enzyme to form acetaldehyde. Further oxidation leads to acetic acid, which can be metabolized by cells.

$$CH_3-CH_2OH \longrightarrow \underset{\text{acetaldehyde}}{CH_3-C\diagup^{\displaystyle O}_{\diagdown H}} \longrightarrow \underset{\text{acetic acid}}{CH_3-C\diagup^{\displaystyle O}_{\diagdown OH}}$$

Oxidation of methanol by enzymatic reactions in the liver produces formaldehyde.

$$CH_3OH \longrightarrow \underset{\text{formaldehyde}}{\overset{\displaystyle H}{\underset{\displaystyle H}{\diagdown \atop \diagup}}C=O}$$

Formaldehyde reacts very rapidly with proteins and causes them to lose their biological function. It is this property that is the reason for the use of formaldehyde in preserving tissue samples in biology. In a living organism the reaction with proteins is disastrous. Enzymes are proteins, and the loss of the function of these biological catalysts causes death. Even in low concentrations, formaldehyde in the blood causes the eye lens to turn opaque and blindness results. Ethanol administered intravenously within a proper time period is an antidote for methanol poisoning. The ethanol competes for the oxidative enzyme and tends to prevent the oxidation of the methanol to formaldehyde. The methanol is then slowly excreted by the body.

Example 16.5

Which of the isomeric $C_4H_{10}O$ alcohols will react with potassium dichromate to produce a ketone, C_4H_8O?

Solution. Inspection of Figure 16.2 reveals that two of the $C_4H_{10}O$ alcohols are primary, one is secondary, and one is tertiary. Only 2-butanol (*sec*-butyl alcohol) can give a ketone.

$$CH_3-\underset{\underset{H}{|}}{\overset{\overset{OH}{|}}{C}}-CH_2-CH_3 \xrightarrow[H_2SO_4]{K_2Cr_2O_7} CH_3-\overset{\overset{O}{\|}}{C}-CH_2-CH_3$$

[Additional examples may be found in **16.43–16.51** at the end of the chapter.]

Solutions of dichromate ion are yellow-orange, whereas solutions of Cr^{3+} are green. Thus, the oxidation of a primary or secondary alcohol by potassium dichromate is accompanied by a color change. Tertiary alcohols do not show a color change. A breath analyzer test for drunk-driving suspects employs a similar chemical change. The degree of the color change gives a measure of the ethyl alcohol content in the suspect's breath.

16.8
Formation of Esters

An alcohol can react with an acid to give an **ester** *and water*. The general reaction with H—O—A as a representation of any acid containing an acidic hydrogen atom bonded to an oxygen atom is given below.

$$\underset{\text{an alcohol}}{R-O-H} + \underset{\text{an acid}}{H-O-A} \longrightarrow \underset{\text{an ester}}{R-O-A} + H_2O$$

Thus, the ester contains both a part of the alcohol and the acid.

Esters of phosphoric acid, pyrophosphoric acid, and triphosphoric acid play important roles in many biological reactions.

$$\underset{\text{phosphoric acid}}{HO-\underset{\underset{OH}{|}}{\overset{\overset{\cdot\cdot}{\overset{O}{\|}}}{P}}-OH} \qquad \underset{\text{pyrophosphoric acid}}{HO-\underset{\underset{OH}{|}}{\overset{\overset{\cdot\cdot}{\overset{O}{\|}}}{P}}-\overset{\cdot\cdot}{O}-\underset{\underset{OH}{|}}{\overset{\overset{\cdot\cdot}{\overset{O}{\|}}}{P}}-OH} \qquad \underset{\text{triphosphoric acid}}{HO-\underset{\underset{OH}{|}}{\overset{\overset{\cdot\cdot}{\overset{O}{\|}}}{P}}-\overset{\cdot\cdot}{O}-\underset{\underset{OH}{|}}{\overset{\overset{\cdot\cdot}{\overset{O}{\|}}}{P}}-\overset{\cdot\cdot}{O}-\underset{\underset{OH}{|}}{\overset{\overset{\cdot\cdot}{\overset{O}{\|}}}{P}}-OH}$$

Several of the esters of these acids are given in Figure 16.4. The metabolism of glucose involves its phosphate ester. An isomerization reaction then produces a phosphate ester of fructose from which, in subsequent steps, esters of two low-molecular-weight compounds, glyceraldehyde and dihydroxyacetone, are formed.

Energy obtained from metabolism of food is stored as potential energy in adenosine triphosphate (ATP) (Figure 16.4). Hydrolysis of adenosine triphosphate to give adenosine diphosphate (ADP) is exothermic. It is this energy that is the source of chemical energy for other biochemical reactions. More details of these reactions will be provided in Chapters 20 and 25 of this text.

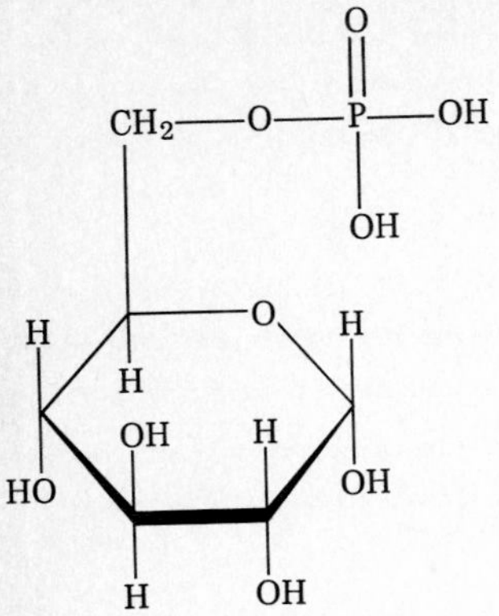

Figure 16.4
Structures of Phosphoric Acid Esters of
Biological Interest.

glucose 6-phosphate

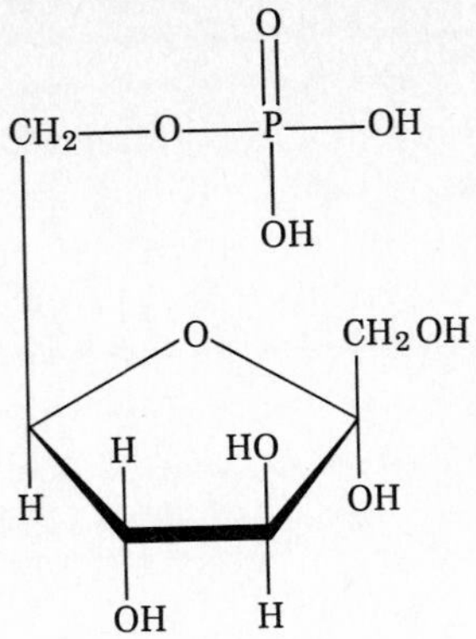

fructose 6-phosphate

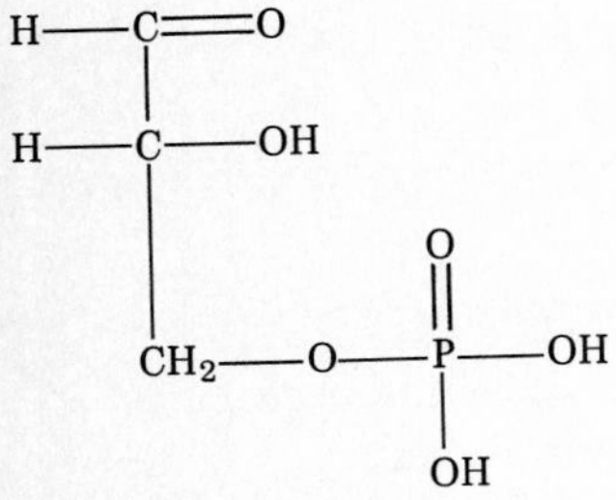

glyceraldehyde 3-phosphate

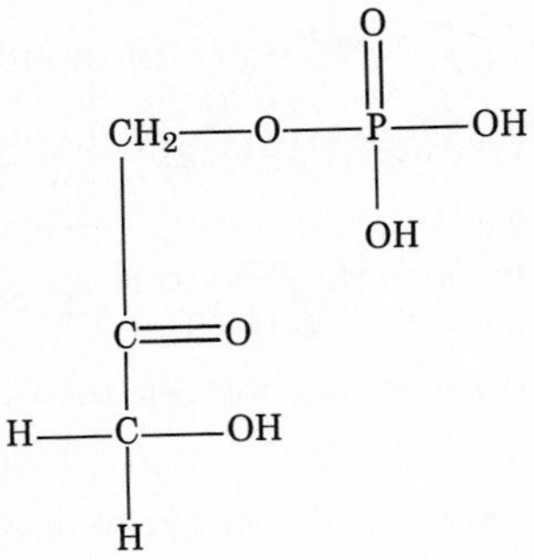

dihydroxyacetone phosphate

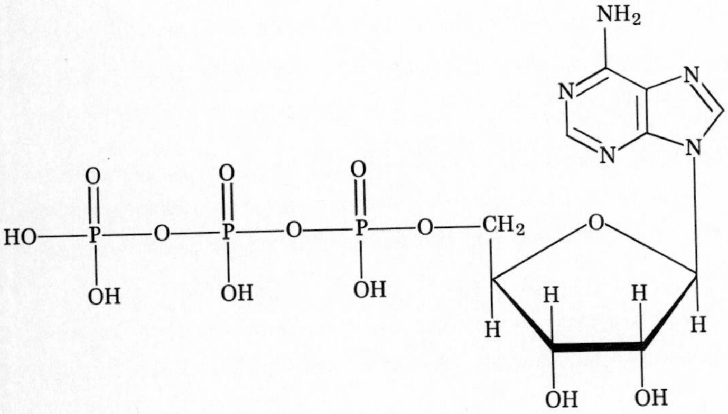

adenosine triphosphate

Esters of nitric acid are called nitrate esters. Many of these esters, such as glyceryl trinitrate (nitroglycerin), are powerful explosives.

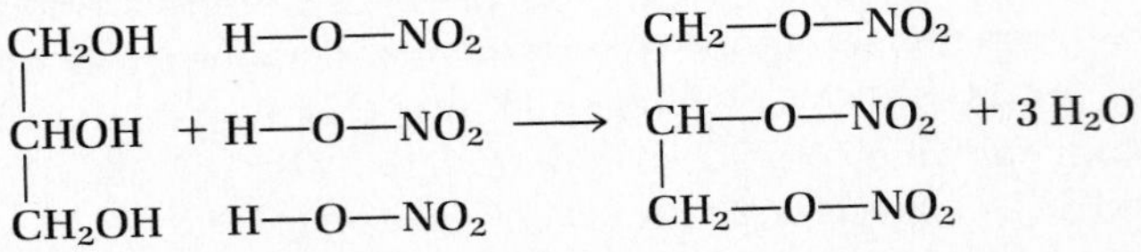

$$\begin{array}{ccc} CH_2OH & H-O-NO_2 & CH_2-O-NO_2 \\ | & & | \\ CHOH & + H-O-NO_2 \longrightarrow & CH-O-NO_2 & + 3\,H_2O \\ | & & | \\ CH_2OH & H-O-NO_2 & CH_2-O-NO_2 \end{array}$$

Glycerol trinitrate is also a smooth-muscle relaxant and vasodilator. It is useful in lowering blood pressure and in treating angina pectoris, a heart disorder.

Esters of sulfuric acid are called *sulfate esters.* The sulfate esters are converted into detergents. The sulfate ester of lauryl alcohol has an acidic proton, which reacts with sodium hydroxide in a neutralization reaction. The sodium salt is the commercial detergent Dreft.

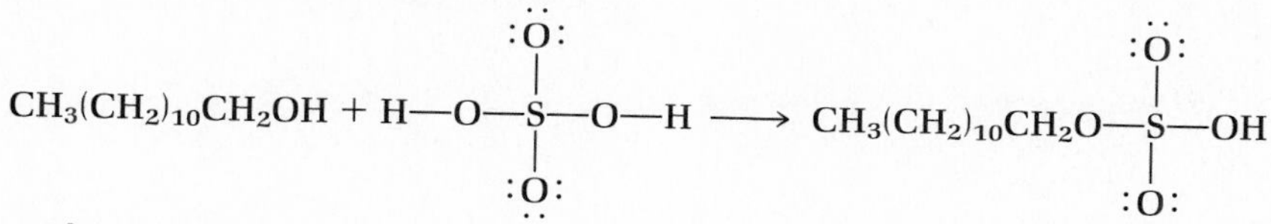

$$CH_3(CH_2)_{10}CH_2OH + H-O-\overset{\overset{\displaystyle :\ddot{O}:}{|}}{\underset{\underset{\displaystyle :\ddot{O}:}{|}}{S}}-O-H \longrightarrow CH_3(CH_2)_{10}CH_2O-\overset{\overset{\displaystyle :\ddot{O}:}{|}}{\underset{\underset{\displaystyle :\ddot{O}:}{|}}{S}}-OH$$

lauryl alcohol   lauryl sulfate

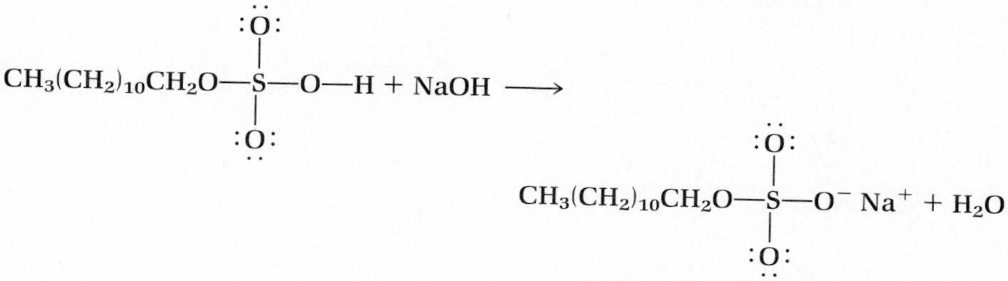

$$CH_3(CH_2)_{10}CH_2O-\overset{\overset{\displaystyle :\ddot{O}:}{|}}{\underset{\underset{\displaystyle :\ddot{O}:}{|}}{S}}-O-H + NaOH \longrightarrow$$

$$CH_3(CH_2)_{10}CH_2O-\overset{\overset{\displaystyle :\ddot{O}:}{|}}{\underset{\underset{\displaystyle :\ddot{O}:}{|}}{S}}-O^-\ Na^+ + H_2O$$

sodium lauryl sulfate

16.9 Phenols

There is a resonance interaction between the orbitals of the nonbonding electrons of oxygen and the orbitals of the multiple bonds of the benzene ring in phenols. Thus, there are some differences between the chemistry of phenols and that of alcohols.

Both alcohols and phenols have in common their reactivity of the hydrogen atom toward active metals such as sodium. Water, alcohols, and phenols all produce hydrogen atom gas when sodium is added.

$$2\,H-OH + 2\,Na \longrightarrow H_2 + 2\,NaOH$$

$$2\,R-OH + 2\,Na \longrightarrow H_2 + 2\,NaOR$$

$$2\,Ar-OH + 2\,Na \longrightarrow H_2 + 2\,NaOAr$$

Phenols are stronger acids than alcohols, although weaker than organic acids (RCO_2H).

$$CH_3C\overset{\displaystyle O}{\underset{\displaystyle O-H}{\diagup}}$$
$$K_a = 1.8 \times 10^{-5}$$

phenol—O—H
$$K_a = 1.3 \times 10^{-10}$$

$$CH_3CH_2O-H$$
$$K_a = 1 \times 10^{-17}$$

As a consequence, phenols will react even with a weak base, such as bicarbonate, whereas alcohols will not. Also, phenols can be dissolved in basic solution, whereas high-molecular-weight alcohols are insoluble.

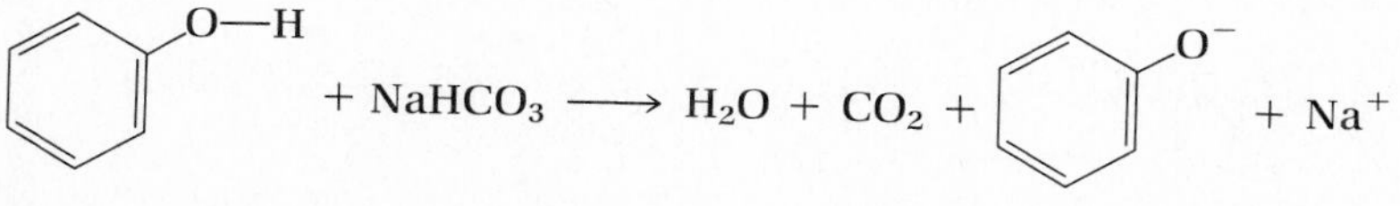

(phenol)O—H $+ NaHCO_3 \longrightarrow H_2O + CO_2 +$ (phenol)O$^-$ $+ Na^+$

$$CH_3(CH_2)_4CH_2OH + NaHCO_3 \longrightarrow \text{no reaction}$$

Figure 16.5
Structures of Phenols.

phenol   thymol   *p*-cresol   urushiol

hexylresorcinol   hexachlorophene

Phenols are active germicides. The structures of some of these phenols are given in Figure 16.5. The English surgeon Joseph Lister used phenol itself as a hospital antiseptic in the late nineteenth century. Because phenol can cause severe burns to skin, it is always used as a dilute solution. A 2% solution can be used to decontaminate medical instruments. The presence of alkyl groups on the aromatic ring causes remarkable changes in biological activity. *p*-Cresol and its ortho and meta isomers are used in Lysol, the commercial disinfectant. Thymol is used by dentists to sterilize a tooth prior to filling it. By contrast, urushiol is the skin irritant of poison ivy.

The efficiency of germicides is measured in units called the phenol coefficient. A 1% solution of a germicide that is as effective as a 10% solution of phenol has a phenol coefficient of 10. If alkyl groups are attached to the aromatic ring, the germicidal action of the compound is increased. Mouthwash solutions usually contain phenolic substances. The antiseptic hexylresorcinol is used in throat

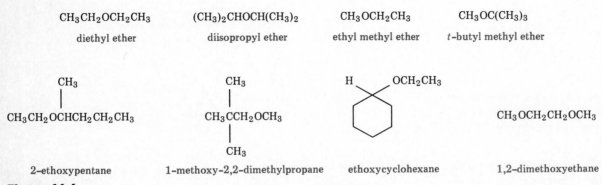

$CH_3CH_2OCH_2CH_3$	$(CH_3)_2CHOCH(CH_3)_2$	$CH_3OCH_2CH_3$	$CH_3OC(CH_3)_3$
diethyl ether	diisopropyl ether	ethyl methyl ether	*t*–butyl methyl ether

$CH_3CH_2OCHCH_2CH_2CH_3$ $CH_3CCH_2OCH_3$ ethoxycyclohexane $CH_3OCH_2CH_2OCH_3$

2-ethoxypentane   1-methoxy-2,2-dimethylpropane   ethoxycyclohexane   1,2-dimethoxyethane

Figure 16.6
Nomenclature of Ethers.

lozenges. Hexachlorophene (phenol coefficient = 120) has been used in some toothpastes, deodorants, and soaps.

It has been discovered that hexachlorophene is absorbed through the skin of newly born monkeys and causes brain damage in these primates. The Food and Drug Administration of the United States has restricted the use of hexachlorophene for this reason, and it is no longer allowed in soaps and skin cleansers that are to be used by infants and young children.

16.10 Ethers

Simple ethers are named according to the alkyl groups attached to the oxygen atom. *For higher-molecular-weight ethers, the smaller alkyl group and the oxygen are called an* **alkoxy group,** and this group is regarded as a substituent on the larger alkane chain. Examples of nomenclature are given in Figure 16.6.

Larger group is parent chain.

$$CH_3CH_2CH_2CHCH_3$$
$$|$$
$$OCH_3$$

Smaller group
is substituent.
2-methoxypentane

Diethyl ether is called simply ethyl ether or ether. It is the ether most familiar as a general anesthetic. Diethyl ether is a colorless liquid with a slightly pungent odor. It boils at 35°C and thus can be administered as a vapor. Ether vapors act as a depressant on the central nervous system and cause unconsciousness. Ether was first used as an anesthetic by Dr. Crawford Long of Atlanta, Georgia, in 1842. Because its high flammability and volatility are hazards in the operating room, it must be administered by highly trained personnel. Diethyl ether is administered along with oxygen, which is a potentially explosive mixture. Precautions must be taken to reduce the possibility of static electricity, which could cause an explosion.

Although diethyl ether is toxic, it is a safe anesthetic, since the amount required to cause unconsciousness is much smaller than the amount that would be lethal. One of the side effects of diethyl ether is nausea and vomiting after surgery. Divinyl ether (vinethene) is less nauseating than diethyl ether and acts more rapidly. However, this substance also can cause an explosion.

$$CH_2{=}CH{-}O{-}CH{=}CH_2$$
divinyl ether

Diethyl ether is a slightly polar liquid and is an excellent solvent for organic compounds. There are few organic compounds that are not soluble in ether. However, it must be used with care in the laboratory because of its high flammability.

The structures of some naturally occurring and synthetically prepared ethers of biological interest are given in Figure 16.7.

16.11 Sulfur Compounds

Sulfur is in the same group of the periodic table as oxygen and forms compounds structurally similar to alcohols. *The —SH group is called the* **sulfhydryl group.** *Compounds containing the —SH group are called* **mercaptans** *or thiols*. These

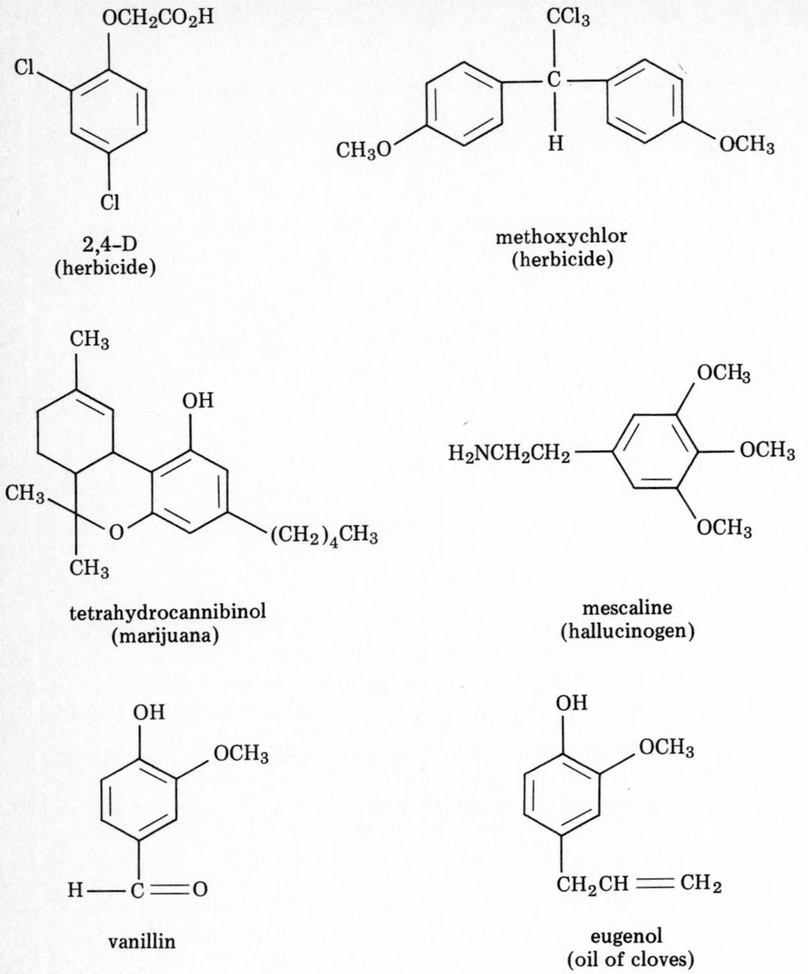

Figure 16.7
Structures of Ethers of Biological Interest.

2,4-D
(herbicide)

methoxychlor
(herbicide)

tetrahydrocannibinol
(marijuana)

mescaline
(hallucinogen)

vanillin

eugenol
(oil of cloves)

compounds, R—S—H, can be considered as derivatives of hydrogen sulfide in which one hydrogen atom is replaced by an alkyl group. Although there are some similarities between the alcohols and mercaptans, there are pronounced differences as well.

Thiols have lower boiling points than their corresponding alcohols. One of the distinguishing properties is their strong, disagreeable odor, which can be detected at the level of parts per billion in air. For this reason, small amounts of thiols are added to natural gas so that leaks can easily be sensed. Some of these interesting compounds are shown in Figure 16.8. Both *trans*-2-butene-1-thiol and 3-methyl-1-butanethiol are responsible for the odor of skunks. 1-Propanethiol is emitted when onions are chopped. Garlic contains allyl mercaptan.

Although they are still classified as weak acids, mercaptans are stronger acids than alcohols.

$$R—S—H + H_2O \longrightarrow R—S^- + H_3O^+ \qquad K_a = 10^{-8}$$

The sulfhydryl group is sufficiently acidic to react with hydroxide ions to form salts.

$$R—S—H + NaOH \longrightarrow R—S^- + Na^+ + H_2O$$

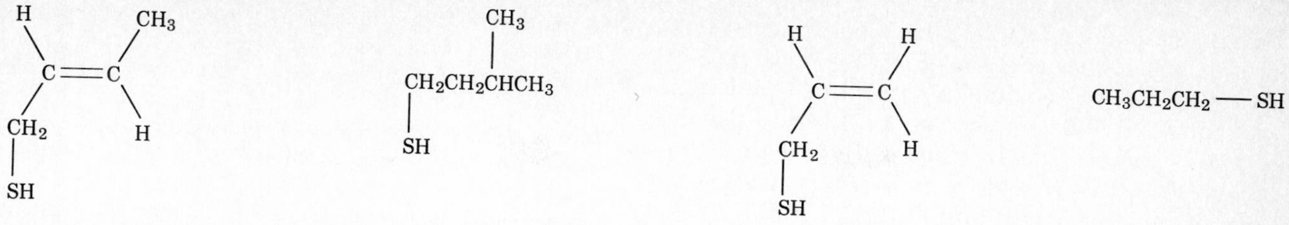

trans-2-butene-1-thiol · 3-methyl-1-butanethiol · 2-propene-1-thiol (allyl mercaptan) · 1-propanethiol

Figure 16.8
Structures of Sulfur Compounds.

Mercaptans are easily oxidized, but yield disulfides rather than the structural analogs of aldehydes and ketones. In the following equation, the symbol [O] represents an unspecified oxidizing agent that removes the hydrogen atoms.

$$2\,R{-}S{-}H \xrightarrow{[O]} \underset{\text{a disulfide}}{R{-}S{-}S{-}R}$$

A biological reaction of great importance involves the oxidation of the —SH group of the amino acid cysteine to yield cystine.

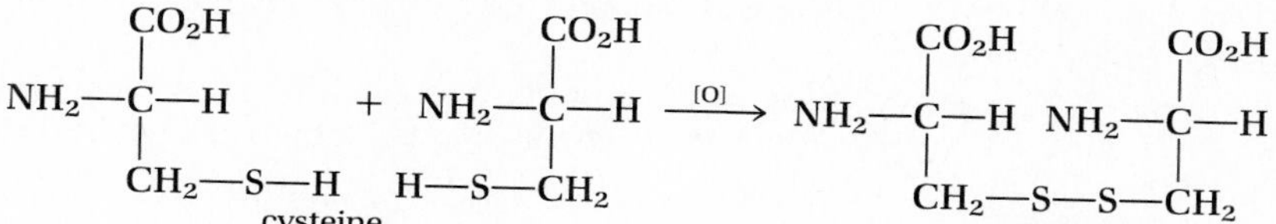

cysteine · cystine

The oxidation of cysteine contained in enzymatic proteins can alter the action of the molecule. Frequently the activity of the enzyme is due to the free sulfhydryl groups. If oxidation involves two sulfhydryl groups, the enzyme may become inactive.

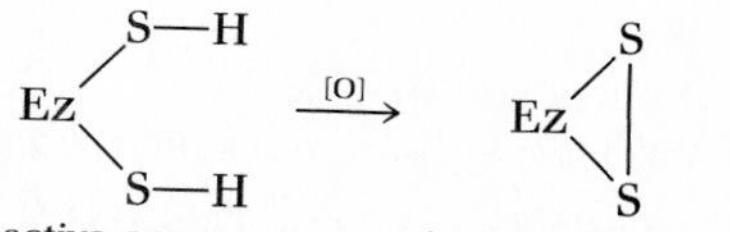

active enzyme · inactive enzyme

Salts of lead and mercury are toxic because of their reaction with the sulfhydryl group of enzymes.

$$2\,Ez{-}S{-}H + Hg^{2+} \longrightarrow Ez{-}S{-}Hg{-}S{-}Ez + 2\,H^+$$

The relationship of disulfide bonds to the structure and activity of enzymes will be presented in Chapter 24.

Summary

Alcohols have the hydroxyl group bonded to a saturated carbon atom. There are three classes of alcohols: primary, secondary, and tertiary. Alcohols have higher boiling points than alkanes of similar molecular weight because of hydrogen bonding between hydroxyl groups in alcohols. Lower-molecular-weight alcohols are soluble in water, but as the size of the alkyl group increases, the solubility decreases.

The oxygen–hydrogen bond of alcohols reacts with active metals to form hydrogen gas. The alcohols are weaker acids than water.

Three important reactions of alcohols are dehydration, oxidation, and the formation of esters. Dehydration of alcohols can form a mixture of alkenes. Oxidation of primary alcohols first produces aldehydes and then acids. Secondary alcohols are oxidized to ketones, whereas tertiary alcohols are not oxidized. Alcohols react with acids to form esters. These reactions of alcohols occur both in the laboratory and in living systems. Reactions in living systems are enzyme catalyzed.

Phenols have the general formula Ar—OH where Ar represents an aryl group derived from an aromatic compound. Phenols are stronger acids than alcohols. Phenols are effective germicides.

Ethers have either alkyl or aryl groups replacing both hydrogen atoms of water. The ethers are slightly polar compounds. Diethyl ether is an excellent solvent for organic compounds.

Thiols or mercaptans are the sulfur analogs of alcohols and have the general formula R—SH. These compounds are more acidic than alcohols. Oxidation of thiols forms disulfides, represented by R—S—S—R. The most distinctive physical property of thiols is their foul odor.

Learning Objectives

As a result of studying Chapter 16 you should be able to

- Draw structural formulas for representative alcohols, phenols, and ethers.
- Name alcohols by the IUPAC method.
- Assign common names for low-molecular-weight alcohols.
- Classify alcohols as primary, secondary, or tertiary from the structural formula.
- Explain how hydrogen bonding influences the physical properties of alcohols.
- List uses for some common alcohols.
- Identify and classify alcohols based on their chemical reactivity.
- Predict the products of dehydration of an alcohol.
- Predict the products of oxidation of alcohols.
- Write the product of reaction of an alcohol with an acid.
- Distinguish between phenols and alcohols.
- Name ethers and write their structures.
- Illustrate the differences in the chemical reactivities of alcohols and phenols.
- Illustrate the differences between alcohols and thiols.

New Terms

An **alcohol** is a compound containing a hydroxyl group bonded to a saturated carbon atom.

An **alkoxide** is a salt formed by the reaction of an alcohol with an active metal.

An **alkoxy group**, represented as RO—, is used to name ethers.

Dehydration is the removal of water from a molecule.

A **disulfide** represented by R—S—S—R is the product of oxidation of a thiol.

An **ester** is a compound formed by the reaction of an alcohol and an acid.

The **hydroxyl group**, represented by —OH, is a functional group contained in alcohols and phenols.

A **mercaptan** is a compound containing the sulfhydryl group, —SH. Mercaptans are also called thiols.

A **phenol** has the hydroxyl group bonded to a carbon atom of an aromatic ring.

A **primary alcohol** has a hydroxyl group bonded to a primary carbon atom.

A **secondary alcohol** has a hydroxyl group bonded to a secondary carbon atom.

The **sulfhydryl group**, represented by —SH, is the functional group in thiols, which are also known as mercaptans.

A **tertiary alcohol** has a hydroxyl group bonded to a tertiary carbon atom.

A **thiol** is a compound containing a sulfhydryl group bonded to a carbon atom. Thiols are also called mercaptans.

Questions and Problems

Terminology

16.1 Explain the differences in structure between an alcohol and a phenol.

16.2 Explain how alcohols are classified.

16.3 How do alcohols differ structurally from ethers?

Nomenclature of Alcohols

16.4 Write the structural formula for each of the following.
- (a) 2-methyl-2-pentanol
- (b) 2-methyl-1-butanol
- (c) 2,3-dimethyl-1-butanol
- (d) 2-methyl-3-pentanol
- (e) 3-ethyl-3-pentanol
- (f) 4-methyl-2-pentanol

16.5 Write the structural formula for each of the following.
- (a) propyl alcohol
- (b) t-butyl alcohol
- (c) n-butyl alcohol
- (d) isopropyl alcohol
- (e) ethyl alcohol
- (f) sec-butyl alcohol
- (g) methyl alcohol
- (h) isobutyl alcohol

16.6 Write the structural formula for each of the following.
- (a) cyclopentanol
- (b) 1-methylcyclobutanol
- (c) trans-2-methylcyclohexanol
- (d) cis-3-ethylcyclopentanol

16.7 Write the structural formula for each of the following.
(a) 1,2-hexanediol
(b) 1,3-propanediol
(c) 1,2,4-butanetriol
(d) 1,2,3,4,5,6-hexanehexol

16.8 Explain why each of the following names is incorrect.
(a) 4-methyl-4-pentanol
(b) 1-ethyl-1-methyl-1-pentanol
(c) 3,4-butanediol

16.9 Name each of the following compounds.

(a) $CH_3-CH_2-\underset{\underset{OH}{|}}{CH}-CH_2-CH_3$

(b) $CH_3-\underset{\underset{OH}{|}}{\overset{\overset{CH_3}{|}}{C}}-CH_2-CH_3$

(c) $CH_3-CH_2-\underset{\overset{|}{OH}}{CH}-CH_3$

(d) $CH_3-\underset{\underset{CH_3}{|}}{\overset{\overset{CH_3}{|}}{C}}-OH$

(e) $CH_3-\underset{\underset{CH_3}{|}}{\overset{\overset{CH_3}{|}}{C}}-CH_2-OH$

(f) $CH_3-\underset{\overset{|}{OH}}{CH}-CH_2-\underset{\overset{|}{CH_3}}{CH}-CH_2-CH_3$

(g) $CH_3-\underset{\overset{|}{OH}}{CH}-CH_2-CH_2-CH_3$

(h) $CH_3-\underset{\underset{CH_3}{|}}{\overset{\overset{CH_3}{|}}{C}}-CH_2-CH_2-OH$

16.10 Name each of the following compounds.

(a) $CH_3-CH_2-CH_2-CH_2-OH$

(b) $CH_3-\underset{\underset{CH_2-OH}{|}}{CH}-CH_2-\overset{\overset{CH_3}{|}}{CH}-CH_3$

(c) $CH_3-CH_2-\underset{\underset{CH_2-OH}{|}}{CH}-CH_3$

(d) $CH_3-CH_2-CH_2-\underset{\underset{CH_2-OH}{|}}{\overset{\overset{CH_3}{|}}{C}}-CH_3$

(e) $CH_3-CH_2-\underset{\underset{CH-OH}{\overset{|}{}}}{\underset{\underset{CH_3}{|}}{CH}}-CH_2-CH_2-CH_3$

(f) $\underset{\underset{CH_2OH}{|}}{CH_3\overset{\overset{CH_3}{|}}{CH}CH}-\overset{\overset{CH_3}{|}}{CH}CH_3$

16.11 Name each of the following compounds.

(a) cyclohexane—OH

(b) CH_3—cyclohexane—OH

(c) cyclopentane—OH with CH_3

(d) cyclopentane with OH

(e) cyclopentane with CH_3 and OH

(f) HO—cyclohexane with CH_3

16.12 Name each of the following compounds.

(a) $\underset{\overset{|}{OH}}{CH_2}-\underset{\overset{|}{OH}}{CH_2}$

(b) $\underset{\overset{|}{OH}}{CH_2}-CH_2-\underset{\overset{|}{OH}}{CH_2}$

(c) $\underset{\overset{|}{OH}}{CH_2}-\underset{\overset{|}{OH}}{CH}-\underset{\overset{|}{OH}}{CH_2}$

(d) $CH_3-\underset{\overset{|}{OH}}{CH}-CH_2-OH$

16.13 Write the structure and give the IUPAC name for the simplest diol.

16.14 Write the structure and give the IUPAC name for the simplest triol.

16.15 Write the structure of each of the following.
(a) cis-1,3-cyclopentanediol
(b) trans-1,4-cyclohexanediol
(c) trans-1,2-cyclohexanediol
(d) cis-1,2-cyclobutanediol

Classification of Alcohols

16.16 Classify each of the compounds in **16.9** as primary, secondary, or tertiary alcohols.

16.17 Classify each of the compounds in **16.10** as primary, secondary, or tertiary alcohols.

16.18 Classify each of the compounds in **16.11** as primary, secondary, or tertiary alcohols.

16.19 There is no compound known as a quaternary alcohol. Explain why.

16.20 Locate and classify the alcohol functional groups in cortisone.

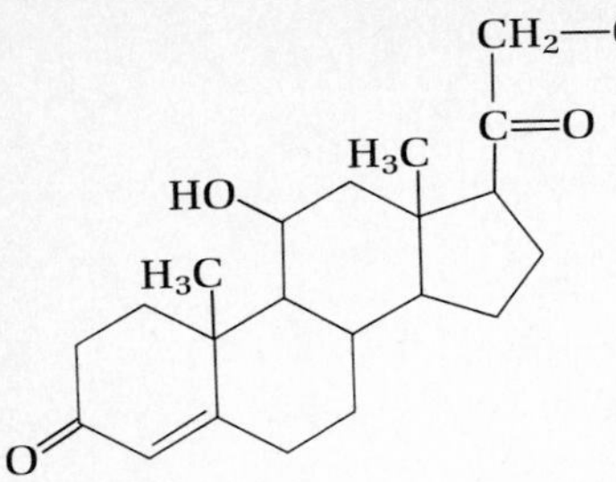

16.21 Locate and classify the alcohol functional groups in the following prostaglandin, which is a smooth-muscle stimulant.

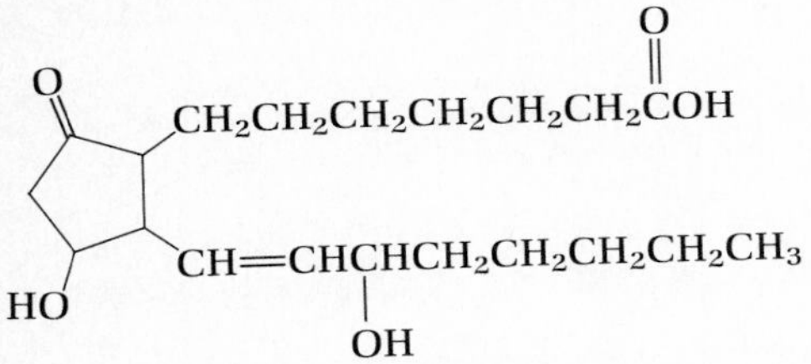

16.22 Classify each of the following alcohols.

(a)

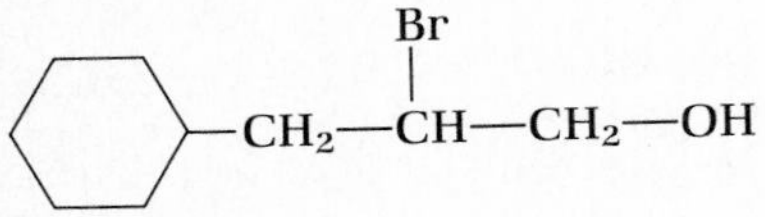

(b)

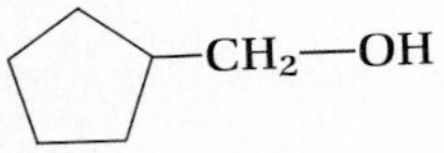

(c) 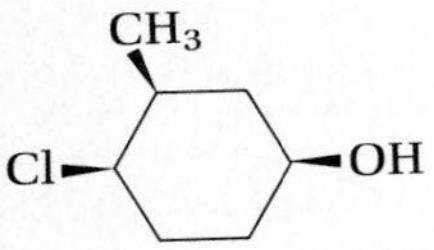 Cl

(d) 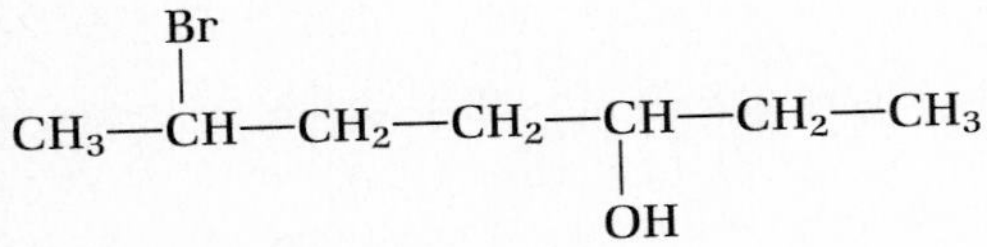 $CH_3-CH-CH_2-CH_2-CH-CH_2-CH_3$
           | |
          Br   OH

(e)

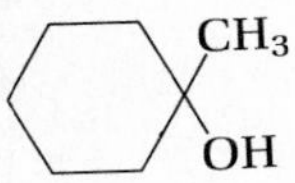

(f)

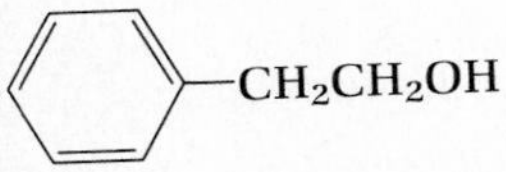

(g)

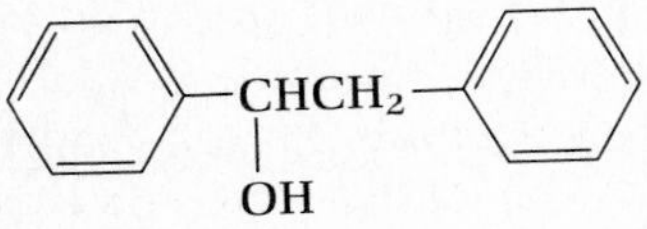

(h)

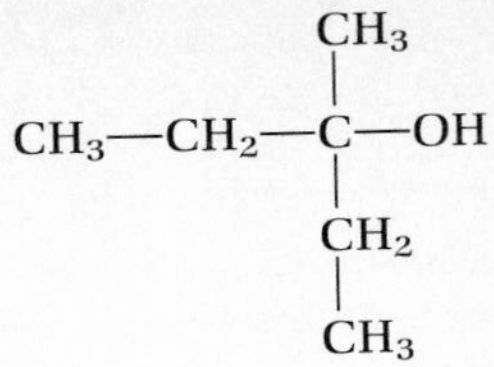

Physical Properties of Alcohols

16.23 1,2-Hexanediol is very soluble in water. Why?

16.24 Ethylene glycol boils at 198°C. Why does this compound have such a high boiling point?

16.25 Give one use for each of the following compounds.
(a) methanol   (b) ethanol
(c) isopropyl alcohol   (d) ethylene glycol
(e) glycerol

16.26 Explain why 1-butanol is less soluble in water than 1-propanol.

16.27 Suggest a reason why 2-methyl-2-propanol is much more soluble in water than 1-butanol.

Reactivity in Alcohols

16.28 Write a balanced equation for the reaction of ethanol with potassium.

16.29 Write a balanced equation for the reaction of methanol with sodium.

16.30 What is the name of the organic product formed from the reaction of ethanol with sodium?

16.31 Write an equation for the reaction of sodium methoxide with hydrogen chloride.

16.32 Write an equation for the reaction of lithium ethoxide with hydrogen bromide.

16.33 Describe a simple visual test of a chemical reaction to distinguish between a sample of octane and a sample of 1-octanol.

Substitution Reactions of Alcohols

16.34 Describe the visual change observed in the reaction of a tertiary alcohol with the Lucas reagent.

16.35 Write the structure of the expected product from each of the following reactions.

(a) $CH_3CH_2CH_2CH_2OH + PBr_3 \longrightarrow$

(b) 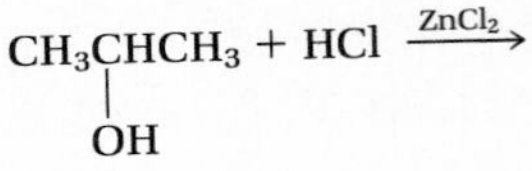 $CH_3CHCH_3 + HCl \xrightarrow{ZnCl_2}$
         |
        OH

(c) 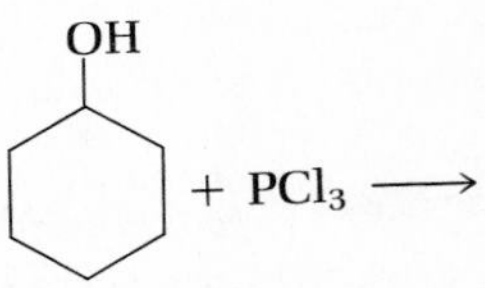
$+ PCl_3 \longrightarrow$

(d) 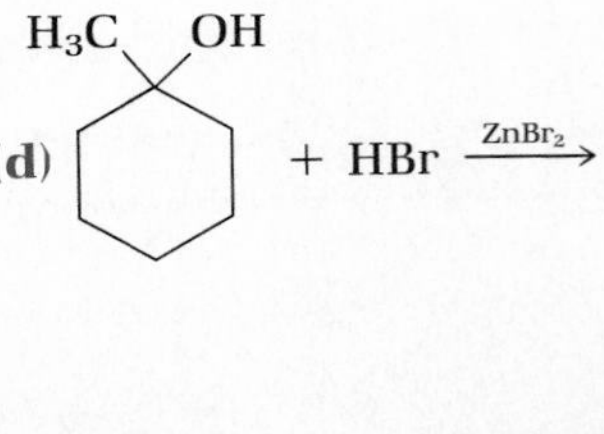
$+ HBr \xrightarrow{ZnBr_2}$

16.36 Describe how the Lucas reagent could be used to distinguish between samples of 3-pentanol and 2-methyl-2-butanol.

16.37 Describe how the following two compounds could be distinguished by using the Lucas reagent.

16.38 Describe how the following three compounds could be distinguished by using the Lucas reagent.

$$CH_3-CH_2-\underset{\underset{CH_3}{|}}{\overset{\overset{CH_3}{|}}{C}}-OH$$

I

$$CH_3-CH_2-\underset{\underset{CH_3}{|}}{CH}-CH_2-OH$$

II

$$CH_3-\underset{\underset{OH}{|}}{CH}-\underset{\underset{H}{|}}{\overset{\overset{CH_3}{|}}{C}}-CH_3$$

III

16.39 Order the following compounds according to their reactivity with the Lucas reagent.

$$CH_3-\underset{\underset{CH_3}{|}}{\overset{\overset{CH_3}{|}}{C}}-CH_2-CH_2-OH$$

I

$$CH_3-\underset{\underset{OH}{|}}{\overset{\overset{CH_2-CH_3}{|}}{C}}-CH_2-CH_3$$

II

$$CH_3-\underset{\underset{CH_3}{|}}{\overset{\overset{H}{|}}{C}}-CH_2-\underset{\underset{OH}{|}}{CH}-CH_3$$

III

Dehydration of Alcohols

16.40 Draw the dehydration product(s) obtained by treating each of the following alcohols with H_2SO_4.
(a) 1-pentanol   (b) 2-pentanol
(c) 2-methyl-2-butanol

16.41 Draw the structures of the dehydration products when each of the following compounds reacts with sulfuric acid.

(a) $CH_3-CH_2-CH_2-OH$

(b) $CH_3-\underset{\underset{CH_3}{|}}{\overset{\overset{CH_3}{|}}{C}}-OH$

(c) $CH_3-CH_2-CH_2-CH_2-CH_2-OH$

(d) $CH_3-CH_2-\underset{\underset{OH}{|}}{CH}-CH_3$

(e) $CH_3-\underset{\underset{\underset{CH_3}{|}}{CH-OH}}{CH}-CH_2-CH_2-CH_3$

(f) $CH_3-CH_2-\underset{\underset{CH_3}{|}}{\overset{\overset{CH_3}{|}}{C}}-OH$

16.42 Draw the structures of the dehydration products from the reaction of each of the following compounds with sulfuric acid.

Oxidation of Alcohols

16.43 What products are formed from the oxidation by $K_2Cr_2O_7$ of each of the compounds in **16.41**?

16.44 The structure of menthol is given below. Classify this alcohol. Draw the structure of the product obtained by oxidation with $K_2Cr_2O_7$.

menthol

16.45 Write the structure of the product formed from the oxidation by K_2CrO_7 of each of the compounds in **16.5**.

16.46 Which of the compounds in **16.41** will give an acid as product when oxidized with potassium dichromate?

16.47 Which of the compounds in **16.41** will give a ketone when oxidized with potassium dichromate?

16.48 Write the structure of the product formed when each of the compounds in **16.42** reacts with sodium dichromate.

16.49 Write the structure of the product formed when each of the compounds in **16.38** reacts with sodium dichromate.

16.50 Ethylene glycol is toxic if ingested. Enzymes in the liver oxidize it to oxalic acid, $C_2H_2O_4$. Oxalic acid then causes kidney damage. Write the structure of oxalic acid.

16.51 1,2-Propanediol is oxidized by enzymes in the liver to pyruvic acid, $C_3H_4O_3$. (Pyruvic acid can be metabolized by the body.) Write the structure of pyruvic acid.

Esters

16.52 Draw an ester produced from each of the following pairs of compounds.
(a) methanol and nitric acid
(b) ethanol and sulfuric acid
(c) 2-propanol and phosphoric acid

16.53 An ester may be produced from the reaction of 2 moles of ethanol and 1 mole of sulfuric acid. Draw the structure.

16.54 Draw the structure of an ester produced from 3 moles of methanol and 1 mole of phosphoric acid.

Phenols

16.55 Give one reaction that can be used to distinguish between cyclohexanol and phenol.

16.56 Phenol can be excreted from the body as the sulfate ester. Draw the structure of the ester.

16.57 How can the following two compounds be chemically distinguished from each other?

16.58 Which of the following compounds is a phenol?

Ethers

16.59 Draw the three isomeric ethers with the molecular formula $C_4H_{10}O$.

16.60 Name each of the following compounds.
(a) CH_3—CH_2—O—CH_2—CH_3
(b) CH_3—O—CH_2—CH_3
(c) CH_3—CH_2—CH_2—O—

16.61 Draw the structure of each of the following ethers.
(a) methyl isobutyl ether
(b) methyl ethyl ether
(c) ethyl phenyl ether

16.62 Ethers are not soluble in water. Explain why.

16.63 Dimethyl ether is a gas at room temperature, whereas ethanol is a liquid. Explain this difference in physical properties for these two isomers.

16.64 Ethers do not react with sodium. Explain why no reaction occurs.

16.65 Explain how diethyl ether and 1-butanol can be distinguished by a visual chemical test.

Sulfur Compounds

16.66 There are two isomeric compounds C_3H_8S with an —S—H group. Draw these compounds.

16.67 How could the following two compounds be distinguished chemically from each other?

$$CH_3(CH_2)_3CH_2—SH$$

$$CH_3CH_2—S—CH_2CH_2CH_3$$

16.68 Low-molecular-weight mercaptans have extremely unpleasant odors. Explain why an aqueous solution of $CH_3CH_2CH_2SH$ has a foul odor and why the addition of NaOH eliminates the odor.

16.69 Thiols have distinctly lower boiling points than the corresponding alcohols. Suggest a reason for this difference.

16.70 Write the product formed when each of the following is oxidized.

(a) CH_3—SH

(b) CH_3CH_2—SH

(c) 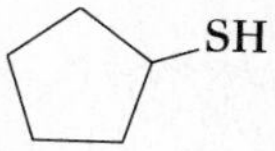 SH

(d) 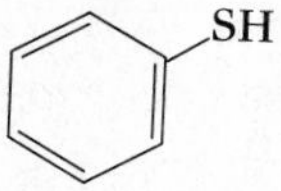 SH

16.71 Although they have not been given, the IUPAC rules for naming thiols may be somewhat obvious based on your previous experience in naming other classes of compounds. Draw the structure of each of the following compounds.

(a) methanethiol

(b) 1-propanethiol

(c) 2-propanethiol

(d) 2-methyl-3-pentanethiol

(e) cyclopentanethiol

(f) *cis*-2-methylcyclopentanethiol

17 Aldehydes and Ketones

We can recognize flowers by both sight and smell. Many flowers release small quantities of ketones, which are responsible for their odor.

In the previous chapter, the compounds of oxygen considered had two single bonds to oxygen. In alcohols, ethers, and phenols, the oxygen is σ bonded to two atoms. Oxygen can also form a covalent double bond, which is the subject of this chapter.

A **carbonyl group** *consists of a carbon atom and an oxygen atom bonded by a double bond*. In the carbonyl group, oxygen shares two of its six valence electrons with carbon. The remaining four electrons of oxygen are present as two nonbonded pairs of electrons. Carbon shares two of its four valence electrons with oxygen, and the remaining two electrons can be shared to form two σ bonds to other atoms.

$$\overset{..}{C}::\overset{..}{O}: \quad \text{or} \quad \overset{.}{C}=\overset{..}{O}:$$

The carbonyl carbon is sp^2 hybridized, and the electron in each of the three hybrid orbitals is shared with one electron of each of the three bonded atoms. The resultant molecule is planar with a bond angle close to 120°. The fourth electron of carbon is in a p orbital, which combines with a p orbital of oxygen to form a π bond. Formaldehyde, CH_2O, is the simplest compound with a carbonyl group. In this case, the two single bonds are to hydrogen atoms (Figure 17.1).

The carbonyl group is extremely important in organic chemistry and biochemistry. It may well be the most prevalent functional group in naturally occurring compounds. Numerous classes of compounds are possible because several different atoms may bond to the carbon atom of the carbonyl group. In this chapter, we will consider only hydrogen or carbon atoms bonded to the carbonyl group. Chapter 18 will deal with acids and esters in which there is another oxygen atom bonded to the carbonyl carbon atom. Amides (Chapter 19) have a nitrogen atom bonded to the carbonyl carbon atom. Other atoms such as sulfur and the halogens may also bond to the carbonyl carbon atom.

Figure 17.1
The Carbonyl Group.
(a) The electron in each p orbital may overlap to form a π bond. (b) The π bond has electrons both above and below the plane of the carbonyl bond. (c) A perspective structure of formaldehyde perpendicular to the plane of the paper. (d) A perspective structure of formaldehyde in the plane of the paper.

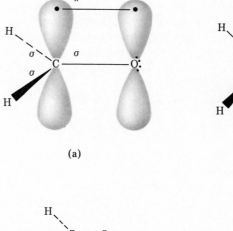

17.2
Representation of Aldehydes and Ketones

Aldehydes *are compounds in which the carbonyl carbon atom is bonded to at least one hydrogen atom.* The other group can be a hydrogen atom, an alkyl group (R), or an aromatic group (Ar). In formaldehyde, the simplest aldehyde, the second group is also a hydrogen atom.

$$\begin{array}{cc} \text{H} & & \text{H} \\ \diagdown & & \diagdown \\ \text{C}=\text{O}: & \text{or} & \text{C}=\text{O}: \\ \diagup & & \diagup \\ \text{R} & & \text{Ar} \end{array}$$

general structural formulas of aldehydes

The condensed formula of an aldehyde is RCHO or ArCHO. Both the H and O following the carbon are understood to be bonded to the carbon. By convention, RCOH is not used, since that formula is used to represent an alcohol.

Ketones *are compounds in which the carbonyl carbon atom is bonded to two other carbon atoms.* The bonded groups may be any combination of alkyl or aromatic groups.

$$\begin{array}{ccc} \text{R} & \text{R} & \text{Ar} \\ \diagdown & \diagdown & \diagdown \\ \text{C}=\text{O}: & \text{C}=\text{O}: & \text{C}=\text{O}: \\ \diagup & \diagup & \diagup \\ \text{R} & \text{Ar} & \text{Ar} \end{array}$$

$$\begin{array}{ccc} \text{or} & \text{or} & \text{or} \\ \\ :\text{O} & :\text{O} & :\text{O} \\ \parallel & \parallel & \parallel \\ \text{R}-\text{C}-\text{R} & \text{R}-\text{C}-\text{Ar} & \text{Ar}-\text{C}-\text{Ar} \end{array}$$

There are two ways to represent the structural formula of a ketone. The structural formula with 120° bond angles represents the actual shape of the molecule. However, for convenience in writing or printing, a linear arrangement of the atoms may be used. This arrangement is also useful to assign names to ketones. In addition, the nonbonded pairs of electrons on the oxygen atom are not displayed but are understood.

The condensed formula of a ketone is RCOR. The oxygen atom is understood to be double bonded to the carbonyl carbon atom at the left of the oxygen atom. The alkyl group on the right is also understood to be bonded to the carbon atom of the carbonyl group and not to the oxygen. This representation should not be confused with ROR used for ethers. In ethers, both R groups are bonded to oxygen.

17.3
Nomenclature of Aldehydes and Ketones

In the IUPAC system, aldehydes are named by the following rules.

1. Select the longest continuous carbon chain with the aldehyde carbon atom.
2. Replace the final -e of the parent hydrocarbon by the ending -al.
3. Number the chain by assigning the number 1 to the aldehyde carbon and numbering the other carbon atoms consecutively. The number 1 is not used in the name to indicate the position of the carbonyl carbon atom, since it is understood to be located at the end of the chain.
4. Determine the identity and location of any branches and indicate this information by prefixing it to the parent name

Examples of the nomenclature of aldehydes are given in Figure 17.2.

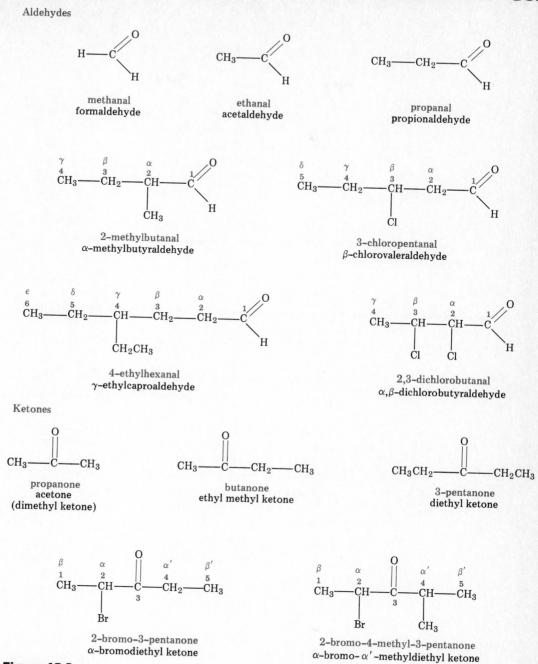

Figure 17.2
Nomenclature of Aldehydes and Ketones.

An earlier nomenclature for aldehydes designates the positions of attached groups by the Greek letters α, β, γ, etc. The α position is the carbon atom next to the carbonyl carbon. This convention was chosen since no substituent can be bonded directly to the carbonyl carbon of an aldehyde. The names of the aldehydes are based on the common names of the structurally related acids (Chapter 18). The common names of some aldehydes are given in Figure 17.2.

Example 17.1

Give the IUPAC name for the following compound.

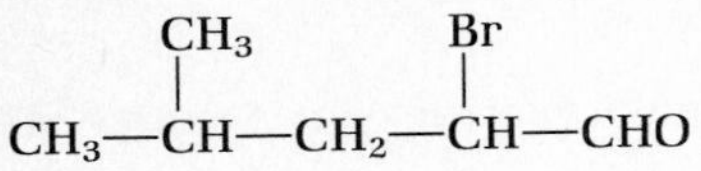

$$CH_3-\overset{\overset{\displaystyle CH_3}{|}}{CH}-CH_2-\overset{\overset{\displaystyle Br}{|}}{CH}-CHO$$

Solution. The aldehyde carbon atom on the right is assigned the number 1, and the five-carbon pentanal is numbered from right to left.

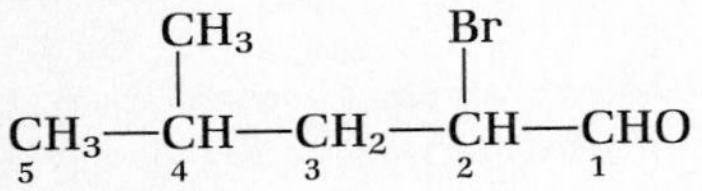

$$\underset{5}{CH_3}-\underset{4}{\overset{\overset{\displaystyle CH_3}{|}}{CH}}-\underset{3}{CH_2}-\underset{2}{\overset{\overset{\displaystyle Br}{|}}{CH}}-\underset{1}{CHO}$$

A methyl group is located at carbon atom number 4 and is part of the hydrocarbon skeleton. The position and identity of the alkyl group are prefixed to the word pentanal to give 4-methylpentanal.

The bromine atom with its position number is then prefixed to the name to give 2-bromo-4-methylpentanal.

[Additional examples may be found in **17.10**, **17.14**, and **17.17** at the end of the chapter.]

Ketones are named by the following IUPAC rules.

1. Select the longest continuous carbon chain containing the ketone carbon atom.
2. Replace the final -e of the parent hydrocarbon by the ending -one.
3. Number the carbon chain so that the carbonyl carbon atom receives the lowest number.
4. Place the number indicating the carbonyl location as a prefix before the parent name.
5. Indicate the identity and location of substituents as a prefix on the parent ketone.

Several examples of IUPAC nomenclature are given in Figure 17.2. Note that in some cases, such as butanone, no number is necessary because the name is unambiguous.

Cyclic ketones are named by assigning the carbonyl carbon atom as number 1. The ring is then numbered to give the lowest number(s) to the atom(s) bearing substituents.

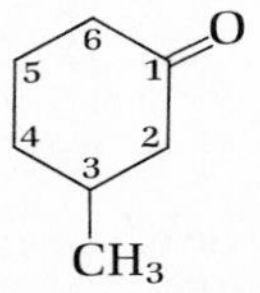

3-methylcyclohexanone 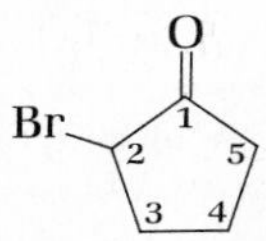   2-bromocyclopentanone

The common names for ketones use the names of the alkyl groups attached to the carbonyl carbon. Examples are given in Figure 17.2. The positions of attached groups are given by Greek letters. Acetone is the most widely used common name for the simplest ketone, CH_3COCH_3.

Example 17.2

Name the following compound.

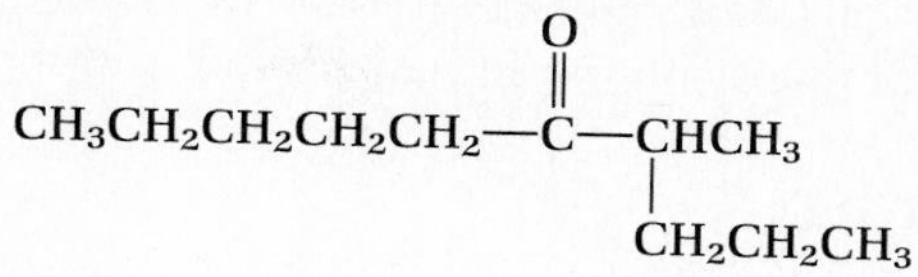

$$CH_3CH_2CH_2CH_2CH_2 \overset{\overset{\displaystyle O}{\|}}{-C} -\underset{\underset{\displaystyle CH_2CH_2CH_3}{|}}{CHCH_3}$$

Solution. First determine that the longest chain of this ketone contains ten carbon atoms. It is then named as a decanone. The chain must be numbered from right to left, so that the carbonyl carbon atom is number 5. Numbering from left to right would give the carbonyl carbon atom the number 6. Since there is a methyl group at position 4, the name is 4-methyl-5-decanone.

The physical properties of aldehydes and ketones are distinctly different from those of either the alkanes or alcohols of similar molecular weight (Table 17.1). Since oxygen is more electronegative than carbon, the shared electrons are pulled toward oxygen, and the carbonyl group is polar. The polarity of the group is shown by use of ↔ or δ+ and δ−.

17.4
Physical Properties of Aldehydes and Ketones

The physical properties of aldehydes and ketones are determined by this polarity.

Aldehydes and ketones have higher boiling points than the alkanes because there are attractive forces between neighboring dipolar molecules. However, these attractive forces are smaller than those of hydrogen bonding. Aldehydes and ketones lack a positive hydrogen atom bonded to an electronegative atom as

Table 17.1

Comparative Boiling Points of Carbonyl Compounds

Compound	Structure	Molecular weight	Boiling point (°C)
ethane	CH_3CH_3	30	−89
methanol	CH_3OH	32	64.6
formaldehyde	HCHO	30	−21
propane	$CH_3CH_2CH_3$	44	−42
ethanol	CH_3CH_2OH	46	78.3
acetaldehyde	CH_3CHO	44	20
butane	$CH_3CH_2CH_2CH_3$	58	−1
1-propanol	$CH_3CH_2CH_2OH$	60	97.1
propionaldehyde	CH_3CH_2CHO	58	48.8
isobutane	$CH_3CH(CH_3)_2$	58	−12
2-propanol	$CH_3CH(OH)CH_3$	60	82.5
acetone	CH_3COCH_3	58	56.1

in the —OH group of alcohols. Therefore, alcohols have higher boiling points than aldehydes and ketones of similar molecular weight.

Aldehydes and ketones can form hydrogen bonds with water. As a consequence, the lower-molecular-weight compounds formaldehyde, acetaldehyde, and acetone are soluble in water in all proportions.

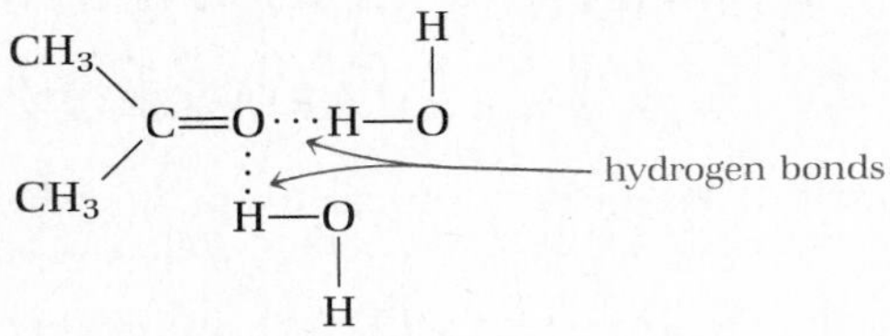

acetone hydrogen bonded to water

As the molecular weights of the carbonyl compounds increase, the differences in physical properties compared with alkanes and alcohols are lessened. The dipole–dipole attractive forces are less important as the chain length increases. As a result, aldehydes and ketones as well as alcohols become more like hydrocarbons. The boiling point differences become smaller, although the order of boiling points is still alcohol > carbonyl compound > alkane. For the same reason, the solubility of carbonyl compounds in water decreases as the chain length increases. The boiling points and solubilities of some aldehydes and ketones are given in Table 17.2.

Table 17.2

Physical Properties of Aldehydes and Ketones

Structure	Common name	IUPAC name	Melting point (°C)	Boiling point (°C)	Solubility (g/100g H$_2$O at 20°C)
Aldehydes					
HCHO	formaldehyde	methanal	−92	−21	miscible
CH$_3$CHO	acetaldehyde	ethanal	−121	20	miscible
CH$_3$CH$_2$CHO	propionaldehyde	propanal	−81	49	16
CH$_3$(CH$_2$)$_2$CHO	butyraldehyde	butanal	−99	76	7
(CH$_3$)$_2$CHCHO	isobutyraldehyde	2-methylpropanal	−66	64	11
CH$_3$(CH$_2$)$_3$CHO	valeraldehyde	pentanal	−91	103	slightly soluble
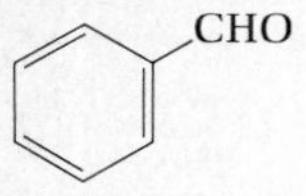 ⬡—CHO	benzaldehyde	benzaldehyde	−26	178	0.3
Ketones					
CH$_3$COCH$_3$	dimethyl ketone (acetone)	propanone	−94	56	miscible
CH$_3$CH$_2$COCH$_3$	methyl ethyl ketone	butanone	−86	80	26
CH$_3$(CH$_2$)$_2$COCH$_3$	methyl propyl ketone	2-pentanone	−78	102	6
CH$_3$CH$_2$COCH$_2$CH$_3$	diethyl ketone	3-pentanone	−41	101	5
CH$_3$(CH$_2$)$_3$COCH$_3$	methyl butyl ketone	2-hexanone	−35	150	2
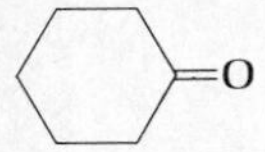 ⬡=O (cyclohexanone)	cyclohexanone	cyclohexanone	−45	157	5
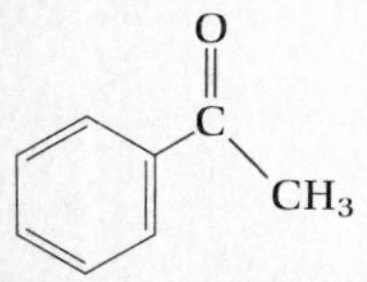 ⬡—CO—CH$_3$	methyl phenyl ketone (acetophenone)	1-phenylethanone	21	202	insoluble

Oxidation of Carbonyl Compounds

In Chapter 16 you learned that primary alcohols are oxidized to aldehydes, which then are easily oxidized to acids.

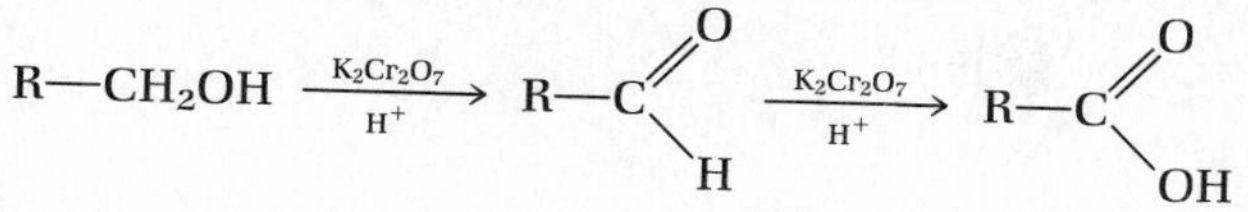

$$R-CH_2OH \xrightarrow[H^+]{K_2Cr_2O_7} R-\overset{O}{\underset{H}{C}} \xrightarrow[H^+]{K_2Cr_2O_7} R-\overset{O}{\underset{OH}{C}}$$

Under the same conditions, secondary alcohols are oxidized to ketones, which are not oxidized further.

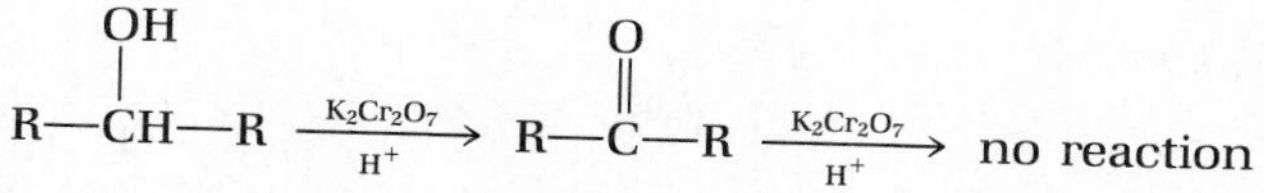

$$R-\overset{OH}{\underset{}{CH}}-R \xrightarrow[H^+]{K_2Cr_2O_7} R-\overset{O}{\underset{}{C}}-R \xrightarrow[H^+]{K_2Cr_2O_7} \text{no reaction}$$

The difference in the reactivities of aldehydes and ketones toward oxidizing agents is useful to distinguish between these two classes of compounds. **Benedict's solution** *contains cupric ion* (Cu^{2+}) *as a complex ion in a basic solution*. The Cu^{2+} will oxidize aldehydes, but not ketones. The aldehyde is oxidized to an acid, and Cu^{2+} is reduced to Cu^+, which precipitates as Cu_2O. The characteristic blue color of Cu^{2+} in solution is diminished as a red precipitate of Cu_2O is formed. Since the solution is basic, a carboxylic acid is not formed as such, but as its conjugate base. An unbalanced equation for this reaction is given below.

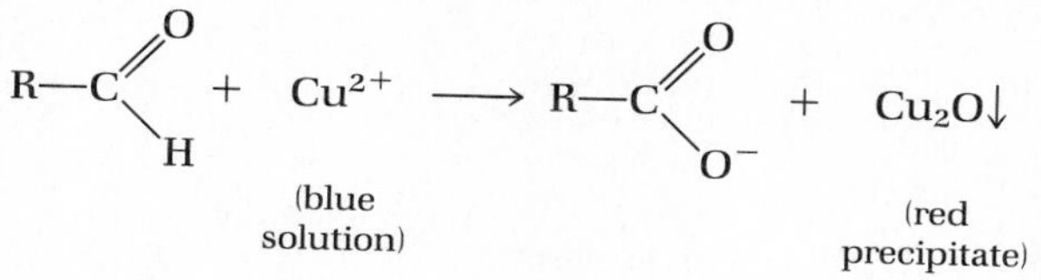

$$R-\overset{O}{\underset{H}{C}} + Cu^{2+} \longrightarrow R-\overset{O}{\underset{O^-}{C}} + Cu_2O\downarrow$$

(blue solution)   (red precipitate)

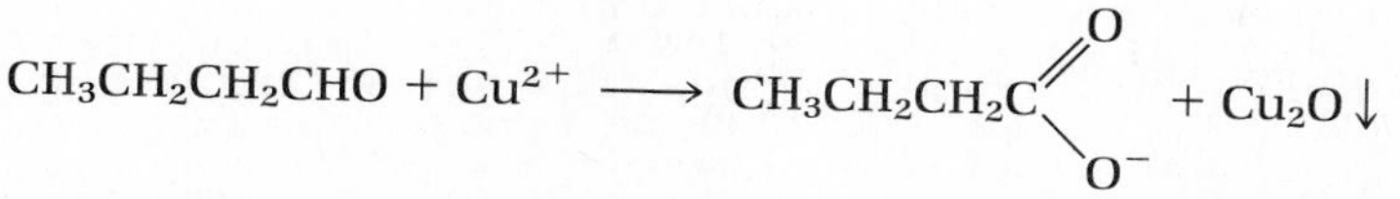

$$CH_3CH_2CH_2CHO + Cu^{2+} \longrightarrow CH_3CH_2CH_2\overset{O}{\underset{O^-}{C}} + Cu_2O\downarrow$$

Fehling's solution gives an identical reaction. The solution *also contains* Cu^{2+}, *but as a different complex ion, in a basic solution*. The presence of glucose, which contains an aldehyde group, in the blood can be detected clinically by using either Benedict's solution or Fehling's solution. Benedict's solution can detect as little as 0.2% glucose in urine. This test is used to monitor glucose levels in the urine of diabetics.

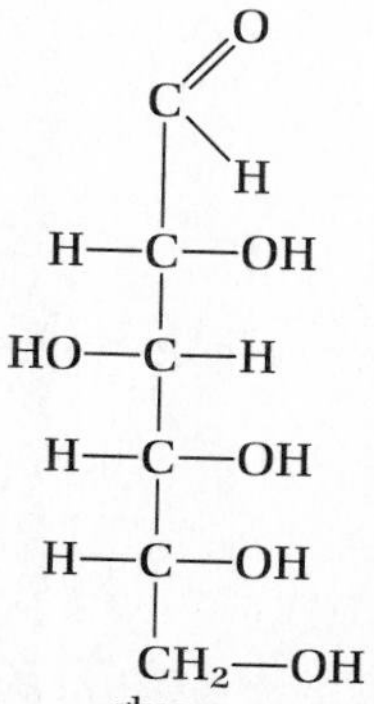

$$\begin{array}{c}
\overset{O}{\underset{}{C}}\diagdown H \\
H-C-OH \\
HO-C-H \\
H-C-OH \\
H-C-OH \\
CH_2-OH
\end{array}$$
glucose

Tollens' reagent also reacts with aldehydes, but not ketones. The reagent *is a basic solution of a silver ammonia complex ion.* When an aldehyde is oxidized by $[Ag(NH_3)_2]^+$, metallic silver is deposited as a mirror on the walls of a test tube. The unbalanced equation is given below.

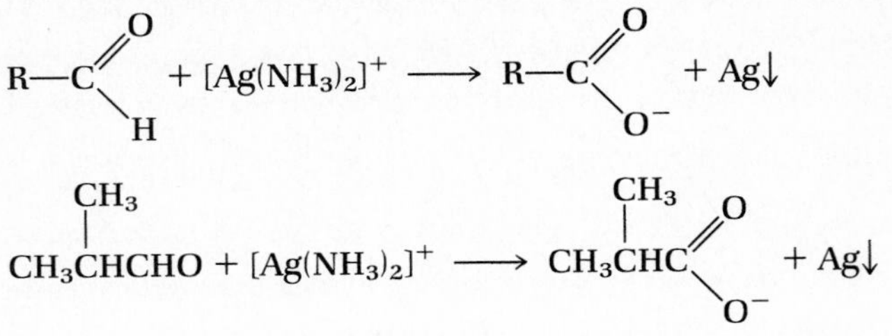

$$R-C\overset{O}{\underset{H}{\diagup}} + [Ag(NH_3)_2]^+ \longrightarrow R-C\overset{O}{\underset{O^-}{\diagup}} + Ag\downarrow$$

$$CH_3\underset{\overset{|}{CH_3}}{CHCHO} + [Ag(NH_3)_2]^+ \longrightarrow CH_3\underset{\overset{|}{CH_3}}{CHC}\overset{O}{\underset{O^-}{\diagup}} + Ag\downarrow$$

The reduction of Ag^+ by formaldehyde is used commercially to apply silver to mirrors.

Example 17.3

How can the isomeric carbonyl-containing compounds of the molecular formula C_4H_8O be distinguished from each other by a simple chemical test?

Solution. There are three isomeric carbonyl-containing compounds, two aldehydes and one ketone.

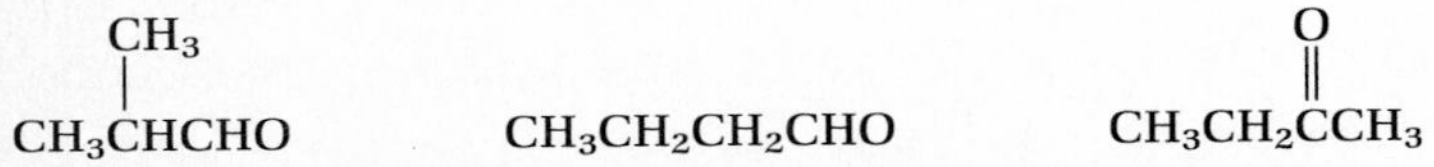

$$\underset{\overset{|}{CH_3}}{CH_3CHCHO} \qquad CH_3CH_2CH_2CHO \qquad CH_3CH_2\overset{O}{\overset{||}{C}}CH_3$$

Both of the aldehydes will react with Tollens' reagent to produce a silver mirror. Therefore these two compounds cannot be distinguished from each other by the use of the reagent. However, the ketone will not react with Tollens' reagent, and therefore the compound that does not yield a silver mirror is butanone.

[Additional examples may be found in **17.27–17.29** and **17.51–17.54** at the end of the chapter.]

Reduction of Carbonyl Compounds

Both aldehydes and ketones can be reduced to alcohols. The reaction with hydrogen gas, called hydrogenation, is catalyzed by nickel, palladium, or platinum.

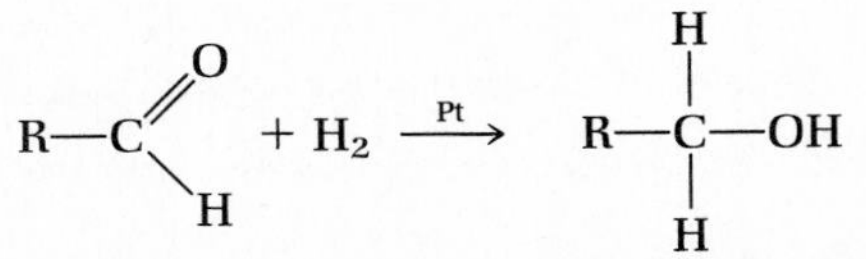

$$R-C\overset{O}{\underset{H}{\diagup}} + H_2 \overset{Pt}{\longrightarrow} R-\overset{H}{\underset{H}{\overset{|}{\underset{|}{C}}}}-OH$$

an aldehyde   a primary alcohol

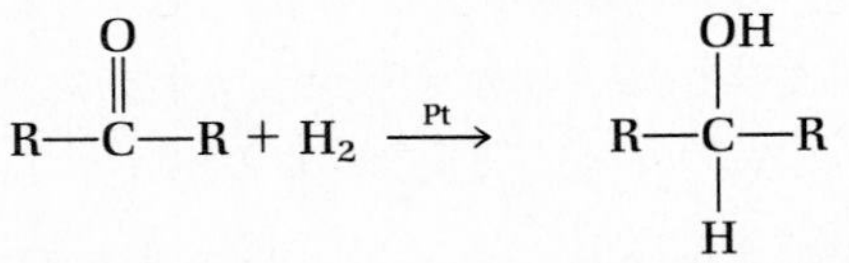

$$R-\overset{O}{\overset{||}{C}}-R + H_2 \overset{Pt}{\longrightarrow} R-\overset{OH}{\underset{H}{\overset{|}{\underset{|}{C}}}}-R$$

a ketone   a secondary alcohol

$$CH_3CH_2CH_2CHO + H_2 \xrightarrow{Pt} CH_3CH_2CH_2CH_2OH$$

butanal   1-butanol

$$\overset{\displaystyle O}{\overset{\|}{CH_3CCH_2CH_2}} + H_2 \xrightarrow{Pt} CH_3\underset{\underset{\displaystyle OH}{|}}{CH}CH_2CH_3$$

butanone   2-butanol

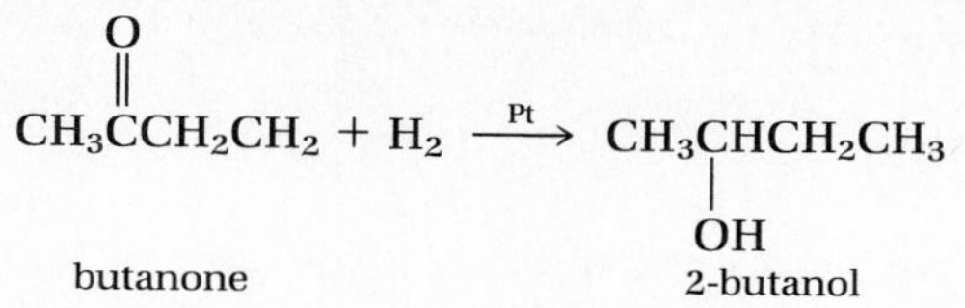

Recall that carbon–carbon double bonds are also reduced by hydrogen (Chapter 15). In any compound containing both carbon–carbon double bonds and carbonyl groups, both are hydrogenated.

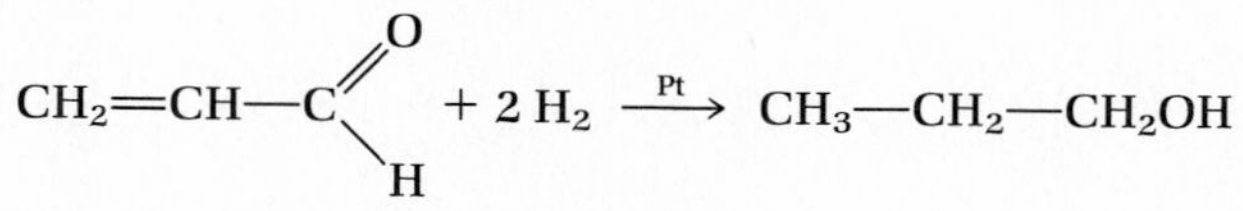

$$CH_2{=}CH{-}\overset{\displaystyle O}{\underset{\displaystyle H}{C}} + 2\,H_2 \xrightarrow{Pt} CH_3{-}CH_2{-}CH_2OH$$

Some reagents will reduce the carbonyl group without affecting the carbon–carbon double bond. One such reagent is lithium aluminum hydride, $LiAlH_4$. The product is obtained from a complex aluminum salt by treating the solution with acid.

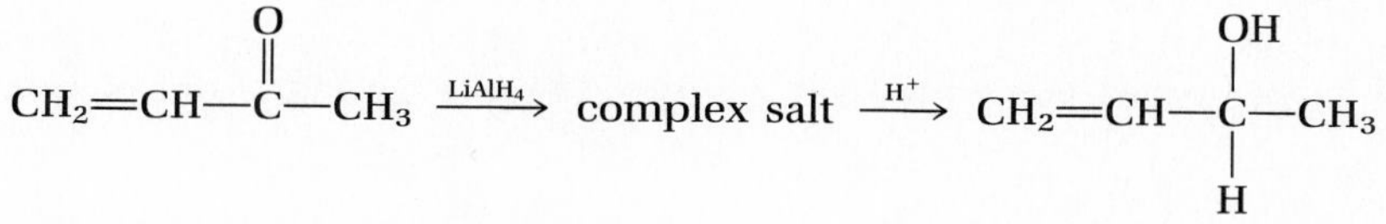

$$CH_2{=}CH{-}\overset{\displaystyle O}{\overset{\|}{C}}{-}CH_3 \xrightarrow{LiAlH_4} \text{complex salt} \xrightarrow{H^+} CH_2{=}CH{-}\underset{\underset{\displaystyle H}{|}}{\overset{\overset{\displaystyle OH}{|}}{C}}{-}CH_3$$

The reduction of the carbonyl group in biological systems occurs with NADH (the reduced form of nicotinamide adenine dinucleotide). Pyruvic acid, produced in muscle tissue, is reduced to lactic acid.

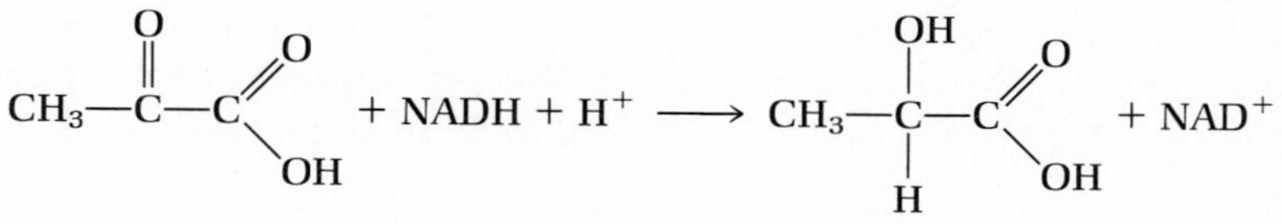

$$CH_3{-}\overset{\displaystyle O}{\overset{\|}{C}}{-}\overset{\displaystyle O}{\underset{\displaystyle OH}{C}} + NADH + H^+ \longrightarrow CH_3{-}\underset{\underset{\displaystyle H}{|}}{\overset{\overset{\displaystyle OH}{|}}{C}}{-}\overset{\displaystyle O}{\underset{\displaystyle OH}{C}} + NAD^+$$

The π bond of carbonyl compounds resembles the π bond of alkenes discussed in Chapter 15. A characteristic reaction of π bonds is the addition of reagents of the type H—A across the double bond where A is some electronegative atom or group of atoms. In the case of unsymmetrical alkenes, a specific order of addition is found. Similarly, for carbonyl compounds only one of two possible isomers is formed.

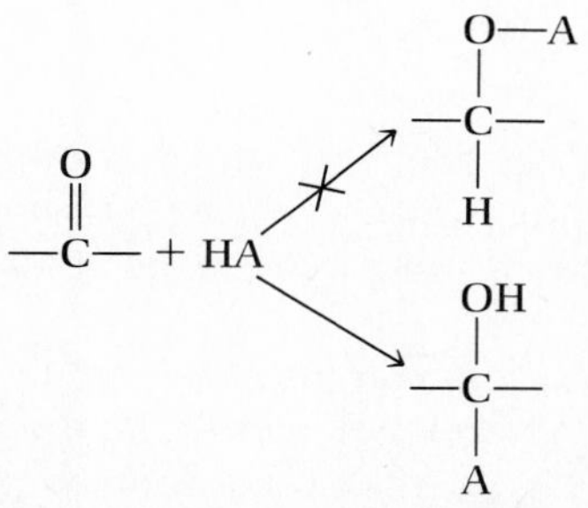

17.6
Addition Reactions of Carbonyl Compounds

$$\overset{\displaystyle O}{\overset{\|}{{-}C{-}}} + HA \;\;\substack{\nearrow \\ \\ \searrow}\;\; \begin{array}{c} \overset{\displaystyle O{-}A}{|} \\ {-}\underset{\underset{\displaystyle H}{|}}{C}{-} \\[4pt] \overset{\displaystyle OH}{|} \\ {-}\underset{\underset{\displaystyle A}{|}}{C}{-} \end{array}$$

You can remember this reaction by noting that the negative oxygen atom of the polar carbonyl group becomes bonded to the positive hydrogen atom of the compound HA. The electronegative group A bonds to the positive carbon atom of the carbonyl group.

Addition of Water

Aldehydes and ketones add water to form hydrates. The proton of water becomes bonded to the oxygen atom of the carbonyl group. The hydroxide ion derived from water adds to the carbon atom.

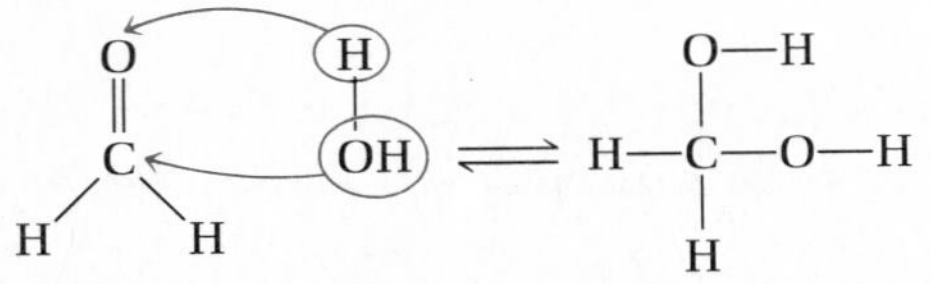

Formalin, used as a preservative of biological specimens, is a 37% by weight solution of formaldehyde in water. Formaldehyde is over 99% hydrated, whereas other aldehydes are substantially less hydrated. Ketones are hydrated to a small extent—usually less than 1%.

The hydrates of aldehydes and ketones cannot be isolated. They exist only in solution, where the large amount of water tends to force the equilibrium to the right. Most hydrates readily lose water when attempts are made to isolate them.

Addition of Alcohol

Alcohols can also add to carbonyl compounds in much the same way as water. However, there is a second subsequent reaction that occurs to give stable compounds of some biological importance. Sugars contain both hydroxyl groups and carbonyl groups that can undergo these reactions.

First let's consider the addition of one mole of an alcohol to one mole of an aldehyde or a ketone. The hydrogen atom of the alcohol adds to the carbonyl oxygen atom and the R'—O portion (alkoxide) adds to the carbon atom.

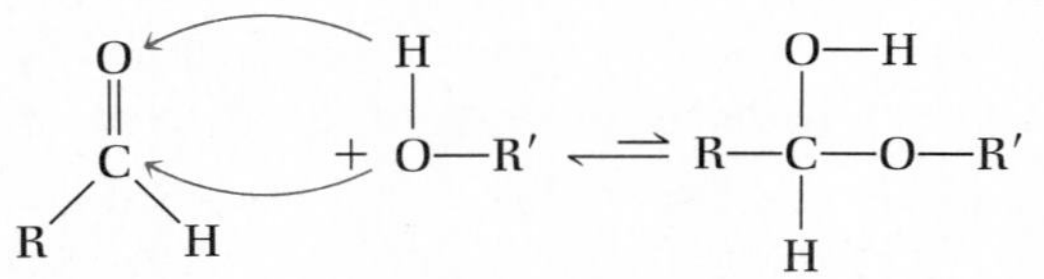

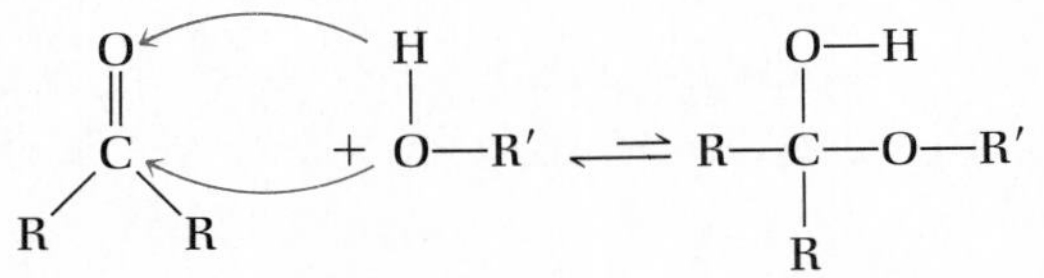

The product is a **hemiacetal** or **hemiketal.** In either case both an OH group and an OR' group are attached to the same carbon atom. However, *the hemiacetal has a hydrogen atom and an alkyl group attached to the carbon, whereas the hemiketal has two alkyl groups attached.*

Hemiacetals and hemiketals generally are unstable compounds. The equilibrium for the above two reactions lies to the left. However, when both the carbonyl group and the alcohol are part of the same molecule, they react to form cyclic products. The equilibrium in this case is favorable because the two functional groups are close to each other. The sugar glucose exists to only a small extent as an open-chain molecule. An intramolecular reaction occurs to form a cyclic hemiacetal (Chapter 21).

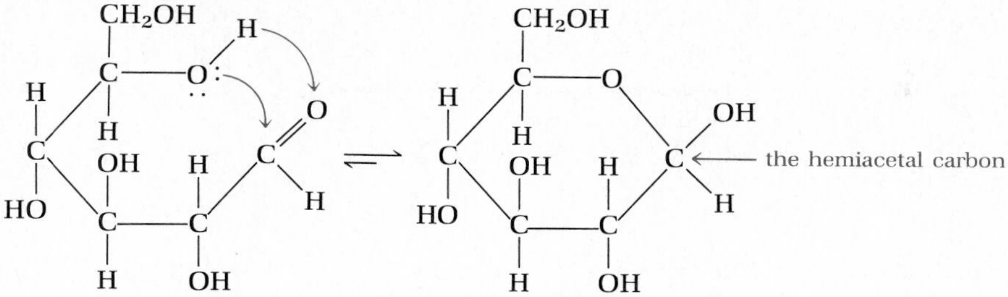

glucose
(open-chain form)

glucose
(a cyclic hemiacetal)

A substitution reaction of OH by OR′ in a hemiacetal or ketal occurs in acidic solution to produce an **acetal** or **ketal.**

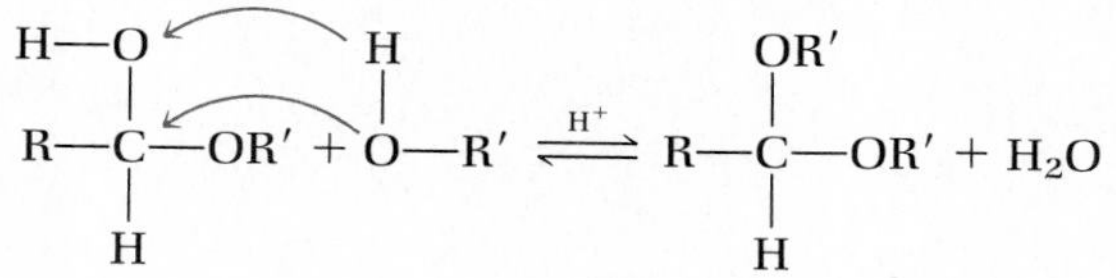

acetal

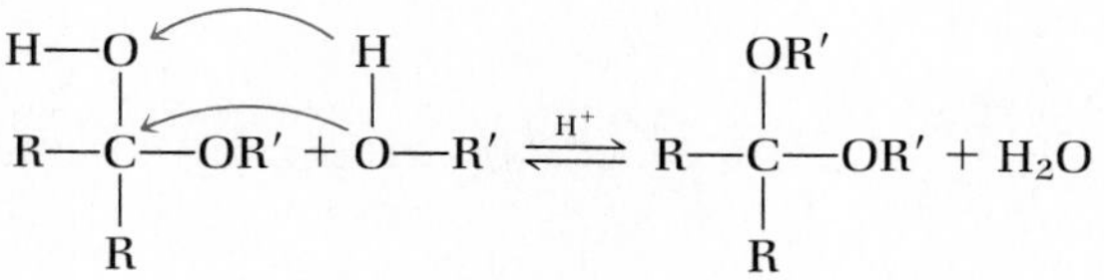

ketal

Note that *both acetals and ketals have two alkoxide groups* (OR′) *attached to the same carbon atom.* The reaction is reversible in acid solution, but formation of the acetal or ketal is favored by removing the water formed in the reaction or by increasing the amount of the alcohol.

hemiacetal + alcohol $\rightleftharpoons$ acetal + water

Adding alcohol "pushes"
equilibrium to the right.

Removing water "pulls"
equilibrium to the right.

The formation of acetals and ketals is catalyzed by acid. *Two moles of alcohol react per mole of carbonyl compound.*

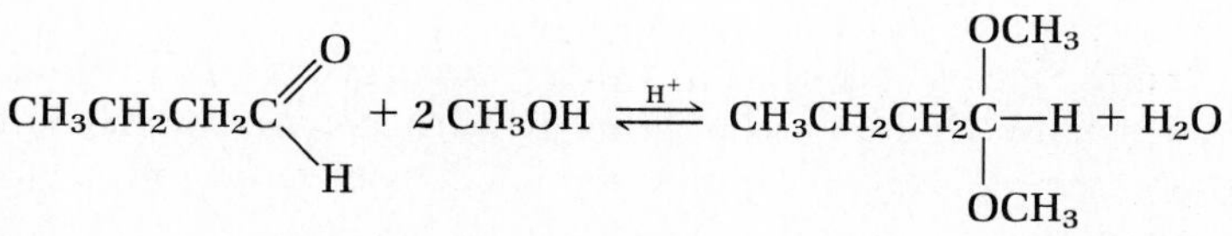

The forward reaction is favored by adding alcohol or removing the water that is formed. The reverse reaction is favored by adding water. Acetals and ketals react with water (a hydrolysis reaction) to give the carbonyl compound and the alcohol.

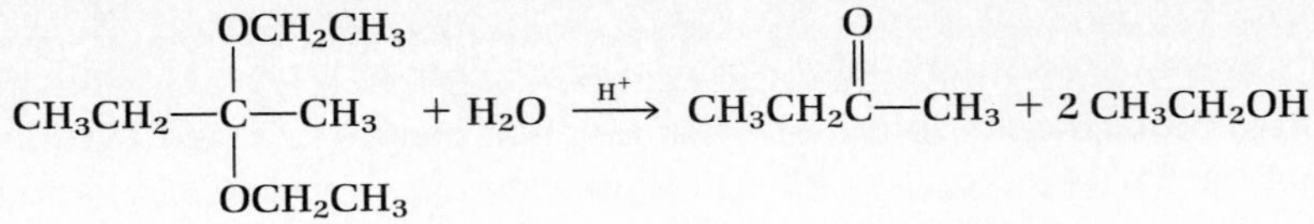

$$CH_3CH_2-\underset{\underset{OCH_2CH_3}{|}}{\overset{\overset{OCH_2CH_3}{|}}{C}}-CH_3 \ + H_2O \ \xrightarrow{H^+} \ CH_3CH_2\overset{\overset{O}{||}}{C}-CH_3 + 2 \ CH_3CH_2OH$$

Acetals and ketals do not react in neutral or basic solution.

Example 17.4

Identify the class to which each of the following compounds belongs.

a. 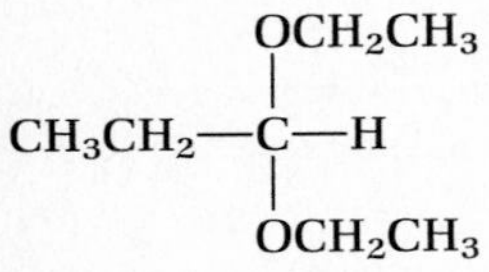 $CH_3CH_2-\underset{\underset{OCH_2CH_3}{|}}{\overset{\overset{OCH_2CH_3}{|}}{C}}-H$

b. 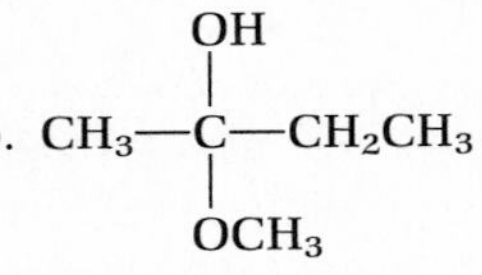 $CH_3-\underset{\underset{OCH_3}{|}}{\overset{\overset{OH}{|}}{C}}-CH_2CH_3$

Solution. a. The carbon atom with two —OCH_2CH_3 groups bonded to it has a bond to hydrogen. The compound is an acetal that can be formed from propanal and ethanol.

b. A carbon atom with one —OCH_3 group and an —OH group bonded to it is the focal point. That carbon atom has two bonds to alkyl groups and the compound is a hemiketal.

[Additional examples may be found in **17.34–17.36** at the end of the chapter.]

Addition of Nitrogen Compounds

The nitrogen–hydrogen bond can add to a carbonyl group in the same way as the oxygen–hydrogen bond adds. For example, ammonia adds as follows.

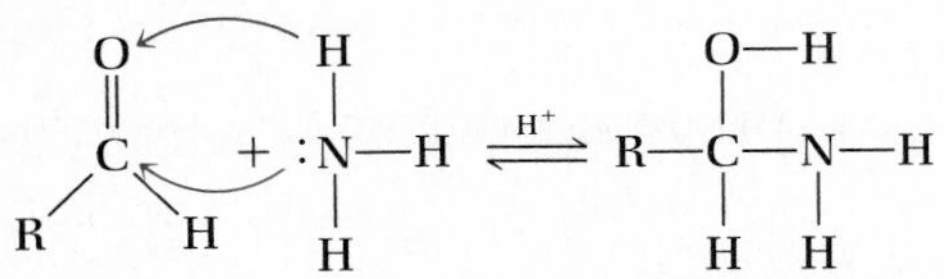

$$\underset{R}{\overset{O}{C}} + \ :\underset{\underset{H}{|}}{N}-H \ \xrightleftharpoons{H^+} \ R-\underset{\underset{H}{|}}{\overset{\overset{O-H}{|}}{C}}-\underset{\underset{H}{|}}{N}-H$$

addition product

The addition product formed is unstable and then eliminates a molecule of water to give an imine.

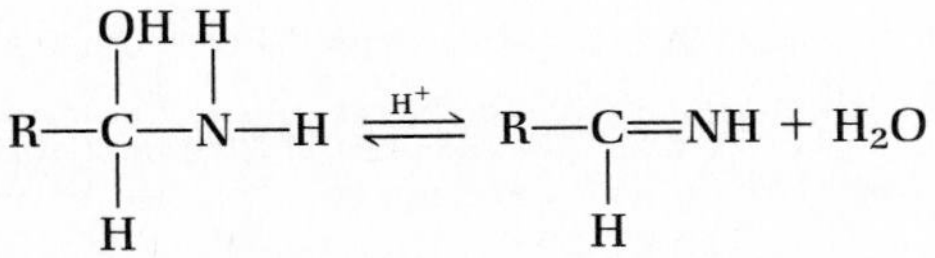

$$R-\underset{\underset{H}{|}}{\overset{\overset{OH \ \ H}{|}}{C}}-N-H \ \xrightleftharpoons{H^+} \ R-\underset{\underset{H}{|}}{C}=NH + H_2O$$

an imine

Imines can be isolated, but they react with water in the reverse of the above two reactions to give the carbonyl compound and the nitrogen compound. Amines, which have carbon groups bonded to nitrogen, also react with carbonyl compounds to give imines.

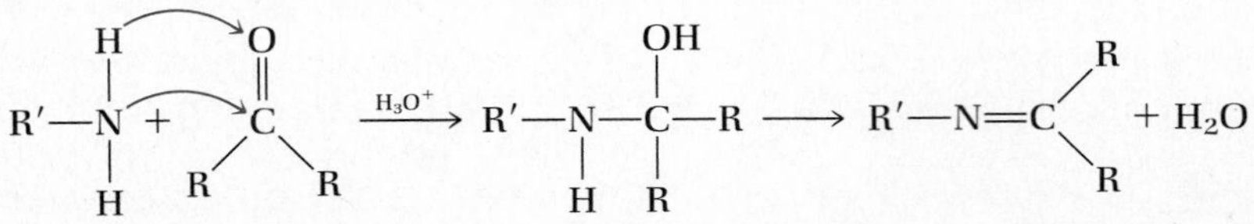

$$R'-\underset{\underset{H}{|}}{\overset{\overset{H}{|}}{N}} + \underset{R}{\overset{O}{\underset{R}{C}}} \ \xrightarrow{H_3O^+} \ R'-\underset{\underset{R}{|}}{\overset{\overset{OH}{|}}{N}}-\underset{\underset{R}{|}}{C}-R \ \longrightarrow \ R'-N=\underset{R}{\overset{R}{C}} + H_2O$$

AN ASIDE

The reaction of carbonyl compounds with amines to form imines is a common biochemical reaction. The reaction of retinal with the protein opsin given on page 316 in Chapter 15 is such an example. Retinal contains an aldehyde group that reacts with a specific amino group in the protein.

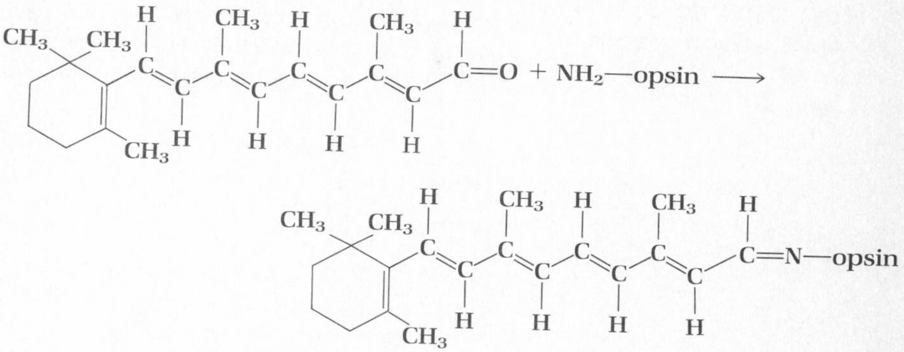

Thus, rhodopsin contains an imine bond.

One of the tests used in hospitals to determine the concentration of glucose in blood is based on the formation of an imine. The aldehyde group of glucose reacts with the amino group of *o*-toluidine, which is the test reagent.

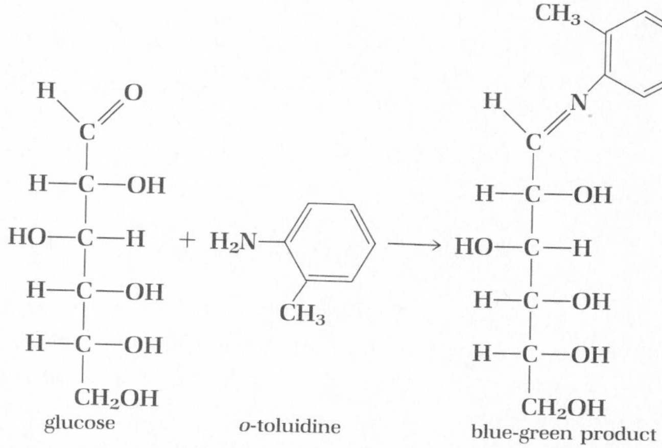

glucose   *o*-toluidine   blue-green product

The product is blue-green, and the color intensity is directly proportional to the amount of glucose present in the blood. This test is used in the diagnosis of diabetes.

Chemists use the addition–elimination reaction of nitrogen compounds to identify aldehydes and ketones. An addition product of a carbonyl compound with 2,4-dinitrophenylhydrazine is easily formed.

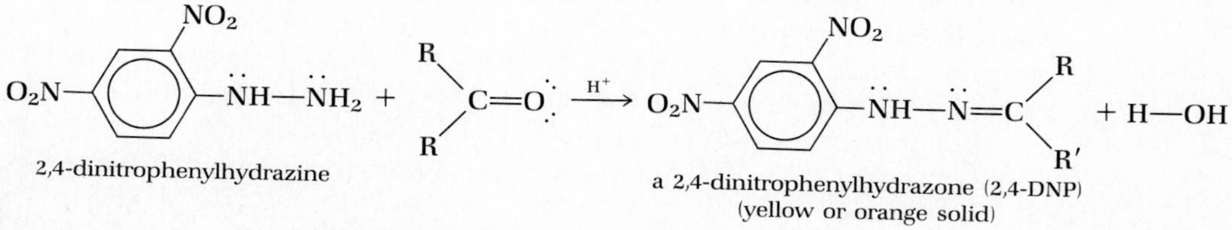

2,4-dinitrophenylhydrazine   a 2,4-dinitrophenylhydrazone (2,4-DNP)
(yellow or orange solid)

The product, which is a bright yellow to orange crystalline solid, is called a 2,4-dinitrophenylhydrazone (2,4-DNP). Its melting point is used to help identify the original aldehyde or ketone.

17.7
Reactions at the Alpha Carbon Atom

To this point we have seen the characteristic reactions of the carbonyl group itself. However, part of the extensive chemistry of carbonyl compounds is due to reactivity of the α-carbon atom. The polar carbonyl group tends to pull electrons in neighboring bonds to itself. The electrons bonding the α-hydrogen atom to the α-carbon atom are to some degree pulled away as shown below.

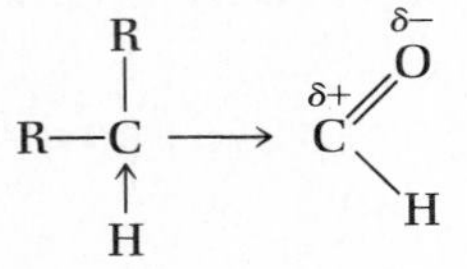

The arrows indicate the direction of electron flow away from the hydrogen atom. As a result of this shift of electrons, the hydrogen atom resembles a proton, and the α-hydrogen atom is somewhat acidic.

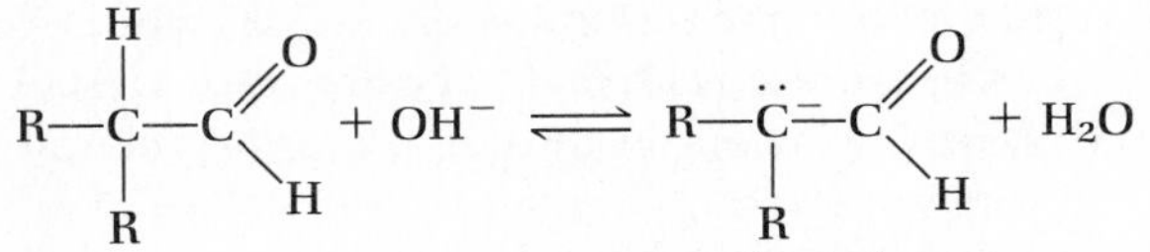

Carbonyl compounds are weak acids, but they are much stronger acids than hydrocarbons. The acid dissociation constant for ethane is estimated to be 10^{-50}, whereas that of acetone is about 10^{-20}.

$$CH_3CH_3 + H_2O \rightleftharpoons CH_3CH_2^- + H_3O^+ \qquad K = 10^{-50}$$

$$CH_3\overset{O}{\overset{\|}{C}}CH_3 + H_2O \rightleftharpoons CH_3\overset{O}{\overset{\|}{C}}CH_2^- + H_3O^+ \qquad K = 10^{-20}$$

Keto–Enol Tautomers

As a result of the acidity of the α-hydrogen atom, aldehydes and ketones can exist as an equilibrium mixture called the keto and the enol forms.

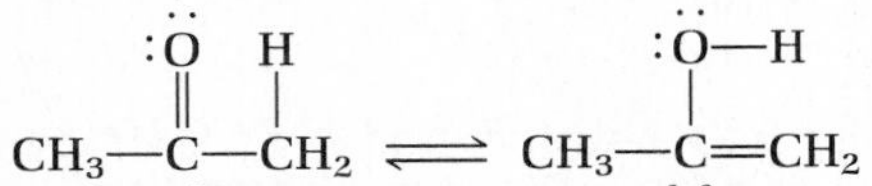

The hydrogen atom bonded to the α-carbon atom is removed and transferred to the carbonyl oxygen atom. In order to maintain the proper number of bonds, the carbon–oxygen double bond becomes a single bond and the carbon–carbon single bond becomes a double bond. The keto and enol forms are, of course, isomers, since their structures differ. *Isomers that differ because of the shift of the position of a hydrogen atom via an equilibrium process are called* **tautomers.** For most aldehydes and ketones, the keto form containing the C=O group predominates. Less than 1% of the isomeric enol form exists. However, this form is often responsible for the reactivity of carbonyl compounds.

The phenomenon of tautomerism is important in the chemistry of carbohydrates (Chapter 21) and in the metabolism of these compounds (Chapter 26). For example, dihydroxyacetone phosphate is isomerized in an enzyme-catalyzed reaction to form glyceraldehyde 3-phosphate. The transfer of a hydrogen atom from the α-carbon atom of dihydroxyacetone phosphate to the carbonyl group yields an enediol intermediate.

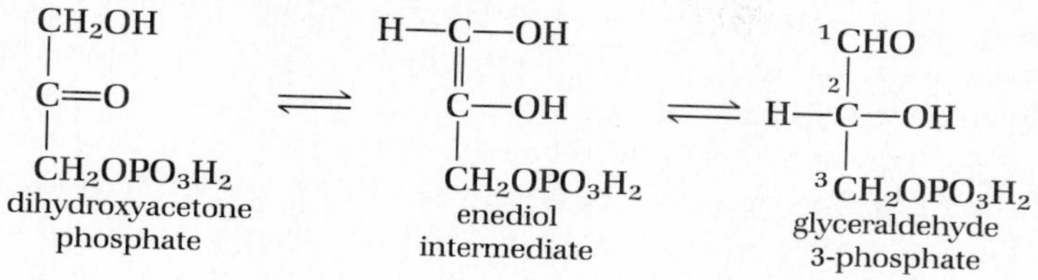

$$\begin{array}{ccc} \text{CH}_2\text{OH} & \text{H---C---OH} & {}^1\text{CHO} \\ | & \| & {}_2| \\ \text{C}{=}\text{O} & \text{C---OH} & \text{H---C---OH} \\ | & | & | \\ \text{CH}_2\text{OPO}_3\text{H}_2 & \text{CH}_2\text{OPO}_3\text{H}_2 & {}^3\text{CH}_2\text{OPO}_3\text{H}_2 \end{array}$$

dihydroxyacetone phosphate   enediol intermediate   glyceraldehyde 3-phosphate

The transfer of a hydrogen atom from the α-carbon atom (position 2) of glyceraldehyde 3-phosphate produces the same enediol intermediate. The isomerization reaction occurs because a common enediol is in equilibrium with the two compounds.

Example 17.5

The following two isomeric compounds can be interconverted. Draw the structure of their common intermediate.

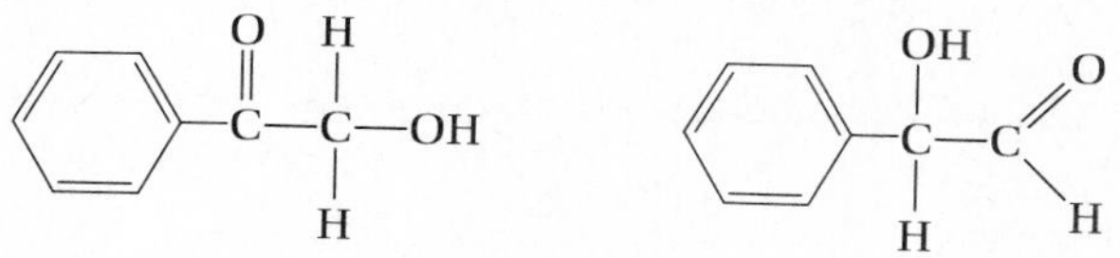

Solution. Both compounds contain a hydroxyl group on a carbon atom α to the carbonyl group. Proton transfer from the α-carbon atom to the carbonyl oxygen atom of either compound gives the same intermediate.

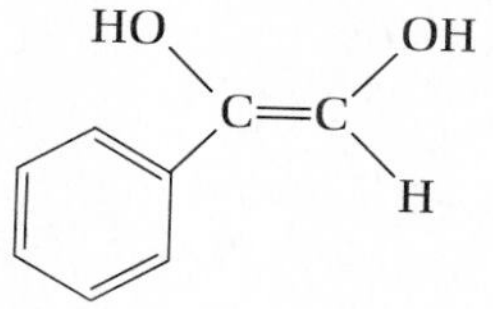

Transfer of a proton from the hydroxyl group nearest the aromatic ring to the terminal carbon atom on the right gives the ketone. Transfer of a proton from the hydroxyl group on the carbon atom on the right to the carbon atom nearest the aromatic ring gives the aldehyde.

[Additional examples may be found in **17.49** and **17.50** at the end of the chapter.]

The Aldol Condensation

A second consequence of the acidity of an α-hydrogen atom is seen in a reaction between two moles of a carbonyl compound

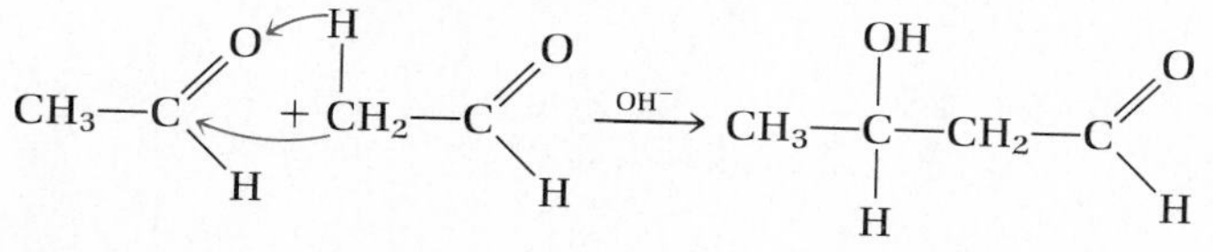

an aldol

One aldehyde adds to the carbonyl group of another molecule of the aldehyde. The acidic α-hydrogen atom adds to the oxygen atom of the carbonyl group. The α-carbon atom adds to the carbon atom of the carbonyl group. Note that the product is called an aldol because it contains both an aldehyde and an alcohol.

The formation of an aldol involves an important structural change—a carbon–carbon bond is formed. Similar types of aldol reactions occur frequently in the biochemical processes that form large molecules. One step in the formation of the sugar fructose is an aldol reaction. The α-hydrogen atom of dihydroxyacetone phosphate adds to the carbonyl oxygen atom of glyceraldehyde 3-phosphate. The α-carbon atom of the ketone adds to the carbonyl carbon atom of the aldehyde.

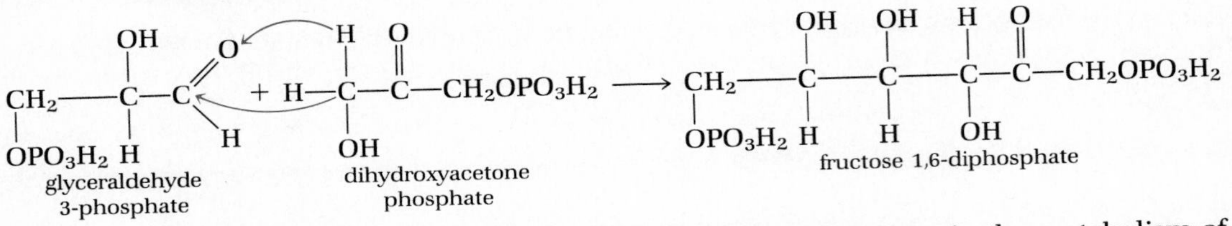

glyceraldehyde
3-phosphate

dihydroxyacetone
phosphate

fructose 1,6-diphosphate

The reverse of this reaction, known as a retro-aldol, occurs in the metabolism of carbohydrates (Chapter 26).

Example 17.6

Draw the product of the aldol condensation of propanal.

Solution. First draw the structural formula of propanal with the carbonyl group arranged toward the right.

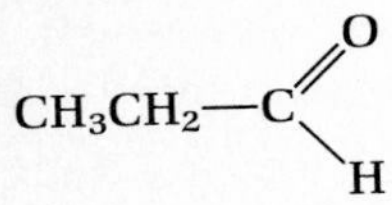

Next, draw a second structural formula of propanal with the α-carbon near the carbonyl carbon atom of the first structure. Arrange the α-hydrogen atom of the second structure so that it is close to the carbonyl oxygen atom of the first molecule.

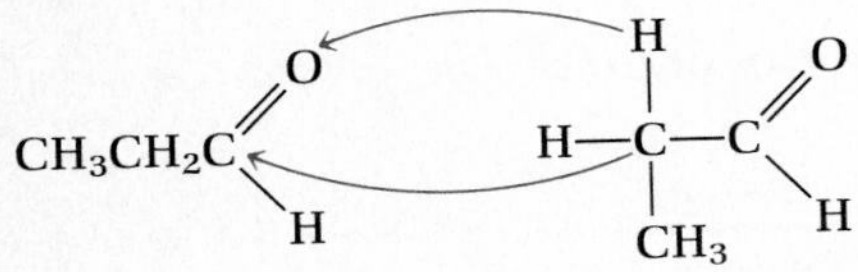

Now form the bond in the indicated direction, and the correct aldol product is predicted.

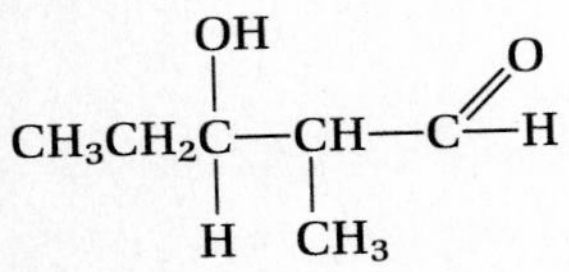

[Additional examples may be found in **17.47** and **17.48** at the end of the chapter.]

Alcohols, ethers, aldehydes, and ketones all contain one oxygen atom per molecule. How, then, can one tell by looking at the molecular formula which type of compound it is? One answer is that the number of hydrogen atoms is different, since the oxidation states are not the same. Consider the following compounds.

$$CH_3CH_2CH_2OH \qquad CH_3CH_2{-}O{-}CH_3 \qquad CH_3CH_2C\!\!\begin{array}{c}O\\ \diagup\diagdown\\ H\end{array} \qquad CH_3\overset{\displaystyle O}{\overset{\|}{C}}CH_3$$

1-propanol   ethyl methyl ether   propanal   propanone

Both ethyl methyl ether and propanol are isomeric and have the molecular formula C_3H_8O. In fact, the general molecular formula for alcohols and ethers is $C_nH_{2n+2}O$.

When an alcohol is oxidized to an aldehyde or ketone, two hydrogen atoms are removed by the oxidizing agent. The general molecular formula for aldehydes and ketones is therefore $C_nH_{2n}O$.

Other structural features can alter the molecular formula of an alcohol, ether, aldehyde, or ketone. Recall that the presence of a double bond or a ring decreases the number of hydrogen atoms by two. A compound containing a ring and an —OH group would then have the general molecular formula $C_nH_{2n}O$. This means that cyclic alcohols are isomeric with aldehydes and ketones.

In order to determine the structure of a compound it is necessary to use chemical tests. You have learned some visual tests that illustrate this method.

1. Alcohols react with sodium to give hydrogen gas.
2. Ethers do not react with sodium.
3. Aldehydes and ketones react with 2,4-dinitrophenylhydrazine, but alcohols and ethers do not.
4. Aldehydes react with Tollens' reagent or Fehling's solution, but ketones, alcohols, and ethers do not.

Consider the compounds given at the start of this section. Propanol is identified by its reactivity with sodium, whereas the isomeric ethyl methyl ether does not react with sodium. The isomeric compounds propanal and propanone react with 2,4-dinitrophenylhydrazine, but only propanal reacts with Tollens' reagent.

Example 17.7 ⎯⎯

A compound $C_6H_{10}O$ reacts with 2,4-dinitrophenylhydrazine, but not with Fehling's solution. The $C_6H_{10}O$ compound reacts with hydrogen in the presence of platinum to form $C_6H_{12}O$. Suggest a structure for the $C_6H_{10}O$ compound.

Solution. The reaction with 2,4-dinitrophenylhydrazine indicates that the compound is an aldehyde or ketone. Since the substance does not react with Fehling's solution, it must be a ketone.

The general molecular formula for a ketone is $C_nH_{2n}O$. A ketone containing six carbon atoms would have the molecular formula $C_6H_{12}O$. Therefore, the unknown compound must have either a carbon–carbon double bond or a ring. Since the reaction with hydrogen gas gives $C_6H_{12}O$, there can be no carbon–carbon double bond. The reduction of the carbonyl group accounts for the two hydrogen atoms added. Only a ring can account for the deficiency of two hydrogens in this ketone. Any $C_6H_{10}O$ structure containing a ring and a ketone group fits the data. Examples include

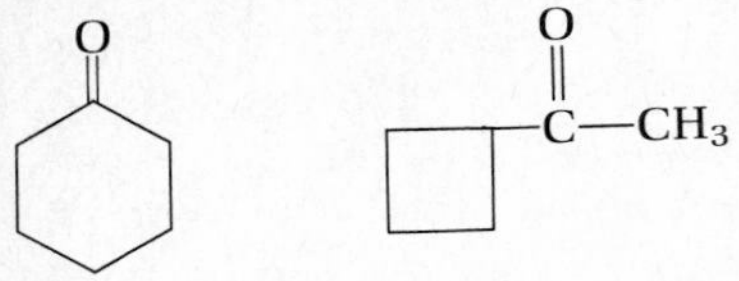

[Additional examples may be found in **17.51–17.54** at the end of the chapter.]

17.9
Naturally Occurring Carbonyl Compounds

As indicated in Section 17.1, carbonyl compounds are perhaps the most abundant organic compounds in nature. A few examples are given in Figure 17.3.

Many carbohydrates (Chapter 21) consist of structures containing aldehydes or ketones. Glucose and fructose are two examples. Note that they are isomers. The polyunsaturated aldehydes 11-*trans*-retinal and 11-*cis*-retinal are in-

Figure 17.3
Structures of Some Biologically Important Carbonyl Compounds.

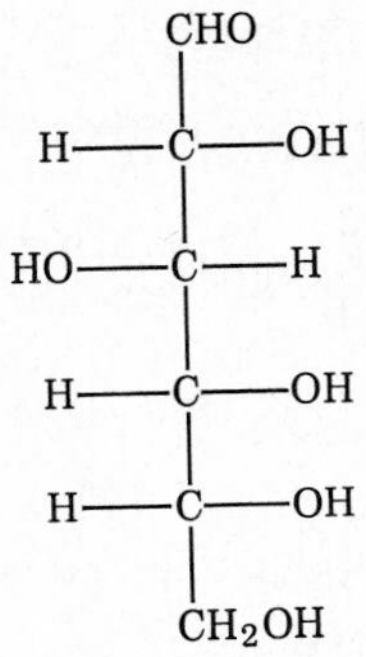

glucose

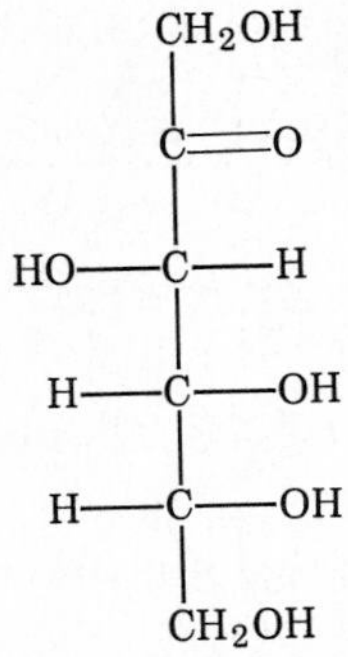

fructose

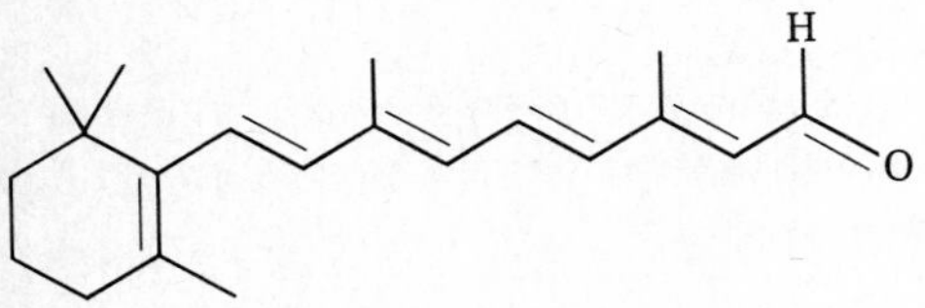

trans–retinal

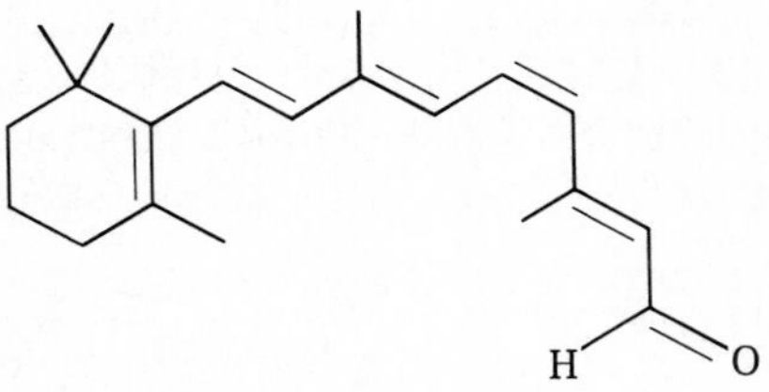

11–*cis*–retinal

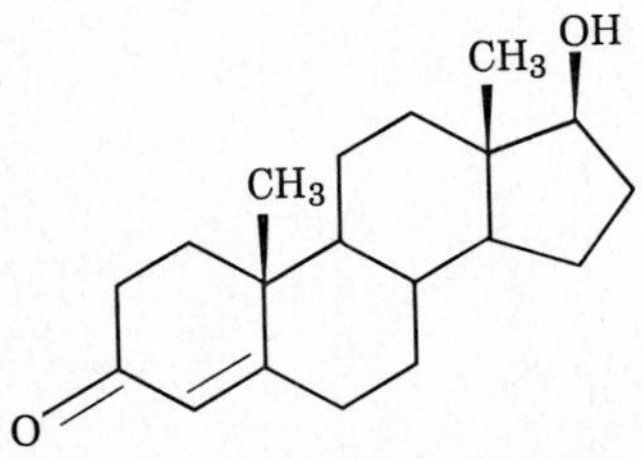

testosterone

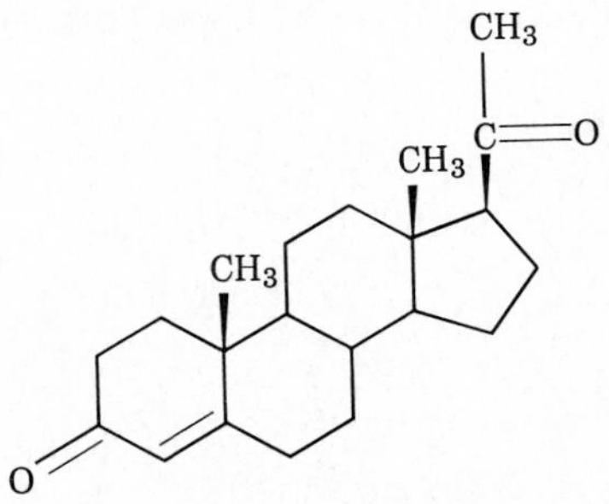

progesterone

volved in the chemistry of sight. The retinals are formed from the oxidation of vitamin A.

The tetracyclic ring structures known as steroids (Chapter 22) are responsible for numerous biochemical functions. Some of the steroids are ketones. These include testosterone, a male sex hormone affecting secondary sexual characteristics, and progesterone, a female sex hormone that prepares the lining of the uterus for a fertilized ovum. The small difference in the structures of these compounds causes remarkably different biological changes.

Summary

Aldehydes and ketones are compounds that have a carbonyl functional group. Aldehydes have one of the bonds from the carbonyl carbon atom to a hydrogen atom and the other bond to an alkyl or aryl group. Ketones have both bonds from the carbonyl group to alkyl or aryl groups.

The carbonyl group is polar because of the electronegativity difference of the carbon atom and the oxygen atom. As a result of this polarity, carbonyl compounds have higher boiling points than hydrocarbons of similar molecular weight. Since carbonyl compounds cannot form hydrogen bonds to themselves, these compounds have lower boiling points than do the structurally related alcohols. Lower-molecular-weight aldehydes and ketones are soluble in water.

In the IUPAC system of nomenclature, aldehydes are designated by dropping the terminal -e of the alkane and adding -al. Ketones are named by replacing the -e of the alkane by -one.

Aldehydes are oxidized by Benedict's, Fehling's, and Tollens' reagents, whereas ketones are not oxidized by these reagents. Both aldehydes and ketones can be reduced to alcohols.

Aldehydes and ketones undergo addition reactions with water, alcohols, ammonia, and other nitrogen compounds. The addition product using 2,4-dinitrophenylhydrazine is called a 2,4-DNP. Carbonyl compounds also undergo a condensation reaction in which the α-carbon atom of one molecule bonds to the carbonyl carbon atom of another molecule. These self-addition reactions are important in building a large molecule from two smaller ones in living systems.

Learning Objectives

As a result of studying Chapter 17 you should be able to

- Identify carbonyl groups and distinguish an aldehyde or a ketone from other compounds.
- Assign IUPAC names to aldehydes and ketones.
- Explain the roles of dipole–dipole attraction and hydrogen bonding in the physical properties of alcohols versus carbonyl compounds.
- Distinguish between aldehydes and ketones by oxidation reactions.
- Write the products of reduction of aldehydes and ketones.
- Identify hemiacetals, hemiketals, acetals, and ketals.
- Write the products of the reaction of carbonyl compounds with nitrogen compounds.
- Compare and explain the acidities of carbonyl compounds and alkanes.
- Write the enol form of a carbonyl compound.
- Write the aldol product of an aldehyde.
- Distinguish among alcohols, ethers, and carbonyl compounds with visual tests.

New Terms

An **acetal** is formed from the reaction of one mole of an aldehyde and two moles of an alcohol. The general formula is $RCH(OR')_2$.

An **aldehyde** is a carbonyl compound whose carbonyl carbon atom is bonded to one hydrogen atom and either an alkyl or aryl group.

Benedict's solution is an alkaline solution of cupric ion as a complex ion. It is a test reagent for aldehydes.

A **carbonyl** group consists of a carbon atom and an oxygen atom bonded by a double bond.

Fehling's solution is an alkaline solution of cupric ion as a complex ion. It is a test reagent for aldehydes.

A **hemiacetal** has the functional grouping
$$R-\overset{\displaystyle H}{\underset{\displaystyle OR'}{C}}-OH$$

A **hemiketal** has the functional grouping
$$R-\overset{\displaystyle R}{\underset{\displaystyle OR'}{C}}-OH$$

A **ketal** is formed from the reaction of one mole of a ketone and two moles of an alcohol. The general formula is $R_2C(OR')_2$.

A **ketone** is a carbonyl compound whose carbonyl carbon atom is bonded to two alkyl or aryl groups or an alkyl and an aryl group.

Tautomers are isomers that differ because of a shift of a hydrogen atom.

Tollens' reagent is an alkaline solution of $[Ag(NH_3)_2]^+$, which is used as a test for aldehydes.

Questions and Problems

Terminology

17.1 Why aren't aldehydes and ketones considered a single class of compounds?

17.2 How does an acetal differ from a ketal?

17.3 How do hemiacetals differ structurally from acetals?

17.4 What are tautomers?

Aldehydes and Ketones

17.5 Describe the geometry of the carbonyl group.

17.6 What structural features of aldehydes distinguish them from ketones?

17.7 Draw the structures of the simplest aldehyde and simplest ketone.

17.8 Identify each of the following as an aldehyde, ketone, or other class of compound.

(a) $CH_3\overset{\overset{\displaystyle O}{\|}}{C}CH_2CH_3$ (b) CH_3CH_2CHO

(c)   (d)

(e)   (f)

Nomenclature

17.9 Write the structure of each of the following compounds.
(a) diethyl ketone
(b) methyl ethyl ketone
(c) acetaldehyde
(d) methyl phenyl ketone
(e) butyraldehyde
(f) α-bromobutyraldehyde
(g) β-chloropropionaldehyde
(h) acetone

17.10 Write the structural formula for each of the following compounds.
(a) 2-methylbutanal   (b) 3-ethylpentanal
(c) 2-bromopentanal   (d) 3,4-dimethyloctanal
(e) 2-ethylhexanal   (f) 2,2-dichloropropanal

17.11 Write the structure of each of the following compounds.
(a) 3-bromo-2-pentanone
(b) 4-methyl-2-pentanone
(c) 2,4-dimethyl-3-pentanone
(d) 3,4-dimethyl-2-pentanone

17.12 Write the structure of each of the following compounds.
(a) cyclopentanone
(b) 2-methylcyclohexanone
(c) 3-bromocyclobutanone
(d) 4,4-dimethylcyclohexanone

17.13 Explain why each of the following is an incorrect IUPAC name.
(a) 5-hexanone
(b) 1-butanone
(c) 5-methylcyclopentanone
(d) 4,4-dimethylcyclobutanone
(e) 3-ethylbutanal
(f) 1-methylpropanal

17.14 Name each of the following compounds by IUPAC nomenclature.

(a) CH_3CHO (b) $CH_3CH_2CH_2CH_2CHO$

(c) $CH_3CH_2CH_2CHO$ (d) $CH_3\overset{\overset{\displaystyle CH_3}{|}}{\underset{\underset{\displaystyle CH_3}{|}}{C}}CH_2CHO$

(e) $CH_3\overset{\overset{\displaystyle CH_3}{|}}{C}HCHO$ (f) $CH_3CH_2\overset{\overset{\displaystyle CH_3}{|}}{C}H\overset{\underset{\displaystyle CH_2CH_3}{|}}{C}HCHO$

(g) $CH_3CH_2\overset{\overset{\displaystyle CH_3}{|}}{C}HCHO$ (h) $CH_3\overset{\overset{\displaystyle CH_3}{|}}{C}HCH_2CH_2CHO$

17.15 Give the IUPAC name for each of the following compounds.

(a) $CH_3CH_2\overset{\overset{\displaystyle O}{\|}}{C}CH_3$ (b) $CH_3CH_2\overset{\overset{\displaystyle O}{\|}}{C}CH_2CH_3$

(c) $CH_3CH_2\overset{\overset{\displaystyle O}{\|}}{C}CH_2CH_3$ (d) $CH_3\overset{\overset{\displaystyle O}{\|}}{C}-\overset{\overset{\displaystyle CH_3}{|}}{\underset{\underset{\displaystyle CH_3}{|}}{C}}CH_3$

(e) $CH_3\overset{\overset{\displaystyle O}{\|}}{\underset{\underset{\displaystyle CH_3}{|}}{C}}HCCH_2CH_3$ (f) $CH_3\overset{\overset{\displaystyle CH_3}{|}}{C}HCH_2\overset{\overset{\displaystyle O}{\|}}{C}CH_2CH_3$

(g)   (h)

17.16 Give the IUPAC name for each of the following compounds.

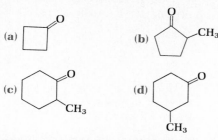

(a)   (b)   (c)   (d)

17.17 Give the IUPAC name for each of the following compounds.

(a) CH_3CHCHO
       |
       Cl

(b) $BrCH_2CH_2CH_2CHO$

(c) $ClCH_2CHCH_2CHO$
            |
            CH_3

(d)

(e) CH_3CHCH_2CHO
         |
         OH

(f) $CH_3CHCHCHO$
         | |
         OH CH_3

(g) $CH_3CCH_2CCH_3$
         | ||
        (OH,O)
         |
         CH_3

(h) $CH_3CCH_2CCH_3$
         | ||
        (OH,O)
         |
         CH_3

17.18 Give common names with Greek letters for compounds (a)–(f) in **17.17**.

Physical Properties

17.19 Aldehydes in general are less soluble in water than the corresponding alcohols of similar molecular weight. Explain why.

17.20 Which compound has the higher boiling point, cyclohexanol or cyclohexanone?

17.21 Arrange the following compounds in order of increasing boiling point.

17.22 Why can't aldehydes and ketones form intermolecular hydrogen bonds with themselves?

17.23 Order hexanal, 1-hexanol, and heptane according to their boiling points.

17.24 The boiling points of butanal and 1-butanol are 76 and 118°C, respectively. The solubilities for the same compounds are 7.0 and 7.9 g/100 mL of water, respectively. Explain why the boiling points are very different but the solubilities are similar.

17.25 Laboratory glassware is rinsed with acetone after washing with water. Considering the solubility and boiling points of acetone, explain why this procedure is used to dry glassware.

Oxidation and Reduction

17.26 Write the structure of the product of each of the following compounds when reacted with sodium dichromate and a strong acid.

(a) 1-propanol   (b) 2-butanol
(c) 1,1-dimethylpropanol (d) cyclohexanol

(e)   (f)

17.27 What observation is made when an aldehyde reacts with Benedict's solution?

17.28 What observation is made when an aldehyde reacts with Tollens' reagent?

17.29 What observation will be made when a ketone is treated with Benedict's solution?

17.30 What is the product when each of the following reacts with hydrogen gas in the presence of a platinum catalyst?

(a) acetone
(b) 3-pentanone
(c) hexanal
(d) cyclohexanone
(e) 4,4-dimethylcyclohexanone
(f) 3,3-dimethylpentanal

17.31 What class of compounds results from the reduction of ketones with hydrogen gas in the presence of a platinum catalyst?

17.32 What class of compounds results from the reduction of aldehydes by hydrogen gas in the presence of a platinum catalyst?

17.33 Write the product of each of the following reactions.

(a) $CH_3CH_2CH{=}CHCH_2CHO \xrightarrow[H_2]{Pt}$

(b) $CH_3CH_2CH{=}CHCH_2CHO \xrightarrow[(2)\ H^+]{(1)\ LiAlH_4}$

(c) $\xrightarrow[H_2]{Pt}$

(d) $\xrightarrow[(2)\ H_2O]{(1)\ LiAlH_4}$

Addition Reactions with Alcohols

17.34 Identify each of the following as a hemiacetal, hemiketal, acetal, or ketal.

(a) $CH_3CH_2CH(OCH_3)_2$ (b) $CH_3CH_2C(OCH_3)_2CH_3$

(c) $CH_3CHOCH_2CH_3$ with OCH_2CH_3 substituent   (d) CH_3CH_2CHOH with OCH_3 substituent

(e) $CH_3CH_2CCH_3$ with OCH_3 and OH substituents   (f) $(CH_3O)_2CHCH_2CH(OCH_3)_2$

(g) cyclohexane with OCH_3 and OCH_3 substituents

(h) cyclohexane with OCH_3 and OH substituents

17.35 What compounds are required to produce each substance in **17.34?**

17.36 Identify each of the following as a hemiacetal, hemiketal, acetal, or ketal.

(a) tetrahydropyran ring with OH

(b) tetrahydropyran ring with OCH_3

(c) tetrahydropyran ring with OCH_3 and CH_3

(d) tetrahydropyran ring with OH and CH_3

(e) sugar ring with CH_2OH, H, OCH_3, HO, OH, H, H, OH

(f) sugar ring with CH_2OH, HO, H, OH, OH, H, H, H, OH

17.37 Would an acetal be stable in gastric juice?

Addition Reactions with Nitrogen Compounds

17.38 Write the structure of the product of the reaction of ethanal with methylamine.

17.39 Write the structure of the product of the reaction of cyclohexanone with 2,4-dinitrophenylhydrazine.

17.40 Explain how the two isomeric substances, 3-penta-

none and cyclopentanol, may be distinguished by use of a single visual test.

17.41 An amine group of alanine reacts with the carbonyl group of pyridoxal. Write the structure of the product.

$CH_3-CH-CO_2H$ with NH_2 substituent

alanine

pyridoxal: pyridine ring with CHO, $HOCH_2$, OH, N, CH_3

Chemistry of the α-Carbon Atom

17.42 Explain why aldehydes and ketones are much more acidic than alkanes.

17.43 Consider the following compound and discuss the acidity of each of the C—H bonds.

$$CH_3CH_2CCH_2CCH_2CH_3$$ with two C=O groups

17.44 Write the enol form of each of the following compounds.

(a) acetone   (b) butanal

(c) cyclohexanone   (d) acetaldehyde

17.45 Formaldehyde does not exist as an enol. Why?

17.46 There are three possible enols of 2-butanone. Explain why and write the structures.

17.47 Write the aldol product expected by treating each of the following with base.

(a) ethanal   (b) propanal   (c) methylpropanal

17.48 What aldehyde is required to produce each of the following compounds by an aldol condensation?

(a) CH_3CHCH_2CHO with OH substituent

(b) $CH_3CH_2CHCHCHO$ with OH and CH_3 substituents

(c) $CH_3CHCH-CCHO$ with OH, CH_3, CH_3 substituents

17.49 Compounds (a) and (b) react with Benedict's solution but (c) and (d) do not. Suggest a reason for this observation.

(a) $CH_3CH_2CCH_2OH$ with C=O group

(b) cyclohexane with CCH_2OH group (C=O)

(c) $CH_3C-CHCH_3$ with C=O and OH groups

(d) cyclohexane with CH_2OH group and C=O

17.50 Glucose reacts with Benedict's solution because it is an aldehyde. Fructose, although a ketone, also reacts with Benedict's solution in aqueous solution. Suggest an isomerization reaction that can account for this observation.

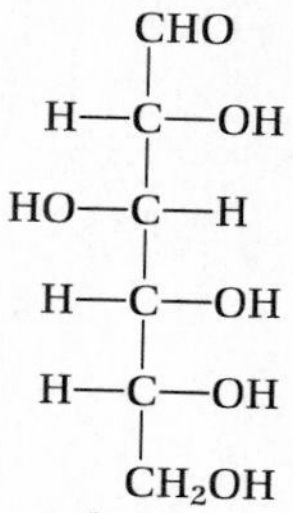

CHO
|
H—C—OH
|
HO—C—H
|
H—C—OH
|
H—C—OH
|
CH₂OH
glucose

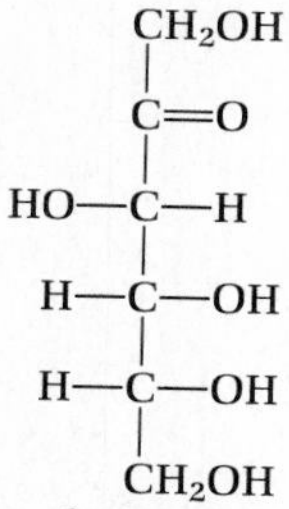

CH₂OH
|
C=O
|
HO—C—H
|
H—C—OH
|
H—C—OH
|
CH₂OH
fructose

Identification of Oxygen Compounds

17.51 Describe a single laboratory test that can be used to distinguish between each member of the following pairs of compounds.

(a) CH_3CHCHO and $CH_3COCH_2CH_3$
   with CH_3 above the second carbon

(b) 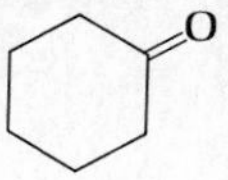 and

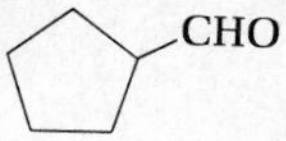

(c) $CH_2=CH—O—CH_2CH_3$ and $CH_3CH_2CH_2CHO$

(d) $CH_3COCH_2CH_3$ and $CH_2=CHCH_2CH_2OH$

17.52 A compound, $C_4H_{10}O$, reacts with sodium dichromate and acid to yield C_4H_8O. The C_4H_8O compound does not react with Benedict's solution. The C_4H_8O compound is reduced by hydrogen gas in the presence of a platinum catalyst to give $C_4H_{10}O$. What are the structures of $C_4H_{10}O$ and C_4H_8O?

17.53 A compound, $C_6H_{10}O$, gives a positive Tollens' test. The compound when reduced by hydrogen gas in the presence of a platinum catalyst yields $C_6H_{12}O$. There are 16 isomeric compounds that fit these data. Draw the structures of at least three of them.

17.54 A compound, C_5H_8O, does not react with Fehling's solution but does react with 2,4-dinitrophenylhydrazine. Reduction of the compound with hydrogen gas in the presence of a platinum catalyst yields $C_5H_{10}O$. Eight possible compounds fit these data. Draw the structures of three of these compounds.

18 Carboxylic Acids and Esters

Dacron is a polyester that is used for a large number of products. The Dacron sails of this boat are light, but have great structural strength.

Carboxylic acids *are compounds containing the carboxyl group.* The carboxyl group is represented in several ways.

$$\begin{array}{c} -\overset{\displaystyle O}{\overset{\|}{C}}-OH \end{array} \quad \text{or} \quad -COOH \quad \text{or} \quad -CO_2H$$

Hydrogen, an alkyl group, or an aromatic group may be bonded to the carboxyl group. The simplest examples of such carboxylic acids are formic acid, acetic acid, and benzoic acid.

$$\underset{\text{formic acid}}{H-\overset{O}{\overset{\|}{C}}-OH} \qquad \underset{\text{acetic acid}}{CH_3-\overset{O}{\overset{\|}{C}}-OH} \qquad \underset{\text{benzoic acid}}{\text{C}_6H_5-\overset{O}{\overset{\|}{C}}-OH}$$

The carboxyl group is so named because it consists of a carbonyl group and a hydroxyl group. However, the chemical reactivity of the carboxyl group is not that of a carbonyl group or of a hydroxyl group. In this chapter, we will see that the component atoms of the carboxyl group must be treated as a whole.

The carboxyl carbon atom is sp^2 hybridized, and three of its electrons form three σ bonds at 120° to each other (Figure 18.1). One of the σ bonds is to hydrogen, to an alkyl group, or to an aromatic group. Each of the other two σ bonds is to an oxygen atom, one to the hydroxyl oxygen atom and the other to the carbonyl oxygen atom. The remaining electron of the carbon atom is in a p orbital that forms a π bond with an electron in a p orbital of the carbonyl oxygen.

Carboxylic acids are widely distributed in nature, and their taste is characterized as tart. Acetic acid, which is formed in the oxidation of ethyl alcohol by bacteria, has been known since biblical times. The vinegar taste that develops in wine oxidized by air is due to acetic acid. The taste of sour milk is due to lactic acid. Oxalic acid is contained in rhubarb, and citric acid is contained in citrus fruits.

Two of the classes of foods, proteins and fats, are derived from acids. Amino acids, which contain both the amino group and the carboxyl group, are the building blocks of proteins. Fats are esters of carboxylic acids and the alcohol glycerol. The citric acid cycle (Chapter 25) is a series of eight enzyme-catalyzed reactions involving acids as both reactants and products.

Figure 18.1
Bonding in Formic Acid.
The atoms bonded to the carboxyl carbon atom of carboxylic acids all lie in a plane with the carboxyl carbon atom. The carboxyl carbon atom is sp^2 hybridized, and there is a π bond between the carbonyl oxygen atom and the carboxyl carbon atom.

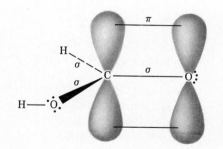

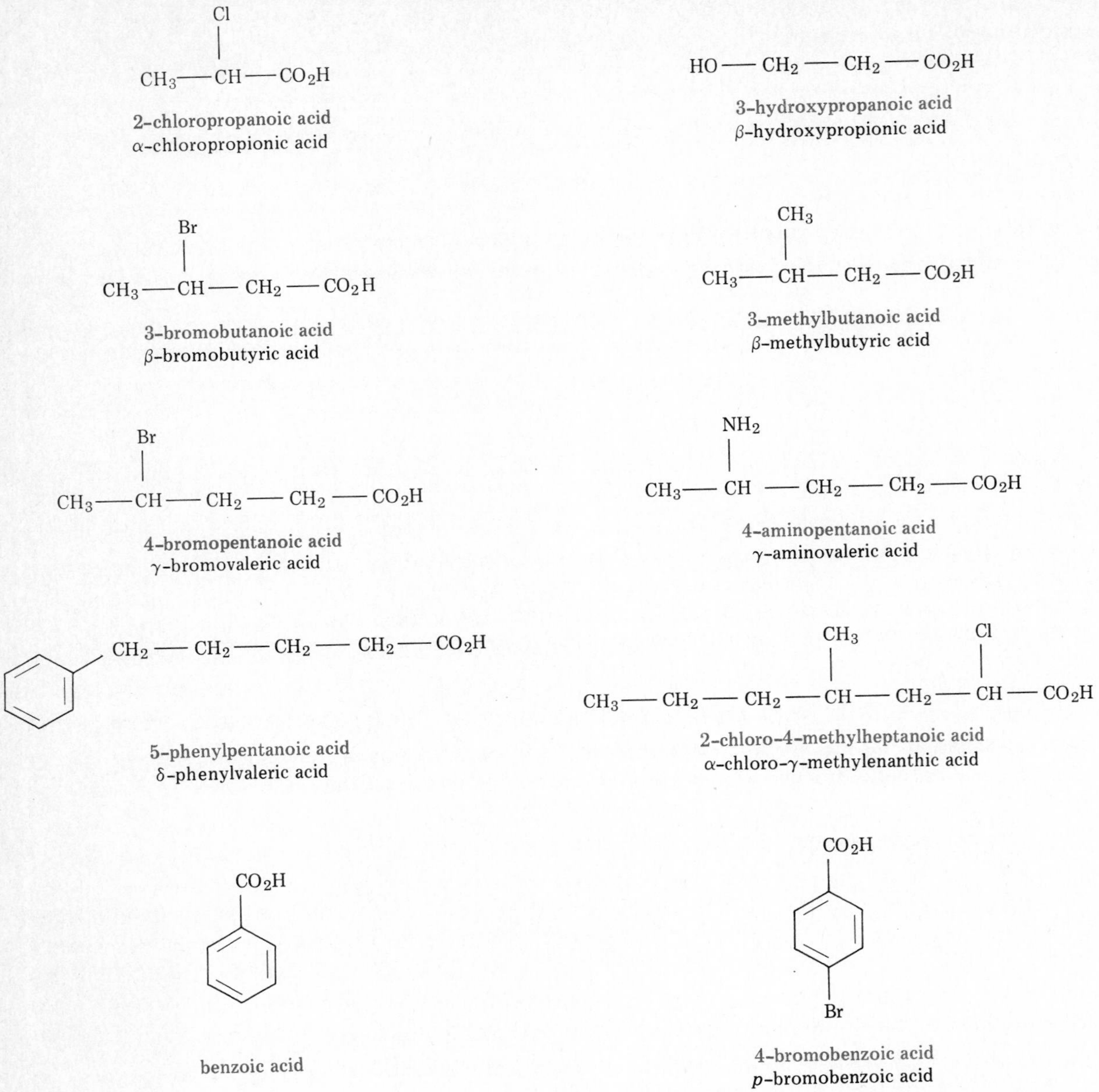

$$CH_3-\underset{\underset{Cl}{|}}{CH}-CO_2H$$

2-chloropropanoic acid
α-chloropropionic acid

$$HO-CH_2-CH_2-CO_2H$$

3-hydroxypropanoic acid
β-hydroxypropionic acid

$$CH_3-\underset{\underset{Br}{|}}{CH}-CH_2-CO_2H$$

3-bromobutanoic acid
β-bromobutyric acid

$$CH_3-\underset{\overset{CH_3}{|}}{CH}-CH_2-CO_2H$$

3-methylbutanoic acid
β-methylbutyric acid

$$CH_3-\underset{\underset{Br}{|}}{CH}-CH_2-CH_2-CO_2H$$

4-bromopentanoic acid
γ-bromovaleric acid

$$CH_3-\underset{\overset{NH_2}{|}}{CH}-CH_2-CH_2-CO_2H$$

4-aminopentanoic acid
γ-aminovaleric acid

$$CH_2-CH_2-CH_2-CH_2-CO_2H$$

5-phenylpentanoic acid
δ-phenylvaleric acid

$$CH_3-CH_2-CH_2-\underset{\overset{CH_3}{|}}{CH}-CH_2-\underset{\overset{Cl}{|}}{CH}-CO_2H$$

2-chloro-4-methylheptanoic acid
α-chloro-γ-methylenanthic acid

benzoic acid

4-bromobenzoic acid
p-bromobenzoic acid

Figure 18.2
Nomenclature of Carboxylic Acids.

18.2
Nomenclature of Carboxylic Acids

Because of the abundance of carboxylic acids in nature and their early discovery, these compounds were long ago given common names, based on their natural sources, that are still widely used (Table 18.1).

In the IUPAC system, the longest continuous chain of carbon atoms containing the carboxyl functional group is selected. The -e ending of the parent alkane is dropped and replaced by the ending -oic acid. The numbering of the chain starts at the carboxyl carbon, but since the carboxyl group must be at the end of

Table 18.1

Nomenclature of Unbranched Carboxylic Acids

Formula	Name	
	Common	IUPAC
HCO_2H	formic acid	methanoic acid
CH_3CO_2H	acetic acid	ethanoic acid
$CH_3CH_2CO_2H$	propionic acid	propanoic acid
$CH_3(CH_2)_2CO_2H$	butyric acid	butanoic acid
$CH_3(CH_2)_3CO_2H$	valeric acid	pentanoic acid
$CH_3(CH_2)_4CO_2H$	caproic acid	hexanoic acid
$CH_3(CH_2)_6CO_2H$	caprylic acid	octanoic acid
$CH_3(CH_2)_8CO_2H$	capric acid	decanoic acid
$CH_3(CH_2)_{10}CO_2H$	lauric acid	dodecanoic acid
$CH_3(CH_2)_{12}CO_2H$	myristic acid	tetradecanoic acid
$CH_3(CH_2)_{14}CO_2H$	palmitic acid	hexadecanoic acid
$CH_3(CH_2)_{16}CO_2H$	stearic acid	octadecanoic acid

the chain, the number 1 is not used in the name. Examples of this nomenclature are given in Figure 18.2.

When common names are used, the positions of groups attached to the parent chain are designated α, β, γ, and so on. The α position is the carbon atom attached to the carboxyl carbon.

Example 18.1

Give both the IUPAC and common names for the following carboxylic acid.

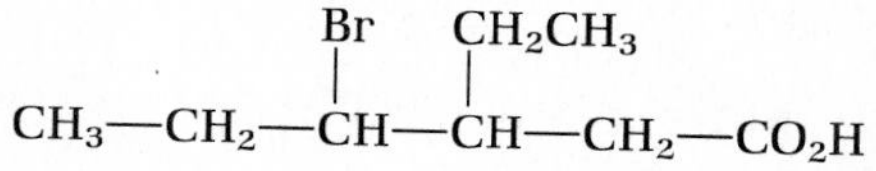

$$\underset{}{CH_3-CH_2-\underset{\overset{|}{Br}}{CH}-\underset{\overset{|}{CH_2CH_3}}{CH}-CH_2-CO_2H}$$

Solution. First determine that the longest continuous chain of the acid contains six carbon atoms. The parent name in the IUPAC system is hexanoic; the common name of the parent acid is caproic acid (Table 18.1).

The substituents are an ethyl group and a bromo group. These groups are on the numbers 3 and 4 carbon atoms, respectively, according to IUPAC rules, but the β and γ carbon atoms according to the older common name system.

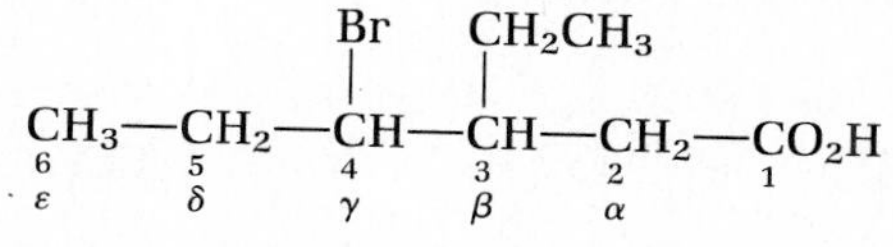

$$\underset{\underset{\varepsilon}{6}}{CH_3}-\underset{\underset{\delta}{5}}{CH_2}-\underset{\underset{\gamma}{4}}{\underset{\overset{|}{Br}}{CH}}-\underset{\underset{\beta}{3}}{\underset{\overset{|}{CH_2CH_3}}{CH}}-\underset{\underset{\alpha}{2}}{CH_2}-\underset{\underset{}{1}}{CO_2H}$$

The ethyl group is part of the hydrocarbon structure and is placed closest to the parent name. The bromo group is prefixed to this name.

4-bromo-3-ethylhexanoic acid
γ-bromo-β-ethylcaproic acid

[Additional examples may be found in **18.5–18.12** at the end of the chapter.]

Dicarboxylic acids are compounds with two carboxyl groups. Some of the simplest of these compounds are known by their common name. Succinic acid and glutaric acid are intermediates in the citric acid cycle.

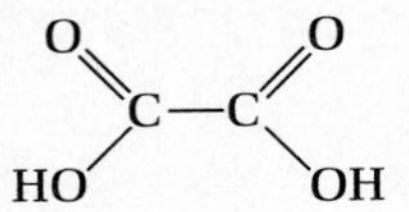

oxalic acid 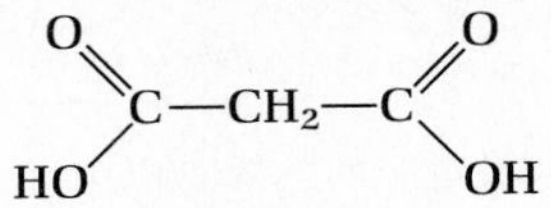   malonic acid 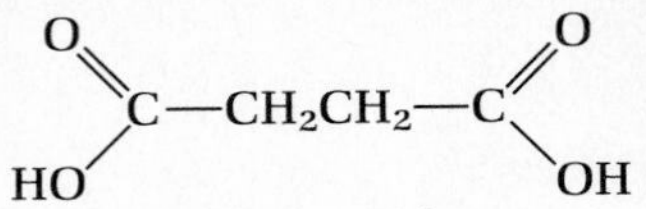   succinic acid

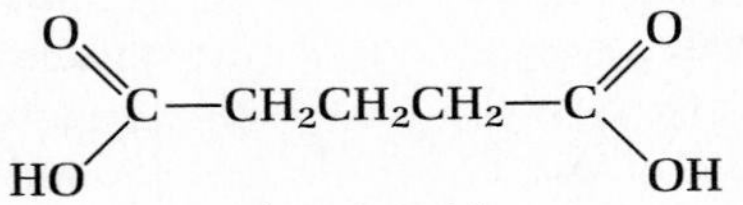

glutaric acid 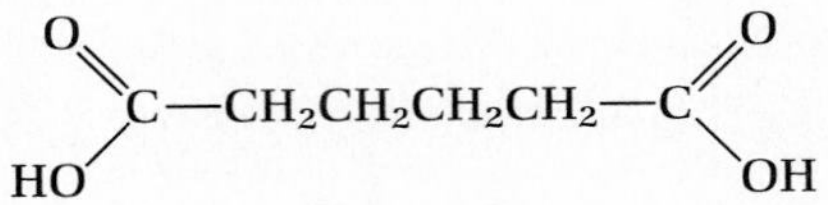   adipic acid

Unsaturated acids of importance include the two geometric isomers maleic acid and fumaric acid. Fumaric acid is an intermediate in the citric acid cycle.

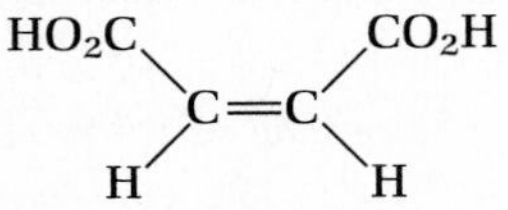

maleic acid 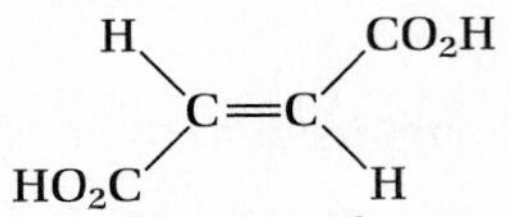   fumaric acid

There are many polyfunctional carboxylic acids that contain functional groups other than the carboxyl group. Hydroxyl groups are common in naturally occurring compounds. Both citric acid and malic acid are intermediates in the citric acid cycle.

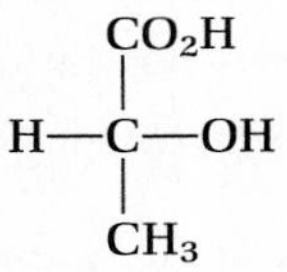

lactic acid 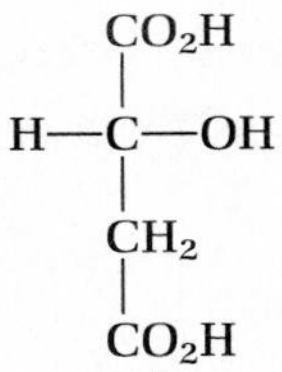   malic acid 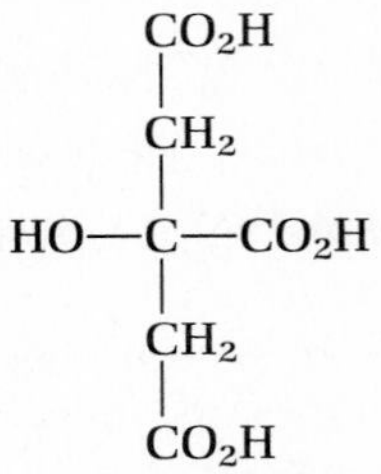   citric acid

Keto acids that are important in the citric acid cycle include oxaloacetic acid and α-ketoglutaric acid. Pyruvic acid is a central metabolic intermediate (Chapter 26).

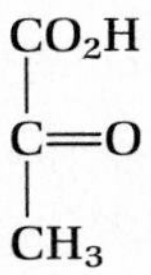

pyruvic acid 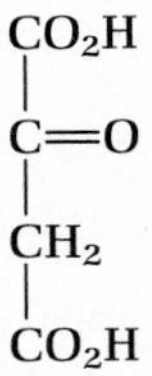   oxaloacetic acid α-ketoglutaric acid

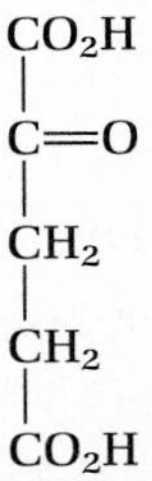

Example 18.2 ──────────────────────────────────

Unsaturated carboxylic acids are named by prefixing the position of the double bond to the parent hydrocarbon name with the ending -ane replaced by -enoic acid. An 18-carbon normal alkane is octadecane. With this information, assign an IUPAC name to oleic acid found in vegetable oils.

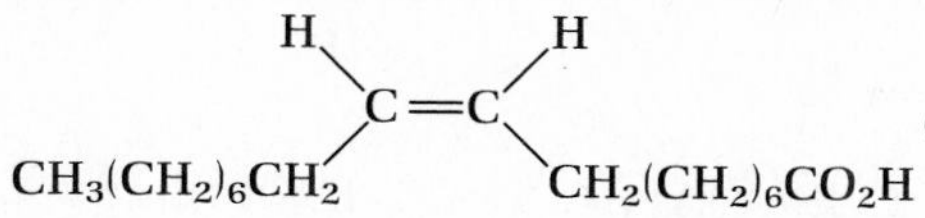

Solution. The double bond is located at carbon atom 9 in the chain numbered from the carboxyl group on the right. Thus, the compound is a 9-octadecenoic acid. The geometry about the double bond is *cis* and thus the complete name is *cis*-9-octadecenoic acid.

The low-molecular-weight carboxylic acids are liquids at room temperature (Table 18.2). The higher-molecular-weight compounds are wax-like solids. Liquid acids have sharp and unpleasant odors. Butyric acid occurs in rancid butter and aged cheese. Caproic, caprylic, and capric acids have the smell of goats. The Latin word for goat is caper and is the source of the common name of these acids.

Low-molecular-weight acids are soluble in water, but the solubility decreases with increasing chain length (Table 18.2). The solubility of these compounds in water is due to hydrogen bonding between water and the carboxyl group.

The high-molecular-weight acids, such as myristic, palmitic, and stearic acids, known as fatty acids, are products of fat digestion by the body. Since these compounds are not appreciably soluble in water, the bile salts produced in the gallbladder act as emulsifying agents to disperse and transport them.

The boiling points of carboxylic acids are abnormally high (Table 18.2). This phenomenon is due to very strong hydrogen bonding, which holds two molecules together as a dimer.

18.3
**Physical Properties
of Carboxylic Acids**

Table 18.2 ──────────────────────────────────
Physical Properties of Carboxylic Acids

IUPAC	Melting point (°C)	Boiling point (°C)	Solubility (g/100 g H$_2$O, at 20°C)
methanoic acid	8	101	miscible
ethanoic acid	17	118	miscible
propanoic acid	−21	141	miscible
butanoic acid	−5	164	miscible
pentanoic acid	−34	186	4.97
hexanoic acid	−3	205	0.96
octanoic acid	17	239	0.068
decanoic acid	32	270	0.015
dodecanoic acid	44	299	0.0055

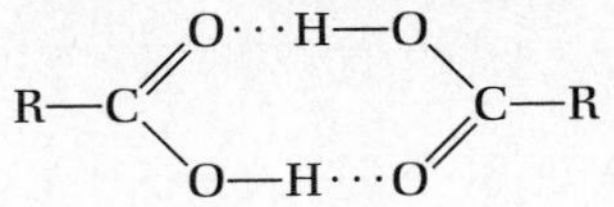

$$R-C\underset{O-H\cdots O}{\overset{O\cdots H-O}{\big\langle\big\rangle}}C-R$$

Even in the vapor phase, carboxylic acids exist as dimers. The aggregates of molecules have a higher boiling point than would a liquid made up of single molecules.

18.4
Acidity of Carboxylic Acids

You learned in Chapter 11 that the ionization of a weak acid such as acetic acid can be expressed by an acid equilibrium constant (K_a).

$$CH_3CO_2H \rightleftharpoons CH_3CO_2^- + H^+$$

$$K_a = \frac{[H^+][CH_3CO_2^-]}{[CH_3CO_2H]} = 1.8 \times 10^{-5} \text{ mole/L}$$

Although acetic acid and other carboxylic acids are more acidic than alcohols or phenols, they are still weak acids. The point of equilibrium for acetic acid and its conjugate base, the acetate ion, lies far to the left. A $1\,M$ solution of acetic acid is only 0.4% ionized.

pK_a Scale

Chemists have devised a pK_a scale to express the acid strength of compounds. This scale, which avoids the use of exponential notation, is defined by

$$pK_a = -\log K_a$$

The definition is analogous to that of pH; the stronger the acid, the smaller is its pK_a.

The relationship between the pK_a of an acid and the pH of a solution is important to understanding many biological processes. The relationship is best seen by rearranging the equilibrium constant expression, taking logarithms of both sides of the equation, and changing signs appropriately.

$$[H^+] = K_a \frac{[CH_3CO_2H]}{[CH_3CO_2^-]}$$

$$\log[H^+] = \log K_a + \log \frac{[CH_3CO_2H]}{[CH_3CO_2^-]}$$

$$-\log[H^+] = -\log K_a - \log \frac{[CH_3CO_2H]}{[CH_3CO_2^-]}$$

$$pH = pK_a - \log \frac{[CH_3CO_2H]}{[CH_3CO_2^-]}$$

Knowing that $\log(x/y) = -\log(y/x)$, we can write the final expression

$$pH = pK_a + \log \frac{[CH_3CO_2^-]}{[CH_3CO_2H]}$$

This equation, which is widely used in biochemistry and organic chemistry, is known as the **Henderson–Hasselbalch equation.** From this equation, *the ratio of the concentrations of an acid and its conjugate base at any pH can be calculated.*

This calculation is important in identifying the exact form in which an acid exists in aqueous solutions in organisms.

The pK_a of an acid can easily be determined by adding 0.5 mole of a strong base to 1.0 mole of an acid. Under these conditions, only one-half of the acid is neutralized, and the concentrations of the remaining acid and its conjugate base are equal. The pH of the resulting solution is then equal to the pK_a.

$$pH = pK_a + \log \frac{[A^-]}{[HA]}$$
$$= pK_a + \log 1$$
$$= pK_a$$

Effect of Structure on Acidity

An important factor in the higher acidity of acids relative to water or alcohols is resonance stabilization of the conjugate base. In the conjugate base, the negative charge is delocalized over a system consisting of orbitals of the carbon atom and the two oxygen atoms. Two structures representing the conjugate base are

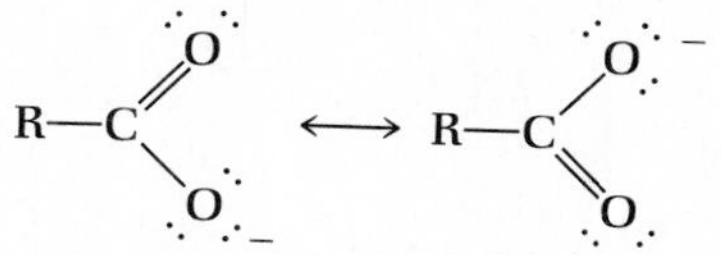

resonance forms of the carboxylate anion

Delocalization of the negative charge over the two oxygen atoms rather than having the charge on one atom results in a more stable structure. In water and alcohols the negative charge of the conjugate bases is on a single oxygen atom.

The structure of the carboxylic acid determines its acidity. Electronegative groups attached to the α-carbon atom increase the acidity. Such groups pull electrons away from the hydrocarbon skeleton and indirectly from the O—H bond. As a consequence, the proton is more easily removed and the K_a is increased (Table 18.3).

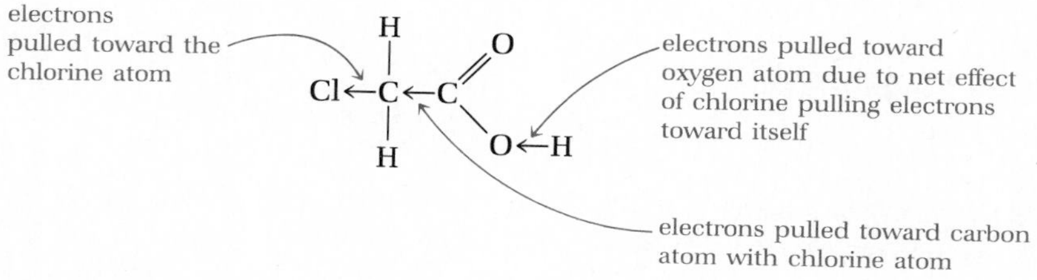

electrons pulled toward the chlorine atom

electrons pulled toward oxygen atom due to net effect of chlorine pulling electrons toward itself

electrons pulled toward carbon atom with chlorine atom

Table 18.3

Effect of Substituents on the Acidity of Carboxylic Acids

Name	Formula	K_a	pK_a
acetic acid	CH_3CO_2H	1.8×10^{-5}	4.74
iodoacetic acid	ICH_2CO_2H	6.7×10^{-4}	3.17
bromoacetic acid	$BrCH_2CO_2H$	1.3×10^{-3}	2.88
chloroacetic acid	$ClCH_2CO_2H$	1.4×10^{-2}	2.85
dichloroacetic acid	Cl_2CHCO_2H	5.0×10^{-2}	1.30
trichloroacetic acid	Cl_3CCO_2H	2.3×10^{-1}	0.64

A variety of acids are involved in biological transformations. Several steps of the citric acid cycle are involved in the conversion of succinic acid into oxaloacetic acid. The acid dissociation constants for the intermediate acids increase markedly. The overall change in acid strength is approximately 300-fold.

$$\underset{\substack{\text{succinic} \\ \text{acid} \\ 6.9 \times 10^{-5}}}{} \longrightarrow \underset{\substack{\text{fumaric} \\ \text{acid} \\ 9.3 \times 10^{-4}}}{} \longrightarrow \underset{\substack{\text{malic} \\ \text{acid} \\ 3.9 \times 10^{-4}}}{} \longrightarrow \underset{\substack{\text{oxaloacetic} \\ \text{acid} \\ 2.0 \times 10^{-2}}}{}$$

As a consequence, the pH of cellular fluid would vary if no buffers were present.

Example 18.3

Compare the acid equilibrium constants of fluoroacetic acid and trifluoroacetic acid to those of chloroacetic acid and trichloroacetic acid.

Solution. Fluorine is the most electronegative of the halogens. As shown in Table 18.3, the acidity increases with increasing electronegativity. Fluoroacetic acid should have a larger K_a than chloroacetic acid. Trifluoroacetic acid should have a larger K_a than trichloroacetic acid.

[Additional examples may be found in **18.19–18.25** at the end of the chapter.]

18.5
Salts of Carboxylic Acids

Although water is too weak a base to convert very much of a carboxylic acid into its conjugate base, the hydroxide ion is a sufficiently strong base to neutralize carboxylic acids.

Carboxylic acids react with bases such as metal hydroxides to produce carboxylate salts. The names of salts are derived from the acid by changing the -ic ending to -ate and preceding this name by the name of the metal.

$$CH_3-\overset{\displaystyle O}{\underset{\displaystyle OH}{C}} + Na^+OH^- \longrightarrow CH_3C\overset{\displaystyle O}{\underset{\displaystyle O^-}{}} Na^+ + H_2O$$

common: sodium acetate
IUPAC: sodium ethanoate

Since carboxylic acids are weak acids, carboxylates are relatively strong bases. They will react with a good proton donor such as hydronium ion to form the carboxylic acid. The reaction of carboxylic acids with hydroxide ion as well as the reaction of carboxylate salts with hydronium ion has some practical applications in separating them from mixtures with nonacidic materials.

The ionic carboxylate salts are more soluble in water than their corresponding carboxylic acids. In the laboratory, carboxylic acids can be separated from other nonpolar organic compounds by adding a solution of sodium hydroxide. For example, octanol is not very soluble in water and does not react with sodium hydroxide. However, octanoic acid reacts with sodium hydroxide and thus will dissolve in the basic solution.

$$CH_3(CH_2)_6CH_2OH + OH^- \longrightarrow \text{no reaction}$$

$$CH_3(CH_2)_6CO_2H + OH^- \longrightarrow CH_3(CH_2)_6CO_2^- + H_2O$$

AN ASIDE

Both carboxylic acids and their carboxylate salts are widely used in industry and in commercial products. The simplest carboxylic acid, formic acid, is used in the processing of leather and textiles. Acetic acid in dilute solution is vinegar; as the pure liquid it is used in the production of some plastics and textiles (acetate rayons).

Lactic acid is used both in tanning leather and in dyeing wool. In addition, lactic acid is added as a preservative in processed cheese and salad dressings. Lactic acid is one of the several acids used to increase the acidity of carbonated beverages.

Citric acid is used in candy and soft drinks. It is also present in practically all plants and is responsible for the sour taste of citrus fruits.

Salicylic acid is both an antipyretic (reduces fever) and an analgesic, known since 1870. However, it irritates the stomach lining and its ester, called acetylsalicylic acid—better known as aspirin—is now preferred for use as a medicine. Approximately 12,000 tons of aspirin is produced annually in the United States.

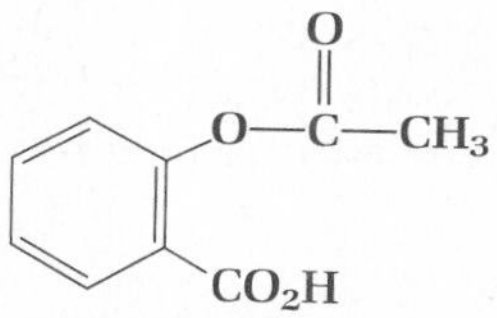

acetylsalicylic acid

The sodium and calcium salts of propanoic acid are used in baked goods to prevent the growth of bacteria and in cheese to inhibit the formation of mold. The sodium and potassium salts of sorbic acid are used to inhibit the growth of molds in fruit drinks, soft drinks, and wines as well as for some meat and fish products.

$$CH_3CH=CHCH=CHCO_2H$$

sorbic acid

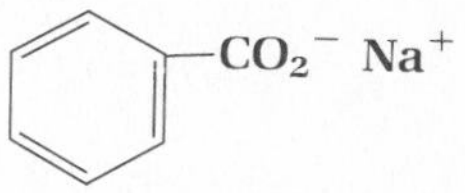

sodium benzoate

Sodium benzoate are used to inhibit the formation of molds in syrups, jams, and jellies as well as margarine, pickles, and pie fillings.

A basic solution can be used to form and dissolve the octanoate salt, while the octanol remains insoluble. After separation, HCl is added to neutralize the basic solution, and insoluble octanoic acid then separates.

$$CH_3(CH_2)_6CO_2^- + H_3O^+ \longrightarrow CH_3(CH_2)_6CO_2H + H_2O$$
(soluble)   (insoluble)

This procedure is very useful in isolating acids from complex mixtures in nature. It is also used to purify acids produced in chemical synthesis in the laboratory.

18.6 Esters

In Section 15.8 you were shown that alcohols react with acids of the general formula H—O—A to form esters. The acids used in that section were all inorganic compounds.

$$\text{R—O—H} + \text{H—O—A} \longrightarrow \underset{\text{an ester}}{\text{R—O—A}} + \text{H}_2\text{O}$$

A carboxylic acid also can react with an alcohol to form an **ester.** The reaction is catalyzed by inorganic acids.

$$\text{R—O—H} + \text{H—O—}\overset{\text{O}}{\overset{\|}{\text{C}}}\text{—R}' \overset{\text{H}^+}{\rightleftharpoons} \text{R—O—}\overset{\text{O}}{\overset{\|}{\text{C}}}\text{—R}' + \text{H}_2\text{O}$$

$$\text{CH}_3\text{CH}_2\text{OH} + \text{H—O—}\overset{\text{O}}{\overset{\|}{\text{C}}}\text{—CH}_3 \overset{\text{H}^+}{\rightleftharpoons} \underset{\text{ethyl acetate}}{\text{CH}_3\text{CH}_2\text{—O—}\overset{\text{O}}{\overset{\|}{\text{C}}}\text{CH}_3} + \text{H}_2\text{O}$$

The equilibrium can be displaced to increase the amount of the ester by distilling the water out of the reaction mixture. If one of the starting reagents is rare or expensive, the yield of ester can be increased by using an excess of the cheaper component of the reaction. Ethyl esters of acids can be obtained by using ethanol as a solvent. Under such conditions, the high concentration of ethanol favors a high conversion of the acid to the ester. Both the removal of water and addition of alcohol result in shifts in the equilibrium predicted by Le Châtelier's principle.

$$\text{alcohol} + \text{acid} \rightleftharpoons \text{ester} + \text{water}$$

Adding alcohol "pushes" reaction to the right

Removal of water "pulls" reaction to the right

Nomenclature of Esters

The esters are named like the salts of acids with the name of the alkyl group of the alcohol replacing the metal. Because there are two systems of naming carboxylic acids, there are two names for the esters. In the examples given in Figure 18.3 the alcohol portion of each molecule is on the right side of the struc-

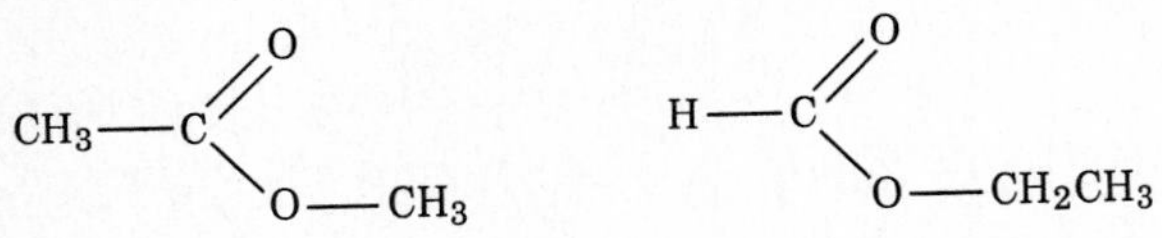

methyl ethanoate
methyl acetate

ethyl methanoate
ethyl formate

ethyl ethanoate
ethyl acetate

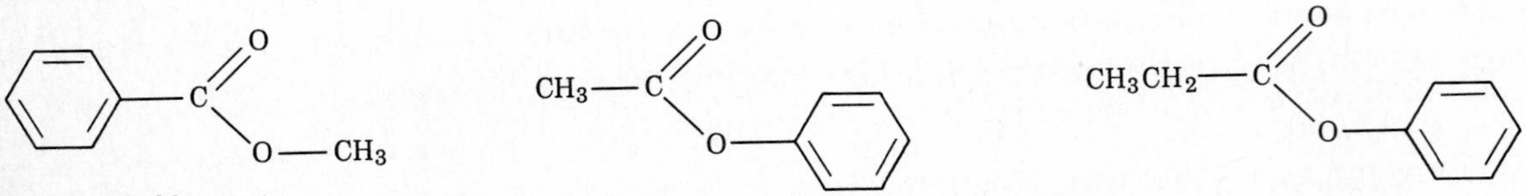

methyl benzoate
methyl benzoate

phenyl ethanoate
phenyl acetate

phenyl propanoate
phenyl propionate

Figure 18.3
Nomenclature of Esters.

ture. However, regardless of how the structure is written, the name of the alcohol is written first in the name of the ester.

The structure R—C$\overset{O}{\diagup}$ that occurs in esters and other derivatives of carboxylic acids is called an **acyl group.** An acyl group is named by replacing the -ic acid of a carboxylic acid by -yl.

$$CH_3-C\overset{O}{\diagup}$$
acetyl group
or
ethanoyl group

$$CH_3CH_2CH_2C\overset{O}{\diagup}$$
butyryl group
or
butanoyl group

Example 18.4

What is the IUPAC name of the following ester?

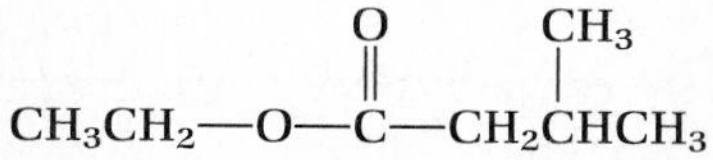

$$\underset{}{CH_3CH_2-O-\overset{\overset{O}{\|}}{C}-CH_2\overset{\overset{CH_3}{|}}{CH}CH_3}$$

Solution. First, identify the alcohol portion of the ester, at the left of the molecule; it contains two carbon atoms. This is an ethyl ester.

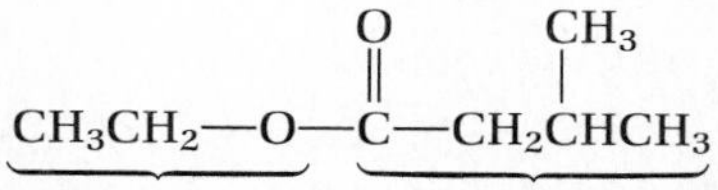

$$\underset{\underset{\text{portion}}{\text{alcohol}}}{\underline{CH_3CH_2-O}}-\underset{\underset{\text{from acid}}{\text{acyl portion}}}{\underline{\overset{\overset{O}{\|}}{C}-CH_2\overset{\overset{CH_3}{|}}{CH}CH_3}}$$

The acyl portion contains a four-carbon chain with a methyl branch. This acid is called 3-methylbutanoic acid.

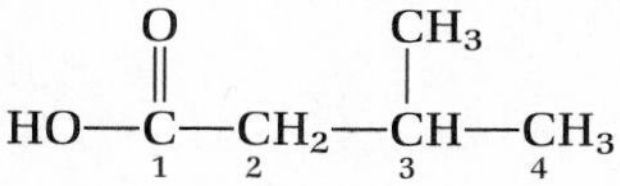

$$HO-\overset{\overset{O}{\|}}{\underset{1}{C}}-\underset{2}{CH_2}-\underset{3}{\overset{\overset{CH_3}{|}}{CH}}-\underset{4}{CH_3}$$

Now change the -ic ending of the acid to -ate and write the name of the alkyl group of the alcohol as a separate word in front of the modified acid name to give ethyl 3-methylbutanoate.

[Additional examples may be found in **18.35–18.41** at the end of the chapter.]

Properties of Esters

Esters are polar molecules, but have lower boiling points than carboxylic acids and alcohols of similar molecular weight. This lower boiling point reflects the absence of hydrogen bonds in esters.

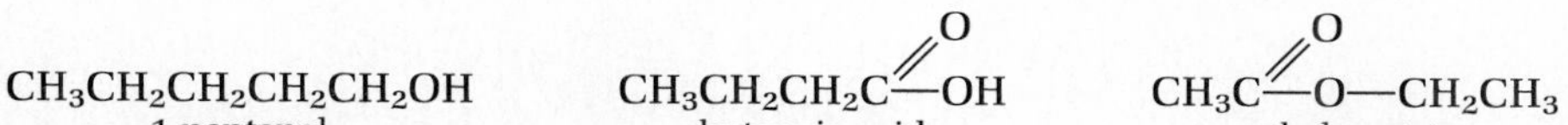

$$CH_3CH_2CH_2CH_2CH_2OH$$
1-pentanol
b.p. 138°C

$$CH_3CH_2CH_2C\overset{O}{\diagup}-OH$$
butanoic acid
b.p. 164°C

$$CH_3C\overset{O}{\diagup}-O-CH_2CH_3$$
ethyl acetate
b.p. 77°C

Table 18.4

Properties of Esters of Carboxylic Acids

Common name	Structure	Boiling point (°C)	Solubility (g/100g H_2O at 20°C)
methyl formate	HCO_2CH_3	32	miscible
methyl acetate	$CH_3CO_2CH_3$	57	24.4
methyl propionate	$CH_3CH_2CO_2CH_3$	99	1.8
methyl butyrate	$CH_3(CH_2)_2CO_2CH_3$	120	0.5
methyl valerate	$CH_3(CH_2)_3CO_2CH_3$	145	0.2
methyl caproate	$CH_3(CH_2)_4CO_2CH_3$	168	0.06
methyl enanthate	$CH_3(CH_2)_5CO_2CH_3$	189	0.03
methyl caprylate	$CH_3(CH_2)_6CO_2CH_3$	208	0.007
methyl acetate	$CH_3CO_2CH_3$	57	24.4
ethyl acetate	$CH_3CO_2CH_2CH_3$	77	7.4
propyl acetate	$CH_3CO_2CH_2CH_2CH_3$	102	1.9
butyl acetate	$CH_3CO_2CH_2CH_2CH_2CH_3$	125	0.9

Esters can form hydrogen bonds via their oxygen atoms with water molecules and thus have some degree of water solubility. However, since esters do not themselves have a hydrogen atom with which to bond to an oxygen atom of water, they are less soluble than carboxylic acids. The solubility and boiling points of esters are listed in Table 18.4.

Common Esters

Many of the odors of fruits are due to esters. For example, ethyl acetate is found in pineapples, isoamyl acetate in apples and bananas, isoamyl isovalerate in apples, and octyl acetate in oranges.

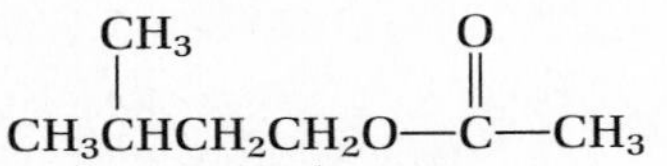

isoamyl acetate 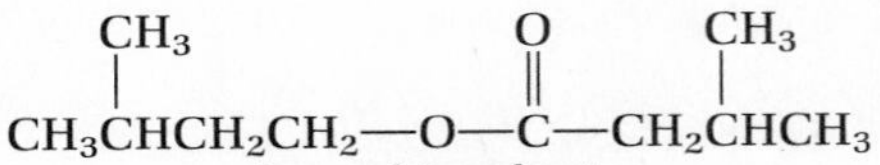   isoamyl isovalerate

In recent years there has been an increased demand for substances that enhance the flavor and aroma of processed fruits and fruit products such as jams and jellies. The esters used in these products are not necessarily the same as those in the natural fruits, but may be others that produce the same odor or taste. A list of some of these flavoring agents is given in Table 18.5.

Salicylic acid can act as a phenol or as an acid. The acetate ester of the phenolic group is acetylsalicylic acid (aspirin). The methyl ester of the carboxyl group of salicylic acid is methyl salicylate (oil of wintergreen) and is used in liniments.

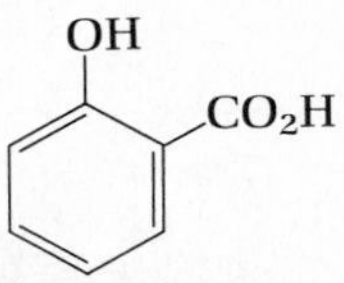

salicylic acid 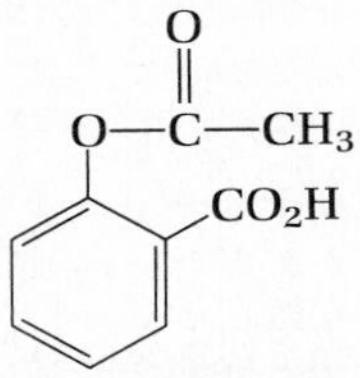   acetylsalicylic acid 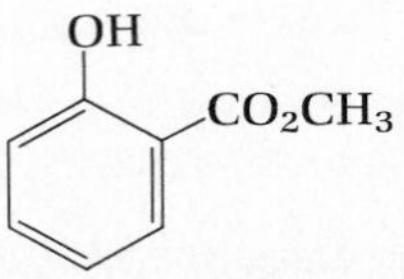   methyl salicylate

Table 18.5

Esters Used as Flavoring Agents

Name	Structure	Flavor
methyl butyrate	$CH_3-O-\overset{\overset{\displaystyle O}{\|\|}}{C}-CH_2CH_2CH_3$	apple
n-pentyl butyrate	$CH_3(CH_2)_4-O-\overset{\overset{\displaystyle O}{\|\|}}{C}-CH_2CH_2CH_3$	apricot
n-pentyl acetate	$CH_3(CH_2)_4-O-\overset{\overset{\displaystyle O}{\|\|}}{C}-CH_3$	banana
n-octyl acetate	$CH_3(CH_2)_7-O-\overset{\overset{\displaystyle O}{\|\|}}{C}-CH_3$	orange
ethyl butyrate	$CH_3CH_2-O-\overset{\overset{\displaystyle O}{\|\|}}{C}-CH_2CH_2CH_3$	pineapple
ethyl formate	$CH_3CH_2-O-\overset{\overset{\displaystyle O}{\|\|}}{C}-H$	rum

Example 18.5

Isobutyl formate has the odor of raspberries. Draw its structural formula.

Solution. The acid portion of the ester is formic acid. The alcohol portion is isobutyl alcohol.

$$H-\overset{\overset{\displaystyle O}{\|\|}}{C}-OH \qquad CH_3-\overset{\overset{\displaystyle CH_3}{\|}}{CH}-CH_2OH$$

The ester can be represented in two ways depending on which component is drawn on the left of the structure. Both structures are identical.

$$H-\overset{\overset{\displaystyle O}{\|\|}}{C}-O-CH_2-\overset{\overset{\displaystyle CH_3}{\|}}{CH}-CH_3 \quad \text{or} \quad CH_3-\overset{\overset{\displaystyle CH_3}{\|}}{CH}-CH_2-O-\overset{\overset{\displaystyle O}{\|\|}}{C}-H$$

[Additional examples may be found in **18.51** at the end of the chapter.]

18.7 Reactions of Esters

Although esters undergo a variety of reactions, only three examples will be presented in this chapter. Two reactions called hydrolysis and saponification are related and involve breaking the ester bond between the acid and alcohol. The second reaction, the Claisen condensation, is like the aldol reaction (Section 17.7), since it involves the reaction of the α-carbon atom and forms a carbon–carbon bond.

Hydrolysis of Esters

The hydrolysis of an ester involves the breaking of the ester bond to produce an acid and an alcohol. The reaction is catalyzed by strong acids. Ester hydrolysis, then, is just the reverse of the esterification reaction.

$$R-\overset{\overset{\textstyle O}{\|}}{C}-O-R' + H_2O \underset{}{\overset{H^+}{\rightleftharpoons}} R-\overset{\overset{\textstyle O}{\|}}{C}-OH + R'OH$$

$$CH_3-\overset{\overset{\textstyle O}{\|}}{C}-O-CH_3 + H_2O \underset{}{\overset{H^+}{\rightleftharpoons}} CH_3-\overset{\overset{\textstyle O}{\|}}{C}-OH + CH_3OH$$

In order to favor the hydrolysis reaction a large excess of water is used. Note that when the ester bond is broken, the water is incorporated into the products.

$$\text{ester} + \text{water} \rightleftharpoons \text{acid} + \text{alcohol}$$

Adding water "pushes" the reaction to the right.

Saponification of Esters

Saponification *is the hydrolysis of an ester bond by a strong base.* An example shows the saponification of methyl acetate.

$$R-\overset{\overset{\textstyle O}{\|}}{C}-OR' + OH^- \rightleftharpoons R-\overset{\overset{\textstyle O}{\|}}{C}-O^- + R'OH$$

$$CH_3-\overset{\overset{\textstyle O}{\|}}{C}-O-CH_3 + OH^- \rightleftharpoons CH_3-\overset{\overset{\textstyle O}{\|}}{C}-O^- + CH_3OH$$

Note that instead of the acetic acid that is produced in the hydrolysis reaction, acetate is formed. Since the reaction mixture is basic, any carboxylic acid formed in the reaction is converted to its conjugate base by loss of a proton to the hydroxide ion.

There is another important distinction between the hydrolysis and the related saponification reactions. Hydrolysis is catalyzed by acid, whereas in the saponification reaction hydroxide is a reagent because equal numbers of moles of hydroxide and ester are required. The hydroxide is consumed in the reaction, and, since it is a strong base, the point of equilibrium is overwhelmingly to the right.

Saponification has long been used to produce soaps from animal fats. The fats are triesters of glycerol and acids containing 12–18 carbon atoms.

$$
\begin{array}{l}
CH_2-O-\overset{\overset{\textstyle O}{\|}}{C}-(CH_2)_{16}CH_3 \\
\ \ \ | \qquad\ \ \overset{\textstyle O}{\|} \\
CH-O-\overset{}{C}-(CH_2)_{16}CH_3 \\
\ \ \ | \qquad\ \ \overset{\textstyle O}{\|} \\
CH_2-O-\overset{}{C}-(CH_2)_{16}CH_3 \\
\qquad\quad \text{tristearin}
\end{array}
\xrightarrow{NaOH,\ H_2O}
\begin{array}{l}
CH_2-OH \\
\ \ | \\
CH-OH \ \ + 3\ CH_3(CH_2)_{16}CO_2^- + 3\ Na^+ \\
\ \ | \\
CH_2-OH \\
\text{glycerol} \qquad\qquad \text{stearate ion}
\end{array}
$$

Example 18.6

Saponification of an extract of apricot followed by acidification gives 1-pentanol and propanoic acid. Write the structure and name of the compound contained in the extract.

Solution. First write the structures of the products. Now in your mind form an ester bond between the oxygen atom of the alcohol and the carbonyl carbon atom of the acid. The other product formed is water.

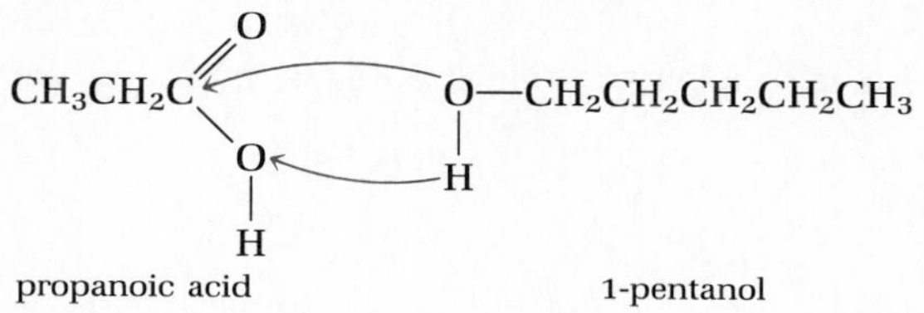

$$CH_3CH_2C \overset{O}{\diagup} \quad O—CH_2CH_2CH_2CH_2CH_3$$

propanoic acid   1-pentanol

$$CH_3CH_2C \overset{O}{\diagup}_{OCH_2(CH_2)_3CH_3}$$

pentyl propanoate

[Additional examples may be found in **18.43–18.50** at the end of the chapter.]

Thioesters

Just as alcohols can be converted into esters, thiols can be converted into thioesters.

$$\underset{\substack{\underbrace{}\\ \text{from} \\ \text{an acid}}}{R—C}\overset{O}{\diagup}\underset{\substack{\underbrace{}\\ \text{from} \\ \text{a thiol}}}{S—R'}$$

Thioesters are extremely reactive, and the thiol group can easily be replaced by an alkoxyl group from an alcohol.

$$R—C\overset{O}{\diagup}S—R' + R''—OH \longrightarrow R—C\overset{O}{\diagup}O—R'' + R'SH$$

One of the most important thioesters is acetyl coenzyme A formed from the thiol group of coenzyme A.

$$CH_3—C\overset{O}{\diagup}S—CoA$$
acetyl coenzyme A

This thioester reacts in certain biological reactions to convert alcohols into acetate esters. For example, choline is acetylated by coenzyme A.

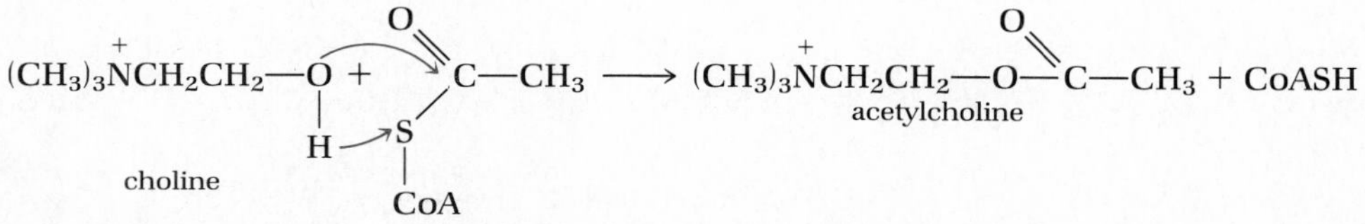

$$(CH_3)_3\overset{+}{N}CH_2CH_2—O + \overset{O}{\diagup}C—CH_3 \longrightarrow (CH_3)_3\overset{+}{N}CH_2CH_2—O\overset{O}{\diagup}C—CH_3 + CoASH$$

choline   acetylcholine

$$\underset{H}{} \quad \underset{CoA}{S}$$

Claisen Condensation

Like aldehydes, esters have an acidic α-hydrogen atom, but their acid dissociation constants are about 10^5 smaller than those of aldehydes. However, sodium ethoxide can remove the proton in an equilibrium reaction.

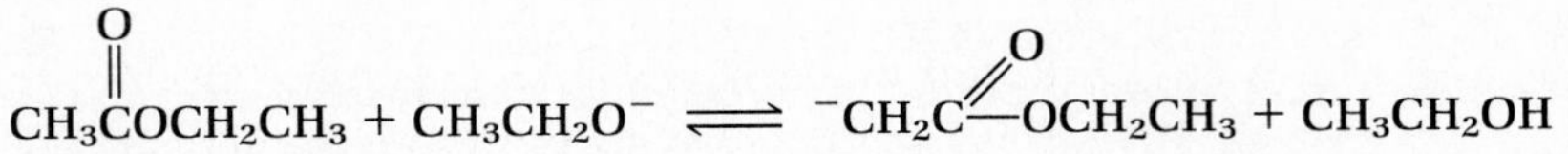

$$CH_3COCH_2CH_3 + CH_3CH_2O^- \rightleftharpoons {}^-CH_2C\!\!-\!\!OCH_2CH_3 + CH_3CH_2OH$$

The conjugate base of the ester can react with another molecule of an ester to form a carbon–carbon bond. However, unlike the aldol product of aldehydes, the addition product of esters is unstable and loses a molecule of alcohol.

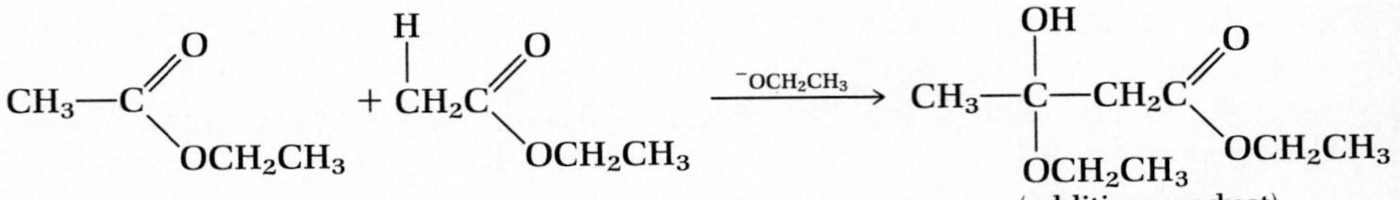

(addition product)

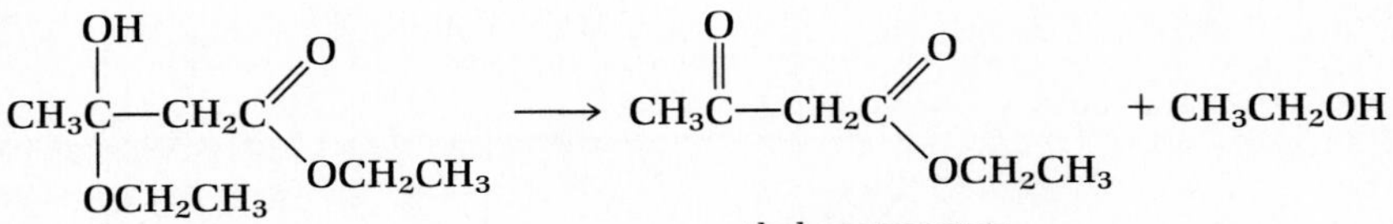

ethyl acetoacetate

Note that since the addition product is a hemiketal, the second step of this reaction sequence occurs because hemiketals are unstable. The addition product, or aldol, formed from an aldehyde is a stable alcohol.

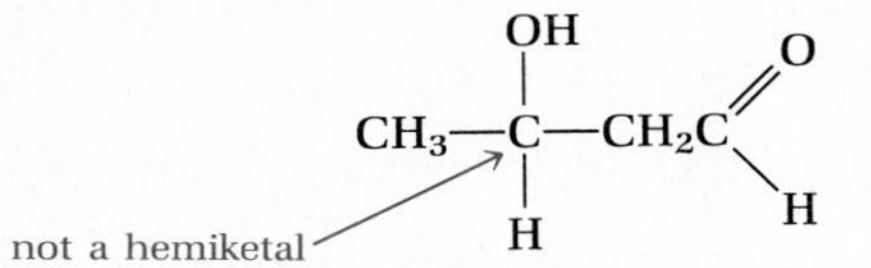

not a hemiketal

The reaction of two moles of an ester to produce a β-keto ester is called the **Claisen condensation.** In nature, Claisen condensations are enzyme catalyzed.

An aldol-type condensation of acetyl coenzyme A is involved in the first step of the citric acid cycle. The acetyl group is transferred to oxaloacetic acid to form a compound that when hydrolyzed yields citric acid.

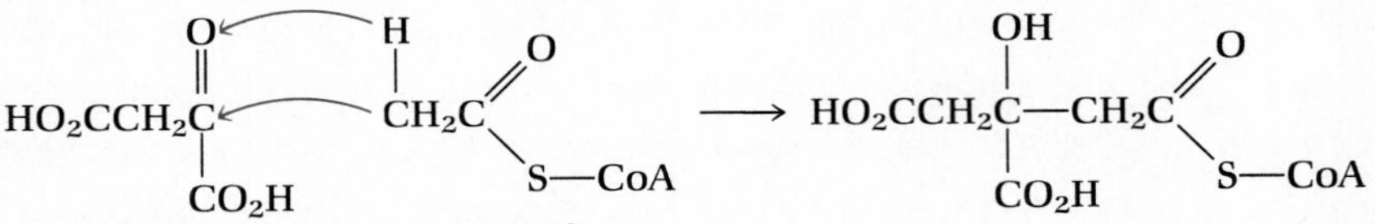

oxaloacetic acid   acetyl coenzyme A

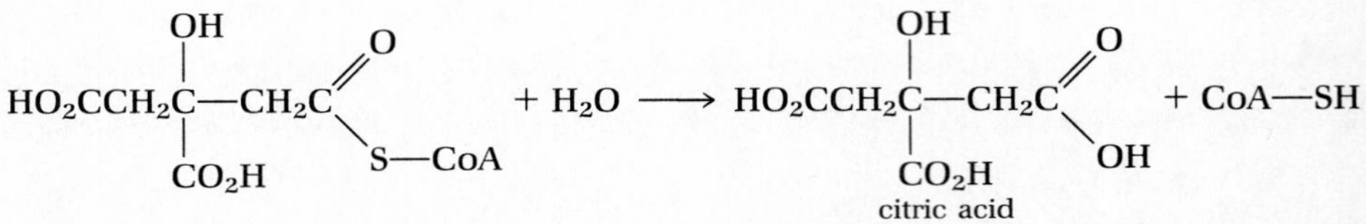

citric acid

In normal metabolism, a large part of acetyl coenzyme A is used to react with oxaloacetic acid to form citric acid. However, in certain illnesses, such as diabetes, the metabolism of fats predominates over the metabolism of carbohydrates (Chapter 27). When there is an insufficient amount of oxaloacetic acid, the acetyl coenzyme A reacts with itself in a Claisen condensation.

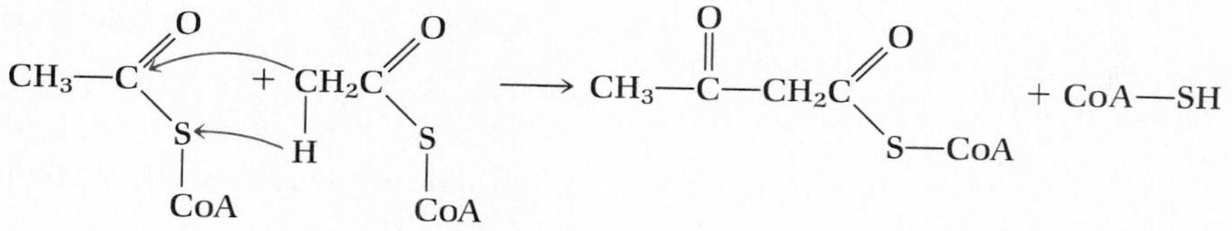

$$CH_3-C\overset{O}{\underset{\underset{CoA}{S}}{\diagup}} + CH_2C\overset{O}{\underset{\underset{CoA}{S}}{\diagup}} \longrightarrow CH_3-\overset{O}{\overset{\|}{C}}-CH_2C\overset{O}{\underset{S-CoA}{\diagup}} + CoA-SH$$

Hydrolysis of the thioester yields acetoacetic acid (3-ketobutanoic acid). Subsequent reactions produce 3-hydroxybutanoic acid and acetone, which are collectively called ketone bodies.

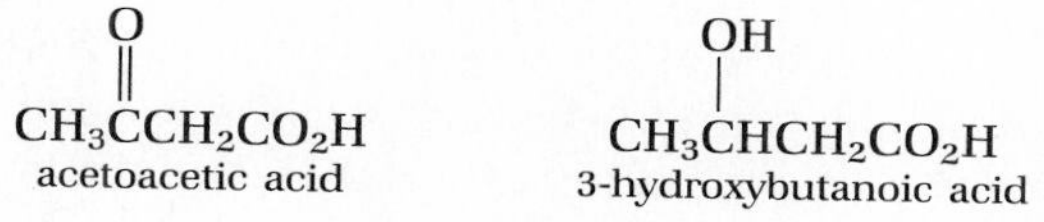

$$\underset{\text{acetoacetic acid}}{CH_3\overset{O}{\overset{\|}{C}}CH_2CO_2H} \qquad \underset{\text{3-hydroxybutanoic acid}}{CH_3\overset{OH}{\overset{\|}{C}H}CH_2CO_2H}$$

There are many commercial products, called condensation polymers, that contain ester bonds. **Condensation reactions** involve the joining of two molecules to form a larger molecule with the elimination of a small molecule such as water. Thus, esterification is a condensation reaction.

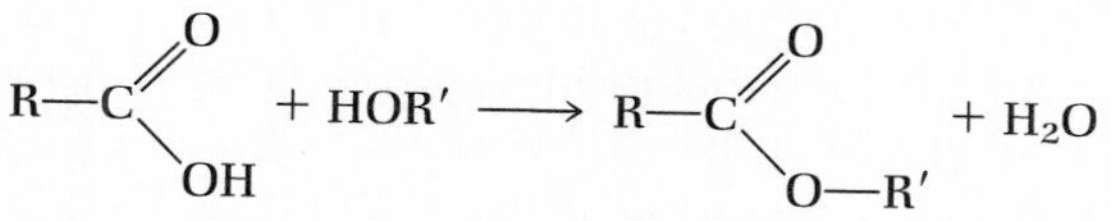

$$R-C\overset{O}{\underset{OH}{\diagup}} + HOR' \longrightarrow R-C\overset{O}{\underset{O-R'}{\diagup}} + H_2O$$

18.8
Polyesters

Condensation polymers *are made by reacting monomers to give large molecules and some small molecule such as water.* Such a polymerization process differs from addition polymerization in which the entire monomer is included in the polymer.

In condensation polymerization to form polyesters, the two monomers are difunctional. Dacron is produced by the reaction of terephthalic acid and ethylene glycol. One step in the reaction sequence is shown below.

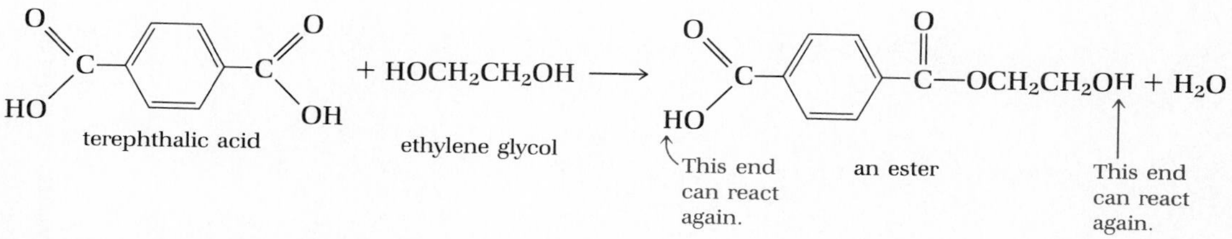

terephthalic acid   ethylene glycol   an ester

This end can react again.   This end can react again.

Although the product looks like a simple ester, it is also both an acid and an alcohol. The acid end of the molecule can react with ethylene glycol, and the alcohol end of the molecule can react with terephthalic acid. After each of these reactions occurs, a larger molecule is formed.

$$HOCH_2CH_2O-\overset{\overset{\displaystyle O}{\|}}{C}-\underset{}{\bigcirc}-\overset{\overset{\displaystyle O}{\|}}{C}-OCH_2CH_2O-\overset{\overset{\displaystyle O}{\|}}{C}-\underset{}{\bigcirc}-\overset{\overset{\displaystyle O}{\|}}{C}\underset{OH}{\diagup} \longrightarrow$$

$$\overset{O}{\underset{HO}{\diagdown}}\overset{\|}{C}-\underset{}{\bigcirc}-\overset{\overset{\displaystyle O}{\|}}{C}-\left[OCH_2CH_2O-\overset{\overset{\displaystyle O}{\|}}{C}-\underset{}{\bigcirc}-\overset{\overset{\displaystyle O}{\|}}{C}\right]_n OCH_2CH_2OH$$

Dacron

This molecule can continue to react to form a high-molecular-weight polyester, which is industrially processed into a fiber called Dacron.

Dacron is used in woven and knitted fabrics in combination with cotton or wool. Because it is physiologically inert, Dacron fabric is also used in the form of a mesh to replace diseased sections of blood vessels.

18.9
Acid Anhydrides

There are three classes of compounds in which an oxygen atom is bonded to a carbonyl carbon atom. Two of these, acids and esters, have already been presented in this chapter. The third type of compound is an acid anhydride.

In acids the carbonyl carbon atom is bonded to —OH, whereas in esters an —OR group is bonded to the carbonyl carbon. **Acid anhydrides** *have the following functional group.*

$$R-\overset{\overset{\displaystyle O}{\|}}{C}-O-\overset{\overset{\displaystyle O}{\|}}{C}-R$$

One carbonyl group is bonded to oxygen, which in turn is bonded to another carbonyl group. The acid anhydride functional group thus is derived from two carboxyl groups.

Acid anhydrides are made by reacting carboxylic acids with P_4O_{10}, a dehydrating agent.

$$R-\overset{\overset{\displaystyle O}{\|}}{C}-OH + HO-\overset{\overset{\displaystyle O}{\|}}{C}-R \xrightarrow{P_4O_{10}} R-\overset{\overset{\displaystyle O}{\|}}{C}-O-\overset{\overset{\displaystyle O}{\|}}{C}-R$$
acid anhydride

$$CH_3-\overset{\overset{\displaystyle O}{\|}}{C}-OH + HO-\overset{\overset{\displaystyle O}{\|}}{C}-CH_3 \xrightarrow{P_4O_{10}} CH_3-\overset{\overset{\displaystyle O}{\|}}{C}-O-\overset{\overset{\displaystyle O}{\|}}{C}-CH_3$$
acetic anhydride

Acid anhydrides react quickly with an alcohol to form an ester and an acid.

$$\underset{\text{cyclohexanol}}{\overset{OH}{\bigcirc}} + CH_3-\overset{\overset{\displaystyle O}{\|}}{C}-O-\overset{\overset{\displaystyle O}{\|}}{C}-CH_3 \longrightarrow \underset{\text{cyclohexyl acetate}}{\overset{O-\overset{\overset{\displaystyle O}{\|}}{C}-CH_3}{\bigcirc}} + CH_3\overset{\overset{\displaystyle O}{\|}}{C}-OH$$

There are counterparts to acid anhydrides of carboxylic acids based on phosphoric acid that are very important in biochemistry. Removal of a molecule of water from two molecules of phosphoric acid forms pyrophosphoric acid.

$$\underset{\text{OH}}{\overset{\text{O}}{\text{HO}-\overset{\|}{\text{P}}-\text{OH}}} + \underset{\text{OH}}{\overset{\text{O}}{\text{HO}-\overset{\|}{\text{P}}-\text{OH}}} \longrightarrow \underset{\text{OH}\quad\text{OH}}{\overset{\text{O}\quad\text{O}}{\text{HO}-\overset{\|}{\text{P}}-\text{O}-\overset{\|}{\text{P}}-\text{OH}}} + \text{H}_2\text{O}$$

Its structure is similar to that of carboxylic acid anhydrides. Three molecules of phosphoric acid can combine to form two anhydride bonds.

anhydride bonds

$$\underset{\underset{\text{triphosphoric acid}}{\text{OH}\quad\text{OH}\quad\text{OH}}}{\overset{\text{O}\quad\text{O}\quad\text{O}}{\text{HO}-\overset{\|}{\text{P}}-\text{O}-\overset{\|}{\text{P}}-\text{O}-\overset{\|}{\text{P}}-\text{OH}}}$$

Phosphoric acid, pyrophosphoric acid, and triphosphoric acid occur in living cells as esters. Some of these structures are shown in Figure 16.4. In Chapter 16, the emphasis was on the chemistry of alcohols and one of their characteristic reactions—that of ester formation. Now we are discussing the same reaction with the emphasis placed on the acid group. Just as alcohols react with carboxylic acids, they also react with phosphoric acid, pyrophosphoric acid, and triphosphoric acid.

ester bond

$$\text{R}-\text{OH} + \underset{\text{OH}}{\overset{\text{O}}{\text{HO}-\overset{\|}{\text{P}}-\text{OH}}} \longrightarrow \underset{\underset{\text{alkyl phosphate}}{\text{OH}}}{\overset{\text{O}}{\text{R}-\text{O}-\overset{\|}{\text{P}}-\text{OH}}} + \text{H}_2\text{O}$$

ester bond

$$\text{R}-\text{OH} + \underset{\text{OH}\quad\text{OH}}{\overset{\text{O}\quad\text{O}}{\text{HO}-\overset{\|}{\text{P}}-\text{O}-\overset{\|}{\text{P}}-\text{OH}}} \longrightarrow \underset{\underset{\text{alkyl diphosphate}}{\text{OH}\quad\text{OH}}}{\overset{\text{O}\quad\text{O}}{\text{R}-\text{O}-\overset{\|}{\text{P}}-\text{O}-\overset{\|}{\text{P}}-\text{OH}}} + \text{H}_2\text{O}$$

ester bond

$$\text{R}-\text{OH} + \underset{\text{OH}\quad\text{OH}\quad\text{OH}}{\overset{\text{O}\quad\text{O}\quad\text{O}}{\text{HO}-\overset{\|}{\text{P}}-\text{O}-\overset{\|}{\text{P}}-\text{O}-\overset{\|}{\text{P}}-\text{OH}}} \longrightarrow \underset{\underset{\text{alkyl triphosphate}}{\text{OH}\quad\text{OH}\quad\text{OH}}}{\overset{\text{O}\quad\text{O}\quad\text{O}}{\text{R}-\text{O}-\overset{\|}{\text{P}}-\text{O}-\overset{\|}{\text{P}}-\text{O}-\overset{\|}{\text{P}}-\text{OH}}} + \text{H}_2\text{O}$$

Note that these esters are also acids. At physiological pH, the protons in the OH groups of these esters are ionized. The phosphate, diphosphate, and triphosphate, which lose 2, 3, and 4 protons, respectively, have -2, -3, and -4 charges, respectively. It is for this reason that these phosphoric acid derivatives are soluble in the aqueous environment of living systems.

Adenosine triphosphate (ATP), a triphosphate ester of adenosine, is produced in the metabolism of fats and carbohydrates and stores some of the energy released in these reactions. When ATP is hydrolyzed in the presence of enzymes, adenosine diphosphate (ADP) is produced and 7.3 kcal of energy is released per mole of ADP produced. These reactions will be discussed in Chapter 25.

$$\text{adenosine} - O - \overset{\overset{\displaystyle O}{\|}}{\underset{\underset{\displaystyle O^-}{|}}{P}} - O - \overset{\overset{\displaystyle O}{\|}}{\underset{\underset{\displaystyle O^-}{|}}{P}} - O - \overset{\overset{\displaystyle O}{\|}}{\underset{\underset{\displaystyle O^-}{|}}{P}} - O^- \ + H_2O \ \longrightarrow$$

ATP

$$\text{adenosine} - O - \overset{\overset{\displaystyle O}{\|}}{\underset{\underset{\displaystyle O^-}{|}}{P}} - O - \overset{\overset{\displaystyle O}{\|}}{\underset{\underset{\displaystyle O^-}{|}}{P}} - OH + H - O - \overset{\overset{\displaystyle O}{\|}}{\underset{\underset{\displaystyle O^-}{|}}{P}} - O^-$$

ADP

Summary

The structural characteristic of carboxylic acids is the carboxyl group, $-CO_2H$. Carboxylic acids are represented as RCO_2H or $RCOOH$. The R group may contain other structural features such as double bonds, hydroxyl groups, or carbonyl groups.

The carboxyl group is very polar and forms hydrogen bonds to itself or other molecules. Strong hydrogen bonds to itself account for the high boiling point compared to ethers, alcohols, aldehydes, and ketones. Since carboxylic acids can hydrogen bond to water, they are very soluble in water. Carboxylic acids react with strong bases to form carboxylate salts. Carboxylic acids are only slightly ionized in water. The equilibrium constant is given the symbol K_a and is called the acid dissociation constant. The pK_a of an acid is equal to $-\log K_a$. Weak acids have small K_a values and large pK_a values.

Esters are formed by the reaction of an acid with an alcohol and have the general formula RCO_2R or RCOOR. Esters are polar compounds, but they cannot form hydrogen bonds to themselves. Consequently, their boiling points are lower than alcohols and acids of similar molecular weight.

The hydrolysis of esters occurs in the presence of a strong acid catalyst to give an equilibrium mixture. The reaction of an ester with a strong base results in saponification to yield an alcohol and a carboxylate salt.

An ester can undergo a condensation reaction. The Claisen condensation results in the formation of a β-keto ester.

Acid anhydrides of phosphoric acid are important in the chemistry of the cell. ATP, a triphosphate ester of adenosine, contains anhydride bonds that store biochemical energy.

Learning Objectives

As a result of studying Chapter 18 you should be able to
- Assign common and IUPAC names to carboxylic acids.
- Explain how hydrogen bonding affects the properties of carboxylic acids.
- Describe the effect of structure on the acidity of carboxylic acids.
- Assign common and IUPAC names to salts of carboxylic acids.
- Assign common and IUPAC names to esters.
- Describe the differences and similarities in the hydrolysis and saponification of esters.
- Write the product of the Claisen condensation of an ester.
- Write the structure of a polyester formed from a diol and a dicarboxylic acid.
- Draw the structures of phosphoric, pyrophosphoric, and triphosphoric acids and their esters.

New Terms

An **acid anhydride** is formed by loss of water in the reaction of two molecules of an acid. Acid anhydrides of carboxylic acids have the functional group

$$-\overset{\overset{\displaystyle O}{\|}}{C} - O - \overset{\overset{\displaystyle O}{\|}}{C} -$$

The **acyl group**, represented by $R-\overset{\diagup O}{C}-$, is contained as part of the structure of acid derivatives such as esters.
A **carboxyl group**, represented by $-CO_2H$, is the functional group of carboxylic acids.

A **carboxylate group** is the anion formed by loss of a proton from a carboxylic acid and is represented by $-CO_2^-$.

A **condensation polymer** is made by reacting monomers to give a polymer and some small molecule such as water.

An **ester** is formed by the reaction of an alcohol and an acid and is represented by RCO_2R.

The **Henderson–Hasselbalch equation** relates the pH of a solution to the pK_a of an acid and the concentration of the acid and its conjugate base.

Saponification is the hydrolysis of an ester bond by a strong base.

Thioesters are esters of thiols and are represented by

$$RC\overset{O}{\underset{}{\diagup}}SR.$$

Questions and Problems

Terminology

18.1 Explain the structural differences among an aldehyde, a ketone, and a carboxylic acid.

18.2 How do an acid and an ester differ structurally?

18.3 What is an acyl group?

18.4 What is an acid anhydride?

Nomenclature of Acids

18.5 Give the common name for each of the following acids.

(a) $CH_3CH_2CO_2H$ (b) HCO_2H

(c) $CH_3CH_2CH_2CO_2H$ (d) CH_3CO_2H

18.6 Give the IUPAC name for each of the acids in **18.5**.

18.7 Give the common name for each of the following acids.

(a) $CH_3(CH_2)_3CO_2H$ (b) $(CH_3)_2CHCO_2H$

(c) $CH_3\underset{\underset{CH_2CH_3}{|}}{C}HCH_2CO_2H$ (d) $CH_3\underset{\underset{CH_3}{|}}{C}HCH_2\underset{\underset{CH_3}{|}}{C}HCO_2H$

(e) $CH_3\underset{\underset{CH_2CH_2CH_3}{|}}{C}HCH_2CO_2H$ (f) $CH_3\underset{\underset{CH_3}{|}}{\overset{\overset{CH_3}{|}}{C}}CH_2\underset{\underset{CH_2}{|}\\\underset{CH_3}{|}}{C}HCH_2CO_2H$

18.8 Give the IUPAC name for each of the acids in **18.7**.

18.9 Give the common name for each of the following acids.

(a) $CH_3\underset{\underset{Cl}{|}}{C}HCH_2CO_2H$ (b) $Br\underset{\underset{CH_3}{|}}{C}HCO_2H$

(c) $CH_3\underset{\underset{Br}{|}}{\overset{\overset{CH_3}{|}}{C}}HCHCH_2CO_2H$ (d) $CH_3\underset{\underset{CH_3}{|}}{\overset{\overset{CH_3}{|}}{C}}CH_2\underset{\underset{}{}}{\overset{\overset{Cl}{|}}{C}}HCO_2H$

18.10 Give the IUPAC name for each of the acids in **18.9**.

18.11 Give the common name for each of the following acids.

(a) $CH_3\underset{\underset{OH}{|}}{C}HCO_2H$ (b) $CH_3CH_2\underset{\underset{OH}{|}}{C}HCO_2H$

(c) $HOCH_2CH_2COOH$

18.12 Give the IUPAC name for each of the acids in **18.11**.

Polyfunctional Acids

18.13 Write a formula for an example of each of the following.

(a) α-hydroxy acid   (b) α-keto acid

(c) β-keto acid   (d) unsaturated acid

(e) dicarboxylic acid

18.14 Name each of the following compounds, which are important in biochemical reactions.

(a) $HO_2CCH_2CH_2CH_2CO_2H$

(b) $HO_2CCH_2CH_2CO_2H$

(c) $CH_3\overset{\overset{O}{\|}}{C}CO_2H$

(d) $HO_2CCH_2\overset{\overset{O}{\|}}{C}CO_2H$

(e) $\underset{H}{\overset{HO_2C}{\diagdown}}C=C\underset{\diagdown CO_2H}{\diagup H}$

(f) $CH_3\underset{\underset{OH}{|}}{C}HCO_2H$

Properties of Acids

18.15 List each group of compounds in order of increasing boiling point.

(a) butanal, 1-butanol, butanoic acid

(b) hexane, hexanoic acid, adipic acid

18.16 What is the structure of the dimer of acetic acid?

18.17 Why is 1-butanol less soluble in water than butanoic acid?

18.18 Adipic acid is much more soluble in water than hexanoic acid. Explain this order of solubilities.

Acidity

18.19 Explain why butyric acid is a much stronger acid than 1-butanol.

18.20 Which is the stronger acid of each pair?

(a) chloroacetic acid or bromoacetic acid

(b) α,α-dichlorobutyric acid or α-chlorobutyric acid

18.21 The acid dissociation constant of α-bromobutyric acid is larger than the acid dissociation constant of butyric acid. However, the acid dissociation constants of butyric acid and γ-bromobutyric acid are very similar. Explain why.

18.22 The acid dissociation constants of β-ketobutyric acid and butyric acid are 2.6×10^{-4} and 1.5×10^{-5}, respectively. Suggest a reason for this order.

18.23 Describe a simple chemical test that could be used to distinguish between octanoic acid and butyl butanoate.

18.24 The K_a values of formic acid and acetic acid are 1.8×10^{-4} and 1.8×10^{-5}, respectively. Which compound is the stronger acid?

18.25 The pK_a values of acetic acid and benzoic acid are 4.74 and 4.19, respectively. Which substance is the stronger acid?

Salts of Carboxylic Acids

18.26 Write an equation for the reaction of each of the following acids with sodium hydroxide.
(a) acetic acid   (b) stearic acid
(c) adipic acid   (d) lactic acid

18.27 Give the common and IUPAC names for each of the following.
(a) $CH_3CH_2CH_2CO_2Na$ (b) $CH_3(CH_2)_{16}CO_2K$
(c) $(CH_3CH_2CO_2)_2Ca$

18.28 Write a balanced equation for the reaction of each of the compounds in **18.27** with hydrochloric acid.

18.29 Carboxylic acids have strong odors, but their corresponding sodium salts are odorless. Why?

18.30 Write the formula of each of the following compounds.
(a) sodium butyrate   (b) potassium propanoate
(c) lithium stearate   (d) calcium acetate

Molecular Formulas

18.31 What is the general molecular formula for carboxylic acids.

18.32 How many isomeric acids have the molecular formula $C_4H_8O_2$?

18.33 How many isomeric esters have the molecular formula $C_4H_8O_2$?

18.34 There are nine isomeric esters of the molecular formula $C_5H_{10}O_2$. Write a structure for each compound.

Nomenclature of Esters

18.35 Write the structure of each of the following compounds.
(a) octyl acetate   (b) t-butyl formate
(c) ethyl butyrate   (d) propyl pentanoate
(e) methyl pyruvate   (f) dimethyl glutarate

18.36 Give the common name for each of the following compounds.

(a) $CH_3CH_2-O-\overset{\overset{\displaystyle O}{\|}}{C}-H$

(b) $CH_3-O-\overset{\overset{\displaystyle O}{\|}}{C}-CH_2CH_2CH_3$

(c) $CH_3(CH_2)_7-O-\overset{\overset{\displaystyle O}{\|}}{C}-CH_3$

(d) $CH_3CH_2-O-\overset{\overset{\displaystyle O}{\|}}{C}-CH_2CH_2CH_3$

(e) $CH_3(CH_2)_4-O-\overset{\overset{\displaystyle O}{\|}}{C}-CH_2CH_2CH_3$

(f) $CH_3(CH_2)_4-O-\overset{\overset{\displaystyle O}{\|}}{C}-CH_3$

18.37 Give the IUPAC name for each of the compounds in **18.36**.

18.38 Give the common name for each of the following compounds.

(a)

(b)

(c)

(d)

(e)

(f)

18.39 Give the IUPAC name for each of the compounds in **18.38**.

18.40 Give the common name for each of the following compounds.
(a) $CH_3CH_2CH_2CO_2CH_2CH_2CH_2CH_3$
(b) $CH_3CO_2CH_2(CH_2)_3CH_3$
(c) $CH_3CO_2CH_2(CH_2)_6CH_3$
(d) $CH_3CH_2CH_2CO_2CH_3$

18.41 Give the IUPAC name for each of the compounds in **18.40**.

18.42 Give the common name of each acyl group.

(a) $CH_3\overset{\overset{\displaystyle O}{\|}}{C}-$ (b) $CH_3CH_2-\overset{\overset{\displaystyle O}{\|}}{C}-$

(c) $H-\overset{O}{\underset{\|}{C}}-$

(d) $\bigcirc-\overset{O}{\underset{\|}{C}}-$

(e) $CH_3CH_2CH_2\overset{O}{\underset{\|}{C}}-$

(f) $-\overset{O}{\underset{\|}{C}}-H$

(g) $-\overset{O}{\underset{\|}{C}}-CH_2CH_2CH_3$

(h) $-\overset{O}{\underset{\|}{C}}-CH_3$

Reactions of Esters

18.43 Write the products of hydrolysis of each of the following esters.

(a) $CH_3-\overset{O}{\underset{\|}{C}}-O-\bigcirc$

(b) $\bigcirc-\overset{O}{\underset{\|}{C}}-O-CH_2CH_3$

(c) $H-\overset{O}{\underset{\|}{C}}-O-CH_2CH_3$

(d) $CH_3-\overset{O}{\underset{\|}{C}}-O-CH_2CH_2CH_3$

(e) $CH_3CH_2CH_2CH_2-\overset{O}{\underset{\|}{C}}-O-CH_2CH_2CH_2CH_3$

(f) $CH_3CH_2-\overset{O}{\underset{\|}{C}}-O-CH_2CH_3$

18.44 Write the products of hydrolysis of each of the following esters.

(a) $CH_3CH_2CH_2-\overset{O}{\underset{\|}{C}}-O-CH_2CH_2CH_2CH_3$

(b) $CH_3CH_2-\overset{O}{\underset{\|}{C}}-O-CH_2CH_2CH_3$

(c) $CH_3-\overset{O}{\underset{\|}{C}}-O-CH_2CH_3$

(d) $CH_3-\overset{O}{\underset{\|}{C}}-O-CH_2CH_2CH_3$

(e) $CH_3\overset{CH_3}{\underset{|}{C}}HCH_2-O-\overset{O}{\underset{\|}{C}}-H$

(f) $CH_3CH_2CH_2-O-\overset{O}{\underset{\|}{C}}-CH_3$

(g) $\bigcirc-CH_2CH_2CH_2-\overset{O}{\underset{\|}{C}}-O-CH_2CH_3$

(h) $\bigcirc-\overset{O}{\underset{\|}{C}}-O-CH_2CH_3$

18.45 Write the products of the saponification of each of the compounds in **18.43** with NaOH.

18.46 Write the products of the saponification of each of the compounds in **18.44** with NaOH.

18.47 Explain why the saponification of an ester is a more complete reaction than the hydrolysis of the same ester.

18.48 Write the products of the saponification of each of the following esters.
(a) dimethyl succinate   (b) methyl lactate
(c) dimethyl fumarate   (d) diethyl adipate

18.49 Write an equation for the reaction of three moles of palmitic acid with glycerol.

18.50 Write an equation for the saponification of glyceryl tristearate.

18.51 What alcohol and acid are required to form each of the following compounds?

(a) $CH_3CH_2-O-\overset{O}{\underset{\|}{C}}-H$

(b) $CH_3CH_2CH_2CH_2CH_2-O-\overset{O}{\underset{\|}{C}}-CH_3$

(c) $CH_3-\overset{O}{\underset{\|}{C}}-O-CH_2CH_3$

(d) $CH_3CH_2CH_2CH_2-O-\overset{O}{\underset{\|}{C}}-CH_2CH_2CH_3$

(e) $CH_3-O-\overset{O}{\underset{\|}{C}}-CH_2CH_2CH_3$

(f) $CH_3CH_2-O-\overset{O}{\underset{\|}{C}}-CH_2CH_2CH_3$

18.52 Write the structure of the ester that can be formed from each acid and alcohol.
(a) $CH_3-CO_2H + CH_3-CH_2-OH$
(b) $CH_3-CH_2-CO_2H + CH_3-CH_2-CH_2-OH$

(c) $\bigcirc-CO_2H + CH_3-OH$

(d) $CH_3-CO_2H + \bigcirc-OH$

Properties of Esters

18.53 Explain why carboxylic acids are more soluble in water than esters of the same molecular weight.

18.54 Explain why esters have lower boiling points than acids of the same molecular weight.

18.55 Which of the following is more soluble in water?

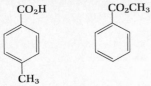

18.56 Arrange the following compounds according to increasing boiling point.
(a) $CH_3CH_2CH_2CH_2OH$ (b) $CH_3CH_2CO_2H$
(c) $CH_3CO_2CH_3$

The Claisen Condensation

18.57 What ester is required to form each of the following by a Claisen condensation?

(a) $CH_3-\overset{\overset{O}{\|}}{C}-CH_2-CO_2CH_3$

(b) $CH_3CH_2\overset{\overset{O}{\|}}{C}CHCO_2CH_2CH_3$
$\qquad\qquad\quad \overset{|}{CH_3}$

(c) $CH_2-\overset{\overset{O}{\|}}{C}-CH-CO_2CH_3$

(d) $CH_3-\overset{\overset{O}{\|}}{C}-CH_2-CO_2CH_2CH_3$

18.58 Write the product of the Claisen condensation of each of the following compounds.
(a) methyl acetate   (b) ethyl propanoate
(c) propyl ethanoate

Polyesters

18.59 List some dicarboxylic acids that could be used to produce polyesters.

18.60 Could glycerol be used to produce polyesters?

18.61 Write the structure of the polymer formed from terephthalic acid and 1,3-propanediol.

18.62 What monomers are needed to produce the following polymer?

$$-O(CH_2)_4O\overset{\overset{O}{\|}}{C}(CH_2)_3C\overset{\overset{O}{\|}}{}O(CH_2)_4O\overset{\overset{O}{\|}}{C}(CH_2)_3\overset{\overset{O}{\|}}{C}-$$

Acid Anhydrides

18.63 Write the structure of the anhydride formed by reacting each of the following acids with P_4O_{10}.
(a) propanoic acid   (b) benzoic acid
(c) butyric acid

18.64 Reaction of succinic acid with P_4O_{10} produces a compound $C_4H_4O_3$. Write the structure of the compound.

18.65 Write the structure of the reaction product of one mole of methanol with each of the following.
(a) phosphoric acid   (b) pyrophosphoric acid
(c) triphosphoric acid

Amines and Amides 19

Nylon is a polyamide of such structural strength that it is used in parachutes.

19.1
Organic Nitrogen Compounds

In the preceding chapters you were introduced to organic compounds composed only of carbon, hydrogen, and oxygen. Now we will consider the organic compounds containing nitrogen, the fourth most abundant element in the human body.

Nitrogen, a member of Group V of the periodic table, has five valence electrons. Three covalent bonds to other atoms are then necessary to achieve a stable octet of electrons about nitrogen. Nitrogen forms strong bonds to both carbon and hydrogen. As a result, there are many different structures containing nitrogen as part of a functional group. Bonding arrangements include (a) three single bonds, (b) a single and a double bond, and (c) a triple bond. Some examples of structures containing these bond types are given in Table 19.1. In this chapter the emphasis will be on compounds containing nitrogen with three single bonds and, to a lesser extent, compounds with a single and a double bond. Nitriles, which contain a carbon–nitrogen triple bond, will not be discussed.

Nitrogen compounds occur widely in nature in both plants and animals. Most are highly active in a physiological sense. Many vitamins contain nitrogen and, in fact, the origin of the name (vital amines) implies the presence of nitrogen as an amine functional group. However, it is now known that not all vitamins contain nitrogen.

The effects of nitrogen compounds on the brain, spinal cord, and nerves can be quite pronounced. Some of the compounds that you may be familiar with include the neurotransmitters dopamine, epinephrine, and serotonin. Parkinson's disease is the result of dopamine deficiency. Epinephrine, commonly called adrenaline, stimulates the conversion of stored glycogen into glucose. Serotonin deficiency is responsible for some forms of mental depression. Hallucinogens, such as mescaline, LSD, and STP, contain nitrogen. Opiates, such as morphine, heroin, and codeine, are also nitrogen compounds.

Nitrogen is contained in high-molecular-weight molecules called proteins, one of the classes of compounds that are the molecular basis of life. The amine functional group of one α-amino acid reacts with the carboxyl group of another amino acid to form an amide bond. This chemistry will be presented in detail in Chapter 23. DNA (deoxyribonucleic acid) and RNA (ribonucleic acid), which are involved in the transmission of genetic information, contain nitrogen (Chapter 29).

Table 19.1

Types of Bonds in Nitrogen Compounds

Functional group	Class of compound
R—NH$_2$	amine
R—C(=O)—NH$_2$	amide
R—C(R)=NH	imine
R—C≡N	nitrile

19.2
Amines

Amines *are organic derivatives of ammonia in which one or more of the hydrogen atoms is replaced by an alkyl or aromatic group.* Thus, amines are organic bases. Amines are called primary, secondary, or tertiary depending on the number of groups attached to nitrogen. Abbreviations for these terms are 1°, 2°, and 3°, respectively.

ammonia   primary amine   secondary amine   tertiary amine

This system of classifying amines is different than that previously used for alcohols. In alcohols, the classification is based on the groups attached to the carbon atom bearing the hydroxyl group. In amines, the focus is on the nitrogen atom

Figure 19.1
Structure of Methylamine.

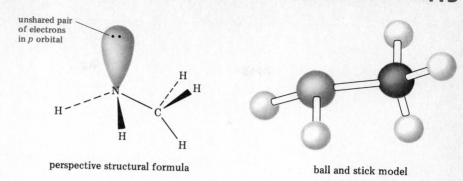

perspective structural formula   ball and stick model

and the number of groups attached to it. For example, *t*-butylamine has a *t*-butyl group, but the amine is primary since it has only one alkyl group bonded to the nitrogen atom. In *t*-butyl alcohol, the alcohol is classified as tertiary because the OH-bearing carbon atom has three alkyl groups.

tertiary
carbon atom   primary amine

$$CH_3-\overset{\overset{\displaystyle CH_3}{|}}{\underset{\underset{\displaystyle CH_3}{|}}{C}}-\overset{\displaystyle}{\underset{\underset{\displaystyle H}{|}}{N}}-H$$

t-butylamine
(a primary amine)

tertiary
carbon atom   tertiary
                           alcohol

$$CH_3-\overset{\overset{\displaystyle CH_3}{|}}{\underset{\underset{\displaystyle CH_3}{|}}{C}}-OH$$

t-butyl alcohol
(a tertiary alcohol)

Recall from Chapter 6 that ammonia is a pyramidal molecule. Nitrogen is sp^3 hybridized, and the three bonds of ammonia are directed to three of the corners of a tetrahedron. The nonbonded electron pair of the nitrogen atom is in a sp^3 orbital directed toward the fourth corner of the tetrahedron.

In the simplest amine, methylamine, CH_3NH_2, one hydrogen atom of ammonia is replaced by a methyl group. The shape of the compound is given in Figure 19.1. The H—N—H and H—N—C bond angles are close to 109.5°. Other primary amines as well as secondary and tertiary amines have similar bond angles.

Amines containing nitrogen as part of a complex ring system are quite common in nature. *Compounds that have one or more atoms in the ring other than carbon are called* **heterocyclic compounds.** One example of a simple nitrogen heterocyclic compound or heterocycle is piperidine.

piperidine
(a secondary amine)

Piperidine, a secondary amine, is structurally similar to cyclohexane. The replacement of an sp^3-hybridized carbon atom by an sp^3-hybridized nitrogen atom results in a compound that also exists in a chair conformation.

Example 19.1 ─────────────────────────────────

The synthetic narcotic analgesic Demerol has the following structure. Is this a primary, secondary, or tertiary amine?

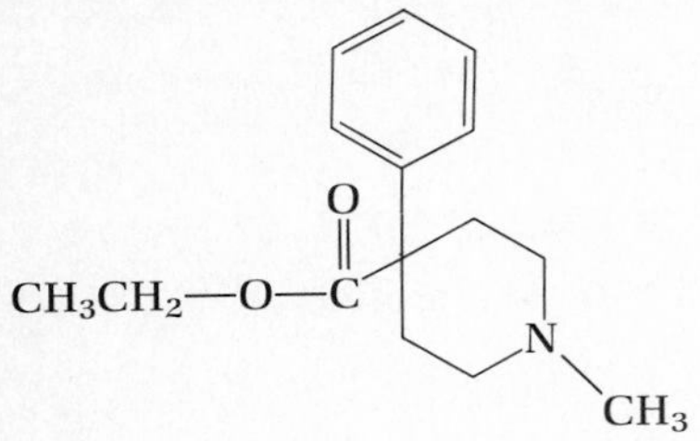

$$CH_3CH_2-O-\overset{\overset{\displaystyle O}{\|}}{C}$$

Solution. There are three carbon atoms bonded to nitrogen. Two of the carbon atoms are in the heterocyclic ring. The third carbon atom is the methyl group. The compound is a tertiary amine. Note that the compound contains a piperidine ring.

─────────────────────────────────

[Additional examples may be found in **19.3–19.12** at the end of the chapter.]

19.3
Nomenclature of Amines

In the common system of nomenclature, amines are named by prefixing the names of the attached alkyl groups to the word amine. The entire name is written as one word. When two or more identical alkyl groups are present, the prefixes di- and tri- are used. Some examples illustrate this method.

$$CH_3CH_2NH_2 \qquad (CH_3)_2NH \qquad CH_3CH_2CH_2\overset{\overset{\displaystyle CH_3}{|}}{N}CH_2CH_3$$

ethylamine   dimethylamine   ethylmethylpropylamine

In the IUPAC system, amines are named as hydrocarbons and the location of the functional group is indicated. The —NH_2 group, called the amino group, appears in primary amines (Figure 19.2). In 2° and 3° amines, the largest alkyl group is considered the parent structure. The name of the N-alkylamino (—NHR) or the N,N-dialkylamino group (—NRR′) is then prefixed to the name of the hydrocarbon. The capital N indicates that the alkyl group is attached to nitrogen and not to the parent hydrocarbon. Some examples are given in Figure 19.2.

Example 19.2 ─────────────────────────────────

Write the common and IUPAC names for the following amine.

$$CH_3\overset{\overset{\displaystyle CH_3}{|}}{CH}CH_2NHCH_3$$

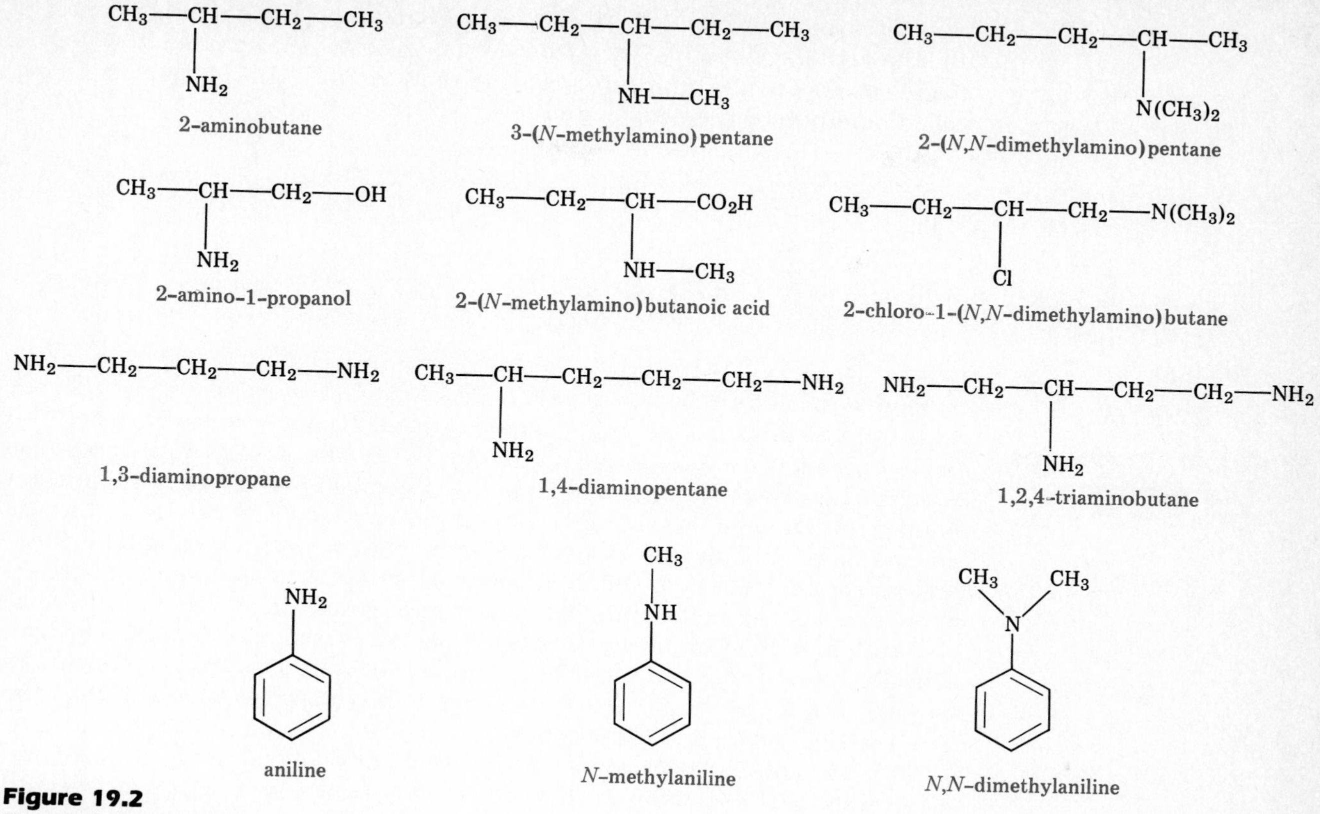

$$CH_3-CH-CH_2-CH_3$$

$$|$$

$$NH_2$$

2-aminobutane

$$CH_3-CH_2-CH-CH_2-CH_3$$

$$|$$

$$NH-CH_3$$

3-(*N*-methylamino)pentane

$$CH_3-CH_2-CH_2-CH-CH_3$$

$$|$$

$$N(CH_3)_2$$

2-(*N,N*-dimethylamino)pentane

$$CH_3-CH-CH_2-OH$$

$$|$$

$$NH_2$$

2-amino-1-propanol

$$CH_3-CH_2-CH-CO_2H$$

$$|$$

$$NH-CH_3$$

2-(*N*-methylamino)butanoic acid

$$CH_3-CH_2-CH-CH_2-N(CH_3)_2$$

$$|$$

$$Cl$$

2-chloro-1-(*N,N*-dimethylamino)butane

$$NH_2-CH_2-CH_2-CH_2-NH_2$$

1,3-diaminopropane

$$CH_3-CH-CH_2-CH_2-CH_2-NH_2$$

$$|$$

$$NH_2$$

1,4-diaminopentane

$$NH_2-CH_2-CH-CH_2-CH_2-NH_2$$

$$|$$

$$NH_2$$

1,2,4-triaminobutane

NH₂ / aniline

CH₃ NH / *N*-methylaniline

CH₃ CH₃ N / *N,N*-dimethylaniline

Figure 19.2
Nomenclature of Amines.

Solution. The two alkyl groups in this secondary amine are methyl and isobutyl.

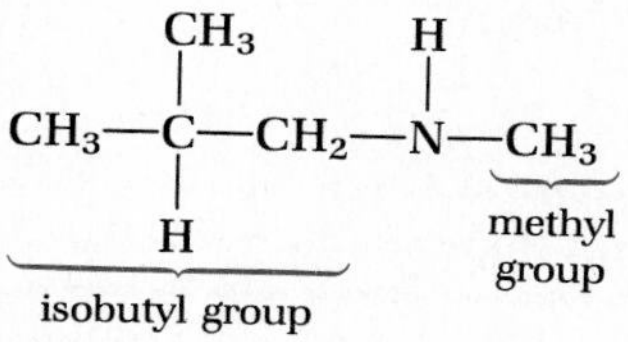

$$CH_3-\underset{\underset{H}{|}}{\overset{\overset{CH_3}{|}}{C}}-CH_2-\underset{\overset{|}{H}}{N}-CH_3$$

isobutyl group   methyl group

The common name is isobutylmethylamine.

The parent hydrocarbon used in the IUPAC name is propane. There are methyl and *N*-methylamino groups substituted on this chain. Proper numbering gives 1-(*N*-methylamino)-2-methylpropane.

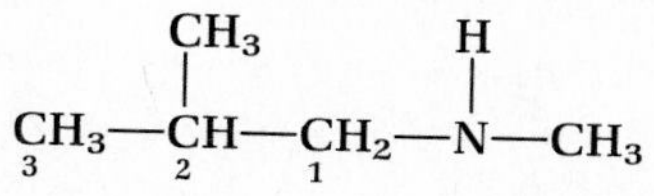

$$\underset{3}{CH_3}-\underset{2}{\overset{\overset{CH_3}{|}}{CH}}-\underset{1}{CH_2}-\underset{\overset{|}{H}}{N}-CH_3$$

[Additional examples may be found in **19.13–19.16** at the end of the chapter.]

There are many naturally occurring compounds in which nitrogen is part of an aromatic ring. In each of the examples shown below, the numbering system is based on assigning a nitrogen atom the number one. These compounds are called heterocyclic aromatic amines.

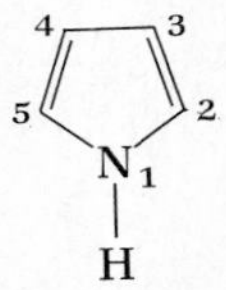

pyrrole

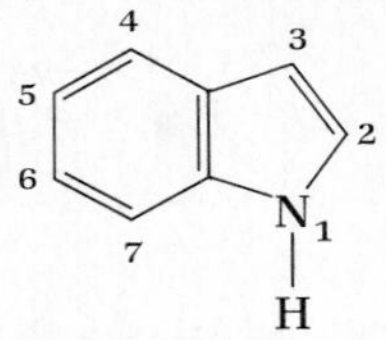

indole

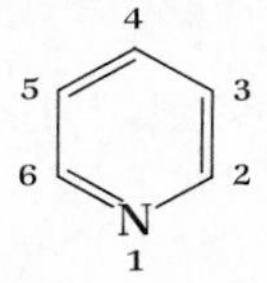

pyridine

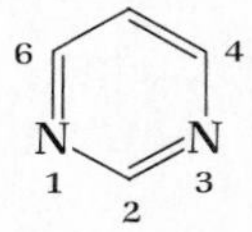

pyrimidine

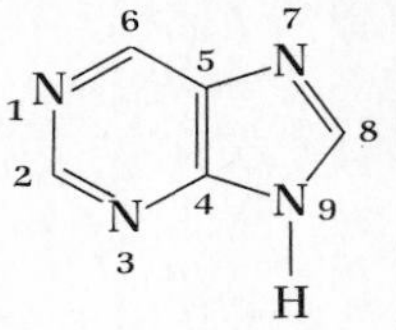

purine

19.4
Physical Properties of Amines

At room temperature the low-molecular-weight amines are gases (Table 19.2), but with increasing molecular weight, amines are liquids and eventually solids. The boiling points of amines are higher than those of alkanes of similar molecular weight. This phenomenon is due in part to the polarity of the carbon–nitrogen bond, but hydrogen bonding is a more important factor (Figure 19.3). Tertiary amines have no hydrogen atoms bonded to the nitrogen atom, and so these amines cannot form hydrogen bonds to each other. As a consequence, tertiary amines have lower boiling points than isomeric primary and secondary amines.

Amines have lower boiling points than alcohols of similar molecular weight. This difference is due to the smaller electronegativity of nitrogen as compared to oxygen. As a result the N—H$\cdots$N hydrogen bond in amines is weaker than the O—H$\cdots$O hydrogen bond in alcohols.

The low-molecular-weight amines with five or fewer carbon atoms are soluble in water. This solubility is due to hydrogen bonds between the amine and water (Figure 19.4). Even tertiary amines are soluble in water because the electron pair of nitrogen may form a hydrogen bond with the hydrogen atoms of water.

Low-molecular-weight amines have sharp penetrating odors similar to ammonia. Higher-molecular-weight compounds smell like decaying fish. In fact, two compounds responsible for the odor of decaying animal tissue are putrescine and cadaverine. These compounds have appropriate common names.

$$NH_2CH_2CH_2CH_2CH_2NH_2$$
putrescine

$$NH_2CH_2CH_2CH_2CH_2CH_2NH_2$$
cadaverine

Table 19.2 _____

Boiling Points of Some Representative Amines

Name	Structure	Boiling point (°C)
ammonia	NH_3	-33
methylamine	CH_3NH_2	-6
ethylamine	$C_2H_5NH_2$	17
propylamine	$n\text{-}C_3H_7NH_2$	49
butylamine	$n\text{-}C_4H_9NH_2$	77
t-butylamine	$t\text{-}C_4H_9NH_2$	44
dimethylamine	$(CH_3)_2NH$	7
trimethylamine	$(CH_3)_3N$	3
aniline	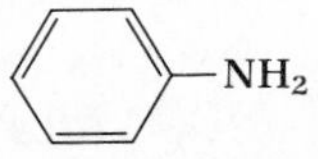 $\langle \rangle$—NH_2	184

Figure 19.3
Hydrogen Bonding in Amines.

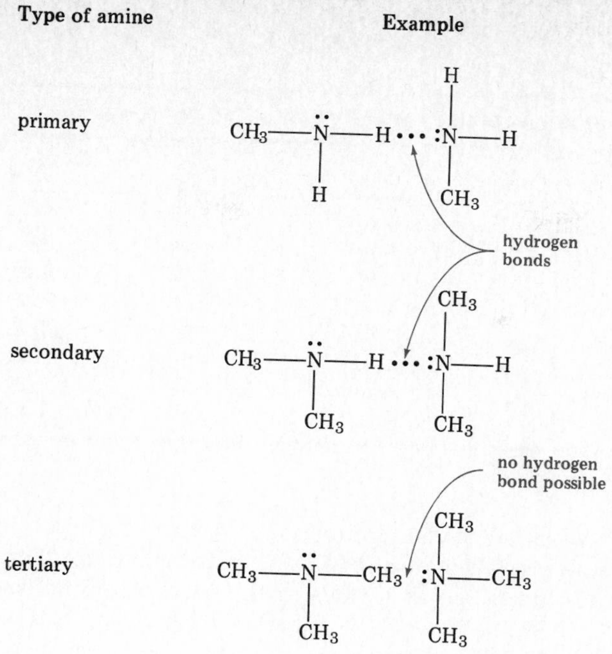

Figure 19.4
Hydrogen Bonds Between Water and Amines.

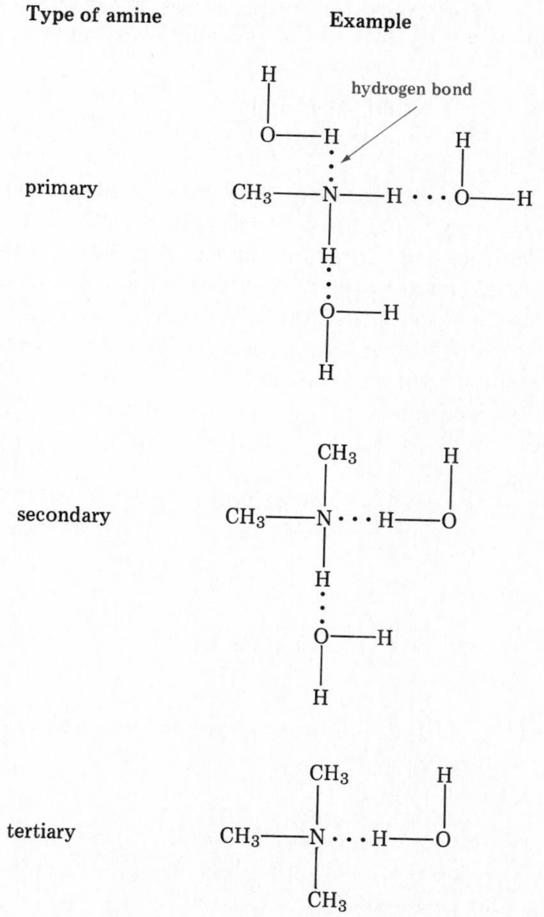

Table 19.3

Basicity of Amines and Acidity of Ammonium Salts

	K_b	K_a	pK_b	pK_a
ammonia	1.8×10^{-5}	5.5×10^{-10}	4.74	9.26
methylamine	4.6×10^{-4}	2.2×10^{-11}	3.34	10.66
ethylamine	4.8×10^{-4}	2.1×10^{-11}	3.20	10.80
dimethylamine	4.7×10^{-4}	2.1×10^{-11}	3.27	10.73
diethylamine	3.1×10^{-4}	3.2×10^{-11}	3.51	10.49
triethylamine	1.0×10^{-3}	1.0×10^{-11}	2.99	11.01
cyclohexylamine	4.6×10^{-4}	2.2×10^{-11}	3.34	10.66
aniline	4.3×10^{-10}	2.3×10^{-5}	9.37	4.63

19.5
Basicity of Amines

Nearly all of the chemical properties of amines reflect the reactivity of the unshared pair of electrons on the nitrogen atom. The electron pair acts as a base in amines much as it does in the base ammonia (see Chapter 11).

Amines are of comparable base strength to ammonia. The position of the following equilibrium for methylamine acting as a base is to the left as indicated by the base ionization constant, K_b.

$$CH_3NH_2 + H_2O \rightleftharpoons \underset{\substack{\text{methylammonium} \\ \text{ion}}}{CH_3NH_3^+} + OH^-$$

$$K_b = \frac{[OH^-][CH_3NH_3^+]}{[CH_3NH_2]} = 4.4 \times 10^{-4}$$

The base ionization constants for several amines are given in Table 19.3. Alkyl-substituted amines are stronger bases than ammonia, as indicated by the larger equilibrium constant for the reaction of these amines. In contrast, anilines are weaker bases than ammonia. The lower basicity of anilines is the result of delocalization of the unshared pair of electrons of nitrogen over the benzene ring. This delocalization makes the unshared electron pair less accessible for bonding with a proton. An excellent illustration of this resonance effect is the fact that the base equilibrium constant for aniline is 10^6 smaller than that for cyclohexylamine.

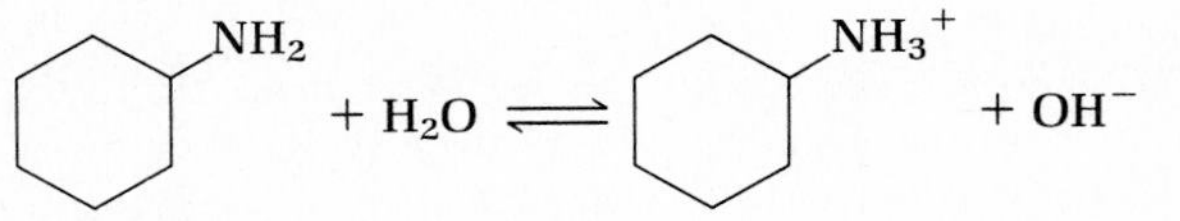

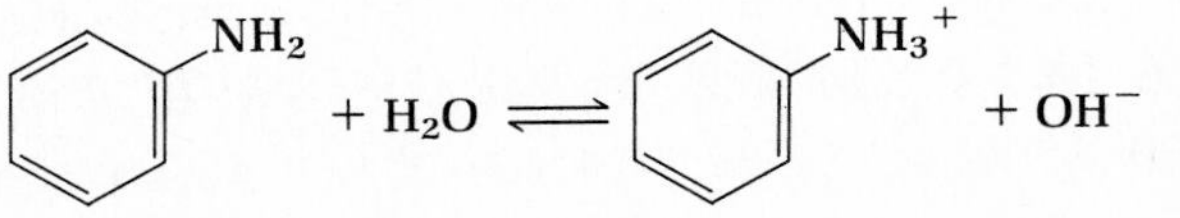

The basicity of amines can also be expressed by pK_b values much like the pK_a values for acids.

$$pK_b = -\log K_b$$

Weak bases have small K_b values and large pK_b. For methylamine $K_b = 4.6 \times 10^{-4}$ and $pK_b = 3.34$.

It is now common to express the strength of a base in terms of the ionization constant, K_a, of its conjugate acid. For methylamine, the reaction is that of the methylammonium ion.

$$\underset{\substack{\text{methylammonium} \\ \text{ion}}}{CH_3NH_3{}^+} + H_2O \rightleftharpoons CH_3NH_2 + H_3O^+$$

$$K_a = \frac{[CH_3NH_2][H_3O^+]}{[CH_3NH_3{}^+]} = 2.2 \times 10^{-11}$$

$$pK_a = 10.66$$

A comparison of the K_b value of methylamine and the K_a value of the methylammonium ion reveals the relationship discussed in Chapter 11 for a base and its conjugate acid. If a base is relatively strong, its conjugate acid is relatively weak. The K_b value for methylamine is relatively large, and the K_a value for the methylammonium ion is relatively small.

Example 19.3

The pK_b values for methylamine and dimethylamine are 3.34 and 3.27, respectively. Which compound is the stronger base?

Solution. The stronger base will have the larger K_b value and hence the smaller pK_b value. Thus, dimethylamine is the stronger base.

Example 19.4

Which of the following is the stronger acid?

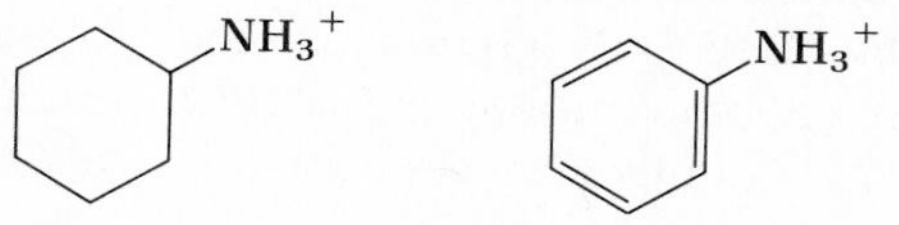

Solution. Aniline is a weaker base than cyclohexylamine. Since weaker bases are conjugate to stronger acids, the anilinium ion is a stronger acid than the cyclohexylammonium ion.

[Additional examples may be found in **19.33–19.37** at the end of the chapter.]

Amines are completely protonated in the reaction with a strong acid such as hydrochloric acid.

$$CH_3NH_2 + HCl \longrightarrow CH_3NH_3{}^+ + Cl^-$$

The ammonium salts thus formed are ionic and are soluble in water if the hydrocarbon portion of the amine is not too large.

Amines may be separated from other substances in naturally occurring material by converting them to ammonium salts. Consider, for example, how one could separate octane from 1-aminooctane. Both compounds are insoluble in water. Addition of HCl converts the 1-aminooctane into its ammonium salt, whereas octane is not affected.

$$CH_3(CH_2)_6CH_2NH_2 + HCl \longrightarrow CH_3(CH_2)_6CH_2NH_3^+ + Cl^-$$

$$CH_3(CH_2)_6CH_3 + HCl \longrightarrow \text{no reaction}$$

The ammonium salt dissolves in the acid solution, whereas the octane does not. After physically separating the octane from the aqueous acid solution, neutralization with sodium hydroxide liberates the amine, which is not very soluble in water.

$$CH_3(CH_2)_6CH_2NH_3^+ + OH^- \longrightarrow CH_3(CH_2)_6CH_2NH_2 + H_2O$$
(soluble in H_2O)   (insoluble in H_2O)

Because ammonium salts are more soluble than amines, drugs containing the amine functional group are often given in the form of salts to improve their solubility in body fluids.

Quaternary ammonium salts *are ammonium salts in which all four groups bonded to nitrogen are organic.* The simplest such compound is tetramethylammonium chloride.

$$(CH_3)_4N^+ \ Cl^-$$
tetramethylammonium chloride

Some detergents, known as invert soaps, contain quaternary ammonium salts. One example of an invert soap is cetyltrimethylammonium chloride.

$$\begin{array}{c} CH_3 \\ | \\ CH_3(CH_2)_{15}\overset{+}{N}-CH_3 Cl^- \\ | \\ CH_3 \end{array}$$

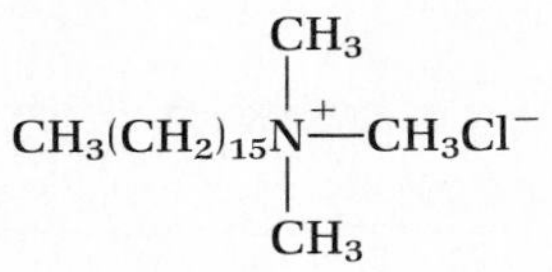

Invert soaps are used as germicides for sterilizing medical instruments and also in some mouthwashes.

Choline and acetylcholine are two important quaternary ammonium salts present in the human body. Choline is combined with other substances in cell membranes (Chapter 22); acetylcholine is involved in nerve impulse transmission.

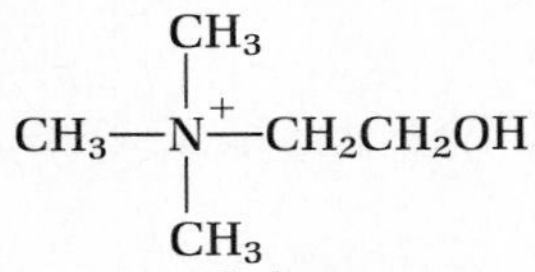

$$\begin{array}{c} CH_3 \\ | \\ CH_3-\overset{+}{N}-CH_2CH_2OH \\ | \\ CH_3 \\ \text{choline} \end{array} \qquad \begin{array}{c} CH_3 \qquad\qquad O \\ | \qquad\qquad\quad || \\ CH_3-\overset{+}{N}-CH_2CH_2-O-CCH_3 \\ | \\ CH_3 \\ \text{acetylcholine} \end{array}$$

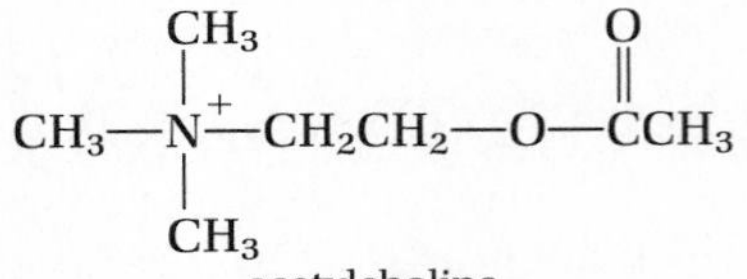

19.6
Reactions of Amines with Nitrous Acid

Nitrous acid is an unstable compound produced by the reaction of a nitrite salt and a strong acid.

$$H_3O^+ + NO_2^- \longrightarrow H_2O + HNO_2$$

Nitrous acid reacts with amines in different ways, depending on whether the amine is primary, secondary, or tertiary. As a result, the reaction is useful in classifying an unknown amine.

Primary amines react with HNO_2 to produce an unstable diazonium salt.

$$R-NH_2 + HNO_2 + H_3O^+ \longrightarrow R-\overset{+}{N}\equiv N + 3\,H_2O$$

In order for this reaction to occur there must be two hydrogen atoms attached to the nitrogen atom. The diazonium salt decomposes to produce nitrogen gas and a mixture of organic products, and the evolution of nitrogen gas is a visual confirmation that the amine is primary.

$$R—\overset{+}{N}\equiv N \longrightarrow N_2 + \text{organic products}$$

Secondary amines react with HNO_2 to form yellow oily compounds called N-nitrosoamines.

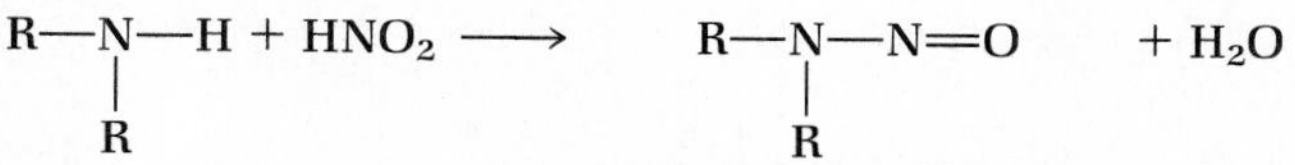

$$R—\underset{\underset{R}{|}}{N}—H + HNO_2 \longrightarrow \qquad R—\underset{\underset{R}{|}}{N}—N{=}O \qquad + H_2O$$

(an N-nitrosodialkylamine)

The yellow oil separates from solution and floats on the solution of the acid used to form the nitrous acid.

Studies of N-nitroso compounds indicate that they are carcinogenic. It has been suggested that nitrites added to bacon, hot dogs, and sandwich meats to retard spoilage may cause stomach cancer. In the stomach, the gastric acid (HCl) can react with nitrite ions to form nitrous acid. The amines present in foods then can react to produce N-nitroso compounds.

The tertiary amines do not react with nitrous acid because they do not have hydrogen atoms bonded to nitrogen. They do not liberate nitrogen gas like primary amines or form N-nitroso compounds like secondary amines. As a result, all three classes of amines are easily distinguished from each other.

Example 19.5 _____

How can a chemist distinguish among the following isomeric compounds?

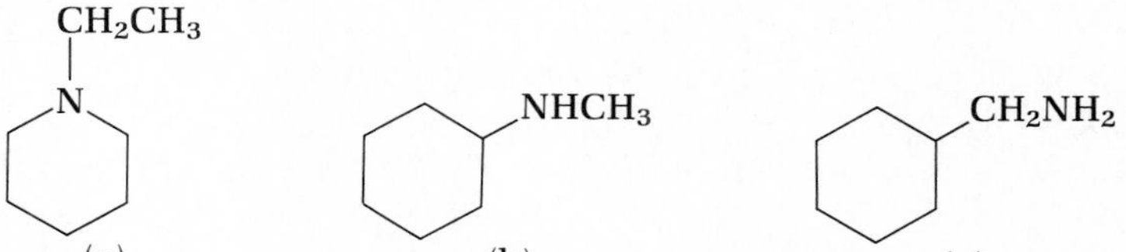

(a)   (b)   (c)

Solution. Compound (c) is a primary amine that will give off nitrogen gas when treated with HNO_2. Compound (b) is a secondary amine that will form a yellow oil with the same reagent. Compound (a) will simply dissolve in the HCl solution used to produce HNO_2.

_____

[Additional examples may be found in questions **19.29–19.32** at the end of the chapter.]

Amides *are compounds in which an amino group or substituted amino group is bonded to a carbonyl carbon atom.* The other two bonds of the nitrogen atom may be to hydrogen atoms, alkyl groups, or aryl groups. The amides are classified on the basis of the combined number of alkyl or aryl groups bonded to the nitrogen atom.

19.7
Amides

an unsubstituted
amide

a monosubstituted
amide

a disubstituted
amide

an unsubstituted
amide

a monosubstituted
amide

a disubstituted
amide

Formamide (Figure 19.5), the simplest amide, has bond angles about the carbon atom close to 120°. Its carbonyl carbon atom is sp^2 hybridized and forms three σ bonds. The electron in the p orbital of the carbon atom forms a π bond with an electron in a p orbital of the oxygen atom.

The structure of formamide resembles that of other carbonyl compounds in that the three atoms bonded to carbon are in the same plane. However, there is an important difference in the bonding between the carbonyl carbon and nitrogen. The nitrogen atom has an unshared pair of electrons that can be delocalized with the π electrons of the carbonyl group. Two resonance structures can be considered to represent formamide.

The bond between carbon and nitrogen is therefore intermediate between a single bond and a double bond. As a consequence, the degree of rotation about the carbon–nitrogen bond is somewhat restricted.

In Chapter 23 we will study the chemistry of proteins, which contain amide bonds. The properties of these compounds reflect the partially restricted rotation about the carbon–nitrogen bond.

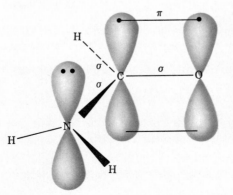

Figure 19.5
Bonding in Formamide.
The geometry of the atoms bonded to the carbon atom of an amide is planar. The carbon atom is sp^2 hybridized, and there is a π bond between the carbon and oxygen atoms. The interaction of the unshared pair of electrons of the nitrogen atom and the carbon–oxygen π bond causes a restriction of the rotation about the carbon–nitrogen bond.

Nomenclature of Amides

In the common system amides are named by dropping the suffix -ic of the related acid and adding the suffix -amide. When there is a substituent on the nitrogen, the prefix N- followed by the name of the group bonded to nitrogen is prefixed to the name. Substituents on the acyl group are designated by α, β, γ, and so on, as in the case for the common names of acids. In the IUPAC system, the final -e of the alkane is replaced by -amide. The substituents on nitrogen are indicated by the same method as in the common system. Numbers are used for substituents on the parent chain in the same way as in acids. Some examples are given below.

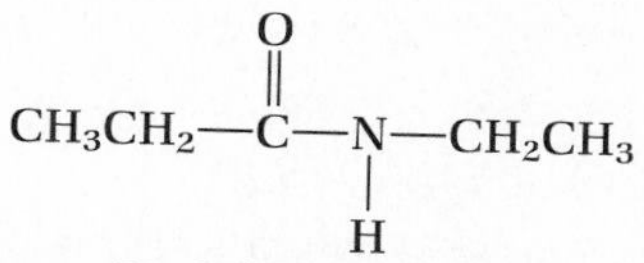

$$CH_3CH_2-\overset{\overset{O}{\|}}{C}-\overset{}{\underset{H}{N}}-CH_2CH_3$$

common: N-ethylpropionamide
IUPAC: N-ethylpropanamide

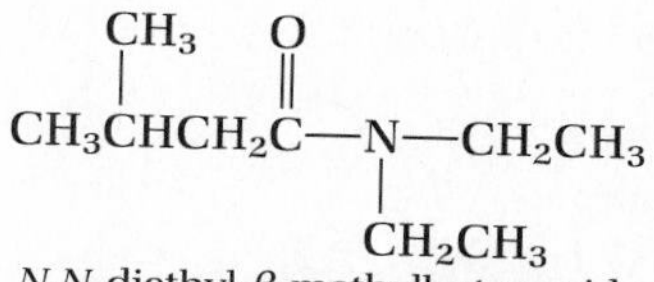

$$CH_3\overset{\overset{CH_3}{|}}{C}HCH_2\overset{\overset{O}{\|}}{C}-\overset{}{\underset{CH_2CH_3}{N}}-CH_2CH_3$$

N,N-diethyl-β-methylbutyramide
N,N-diethyl-3-methylbutanamide

Example 19.6

Give the IUPAC name for the following amide.

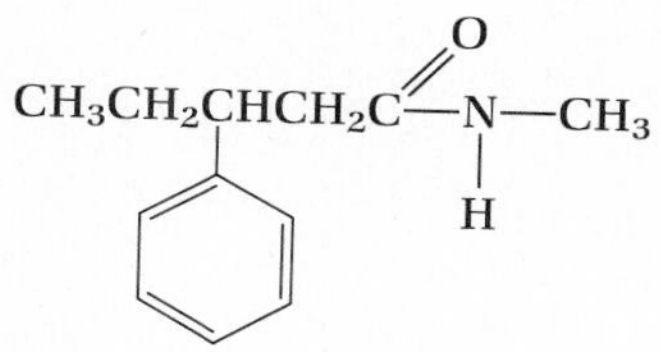

$$CH_3CH_2CHCH_2\overset{\overset{O}{\|}}{C}-\overset{}{\underset{H}{N}}-CH_3$$

Solution. The longest continuous carbon chain contains five carbon atoms. The parent name is pentanamide.

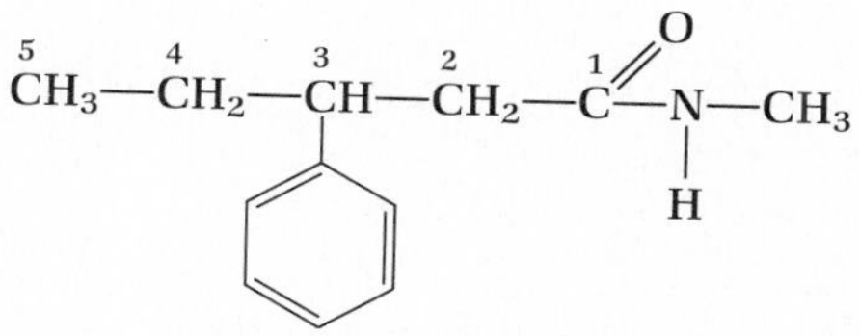

$$\overset{5}{CH_3}-\overset{4}{CH_2}-\overset{3}{CH}-\overset{2}{CH_2}-\overset{1}{C}\overset{O}{\diagup}-\overset{}{\underset{H}{N}}-CH_3$$

A phenyl group is located on carbon atom 3. If we disregard the substituent on the nitrogen atom, we would name the compound 3-phenylpentanamide. The methyl group on the nitrogen atom is designated N-methyl and the complete name is N-methyl-3-phenylpentanamide.

[Additional examples may be found in **19.40–19.44** at the end of the chapter.]

Properties of Amides

Although amides contain a nitrogen atom with an unshared pair of electrons, they are not basic. This difference in basicity compared to amines is due to the effect of the carbonyl group. The carbonyl group draws electrons away from the nitrogen atom, and the resulting dispersion of these electrons decreases their availability to accept a proton.

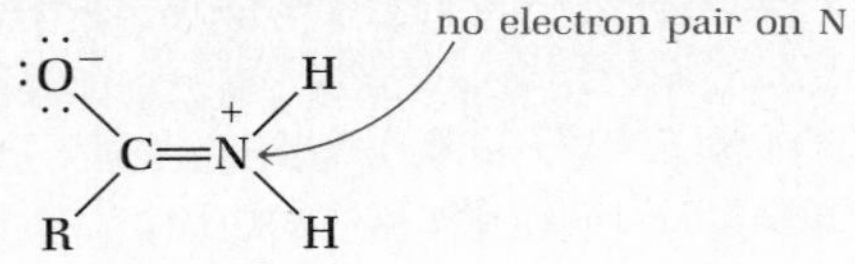

(a resonance form)

With the exception of formamide, the unsubstituted amides are solids at room temperature. Intermolecular hydrogen bonding between the amide hydrogen atom and the carbonyl oxygen atom is strong and accounts for the high melting and boiling points of unsubstituted amides. The consequences of hydrogen bonding of the amides in proteins will be discussed in Chapter 23. Substitution of the hydrogen atoms on the nitrogen atom by alkyl or aryl groups reduces

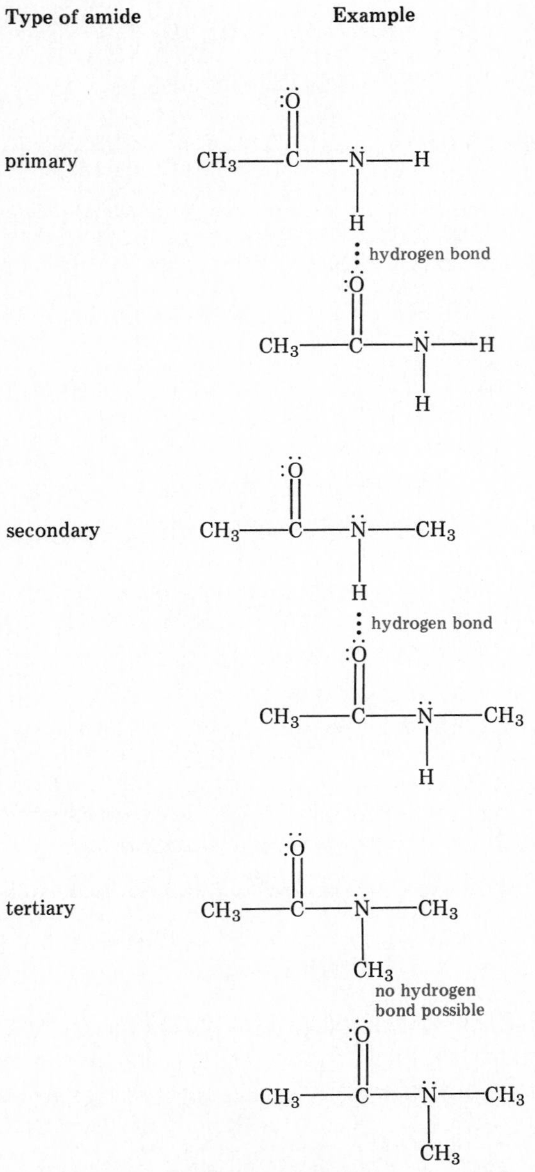

Type of amide	Example

Figure 19.6
Hydrogen Bonds in Amides.

the number of possible intermolecular hydrogen bonds and lowers the melting and boiling points (Figure 19.6).

Low-molecular-weight amides are soluble in water. The solubility is due to intermolecular hydrogen bonds of the amide with water.

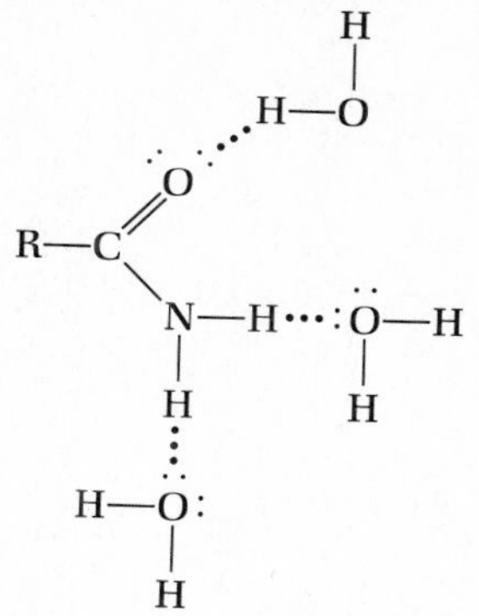

Even low-molecular-weight disubstituted amides are water soluble because the carbonyl oxygen atom can hydrogen bond to the hydrogen atoms of water.

When carboxylic acids are heated with ammonia, a primary amine, or a secondary amine, the products are an amide and water.

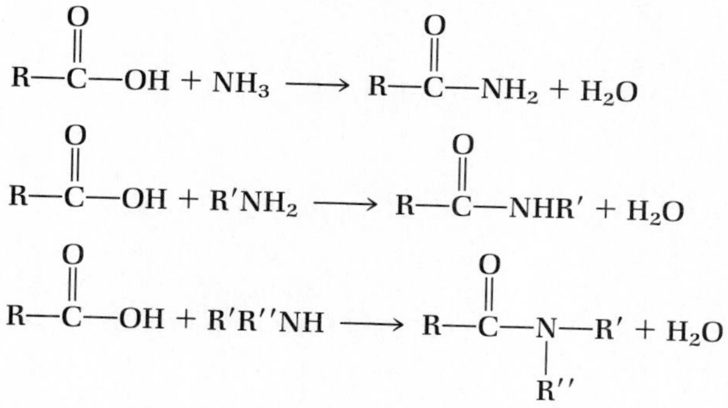

$$R-\overset{\overset{\textstyle O}{\|}}{C}-OH + NH_3 \longrightarrow R-\overset{\overset{\textstyle O}{\|}}{C}-NH_2 + H_2O$$

$$R-\overset{\overset{\textstyle O}{\|}}{C}-OH + R'NH_2 \longrightarrow R-\overset{\overset{\textstyle O}{\|}}{C}-NHR' + H_2O$$

$$R-\overset{\overset{\textstyle O}{\|}}{C}-OH + R'R''NH \longrightarrow R-\overset{\overset{\textstyle O}{\|}}{C}-\underset{\underset{\textstyle R''}{|}}{N}-R' + H_2O$$

19.8
Formation of Amides

Tertiary amines do not react to form amides, since they have no hydrogen atom bonded to the nitrogen atom.

Proteins *are polyamides of α-amino acids.* The amino group of one amino acid reacts with the acid group of another amino acid in an enzyme-catalyzed reaction.

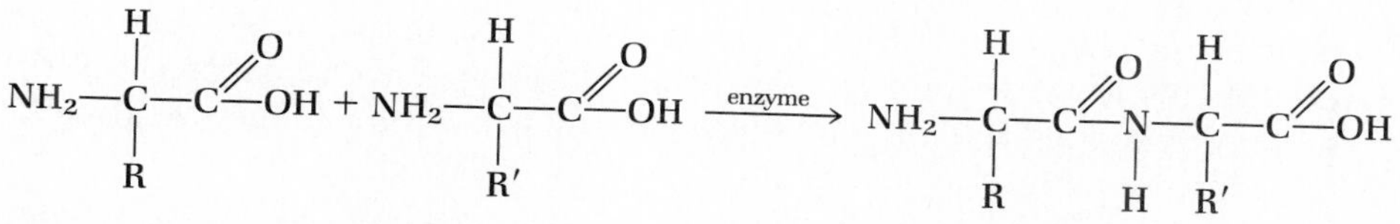

$$NH_2-\underset{\underset{\textstyle R}{|}}{\overset{\overset{\textstyle H}{|}}{C}}-\overset{\overset{\textstyle O}{\diagup}}{C}-OH + NH_2-\underset{\underset{\textstyle R'}{|}}{\overset{\overset{\textstyle H}{|}}{C}}-\overset{\overset{\textstyle O}{\diagup}}{C}-OH \xrightarrow{enzyme} NH_2-\underset{\underset{\textstyle R}{|}}{\overset{\overset{\textstyle H}{|}}{C}}-\overset{\overset{\textstyle O}{\diagup}}{C}-\underset{\underset{\textstyle H}{|}}{N}-\underset{\underset{\textstyle R'}{|}}{\overset{\overset{\textstyle H}{|}}{C}}-\overset{\overset{\textstyle O}{\diagup}}{C}-OH$$

The resultant compound has both an amine and a carboxyl functional group. Continued reaction with these same or different amino acids results in a polypeptide.

Synthetic polyamides can be produced from diamines and dicarboxylic acids by condensation polymerization. Adipic acid and hexamethylenediamine (1,6-diaminohexane) can condense to form an amide.

This end can react with an amine.

This end can react with an acid.

$$\underset{\text{adipic acid}}{\overset{\text{O}}{\underset{\text{HO}}{\parallel}}\text{C(CH}_2)_4\text{C}\overset{\text{O}}{\underset{\text{OH}}{\parallel}}} + \underset{\text{hexamethylenediamine}}{\overset{\text{N}}{\underset{\text{H}}{|}}\text{N(CH}_2)_6\text{N}\overset{\text{H}}{\underset{\text{H}}{|}}} \longrightarrow \underset{\text{an amide}}{\overset{\text{O}}{\underset{\text{HO}}{\parallel}}\text{C(CH}_2)_4\text{C}\text{—N(CH}_2)_6\text{N}\overset{\text{H}}{\underset{\text{H}}{|}}} + \text{H}_2\text{O}$$

The amide is also an amine and a carboxylic acid. Reaction of the amine end with another molecule of adipic acid produces another amide linkage. Reaction of the acid end with hexamethylenediamine produces another amide linkage.

This end can react with an acid. ⟶

⟵ This end can react with an amine.

$$\underset{\text{H}}{\overset{\text{H}}{|}}\text{N(CH}_2)_6\text{N}\text{—C(CH}_2)_4\text{C}\text{—N(CH}_2)_6\text{N}\text{—C(CH}_2)_4\text{C}\overset{\text{O}}{\underset{\text{OH}}{\parallel}}$$

This sequence of reactions can occur again and again to produce a polymer that is a polyamide. The specific polymer formed from adipic acid and hexamethyl-enediamine is called nylon 66 or more commonly nylon. The 66 refers to the six-carbon diacid and the six-carbon diamine.

$$\underset{\text{HO}}{\overset{\text{O}}{\parallel}}\text{C(CH}_2)_4\text{C}\left[\text{N(CH}_2)_6\text{N—C(CH}_2)_4\text{C}\right]_n\text{N(CH}_2)_6\text{N}\overset{\text{H}}{\underset{\text{H}}{}}$$

nylon 66

Nylon can form hydrogen bonds just like any simple amide, and nylon forms many hydrogen bonds between one polymer chain and adjacent chains (Figure 19.7). The strength of nylon is due to the many hydrogen bonds that hold the chains together in fibers. Protein structure is also strongly influenced by both intermolecular and intramolecular hydrogen bonds (Chapter 23).

19.9
Hydrolysis of Amides

Amides can react with water to break the carbon–nitrogen bond and form an acid and ammonia or an amine. The process corresponds to the reverse of the method used to prepare amides. A similar relationship was presented for esters in the preceding chapter. There are, however, important differences. Esterifica-tion results in an equilibrium mixture, but the equilibrium constant for the for-mation of amides is large. The hydrolysis of esters then occurs relatively easily, whereas the amides are quite resistant to hydrolysis.

Amides may be hydrolyzed only by using a strong acid or strong base.

$$\overset{\text{O}}{\overset{\parallel}{\text{C}}}\text{—NHCH}_3 + \text{OH}^- \longrightarrow \overset{\text{O}}{\overset{\parallel}{\text{C}}}\text{—O}^- + \text{CH}_3\text{NH}_2$$

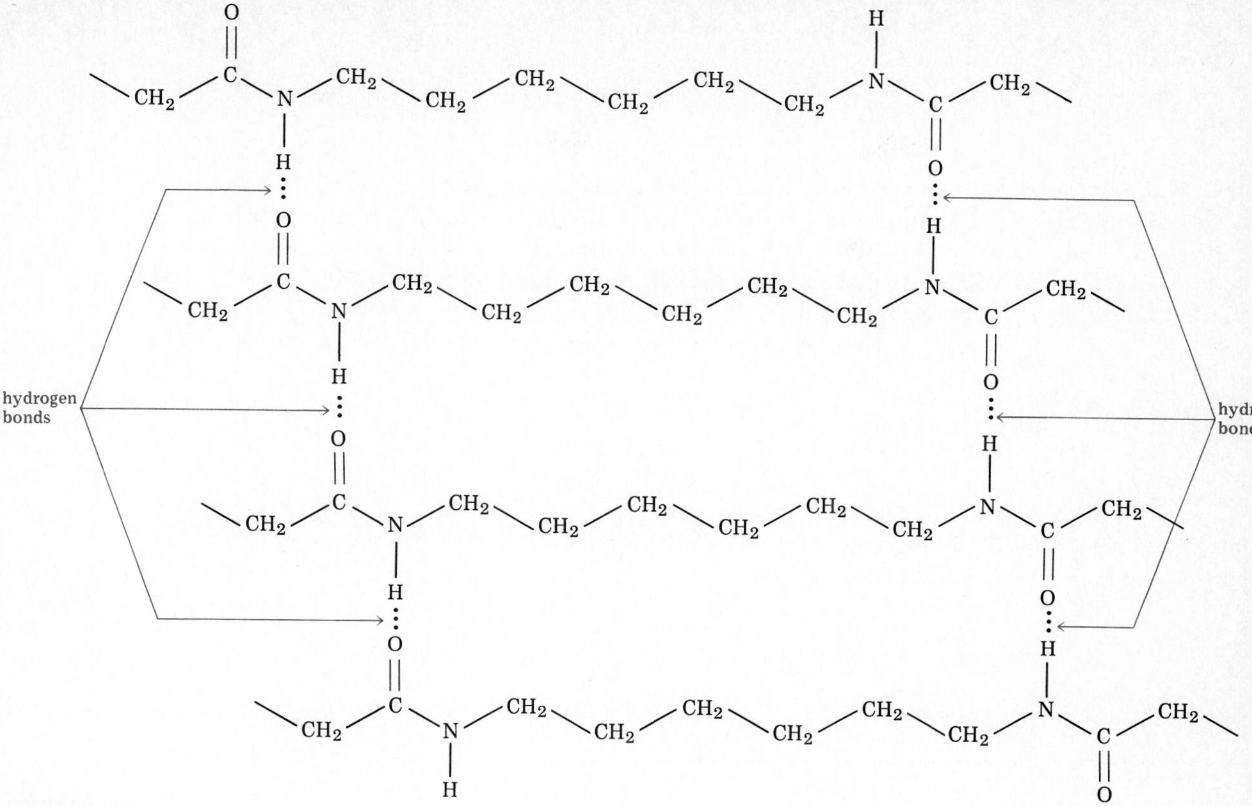

Figure 19.7
Intermolecular Hydrogen Bonding in Nylon.

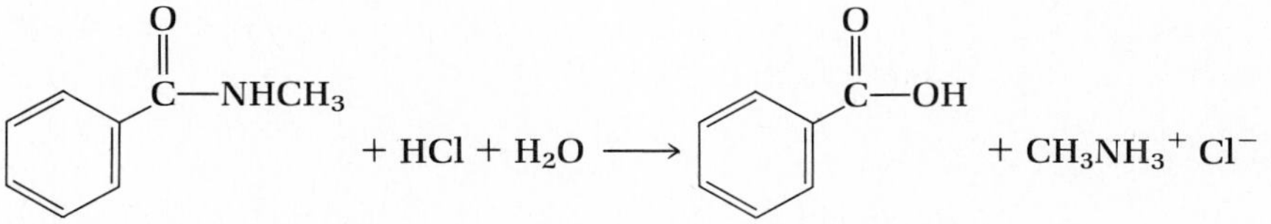

Under basic conditions, the salt of the carboxylic acid is formed, and a mole of base per mole of amide is required. Under acidic conditions, the ammonium salt of the amine is formed, and a mole of acid per mole of amide is required.

Amide hydrolysis requires heating for hours with a strong acid or base, whereas ester hydrolysis usually requires less severe conditions. The great stability of amides toward hydrolysis then means that proteins will not react at physiological pH and at body temperature. However, in the presence of specific enzymes, the hydrolysis of amides is rapid. These reactions will be discussed in Chapter 23.

Example 19.7 _____

What are the products of the hydrolysis of phenacetin by a base? Phenacetin was formerly used in so-called APC tablets consisting of aspirin, phenacetin, and caffeine.

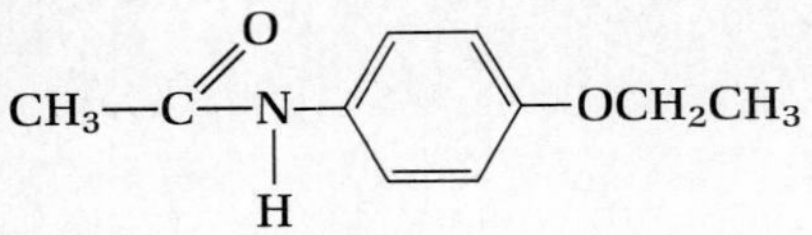

$$CH_3-\overset{\overset{\displaystyle O}{\parallel}}{C}-\underset{\underset{\displaystyle H}{\mid}}{N}-\text{\hexagon}-OCH_2CH_3$$

phenacetin

Solution. The functional group on the right side of the benzene ring is an ether that is very stable. The functional group on the left is an amide.

Hydrolysis of an amide breaks the bond between the nitrogen atom and the carbonyl group. The acid fragment will be acetic acid. The amine fragment will be an aniline containing an ether substituent.

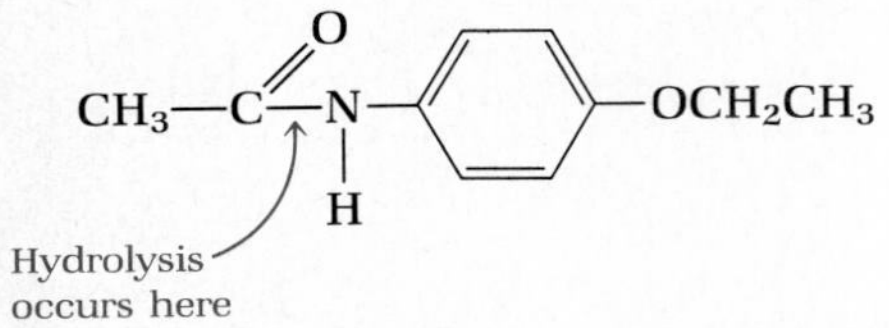

$$CH_3-\overset{\overset{\displaystyle O}{\parallel}}{C}-\underset{\underset{\displaystyle H}{\mid}}{N}-\text{\hexagon}-OCH_2CH_3$$

Hydrolysis
occurs here

Because a base is used in the hydrolysis, the acid product will exist as the acetate ion. The amine is *p*-ethoxyaniline.

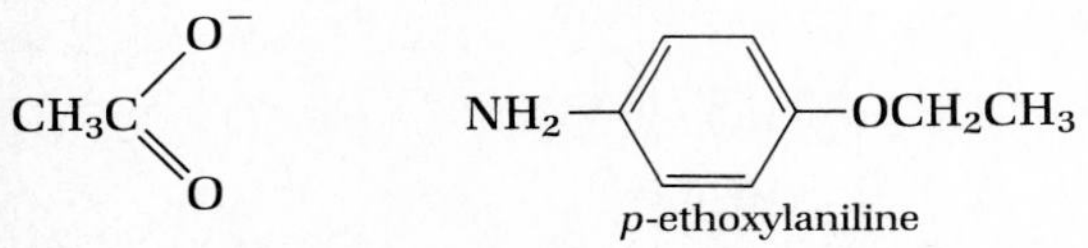

$$CH_3C\overset{\displaystyle O^-}{\underset{\displaystyle O}{\diagup}} \qquad NH_2-\text{\hexagon}-OCH_2CH_3$$

p-ethoxylaniline

[Additional examples may be found in **19.49** and **19.50** at the end of the chapter.]

19.10
Important Heterocyclic Nitrogen Compounds

A number of heterocyclic nitrogen compounds were listed in Section 19.3. Those ring systems occur in a variety of compounds important to life. Among these are vitamins and the compounds present in DNA.

Pyrroles and Indoles

The pyrrole ring occurs in several compounds of biological importance. Four of these rings are incorporated in chlorophyll, hemoglobin (Figure 19.8), and vitamin B_{12}. In each molecule there is a metal surrounded by four nitrogen atoms contained in pyrrole rings.

Indole rings occur in both tryptophan, an amino acid, and serotonin, a hormone that is important in the chemistry of the brain.

Pyridines

The pyridine ring is contained in the deadly poison nicotine (Figure 19.8). Gardeners who have improperly handled nicotine as an insecticide have died from respiratory failure in a matter of minutes. The tolerance of nonsmokers for nicotine is as little as 4 mg, but smokers tend to build some tolerance for nicotine. In contrast to the toxic effects of nicotine, consider the benefits of nicotinamide, which is vitamin B_3 (niacin). This substance has been assigned a newer name, niacinamide.

Vitamin B_6, which is actually two substances, also contains a pyridine ring.

Pyrroles and Indoles

Figure 19.8
Important Heterocyclic Nitrogen Compounds.

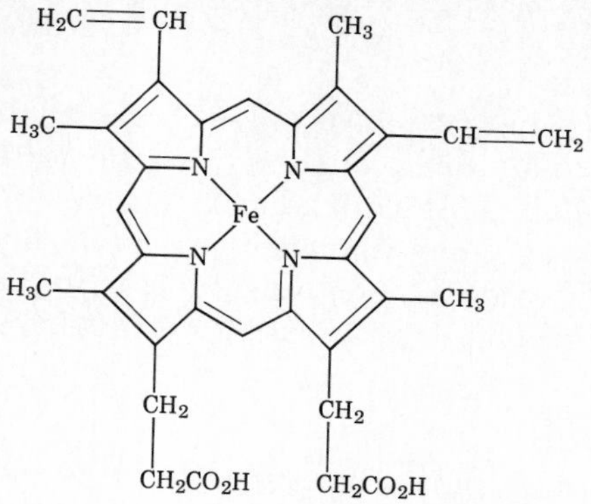

heme

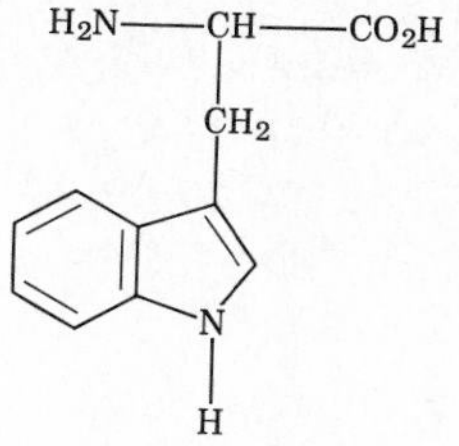

tryptophan

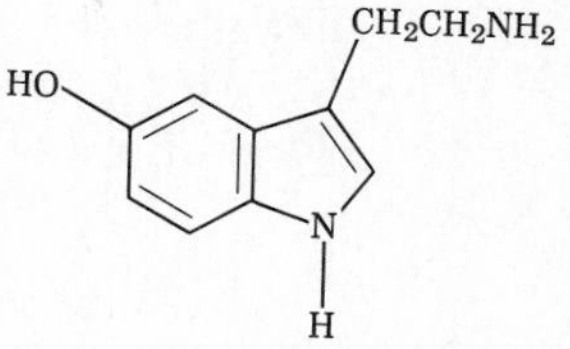

serotonin

Pyridines

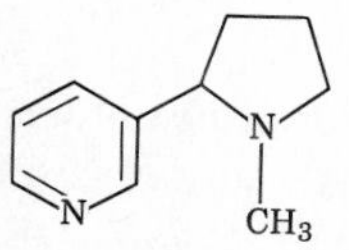

nicotine

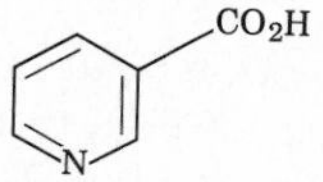

nicotinic acid

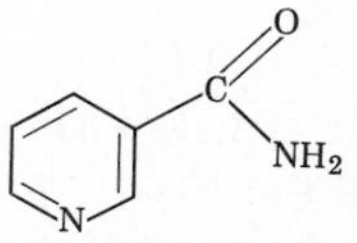

nicotinamide

the B_3 vitamins

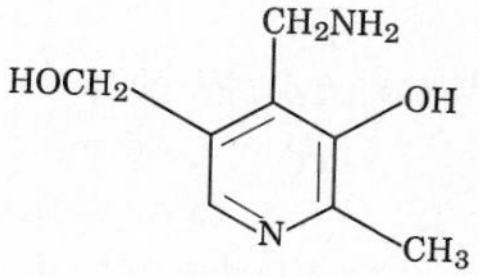

pyridoxamine

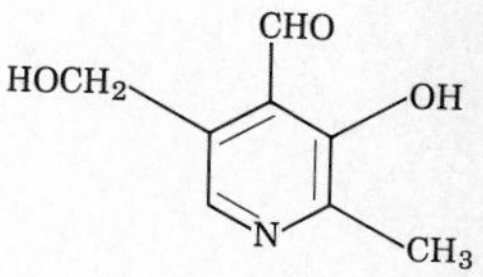

pyridoxal

the B_6 vitamins

Pyrimidines

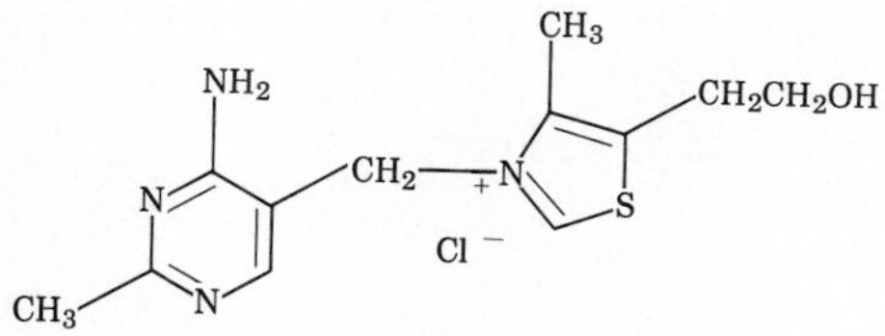

thiamine chloride

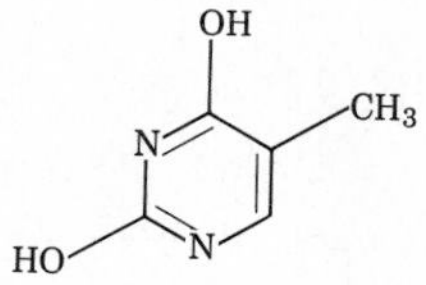

thymine

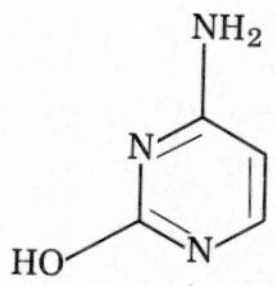

cytosine

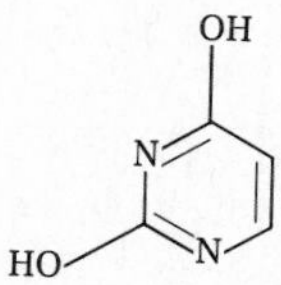

uracil

Purines

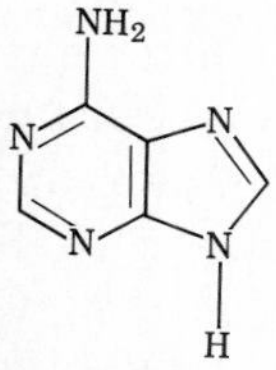

adenine

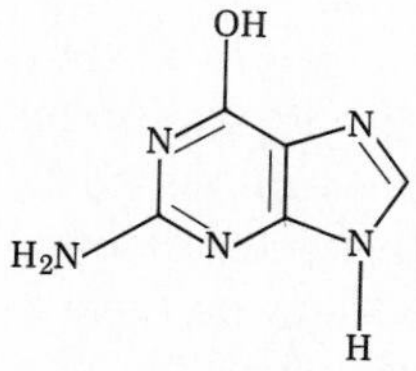

guanine

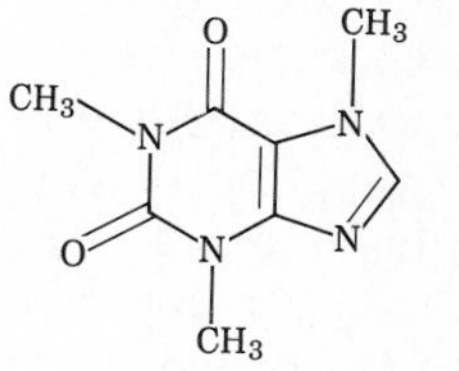

caffeine

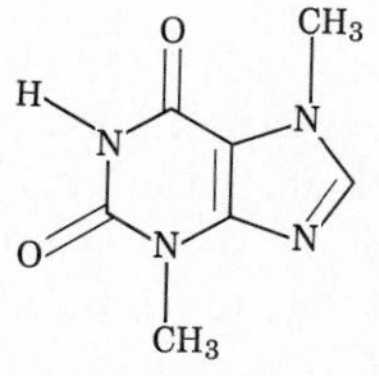

theobromine

Pyrimidines

Thiamine (vitamin B_1) contains a pyrimidine ring (Figure 19.8). Only 1 mg of thiamine a day is necessary for humans. A deficiency causes disorders of the nervous system.

Thymine, cytosine, and uracil are pyrimidines that are incorporated in the nucleic acids DNA and RNA. The nucleic acids are responsible for the transmission of hereditary information and the control of protein synthesis.

Purines

The substituted purines adenine and guanine (Figure 19.8) are contained in nucleic acids. Adenine is contained in ATP, the biological energy storage compound, as well as in NAD^+ and FAD, two biological oxidizing agents. Caffeine and the related theobromine are both present in tea and cocoa. There is a possibility that these compounds may interfere with the synthesis and repair of DNA in the body (Chapter 29). Both compounds resemble adenine and guanine and could be incorporated in the DNA molecule.

19.11
Drugs Containing Nitrogen

In the first section of this chapter it was pointed out that naturally occurring and synthetic nitrogen compounds are very physiologically active. In this section, several of the various classes of nitrogen-containing drugs will be presented. However, this material is only a miniscule fraction of the known chemistry of nitrogen compounds.

Acetanilides

Acetanilide is an aromatic amide once used as an antipyretic (reduces fever) and analgesic. Because of the toxicity of acetanilide, phenacetin and acetaminophen replaced it as over-the-counter medicines. Phenacetin formerly was combined with aspirin and caffeine and sold as APC tablets. Acetaminophen is the active ingredient in Tylenol.

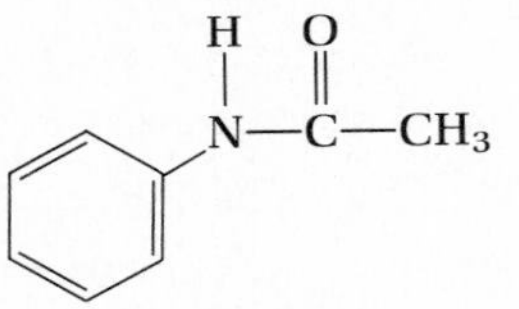

acetanilide 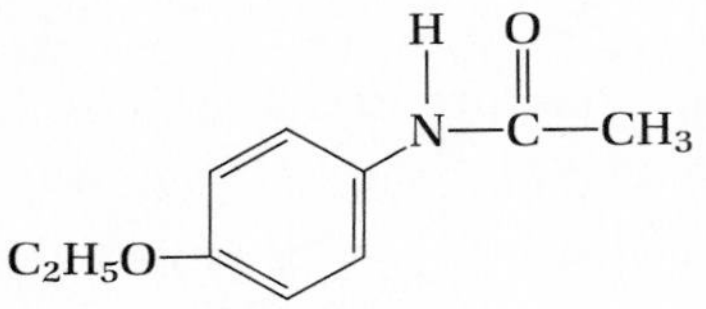   phenacetin 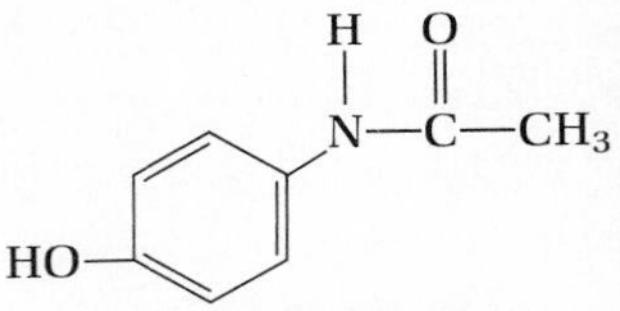   acetaminophen

Amphetamines

Amphetamines are drugs that mimic the action of epinephrine (adrenaline) and norepinephrine (Figure 19.9). Both epinephrine and norepinephrine are produced by the adrenal medulla. Norepinephrine maintains the muscle tone of blood vessels and hence controls blood pressure.

Amphetamine (also known as Benzedrine) has a structure similar to epinephrine. It is a moderate appetite depressant and stimulates the cortex of the brain, which effectively counters fatigue. Amphetamines are known as pep pills or uppers. Methamphetamine is "speed," and methoxyamphetamine is STP. Amphetamine and its structurally related compounds have been used as illicit drugs. They can produce severe physiological reactions. In addition, once the drug wears off the user tends to "crash" into a state of physical and mental exhaustion.

Barbiturates

Barbiturates, derivatives of barbituric acid (Figure 19.9), are among the most widely abused drugs. Not only are they habit forming but the user tends to build a tolerance for them. As a consequence, larger amounts are taken to achieve the same effect.

Figure 19.9
Nitrogen-Containing Drugs.

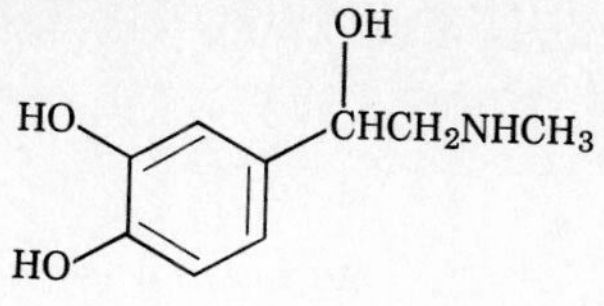

epinephrine

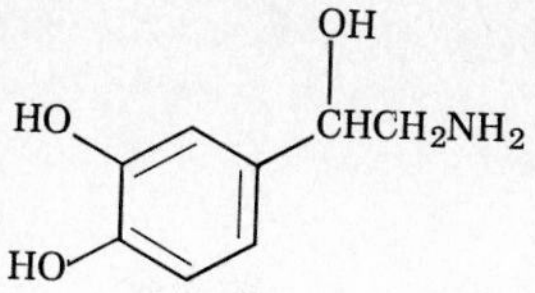

norepinephrine

Amphetamines

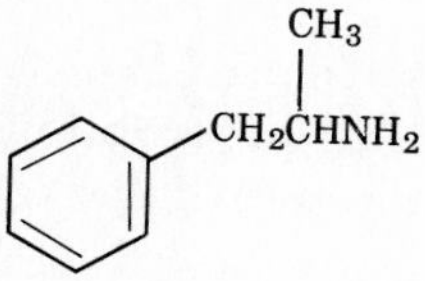

amphetamine

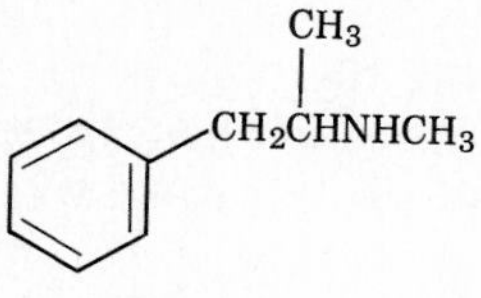

methamphetamine

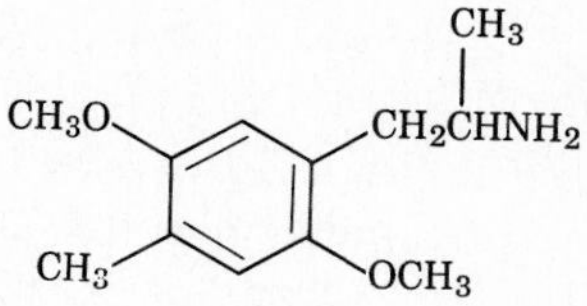

methoxyamphetamine

Barbiturates

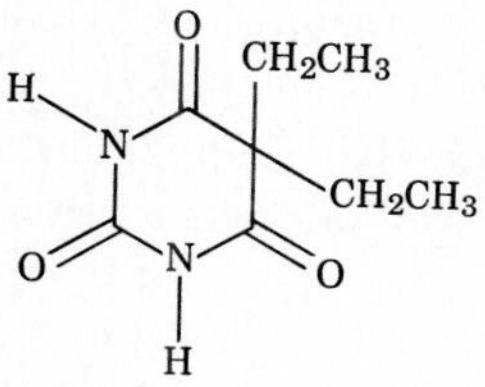

barbital

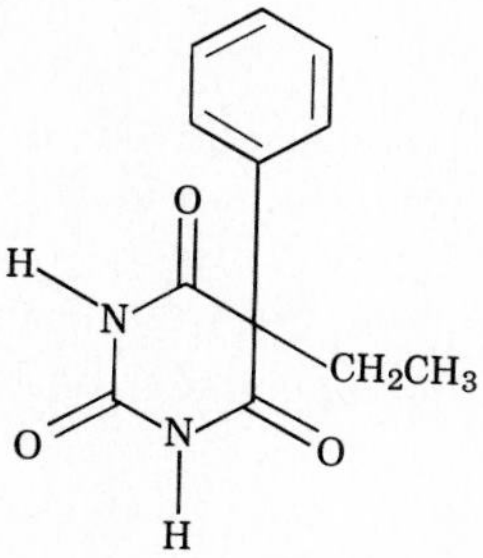

phenobarbital

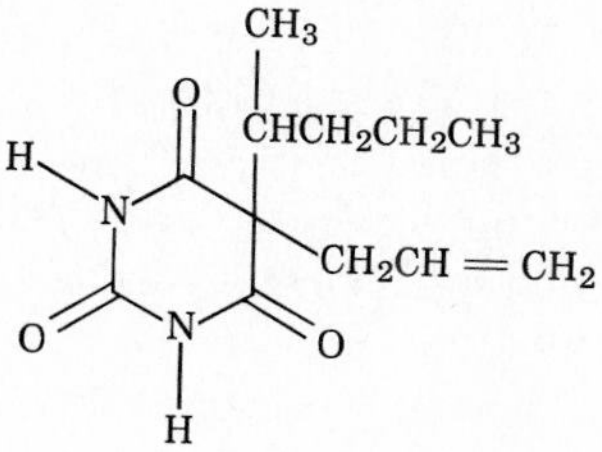

secobarbital

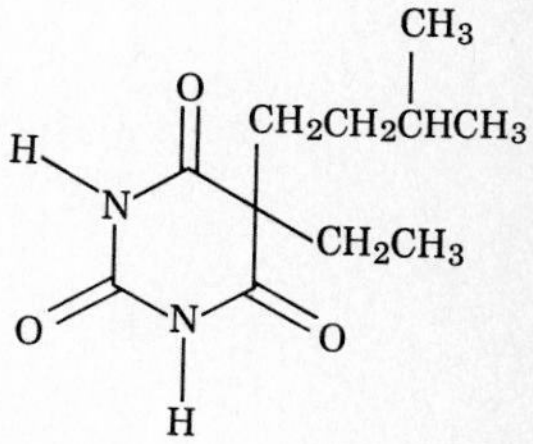

amobarbital

Narcotics

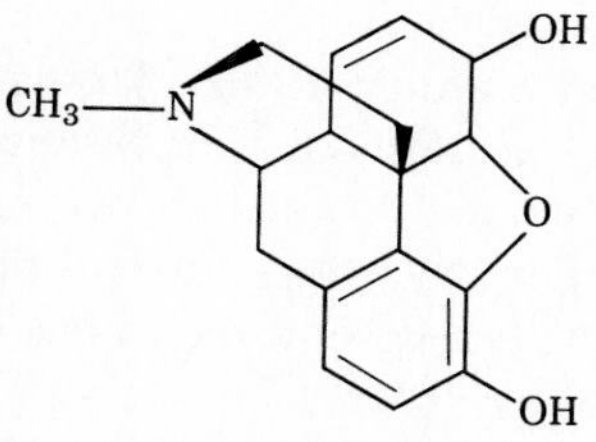

morphine

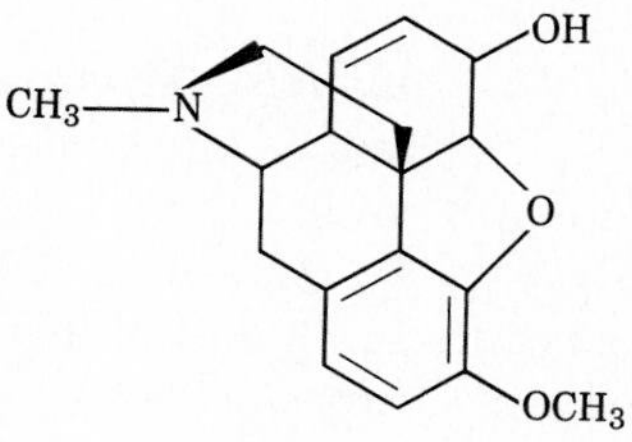

codeine

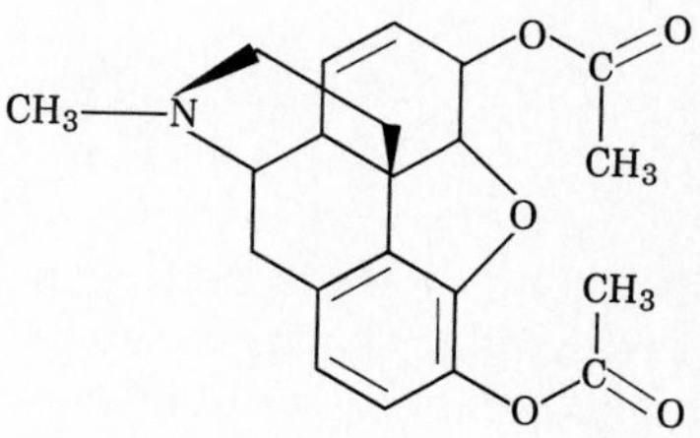

heroin

Hallucinogens

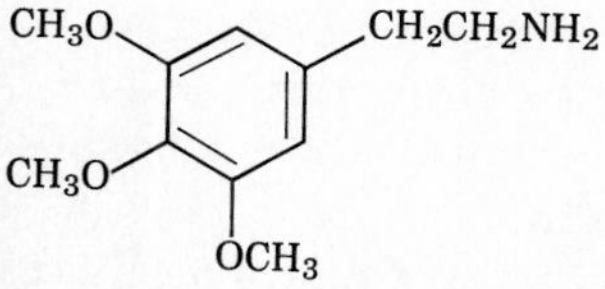

mescaline

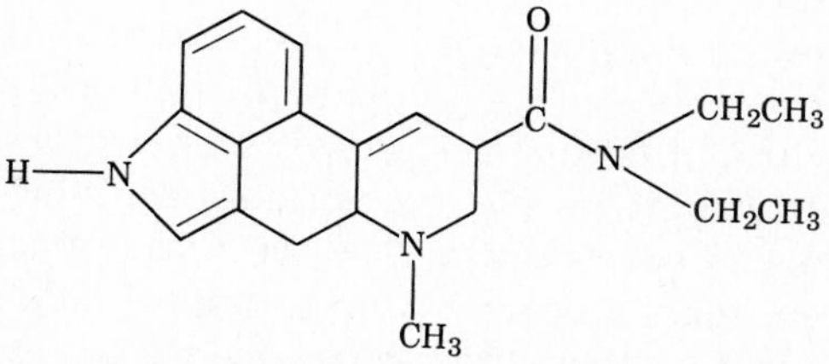

lysergic acid diethylamide

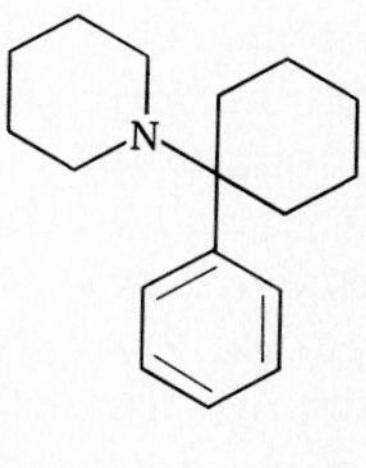

phencyclidine (PCP)

Barbiturates, sometimes used as sedatives, are more pronounced in their effect when the body also contains alcohol. Therefore, sleeping pills taken after drinking alcoholic beverages can be hazardous and even cause death.

Narcotics

Unfortunately, narcotics are the most powerful painkillers and also the most addictive of drugs. For this reason the use of narcotics for severe pain is carefully controlled.

Morphine (Figure 19.9) is obtained from the milky syrup of the poppy. As little as 10 mg relieves severe pain for about 4 hours.

Codeine is also obtained from the poppy. It is not as effective a painkiller as morphine, but it is less addictive. Codeine is an ingredient of some prescription pain-relief drugs and cough syrups. By weight, codeine is about ten times as potent a painkiller as aspirin.

Heroin is produced synthetically from morphine. It is extremely addictive, and dependence will result from the very outset of its use. This substance is not used for any medical reasons, and its possession is illegal in the United States.

Hallucinogens

Hallucinogens, also known as psychedelic drugs, drastically alter one's perception. The resulting impaired judgment can lead to serious harm to the user and others.

One of the most powerful hallucinogens is lysergic acid diethylamide (Figure 19.9) or LSD. Oral doses as low as 0.05 mg cause varied unpredictable reactions in a user.

Mescaline (Figure 19.9), an illegal drug, is a hallucinogen contained in peyote cactus. It has been used by some American Indian tribes in religious ceremonies.

Phencyclidine (Figure 19.9) or PCP is a synthetic hallucinogen made illegally in home laboratories.

Summary

Amines can be viewed as derived from ammonia in which one, two, or three hydrogen atoms are replaced by hydrocarbon groups. Amines are classified as primary, secondary, or tertiary depending on the number of hydrocarbon groups replacing hydrogen atoms. If the nitrogen atom of an amine is one of the atoms of a ring, the compound is called a heterocyclic amine. Heterocyclic aromatic amines, which contain a nitrogen atom replacing a carbon atom, are important biological compounds.

Primary and secondary amines form intermolecular hydrogen bonds, but because nitrogen is less electronegative than oxygen, amines form weaker hydrogen bonds than alcohols. The primary and secondary amines have higher boiling points than the nonpolar hydrocarbons, but lower boiling points than alcohols. Tertiary amines cannot form intermolecular hydrogen bonds with themselves and have low boiling points.

Amines containing a small number of carbon atoms are soluble in water. These compounds have a fishy odor.

Amines are weak bases and form weakly basic solutions. The reaction of a strong acid with an amine forms an ammonium salt, which is much more soluble in water than the amine.

The reaction of amines with nitrous acid gives distinctly different products. Primary amines form nitrogen gas; secondary amines form oily N-nitroso compounds; tertiary amines dissolve in the acid solution with no apparent reaction.

Amides are derived from a carboxylic acid by replacing the hydroxyl group with a NH_2, NHR, or NR_2 group. The resulting carbon–nitrogen bond has double-bond character, and the restricted rotation is an important factor in proteins, which consist of amino acids bonded via amide bonds.

Unsubstituted amides have high melting points and boiling points owing to intermolecular hydrogen bonds. Substitution of hydrocarbon groups for hydrogen atoms on nitrogen reduces the number of intermolecular hydrogen bonds and lowers the melting point and boiling point.

Amide bonds may be hydrolyzed by acid or base, but under more vigorous conditions than for the hydrolysis of esters.

Learning Objectives

As a result of studying Chapter 19 you should be able to

- Classify primary, secondary, and tertiary amines.
- Assign common and IUPAC names to amines.
- Relate the physical properties of amines and their structure.
- Distinguish among primary, secondary, and tertiary amines by a chemical test.
- Recognize the common heterocyclic aromatic rings.
- Assign common and IUPAC names to amides.
- Relate the physical properties of amides and their structures.
- Describe the formation of a polyamide in a condensation polymerization reaction.
- Describe the classes of the nitrogen-containing drugs, amphetamines, barbiturates, narcotics, and hallucinogens.

New Terms

An **amide** is the functional group $-\overset{\overset{\displaystyle O}{\|}}{C}-\overset{\displaystyle |}{N}-$. Amides may
be classified as unsubstituted, monosubstituted, or disubstituted based on the groups bonded to the nitrogen atom.
An **amine** is a derivative of ammonia in which one or more hydrogen atoms is replaced by an alkyl or aryl group.
An **amino** group is $-NH_2$.
Heterocyclic compounds have one or more atoms in the ring other than carbon.
The **pK_b** of a base is equal to $-\log K_b$.
Primary amines have a single hydrocarbon group replacing one hydrogen atom of ammonia.
Proteins are polyamides of α-amino acids.
A **quaternary ammonium ion** has four hydrocarbon groups bonded to nitrogen and bears a positive charge.

A **secondary amine** has two hydrogen atoms of ammonia replaced by hydrocarbon groups.
A **tertiary amine** has all three hydrogen atoms of ammonia replaced by hydrocarbon groups.

Questions and Problems

Terminology

19.1 What are the structural differences between an amide and an amine?

19.2 Explain the classification system of amines, and contrast it with the classification system of alcohols.

Classification of Amines

19.3 Coniine, the poison in hemlock, has the following structure. Classify this amine.

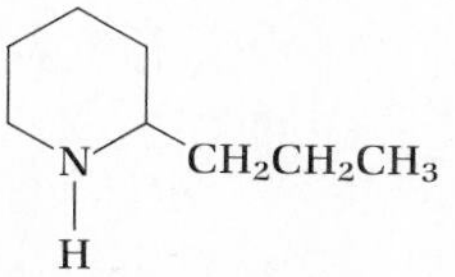

$CH_2CH_2CH_3$

19.4 Classify the amine sites in the amino acid tryptophan.

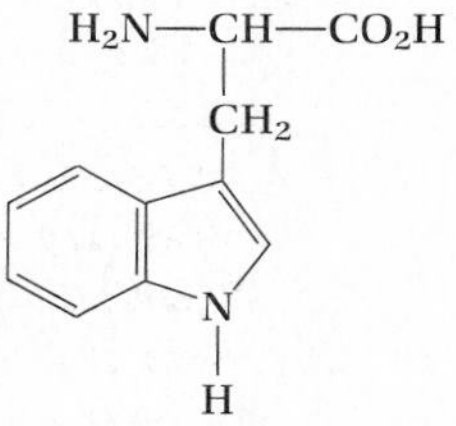

$H_2N-CH-CO_2H$
CH_2

19.5 Classify the amine sites in novocaine.

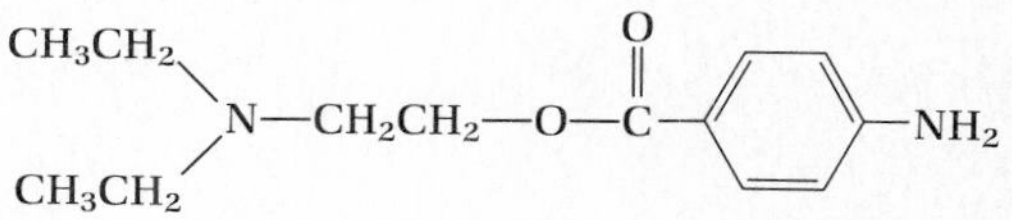

CH_3CH_2
CH_3CH_2
$N-CH_2CH_2-O-\overset{\overset{\displaystyle O}{\|}}{C}-$
$-NH_2$

19.6 Classify the amine sites in histidine.

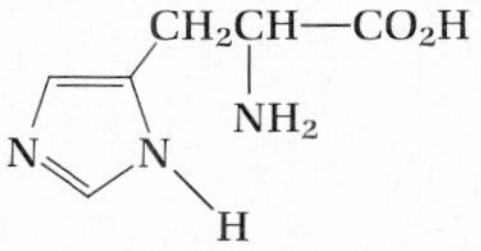

CH_2CH-CO_2H
NH_2

19.7 Classify arecoline as an amine. The compound is a narcotic found in the nut of the betel palm.

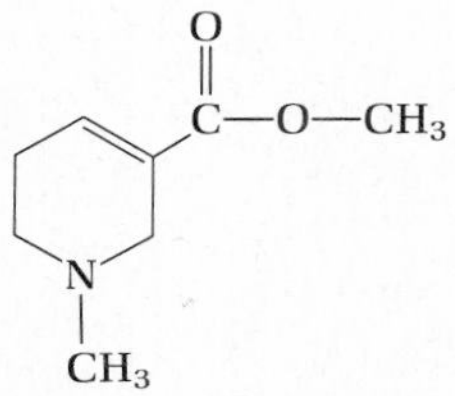

$\overset{\overset{\displaystyle O}{\|}}{C}-O-CH_3$
CH_3

19.8 Classify the amine sites in nicotine.

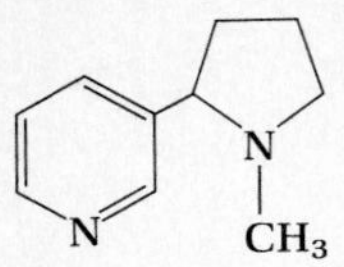

19.9 Classify each of the following amines.

(a) $CH_3-\overset{\overset{\displaystyle H}{|}}{N}-CH_2CH_3$

(b) $CH_3CH_2-\overset{\overset{\displaystyle CH_2CH_3}{|}}{N}-CH_2CH_3$

(c) $CH_3CH_2-\overset{\overset{\displaystyle H}{|}}{N}-\underset{\underset{\displaystyle CH_3}{|}}{C}HCH_3$

(d) $CH_3CHCH_2-NH_2$

(e) $CH_3CH_2-\overset{\overset{\displaystyle CH_3}{|}}{N}-CH_3$

(f) $CH_3\overset{\overset{\displaystyle CH_3}{|}}{C}H-\overset{\overset{\displaystyle H}{|}}{N}-CH_2CH_3$

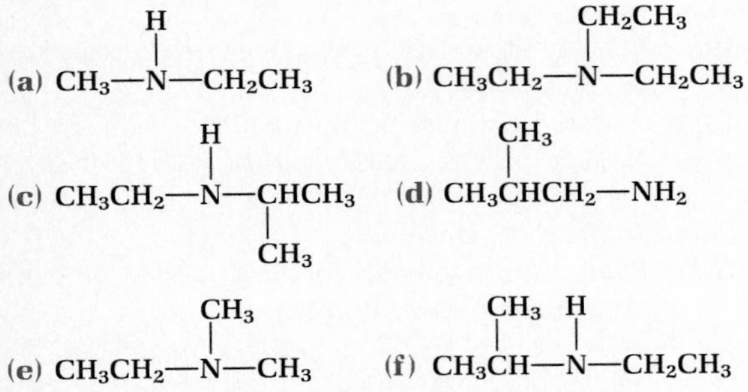

19.10 Classify each of the following amines.

(a)

(b)

(c)

(d)

(e)

(f)

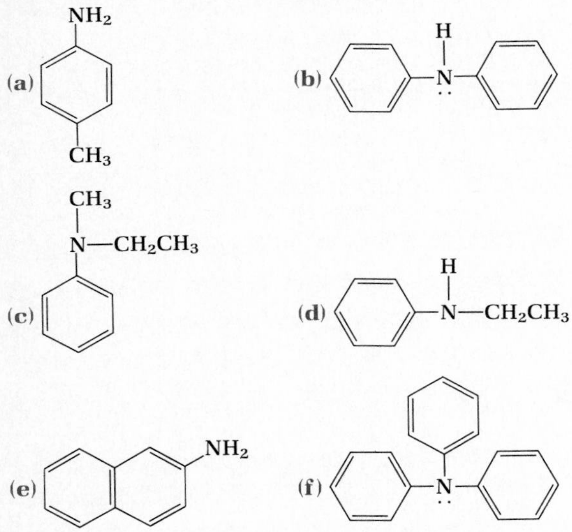

19.11 Classify each of the following amines.

(a)

(b)

(c)

(d)

(e)

(f)

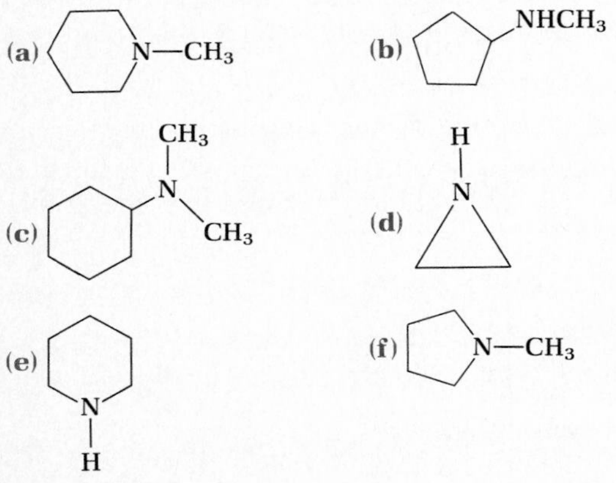

19.12 Classify each of the following amines.

(a)

(b)

(c)

(d)

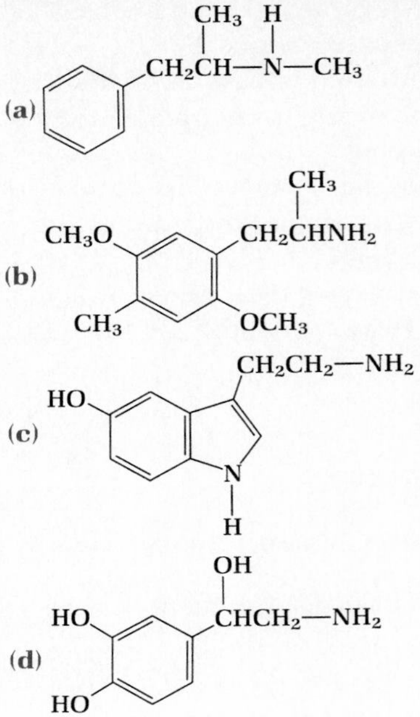

Nomenclature

19.13 Give the common name of each of the compounds in **19.9.**

19.14 Give the IUPAC name of each of the following compounds.

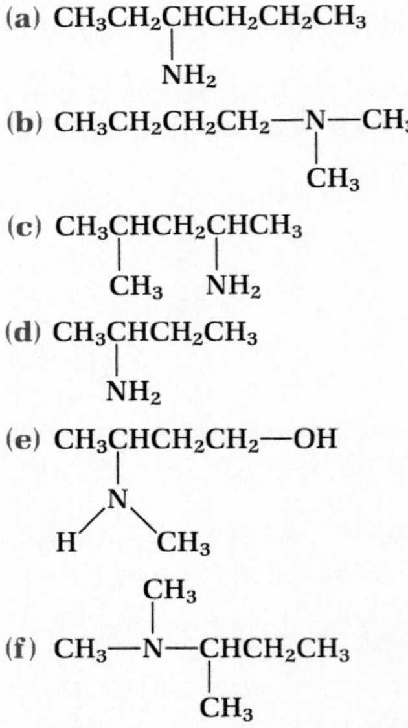

(a) $CH_3CH_2\overset{\overset{\displaystyle }{}}{C}HCH_2CH_2CH_3$ with NH_2

(b) $CH_3CH_2CH_2CH_2-\overset{\overset{\displaystyle }{}}{N}-CH_3$ with CH_3

(c) $CH_3\overset{\overset{\displaystyle }{}}{C}HCH_2\overset{\overset{\displaystyle }{}}{C}HCH_3$ with CH_3 and NH_2

(d) $CH_3\overset{\overset{\displaystyle }{}}{C}HCH_2CH_3$ with NH_2

(e) $CH_3\overset{\overset{\displaystyle }{}}{C}HCH_2CH_2-OH$ with $\underset{\underset{\displaystyle H \quad CH_3}{}}{N}$

(f) $CH_3-\overset{\overset{\displaystyle }{}}{N}-\overset{\overset{\displaystyle CH_3}{}}{C}HCH_2CH_3$ with CH_3

19.15 Give the IUPAC name of each of the following diamines.

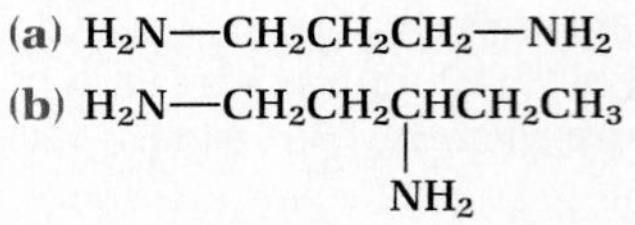

(a) $H_2N-CH_2CH_2CH_2-NH_2$

(b) $H_2N-CH_2CH_2\overset{\overset{\displaystyle }{}}{C}HCH_2CH_3$ with NH_2

(c) $H_2N—CH_2(CH_2)_3CH_2—NH_2$

(d) $CH_3CHCH_2CHCH_3$
      | |
     NH_2 NH_2

19.16 Give the IUPAC name of each of the following compounds.

(a) $H_2N—CH_2—CO_2H$ (b) $CH_3CH—CO_2H$
                                                        with NH_2 above

(c) $CH_3CHCH_2CH—CO_2H$ (d) $CH_3CHCH_2CH_2—CHO$
         with NH_2 and Cl with $N(CH_3)_2$

(e) $CH_3CHCHCH_2CH_3$
       with HO and NH_2

Molecular Formulas

19.17 Determine the general molecular formula for a saturated amine.

19.18 Determine the general molecular formula for a saturated diamine.

19.19 Determine the general molecular formula for a saturated cyclic amine.

19.20 Write the molecular formula for each amine in **19.11**.

Isomers of Amines

19.21 How many isomers are possible with the molecular formula C_2H_7N?

19.22 How many amines are possible with the molecular formula C_3H_9N?

19.23 There are four primary, three secondary, and one tertiary amines with the molecular formula $C_4H_{11}N$. Write their structures.

Properties of Amines

19.24 The boiling points of the isomeric compounds propylamine and trimethylamine are 49 and 3.5°C, respectively. Explain this large difference.

19.25 The fishy odor of an aqueous solution of propylamine is eliminated when an equimolar amount of HCl is added. Explain why.

19.26 The boiling point of 1,2-diaminoethane is 116°C. Explain why this compound boils at a much higher temperature than propylamine (49°C), which has a similar molecular weight.

19.27 Explain why lemon juice, which contains citric acid, is put on fish dishes.

19.28 Why is morphine administered to patients as morphine sulfate?

Classification of Amines

19.29 A compound, $C_4H_{11}N$, reacts with nitrous acid to yield a yellow oil. What is the molecular formula of the oil? What structures are possible for $C_4H_{11}N$?

19.30 A substance, $C_4H_{11}N$, reacts with nitrous acid to form

nitrogen gas. What structures are possible for $C_4H_{11}N$?

19.31 What reactions would occur when each of the compounds in **19.9** reacts with nitrous acid?

19.32 What reactions would occur when each of the compounds in **19.11** reacts with nitrous acid?

Basicity of Amines

19.33 Estimate the K_b for each of the following compounds.

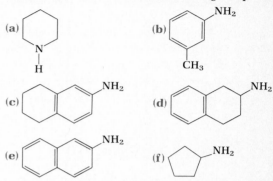

19.34 The K_b values for diethylamine and triethylamine are 3.51 and 2.99, respectively. Which compound is the stronger base?

19.35 The K_b values for dimethylamine and diethylamine are 4.7×10^{-4} and 3.1×10^{-4}, respectively. Which compound is the stronger base?

19.36 Pyrrole is a very weak base. Suggest a reason for this fact.

19.37 Write a balanced equation for each of the following reactions.

(a) ethylamine + HCl $\longrightarrow$

(b) aniline + HBr $\longrightarrow$

(c) dimethylamine + HCl $\longrightarrow$

(d) trimethylamine + HBr $\longrightarrow$

(e) ethylamine + H_2SO_4 $\longrightarrow$

(f) 1,4-diaminobutane + H_2SO_4 $\longrightarrow$

Heterocyclic Aromatic Rings

19.38 What is the molecular formula of each of the following compounds?

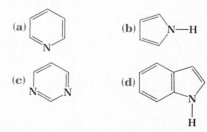

19.39 Draw the structure of each of the following compounds.

(a) 2-ethylpyrrole   (b) 3-bromopyridine

(c) 2,5-dimethylpyrimidine   (d) 2,6,8-trimethylpurine

Nomenclature of Amides

19.40 Give the common name of each of the following amides.

(a) $CH_3CH_2\overset{\text{O}}{\underset{}{C}}-NH_2$ (b) $CH_3CH_2\overset{\text{O}}{\underset{}{C}}-NHCH_2CH_3$

(c) $CH_3CH_2\underset{\underset{CH_3}{|}}{CH}\overset{\text{O}}{\underset{}{C}}-NH_2$ (d) $CH_3CH_2CH_2\overset{\text{O}}{\underset{}{C}}-N(CH_3)_2$

19.41 Give the IUPAC name of each of the amides in **19.40.**

19.42 Give the IUPAC name of each of the following amides.

(a) $CH_3CH_2\overset{\text{O}}{\underset{}{C}}-\underset{\underset{CH_3}{|}}{N}-CH_3$ (b) $H-\overset{\text{O}}{\underset{}{C}}-\underset{\underset{CH_3}{|}}{N}-CH_2CH_3$

(c) $CH_3\underset{\underset{CH_3}{|}}{CH}CH_2\overset{\text{O}}{\underset{}{C}}-\underset{\underset{H}{|}}{N}-CH_3$

(d) 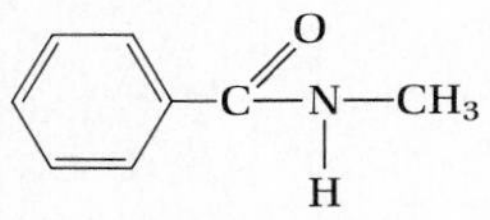 $C_6H_5-\overset{\text{O}}{\underset{}{C}}-\underset{\underset{H}{|}}{N}-CH_3$

19.43 Write the structural formula for each of the following compounds.
 (a) N-ethylpropanamide
 (b) N,N-dimethylbutanamide
 (c) N-phenylethanamide
 (d) pentanamide

19.44 Write the structural formula for each of the following compounds.
 (a) 2-methylbutanamide
 (b) 3-chlorobutanamide
 (c) N-methyl-4-bromobutanamide
 (d) N,N-dimethyl-2-ethylbutanamide

Properties of Amides

19.45 Amides contain a nitrogen atom, but yet are not basic. Explain why.

19.46 How many sites in acetamide can hydrogen bond to water?

19.47 Explain why N,N-dimethylformamide is soluble in water even though it has no nitrogen–hydrogen bonds.

19.48 Predict the order of boiling points of the following compounds.

(a) $CH_3CH_2\overset{\text{O}}{\underset{}{C}}-NH_2$ (b) $CH_3\overset{\text{O}}{\underset{}{C}}-NHCH_3$

(c) $H-\overset{\text{O}}{\underset{}{C}}-N(CH_3)_2$

Reactions of Amides

19.49 Write the products of each of the following reactions.

(a) 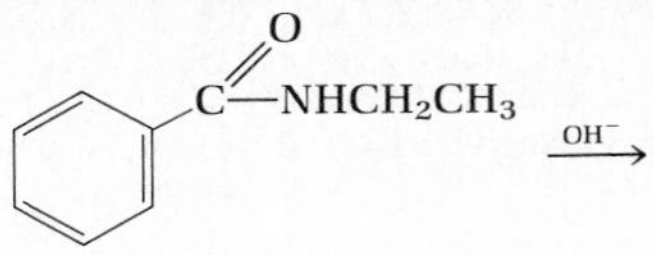 $C_6H_5-\overset{\text{O}}{\underset{}{C}}-NHCH_2CH_3 \xrightarrow{OH^-}$

(b) $CH_3CH_2\overset{\text{O}}{\underset{}{C}}-N(CH_3)_2 \xrightarrow{HCl}$

(c) 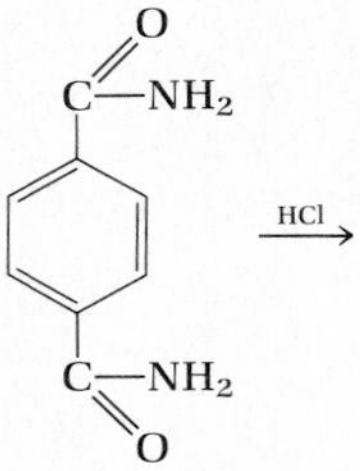 para-disubstituted benzene with $\overset{\text{O}}{\underset{}{C}}-NH_2$ and $\overset{}{\underset{\text{O}}{C}}-NH_2$ groups $\xrightarrow{HCl}$

(d) nicotinamide $\xrightarrow{OH^-}$

19.50 Write the products of each of the following reactions.

(a) $CH_3\overset{\text{O}}{\underset{}{C}}-NHCH_2CH_3 + OH^- \longrightarrow$

(b) $CH_3CH_2\overset{\text{O}}{\underset{}{C}}-NHCH_3 + H_3O^+ \longrightarrow$

(c) 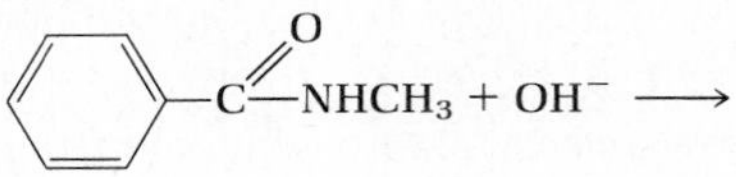 $C_6H_5-\overset{\text{O}}{\underset{}{C}}-NHCH_3 + OH^- \longrightarrow$

(d) 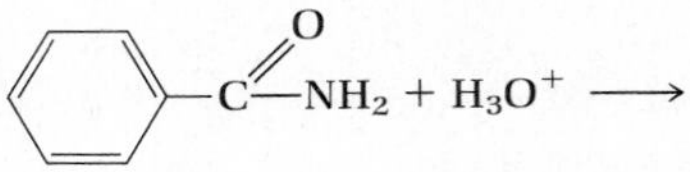 $C_6H_5-\overset{\text{O}}{\underset{}{C}}-NH_2 + H_3O^+ \longrightarrow$

Polyamides

19.51 Nylon is resistant to dilute acid or base, but polyesters are damaged by acid or base. Explain this difference.

19.52 Draw a representation of the condensation polymer formed by each of the following pairs of compounds.
 (a) terephthalic acid and 1,2-diaminoethane
 (b) adipic acid and 1,3-diaminopropane
 (c) succinic acid and hexamethylenediamine

Physiologically Active Nitrogen Compounds

19.53 Draw two aromatic heterocyclic rings found in some vitamins.

19.54 Draw two aromatic heterocyclic rings found in RNA and DNA.

19.55 What is the difference in the physiological effects of barbiturates and amphetamines?

19.56 What is the structural difference between morphine and heroin?

19.57 What is the structural difference between codeine and morphine?

19.58 Classify each type of nitrogen atom in LSD.

Introduction to Biochemistry

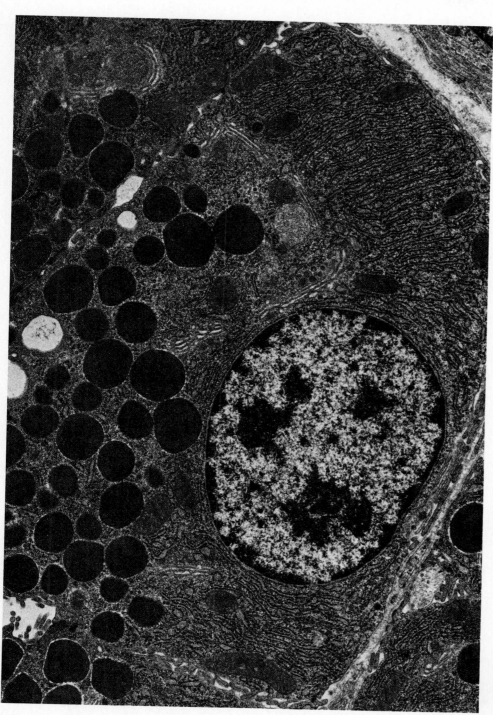

The cell is a complex structure composed of carbohydrates, lipids, proteins, and nucleic acids.

20.1
What Is Biochemistry?

Biochemistry *is the study of the composition, structure, and reactions of substances in living systems*. As you may recall from Chapter 1, the definition of chemistry is very similar to this definition. The only difference is the restriction to substances found in living systems. An alternative definition of biochemistry is the study of the molecular basis of living organisms.

Louis Pasteur might have been the first biochemist, since he had interest in the reactions that occur in living systems. In his classic experiments, he studied growing yeast cells. These cells cause the fermentation of grape sugar, which is converted into alcohol and a number of other products. Pasteur showed that the products of fermentation depend on the type of yeast cells used.

Hans and Eduard Büchner showed in 1897 that the yeast cells are not necessary for fermentation but only some of the chemicals contained in them. Cell-free yeast juices also ferment grapes! These juices contains catalysts called enzymes, which are proteins. All living cells contain thousands of types of enzymes, and each type catalyzes a different reaction in the cell. However, enzymes can be isolated from cells and purified like other chemicals. The enzymes are not living and can catalyze reactions outside a cell.

In 1905, it was shown by other scientists that phosphate, considered to be an inorganic substance, is also involved in fermentation. When phosphate ions are added to yeast juice, the rate of fermentation is increased. Thus, ordinary chemicals are also responsible for biochemical reactions.

Living systems, as compared to inanimate matter, have long fascinated chemists. However, in spite of the obvious remarkable differences, living matter obeys the same physical and chemical laws as any other type of matter. This and the following chapters of the text will show that while there are many series of reactions involved in any life process, each reaction can be understood based on material presented in earlier chapters. In essence, biochemistry is still chemistry.

One characteristic of living systems is great structural complexity. Even the simplest life forms such as the bacterium *Escherichia coli* (*E. coli*) contain many kinds of molecules that act in concert with each other. There are about 5000 different compounds in this bacterium. In humans there are probably millions of compounds. Each molecule has a specific function and usually cannot be substituted for by any other molecule. The broadest question of biochemistry, then, is how do individual "nonliving" molecules contribute to the maintenance, growth, and reproduction of life forms?

In this chapter we will define several classes of biomolecules that will then be discussed in detail in Chapters 21–23 and 29. In each class of biomolecules you will find familiar functional groups. Thus, you can anticipate the reactions that these molecules can undergo. The role of enzymes in the catalysis of biochemical reactions is presented in Chapter 24.

Living organisms extract and use energy from food to maintain their structure, to grow, and to reproduce. The method of efficiently using energy for these processes is a distinctive characteristic of living organisms. Inanimate matter tends to react to give products of lower chemical energy by spontaneous processes. Water flowing downhill is a physical example of this natural tendency to seek lower energy states. The methods by which energy is extracted, stored, and used will be discussed in Chapter 25, which is devoted to biochemical energy. Chapters 26–28 integrate the concepts developed in Chapter 25 with the specific reactions of metabolism of carbohydrates, fats, and amino acids.

Perhaps the most fascinating aspect of biochemistry is the study of the self-replication of organisms. Inanimate matter does not have the capacity to reproduce and transmit structural and chemical characteristics. Much of the excitement of biochemistry in the 1960s was in the discovery of the structure of deoxyribonucleic acid (DNA) and the evaluation of the genetic code. Chapter 29 is devoted to this subject.

There are over a million species of living organisms, but in spite of this diversity there are repeated patterns of both molecular structure and associated chemical reactivity. The same classes of compounds are used to create the structure of all organisms. Energy is extracted, stored, and used in the form of chemical energy contained in adenosine triphosphate (ATP) regardless of the organism. All forms of life, from the simplest bacterium to the highest life forms, store and transmit genetic information in the form of DNA. Thus, biochemistry really is the chemistry of a few classes of compounds and reactions regardless of the organism considered.

Biochemistry, biology, and ultimately medicine are strongly related. Discoveries in biochemistry often directly result in clinical application in medicine. For example, the presence in the blood of an enzyme that is normally in a specific organ may be the result of damage to the organ, which then leads to release of the enzyme into the blood. Both liver damage and myocardial infarction can be diagnosed by analysis of blood serum enzymes (Chapter 24).

Biochemical reactions are many times done outside of living systems. Such experiments are called **in vitro,** meaning in glass. Experiments and reactions studied in the living animal or plant are called **in vivo.** Of course, biochemists have to be careful to prove that in vitro experiments correspond to what actually occurs in vivo.

Experiments done in vitro have the advantage that selected reactions may be more easily studied. Each of the possible chemicals can be changed and the conditions of the reaction controlled. An in vivo experiment is subject to far more variables. However, ultimately the chemistry of living organisms is the interest of biochemists.

The pursuit of biochemical knowledge does not have, nor is it approaching, an end. Some biochemical processes such as metabolism are reasonably well understood. Great strides have been made in developing a theory of transmission of the genetic code. However, with the advance of knowledge come still more questions of even greater complexity. Furthermore, with the development of more sophisticated technology, there are more areas of biochemistry that can be probed.

20.2 Composition of Biochemicals

Of the 105 elements, only ten are contained in amounts greater than 0.1% of the human body (Table 20.1). Carbon, hydrogen, and oxygen, which constitute 93% of the compounds of the body, are present in virtually all biochemical compounds. Nitrogen, phosphorus, and sulfur are present in covalent molecules, while sodium, potassium, calcium, and chlorine are found as ions in "inorganic" substances.

Abundance alone does not necessarily indicate the importance of the elements to life. Iron is a part of the protein hemoglobin, which is present in the blood and is responsible for carrying oxygen. Iodine is vital to the proper functioning of the thyroid. Cobalt is at the center of the very complex vitamin B_{12}. Metals such as zinc, copper, and manganese are present in small amounts, but are essential for some enzymes to be functional. In each enzyme there are thousands of carbon, hydrogen, oxygen, and nitrogen atoms for every metal atom.

Nevertheless, without the metal the enzyme doesn't function in catalyzing the conversion of reactants into products.

Example 20.1

The molecular weight of hemoglobin $(C_{2952}H_{4664}O_{832}N_{812}S_8Fe_4)$ is approximately 65,000. What is the percentage of iron in hemoglobin?

Solution. By the definition of percentage composition developed in Chapter 1, the percentage of iron is equal to the mass of iron in the molecule divided by the mass of the molecule with the resulting quotient multiplied by 100. The mass of four iron atoms is four times the atomic weight of iron.

$$4 \times 55.8 \text{ amu} = 223 \text{ amu}$$

The percentage of iron is given by

$$\frac{223 \text{ amu}}{65,000 \text{ amu}} \times 100 = 0.34\%$$

In order to understand biochemistry we must be concerned with exactly how the atoms are combined in compounds. More specifically, it is convenient as always to organize the many compounds into classes. Then the physical and chemical properties of each class may be understood based on structure.

Water is the most abundant compound in living organisms. Human tissue is about 75% water, whereas human blood is about 80% water. The *E. coli* cell is about 70% water (Table 20.2). While water is very important to life, it is usually not considered a biochemical. After water, the four most abundant classes of compounds in most life forms are proteins, carbohydrates, lipids, and nucleic acids. The approximate percentages for the *E. coli* cell are given in Table 20.2.

Although proteins and carbohydrates are familiar to you, the terms *lipids* and *nucleic acids* are less well known. Lipids are a conglomerate class of compounds that include fats and oils as well as several other types of compounds. Note then that the three classes of compounds discussed in diets—proteins, carbohydrates, and fats—are also important classes of compounds in biochemistry. Nucleic acids are better known by the abbreviations for two members of this class, DNA (deoxyribonucleic acid) and RNA (ribonucleic acid).

Table 20.1

Composition of the Human Body

Element	% by weight	Element	% by weight
oxygen	65	sulfur	0.2
carbon	18	sodium	0.1
hydrogen	10	chlorine	0.1
nitrogen	3	magnesium	0.05
calcium	2	iron	trace
phosphorus	1.1	iodine	trace
potassium	0.3	bromine	trace

Table 20.2 ――――――――
Composition of an E. coli Cell

Component	% by weight	Number of types of component
water	70	1
proteins	15	3000
carbohydrates	3	50
lipids	2	40
nucleic acids		
DNA	1	1
RNA	6	1000
simple organic compounds	2	500
inorganic ions	1	12

Carbohydrates *are polyhydroxy aldehydes or ketones and compounds that can be hydrolyzed to produce polyhydroxy aldehydes and ketones.* Glyceraldehyde is the simplest example of a carbohydrate.

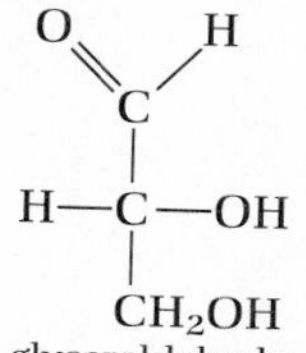

glyceraldehyde

You already know that alcohols can be oxidized and can form esters and that aldehydes may be oxidized or reduced and can form hemiacetals and acetals. Thus, based on Chapters 16 and 17, you are prepared for the study of carbohydrates presented in Chapter 21.

Proteins *are polyamides of a variety of amino acids, which are molecules that contain both an acid group and an amino group.* Alanine is one of the amino acids found in proteins.

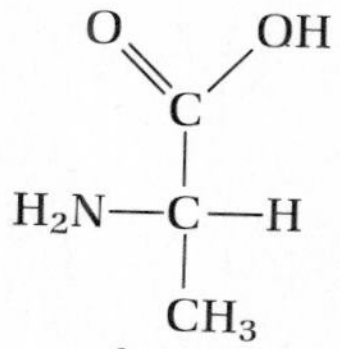

alanine

The chemistry of acids was given in Chapter 18, whereas that of amines was in Chapter 19. Acids and amines combine to give amides, as previously indicated in Chapter 19. The chemistry of proteins, which is in part that of the amide functional group, will be presented in Chapter 23. Some amino acids contain other functional groups such as hydroxyl and sulfhydryl, which also play a role in the chemistry of proteins.

Fats *of the lipid class are esters of glycerol and long-chain fatty acids, such as stearic acid.* The chemistry of such substances is based on the reactions given in

Chapter 18. Phosphoric acid also forms esters with glycerol as part of the structure of phospholipids. The phospholipids are important in formation of cell membranes.

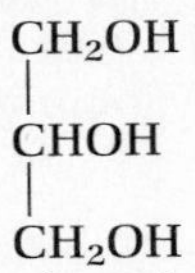

$$CH_2OH$$
$$|$$
$$CHOH \qquad CH_3(CH_2)_{16}CO_2H$$
$$|$$
$$CH_2OH$$
glycerol   stearic acid

$$\begin{array}{c} O \\ \parallel \\ HO-P-OH \\ | \\ OH \end{array}$$
phosphoric acid

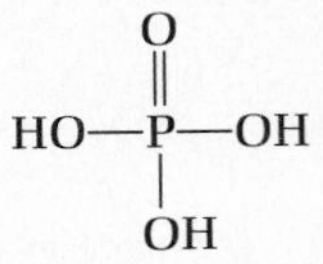

There are even more complex compounds that contain combinations of carbohydrates with either proteins or lipids. These compounds are called glycoproteins and glycolipids, respectively. The basis of these names will be discussed in Chapter 21 when the chemistry of carbohydrates is considered.

Nucleic acids *are composed of some pyrimidine and purine bases* (Chapter 19), *the carbohydrates ribose and deoxyribose, and the inorganic acid phosphoric acid.* These substances will be discussed in Chapter 29.

Example 20.2

Based on their structural formulas, classify ribose and serine.

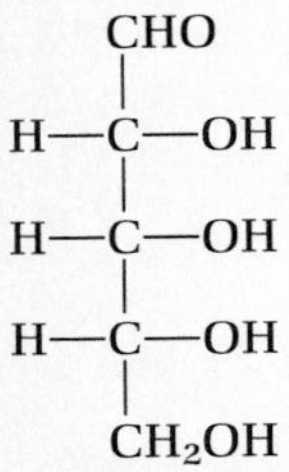

$$\begin{array}{c} CHO \\ | \\ H-C-OH \\ | \\ H-C-OH \\ | \\ H-C-OH \\ | \\ CH_2OH \end{array}$$
ribose

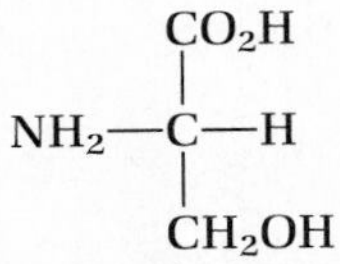

$$\begin{array}{c} CO_2H \\ | \\ NH_2-C-H \\ | \\ CH_2OH \end{array}$$
serine

Solution. Ribose contains an aldehyde functional group and four hydroxyl groups. Thus, ribose is a polyhydroxy aldehyde and is classified as a carbohydrate.

Serine contains a carboxyl group, an amino group, and a hydroxyl group. The compound is classified as an amino acid based on the presence of the amino group and the carboxyl group. Note that the compound structurally resembles alanine, an amino acid.

[Additional examples may be found in **20.8–20.10** at the end of the chapter.]

20.3 Biochemical Reactions

Biochemical reactions are governed by the same chemical principles as any other reaction. Thus, kinetics and equilibria (Chapter 10) must be considered in the study of biochemical reactions. There are some differences in studying biochemical reactions as compared to other reactions. Biochemical processes involve many separate but interdependent reactions.

Biochemists use the general term *metabolism* to describe all of the reactions that occur in a living organism. The molecules that are absorbed by the body may be either built into more complex molecules or broken into simpler molecules.

Anabolic reactions (anabolism) *are metabolic reactions that build (synthesize) more complex molecules.* **Catabolic reactions (catabolism)** *are reactions that break down (degrade) larger molecules into smaller molecules.* Both anabolic and catabolic reactions are constantly occurring in living organisms, which integrate these reactions to maintain cellular functions.

Most anabolic and catabolic reactions are catalyzed by enzymes. Many enzymes are entirely protein, while others consist of protein and another substance, called a coenzyme. Enzymes are needed for a reaction to go fast enough to support the living cell. As a result of the enzymes, organisms do not need to resort to either the high temperatures or strongly acidic or basic conditions that must be used in the laboratory to increase the rate of a reaction. Enzymes will be discussed in Chapter 24.

Anabolic reactions require energy; catabolic reactions release energy. *The amount of energy required or released in a chemical reaction is called the* **change in free energy** *and is symbolized by* $\Delta G°$. When energy is released by a reaction, the $\Delta G°$ is negative. Conversely, a positive $\Delta G°$ means that energy must be supplied for the reaction to occur. The units used for $\Delta G°$ are kcal/mole. Recall from Section 10.6 that free energy changes represent both enthalpy and entropy changes.

A compound contains a certain amount of stored free energy in its bonds (Section 10.6). When a reaction occurs, the product may contain more or less stored free energy since its bonds are different. *If the product has lower free energy, the reaction is* **exergonic.** *If the product has higher chemical energy, the reaction must be* **endergonic.**

The energy balance of living organisms is provided by ATP (Chapter 25) and its related compound, ADP. Energy released in catabolic reactions has to be stored to provide energy for anabolic reactions. This is accomplished by converting ADP into ATP, which is an endergonic process. The energy released by the exergonic catabolic reaction is saved in ATP, which is produced from ADP, much as money earned is placed in a bank until needed. The ATP level in the organism (bank account) is then increased.

When energy is required for an anabolic reaction, it is supplied by converting ATP into ADP.

$$\text{ATP} + \text{H}_2\text{O} \longrightarrow \text{ADP} + \text{phosphate} + 7.3 \text{ kcal/mole}$$

The bank account of energy is then depleted on demand of the organism. The energy released by the hydrolysis of ATP is 7.3 kcal/mole. In terms of the change of free energy $\Delta G° = -7.3$ kcal/mole.

There are several other phosphoric acid derivatives that store and provide energy in metabolic reactions. Phosphoenolpyruvate releases 14.8 kcal/mole, whereas creatine phosphate releases 10.3 kcal/mole.

phosphoenolpyruvate   creatine phosphate

Further details of biochemical energy will be given in Chapter 25. However, it should be noted at this point that energy is constantly being produced and used in living systems. Endergonic reactions occur only because exergonic reactions also occur.

Example 20.3

The hydrolysis of phosphoenolpyruvate has $\Delta G° = -14.8$ kcal/mole. What is the $\Delta G°$ for the reverse of this reaction in which phosphoenolpyruvate is formed?

$$\begin{array}{ccc} CO_2H & O & \\ | & \| & \\ C-O-P-OH + H_2O & \longrightarrow & \\ \| & | & \\ CH_2 & OH & \end{array} \quad \begin{array}{c} CO_2H \\ | \\ C=O + H_3PO_4 \\ | \\ CH_3 \end{array}$$

Solution. Since energy can neither be created nor destroyed in ordinary chemical reactions, the energy released by a reaction must be gained by the reverse of that process. Thus, the formation of phosphoenolpyruvate requires energy and $\Delta G° = +14.8$ kcal/mole.

[Additional examples may be found in **20.12–20.14** at the end of the chapter.]

20.4
The Cell

In our discussions of the reactions of biomolecules we will relate both structure and reactivity to cells and their components. Thus, it is useful to describe at this point some of the structural features of cells. Cells are divided into two classes, eukaryotic and prokaryotic. Eukaryotic comes from the Greek *eu* for true and *karyon* for nucleus. **Eukaryotes** *are cells that contain nuclei, whereas* **prokaryotes** *don't.* Eukaryotic cells are found in higher classes of organisms and some complex microorganisms such as yeast. Perhaps the most widely studied eukaryotic cell is the liver parenchymal cell. Rat livers are frequently used in biochemical studies because of their availability and the ease with which they can be separated into cell components. The bacterium *Escherichia coli* (*E. coli*) is the best-studied example of a prokaryotic cell.

Eukaryotic cells contain many distinctive structures in addition to the nucleus. *The components of a cell are called* **organelles.** Some of these organelles are the mitochondria, Golgi complex, endoplasmic reticulum, and lysosomes (Figure 20.1). These organelles provide for the separation of the many reactions that must simultaneously occur within a single cell. Thus, both catabolic and anabolic reactions can take place within a cell. For example, in one part of the cell, glucose may be converted into glycogen, one of the energy-storing compounds. In another part of the cell, glucose may be oxidized to carbon dioxide and water. The relationship of the organelles to the cell, the macromolecules, and simpler biochemicals is shown in Figure 20.2.

Membranes

Membranes separate cells from their environment as well as organelles from each other within the cell. Membranes are indispensable for the organization and proper operation of the cell. They have selective permeability—that is, they allow

Figure 20.1
The Cell.

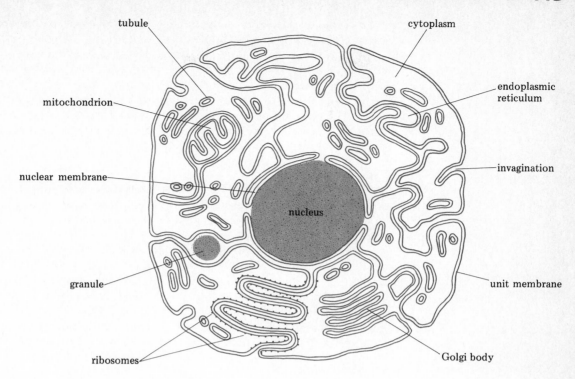

certain molecular and ionic species to pass through them. Internal membranes form the boundaries of the organelles.

While membranes differ widely both in structure and function, there are some common features. Membranes are usually only a few molecules thick and are between 60 and 100 Å. The membranes mainly consist of lipids and proteins in ratios varying from 1:4 to 4:1. Carbohydrates are bonded to some lipids to form glycolipids and to some proteins to form glycoproteins.

Passage of molecules through a membrane depends on the components of the membrane. In a sense the control is similar to the rule of solubilities that likes dissolve likes. Lipids allow the passage of some nonpolar molecules and restrict

Figure 20.2
Molecular Organization of the
Components of a Cell.

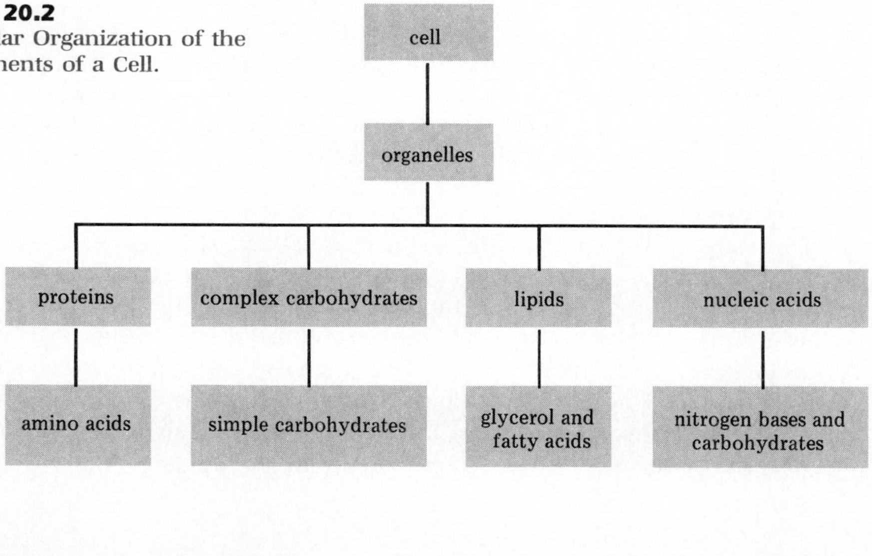

cell

organelles

proteins — amino acids

complex carbohydrates — simple carbohydrates

lipids — glycerol and fatty acids

nucleic acids — nitrogen bases and carbohydrates

polar molecules. Specific proteins with polar sites are responsible for passage and transport of polar molecules and ions.

The proteins and lipids are held together in the membrane by noncovalent interactions that are not sufficiently strong to hold every molecule rigidly in place. The structure is fluid and allows the movement of lipids in the plane of the membrane. Proteins can move as well, but are more restricted by interactions within the membrane. Membranes have two surfaces, and the inside and outside surfaces have different compositions and structures. Further details of membrane structure will be given in Chapter 22.

The Nucleus

The **nucleus** *is about 5 μm in diameter and is surrounded by a double membrane with a space in between.* The details of the transport of material across the nuclear membrane are still not well understood, but it is known that large pores exist in the membrane.

Pairs of chromosomes are found in nuclei of eukaryotic cells. In human cells there are 22 pairs of chromosomes plus a pair of X chromosomes in females or an X and a Y chromosome in males. The DNA of these chromosomes is much larger than the DNA contained in the cytoplasm of prokaryotic cells.

The nucleus of the cell is a control center. It is responsible for directing the chemical reactions in the cell as well as the reproduction of the cell. DNA replication and formation of RNA (Chapter 29) occur in the nucleus.

Mitochondria

Mitochondria *are oval-shaped and are about 2 μm in length and 0.5 μm in diameter.* A cell may contain several hundred to a thousand mitochondria depending on the function of the cell. These organelles provide the site at which many catabolic reactions occur. As a result, the energy released by the oxidation of food is stored as ATP in the mitochondria. For this reason the mitochondria are often called the power packs of the cell. Mitochondria are surrounded by two membranes, an outer and an inner membrane. These two membranes differ in composition, permeability, and enzymatic activity. Further details of the structure of the mitochondria will be given in Chapter 25.

Endoplasmic Reticulum

The **endoplasmic reticulum** *is a network of membrane-bound channels.* The fluid within the endoplasmic reticulum is different from the fluid outside the endoplasmic reticulum. For example, this fluid in liver cells contains enzymes that catalyze the formation of lipids. The surface of the endoplasmic reticulum is studded with structures called ribosomes, which are the sites of protein synthesis (Chapter 29). Proteins synthesized at the ribosomes can cross the membrane and pass through the channels to the periphery of the cell.

Golgi Complex

The **Golgi complex** *or* **Golgi apparatus** *is a stack of flattened membrane sacs.* The proteins formed at the endoplasmic reticulum are transported as glycoproteins via vesicles to the Golgi complex. In the Golgi complex, the temporarily stored glycoproteins are chemically altered. Some are converted into different glycoproteins for further transport. Vesicles then can carry some proteins to other parts of the cell while other proteins are secreted from the cell by the Golgi complex. In recent years evidence has appeared that not all of the proteins need to form glycoproteins in order to be secreted or used inside the cell.

Lysosomes

Lysosomes *are organelles, surrounded by a single membrane of protein and lipids, that contain granules of protein enzymes.* These aggregates of enzymes can hydrolytically break down other molecules. In the lysosomes glycogen is hydrolyzed into glucose and proteins are hydrolyzed into amino acids. The function of the lysosomes is to digest material brought into the cell. The membrane surrounding the lysosome prevents the contents from contacting other parts of the cell. Without this protection, the cell itself would be digested.

Example 20.4 ─────────────────────────────────

To gain a better idea of the length of mitochondria, express 2 μm in centimeters.

Solution. First, use the conversion factor to express 2 μm in meters.

$$2 \, \mu m \times \frac{1 \, m}{10^6 \, \mu m} = 2 \times 10^{-6} \, m$$

Next use the conversion factor to express 2×10^{-6} m in centimeters.

$$2 \times 10^{-6} \, m \times \frac{100 \, cm}{1 \, m} = 2 \times 10^{-4} \, cm$$

The structure of a molecule refers to the order in which atoms are bonded to each other and the types of bonds between them. In addition to structure, configuration is important to a complete description of molecules. **Configuration** *refers to the spatial arrangement of atoms.* We saw that *cis-trans* isomerism is possible when there is restricted rotation about carbon–carbon bonds as in cycloalkanes and alkenes. These isomers are called geometric isomers. However, such isomers are but one type of a more general class of stereoisomers. **Stereoisomers** *have the same structure but different configurations.* One type of stereoisomerism that is very important in biochemistry is called optical isomerism. Optical isomers are the result of the asymmetry of molecules.

Many molecules are symmetrical. A molecule is symmetrical if it can be reoriented in space so that it is indistinguishable from the original orientation. Cyclopentane (Figure 20.3) can be rotated about the axis perpendicular to the plane of the molecule. For every 360°/5 or 72° of rotation, a new orientation results

20.5
Structure and Configuration of Biochemicals

Figure 20.3
Axis and Plane of Symmetry of Molecules.

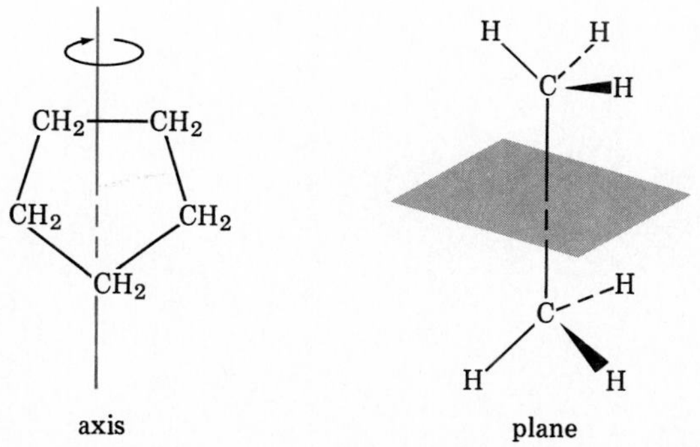

axis

plane

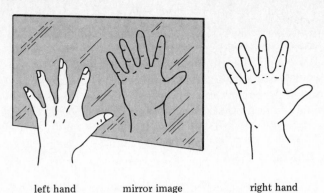

Figure 20.4
Mirror-Image Relationships.

left hand   mirror image   right hand
of left hand

that is identical to the original. *Molecules that can be rotated about an axis to yield an equivalent structure indistinguishable from the original have an axis of symmetry.*

If a plane can be passed through a molecule to produce two halves that are mirror reflections of each other, the molecule has a **plane of symmetry.** Ethane has a plane of symmetry, illustrated in Figure 20.3, which divides it in half. Each half is the mirror image of the other.

Molecules with an axis or plane of symmetry have mirror images that are superimposable on or identical to themselves. **Asymmetric** *molecules lack a center or plane of symmetry and have mirror images that are not superimposable on the original molecule.* In order to see this feature of molecules better, let's first consider human hands. A person's hands are asymmetric. The left hand is the mirror image of the right hand and vice versa (Figure 20.4). The left hand is not superimposable on (identical to) its mirror image (right hand).

In Figure 20.5 a ball and stick model of bromochloromethane is illustrated. The mirror image of bromochloromethane can be rotated 180° about the carbon–chlorine bond to yield a structure that is identical to the original. The two models can be brought together so that each sphere in one model meets a corresponding sphere in the other model. Bromochloromethane has a plane of symmetry that is the plane of the page. Any time that a tetrahedral carbon atom has two or more identical atoms or groups attached, the molecule will have a plane of symmetry. As a consequence the molecule and its mirror image are superimposable.

When four different atoms or groups of atoms are attached to a tetrahedral carbon atom, the molecule is asymmetric. There are two possible

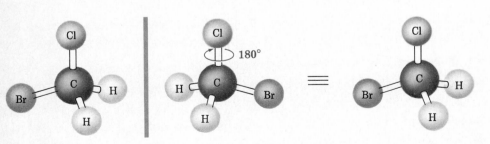

Figure 20.5
Equivalence of Mirror Images in a Symmetric Molecule.

mirror

Table 20.3

Properties of Enantiomers

Compound	Melting point (°C)	Solubility (g/100 mL)	$[\alpha]_D^{25°}$
L-lactic acid	53	∞	+3.8
D-lactic acid	53	∞	−3.8
L-alanine	297	16.5	+2.7
D-alanine	297	16.5	−2.7

arrangements in space for the groups of atoms. Note that the mirror images of bromochlorofluoromethane, illustrated in Figure 20.6, cannot be completely superimposed.

Stereoisomers that are nonsuperimposable mirror images of each other are **enantiomers.** Enantiomers of substances such as lactic acid and alanine have identical physical properties and chemical properties in a symmetrical environment. The physical properties of two substances, lactic acid and alanine, that can exist as enantiomeric pairs are listed in Table 20.3. The use of the prefix D or L to represent enantiomers will be discussed in Section 20.7.

$$\overset{\text{OH}}{\underset{\text{lactic acid}}{\text{CH}_3\text{CHCO}_2\text{H}}} \qquad \overset{\text{NH}_2}{\underset{\text{alanine}}{\text{CH}_3\text{CHCO}_2\text{H}}}$$

Enantiomers are distinguishable in an asymmetric environment. Thus, reactions with other asymmetric molecules will proceed differently. It is this difference that is the basis for life processes. In the next three chapters the exclusive

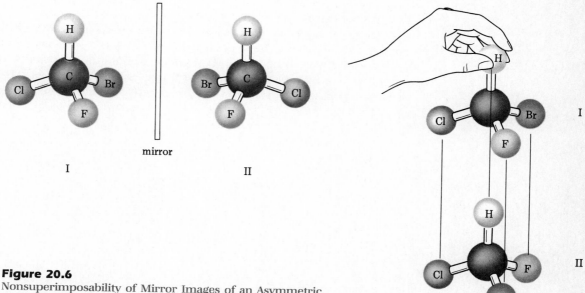

Figure 20.6
Nonsuperimposability of Mirror Images of an Asymmetric Molecule.

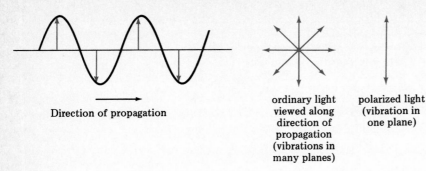

Figure 20.7
Representation of Light as a Wave Motion.

Direction of propagation

ordinary light viewed along direction of propagation (vibrations in many planes)

polarized light (vibration in one plane)

formation and reaction of one isomer rather than its enantiomer will be illustrated many times.

Enantiomers do behave differently toward plane-polarized light. It is this one physical property that distinguishes one enantiomer from the other in a pair of enantiomers. **Enantiomers rotate the plane of polarized light and are said to be optically active.**

20.6
Optical Activity

Plane-Polarized Light

White light consists of a wave motion in which the waves have a variety of lengths (which correspond to different colors) and are vibrating in all possible planes. Light can be monochromatic (one color) if filters are used to eliminate the other colors or if it is generated by a special source. A sodium lamp, a monochromatic source of yellow light, is often used in the study of optical activity. Although the light consists of waves of only one wavelength, the individual light waves are vibrating in all possible planes (Figure 20.7).

A Nicol prism has the property of a screen to allow the passage of light waves vibrating in a single plane. *Light consisting of waves vibrating in a single plane is called* **plane-polarized light.** Figure 20.8 shows the action of Nicol prisms.

The Polarimeter

An instrument called a polarimeter is used to measure the optical activity of compounds. It has two Nicol prisms, one of which is a polarizer and the other an analyzer. Between these two portions of the apparatus is a polarimeter tube that contains a solution of the compound to be examined. Compounds that do not affect the passage of the plane-polarized light are said to be optically inactive.

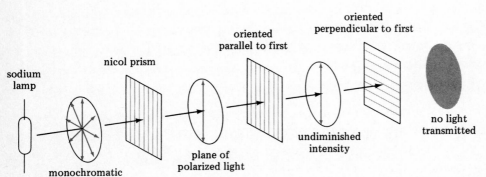

sodium lamp

nicol prism

monochromatic light

plane of polarized light

oriented parallel to first

undiminished intensity

oriented perpendicular to first

no light transmitted

Figure 20.8
Effect of Nicol Prism on Light Transmission.

Figure 20.9
Representation of a Polarimeter.

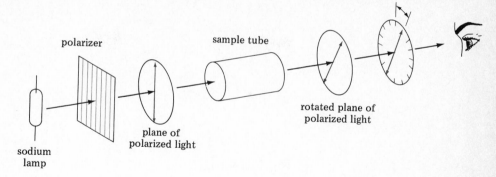

Optically active compounds alter the plane of polarized light, and it is necessary to rotate the analyzer in order to allow the light to emerge. The number of degrees and the direction of rotation of the analyzer correspond to the angle that the light is rotated by the compound (Figure 20.9).

Specific Rotation

The specific rotation, symbolized by $[\alpha]$, is the number of degrees of rotation of a solution at a concentration of 1 g/mL in a 1 decimeter (dm) long tube. The specific rotation depends on the temperature and the light source used. The specific rotation for experiments carried out at 25°C with the yellow D line of the spectrum of sodium is symbolized by $[\alpha]_D^{25°}$. *Substances that rotate the plane of polarized light in a clockwise direction are* **dextrorotatory.** *Those that rotate light in a counterclockwise direction are* **levorotatory.** Dextrorotatory and levorotatory are symbolized by the use of positive and negative signs, respectively. Thus, (+)-glucose is the dextrorotatory enantiomer.

Enantiomers rotate polarized light the same number of degrees but in the opposite direction (Table 20.3). In nature, enantiomers do not occur in pairs like right and left hands. Usually only one member of an enantiomeric pair is produced or can be used by an organism. One exception is lactic acid. The isomer with the positive rotation occurs in human muscles, whereas its enantiomer with a negative rotation occurs in some bacteria.

Lactic acid, $CH_3CH(OH)CO_2H$, contains a carbon atom with four different groups about it and exists as two enantiomeric forms. These are illustrated in Figure 20.10 in which their mirror-image relationship is shown. In order to write two-dimensional formulas that represent these three-dimensional molecules, a convention is used. The carboxyl group, the methyl group, and the asymmetric carbon atom are arranged vertically and the asymmetric carbon atom is placed in the plane of the paper. Both the carboxyl group and the methyl group are behind the plane of the page, whereas the hydrogen atom and the hydroxyl group are above the plane. The projection of this arrangement on a plane is the standard two-dimensional representation of lactic acid and is called a Fischer projection formula.

The two enantiomeric lactic acid molecules present an interpretational problem. In projecting the atoms into a two-dimensional representation, the es-

20.7
Representing Enantiomers

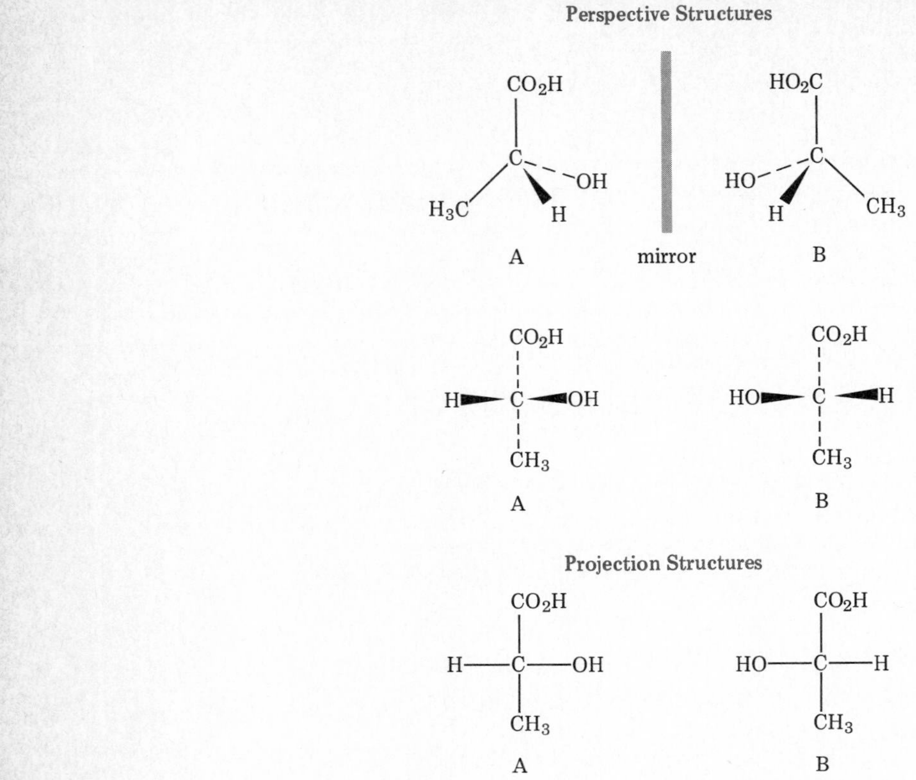

Perspective Structures

Figure 20.10
Enantiomers of Lactic Acid.

A   mirror   B

A   B

Projection Structures

A   B

sence of the asymmetry, the three-dimensionality, has been lost. Of itself, a two-dimensional structure can never be asymmetric, since it has a plane of symmetry.

If one representation were lifted out of the plane and flipped over, it would become identical to the second. An error results from ignoring the three-dimensionality of molecules. If molecule A were flipped over, the carboxyl group and methyl group, originally behind the plane, would be in front of the plane. These groups do not occupy identical positions with respect to the carboxyl group and methyl group of molecule B, which are behind the plane. In order to avoid the error of apparently achieving a two-dimensional equivalence of nonequivalent three-dimensional molecules, it is important not to lift two-dimensional representations out of the plane of the paper.

Two projection formulas can be drawn to depict any pair of enantiomers. Early chemists did not know which representation corresponded to which compound because they were unable to see the arrangement of the atoms about carbon in space. Therefore it was necessary to assign arbitrarily a relative configuration to one of the members of the enantiomeric pair.

As a standard, the enantiomer of glyceraldehyde that rotates plane-polarized light in a clockwise direction was assigned the configuration with the hydroxyl group on the right side in the projection formula.

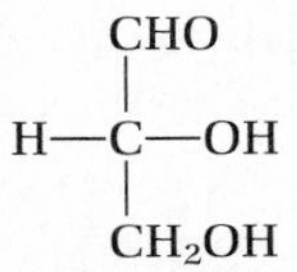

D-(+)-glyceraldehyde

The prefix D indicates the configuration at the asymmetric carbon; the symbol (+) indicates that the compound is dextrorotatory. The mirror image compound is L-(−)-glyceraldehyde and corresponds to the structure in which the hydroxyl group is on the left in the projection formula.

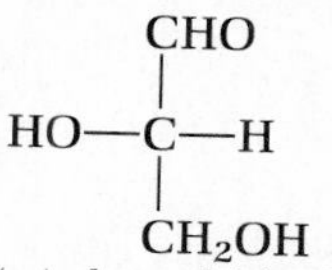

$$\begin{array}{c} CHO \\ | \\ HO{-}C{-}H \\ | \\ CH_2OH \end{array}$$
L-(−)-glyceraldehyde

It rotates plane-polarized light in a counterclockwise direction. The designations D and L refer to configuration only and are not to be interpreted as dextrorotatory and levorotatory. The direction of rotation of light is symbolized by the plus and minus signs.

The D or L configuration of all other asymmetric compounds can be determined by chemically relating them to either D- or L-glyceraldehyde. If D-(+)-glyceraldehyde is converted into lactic acid by converting the CH_2OH group into CH_3 and the CHO group into CO_2H, then the resulting lactic acid must have the same relative configuration as D-(+)-glyceraldehyde. The lactic acid that results is levorotatory and is designated D-(−)-lactic acid. This one transformation illustrates that the D and L notations do not refer to the direction of rotation of plane-polarized light.

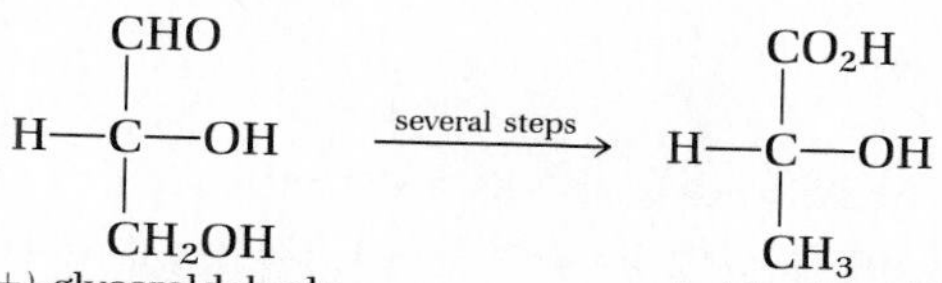

$$\begin{array}{c} CHO \\ | \\ H{-}C{-}OH \\ | \\ CH_2OH \end{array} \xrightarrow{\text{several steps}} \begin{array}{c} CO_2H \\ | \\ H{-}C{-}OH \\ | \\ CH_3 \end{array}$$
D-(+)-glyceraldehyde   D-(−)-lactic acid

A similar conversion of L-(−)-glyceraldehyde yields L-(+)-lactic acid.

In 1950, the actual or absolute configuration of an optically active substance was finally determined by x-ray analysis. The absolute arrangement of the atoms corresponded to those assigned based on the relative configuration of glyceraldehyde. Therefore the original guess was correct.

Example 20.5

What is the configuration of alanine shown as a Fischer projection in Section 20.2? What is the sign of its optical rotation?

Solution. The carbon chain is written vertically and the amino group is on the left. Thus, the compound is of the L configuration.

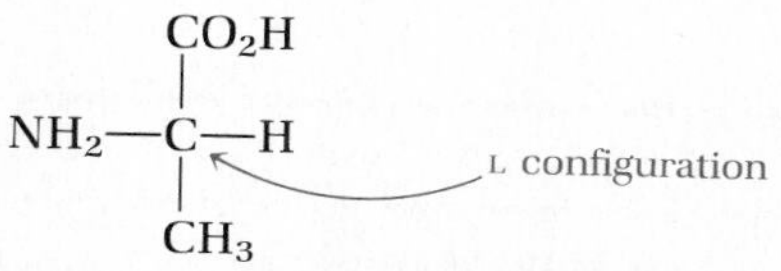

$$\begin{array}{c} CO_2H \\ | \\ NH_2{-}C{-}H \\ | \\ CH_3 \end{array}$$ L configuration

The sign of the rotation is not related to configuration in any way and therefore cannot be predicted.

[Additional examples may be found in **20.23–20.28** at the end of the chapter.]

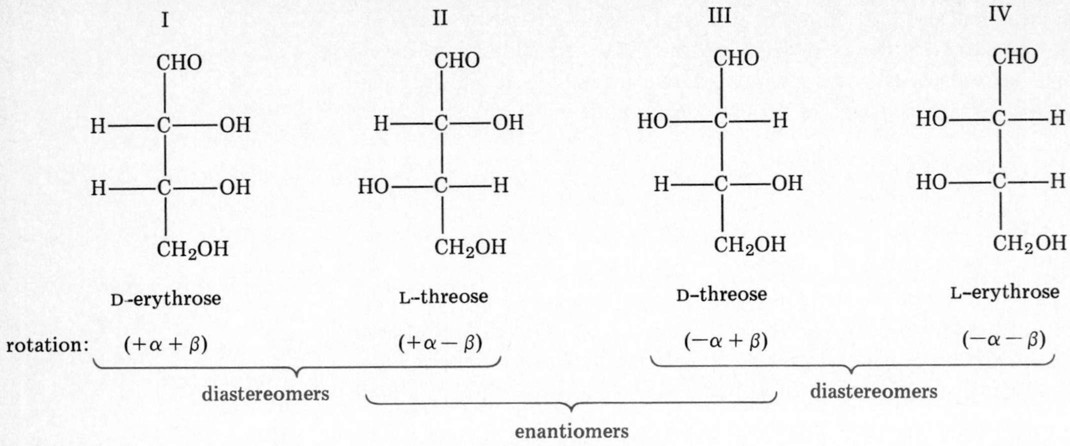

I	II	III	IV

Figure 20.11
Enantiomers and
Diastereomers.

CHO	CHO	CHO	CHO
H—C—OH	H—C—OH	HO—C—H	HO—C—H
H—C—OH	HO—C—H	H—C—OH	HO—C—H
CH$_2$OH	CH$_2$OH	CH$_2$OH	CH$_2$OH
D-erythrose	L-threose	D-threose	L-erythrose

rotation: $(+\alpha + \beta)$ $(+\alpha - \beta)$ $(-\alpha + \beta)$ $(-\alpha - \beta)$

diastereomers   diastereomers

enantiomers

20.8
Compounds with Multiple Nonequivalent Asymmetric Centers

The number of optically active isomers for a molecule containing n nonequivalent asymmetric carbon atoms is 2^n. The term *nonequivalent* means that the asymmetric carbon atoms do not contain identical sets of substituents. Consider the following formula.

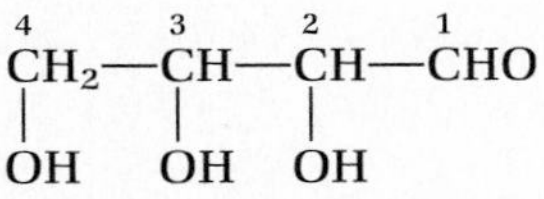

$$\overset{4}{CH_2}-\overset{3}{CH}-\overset{2}{CH}-\overset{1}{CHO}$$
$$\;\;\;|\quad\;\;\;|\quad\;\;\;|$$
$$\;\;OH\quad OH\quad OH$$

The two internal carbon atoms numbered 2 and 3 are both asymmetric and nonequivalent. Both asymmetric centers contribute to the total rotation of the molecule, but their contributions are not identical since the bonds from each atom are not identical. If the contribution to the rotation for carbon 2 is either $+\alpha$ or $-\alpha$, depending on its configuration, and if that for carbon 3 is $+\beta$ or $-\beta$, then four possible total rotations can be written: $+\alpha + \beta$ and $+\alpha - \beta$; $-\alpha + \beta$ and $-\alpha - \beta$. Figure 20.11 shows two-dimensional representations of the four isomers.

Structures II and III are mirror images, are not superimposable on each other, and are classified as enantiomers. This can be verified in two dimensions by imagining a mirror placed between II and III. Neglecting the fact that reversal of the printed letter occurs and remembering that the letters actually represent atoms, we see that the mirror image of any portion of molecule II is equivalent to the same region in molecule III. The sum of the rotatory contributions of II $(+\alpha - \beta)$ is equal but opposite in sign to that of III $(-\alpha + \beta)$.

Structures I and IV are mirror images and are nonsuperimposable. These enantiomers rotate light in opposite directions but by the same absolute value.

Table 20.4

Properties of Diastereomers

Compound	Melting point (°C)	Solubility in ethanol	$[\alpha]_D^{25°}$
D-erythrose	(liquid)	very	−21.5
L-erythrose	(liquid)	very	+21.5
D-threose	130	slightly	+29.1
L-threose	130	slightly	−29.1

Structures I and II are called diastereomers. **Diastereomers** *are pairs of stereoisomers that are not enantiomers.* The pairs II and IV, I and III, and III and IV are all diastereomeric pairs. The physical properties of enantiomers are the same, but because diastereomers are molecules with different internal relative configurations, they have different physical properties (Table 20.4).

Example 20.6 _____

ʟ-Threonine, an amino acid, has the following condensed molecular formula. Write the Fischer projections of the possible stereoisomers.

$$CH_3CH(OH)CH(NH_2)CO_2H$$

Solution. Both carbon 2 and carbon 3 are asymmetric. Thus, four structures are possible. The Fischer projections are written by placing the carboxyl group at the top of the vertical chain. The amino and hydroxyl groups may be on the right or left side of the projection formulas.

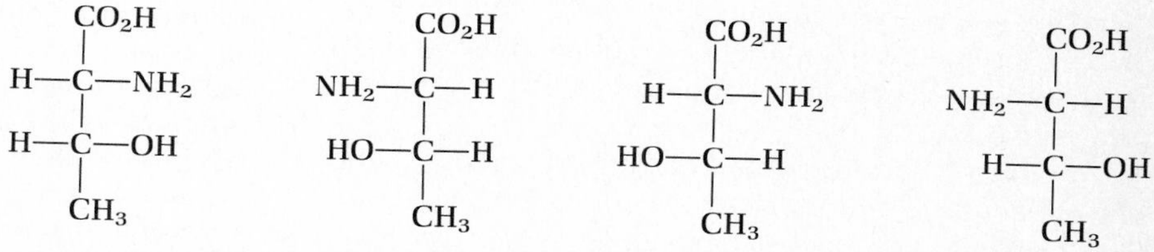

The actual structure of ʟ-threonine in proteins is given by the Fischer projection at the right.

20.9
Synthesis of Stereoisomers

Any laboratory synthesis starting from compounds that are not asymmetric, which results in an asymmetric compound, always yields an equimolar mixture of ᴅ and ʟ forms. *An equimolar mixture of enantiomers is called a* **racemic mixture.** The individual molecules of the mixture are optically active, but any rotation of plane-polarized light by the ᴅ form is canceled by the opposite rotation of the ʟ form.

Chemical reactions carried out in living systems lead to the formation of optically active substances. The compounds of the living system include enzymes, hormones, carbohydrates, lipids, proteins, and nucleic acids, which are all optically active. When a compound containing an asymmetric center is produced in an asymmetric environment, only one enantiomer results.

The simplest analogy for the formation of one isomer in biological systems that are themselves optically active involves the use of gloves and hands. Both gloves and hands are asymmetric. Only the right hand will fit properly into the right glove. The left hand does not fit in the right glove. If one were to make a glove for the right hand, it would have to be constructed in one specific way. In biological systems, only one enantiomer will react because it physically and chemically fits with the asymmetric biological molecules present in the organism. Similarly, when a molecule with an asymmetric center is produced by an organism, only one enantiomer results.

Summary

The cell is the functional unit of biological activity. The two basic types of cells are the eukaryotes, which contain nuclei, and the prokaryotes, which do not. The principal components of eukaryotes are the membrane and the organelles: the nucleus, mitochondria, endoplasmic reticulum, Golgi complex, and lysosomes.

Although cells have great structural complexity, the compounds that make up the cell are of four types: carbohydrates, lipids, proteins, and nucleic acids. These classes are used to create the structure of all organisms.

Metabolism is a series of chemical reactions to obtain energy from food and to construct cellular material. Catabolic reactions degrade larger molecules into smaller molecules. Anabolic reactions build complex molecules from smaller molecules. The amount of energy released in catabolic reactions and required in anabolic reactions is symbolized by the change in free energy, $\Delta G°$.

Configuration, the spatial arrangement of atoms in a molecule, is important to an understanding of the biomolecules. Asymmetric molecules lack a plane or center of symmetry and can exist in enantiomeric forms. Enantiomers rotate the plane of polarized light. Compounds that rotate the plane of polarized light in a clockwise direction are dextrorotatory, whereas those that rotate the plane of polarized light in a counterclockwise direction are levorotatory.

Glyceraldehyde is used as a standard to define the configuration of optically active compounds. Compounds with multiple asymmetric carbon atoms can exist as 2^n stereoisomers. Stereoisomers that are not enantiomers are diastereomers.

Learning Objectives

As a result of studying Chapter 20 you should be able to

- Explain the terms in vitro and in vivo.
- List the elements important to living organisms.
- Describe the major classes of biochemicals.
- Distinguish between enantiomers and diastereomers.
- Interpret two-dimensional projection formulas as three-dimensional structures.
- Explain the difference between catabolic and anabolic reactions.
- Outline how energy is saved and used by a cell.
- Describe the organelles of a cell and their function.

New Terms

Anabolic reactions are metabolic reactions that require energy to build complex molecules from smaller molecules.

Asymmetric molecules lack a center or plane of symmetry and have mirror images that are not superimposable on the original molecule.

Biochemistry is the study of the composition, structure, and reactions of substances in living systems.

Carbohydrates are polydroxy aldehydes or ketones or compounds that can be hydrolyzed to produce polyhydroxy aldehydes or ketones.

Catabolic reactions release energy and degrade larger molecules into smaller molecules.

Configuration refers to the spatial arrangement of atoms.

Dextrorotatory compounds rotate the plane of polarized light in a clockwise direction.

Diastereomers are stereoisomers that are not enantiomers (are not mirror images of each other).

Enantiomers are stereoisomers that are mirror images of each other.

An **endergonic** reaction has products of higher free energy.

The **endoplasmic reticulum** is a network of membrane-bound channels in a cell.

Eukaryotes are cells that contain nuclei.

An **exergonic** reaction has products of lower free energy.

Fats are esters of glycerol and long-chain saturated carboxylic acids.

The **Golgi complex** is a stack of flattened membrane sacs in a cell.

Levorotatory compounds rotate the plane of polarized light in a counterclockwise direction.

Lipids are a class of biomolecules that include fats.

Lysosomes are organelles that contain granules of protein enzymes that can hydrolyze molecules.

Membranes separate cells from their environment as well as organelles from each other.

The **mitochondria** are oval-shaped organelles that store chemical energy by producing ATP.

The **nucleus** of a cell is the control center of a cell, directing chemical reactions of the cell as well as its reproduction.

Nucleic acids are composed of pyrimidine and purine bases, ribose or deoxyribose, and phosphoric acid.

Organelles are distinctive structural components of a cell.

Plane-polarized light consists of light waves vibrating in a single plane.

Prokaryotes are cells that do not contain a nucleus.

Proteins are polyamides of amino acids.

A **racemic mixture** is an equimolar mixture of enantiomers.

Stereoisomers are isomers with the same structure but different configurations.

An **in vitro** biochemical reaction is one done outside a living system.

An **in vivo** biochemical reaction is one that occurs in a living animal or plant.

Questions and Problems

Terminology

20.1 Distinguish between the terms in vivo and in vitro.

20.2 How are anabolic and catabolic reactions related to each other in living organisms?

20.3 Explain the difference between the terms enantiomer and diastereomer.

20.4 Describe how a eukaryotic cell and a prokaryotic cell differ.

Composition of Biochemicals

20.5 What are the three most abundant elements in the human body?

20.6 List three elements that exist in covalent biochemicals in less than 10% abundance.

20.7 List some inorganic ions that occur in significant amounts in the body.

20.8 What are the major classes of biochemicals?

20.9 Draw representative structures of each of the following biochemicals.

    (a) amino acid   **(b)** carbohydrate   **(c)** lipid

20.10 Classify each of the following biochemicals.

(a) 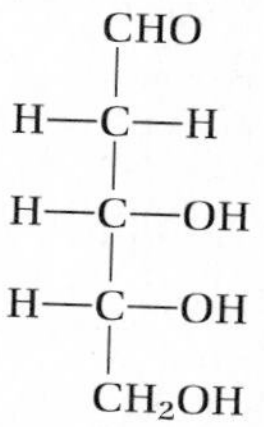
$$\begin{array}{c} CHO \\ | \\ H-C-H \\ | \\ H-C-OH \\ | \\ H-C-OH \\ | \\ CH_2OH \end{array}$$

(b)
$$\begin{array}{c} CO_2H \\ | \\ NH_2-C-H \\ | \\ CH_2 \\ | \\ \end{array}$$

(c)
$$\begin{array}{c} CH_2OH \\ | \\ C=O \\ | \\ H-C-OH \\ | \\ H-C-OH \\ | \\ CH_2OH \end{array}$$

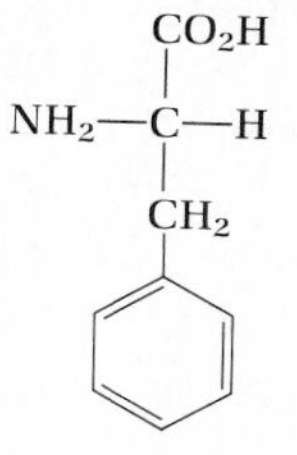

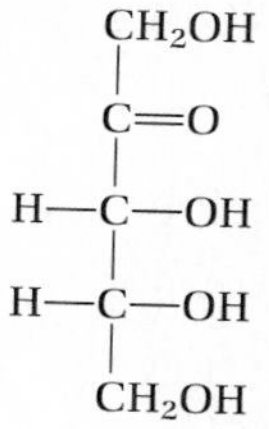

(d) 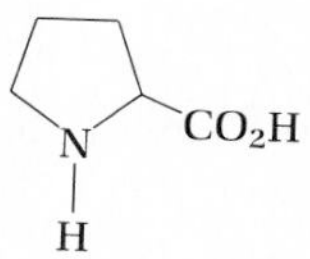
$$\begin{array}{c} \\ N-CO_2H \\ | \\ H \end{array}$$

(e) 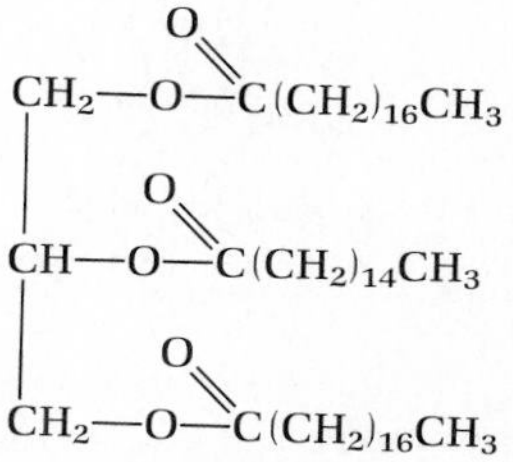
$$\begin{array}{c} O \\ \| \\ CH_2-O-C(CH_2)_{16}CH_3 \\ | \quad O \\ \quad \| \\ CH-O-C(CH_2)_{14}CH_3 \\ | \quad O \\ \quad \| \\ CH_2-O-C(CH_2)_{16}CH_3 \end{array}$$

(f) 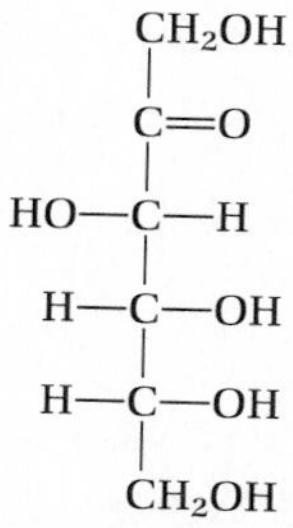
$$\begin{array}{c} CH_2OH \\ | \\ C=O \\ | \\ HO-C-H \\ | \\ H-C-OH \\ | \\ H-C-OH \\ | \\ CH_2OH \end{array}$$

20.11 Calculate the percentage of sulfur in hemoglobin (see Example 20.1).

Energy and Biochemistry

20.12 What is the sign of $\Delta G°$ for an anabolic reaction?

20.13 A reaction has a negative $\Delta G°$. Is the reaction exergonic or endergonic?

20.14 The hydrolysis of creatine phosphate is exergonic by 10.3 kcal/mole. What is the sign of $\Delta G°$ for the reaction?

20.15 What compound provides the energy for anabolic reactions?

Symmetry and Asymmetry

20.16 Which of the following structures can exist in optically active forms?

    (a) $CH_3CH_2CHBrCH_3$ **(b)** $CH_3CH_2CBr_2CH_3$

    (c) $CH_3CH_2CHBrCH_2CH_3$ **(d)** $CH_3CH_2CH_2CHBrCH_3$

    (e) $\underset{\underset{OH}{|}}{CH_3CHCO_2CH_3}$ **(f)** $HOCH_2CH_2CO_2CH_3$

    (g) $CH_3\overset{\overset{O}{\|}}{C}CO_2H$ **(h)** $CH_3\underset{\underset{OH}{|}}{CH}CH_2CO_2H$

20.17 Which of the following compounds has a plane of symmetry?

    (a) *cis*-1,2-dibromocyclobutane

    (b) *trans*-1,2-dibromocyclobutane

    (c) *cis*-1,3-dibromocyclobutane

    (d) *trans*-1,3-dibromocyclobutane

20.18 Ephedrine, a bronchodilator, has the following structural formula. How many asymmetric centers are in the molecule?

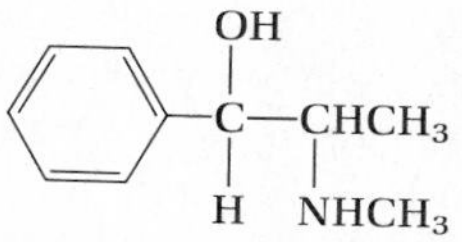

$$\begin{array}{c} OH \\ | \\ C-CHCH_3 \\ | \quad | \\ H \quad NHCH_3 \end{array}$$

20.19 Determine the number of asymmetric centers in the male sex hormone testosterone.

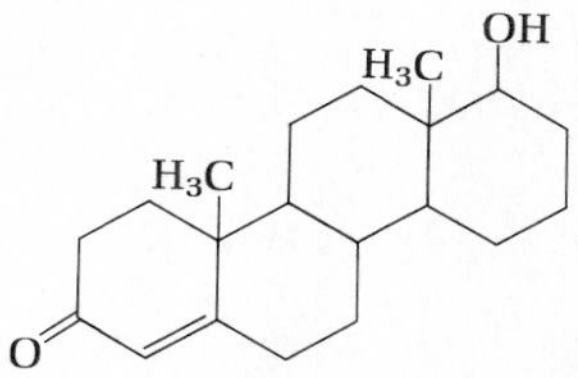

20.20 Determine the number of asymmetric centers in estradiol, one of the female sex hormones.

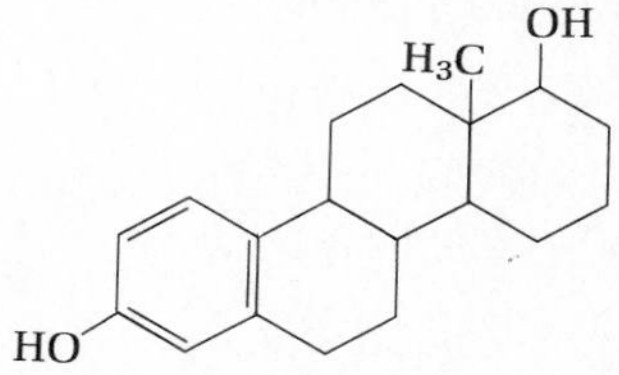

20.21 Citric acid is the central compound of the citric acid cycle, one of the processes to produce ATP for the body. How many asymmetric centers are in citric acid?

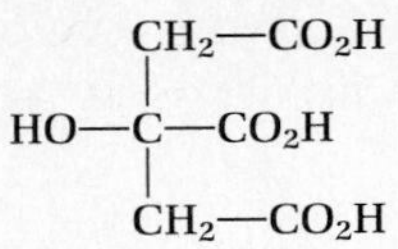

$$CH_2-CO_2H$$
$$HO-C-CO_2H$$
$$CH_2-CO_2H$$

20.22 D-Fructose, a carbohydrate that is an important metabolic intermediate, has the following structure. How many asymmetric centers are in the molecule?

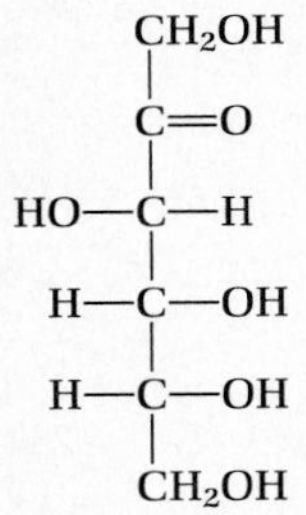

$$CH_2OH$$
$$C=O$$
$$HO-C-H$$
$$H-C-OH$$
$$H-C-OH$$
$$CH_2OH$$

Optical Activity

20.23 The sugar D-glucose has a specific rotation of +53°. What is the specific rotation of L-glucose?

20.24 Explain the meaning of the prefixes in D-(−)-glyceraldehyde and D-(−)-lactic acid.

20.25 D-(−)-Lactic acid is converted into a methyl ester by reacting it with methanol. What is the configuration of the ester? Can you predict its sign of rotation?

20.26 The amino acid L-threonine has a specific rotation of 28.3°. What is the specific rotation of D-threonine?

20.27 A synthetic amino acid, L-allothreonine, has a specific rotation of +9.6°. Explain how a compound of the L configuration can have a positive specific rotation.

20.28 The substance D-tartaric acid has a specific rotation of −12°. What is the specific rotation of L-tartaric acid?

Projection Formulas

20.29 Convert the following into two-dimensional projection formulas.

(a) 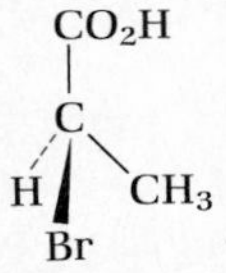
$$CO_2H$$
$$C$$
$$H \quad CH_3$$
$$Br$$

(b) 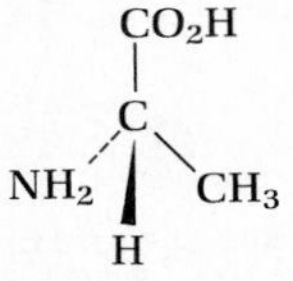
$$CO_2H$$
$$C$$
$$NH_2 \quad CH_3$$
$$H$$

20.30 Draw three-dimensional structures for each of the following two-dimensional formulas.

(a) 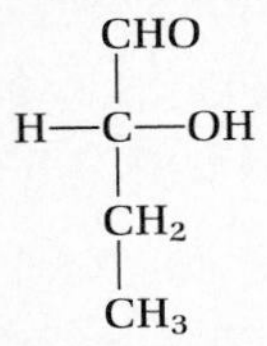
$$CHO$$
$$H-C-OH$$
$$CH_2$$
$$CH_3$$

(b) 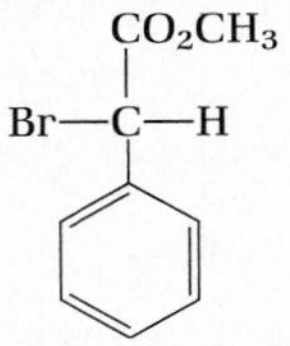
$$CO_2CH_3$$
$$Br-C-H$$

20.31 Draw the projection formula for the amino acid L-serine from the following three-dimensional structure.

$$CO_2H$$
$$C$$
$$HOCH_2 \quad H$$
$$NH_2$$

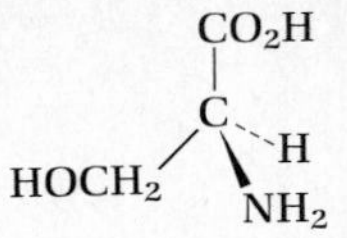

20.32 Draw the projection formula for the amino acid L-threonine from the following three-dimensional structure.

$$H \quad NH_2$$
$$HO \quad C-C \quad H$$
$$CH_3 \quad CO_2H$$

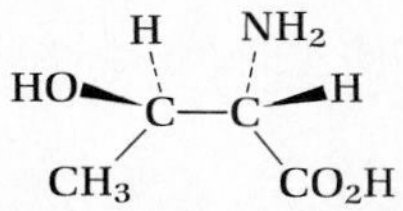

20.33 The projection formula for the synthetic amino acid L-allothreonine is given below. Draw the projection formula for D-allothreonine.

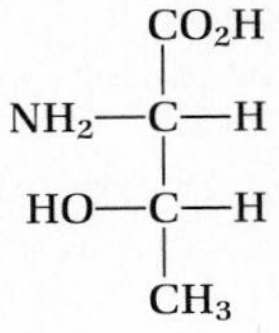

$$CO_2H$$
$$NH_2-C-H$$
$$HO-C-H$$
$$CH_3$$

Diastereomers

20.34 How many optically active isomers are there of $CH_3CH(OH)CH(OH)CH(OH)CO_2H$?

20.35 Draw planar representations of the optically active isomers of **20.16.**

20.36 Examine the four structures listed in Example 20.6. What stereochemical relationships exist between them?

20.37 Consider the following four projection formulas. Determine the two missing optical rotations.

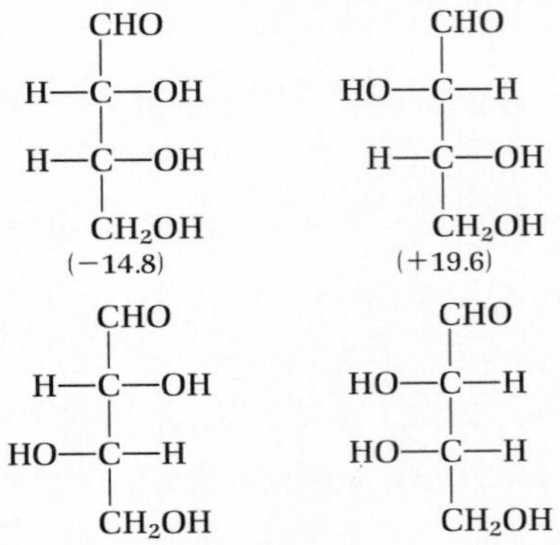

$$CHO$$
$$H-C-OH$$
$$H-C-OH$$
$$CH_2OH$$
$$(-14.8)$$

$$CHO$$
$$HO-C-H$$
$$H-C-OH$$
$$CH_2OH$$
$$(+19.6)$$

$$CHO$$
$$H-C-OH$$
$$HO-C-H$$
$$CH_2OH$$

$$CHO$$
$$HO-C-H$$
$$HO-C-H$$
$$CH_2OH$$

Stereoisomers in Biochemistry

20.38 Why do biochemical reactions produce one member of a pair of possible enantiomeric compounds?

20.39 D-Glucose is a sugar that the body can metabolize. Suggest what would occur if one would eat L-glucose.

20.40 The mold *Penicillium glaucum* will metabolize D-tartaric acid. Explain what will occur when a racemic mixture of tartaric acid is "fed" to the mold.

20.41 Natural adrenaline is levorotatory. The enantiomer has about 5% of the biological activity of the natural compound. Explain why.

20.42 Pantothenic acid (once called vitamin B_3) has the following structure. Only the enantiomer with $[\alpha]_D^{25°} = +37°$ can be used by the body. Explain why the levorotatory isomer cannot be used by the body.

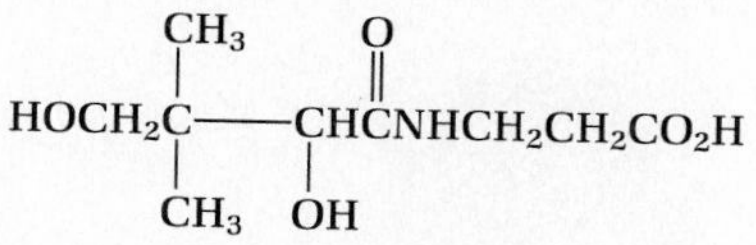

$$\text{HOCH}_2\overset{\overset{\text{CH}_3}{|}}{\underset{\underset{\text{CH}_3}{|}}{\text{C}}}\text{---}\overset{}{\underset{\underset{\text{OH}}{|}}{\text{CH}}}\overset{\overset{\text{O}}{\|}}{\text{C}}\text{NHCH}_2\text{CH}_2\text{CO}_2\text{H}$$

The Cell

20.43 Give a function of each of the following.
  (a) mitochondria   (b) nucleus
  (c) endoplasmic reticulum   (d) lysosomes

20.44 What are two types of biochemicals found in cell membranes?

20.45 What types of forces hold the components together in a cell membrane?

20.46 Explain why the compositions of various cell membranes differ.

21 Carbohydrates

Cellulose is the most abundant carbohydrate. This birch tree in Vermont, as well as the other trees in the mountains, are principally cellulose.

Carbohydrates are one of the four major classes of biochemicals. The name carbohydrate originated in the nineteenth century from the French term *hydrate de carbon*, which indicated these substances could be represented by the general formula $C_n(H_2O)_n$. For example, glucose has a molecular formula $C_6H_{12}O_6$ or $C_6(H_2O)_6$, and ribose has a molecular formula $C_5H_{10}O_5$ or $C_5(H_2O)_5$. The general formula $C_n(H_2O)_n$ is not unique to carbohydrates and thus cannot be used as a criterion for identifying this class of compounds. For example, formaldehyde, CH_2O, and acetic acid, $C_2H_4O_2$, both fit the same general formula, but are not carbohydrates. Furthermore, there are many substances now classified as carbohydrates that have other ratios of carbon, hydrogen, and oxygen. Deoxyribose, $C_5H_{10}O_4$, and rhamnose, $C_6H_{12}O_5$, do not conform to the original general formula for carbohydrates.

The modern definition of carbohydrates is compounds that are polyhydroxy aldehydes or ketones or compounds that can be hydrolyzed to these compounds. Carbohydrates include sugars, starches, cellulose, and other substances found in the roots, stems, and leaves of all plants. These compounds formed by photosynthesis are an important source of energy for animals.

The sun pours about 5×10^{21} kcal of energy into the earth each year. One third of this energy is reflected back into space by dust and clouds. The remaining energy eventually is radiated back into space. If the energy were not returned into space, our planet would become intolerably hot. Only a small amount of solar energy is retained, and it maintains the average temperature of the planet.

Less than 0.05% of the solar energy reaching the earth is used for photosynthesis in plants. This small fraction of the solar energy provides the earth with its plants, which produce oxygen. Some of these plants provide food for animals. During photosynthesis, plants convert carbon dioxide and water into carbohydrates and oxygen.

$$n\,CO_2 + n\,H_2O \longrightarrow (CH_2O)_n + n\,O_2$$

The two major carbohydrates produced in plants are starch and cellulose. Starch forms the nutritional reservoir of plants, and cellulose serves a structural role. Although it is not generally known, the major photosynthetic source of carbohydrates and oxygen actually is the phytoplankton of the oceans. They convert about 80 trillion pounds of carbon dioxide each year.

The importance of photosynthesis cannot be overstated. By this process solar energy is captured and changed into chemical energy. In photosynthesis, the carbon dioxide is converted into hydrogen-containing compounds, which are more reduced compounds of carbon. These reduced compounds contain the stored chemical energy in the carbon–hydrogen bonds. Animals that eat the reduced carbon compounds obtain the stored energy by oxidative metabolic reactions

$$(CH_2O)_n + n\,O_2 \longrightarrow n\,CO_2 + n\,H_2O$$

Carbohydrates are subdivided into three major classes—monosaccharides, oligosaccharides (from the Greek *oligos*, a few), and polysaccharides. **Monosaccharides** *are carbohydrates that cannot be hydrolyzed into simpler carbohydrates.* Although monosaccharides can contain as few as three carbon atoms, the five- and six-carbon monosaccharides are the most numerous and important.

Monosaccharides are classified by the number of carbon atoms and the type

21.1
Carbohydrates— Sources of Energy

21.2
Classification of Carbohydrates

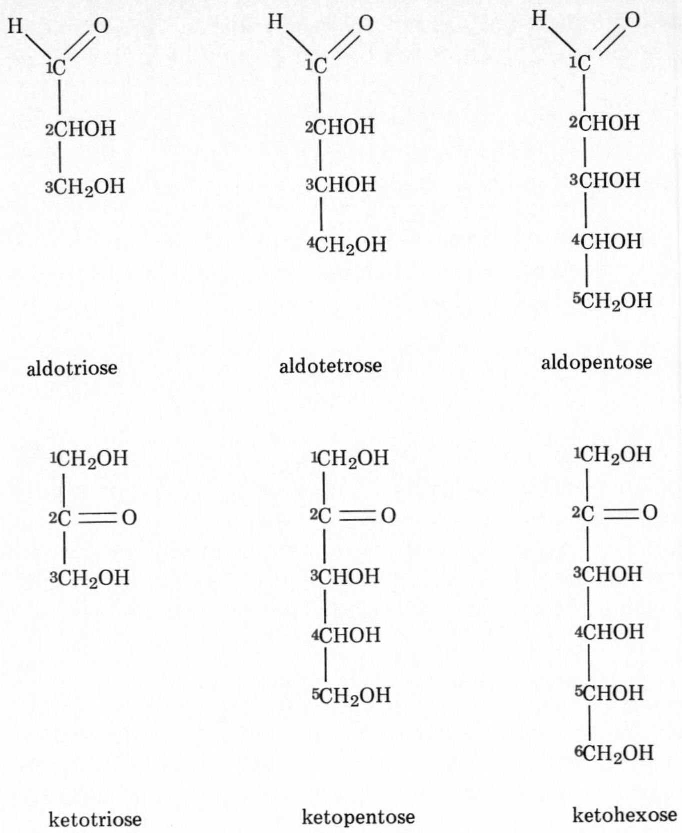

Figure 21.1
Classification of Mono-saccharides.

$$
\begin{array}{ccc}
\underset{1C}{H}\diagdown^O & \underset{1C}{H}\diagdown^O & \underset{1C}{H}\diagdown^O \\
| & | & | \\
2CHOH & 2CHOH & 2CHOH \\
| & | & | \\
3CH_2OH & 3CHOH & 3CHOH \\
 & | & | \\
 & 4CH_2OH & 4CHOH \\
 & & | \\
 & & 5CH_2OH
\end{array}
$$

aldotriose   aldotetrose   aldopentose

$$
\begin{array}{ccc}
1CH_2OH & 1CH_2OH & 1CH_2OH \\
| & | & | \\
2C{=}O & 2C{=}O & 2C{=}O \\
| & | & | \\
3CH_2OH & 3CHOH & 3CHOH \\
 & | & | \\
 & 4CHOH & 4CHOH \\
 & | & | \\
 & 5CH_2OH & 5CHOH \\
 & & | \\
 & & 6CH_2OH
\end{array}
$$

ketotriose   ketopentose   ketohexose

of carbonyl group. An ending -ose is used to indicate that a compound is a carbohydrate. A prefix aldo or keto is used if the compound is an aldehyde or ketone, respectively. The number of carbon atoms is given by tri, tetr, pent, and hex. Examples of this classification are given in Figure 21.1. Aldoses are numbered from the carbonyl carbon atom, whereas ketoses are numbered from the end of the carbon chain closest to the carbonyl carbon atom.

Common names are used for the monosaccharides. Chemists must learn each name as well as the configuration at each carbon atom. However, you should learn only the names and configurations of the monosaccharides of biological importance.

Oligosaccharides *contain a small number of monosaccharide units joined by acetal bonds* (Section 17.6), *which are called glycosidic linkages in carbohydrate chemistry.* These bonds are between the aldehyde or ketone site of one monosaccharide and a hydroxyl group of another monosaccharide. Hydrolysis of this glycosidic linkage produces the component monosaccharides. Two monosaccharides are combined in disaccharides, three in trisaccharides, and so on. The important disaccharides include maltose, lactose, and sucrose (Section 21.6).

A number of monosaccharides and disaccharides taste sweet. The most common one is sucrose, which is table sugar. For this reason, the term sugar is used in discussing the lower molecular weight carbohydrates. The relative sweetness of some monosaccharides and disaccharides is given in Table 21.1.

Polysaccharides *contain hundreds to thousands of monosaccharide units joined by a series of glycosidic linkages.* These molecules are naturally occurring

Table 21.1

Relative Sweetness of Some Sugars

Sugar	Relative sweetness
lactose (disaccharide)	16
galactose (monosaccharide)	32
glucose (monosaccharide)	75
sucrose (disaccharide)	100
fructose (monosaccharide)	170

polymers. Polysaccharides that contain only one type of monosaccharide are *homopolysaccharides. Heteropolysaccharides* are polysaccharides that contain more than one type of monosaccharide. Starch and cellulose in plants are homopolysaccharides, since they yield only glucose when hydrolyzed. *Animals store glucose as a homopolysaccharide called* **glycogen.** Thus, in all life forms, glucose and its polymers play a central role.

Example 21.1

Classify the carbohydrate D-xylulose.

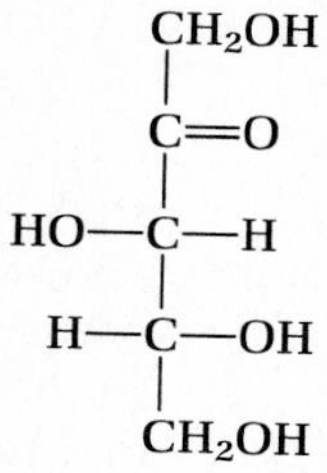

$$
\begin{array}{c}
CH_2OH \\
| \\
C{=}O \\
| \\
HO{-}C{-}H \\
| \\
H{-}C{-}OH \\
| \\
CH_2OH
\end{array}
$$

Solution. The compound contains five carbon atoms and is classified as a pentose. Since the carbonyl group is a ketone, this compound is a ketopentose.

[Additional examples may be found in **21.5–21.9** at the end of the chapter.]

In this section the monosaccharides containing from three to six carbon atoms will be considered. Projection formulas (Section 20.7) are commonly used to represent the structure of these compounds with multiple asymmetric carbon atoms. The vertical lines represent bonds projecting away from you, whereas the horizontal lines represent bonds projecting toward you.

21.3
Monosaccharides

Aldotrioses

The simplest aldose is the triose glyceraldehyde, which can exist in two optically active forms.

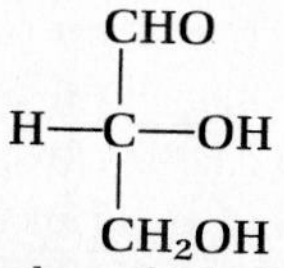

$$
\begin{array}{c}
CHO \\
| \\
H{-}C{-}OH \\
| \\
CH_2OH
\end{array}
\qquad
\begin{array}{c}
CHO \\
| \\
HO{-}C{-}H \\
| \\
CH_2OH
\end{array}
$$

D-glyceraldehyde 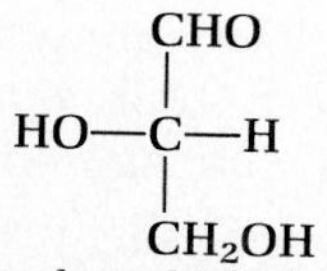   L-glyceraldehyde

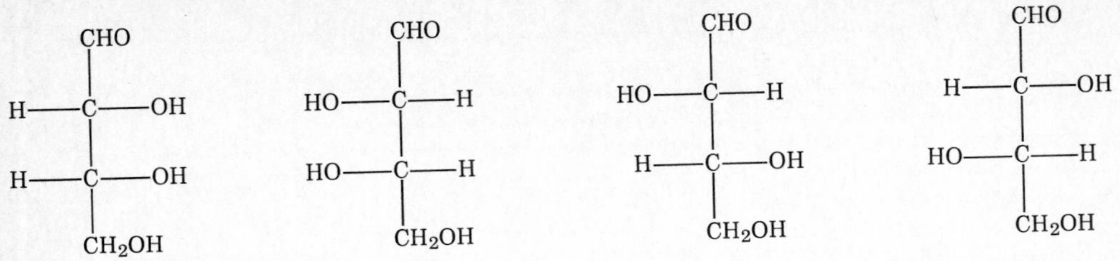

Figure 21.2
The Aldotetroses.

Other monosaccharides may be mentally derived from glyceraldehyde by lengthening the carbon chain at the carbonyl carbon atom. The monosaccharides then are designated as D or L based on their configuration at the asymmetric carbon atom bonded to the CH_2OH group. This carbon atom is the original one bonded to the secondary hydroxyl group in glyceraldehyde. Most of the important naturally occurring monosaccharides are of the D configuration.

Aldotetroses

An aldotetrose contains two nonequivalent asymmetric centers, making possible the four isomers illustrated in Figure 21.2. There are only four possible right–left combinations for the location of the hydroxyl groups (or hydrogen atoms) in the projection formula. Note that the configuration at the third carbon atom of D-erythrose and D-threose is identical to the configuration of the second carbon atom in D-glyceraldehyde. Both D-erythrose and D-threose can be mentally derived from D-glyceraldehyde by extending the carbon chain from the carbonyl carbon. The new H—C—OH group can be positioned in either of two ways to yield the tetroses D-erythrose and D-threose. In a similar manner the isomers L-erythrose and L-threose can be mentally derived from L-glyceraldehyde.

The molecules D-erythrose and D-threose are designated D because of the location of the hydroxyl group on the right side of the asymmetric center closest to the CH_2OH group and therefore farthest away from the aldehyde. D-Erythrose and L-erythrose are enantiomers (Section 20.5), as are D-threose and L-threose. Other combinations such as D-erythrose and D-threose, D-erythrose and L-threose, and L-erythrose and D-threose are diastereomers. Recall that enantiomers possess the same physical properties; diastereomers do not.

Aldopentoses

There are eight isomeric aldopentoses because there are three nonequivalent asymmetric centers. The eight isomers can be considered as four pairs of enantiomers. In other words, there are four aldopentoses of the D configuration and four of the L configuration. The four compounds of the D configuration shown in Figure 21.3 are diastereomers and have different properties. Ribose, which is incorporated in RNA (ribonucleic acid), is produced in the pentose phosphate pathway (Section 21.4).

Aldohexoses

There are 16 isomeric aldohexoses. The eight diastereomeric compounds of the D configuration are given in Figure 21.3. Of course, there are eight mirror-image compounds or enantiomers of the L configuration, one for each compound shown. Glucose, mannose, and galactose are the most important aldohexoses in biochemical reactions.

Figure 21.3
Structures of the Aldoses.

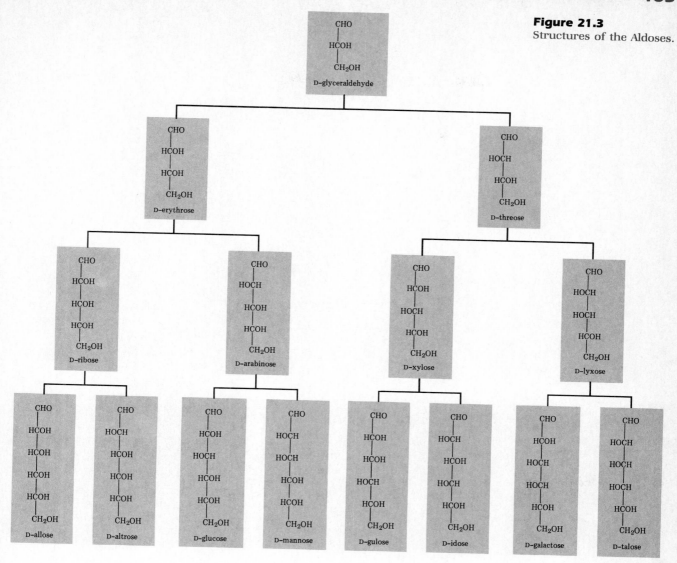

Ketoses

The simplest ketose, dihydroxyacetone, does not contain any asymmetric carbon atoms (Figure 21.4). This ketose is produced in the metabolism of glucose (Chapter 26). Other important ketoses are fructose, which is produced by isomerization of glucose, and the ketopentoses ribulose and xylulose (Figure 21.4), which are both intermediates in the pentose phosphate pathway (Section 21.4).

Example 21.2 _____

What is the structure of L-glucose?

Solution. L-Glucose is the enantiomer of D-glucose—its mirror image. Therefore, the isomer can be drawn by reflecting the planar projection formula in an imagined mirror perpendicular to the plane of the page and parallel to the carbon

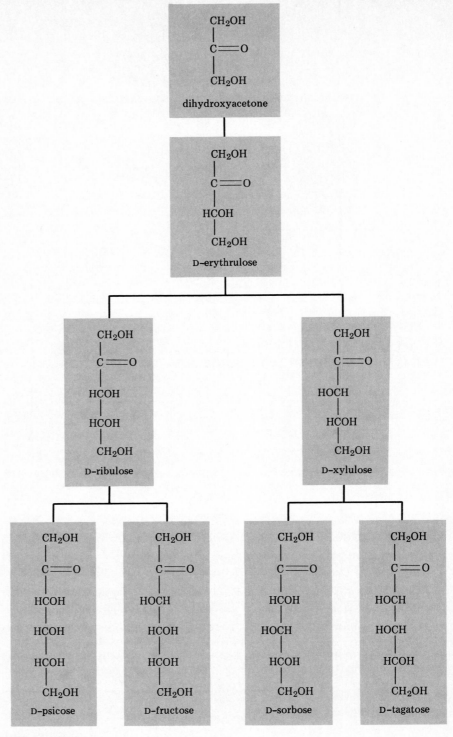

Figure 21.4
Structures of the Ketoses.

AN ASIDE

Although carbohydrates both provide energy and serve as a source of atoms for biochemical reactions in our body, they are not absolutely required in the diet. Even in diets lacking carbohydrates, the body's metabolic reactions produce carbohydrates from other dietary chemicals such as proteins. Details of these reactions will be discussed later in this and subsequent chapters.

In many diets, carbohydrates provide about 50% of our dietary calories. A minimum of about 75 g of carbohydrates per day is sufficient to allow the citric acid cycle (Chapter 25) to operate normally.

The type of carbohydrate in dietary sources is considered important. The Food and Nutrition Board of the National Research Council of the National Academy of Science recommends that complex carbohydrates be the major fraction of this component of a diet. Refined sugar and sugared foods should be 10% or less of the calorie intake. Refined sugar contains approximately the same number of calories as naturally occurring complex carbohydrates. However, refined sugars contain "empty" calories, since they do not have other nutrients required for good health. Our bodies have energy or calorie requirements, but also need a variety of nutrients with which to build and maintain cells and tissues.

---

chain. Each hydroxyl group that is on the right in D-glucose is on the left in L-glucose. The hydroxyl group on the left in D-glucose is on the right in L-glucose.

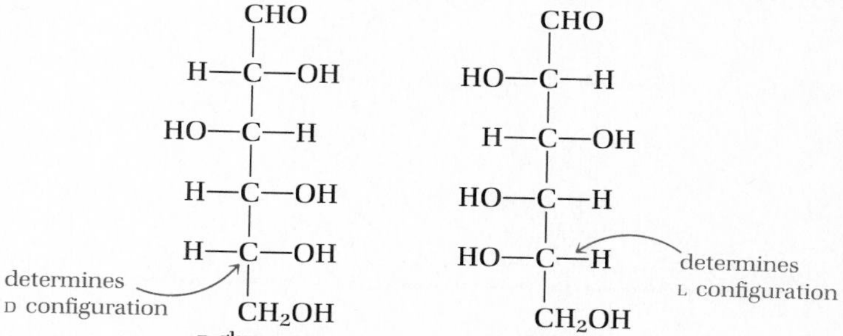

D-glucose / L-glucose Fischer projections, with "determines D configuration" pointing to the bottom stereocenter of D-glucose and "determines L configuration" pointing to the bottom stereocenter of L-glucose.

Example 21.3

What relationship exists between each possible pair of these compounds?

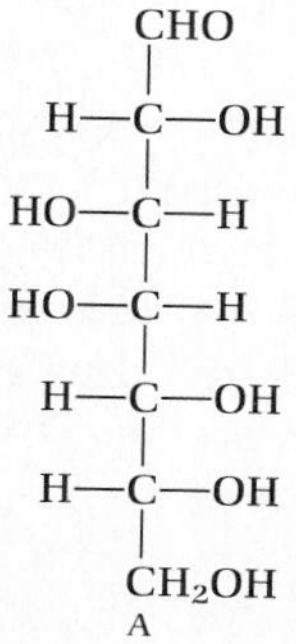

Three Fischer projections labeled A, B, and C.

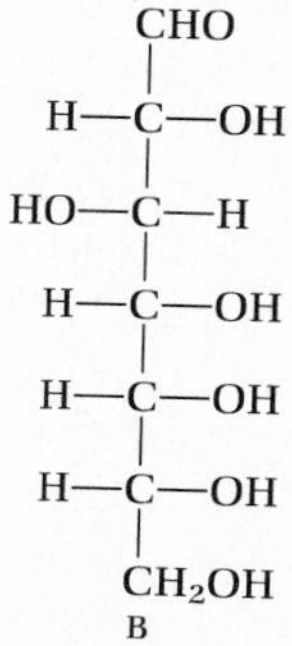

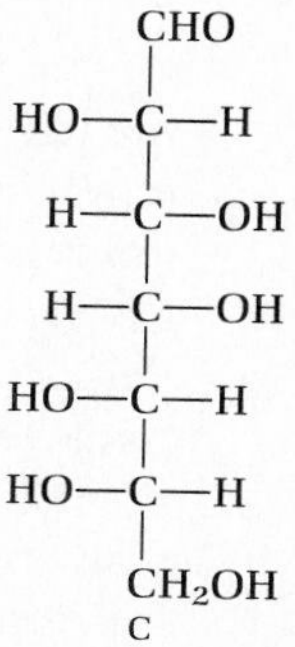

Solution. A and B represent compounds of the D series of aldoheptoses. They are diastereomers because they cannot be enantiomers. Enantiomers are always of opposite configuration and have to be in different series, that is, D and L. Structure C is of the L series and is the mirror image of A and therefore the enantiomer of A. Structures B and C are of different series, but since they cannot be mirror images (only one mirror image is possible per compound and that has been indicated as A with C), they must be diastereomers.

Example 21.4

What aldohexoses correspond to D-fructose in configuration at the asymmetric centers on carbons 3 through 5?

Solution. A search of Figure 21.3, which contains representations of the D-aldohexoses, reveals that only two compounds have identical configurations with D-fructose at asymmetric centers on carbons 3 through 5. This should have been expected because the aldohexoses have one more asymmetric center than fructose. There can be only two possible different configurations at that one center.

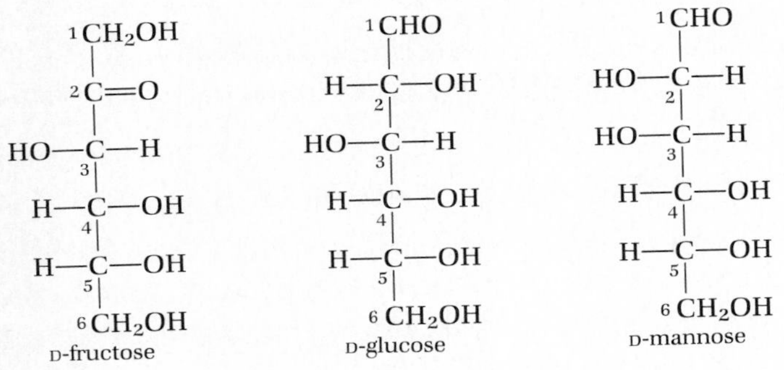

| D-fructose | D-glucose | D-mannose |

[Additional examples may be found in **21.10–21.12** at the end of the chapter.]

21.4
Glucose—A Key Compound

Glucose in the form of a phosphate ester, glucose 6-phosphate, is one of three key compounds in metabolic reactions. The other two substances, pyruvic acid and acetyl coenzyme A, will be considered in Chapter 26.

When glucose enters a cell from the blood, it is rapidly phosphorylated to produce glucose 6-phosphate (Figure 21.5). It is this phosphate ester that is really the key compound in the integration of a number of metabolic pathways. In order to release glucose from a cell and return it to the blood, it is necessary to have the enzyme glucose 6-phosphatase to remove the phosphate group. Organs such as the liver and kidneys contain this enzyme and therefore serve as exporters of glucose to the blood and eventually to other organs. The brain and muscles lack glucose 6-phosphatase and therefore use the glucose that enters their cells.

Glucose is essentially the only fuel for the human brain, except under severe starvation conditions. The brain requires a continuous source of glucose because it does not have any fuel storage capacity. The brain consumes about 120 g of

Figure 21.5
Metabolic Pathways of Glucose.

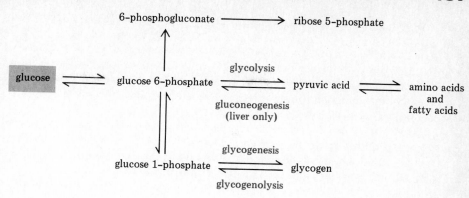

glucose daily, which corresponds to about 400 kcal of chemical energy. Once glucose enters the brain, phosphorylation occurs rapidly and the glucose is used as a source of chemical energy.

Now let's consider the central role of glucose 6-phosphate in cells of other organs such as the muscles and the liver. When glucose 6-phosphate is abundant in a cell, or if the ATP level is high, glycogenesis (glycogen creation) occurs. *In* **glycogenesis,** *the glucose 6-phosphate is isomerized to glucose 1-phosphate and then is converted to* **glycogen.** *The reverse of this process, conversion of glycogen into glucose 6-phosphate, is called* **glycogenolysis** *(glycogen hydrolysis).* Glycogenolysis occurs when glucose 6-phosphate and ATP are required by the muscles for a quick burst of energy. Glycogenolysis also occurs in the liver in order to export glucose to other organs via the blood.

Glycolysis *is a sequence of ten reactions converting glucose into pyruvic acid and simultaneously forming ATP.* In glycolysis glucose 6-phosphate is first isomerized to fructose 6-phosphate. This enzyme-catalyzed process occurs via an enediol intermediate (Section 17.7).

glucose 6-phosphate   fructose 6-phosphate

phosphoglucose isomerase

In the last step of the glycolysis sequence, pyruvic acid is produced, which may be converted into the acetyl group of acetyl coenzyme A. Acetyl coenzyme A fuels the citric acid cycle, which serves to release more stored chemical energy and results in the formation of more ATP. Thus, glycolysis occurs when ATP is required by the organism or when pyruvic acid is needed as a source of carbon compounds in anabolic processes.

Glucose and ultimately glycogen can be synthesized from pyruvic acid. *Conversion of pyruvic acid into glucose is called* **gluconeogenesis.** Compounds such

AN ASIDE

For the body to function properly the concentration of glucose in the blood should be about 65–95 mg/100 mL. (The units used in clinical situations are mg/dL.) The blood sugar level may decrease temporarily during strenuous exercise because the glucose may not be replenished rapidly enough from liver glycogen or by gluconeogenesis.

A condition of low blood sugar is called **hypoglycemia.** It may be characterized by a rapid heartbeat, general weakness, trembling, perspiration, whitening of the skin, and loss of consciousness. The loss of consciousness is the result of deprivation of the necessary glucose for the brain cells.

A condition of high blood sugar is called **hyperglycemia.** The digestion of carbohydrates may result in absorption of glucose into the blood faster than glycogen can be formed by the process of glycogenesis. As the blood sugar level increases, the body can transform glucose into fat and store the fat in adipose tissue.

When glucose reaches the level of 140–160 mg/100 mL of blood, neither glycogen nor fat can be formed rapidly enough to decrease the glucose level. *The condition in which glucose then is excreted by the kidneys and is eliminated in the urine is called* **glucosuria.**

Permanent hyperglycemia occurs in diabetes mellitus, a disease caused by an abnormally low blood insulin level. Insulin helps glucose to enter cells and also promotes the conversion of glucose into glycogen. As a result of low blood insulin levels, glucose accumulates in the blood and is eventually eliminated in the urine where it may be detected clinically.

Diabetics are treated with supplemental insulin to decrease the concentration of glucose in the blood. Careful control of the dosage is required because if

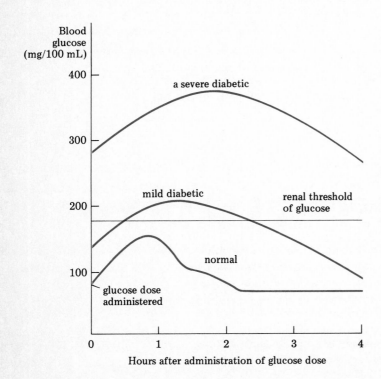

Figure 21.6
Glucose Tolerance Test.
The blood glucose level of a normal individual rises but is brought in control without excretion of glucose in the urine. A mild diabetic will experience a period of glycosuria in which the blood glucose exceeds the renal threshold. Severe diabetics may experience constant glycosuria.

a diabetic takes too much insulin, hyperinsulinism results. Excessive amounts of glucose are then removed from the blood and hypoglycemia occurs. Severe hypoglycemia from too much insulin can cut the supply of glucose to the brain, a condition called insulin shock.

The ability of the body to handle dietary glucose is called **glucose tolerance** and is evaluated by a glucose tolerance test. A person is given a glucose solution to drink in the amount of 1.75 g per kilogram of body weight. The blood sugar is then checked at intervals of a few hours. Typical data are given in Figure 21.6. The blood sugar level increases rapidly for all individuals. However, a nondiabetic individual manages to decrease this level by a variety of metabolic processes. A diabetic person cannot secrete insulin, and so the blood glucose level remains high.

---

as glycerol from fats, lactic acid, and alanine can yield pyruvic acid for gluconeo-genesis.

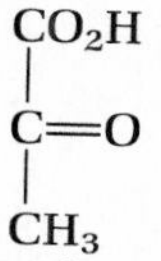

pyruvic acid	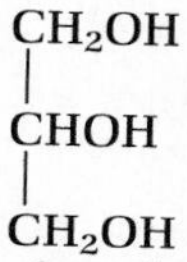 glycerol	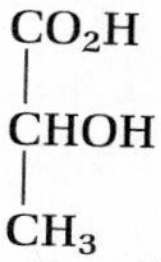 lactic acid	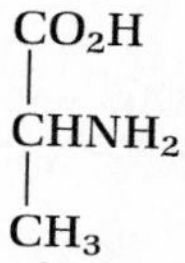 alanine
CO_2H	CH_2OH	CO_2H	CO_2H
$C{=}O$	$CHOH$	$CHOH$	$CHNH_2$
CH_3	CH_2OH	CH_3	CH_3

The major site of gluconeogenesis is the liver, but it also occurs in the kidneys. Very little gluconeogenesis occurs in the brain, skeletal muscle, or heart muscle. Thus, gluconeogenesis in the liver is an important process to produce glucose to maintain the glucose level in the blood for the use of other organs.

The conversion of glucose 6-phosphate by the cell into ribulose 5-phosphate occurs in the **pentose phosphate pathway.** This process is responsible for the

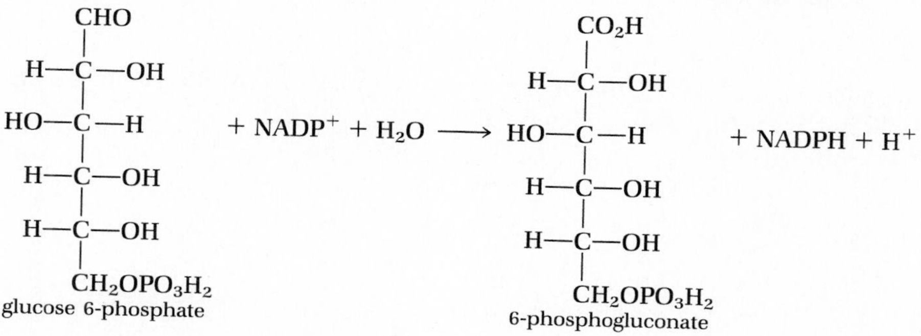

glucose 6-phosphate
$$\begin{array}{c} CHO \\ H{-}C{-}OH \\ HO{-}C{-}H \\ H{-}C{-}OH \\ H{-}C{-}OH \\ CH_2OPO_3H_2 \end{array} + NADP^+ + H_2O \longrightarrow \begin{array}{c} CO_2H \\ H{-}C{-}OH \\ HO{-}C{-}H \\ H{-}C{-}OH \\ H{-}C{-}OH \\ CH_2OPO_3H_2 \end{array} + NADPH + H^+$$
6-phosphogluconate

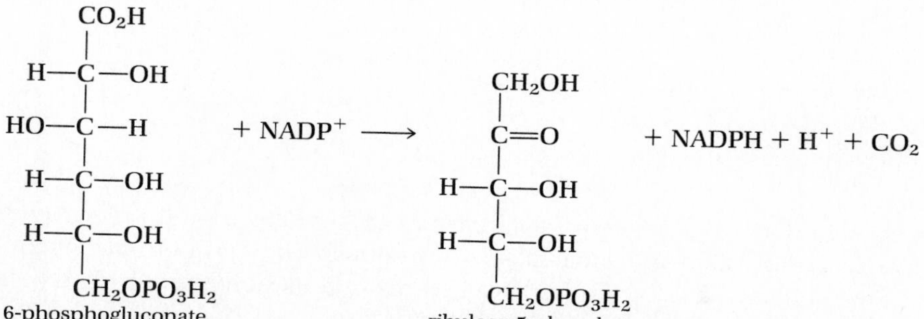

6-phosphogluconate
$$\begin{array}{c} CO_2H \\ H{-}C{-}OH \\ HO{-}C{-}H \\ H{-}C{-}OH \\ H{-}C{-}OH \\ CH_2OPO_3H_2 \end{array} + NADP^+ \longrightarrow \begin{array}{c} CH_2OH \\ C{=}O \\ H{-}C{-}OH \\ H{-}C{-}OH \\ CH_2OPO_3H_2 \end{array} + NADPH + H^+ + CO_2$$
ribulose 5-phosphate

formation of nicotinamide dinucleotide phosphate in its reduced form, NADPH. The reducing characteristics of NADPH are different from those of NADH. NADH is oxidized in catabolic reactions, which form ATP and result in the storage of chemical energy. NADPH is oxidized and reduces biochemicals in anabolic reactions.

Ribulose 5-phosphate is converted into ribose 5-phosphate via an enediol intermediate. Ribose 5-phosphate is required for synthesis of ribonucleic acids.

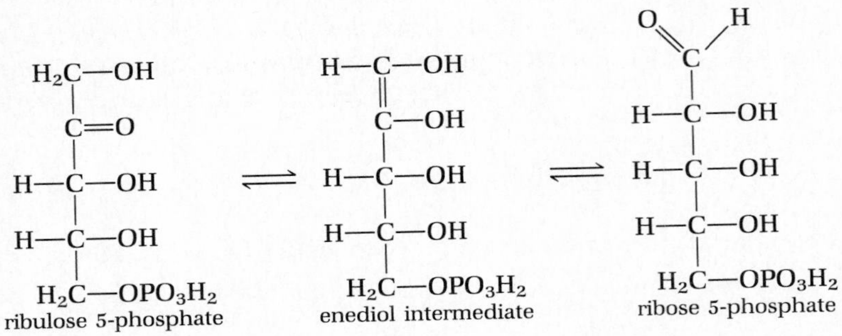

ribulose 5-phosphate ⇌ enediol intermediate ⇌ ribose 5-phosphate

In one of the other steps in the pentose phosphate pathway, ribulose 5-phosphate is isomerized to xylulose 5-phosphate.

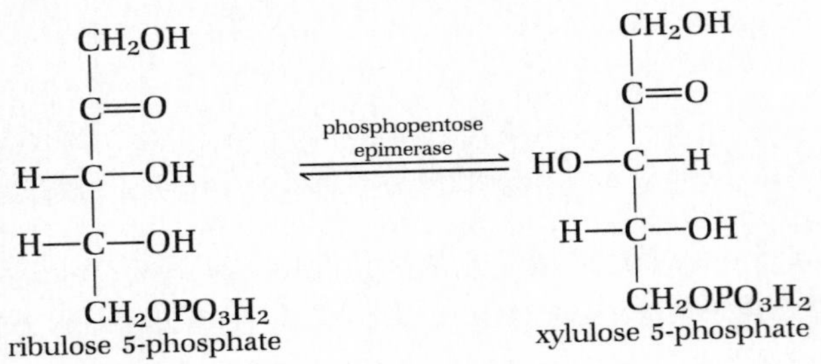

ribulose 5-phosphate ⇌ (phosphopentose epimerase) ⇌ xylulose 5-phosphate

21.5
Hemiacetals and Hemiketals

Recall from Section 17.6 that aldehydes or ketones can react reversibly with alcohols to form hemiacetals or hemiketals.

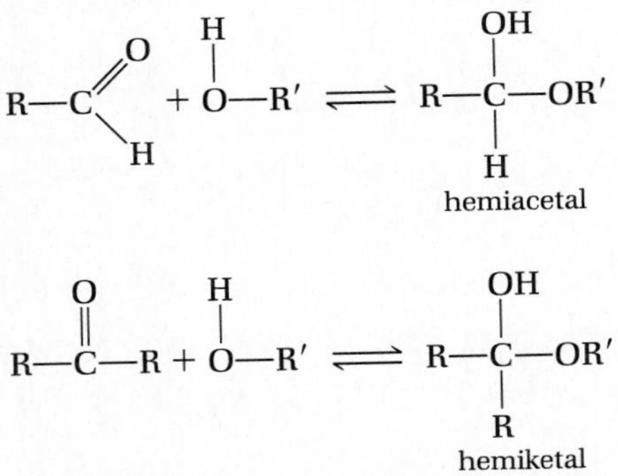

hemiacetal

hemiketal

Because aldoses and ketoses have both a carbonyl group and several hydroxyl groups within the same molecule, an intramolecular reaction can occur. The intramolecular addition of a hydroxyl group to a carbonyl group results in the formation of a cyclic hemiacetal or hemiketal.

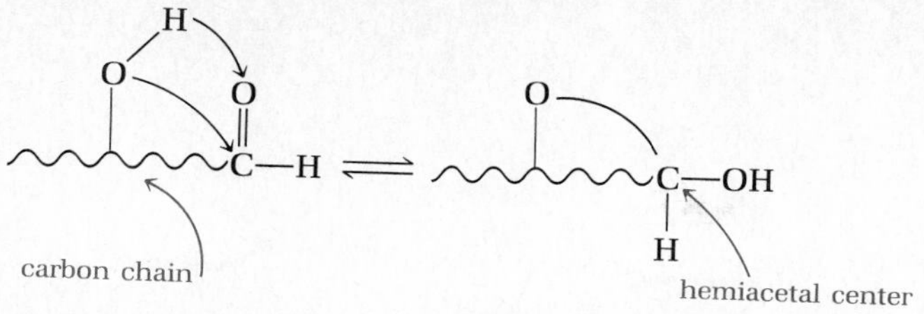

The probability of an intramolecular addition reaction to form a cyclic hemiacetal or hemiketal is more than that of an intermolecular reaction between separate alcohol and carbonyl compounds. The formation of unstrained five- and six-membered rings is favorable because of the proximity of the two functional groups.

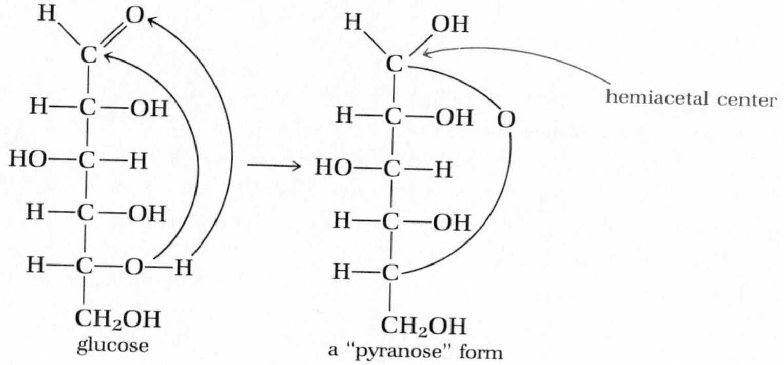

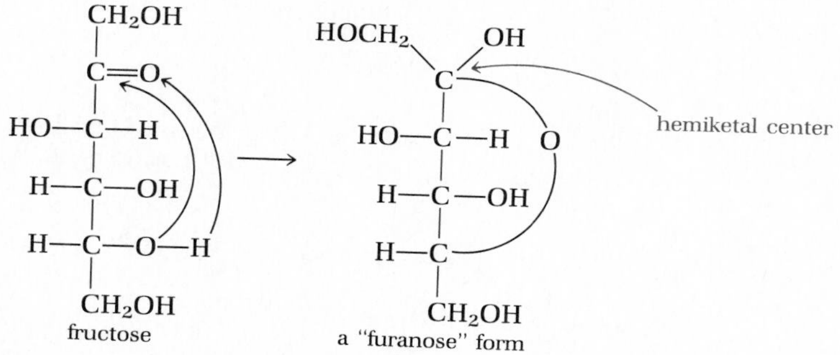

Aldoses with more than four carbon atoms and ketoses with more than five carbon atoms tend not to exist in the open-chain structures shown in Section 21.3. These aldoses and ketoses react intramolecularly to form hemiacetals and hemiketals. Both glucose and fructose react via the hydroxyl group at carbon 5. In the case of glucose the resulting compound has a six-membered ring and is called a pyranose. For fructose the cyclic hemiketal contains a five-membered ring and is called a furanose.

Haworth Projection Formula

In order to represent carbohydrate structures as they actually exist, a convention known as the Haworth projection formula is used. With this convention, all of the carbohydrates are represented consistently so that similarities and differences in structures may be seen more readily. We will examine this convention using glucose as an example.

The Fischer projection formula of glucose can be written in an open form

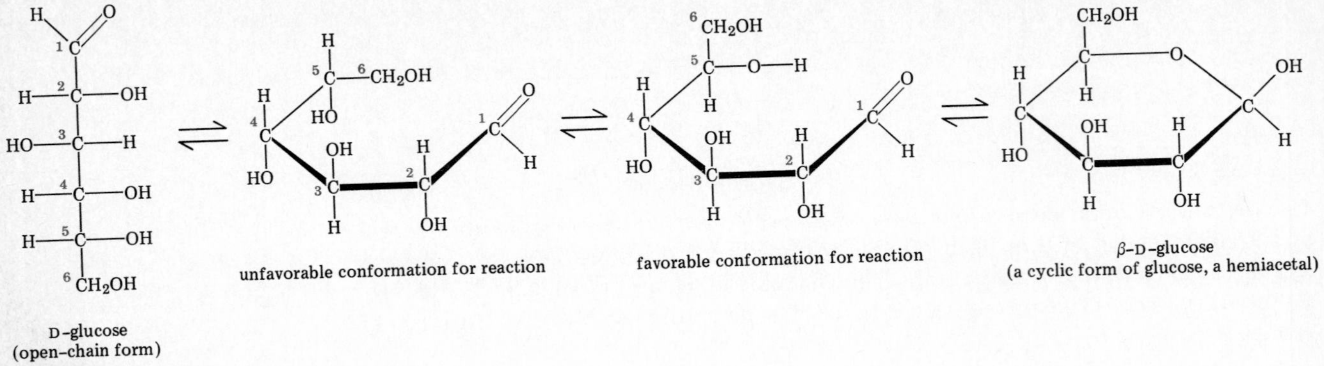

Figure 21.7
Formation of a Cyclic Hemiacetal.

shaped like a C (Figure 21.7). Recall that the carbon chain in a projection formula is directed away from the reader (Section 20.7). Allow this curved chain to "fall down" toward the right. Groups on the right in the Fischer projection are then directed downward, while groups on the left are directed upward. Note that in this conformation the 5–OH group is not near enough to the carbonyl carbon to form a ring. Rotation about the bond between the numbers 4 and 5 carbon atoms produces a conformation suitable for reaction to form a hemiacetal. Addition of the hydroxyl group to the carbonyl group results in a ring structure. In the Haworth projection the ring is viewed as planar and perpendicular to the plane of the paper. The heavy lines directed toward the reader are to atoms in front of the plane of the page.

Figure 21.8
Equilibrium Between Anomeric Forms of Glucose.

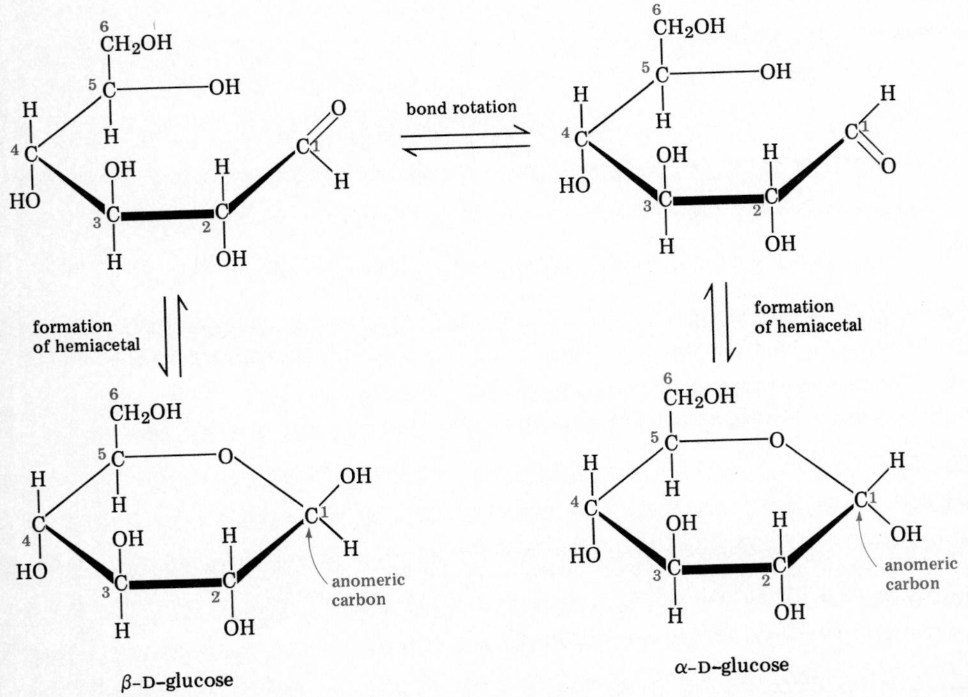

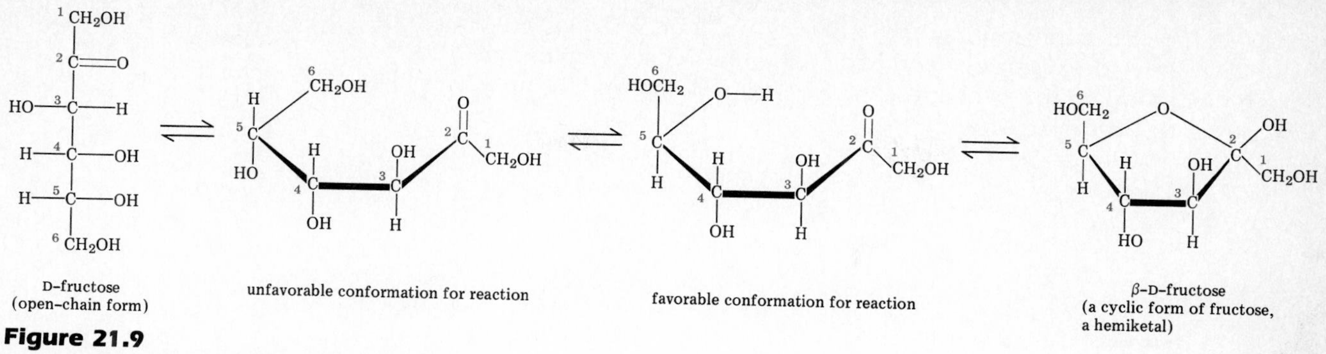

Figure 21.9

Formation of Cyclic Hemiketal of D-Fructose.

When glucose forms a cyclic hemiacetal, a new asymmetric center is created at the original carbonyl carbon atom. There are now four different groups bonded to it. As a result, two configurations are possible at that center. The configuration formed depends on the conformation of the carbonyl group when the bond forms (Figure 21.8). If the hydroxyl group of the hemiacetal is directed below the plane, the compound is a α-D-glucose. The hydroxyl group is above the plane of the ring in β-D-glucose. In both isomers the CH_2OH group is above the plane of the ring. All carbohydrates of the D configuration have the CH_2OH group located in this position in the Haworth projection.

The α and β forms of D-glucose are diastereomers that differ in configuration at one center. *Compounds whose configurations differ only at one center are called* **anomers.** The asymmetric carbon at the hemiacetal center that forms in the cyclization reaction is called the anomeric carbon atom.

Fructose reacts to form a cyclic hemiketal between the 5—OH group and the ketone. A ring of four carbon atoms and one oxygen atom results (Figure 21.9). The anomeric carbon atom is the second of the open-chain form. Again the α and β designate configuration at the anomeric carbon atom and refer to the location of the hydroxyl group.

Example 21.5 _____

Draw the Haworth projection of the α anomer of the pyranose form of D-galactose.

Solution. First write galactose in the Fischer projection and then draw the six-membered ring consisting of five carbon atoms and one oxygen atom.

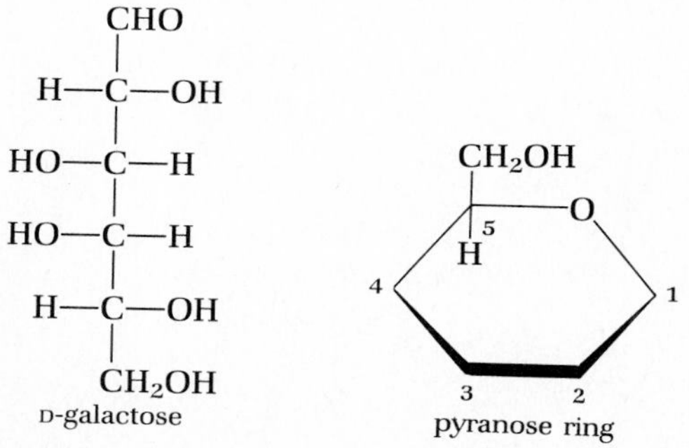

Note that for the D configuration, the CH_2OH group is above the plane of the ring. Now enter the hydroxyl groups and hydrogen atoms at carbon atoms 2, 3, and 4. An atom or group on the right in the Fischer projection is on the "bottom" of the ring of the Haworth projection.

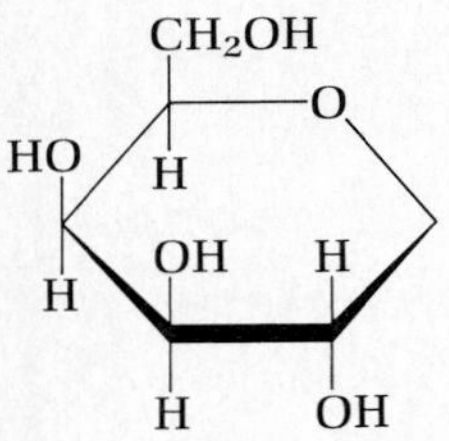

Finally, the α anomer must have a hydroxyl group below the plane of the ring at the anomeric carbon atom, which is the number 1 carbon atom in this case.

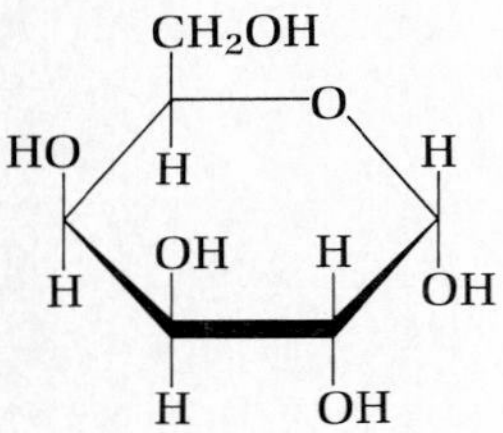

[Additional examples may be found in **21.19–21.27** at the end of the chapter.]

Chair Conformations

Six-membered rings of σ-bonded atoms are not planar, but exist in a chair conformation (Section 14.8). In the chair conformation the bonds from the ring are either equatorial or axial. The chair conformations of α- and β-D-glucose are shown in Figure 21.10. In β-D-glucose all of the groups are equatorial. This arrangement keeps groups away from each other. In α-D-glucose the anomeric hydroxyl group is in a crowded environment. The β form is then the more stable in an equilibrium mixture of the two anomers. Of all the aldohexoses, glucose is the only substance that can have all substituents in equatorial positions in one of the two possible anomers. This fact may explain the uniqueness of glucose in nature.

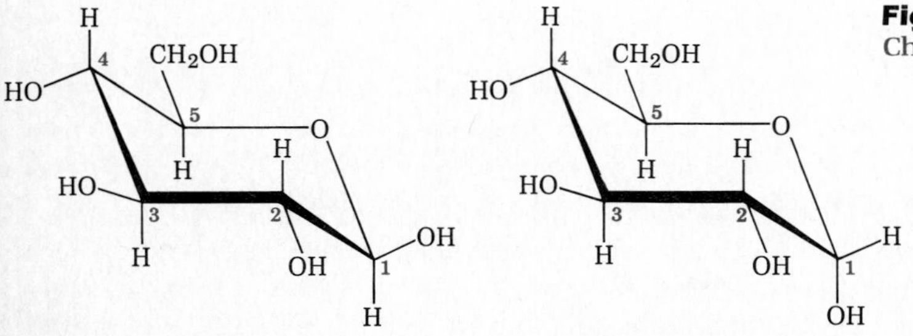

β-D-glucose   α-D-glucose

Figure 21.10
Chair Conformation of Anomeric Forms of D-Glucose.

Mutarotation

Crystalline α-D-glucose melts at 146°C and has $[\alpha]_D = +19°$. The β anomer melts at 150°C and has $[\alpha]_D = +113°$. If either anomer is dissolved in water, the rotation of the solution slowly changes to an equilibrium value of +52.7°. *A gradual change in rotation to an equilibrium point is known as* **mutarotation.** Mutarotation is the result of the cyclic hemiacetals giving the open-chain form in solution followed by closing of the ring again. Both anomers then exist in equilibrium.

$$\alpha\text{-D-glucose} \rightleftharpoons \text{open chain} \rightleftharpoons \beta\text{-D-glucose}$$

If both anomers were of equal stability, a 50:50 mixture would result in a rotation that is the average of the two components.

$$\frac{113° + 19°}{2} = 65.5°$$

However, at equilibrium there are 63% of the β anomer and 37% of the α anomer and the rotation is 52.7°. Less than 0.01% of the open chain is present.

In Section 17.6 you saw that the hemiacetals and hemiketals can react with alcohols to yield acetals and ketals, respectively. The reaction is acid catalyzed, and the equilibrium may be shifted to the right by addition of excess alcohol or removal of the water formed.

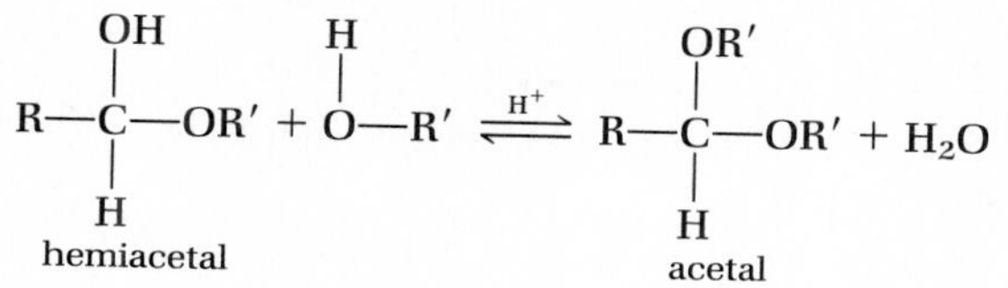

21.6
Reactions with Alcohols

hemiacetal   acetal

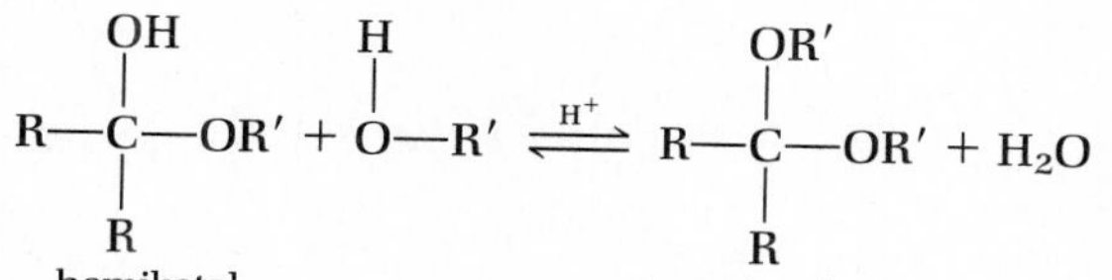

hemiketal   ketal

Monosaccharides that exist as hemiacetals and hemiketals can also form acetals and ketals in a reaction with an alcohol. *The acetal and ketal of carbohydrates are called* **glycosides,** *and the carbon–oxygen bond is called a glycosidic bond.* Since hemiacetals or hemiketals exist in equilibrium, there are two possible glycosides that may be formed (Figure 21.11).

Like acetals and ketals, glycosides are stable enough to be isolated, and each anomer has different physical properties. Glycosides are stable in neutral or basic solution, and thus mutarotation cannot occur. In acid solution, glycosides are hydrolyzed, which is the reverse of the reaction shown in Figure 21.10. Then mutarotation can occur.

Example 21.6

Examine the following molecule to determine its component functional groups. From what compounds may the substance be formed?

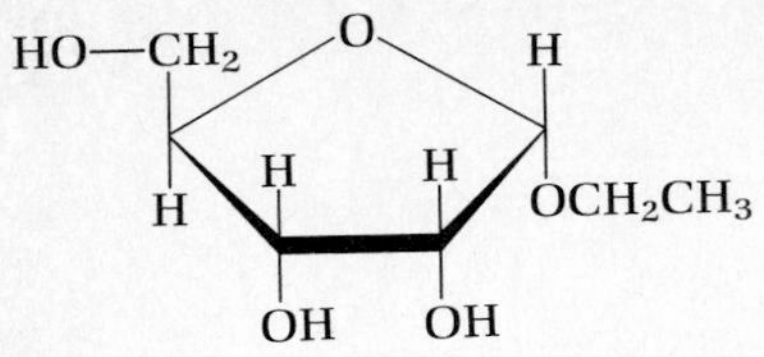

Solution. The compound is a furanose form of a carbohydrate, since there are four carbon atoms and one oxygen atom in a five-membered ring. The ring carbon atom on the right is an acetal because there are one hydrogen atom and two OR groups.

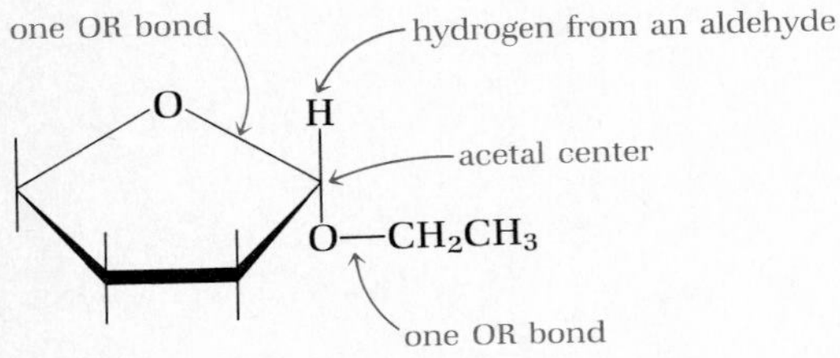

The acetal has the α configuration, and the alcohol used to form the acetal is ethanol.

The carbohydrate has the D configuration because the CH_2OH group is "up" in the Haworth projection. The other two asymmetric carbon atoms of the pentose have hydroxyl groups "down," which corresponds to the right in the Fischer projection. The compound is ribose.

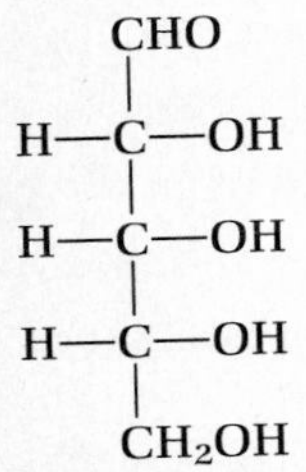

[Additional examples may be found in **21.36–21.39** at the end of the chapter.]

21.7
Disaccharides

Disaccharides *are glycosides formed from two monosaccharides, one acting as the hemiacetal or hemiketal and the other as the alcohol.* The most common bonding links the number 1 carbon atom of the hemiacetal to the number 4 carbon atom of the monosaccharide, which serves as the alcohol. Such bonds are designated $\alpha(1\rightarrow4)$ or $\beta(1\rightarrow4)$ depending on the configuration of oxygen of the glycosidic bond. Maltose, cellobiose, and lactose all have $1\rightarrow4$ disaccharide bonds. Sucrose has a $1\rightarrow2$ disaccharide bond.

Maltose

Maltose (Figure 21.12) is composed of two molecules of D-glucose. The glycosidic oxygen of one glucose is α, and it is bonded to the number 4 carbon atom of another glucose. Maltose then is $\alpha(1\rightarrow4)$ linked. The right ring of maltose is a

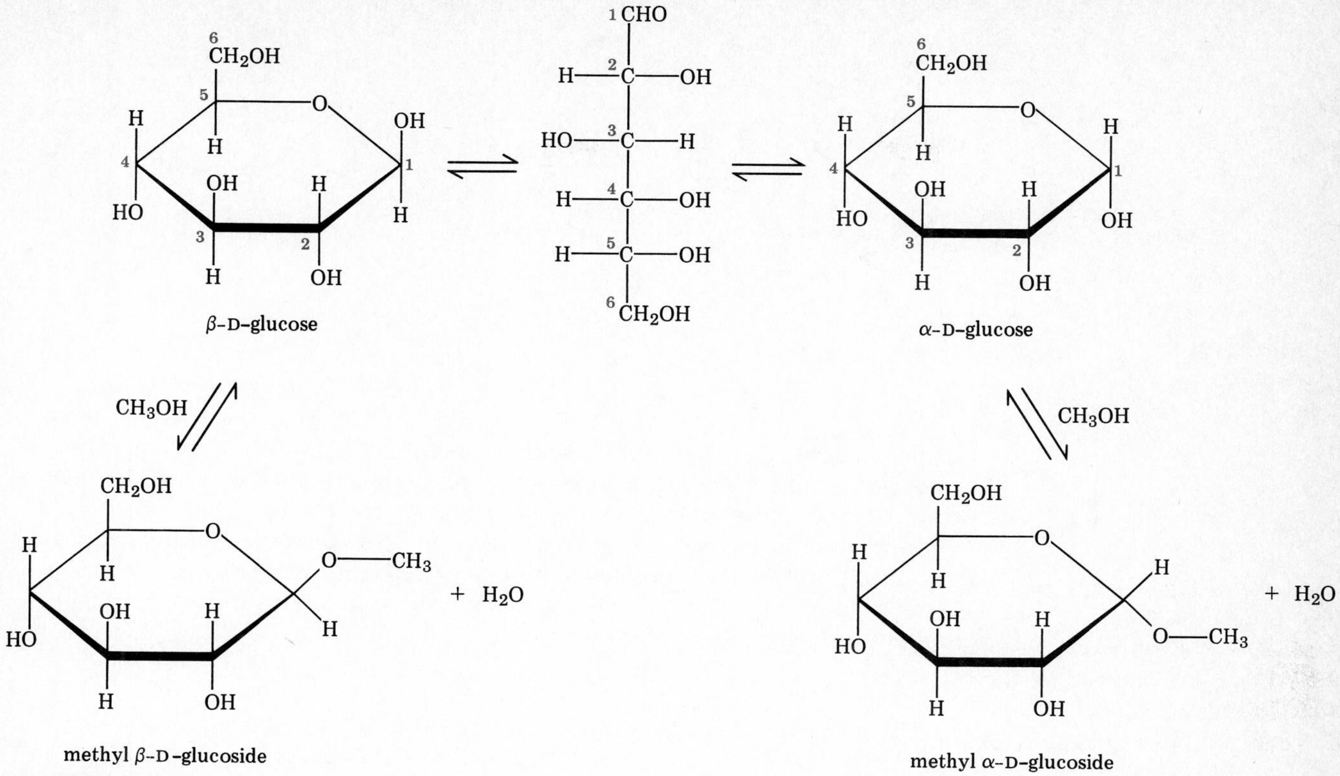

Figure 21.11
Formation of Anomeric Glycosides of Glucose.

hemiacetal and can be either an α- or a β-hemiacetal. In solution both forms of maltose exist in equilibrium. The prefixes α and β refer to the configuration at the hemiacetal center. This designation should not be confused with the acetal center or glycosidic bond, which is always α in maltose.

Maltose is produced by hydrolysis of starch (a homopolysaccharide) by the enzyme amylase. Further hydrolysis, catalyzed by the enzyme maltase, results in two molecules of D-glucose.

Cellobiose

Cellobiose (Figure 21.13) is also composed of two molecules of D-glucose. However, the glycosidic linkage is $\beta(1\rightarrow4)$. The right ring is a hemiacetal, which

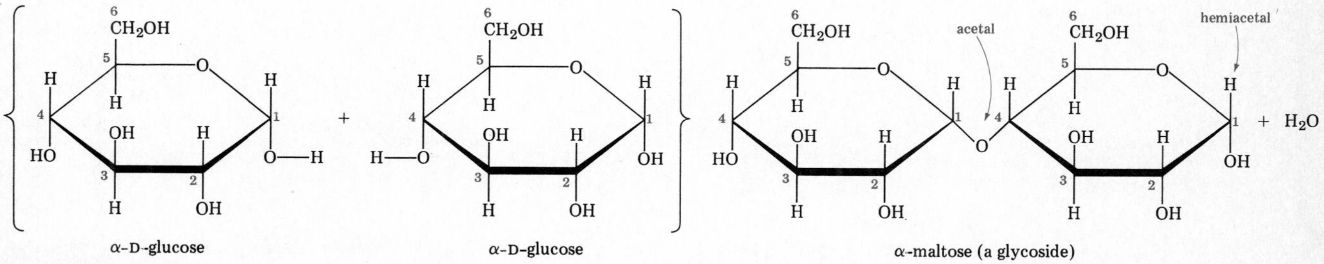

Figure 21.12
Structure of Maltose.

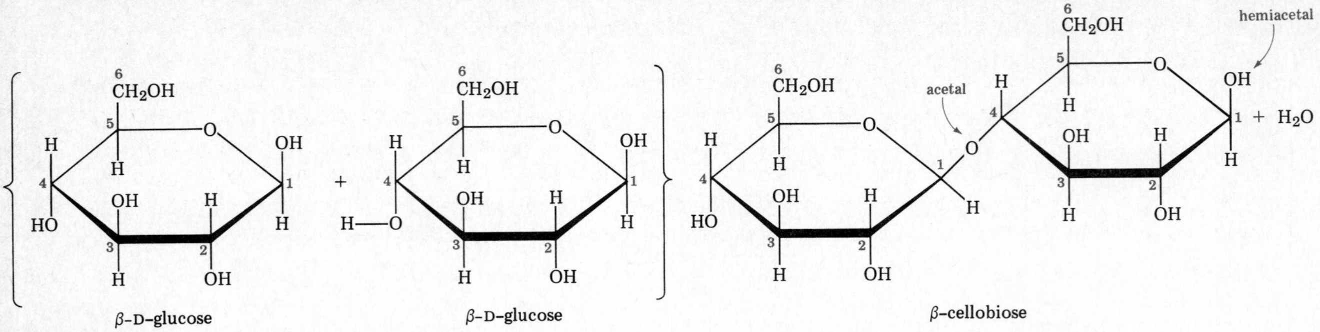

Figure 21.13
Structure of Cellobiose.

can be either α or β. In solution both forms of cellobiose exist in equilibrium. Again, you are cautioned not to confuse the hemiacetal center with the acetal center or glycosidic bond, which is always β in cellobiose.

Cellobiose is produced by hydrolysis of the homopolysaccharide cellulose. In contrast to maltose, humans do not have enzymes to hydrolyze cellobiose. This apparently simple difference in configuration illustrates the remarkable difference in the specificity of enzymes.

Lactose

The disaccharide lactose is the major sugar found in milk and is often called milk sugar (Figure 21.14). The lactose content of milk varies with species; cow's milk contains about 5% lactose, whereas human milk contains about 7% lactose. Lactose has the number 1 carbon atom of galactose joined by a β-glycosidic linkage to the number 4 carbon atom of glucose. The enzyme lactase, which is present in the small intestine, catalyzes the hydrolysis. Galactose must then be isomerized into glucose needed for metabolic reactions in a reaction catalyzed by the enzyme UDP-galactose-4-epimerase.

Some infants inherit a genetic disease called **galactosemia.** These children lack the enzyme necessary to convert galactose into glucose. If the disease is not detected, milk or other foods containing galactose cause vomiting, diarrhea, cataracts, and even mental retardation. Exclusion of galactose from all sources in a strict diet is the only avenue open to such children.

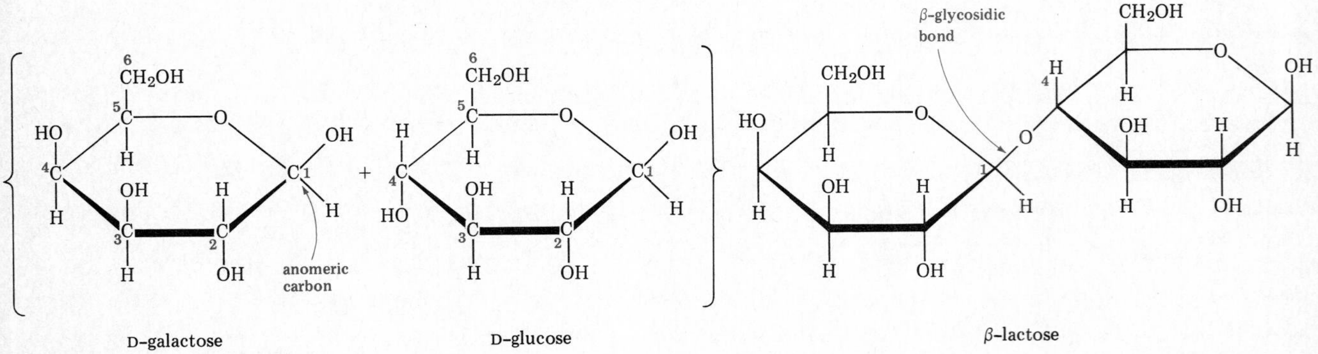

Figure 21.14
Structure of Lactose.

AN ASIDE _____

   Lactose intolerance *is a disease in which people cannot eat any food containing lactose.* These individuals lack the enzyme lactase (also known as β-galactosidase) needed to hydrolyze lactose. Although the disease is not serious, the high level of unhydrolyzed lactose in intestinal juice draws water out of tissues by osmosis. The result is abdominal distention, cramping, and diarrhea.

   The level of the enzyme lactase in humans varies with both age and race. Most humans have sufficient lactase for the early years of life when milk is part of their diet. However, as adults the enzyme level decreases and lactose intolerance results. This trait shows remarkable genetic variations that have been examined in some homogeneous societies. For example, the majority of northern Europeans have high lactase levels as do several nomadic pastoral tribes in Africa (Figure 21.15). This ability to digest milk as adults may be the result of an evolutionary process in societies that milked animals. Those individuals with the enzyme necessary to digest milk may have had an adaptational advantage.

   Some societies such as the Thais and the Chinese have a high lactose intolerance. Similarly, both the Ibo and Yoruba of Nigeria cannot tolerate lactose as adults. The Fula and Hausa of the Sudan have quite different lactose intolerances. The Fula raise and milk a breed of cattle called fulani, whereas the Hausa, who show lactose intolerance, do not raise cattle. The Tussi, a cattle owner class of the Rundi of east Africa, also can tolerate lactose.

Figure 21.15
Lactose Intolerance.

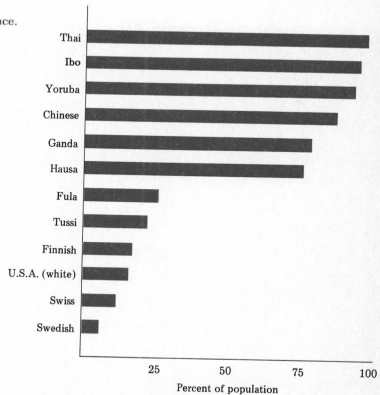

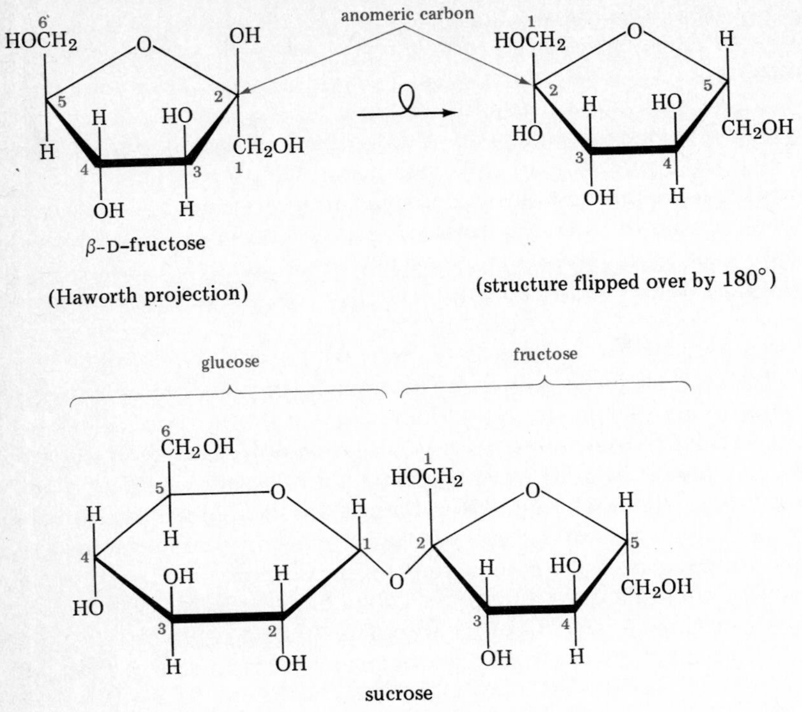

Figure 21.16
Structure of Sucrose.

β-D-fructose

(Haworth projection)

(structure flipped over by 180°)

glucose

fructose

sucrose

Sucrose

Sucrose, common table sugar, is a quite different disaccharide from those discussed to this point. Sucrose (Figure 21.16) is composed of glucose and fructose. The glycosidic linkage is α at the number 1 carbon of glucose and β at the number 2 carbon of fructose. As a result, there are both an acetal and a ketal functional group in sucrose. Since there is no hemiacetal center, sucrose can exist only as a single compound.

Example 21.7

Describe the structure of the following disaccharide.

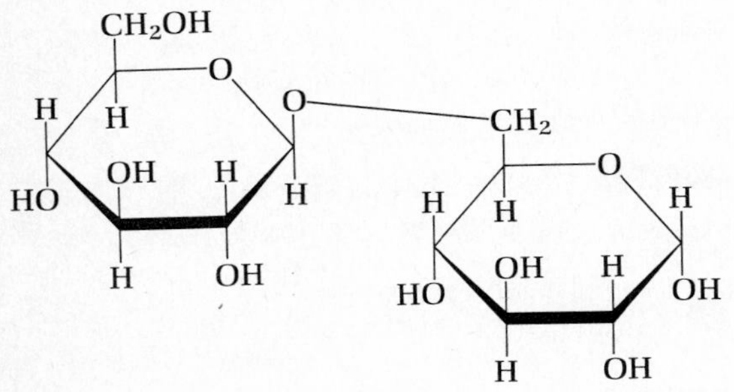

Solution. The hemiacetal center located on the ring at the right has a hydroxyl group below the plane of the ring and has the α configuration.
The glycosidic bond is from the number 1 carbon atom of the ring on the left

to the number 6 carbon atom of the ring on the right. Furthermore, the oxygen bridge is β to the ring on the left. The bridge is $\beta(1\rightarrow6)$. Both monosaccharide components are glucose.

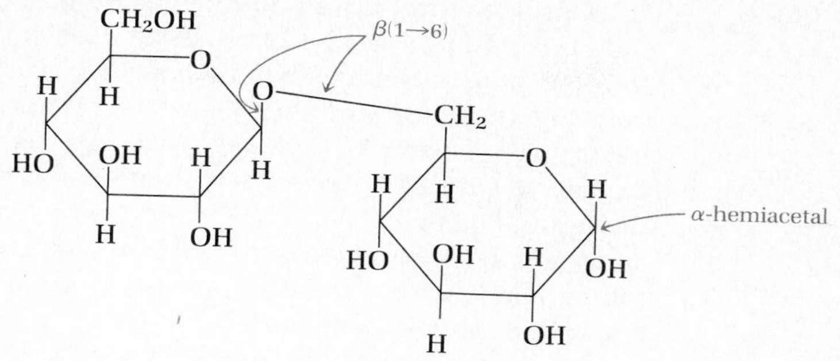

[Additional examples may be found in **21.40–21.45** at the end of the chapter.]

Monosaccharides can form long chains linked to each other by glycosidic bonds similar to those described for disaccharides. These high-molecular-weight substances are called polysaccharides. Starch and cellulose are homopolysaccharides containing only glucose. Among the heteropolysaccharides are hyaluronic acid, found in the vitreous humor of the eye, heparin, an anticoagulant in blood,

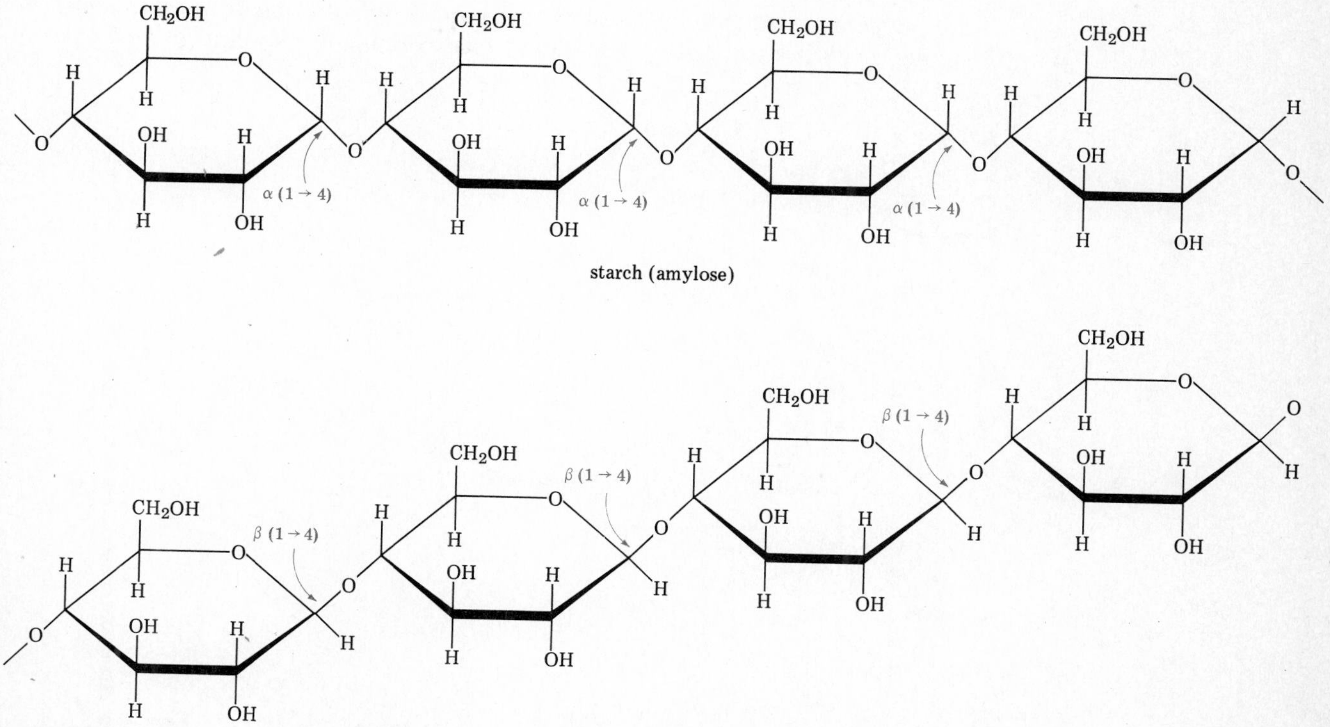

21.8
Polysaccharides

Figure 21.17
Structures of Starch (Amylose) and Cellulose.

and chondroitin, a component of cartilage and tendons. Since the structures of heteropolysaccharides are more complex than homopolysaccharides, only homopolysaccharides will be considered in this book.

One structural feature of polysaccharides is of great biological importance. Only α-linked polysaccharides can be digested by the majority of animals. The β-linked polysaccharides can be digested by cattle and other herbivores. However, digestion occurs only because microorganisms, present in their digestive tracts, have enzymes to hydrolyze β-glycosides. Starches are α-linked polymers of glucose (Figure 21.17) and serve as the major source of food for some animals. Cellulose (Figure 21.17) is a β-linked polymer of glucose. Cellulose accounts for approximately half of all of the carbon atoms contained in the plant kingdom. Unlike starch, which is soft and has no structural stability, cellulose is tough and forms the support material for plants.

Starch is available in potatoes, rice, wheat, and cereal grains. The exact molecular weight and structure of starch depend on its source. Two types of starch are known, amylose and amylopectin. Amylose is a linear polymer (Figure 21.17). The molecular weight of amylose may be between 40,000 and 400,000. These molecular weights correspond to between 200 and 2000 glucose units. Amylopectin contains chains similar to those in amylose, but only about 25 glucose units occur per chain. Branches of glucose-containing chains are interconnected by a glycosidic linkage between the hydroxyl group of carbon atom number 6 of one chain and the number 1 carbon atom of another glucose chain (Figure 21.18). The molecular weight of amylopectin may be as high as 1 million. Since each chain has an average molecular weight of 3000, there may be as many as 300 interconnected chains.

Glycogen, which is synthesized by animals as a storage form of glucose, is similar in structure to amylopectin. It has a molecular weight greater than 3

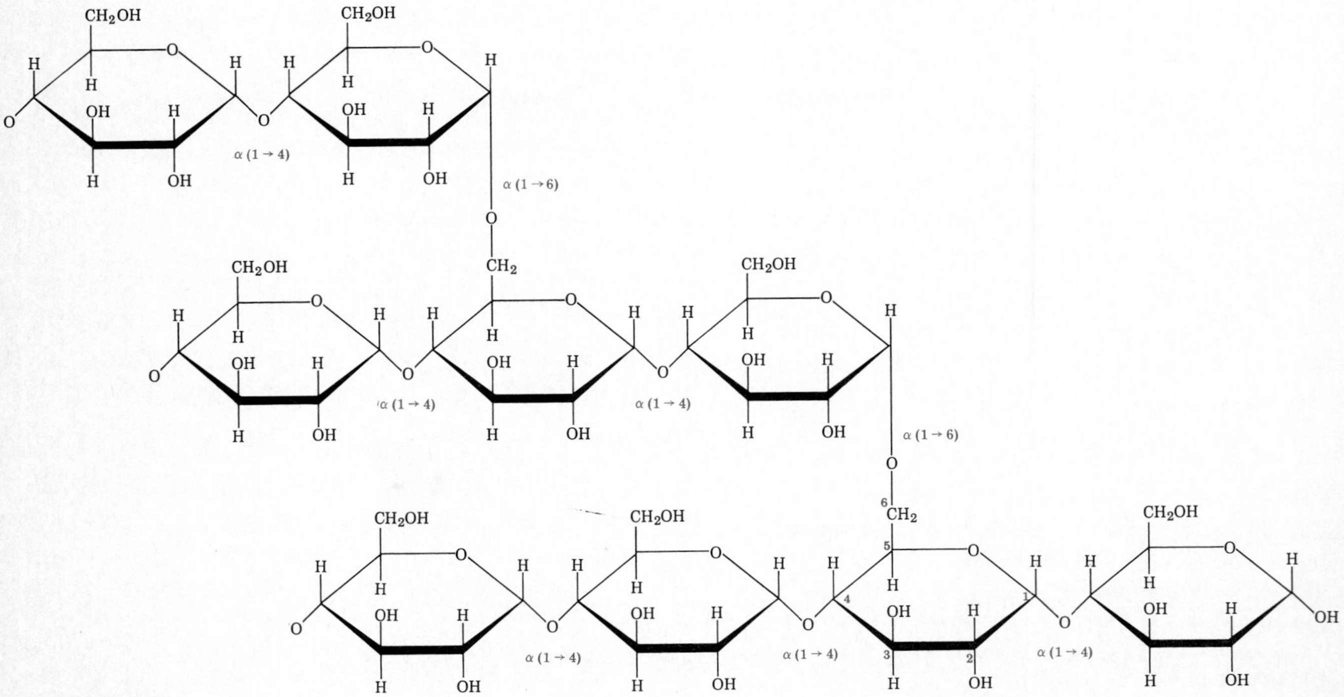

Figure 21.18
Branched Polysaccharide Structure of Amylopectin and Glycogen.

million. Although this molecular weight is larger than that of amylopectin, the weight is due not to longer chains but rather to more branching of chains. The average chain in glycogen is 12 glucose units.

During periods of diminished food intake, animals draw on glycogen stores as one of the sources of energy. Although glycogen is found throughout the body, it appears in the largest amounts in the liver. An average adult has enough glycogen for normal activity for about 15 hours. Thus, we need not eat frequently to maintain our energy.

In Section 17.5 you saw that aldehydes are easily oxidized to carboxylic acids. Oxidizing agents in Tollens' solution, Benedict's solution, and Fehling's solution are used to test for the presence of an aldehyde. These same solutions are used to detect aldehyde groups in carbohydrates. For example, Benedict's solution is used to detect glucose in urine. If no glucose is present in urine, the Benedict's solution remains blue, but the presence of glucose results in a red precipitate. Based on the amount of sugar present, the mixture of the precipitate and solution may vary from green to yellow to orange to red with increasing glucose concentration.

Oxidation of an aldose by Benedict's solution forms a glyconic acid. In the specific case of D-glucose, the glyconic acid is called D-gluconic acid. In the basic solution, the conjugate base is formed.

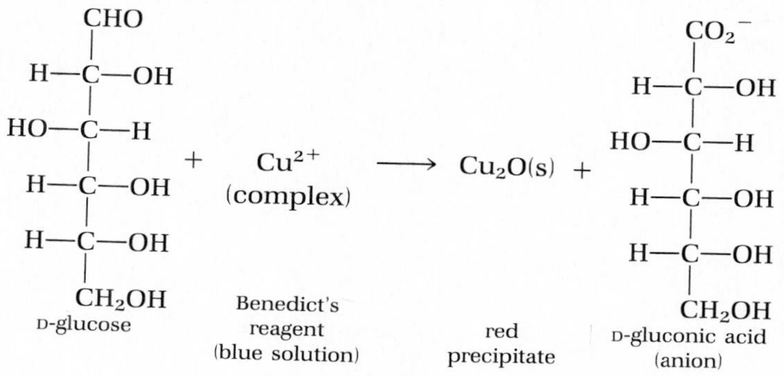

Although the aldoses exist as hemiacetals, there is a small amount of the open-chain form containing an aldehyde group. As the aldehyde reacts, more of the open-chain form is produced as the equilibrium shifts as expected based on Le Châtelier's principle.

Glyconic acids do not actually exist in the open-chain form. A reaction occurs between the carboxylic acid and the 5–OH group to produce a cyclic ester. Cyclic esters are called lactones.

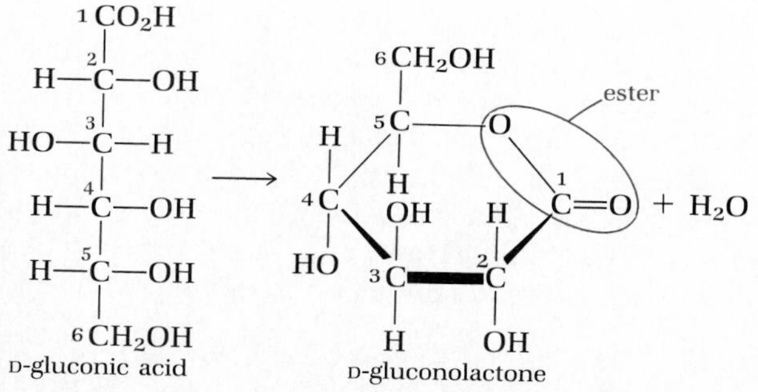

Oxidations of carbohydrates in living cells are catalyzed by enzymes. Processes occur to oxidize the D-glucose at the aldehyde to give D-gluconic acid as well as to oxidize the primary alcohol to give D-glucuronic acid.

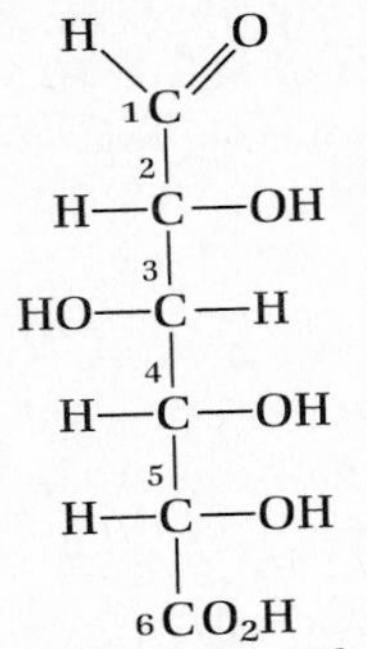

$$
\begin{array}{c}
\overset{\displaystyle H}{\underset{\displaystyle 1C}{\diagdown}} \diagup \overset{\displaystyle O}{} \\
\end{array}
$$

D-glucuronic acid

D-Gluconic acid as a phosphate ester is involved in metabolism. D-Gluconic acid is combined in the coating of some cells and in chondroitin of cartilage.

Ketoses also react with Benedict's solution! You would not expect this observation because ketones are not oxidized by Benedict's solution. The explanation for this apparent inconsistency is that ketoses are α-hydroxy ketones, which can be isomerized to tautomers (Section 17.6). A ketose can form a tautomer that is an enediol. While this enediol is derived from an α-hydroxy ketone, it may not only revert back to the α-hydroxy ketone but also form an isomeric α-hydroxy aldehyde.

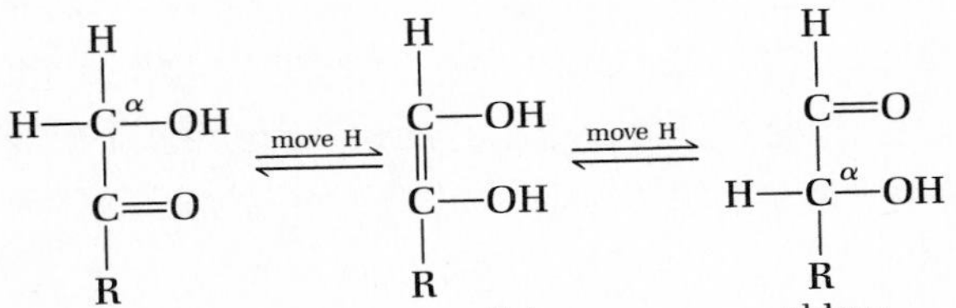

Movement of hydrogen to the number 1 carbon regenerates the original ketose. However, movement of hydrogen to the number 2 carbon forms an aldose. In basic solution, then, some aldose is formed from the ketose. As the aldose reacts with Benedict's solution, more ketose is converted into aldose as the equilibrium shifts as predicted by Le Châtelier's principle.

Carbohydrates that react with Benedict's solution are called **reducing sugars.** The term reducing refers to the effect of the carbohydrate on Benedict's solution. The carbohydrate is oxidized, but the Benedict's solution is reduced. Both aldoses and ketoses are reducing sugars, but glycosides formed from a monosaccharide and an alcohol are not reducing sugars. Glycosides are acetals that are not in equilibrium with an aldehyde. Disaccharides may be reducing sugars because the open-chain form in equilibrium with a hemiacetal center contains an aldehyde. It is this aldehyde that reacts with Benedict's solution; the glycosidic linkage is not affected. Both maltose and lactose are reducing sugars, but sucrose is not. Sucrose does not have a hemiacetal or hemiketal center and cannot produce a carbonyl group.

Example 21.8 _____

Is ribulose a reducing sugar?

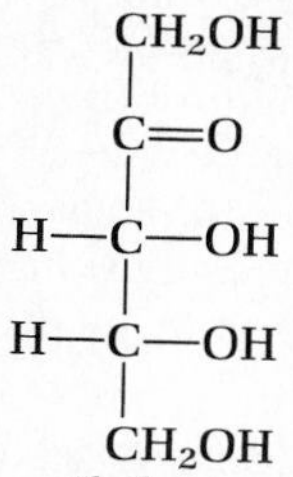

$$
\begin{array}{c}
CH_2OH \\
| \\
C{=}O \\
| \\
H{-}C{-}OH \\
| \\
H{-}C{-}OH \\
| \\
CH_2OH
\end{array}
$$
ribulose

Solution. This ketose will exist in equilibrium with an enediol intermediate, which is also in equilibrium with ribose. Since ribose contains an aldehyde group, ribulose will give a reaction with Benedict's solution and is classified as a reducing sugar.

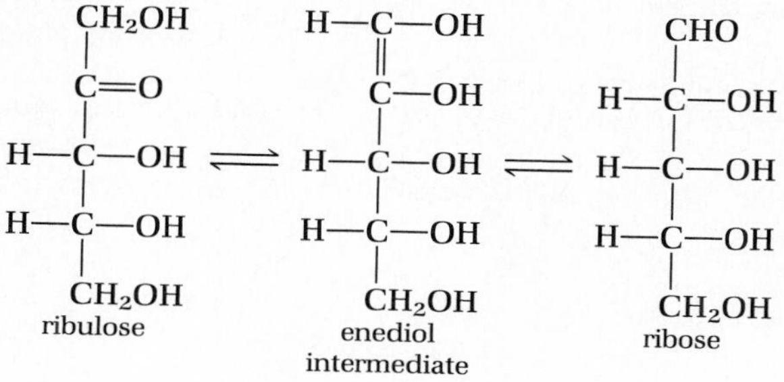

$$
\begin{array}{ccc}
CH_2OH & H{-}C{-}OH & CHO \\
| & || & | \\
C{=}O & C{-}OH & H{-}C{-}OH \\
| & | & | \\
H{-}C{-}OH \rightleftharpoons & H{-}C{-}OH \rightleftharpoons & H{-}C{-}OH \\
| & | & | \\
H{-}C{-}OH & H{-}C{-}OH & H{-}C{-}OH \\
| & | & | \\
CH_2OH & CH_2OH & CH_2OH \\
\text{ribulose} & \text{enediol} & \text{ribose} \\
& \text{intermediate} &
\end{array}
$$

[Additional examples may be found in **21.55–21.62** at the end of the chapter.]

Summary

The simplest carbohydrates, known as monosaccharides, are polyhydroxy aldehydes and polyhydroxy ketones. These compounds, which taste sweet, are called sugars. The aldehydes are called aldoses, whereas the ketones are called ketoses.

Configuration at the several asymmetric centers is an important aspect to the chemistry of monosaccharides. The configuration at the asymmetric carbon atom farthest from the aldehyde or ketone group determines the D or L configuration of the compound.

Aldoses and ketoses form cyclic hemiacetals and hemiketals. The five-membered rings are furanoses, whereas the six-membered rings are pyranoses. A Haworth projection is used to represent the hemiacetals and hemiketals. The additional asymmetric center formed in the hemiacetal or hemiketal is called the anomeric center. This center is responsible for the phenomenon of mutarotation in which the optical rotation of a mixture of anomers attains an equilibrium value.

Monosaccharides can react with alcohols to form acetals and ketals, which are called glycosides. If the alcohol is another monosaccharide, a disaccharide results.

Polysaccharides contain many monosaccharide units bonded by glycosidic bonds. The most common polysaccharides in plants are starch and cellulose. In starch there are α-1,4-glycosidic bonds between glucose units, whereas in cellulose there are β-1,4-glycosidic bonds. In animals, glycogen is a branched polymer of glucose containing α-1,4-glycosidic and α-1,6-glycosidic bonds.

Reducing sugars react with Benedict's solution to give a red precipitate (and with Tollens' reagent to give a silver mirror). Most monosaccharides are reducing sugars. Disaccharides with a hemiacetal or hemiketal center are also reducing sugars.

Learning Objectives

As a result of studying Chapter 21 you should be able to

- Classify a carbohydrate as a monosaccharide, disaccharide, or polysaccharide.
- Classify monosaccharides based on the carbonyl group and the number of carbon atoms.
- Outline the central role of glucose in several metabolic reactions.
- Draw Haworth projection formulas for the common monosaccharides.
- Explain how a monosaccharide mutarotates.
- Distinguish between hemiacetal and acetal centers in carbohydrates.
- Draw the structures of the common disaccharides.
- Describe the biological significance of α- and β-glycosidic bonds.
- Predict whether a carbohydrate is a reducing sugar based on its structure.

New Terms

An **aldose** is a carbohydrate containing an aldehyde group.

Anomers are stereoisomers that differ in configuration at one asymmetric center.

A **disaccharide** is a carbohydrate that can be hydrolyzed to two monosaccharides.

Galactosemia is a genetic disease preventing an individual from converting galactose into glucose.

Glucose tolerance is a measure of the ability of the body to handle dietary glucose.

Glucosuria is a condition of a blood sugar level above the renal threshold.

Gluconeogenesis is the conversion of pyruvic acid into glucose by the liver.

Glycogen is a storage form of glucose in animals.

Glycogenesis is the conversion of glucose into glycogen.

Glycogenolysis is the conversion of glycogen into glucose.

Glycolysis is the conversion of glucose into pyruvic acid.

A **glycoside** is an acetal or ketal of a carbohydrate.

Hypoglycemia is a condition of low blood sugar.

Hyperglycemia is a condition of high blood sugar.

A **ketose** is a carbohydrate having a ketone functional group.

Lactose intolerance is a disease in which individuals cannot hydrolyze lactose for further metabolic reactions.

A **monosaccharide** is a simple carbohydrate that cannot be further hydrolyzed.

Mutarotation is the change in optical rotation due to equilibrium between anomeric forms.

Oligosaccharides contain a small number of monosaccharide units joined by acetal bonds.

The **pentose phosphate pathway** converts glucose 6-phosphate into ribulose 5-phosphate in the cell.

A **polysaccharide** is a polymer consisting of many monosaccharides linked by glycosidic bonds.

A **reducing sugar** causes the reduction of Benedict's solution (or Tollens' reagent).

Questions and Problems

Terminology

21.1 Describe an aldose and a ketose.

21.2 What is a glycosidic bond?

21.3 To what carbon atom do the letters D and L refer in monosaccharides?

21.4 Is a reducing sugar a substance that can be used to lose weight? Explain.

Classification of Monosaccharides

21.5 Draw perspective formulas of D-glyceraldehyde and L-glyceraldehyde. Explain the meaning of the designations D and L.

21.6 Write a structure of any compound that is an aldopentose.

21.7 Write a structure of any compound that is a ketohexose.

21.8 Draw the four stereoisomeric aldotetroses. Which are enantiomers and which are diastereomers?

21.9 Classify each of the following monosaccharides.

21.10 Draw open-chain projection formulas for each of the following compounds.
(a) D-ribose (b) D-fructose (c) D-glucose
(d) D-mannose

21.11 Classify each of the following monosaccharides as D or L.

(a) 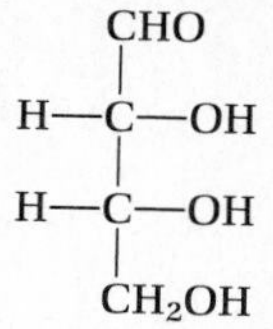
$$
\begin{array}{c}
\text{CHO} \\
\text{H—C—OH} \\
\text{H—C—OH} \\
\text{CH}_2\text{OH}
\end{array}
$$

(b) 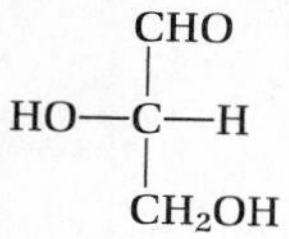
$$
\begin{array}{c}
\text{CHO} \\
\text{HO—C—H} \\
\text{CH}_2\text{OH}
\end{array}
$$

(c)
$$
\begin{array}{c}
\text{CH}_2\text{OH} \\
\text{C=O} \\
\text{HO—C—H} \\
\text{H—C—OH} \\
\text{H—C—OH} \\
\text{CH}_2\text{OH}
\end{array}
$$

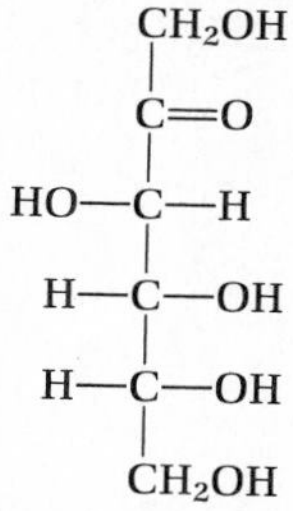

(d) 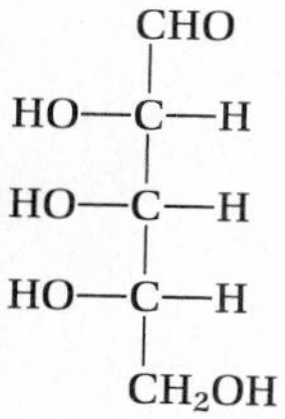
$$
\begin{array}{c}
\text{CHO} \\
\text{HO—C—H} \\
\text{HO—C—H} \\
\text{HO—C—H} \\
\text{CH}_2\text{OH}
\end{array}
$$

21.12 Examine the structure of 2-deoxy-D-ribose. How does it differ from ribose? Why is the compound considered a carbohydrate even though its molecular formula is $C_5H_{10}O_4$?

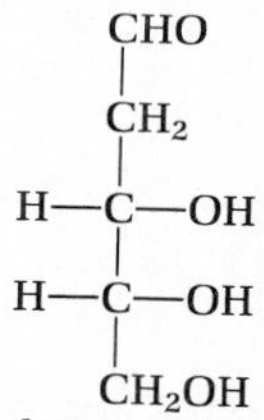

$$
\begin{array}{c}
\text{CHO} \\
\text{CH}_2 \\
\text{H—C—OH} \\
\text{H—C—OH} \\
\text{CH}_2\text{OH}
\end{array}
$$
2-deoxy-D-ribose

Glucose and Metabolism

21.13 What is the role of glucose 6-phosphatase in the cell? Which organs contain this enzyme?

21.14 Distinguish between glycogenolysis and gluconeogenesis.

21.15 Consider the brain, muscles, and the liver. Which organs can store glucose? Which can export glucose?

21.16 In what ways can the liver form glucose for release to other organs?

21.17 Calculate the molarity of glucose in the blood when the concentration is 80 mg/100 mL.

21.18 Using the data in **21.17,** calculate the number of grams of glucose in the blood assuming a blood volume of 5 L.

Haworth Projection Formulas

21.19 What is meant by the terms pyranose and furanose?

21.20 Draw a Haworth projection of an anomer of the hemiacetal of each of the following compounds.
(a) 5-hydroxyhexanal   (b) 4-hydroxyhexanal
(c) 5-hydroxy-2-hexanone (d) 6-hydroxy-2-hexanone

21.21 Draw the Haworth projection formula of the pyranose form of each of the following compounds.

(a) α-D-mannose   (b) β-D-galactose
(c) α-D-glucose   (d) α-D-galactose

21.22 Draw the Haworth projection formula of the furanose form of each of the following compounds.
(a) α-D-fructose   (b) β-D-fructose
(c) α-D-ribulose   (d) β-D-xylulose

21.23 Draw the Haworth projection formula of the pyranose form of each of the following compounds.
(a) α-D-fructose   (b) β-D-sorbose

21.24 Identify each of the following structures by the name of the monosaccharide.

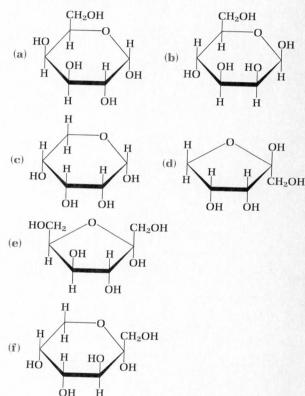

21.25 Which of the compounds in **21.24** are pyranoses and which are furanoses?

21.26 What is the configuration at the anomeric carbon atom of each compound in **21.24?**

21.27 Ribose exists in equilibrium in solution as four isomers. These are 20% of the α-pyranose, 56% of the β-pyranose, 6% of the α-furanose, and 18% of the β-furanose. Draw the Haworth projection of each of these isomers.

Chair Conformations

21.28 Draw the chair conformation of the β-pyranose of glucose.

21.29 Draw the chair conformation of the β-pyranose of mannose. Explain how this structure differs from the β-pyranose of glucose.

21.30 Draw the chair conformations of the two anomeric pyranose forms of galactose. Which of the two forms is more stable?

Mutarotation

21.31 Can all aldopentoses and aldohexoses mutarotate?

21.32 Will L-glucose mutarotate?

21.33 Name two disaccharides that will mutarotate.

21.34 Which of the following compounds can mutarotate?

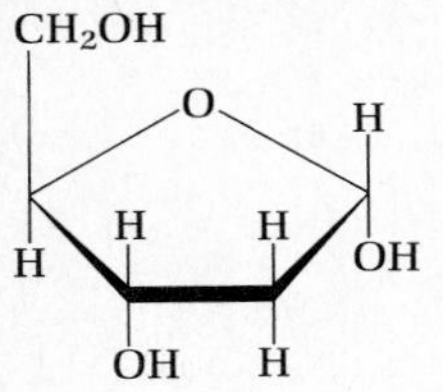

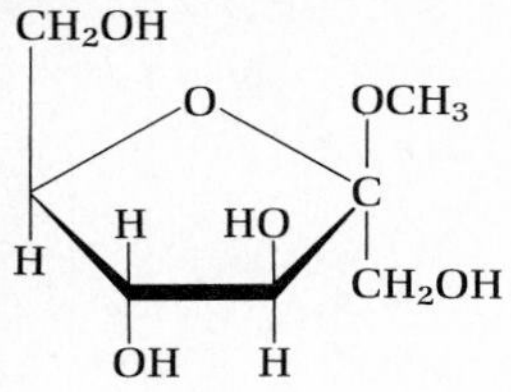

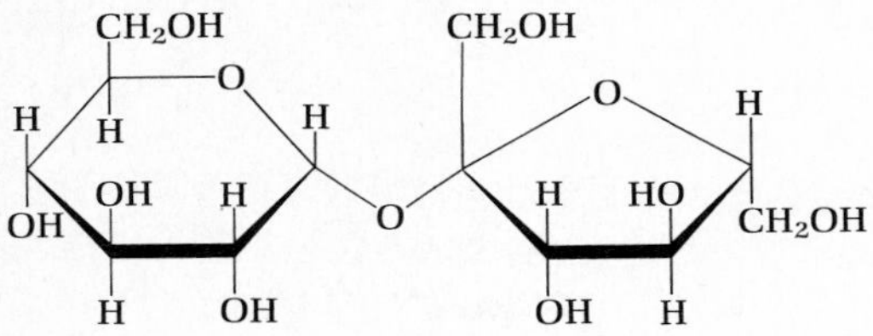

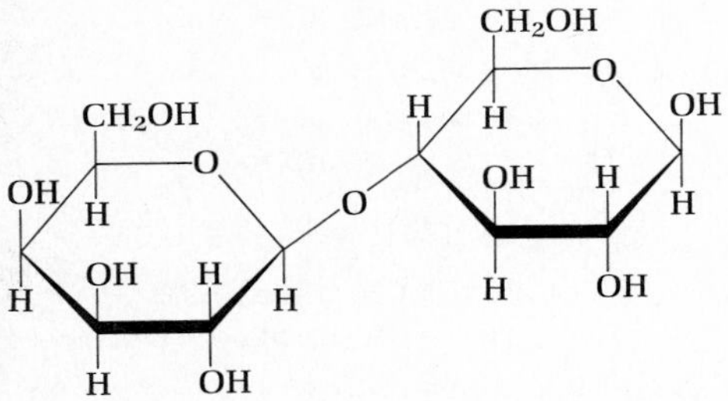

21.35 Draw the two pyranose forms of D-galactose and explain what occurs when this compound mutarotates.

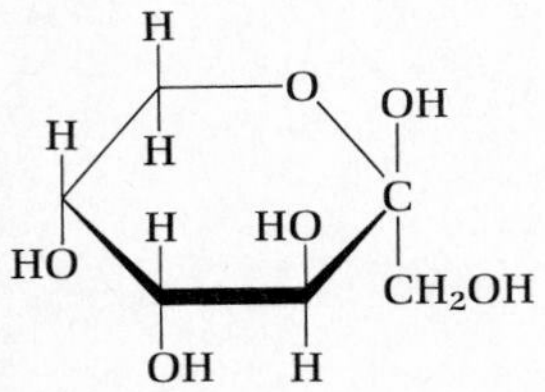

Glycosides

21.36 Write the Haworth projection formulas of the two glycosides formed from the pyranose forms of glucose and methyl alcohol.

21.37 Explain why the methyl glycoside of the α-pyranose form of glucose does not mutarotate.

21.38 Write the Haworth projection formulas of the two glycosides formed from the furanose forms of fructose and methyl alcohol.

21.39 Identify the following compounds by their components, configuration at the acetal or ketal carbon atom, and the ring form.

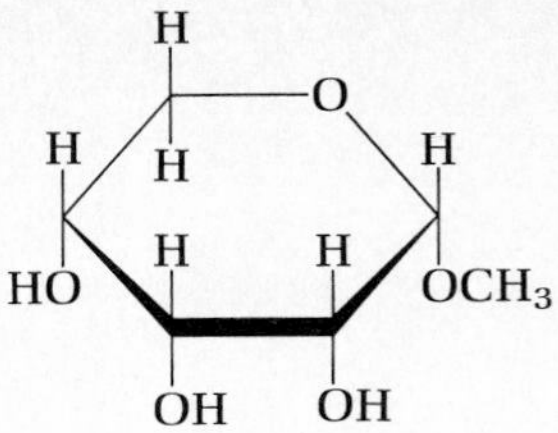

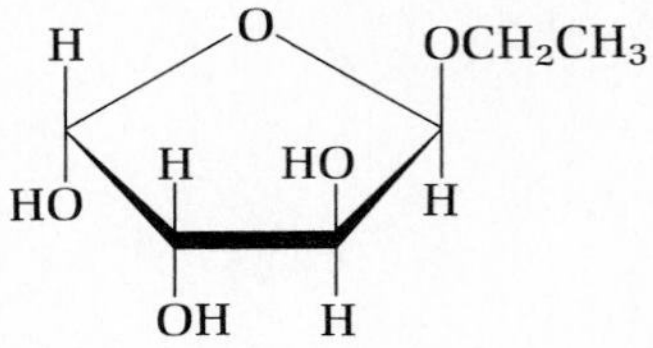

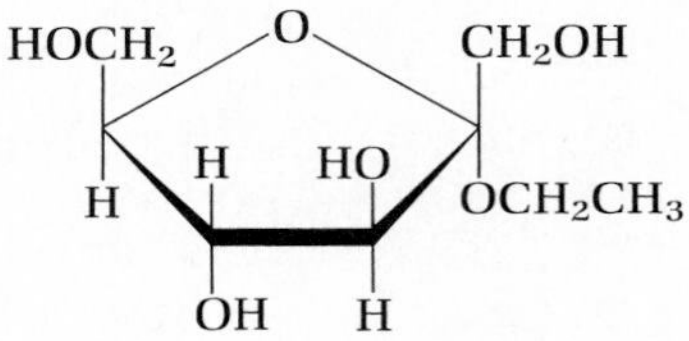

Disaccharides

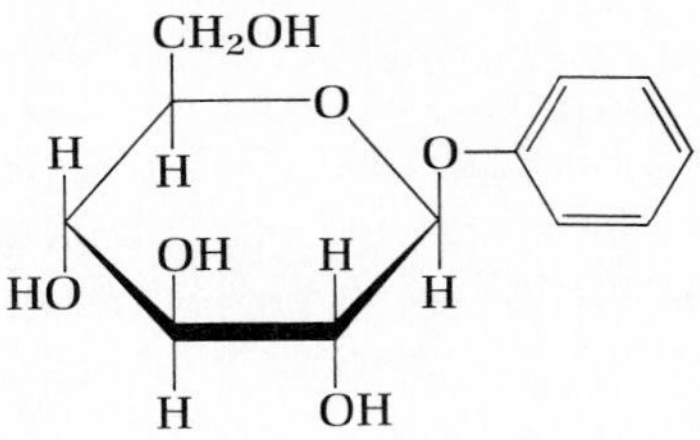

21.40 What are the general structures of acetals and ketals?

21.41 What part of a disaccharide is an acetal or a ketal?

21.42 What monosaccharides are produced from the hydrolysis of each of the following compounds?
(a) lactose   (b) maltose   (c) sucrose
(d) cellobiose

21.43 Describe the types of glycosidic linkages in each of the following compounds.

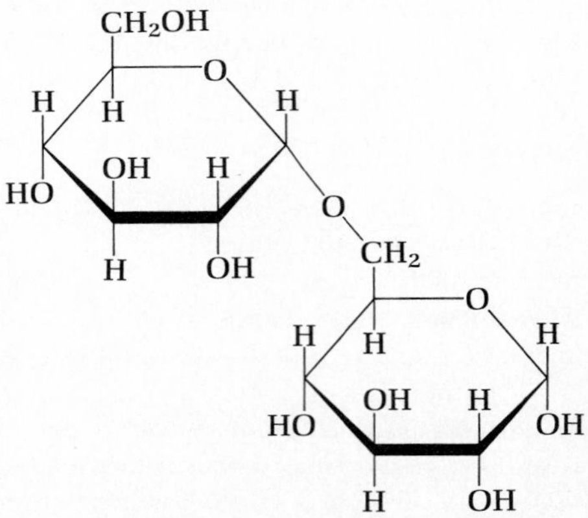

CH₂OH

(b)

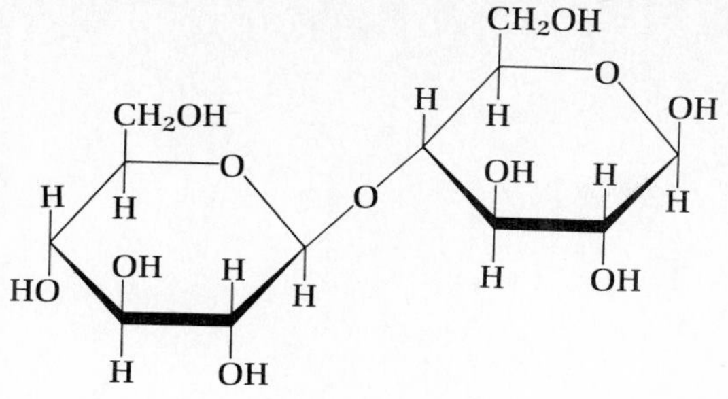

CH₂OH

HO OH H

H OH

CH₂OH

OH

OH H

H OH

(c)

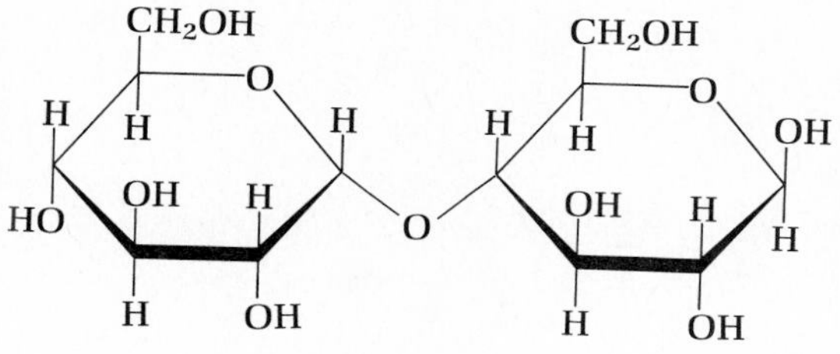

CH₂OH

HO OH H

H OH

CH₂OH

OH H

H OH

(d)

CH₂OH

HO OH H

H OH

CH₂OH

OH H

OH H

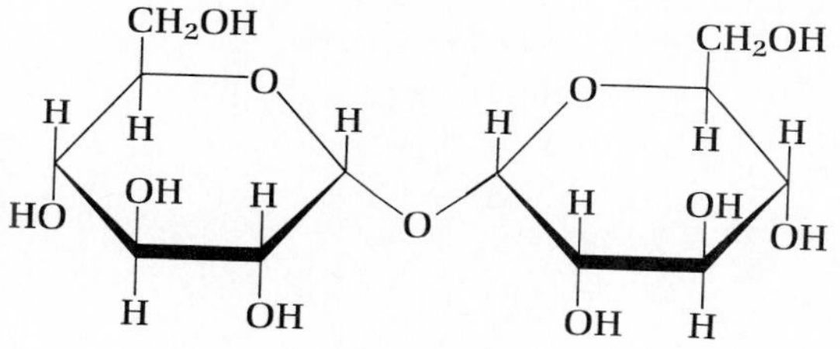

21.44 Can the disaccharide given in Example 21.7 muta-rotate?

21.45 Trehalose, a disaccharide found in mushrooms, has the following structure. What are the constituent monosaccharides? Describe the glycosidic bond.

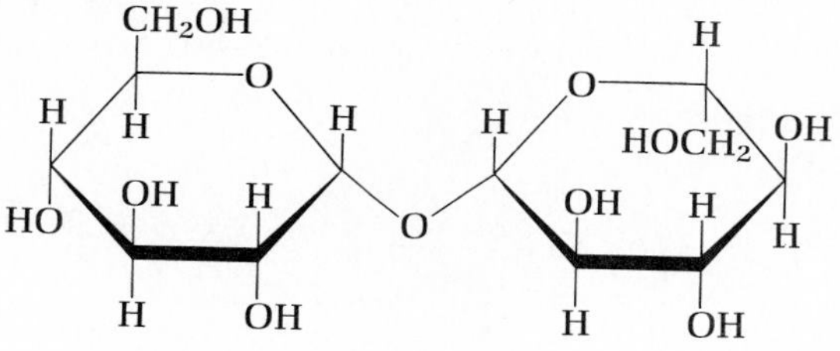

CH₂OH

HO OH H

H OH

HOCH₂ OH

OH H

H OH

Polysaccharides

21.46 What are the structural characteristics of each of the following substances?

(a) amylose   (b) glycogen   (c) amylopectin
(d) cellulose

21.47 What are the main structural differences between amylose and amylopectin?

21.48 Why can't humans digest cellulose?

21.49 How do amylopectin and glycogen resemble each other? How do they differ?

21.50 Describe the biological importance of glycogen.

Isomerization of Monosaccharides

21.51 Treatment of D-glucose with dilute base at room temperature produces an equilibrium mixture of D-glucose, D-mannose, and D-fructose. Explain how this isomerization occurs.

21.52 Write the structure of one aldose and one ketose that can exist in equilibrium with D-ribose in basic solution.

21.53 Explain how dihydroxyacetone and glyceraldehyde may exist in equilibrium with each other in basic solution.

21.54 What aldose and ketose can exist in equilibrium with D-galactose in basic solution?

Oxidation of Carbohydrates

21.55 Ketones do not react with Benedict's solution, but fructose gives a red precipitate with Benedict's solution. Explain this observation.

21.56 Are all aldoses reducing sugars?

21.57 Will 2-deoxy-D-ribose react with Benedict's solution (see **21.12**)?

21.58 Explain why D-ribulose, although classified as a ketose, is a reducing sugar.

21.59 Which of the compounds in **21.34** will react with Benedict's solution?

21.60 Which of the compounds in **21.43** are reducing sugars?

21.61 Which common disaccharides are reducing sugars?

21.62 Explain why sucrose is not a reducing sugar.

21.63 Draw the open-chain formula for each of the following compounds.

(a) D-glucuronic acid   (b) D-galacturonic acid
(c) D-mannouronic acid

21.64 Alginic acid, a polymer of the following compound, is used as a thickening agent in foods. Hydrolysis of alginic acid gives the following compound. What type of compound is the substance and from what monosaccharide is it derived?

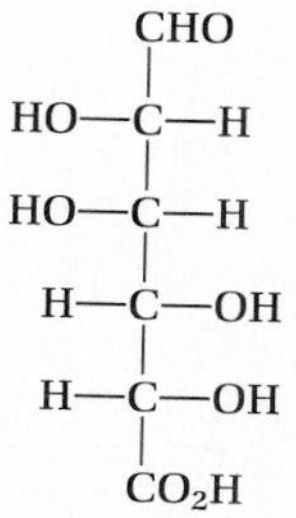

CHO

HO—C—H

HO—C—H

H—C—OH

H—C—OH

CO₂H

21.65 Draw the open-chain form of the compound obtained by oxidizing mannose with Tollen's reagent. How does this compound differ from the compound given in **21.64?**

21.66 Pectic acid, used to form jellies from fruits, is a polymer of the following compound. What type of compound is this substance and from what monosaccharide is it derived?

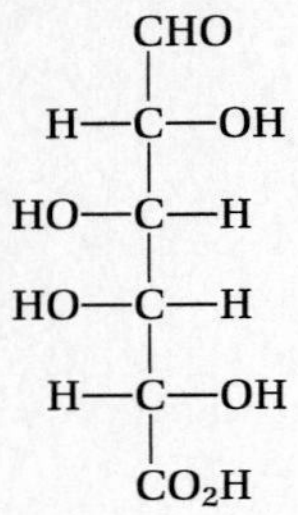

```
        CHO
         |
   H —— C —— OH
         |
  HO —— C —— H
         |
  HO —— C —— H
         |
   H —— C —— OH
         |
        CO₂H
```

Lipids

The feathers of this duck are coated with a wax that prevents them from being wetted. Without this coating the duck would be less buoyant.

The structures of carbohydrates and proteins are easily defined in terms of functional groups. The chemistry of carbohydrates is that of polyhydroxy aldehydes and ketones, whereas the chemistry of proteins is that of amino acids. However, lipids are not a single type of compound. They are a conglomerate class of compounds with a variety of functional groups.

Lipids, which are relatively nonpolar, can be separated from the other more polar biochemicals by dissolving them in nonpolar organic solvents like chloroform and diethyl ether. In fact, the classical definition is *lipids are biochemical compounds that are soluble in nonpolar organic solvents.* Lipid is sometimes used as a synonym for fat because fats are soluble in nonpolar organic solvents. However, fat is only one of the various types of lipids.

Lipids are broadly classified into two groups based on their characteristic hydrolysis reactions (Figure 22.1).

    I. *Lipids that are not hydrolyzed by base are called* **simple lipids.** These include terpenes, which are produced in plants, and steroids, which are important hormones in animals.

    II. *Lipids that are hydrolyzed by base are called* **complex lipids.** The most common hydrolysis products are long-chain carboxylic acids called fatty acids.

The fatty acids in complex lipids have an even number of carbon atoms. Some of the acids are saturated, but other acids have one or more double bonds. The other components obtained in the hydrolysis of complex lipids vary considerably and determine the subclasses of complex lipids.

    1. **Waxes** *are esters of long-chain alcohols and fatty acids.*
    2. **Triglycerides,** *also known as* **triacylglycerols,** *are esters of glycerol and long-chain fatty acids.* Triglycerides that are relatively saturated are fats, whereas the more unsaturated compounds are oils.
    3. **Phosphoglycerides** *consist of glycerol, phosphoric acid, long-chain fatty*

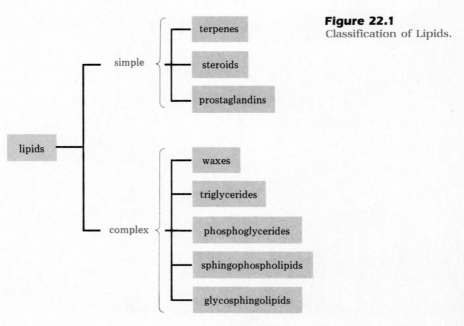

Figure 22.1
Classification of Lipids.

acids, and certain low-molecular-weight alcohols. Phosphoglycerides are important constituents of cell membranes.

4. **Sphingophospholipids** *consist of sphingosine, phosphoric acid, long-chain fatty acids, and choline.* Sphingophospholipids are part of cell membranes found in the brain and in the protective coat of nerves.

5. **Glycosphingolipids** *consist of sphingosine, fatty acids, and a carbohydrate that may be a monosaccharide or an oligosaccharide.* Glycosphingolipids containing only glucose or galactose are called cerebrosides, whereas those containing oligosaccharides are called gangliosides.

Terpenes, which are abundant in the oils of plants and flowers, have distinctive odors and flavors. They are responsible for the odors of pine trees and some flowers, such as geraniums, and for the color of carrots and tomatoes. Although terpenes are of relatively minor biochemical significance in man, vitamins A, E, and K are terpenes. Terpenes are considered as consisting of isoprene units (Figure 22.2). The isoprene units are joined together, usually head-to-tail, but occasionally tail-to-tail. Terpenes may have different degrees of unsaturation, and a variety of functional groups may be present.

Terpenes are classified according to the number of isoprene units they contain. The simplest terpene class, the monoterpenes, contain two isoprene units. Examples include geraniol, limonene, and carvone. The colored lines separate the hypothetical isoprene units.

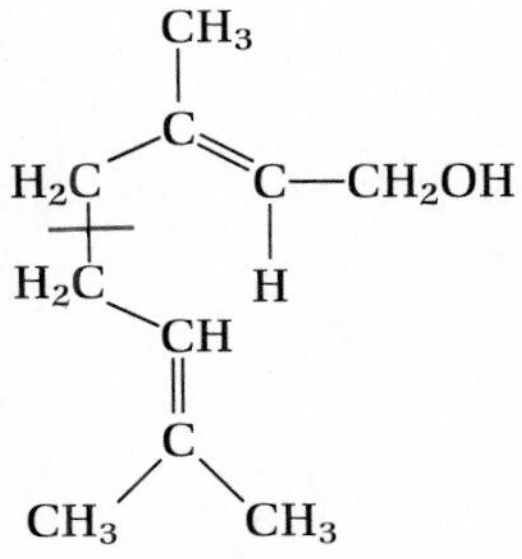

22.2
Terpenes

geraniol
(geranium oil)

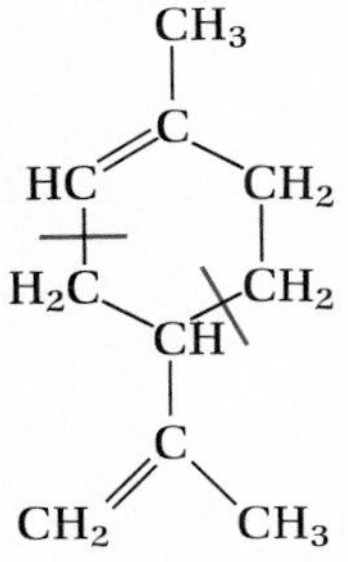

limonene
(citrus oils)

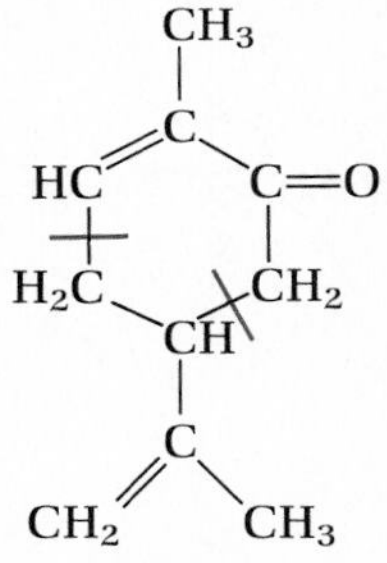

carvone
(oil of caraway)

Sesquiterpenes consist of three isoprene units. Several of these structures are shown below in condensed formulas.

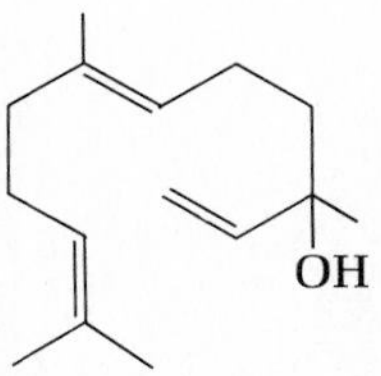

nerolidol
(oil of neroli)

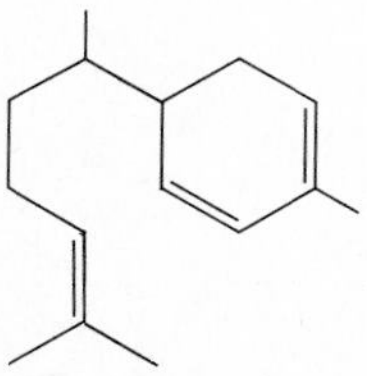

zingiberene
(oil of ginger)

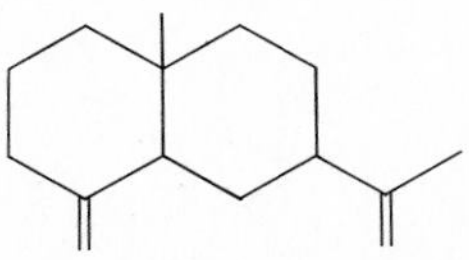

β-selinene
(oil of celery)

Diterpenes consist of four isoprene units. The saturated alcohol phytol,

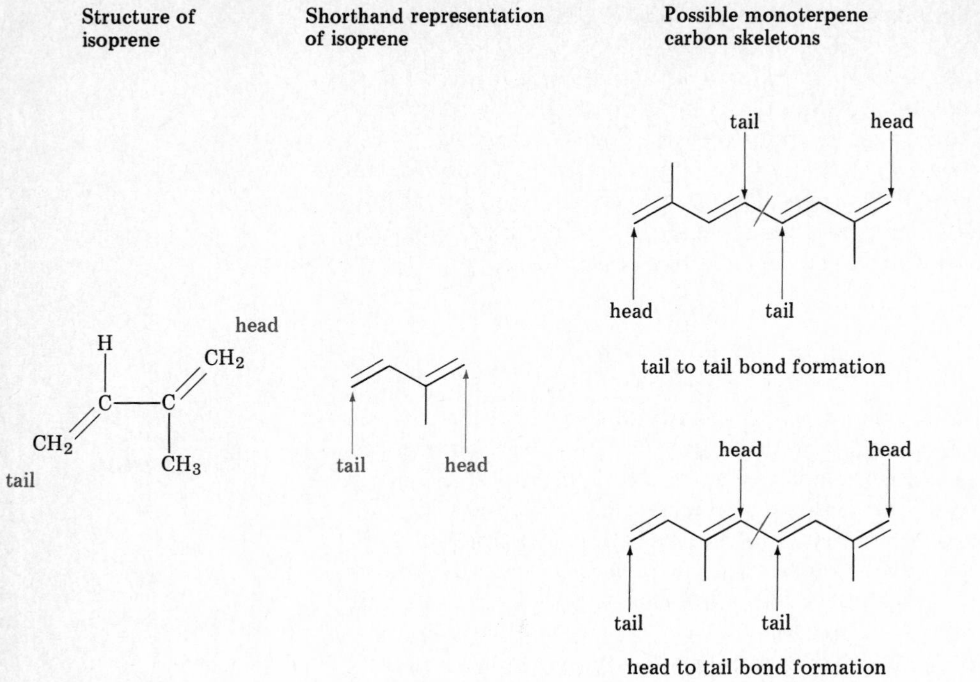
Structure of isoprene
Shorthand representation of isoprene
Possible monoterpene carbon skeletons
head
H
CH_2
C
C
CH_2
CH_3
tail
tail
head
tail
head
head
tail
tail to tail bond formation
head
head
tail
tail
head to tail bond formation

Figure 22.2
Derivation of Terpenes from Isoprene.

which occurs as an ester in chlorophyll, is a diterpene. Vitamin A is also a diterpene.

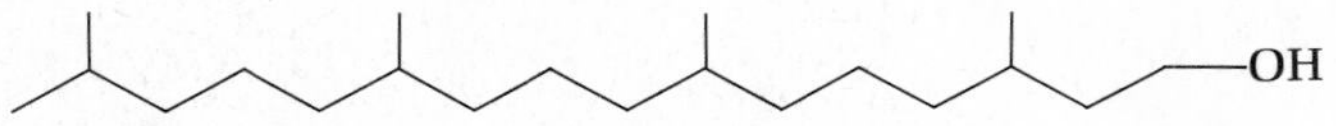
OH

phytol

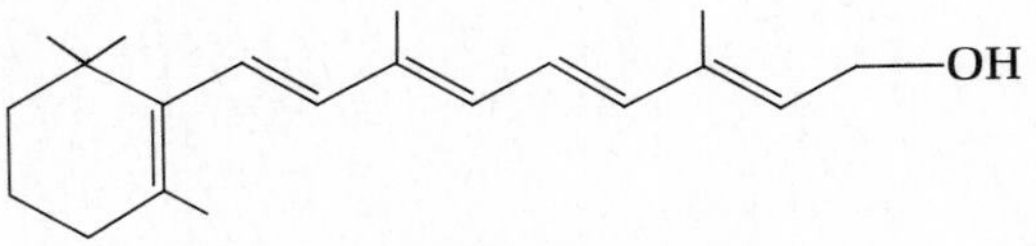
OH

vitamin A

Triterpenes consist of six isoprene units. Squalene, a triterpene, has the unusual tail-to-tail arrangement in the center of the molecule. This terpene reacts to form lanosterol found in wool fat.

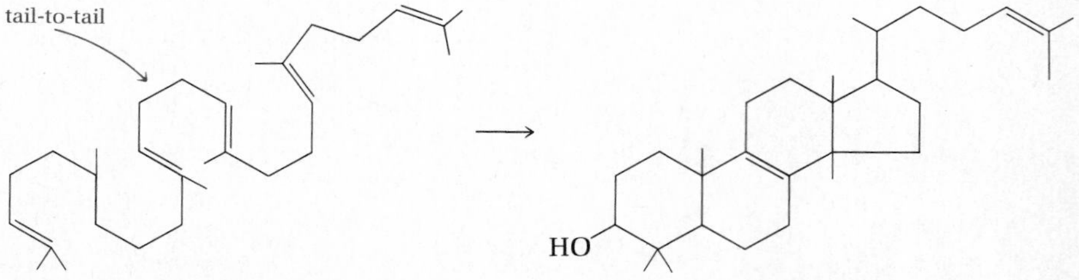
tail-to-tail
HO

squalene lanosterol

Example 22.1 ───────────────────────────────────

Classify the following terpene and indicate the division into isoprene units.

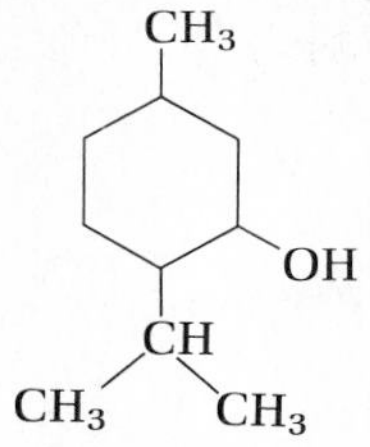

Solution. The compound contains ten carbon atoms and contains two isoprene units. The compound is a monoterpene.

The isopropyl group provides three of the necessary carbon atoms for one isoprene unit. Only two other carbon atoms are required. One carbon atom is the atom to which the isopropyl group is attached. The remaining carbon atom may be either of the carbon atoms bonded to this carbon atom.

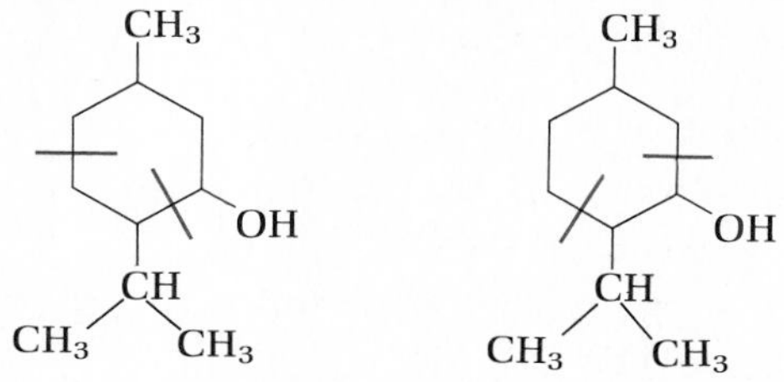

The two pairs of colored lines indicate the possible divisions into isoprene units.

───────────────────────────────────

[Additional examples may be found in **22.10–22.12** at the end of the chapter.]

Steroids *are tetracyclic compounds* (Figure 22.3) *containing three six-membered rings and a five-membered ring.* Each ring is numbered by a standard system. Steroids contain a variety of functional groups such as hydroxyl, carbonyl, and carbon–carbon double bonds.

22.3
Steroids

Cholesterol

Cholesterol is a well-known steroid that is synthesized in most animals from squalene. It occurs in amounts up to 25% in cell membranes and affects their flexibility. The higher the percentage of cholesterol, the more rigid is the cell membrane. In humans the central nervous system has a high cholesterol content, and the brain is about 10% cholesterol on a dry weight basis. Human plasma contains about 50 mg of free cholesterol per 100 mL and about 170 mg of cholesterol esterified with various fatty acids. Cholesterol plays a vital biological role in the formation of steroid hormones, vitamin D, and bile acids.

Corticosteroids

Steroids produced in the adrenal cortex are called **corticosteroids.** The corticosteroids are of two types: glucocorticoid and mineralocorticoid. Both classes

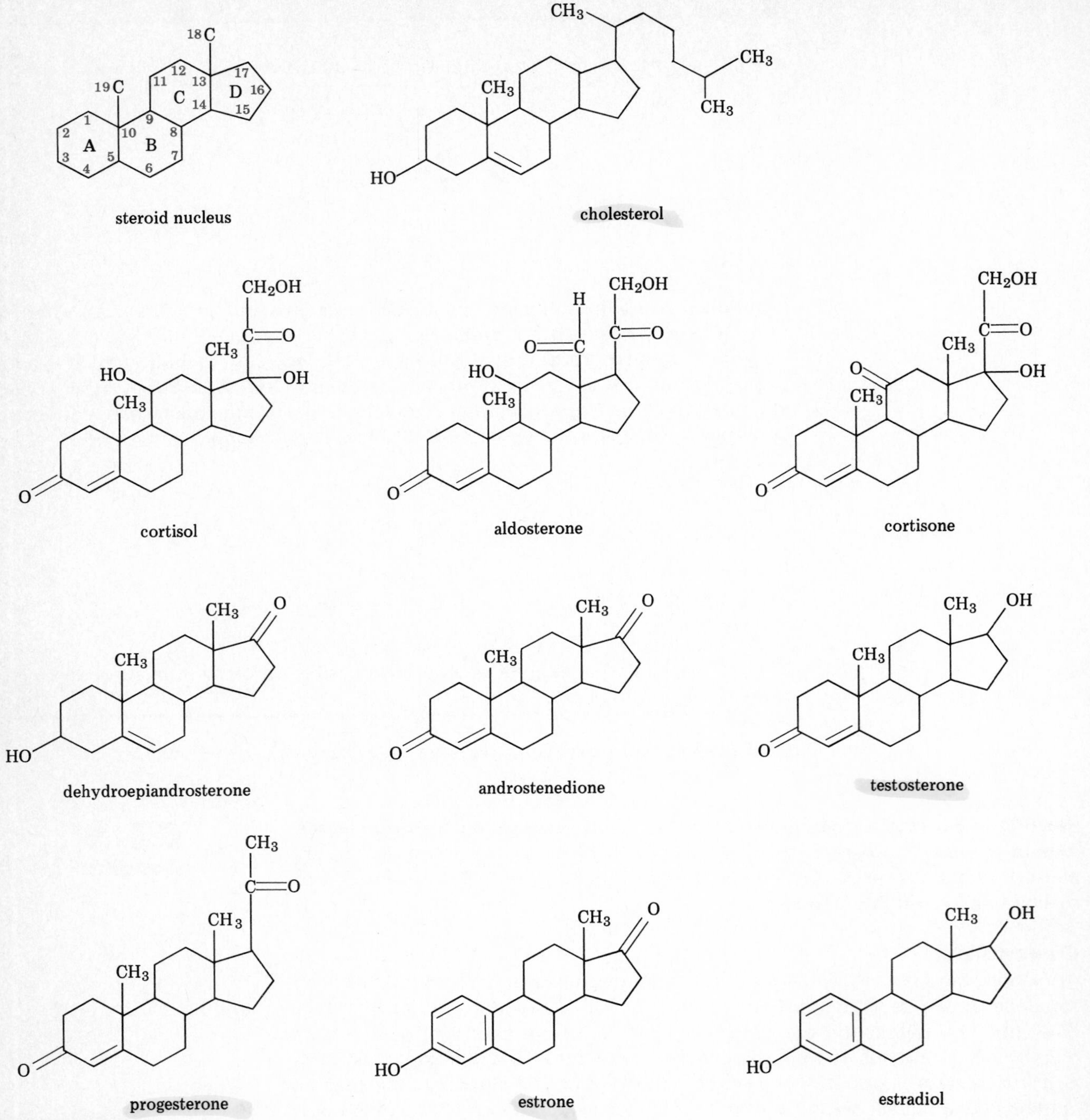

Figure 22.3
Structures of Some Steroids.

of compounds are hormones. **Hormones** *provide a communication pathway between various tissues in multicellular organisms.*

Glucocorticoids *are involved, along with insulin, in controlling the glucose balance in the body.* Glucocorticoids such as cortisol (Figure 22.3) promote gluconeogenesis and the formation of glycogen, whereas insulin facilitates the use of

AN ASIDE

There has been considerable debate about the role of dietary sources of cholesterol in heart disease. However, cholesterol is synthesized by the liver and to some extent by the intestine. About 800 mg of cholesterol per day is produced by adults on a low-cholesterol diet. When cholesterol is absorbed from dietary sources, the synthesis process is repressed.

Arteriosclerosis or "hardening of the arteries" is one of the medical problems of aging. When accompanied by a buildup of cholesterol deposits on the inner surfaces of arteries, the condition is called *atherosclerosis.* Although all of the reasons for the deposition of cholesterol are not known, one is thought to be the decreased ability to metabolize fat with increasing age. Therefore, the cholesterol content in the tissues and blood increases.

Atherosclerosis is a condition in which the diameter of the arteries decreases and there is increased resistance to blood flow. As a result the individual's blood pressure increases. If any blood clots form within the restricted arteries, blood flow may cease and the cells will die from oxygen deprivation. The death of tissue in this manner is called infarction. Although infarction may occur in different tissue, the best known and most serious is myocardial infarction, which involves the arteries of the heart.

---

glucose by cells. These two substances must be in balance in order for the body's sugar to be properly metabolized to provide energy or to be stored as glycogen.

The adrenal cortex secretes 25 mg of cortisol per day. This steroid depresses the synthesis of protein in muscle tissue. As a consequence, amino acids become available for conversion to pyruvic acid and then to glucose by gluconeogenesis and glycogen by glycogenesis. Thus, cortisol increases the supply of carbohydrates at the expense of proteins.

Mineralocorticoids *affect the electrolyte balance of body fluids and hence the water balance.* Aldosterone, secreted by the adrenal cortex, is the most active of the mineralocorticoids. This hormone causes the kidney tubules to reabsorb Na^+, Cl^-, and HCO_3^-. While only about 0.2 mg of aldosterone is secreted each day by a normal adult, the water balance of the body is essentially under the control of this hormone.

Cortisone (Figure 22.3) is the best known of the corticosteroids because of its use in medicine. Cortisone is an effective antiinflammatory agent that is used in treatment of a host of diseases, including rheumatoid arthritis and bronchial asthma.

Sex Hormones

The testes of the male and the ovaries of the female produce steroids that are sex hormones. These hormones control growth and development of reproductive organs, the development of secondary sex characteristics, and the reproductive cycle.

Male sex hormones are **androgens.** Testosterone, the most important androgen, is produced in the testes. Its function is to stimulate production of sperm by the testes and to promote the growth of the male sex organs. Testosterone also is responsible for muscle development. *The* **anabolic steroids** *are synthetic substances related to testosterone. These steroids are used by some athletes to promote muscle development,* but their use is banned by most athletic unions.

The androgens dehydroepiandrosterone and androstenedione (Figure 22.3) are also produced in the adrenal cortex. These steroids are responsible for such secondary sex characteristics in males as deep voices and beards. Excessive androgens in females cause masculinization. **Estrogens** *are female sex hormones.* They are produced in the adrenal cortex as well. However, the major source of male and female sex hormones is the gonads.

Progesterone and two estrogens, estrone and estradiol, are important female sex hormones (Figure 22.3) produced in the ovaries that control the menstrual cycle. After menstrual flow, estrogen secretion causes growth of the lining of the uterus and the ripening of the ovum. Progesterone secretion acts to prevent other ova from ripening after ovulation and aids in maintaining the fertilized egg after implantation. If production of progesterone should decrease, spontaneous abortion would occur. This condition, known as progesterone deficiency, is responsible for patterns of repeated miscarriages.

The estrogens cause the growth of tissues of female sexual organs. Although estrogens are secreted in childhood, the rate increases by 20-fold after puberty. The fallopian tubes, uterus, and vagina all increase in size. The estrogens also initiate growth of the breasts and the breast's milk-producing ductile system.

Example 22.2

Examine the structures of estrone and estradiol and determine the similarities and differences of these compounds.

Solution. Both compounds (see Figure 22.3) have a phenolic ring as the A ring of the steroid. The remaining three rings—B, C, and D—are identical, and there is a methyl group at position 13. The only structural difference is the ketone at position 17 in estrone and the secondary alcohol at position 17 in estradiol. Oxidation of estradiol would yield estrone.

Example 22.3

Norethindrone is a synthetic substance used in some contraceptive pills. Examine its structure and suggest a mechanism for its action.

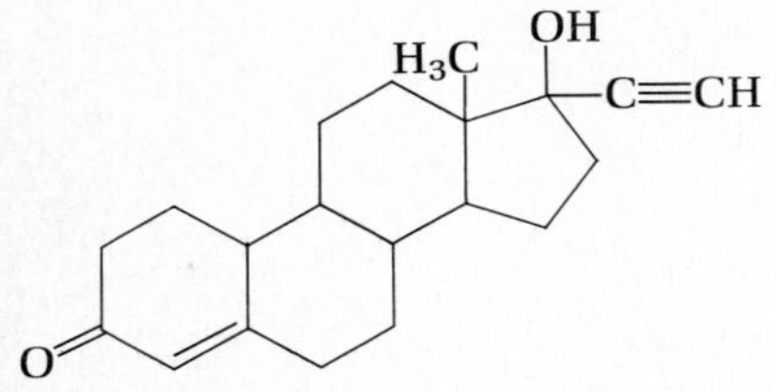

Solution. The compound resembles progesterone. High levels of progesterone prevent maturation and release of an egg. Thus, norethindrone may mimic the action of progesterone and make the female body behave as if it is pregnant.

[Additional examples may be found in **22.13–22.19** at the end of the chapter.]

Figure 22.4
Structure of Some Waxes.

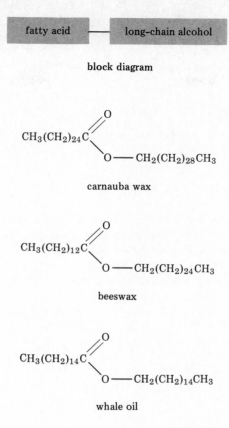

block diagram

carnauba wax

beeswax

whale oil

Waxes (Figure 22.4) are esters of fatty acids and long-chain alcohols, both of which contain an even number of carbon atoms. Waxes are low-melting solids that coat the surface of plant leaves and fruits and also coat hair and feathers to provide a water barrier. When the wax is removed from the feathers of an aquatic bird as a result of an oil spill, the feathers become wet and the bird cannot maintain its buoyancy.

Carnauba wax, widely used in floor polish and car wax, coats the leaves of palm trees. Beeswax is secreted by bees and is used as a structural material for the beehive. Whale oil is actually a wax. For each ester we may represent the structure as a block diagram. Upon hydrolysis with base, a molecule of a wax yields an alcohol and a fatty acid as a carboxylate salt.

22.4
Waxes

Example 22.4

Half of the dry weight of a copepod that lives in the waters of British Columbia is a compound $C_{36}H_{62}O_2$. Hydrolysis of the compound yields an unbranched acid, $C_{20}H_{30}O_2$, and a straight-chain alcohol, $C_{16}H_{34}O$. Hydrogenation of the acid yields $C_{20}H_{40}O_2$. Describe the structure of the $C_{36}H_{62}O_2$ compound.

Solution. Since the compound yields an acid and an alcohol when hydrolyzed, it must be an ester. Esters of long-chain alcohols and fatty acids are classified as waxes.

The formula $C_{16}H_{34}O$ corresponds to a saturated alcohol. The acid, $C_{20}H_{30}O_2$, must contain five double bonds, since hydrogenation results in the addition of five moles of hydrogen gas. The location of the double bonds cannot be determined from the information given.

A general representation of the wax is

$$C_{19}H_{30}C \overset{O}{\diagup} O-(CH_2)_{15}CH_3$$

[Additional examples may be found in **22.20–22.24** at the end of the chapter.]

22.5
Fatty Acids

Triglycerides, phosphoglycerides, sphingophospholipids, and glycosphingolipids upon hydrolysis all yield fatty acids. **Fatty acids** *are long-chain carboxylic acids containing an even number of carbon atoms.* These acids usually range from 14 to 22 carbon atoms, but the 16- and 18-carbon acids are the most abundant. Oleic acid, an 18-carbon acid, which is the most abundant unsaturated fatty acid, has the *cis* configuration. The double bonds of most naturally occurring unsaturated fatty acids are *cis*. Mammals cannot synthesize unsaturated fatty acids containing more than one double bond. For this reason polyunsaturated fatty acids are required in the diet. Linoleic and linolenic acids are then essential fatty acids.

The structures and melting points of some fatty acids are given in Table 22.1. The melting points increase with increasing chain length. This is as expected, since the London forces increase with increasing chain length. Increasing the number of *cis* double bonds strongly decreases the melting point. Unsaturated acids of a *cis* configuration are "bent" molecules because of the presence of the double bonds. The space-filling models of oleic and linoleic acids and of stearic acid are shown in Figure 22.5. At physiological pH the fatty acids exist as anions.

The hydrocarbon chains of saturated acids can pack together tightly in the solid. The London forces can thus hold neighboring chains close together, whereas the "bends" in unsaturated acids hinder efficient packing and the London forces are weaker. As a result, unsaturated acids have lower melting points than saturated acids.

Table 22.1

Structures and Melting Points of Common Fatty Acids

Name	Formula	Melting point (°C)
Saturated		
Lauric	$CH_3(CH_2)_{10}CO_2H$	44
Myristic	$CH_3(CH_2)_{12}CO_2H$	54
Palmitic	$CH_3(CH_2)_{14}CO_2H$	63
Stearic	$CH_3(CH_2)_{16}CO_2H$	70
Arachidic	$CH_3(CH_2)_{18}CO_2H$	76
Unsaturated		
Oleic	$CH_3(CH_2)_7CH{=}CH(CH_2)_7CO_2H$	13
Linoleic	$CH_3(CH_2)_4CH{=}CHCH_2CH{=}CH(CH_2)_7CO_2H$	−5
Linolenic	$CH_3CH_2CH{=}CHCH_2CH{=}CHCH_2CH{=}CH(CH_2)_7CO_2H$	−11
Arachidonic	$CH_3(CH_2)_4CH{=}CHCH_2CH{=}CHCH_2CH{=}CHCH_2CH{=}CH(CH_2)_3CO_2H$	−50

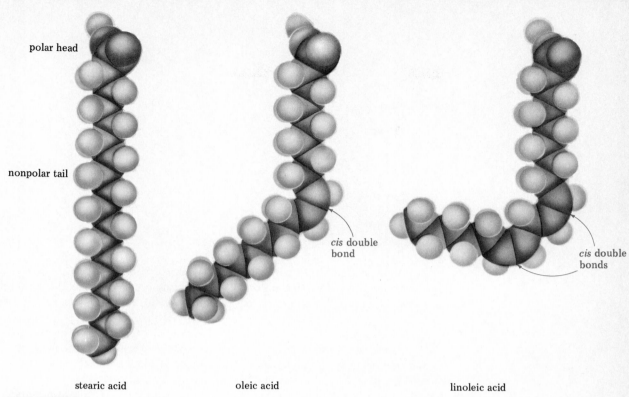

stearic acid   oleic acid   linoleic acid

Figure 22.5
Space-Filling Models of Fatty Acids.

Example 22.5 ————————————————————————————

The melting point of palmitoleic acid (*cis*-9-hexadecenoic acid) is −1°C. Compare this melting point to the melting point of palmitic acid and explain the difference.

$$CH_3(CH_2)_5 \quad \quad (CH_2)_7CO_2H$$
$$C{=}C$$
$$H \quad \quad \quad H$$

palmitoleic acid

Solution. The melting point of palmitic acid is 63°C, which is 64 degrees higher than the melting point of palmitoleic acid. This difference is due to the presence of a *cis* double bond in palmitoleic acid, which prevents close approach of the hydrocarbon chains and results in lower London forces.

——

[Additional examples may be found in **22.27–22.33** at the end of the chapter.]

Soaps

Salts of long-chain fatty acids are soaps (Section 18.6). The best soaps are made from saturated acids that have 14–18 carbon atoms. Sodium salts are com-

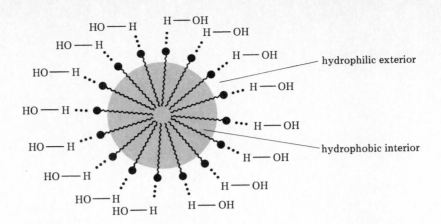

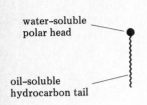

water-soluble
polar head

oil-soluble
hydrocarbon tail

Figure 22.6
Micelle of Soap.

monly used in cake soaps, whereas the softer potassium salts are used in shaving creams.

Because of the long nonpolar hydrocarbon chain, the carboxylate salts of fatty acids do not form a true solution, but are dispersed as micelles. *A **micelle** is a submicroscopic aggregation of molecules or ions.* In the micelle of carboxylate salts, the nonpolar hydrocarbon chains are within the sphere. The negatively charged carboxylate groups are on the outside of the sphere (Figure 22.6).

The nonpolar hydrocarbon chain is **hydrophobic** (*water-repelling*), whereas the carboxylate group is **hydrophilic** (*water-attracting*) and forms hydrogen bonds to water. The micelle is held together by hydrophobic bonds. **Hydrophobic bonds** *are London forces between hydrophobic hydrocarbon chains.* The surface of the micelle, which may contain several hundred carboxylate groups, has a high negative charge. As a result, individual micelles repel each other and they remain suspended in the water.

Grease will not dissolve in water because it is nonpolar. However, grease will dissolve in the hydrocarbon region of a micelle. The hydrocarbon chains penetrate the grease, remove molecules of the grease, and dissolve the grease within the micelle. This process accounts for the cleansing action of a soap.

22.6
Triglycerides

Fatty acids occur in a free form in only small amounts in nature. However, they are common in complex lipids such as triglycerides. Triglycerides, triesters of glycerol and fatty acids, are also known as fats and oils. Triglycerides may be represented by the following general formula and block diagram.

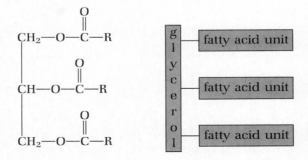

AN ASIDE

Prostaglandins are one of the most biologically potent classes of compounds recently discovered. **Prostaglandins** *are biological derivatives of arachidonic acid* and are formed via prostanoic acid (Figure 22.7). Prostaglandins are 20-carbon fatty acids containing a *trans*-substituted five-membered ring. They are classified according to the number and arrangement of double bonds and hydroxyl and ketone groups.

The prostaglandins, which were originally isolated from the prostate gland, occur in many body tissues and are physiologically active at very low concentrations. The molecular basis of prostaglandin activity is still not well understood. However, there is considerable clinical interest in these compounds. Regulation of menstruation, control of fertility, contraception, control of blood pressure, and blood clot prevention are all potential uses of prostaglandins. For this reason, many prostaglandins have been synthesized in the laboratory for pharmacological use.

Prostaglandin E_1 inhibits the breakdown of fat in adipose tissue. Prostaglandin E_2 induces labor in pregnant women when administered at intravenous dosages of 4 μg (4×10^{-6} g) per minute. The chemical stimulates uterine contractions in women whose supply is insufficient for normal labor.

Figure 22.7
Structure of Prostaglandins.

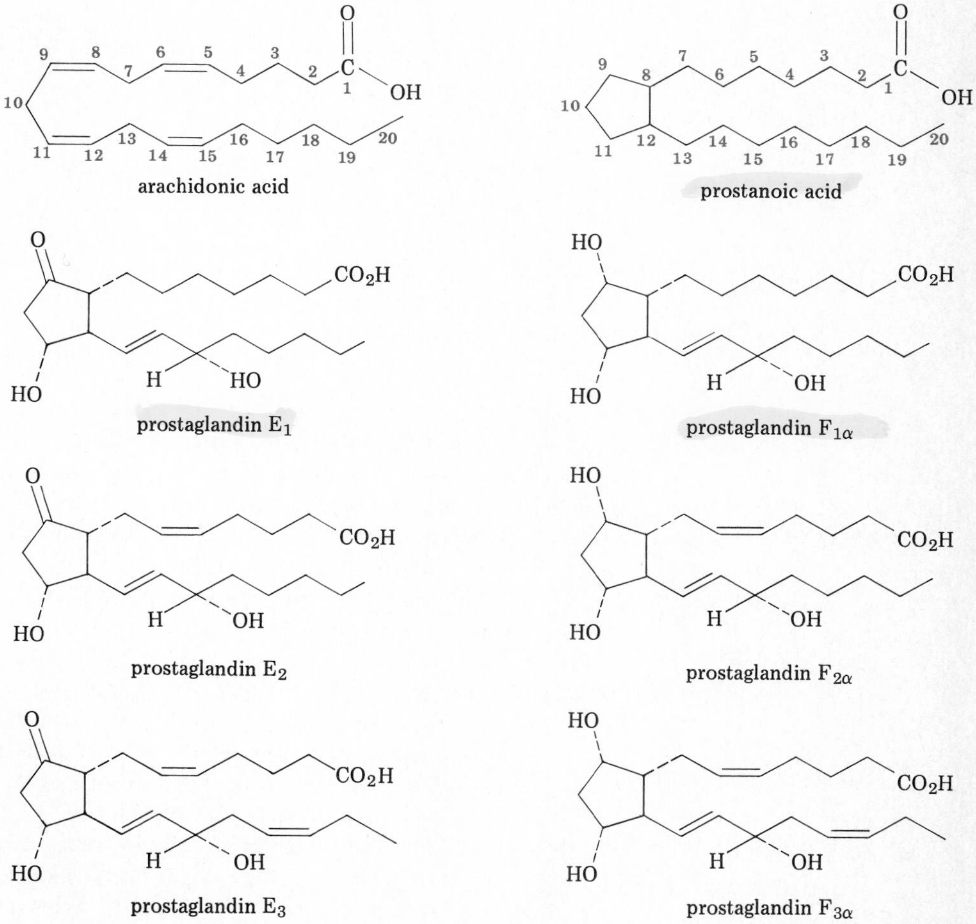

Some men are infertile because of a low prostaglandin level in their sperm. When prostaglandins are mixed with their sperm and then used in artificial insemination, the likelihood of pregnancy is increased. The prostaglandins cause uterine contractions that move the sperm toward the ovum in the fallopian tube.

Much research is being done to develop a variety of synthetic prostaglandins for use as therapeutic drugs. Natural prostaglandins cannot be taken orally because they are rapidly degraded and do not survive sufficiently long for effective action. Thus, one of the goals of researchers is to develop modified prostaglandins that can be administered orally.

Fats and oils are not simple compounds but rather mixtures in which the acid fraction of the molecule may vary in chain length and degree of unsaturation. A single molecule of a fat or oil may contain three different acid residues.

Fats *have a high percentage of saturated acids, whereas oils have a high percentage of unsaturated acids.* They are solids or semisolids and are usually obtained from animals. The important acids found in these sources are lauric, palmitic, and stearic acids (Table 22.1). An example of a fat molecule containing all three of these saturated acids is given below.

$$
\begin{array}{c}
\overset{\displaystyle H}{\underset{\displaystyle |}{}} \quad \overset{\displaystyle O}{} \\
H-\overset{|}{\underset{|}{C}}-O-\overset{\parallel}{C}-(CH_2)_{16}CH_3 \\
\\
\overset{\displaystyle O}{} \\
H-\overset{|}{\underset{|}{C}}-O-\overset{\parallel}{C}-(CH_2)_{14}CH_3 \\
\\
\overset{\displaystyle O}{} \\
H-\overset{|}{\underset{|}{C}}-O-\overset{\parallel}{C}-(CH_2)_{10}CH_3 \\
\overset{\displaystyle |}{H}
\end{array}
$$

Oils that are derived from vegetable sources such as olives, corn, and soybeans are similar to fats, except that the acid residues tend to be more unsaturated. The unsaturated acid residues in the molecules of oils lower their melting points, and they are usually liquids. Among the unsaturated acids found in oils are oleic, linoleic, and linolenic acid. These acids all contain 18 carbon atoms, but differ in their degree of unsaturation.

Animals accumulate fat, adipose tissue, when their intake of food exceeds their energy output requirements. Vital organs such as kidneys are enclosed in adipose tissue, which prevents damage when they are subjected to a blow. A subcutaneous layer of fat helps insulate the animal against heat loss. Although plants do not generally store fats and oils for energy requirements, some (such as peanuts and olives) produce triglycerides in abundance.

Triglycerides can be separated from natural sources by simple physical methods. Oils can be obtained from vegetables such as olives, corn, or soybeans by pressing. Animal fats can be separated from other tissues by melting the fat and removing the liquid in a process much like cooking bacon.

The exact composition of triglycerides varies, as indicated in Table 22.2. The

Table 22.2

Composition of Fats and Oils

| | Melting point (°C) | Percent composition of fatty acids | | | | | | |
| | | Saturated | | | | Unsaturated | | |
		Myristic	Palmitic	Stearic	Arachidic	Oleic	Linoleic	Linolenic
Animal fats								
Butter	32	11	29	9	2	27	4	—
Lard	30	1	28	12	—	48	6	—
Human fat	15	3	24	8	—	47	10	—
Plant oils								
Corn	−20	1	10	3	—	50	34	—
Cottonseed	−1	1	23	1	1	23	48	—
Linseed	−24	—	6	2	1	19	24	47
Olive	−6	—	7	2	—	84	5	—
Peanut	3	—	8	3	2	56	26	—
Soybean	−16	—	10	2	—	29	51	6

conditions in which vegetables are grown affect the degree of unsaturation; for example, linseed oil obtained from flaxseed grown in warm climates may contain up to twice the degree of unsaturation found in oil obtained from seed grown in cold climates. Similarly, the composition of lard from hogs depends on their diet; the fat of corn-fed hogs has less unsaturation than that of peanut-fed hogs.

The relationship between human consumption of saturated fats and arterial disease has been the object of extensive medical research. Unsaturated fats are believed to be beneficial in preventing arterial deposits. Safflower oil, because of its high content of unsaturated material, is now a popular product.

Example 22.6

Olive oil contains 84% oleic acid. Draw a structure for one of the possible components of olive oil.

Solution. Although all oils are mixtures of triglycerides, the large percentage of oleic acid means that there must be a large amount of a triglyceride containing only oleic acid.

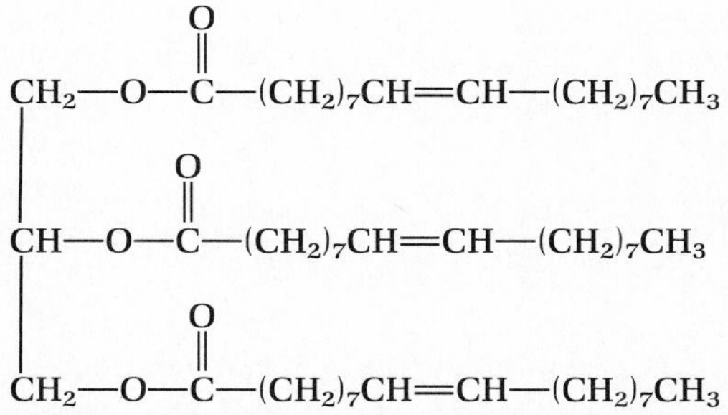

$$
\begin{array}{l}
\text{CH}_2\text{—O—}\overset{\overset{\displaystyle O}{\|}}{\text{C}}\text{—(CH}_2)_7\text{CH=CH—(CH}_2)_7\text{CH}_3 \\[6pt]
\text{CH—O—}\overset{\overset{\displaystyle O}{\|}}{\text{C}}\text{—(CH}_2)_7\text{CH=CH—(CH}_2)_7\text{CH}_3 \\[6pt]
\text{CH}_2\text{—O—}\overset{\overset{\displaystyle O}{\|}}{\text{C}}\text{—(CH}_2)_7\text{CH=CH—(CH}_2)_7\text{CH}_3
\end{array}
$$

[Additional examples may be found in **22.39–22.48** at the end of the chapter.]

22.7
Phosphoglycerides

In triglycerides all three hydroxyl groups of glycerol are esterified with carboxylic acids. If two hydroxyl groups are esterified with carboxylic acids and one hydroxyl group with phosphoric acid, a phosphatidic acid results. At physiological pH phosphatidic acid exists in an ionized phosphatidate form.

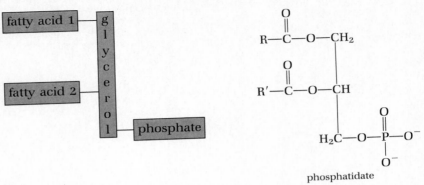

phosphatidate

A phosphatidic acid contains an asymmetric carbon atom. In naturally occurring materials only the L configuration occurs.

Phosphatidic acid is only a minor component of phosphoglycerides; they are usually esterified with an alcohol to form diesters called phosphatides. The phosphatides contain the alcohol choline, ethanolamine, or inositol or the amino acid serine.

$$HO-CH_2-CH_2-NH_2$$
ethanolamine

$$HO-CH_2-CH_2-\overset{+}{N}(CH_3)_3$$
choline

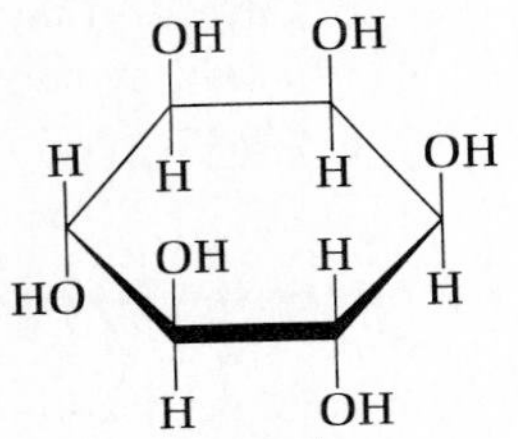

inositol

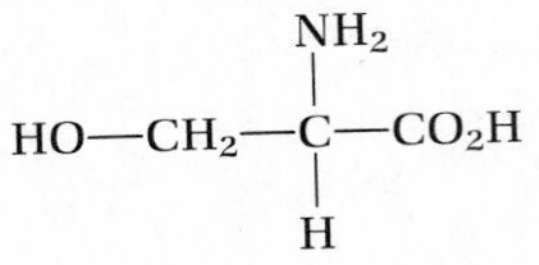

$$HO-CH_2-\overset{NH_2}{\underset{H}{C}}-CO_2H$$
serine

A block diagram of a phosphatide and general formulas for the three types of phosphatides are given in Figure 22.8. The phosphatides are shown as anions since they exist in this form at physiological pH.

The dissociation of the proton on the phosphate oxygen in phosphoglycerides is essentially complete at pH 7.0, as the acid dissociation constant is in the range of 10^{-1} to 10^{-2}. Phosphatidyl choline has a positive charge at nitrogen since it is a quaternary ammonium ion. Phosphatidyl ethanolamine is protonated at nitrogen at pH 7.0. As a result, both phosphatidyl choline and phosphatidyl ethanolamine are dipolar yet have no net charge. Phosphatidyl serine has three charged sites at pH 7.0. The phosphate oxygen bears a negative charge, since the proton is dissociated at pH 7.0. In addition, the carboxyl group of serine exists as the carboxylate ion, and the amine group of serine is protonated. In total, phosphatidyl serine bears a net negative charge.

The older name of phosphatidyl cholines is **lecithins.** The phosphatidyl cholines are found in the protoplasm of body cells, but they are not dissolved in water, rather in micelles. These micelles are excellent emulsifying agents and serve to transport fat molecules from one tissue to another. The nonpolar fat

Figure 22.8
Phosphoglycerides.

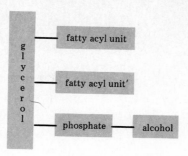

block diagram

phosphatidyl ethanolamine

phosphatidyl choline

phosphatidyl serine

phosphatidyl inositol

molecules dissolve within the micelle much as grease does in a soap micelle.

Phosphatidyl ethanolamines used to be called **cephalins.** These lipids are found in the heart and liver and in high concentration in brain tissue. Phosphatidyl ethanolamines are also essential in the blood-clotting mechanism.

The structures of the phosphoglycerides can be represented by a simplified symbol shown in Figure 22.9. All phosphoglycerides have polar sites, which can be represented as a circle. The two nonpolar hydrocarbon chains of the fatty acids are represented as wavy lines or "tails" attached to the polar "head." Thus, structural features of a complex compound such as phosphatidyl choline (Figure 22.8) can be conveniently represented with the "head-and-tail" model in considering the structure of cell membranes (Section 22.10).

Like phosphoglycerides, sphingophospholipids contain a phosphate group and choline. However, there is a substantial difference because sphingophospholipids contain a long-chain unsaturated amino alcohol called sphingosine rather

22.8
Sphingophospholipids

AN ASIDE

In Chapter 21 you learned that some chemical energy is stored as glycogen in the liver. Although this chemical energy is readily available during strenuous exercise, it is limited in amount. Most of the energy reserves of the body are in the form of triglycerides.

A person weighing 150 lb has chemical energy available in the following amounts.

blood sugar	40 kcal
glycogen	600 kcal
triglycerides	80,000 kcal

Blood sugar can sustain metabolic activity for only a few minutes. Glycogen, a readily available storage form of glucose, would be expended in a few hours of moderate activity. Therefore, larger reserves of chemical energy are needed for survival. One such reserve is unabsorbed food in the digestive tract, which can be converted into blood sugar and glycogen. However, the most important energy reserves are the triglycerides in adipose tissue, which constitutes about 15% of body weight. The lipid reserve is sufficient to maintain life for about 40 days providing water is available.

Triglycerides are a concentrated store of metabolic energy compared to carbohydrates. Oxidation of fatty acids yields about 9 kcal/g, whereas carbohydrates yield about 4 kcal/g. This difference, a factor of about 2, is due to the more reduced state of fatty acids compared to carbohydrates. There is more stored chemical energy in reduced molecules than in oxidized molecules.

Triglycerides, which are nonpolar, are stored in a nearly anhydrous form. Glycogen, which contains many hydroxyl groups, is very polar and binds about 2 g of water per gram of compound. Thus, 1 g of anhydrous fat stores about six times as much energy as a gram of hydrated glycogen.

Migratory birds provide an excellent example of the efficient storage of chemical energy in the form of triglycerides. Birds that must travel for several days over water routes must store fat as a source of energy. If the energy had to be provided from glycogen, the bird would have to carry six times as much weight for fuel!

than glycerol (Figure 22.10). Conversion of the amine of sphingosine into an amide with a fatty acid yields a ceramide. Esterification of the primary alcohol of a ceramide with phosphoric acid and choline results in a sphingolipid.

Compare the structure of a sphingophospholipid and phosphoglyceride shown in Figure 22.9. Although the components are different, the overall structures are similar. Both compounds consist of a polar head and two nonpolar tails. However, chemically there are significant differences between the two types of compounds. Sphingophospholipids, which have a single amide group, are more stable to hydrolysis than phosphoglycerides, which have two carboxylic esters.

Sphingomyelins, compounds found in the myelin sheath surrounding nerve fibers, are sphingophospholipids. Sphingomyelins have acid residues that are 20–26 carbon atoms long. These long chains intertwine and wrap around each

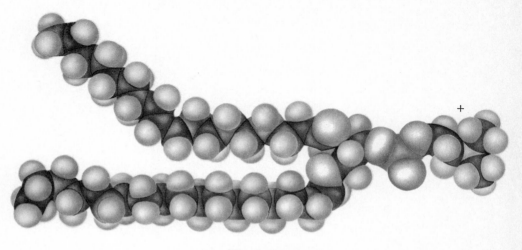

1-stearoyl-2-oleolyl-phosphatidyl choline

space-filling molecular model

nonpolar chains — polar head

simplified symbol

Figure 22.9
Representations of Phosphoglycerides.

other to form a very stable coat for the sensitive nerve fibers. In individuals with some genetic diseases, the carbon chains are shorter and defects result in the myelin sheath. Gaucher's disease, Niemann–Pick disease, multiple sclerosis, and leukodystrophy are all the result of unstable myelin membranes.

Glycosphingolipids are similar to sphingophospholipids since they contain sphingosine and a fatty acid residue bonded as an amide in a ceramide. However, glycosphingolipids contain no phosphate. A carbohydrate is bonded via a glycosidic linkage to the primary alcohol oxygen atom of the ceramide.

22.9
Glycosphingolipids

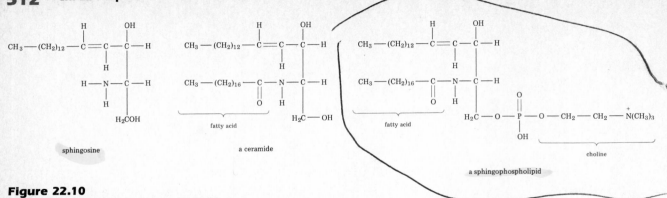

Figure 22.10
Sphingosine and Sphingophospholipids.
A ceramide contains an amide group formed from the amine of sphingosine and the carboxyl group of an acid. A sphingophospholipid contains a phosphate ester of the ceramide, which in turn is an ester of the alcohol choline.

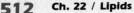

Cerebrosides and gangliosides are two of the several types of glycosphingo-lipids. Cerebrosides usually contain only glucose or galactose, whereas ganglio-sides contain an oligosaccharide. Both are found in the myelin sheath. In addi-tion, cerebrosides are in the white matter of the brain, whereas gangliosides occur in the gray matter.

Gangliosides are synthesized or degraded by sequential addition or removal of monosaccharide units. The degradation of a ganglioside occurs inside lyso-somes, which contain the necessary enzymes for the reaction. Tay–Sachs disease is a genetic disease that results from the abnormal accumulation of gangliosides in the brain. The elevated level of gangliosides is the result of a deficiency of the enzymes necessary for its degradation. Retarded development, blindness, and death in a few years are the consequences of this enzyme deficiency. Tay–Sachs disease occurs in American Jews at an incidence 100 times higher than for other Americans.

Example 22.7

Identify the components of the following compound and classify the com-pound.

$$CH_3(CH_2)_7—CH=CH(CH_2)_7—\overset{\overset{\textstyle O}{\|}}{C}—O—\underset{\underset{\textstyle CH_2—O—\underset{\underset{\textstyle OH}{|}}{\overset{\overset{\textstyle O}{\|}}{P}}—O—CH_2CH_2NH_2}{|}}{\overset{\overset{\textstyle CH_2—O—\overset{\overset{\textstyle O}{\|}}{C}—(CH_2)_{14}—CH_3}{|}}{CH}}$$

Solution. The carboxylic acid residue at the top right of the structure is palmitic acid, and the carboxylic acid residue at the left is oleic acid. A phosphate ester of ethanolamine appears at the bottom right. The center portion of the structure is derived from glycerol. This structure represents a phosphatidyl ethanolamine or cephalin.

[Additional examples may be found in **22.49–22.66** at the end of the chapter.]

Biological membranes separate cells from their environment, as well as organelles from each other within the cytoplasm of the cell. However, membranes are more than just sacks to contain the cells and organelles. Oxygen for metabolic reactions must be taken into the cell, while carbon dioxide and some aqueous solutions must be discharged from the cell. Cell membranes must also allow particular nutrients to enter the cell as well as allow substances produced within the cell to get out. The various membranes control the flow of chemicals into and out of the cell and its organelles and have highly selective permeability.

22.10 Biological Membranes

Composition of Membranes

Cell membranes are about 75 Å thick and predominately consist of phospholipids and proteins with some smaller amounts of carbohydrates. The fraction of each component is related to the function of the membrane. The myelin covering of the nerve fibers is about 80% lipid. This covering is very nonpolar and serves a protective function. Most other cell membranes must allow certain ions and polar molecules to pass in and out of the cells and their lipid content is about 50%. It is the more polar proteins of these cell membranes that participate in the transport process that occurs across the cell membrane.

Mitochondrial membranes have only about 20% lipid. Proteins make up the major part of the cell membrane in the mitochondria. The organelle plays an important role in energy conversions within cells and many molecules must cross the mitochondrial membrane.

Structure of Membranes

Each type of lipid in a membrane has a polar or ionic "head" and two nonpolar "tails" (Figure 22.11). The phospholipids form bilayers in aqueous solutions so that each side of the membrane has polar "heads" toward the exterior and nonpolar "tails" toward the interior. The polar "heads" are hydrophilic and are exposed to the fluid surrounding and contained in cells. The hydrophobic "tails" are kept away from water molecules and intermingle within the bilayer. The

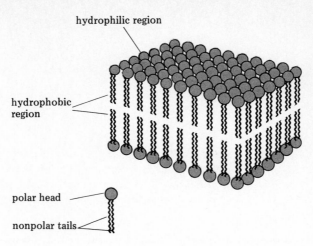

hydrophilic region

hydrophobic region

polar head

nonpolar tails

Figure 22.11
Fluid Model of a Membrane. The lipids may move about one side of the bilayer and keep their hydrophobic tails in the interior while the hydrophilic heads are in contact with surrounding aqueous solution.

properties of the membrane are determined by the type of fatty acid and the kind of polar group in the phospholipids.

The components and organization of membranes differ on the inner and outer surface. For example, phosphatidyl choline and sphingomyelin tend to be located in the outer surface, whereas phosphatidyl ethanolamine and phosphatidyl serine are usually in the inner surface. While the reasons for these different locations are not understood, they must be related to the controlled permeability of membranes.

Proteins in Membranes

Protein molecules are present in the bilayer in two different ways (Figure 22.12). *Proteins extending from the edge of the bilayer into the interior of the membrane are* **integral proteins.** Integral proteins have extensive interaction with the interior of the membrane. These proteins may be imbedded in only one side or may extend to the other side. *Proteins that extend across the membrane are called* **transmembrane proteins.** It is thought that integral proteins interact with the interior of the membrane by hydrophobic forces between certain nonpolar portions of a protein and the "tails" of the lipid.

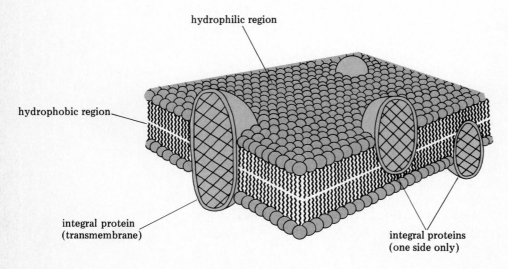

hydrophilic region

hydrophobic region

integral protein (transmembrane)

integral proteins (one side only)

Figure 22.12
Integral Proteins in a Biological Membrane.
The proteins may be imbedded in only one side of the membrane or extend across both sides of the membrane.

*Proteins that are associated with the surface of the bilayer are called **peripheral proteins.*** Peripheral proteins may be bound to membranes by electrostatic forces or hydrogen bonds. It appears that peripheral proteins are associated at the site of the integral proteins.

Carbohydrates in Membranes

Membranes contain carbohydrates combined with lipids in glycolipids and also combined with proteins in glycoproteins. These glycolipids and glycoproteins are always located on the outer surface of membranes of mammalian cells. These carbohydrate groups have two functions. One of the purposes is to orient the protein portion of a glycoprotein properly in the membrane. The carbohydrate portion is hydrophilic and remains directed toward the water in the external environment. As a result, the protein is maintained back toward and anchored in the membrane. The second function is apparently to provide some intercellular recognition. This allows not only the grouping of cells to form tissue, but also the recognition of foreign cells as part of the immune system.

Membrane Fluidity

There are no covalent bonds holding the neighboring lipid molecules together in the bilayer. The forces of attraction in membranes are of the London type. These forces depend on both the length and the geometry of the hydrocarbon chain. The longer the chain, the stronger are the London forces and the more rigid is the membrane. The degree of saturation affects the flexibility of the membrane. Unsaturated fatty acids have bends in the chain and as a result do not pack together as well as saturated acids in the bilayer. Such membranes are more flexible.

The lipids and proteins in membranes can move laterally (around one side) in the membrane. A phospholipid can move about 10^{-4} cm in 1 second, but the lateral mobility of proteins varies considerably. Some proteins are almost as mobile as lipids, whereas others are essentially immobile. The reason why some proteins are "anchored" is not well understood.

Lipid movement from one side of a membrane to the other side is very slow. Such transverse diffusion or "flip-flop" of the polar head from one side to the other occurs at a rate 10^9 times slower than lateral movement. Proteins, which are much more polar than lipids, do not undergo transverse diffusion.

There are several ways in which molecules and ions can pass through a biological membrane. These include diffusion, passive transport, and active transport. Passive transport is also called facilitated diffusion.

Small molecules can move through or diffuse across the membrane. No cellular energy is required for this process, which depends only on concentration differences on each side of the membrane. Diffusion of material is controlled by differences in concentration as discussed in Section 9.7 dealing with osmotic pressure. However, this process is a minor pathway in biological membranes.

Facilitated diffusion *occurs without cellular energy requirements and in a direction dictated by concentration* (Figure 22.13). However, the membrane is postulated to contain a carrier within it that facilitates the diffusion process. These carriers are proteins with molecular weights in the range of 9000–40,000. There are a variety of carriers, each specifically designed for certain molecules. Entry of glucose into cells occurs in this manner. The carrier meets a specific molecule at

22.11
Diffusion and Transport

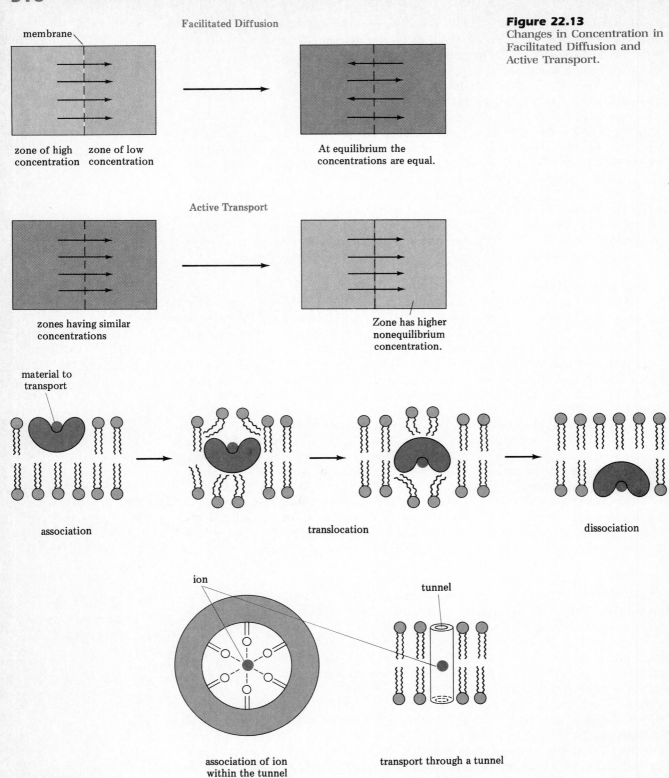

Facilitated Diffusion

membrane

zone of high   zone of low
concentration   concentration

At equilibrium the
concentrations are equal.

Active Transport

zones having similar
concentrations

Zone has higher
nonequilibrium
concentration.

material to
transport

association

translocation

dissociation

ion

tunnel

association of ion
within the tunnel

transport through a tunnel

Figure 22.13
Changes in Concentration in
Facilitated Diffusion and
Active Transport.

Figure 22.14
Transport Models for Biological Membranes.

one surface of the membrane and forms a complex (Figure 22.14). The complex then moves to the other side of the membrane and discharges the molecule. *The processes of facilitated diffusion are called association, translocation, and dissociation.*

Active transport is similar to facilitated diffusion in that a specific interaction is required between a component of the cell and the molecule to be transported. The difference is that **active transport** *occurs against the natural flow expected by concentration differences, and material moves from a region of low concentration to one of high concentration* (Figure 22.12). In order for active transport to occur, cellular ATP must be used to provide energy.

The sodium ion–potassium ion transport system is known as the "sodium pump." Sodium ions are pumped out of the cell to maintain a concentration of $0.1\,M$, while the extracellular sodium ion concentration is $0.14\,M$. At the same time, potassium ions are pumped into the cell from an extracellular concentration of $0.005\,M$ to provide an intracellular concentration of about $0.15\,M$.

Summary

Lipids are a broad class of relatively nonpolar materials that are grouped together on the basis of their solubility in organic solvents and low solubility in water. Simple lipids, which include terpenes and steroids, are not hydrolyzed by base. Complex lipids, which include waxes, triglycerides (triacylglycerols), phosphoglycerides, sphingophospholipids, and glycosphingolipids, are hydrolyzed to yield fatty acids and other components individually characteristic of the class of lipids.

Terpenes contain multiple isoprene units, whereas steroids contain a fused set of four rings. Each class of compounds may contain a variety of functional groups. Two important classes of steroids are the corticosteroids, produced by the adrenal cortex, and the sex hormones, produced by the gonads.

Waxes are esters of fatty acids and long-chain alcohols, both of which contain an even number of carbon atoms. Triglycerides or triacylglycerols are esters of glycerol and fatty acids. The degree of unsaturation determines the classification of a triglyceride as a fat or oil.

Phosphate groups occur in both phosphoglycerides and sphingophospholipids. The phosphoglycerides consist of one unit of glycerol, phosphate, and an alcohol and two units of a fatty acid. Sphingophospholipids have one unit of sphingosine, phosphate, choline, and a fatty acid.

Glycosphingolipids contain one unit of sphingosine, a fatty acid as an amide, and a sugar. Glycosphingolipids contain no phosphate. Cerebrosides contain glucose or galactose as the sugar, whereas gangliosides contain an oligosaccharide.

When placed in water, the phosphate-containing lipids form a lipid bilayer. The assembly is maintained by hydrophobic forces between nonpolar hydrocarbon chains in the middle of the bilayer. The polar heads of the phospholipids interact with water on both sides of the bilayer. Other components such as proteins, glycolipids, and glycoproteins associate within the bilayer. These components provide each membrane with its unique control of the passage of nutrients into the cell.

Molecules and ions may pass through a biological membrane by passive transport (facilitated diffusion) or active transport. The latter process requires energy from the hydrolysis of ATP. The sodium pump is one of the important active transport systems to maintain the proper concentration of ions in a cell.

Learning Objectives

As a result of studying Chapter 22 you should be able to

- Classify lipids as simple or complex.
- Draw examples of the two classes of simple lipids.
- Describe the components of each of the classes of complex lipids.
- Recognize and name the saturated and unsaturated fatty acids.
- State the function of several steroid hormones—cortisone, aldosterone, the androgens, testosterone, progesterone, and the estrogens.
- List some of the medical uses of prostaglandins.

- Draw the structures of a fat and an oil.
- Sketch a model of a bilayer showing hydrophobic and hydrophilic regions.
- Explain the movement of molecules across a cell membrane.

New Terms

Active transport moves material across membranes from low concentration to high concentration.

Anabolic steroids are synthetic substances related to testosterone that are used by some athletes to promote muscle development.

Androgens are male sex hormones; testosterone is one example.

A **cephalin** is a phosphatidyl ethanolamine.

Cerebrosides are glycosphingolipids containing only glucose or galactose.

Complex lipids can be hydrolyzed by base.

Corticosteroids are steroids produced by the adrenal cortex and include glucocorticoids and mineralocorticoids.

Estrogens are female sex hormones; examples are progesterone, estrone, and estradiol.

Facilitated diffusion occurs without cellular energy requirements.

A **fat** is a triglyceride with a high percentage of saturated fatty acids.

Fatty acids are long-chain carboxylic acids containing an even number of carbon atoms.

Gangliosides are glycosphingolipids containing an oligosaccharide as the sugar unit.

Glucocorticoids, which are produced by the adrenal cortex, are involved in the control of glucose levels in the body.

A **glycolipid** is a covalent molecule containing both a sugar and a lipid unit.

Glycosphingolipids consist of sphingosine, fatty acids, and a carbohydrate.

Hydrophilic, which means water attracting, is used to describe the surface of lipid bilayers and micelles.

Hydrophobic, which means water repelling, is used to describe the interior of lipid bilayers and micelles.

Hydrophobic bonding is a term describing the London forces between nonpolar hydrocarbon chains.

Integral proteins are proteins extending from the surface into the interior of a membrane.

A **lecithin** is a phosphatidyl choline.

The **lipid bilayer** is used to describe the structure of cell membranes.

A **micelle** is an aggregate of molecules or ions assembled so that hydrophobic portions are in the interior and hydrophilic portions are on the surface.

Mineralocorticoids, which are produced by the adrenal cortex, are involved in the control of ions in the body, which in turn affects water balance.

An **oil** is a triglyceride containing a high percentage of unsaturated fatty acids.

Peripheral proteins are attached by ionic forces to the surface of a bilayer.

A **phosphoglyceride** consists of one unit each of glycerol, phosphate, and alcohol and two units of fatty acids.

Progesterone is a female sex hormone responsible for maintaining pregnancy.

Prostaglandins are biological derivatives of arachidonic acid that occur in low concentrations in body tissue and have high physiological activity.

Sex hormones are steroids produced predominately by the gonads (ovaries in women and testes in men).

Simple lipids cannot be hydrolyzed by base.

Sphingophospholipids consist of one unit each of sphingosine, a fatty acid as an amide, phosphate, and choline.

Steroids are lipids containing a characteristic system of four fused rings of carbon atoms.

Testosterone is a male sex hormone.

Transmembrane proteins extend across the membrane.

Triacylglycerols is a newer name for triglycerides.

Triglycerides are esters of glycerol and fatty acids.

A **wax** is an ester of a fatty acid and a long-chain alcohol.

Questions and Problems

Terminology

22.1 Distinguish between the terms hydrophilic and hydrophobic.

22.2 Distinguish between facilitated diffusion and active transport.

22.3 What do the terms glycolipid and glycoprotein mean?

Classification of Lipids

22.4 Which of the following lipids do not undergo hydrolysis reactions?
   (a) waxes   (b) steroids
   (c) terpenes   (d) cephalins
   (e) oils   (f) glycosphingolipids

22.5 Which of the following do not contain a fatty acid as part of their structure?
   (a) terpenes   (b) waxes
   (c) steroids   (d) oils
   (e) fats   (f) phosphoglycerides

22.6 How do fats and oils differ in composition?

22.7 Explain why cholesterol, which has no ester functional group, is considered a lipid.

22.8 What are the differences and similarities in the structures of phosphoglycerides and sphingophospholipids?

22.9 What are the differences and similarities in the structures of sphingophospholipids and glycosphingolipids?

Terpenes

22.10 Classify each of the following terpenes.

(a)

(b)

(c)

22.11 Divide each compound of **22.10** into isoprene units.

22.12 Which terpene in **22.10** has a tail-to-tail arrangement of isoprene units?

Steroids

22.13 Name one male sex hormone produced in the testes and one produced in the adrenal cortex.

22.14 Name two female sex hormones produced in the ovaries and describe their biological function.

22.15 Mestranol is one of the numerous compounds used as an oral contraceptive. Determine each of the functional groups and compare it to the female sex hormones.

22.16 What roles do corticosteroids play in the human body?

22.17 Name the functional groups in cortisol.

22.18 Name the functional groups in aldosterone.

22.19 How do the structures of androstenedione and testosterone differ?

Waxes

22.20 What are the structures and molecular formulas of the products of hydrolysis of carnauba wax?

22.21 Describe how whale oil could be converted into a soap.

22.22 What structural features tend to make waxes solid compounds?

22.23 Which of the following structures could not be naturally occurring waxes?
(a) $CH_3(CH_2)_{30}CO_2CH_2CH_2CH_3$
(b) $CH_3(CH_2)_{28}CO_2CH_2(CH_2)_{19}CH_3$
(c) $CH_3(CH_2)_{27}CO_2CH_2(CH_2)_{18}CH_3$

22.24 The wax of the copepod described in Example 22.4 is unsaturated. This species lives in cold water and uses the wax as a source of metabolic energy. Explain the benefit of the unsaturation in the acid portion of this ester.

Fatty Acids

22.25 What do stearic acid and oleic acid have in common? How do they differ?

22.26 Cod liver oil is a triglyceride containing palmitoleic acid. Based on the name, suggest a structure for the acid.

22.27 Predict the melting point of $CH_3(CH_2)_{20}CO_2H$.

22.28 Why does linoleic acid have a lower melting point than oleic acid?

22.29 Write the structures of four naturally occurring carboxylic acids having eighteen carbon atoms.

22.30 Steareolic acid is named 9-octadecynoic acid by the IUPAC method. The molecular formula is $C_{18}H_{32}O_2$. Write its structure.

22.31 The melting point of elaidic acid (*trans*-9-octadecenoic acid) is 45°C. Compare this value to the melting points of stearic acid and oleic acid and explain the differences.

22.32 The melting point of the following compound is 48°C. Compare this value to the melting points of stearic and oleic acids and explain the differences.

$$CH_3(CH_2)_7CH\!=\!CH(CH_2)_7CO_2H$$

22.33 A compound called hypogeic acid is prepared in the laboratory and is now named 7-hexadecenoic acid. Its melting point is 33°C. What is the geometry at the double bond?

22.34 Which of the following is unlikely to be found as an ester in a naturally occurring compound?
(a) $CH_3(CH_2)_{15}CO_2H$ (b) $CH_3(CH_2)_{20}CO_2H$
(c) $(CH_3)_2CH(CH_2)_{14}CO_2H$

22.35 Explain the cleansing action of soap.

22.36 Examine the structure of the following invert soap (Section 19.5) and explain its cleansing action. How does the material differ from soap and what are the similarities?

$$CH_3(CH_2)_{15}\!-\!\overset{\overset{\displaystyle CH_3}{|}}{\underset{\underset{\displaystyle CH_3}{|}}{N^+}}\!-\!CH_3$$

22.37 Examine the structure of the following detergent (Section 16.8) and explain its cleansing action. How does the material differ from soap and what are the similarities?

$$CH_3(CH_2)_{10}CH_2\!-\!OSO_3^-\ Na^+$$

Triglycerides

22.38 Write a balanced equation for the hydrolysis of a fat molecule using a base.

22.39 A sample of one oil is hydrolyzed to produce 50% oleic acid and 35% linoleic acid. A second oil produces 25% oleic and 50% linoleic acid. Which oil is more unsaturated?

22.40 What functions do fats serve in animals?

22.41 Identify the following compound as a fat or an oil.

$$CH_2-O-\overset{\displaystyle O}{\overset{\|}{C}}-(CH_2)_{14}CH_3$$
$$CH-O-\overset{\displaystyle O}{\overset{\|}{C}}-(CH_2)_{10}CH_3$$
$$CH_2-O-\overset{\displaystyle O}{\overset{\|}{C}}-(CH_2)_7CH=CH(CH_2)_7CH_3$$

22.42 Identify the fatty acids in the compound of **22.41.**

22.43 Identify the following compound as a fat or an oil.

$$CH_2-O-\overset{\displaystyle O}{\overset{\|}{C}}(CH_2)_7CH=CHCH_2CH=CH(CH_2)_4CH_3$$
$$CH-O-\overset{\displaystyle O}{\overset{\|}{C}}(CH_2)_7CH=CH(CH_2)_7CH_3$$
$$CH_2-O-\overset{\displaystyle O}{\overset{\|}{C}}(CH_2)_7CH=CHCH_2CH=CHCH_2CH=CHCH_2CH_3$$

22.44 Identify the fatty acids in the compound of **22.43.**

22.45 Write the structure of a triglyceride containing palmitic acid as an ester at the secondary carbon atom and stearic acid as esters at the two primary carbon atoms of glycerol. Can this compound exist in optically active form?

22.46 Can the following compound be optically active?

$$CH_2-O-\overset{\displaystyle O}{\overset{\|}{C}}-(CH_2)_{16}CH_3$$
$$CH_2-O-\overset{\displaystyle O}{\overset{\|}{C}}-(CH_2)_{16}CH_3$$
$$CH_2-O-\overset{\displaystyle O}{\overset{\|}{C}}-(CH_2)_{14}CH_3$$

22.47 Hydrolysis of an optically active triglyceride gives one mole each of glycerol and oleic acid and two moles of stearic acid. Write a structure for the triglyceride.

22.48 How many moles of hydrogen gas will react with the oil in Example 22.6?

Phosphoglycerides

22.49 What products result from hydrolysis of phosphoglycerides?

22.50 Draw the structures of alcohols found in phosphoglycerides.

22.51 What charges exist on the polar head of each of the types of phosphoglycerides at physiological pH?

22.52 Identify the components of the following phosphoglyceride.

$$CH_3(CH_2)_{16}\overset{\displaystyle O}{\overset{\|}{C}}-O-CH_2$$
$$CH_3(CH_2)_7C=C(CH_2)_7\overset{\displaystyle O}{\overset{\|}{C}}-O-CH$$
$$\underset{H\ \ H}{} \qquad H_2C-O-\overset{\displaystyle O}{\underset{\underset{O^-}{\|}}{P}}-OCH_2CH_2\overset{+}{N}(CH_3)_3$$

22.53 What are the hydrolysis products of the following phosphoglyceride?

$$CH_3(CH_2)_{14}\overset{\displaystyle O}{\overset{\|}{C}}-O-CH_2$$
$$CH_3(CH_2)_7C=C(CH_2)_7\overset{\displaystyle O}{\overset{\|}{C}}-O-CH$$
$$\underset{H\ \ H}{} \qquad H_2C-O-\overset{\displaystyle O}{\underset{\underset{O^-}{\|}}{P}}-OCH_2\overset{\overset{+}{N}H_3}{\underset{H}{C}}-CO_2^-$$

22.54 What is a phosphatidic acid and how is it converted into a phosphoglyceride?

22.55 How many ionizable hydrogen atoms are there in a phosphatic acid?

22.56 How do cephalins differ structurally from lecithins?

22.57 What is the common name for phosphatidyl choline?

Sphingophospholipids

22.58 How do sphingophospholipids differ from phosphoglycerides?

22.59 Why are sphingophospholipids not hydrolyzed as readily as phosphoglycerides?

22.60 Sphingophospholipids are said to have two nonpolar tails. One is a fatty acid residue. What is the structure of the second chain?

22.61 Identify the components of the following sphingophospholipid.

$$HO-CHCH=CH(CH_2)_{12}CH_3$$
$$CH-N-\overset{\displaystyle O}{\overset{\|}{C}}-(CH_2)_{12}CH_3$$
$$\underset{H}{}$$
$$(CH_3)_3\overset{+}{N}CH_2CH_2-O-\overset{\displaystyle O}{\underset{\underset{O^-}{\|}}{P}}-O-CH_2$$

22.62 What are sphinogomyelins? What structural feature allows them to serve a uniquely important biological function?

Glycosphingolipids

22.63 How are glycosphingolipids similar to sphingophospholipids? In what ways do the two compounds differ?

22.64 What is the difference between a cerebroside and a ganglioside? Where are these compounds found?

22.65 What type of bond joins the sugar units to the sphingosine part of a glycosphingolipid?

22.66 Based on the structure of glycosphingolipids, predict whether these molecules are more stable in acidic or basic solution.

Biological Membranes

22.67 How does the structure of the fatty acid affect the rigidity of a cell membrane?

22.68 How are proteins incorporated in a cell membrane?

22.69 Explain how cholesterol can be incorporated in a cell membrane.

22.70 What kind of forces hold a cell membrane together?

22.71 Besides lipids, what other classes of biochemicals are found in membranes?

22.72 Peripheral proteins may be removed by washing with a detergent solution. Explain why.

22.73 Explain why inhaling hydrocarbon vapors can damage the cell membranes in your lungs.

22.74 What relationship exists between the protein content and the permeability of a cell membrane?

22.75 Where is the sugar portion of a glycolipid located in a cell membrane?

22.76 Describe each of the following.
(a) peripheral protein
(b) integral protein
(c) transmembrane protein

22.77 What role do sugar residues of glycolipids and glycoproteins play in the cell?

22.78 Explain why it is unlikely that a glycoprotein would undergo transverse diffusion.

Diffusion and Transport

22.79 Describe two methods by which materials cross cell membranes.

22.80 Why is an active transport system necessary to maintain the potassium ion content of a cell?

22.81 What substance provides energy for the sodium pump?

23 Amino Acids and Proteins

A weight lifter must develop muscles that consist of proteins.

All living cells contain high-molecular-weight substances called proteins. The name protein is derived from the Greek proteios meaning preeminence or holding first place. Recall that, after water, proteins are the most abundant component of the organelles (Sec. 20.4). In addition, no other biochemical serves so many purposes. As indicated in the preceding chapter, proteins are important in the transport of substances across cell membranes. Proteins are the major structural substances of skin, blood, muscle, hair, and other tissues of the body. Enzymes, the biological catalysts, are proteins, as are some hormones that regulate metabolic processes. Antibodies, which resist the effect of foreign substances that enter the body, are also proteins.

Proteins are formed from simpler organic molecules called amino acids. The amino acids contain both amino (Chapter 19) and carboxylic acid (Chapter 18) functional groups. Each amino acid is bonded to another by amide bonds, which in proteins are called peptide bonds.

peptide bonds

Proteins *are polymers of amino acids linked together by* **peptide bonds** *formed between the carboxyl group of one amino acid and the amino group of another amino acid.* The molecular weights of proteins range from about 10,000 to over a million. Hemoglobin, $C_{2952}H_{4664}O_{832}N_{812}S_8Fe_4$, for example, has a molecular weight of 65,000.

Classification of Proteins

Proteins are very complex molecules, and it is difficult to classify them unambiguously. One of the earliest classification methods was based on solubility. Two major classes are globular proteins, which are soluble in water or salt solutions, and fibrous proteins, which are insoluble in water.

Globular proteins *have tightly folded peptide chains that are roughly spherical in shape.* These proteins usually serve mobile or dynamic functions in a living organism. Albumins and globulins are two types of globular proteins. Ovalbumin in egg white, lactalbumin in milk, and serum albumin in the blood are examples of albumins. Albumins in blood serve as carriers of some molecules that otherwise would not be soluble in blood. Globulins also occur in blood serum and are part of the body's defense against infectious diseases.

Fibrous proteins *contain polypeptide chains arranged in extended, parallel layers, or sheet-like structures.* They usually have connective, structural, or protective functions. Examples of classes of fibrous proteins are collagens, elastins, myosins, and keratins. The connective tissues of animals such as skin and tendons contain collagen. The elastic tissues in ligaments and arteries contain elastins. Myosins are involved in the contraction and extension of muscles. Keratins occur in hair, wool, nails, and hoofs.

Proteins can also be classified according to composition. *Proteins that when hydrolyzed yield compounds in addition to amino acids are* **conjugated proteins.** *The nonamino acid unit is called a* **prosthetic group.** There are five types of conjugated proteins.

1. When the prosthetic group is a carbohydrate, the protein is called a *glycoprotein*. Heparin, which inhibits the clotting of blood, and mucin, a component of saliva, are both glycoproteins.
2. Proteins combined with nucleic acids in ribosomes and some viruses (Chapter 29) are *nucleoproteins*.
3. *Phosphoproteins* contain phosphoric acid bonded to the protein by ester groups. Vitellin of egg yolk and casein of milk are both examples of phosphoproteins.
4. When the prosthetic group is colored, the complex protein is called a *chromoprotein*. The red pigment heme occurs in the hemoglobin of the blood.
5. *Lipoproteins* are proteins containing lipid molecules.

Protein Functions

A newer system of protein classification is based on the protein function in the body. The various functions are described below and are summarized with examples in Table 23.1.

1. *Structural proteins* provide mechanical support and are a major part of connective tissue and bones. Both keratins and elastins are structural proteins.
2. *Contractile proteins* are responsible for coordination of motion. They can stretch and contract and are found in muscle. Other examples include

AN ASIDE

The role of lipoproteins in the blood and their relationship to serum cholesterol level has only recently been determined. Blood serum is about 0.5–1.0% lipoprotein. The lipoproteins are classified into categories according to their density, which is related to the fractions of lipid and protein in the molecules.

Very-low-density lipoproteins (**VLDL**) have densities in the 1.006–1.019 g/mL range and consist of about 5% protein. **Low-density lipoproteins** (**LDL**) have densities in the range 1.019–1.063 g/mL and consist of about 25% protein. The **high-density lipoproteins** (**HDL**) have densities in the range of 1.063–1.21 g/mL and consist of about 50% protein.

The VLDL are the principal carriers of triglycerides, whereas LDL are the carriers of 80% of the cholesterol. HDL carry the remaining cholesterol. However, the functions of LDL and HDL in carrying cholesterol are quite different. LDL carry cholesterol to cells for their use, whereas HDL carry excess cholesterol away from cells to the liver for processing and excretion from the body.

If the cholesterol carried to the cells exceeds the cell's requirements and the HDL level is too low to remove the excess cholesterol, the cholesterol will be left as a deposit. Individuals with high HDL levels thus have an efficient means of removing unneeded cholesterol and avoiding heart disease.

The average concentration of HDL is 45 mg/100 mL for men and 55 mg/100 mL for women. Male long-distance runners may have as much as 75 mg/100 mL HDL. Thus some scientists feel that the HDL level explains why proportionately fewer women have heart attacks than do men. Exercising is also thought to increase the HDL concentration.

Table 23.1

Biological Functions of Proteins

Protein function	Role
structural	
collagen	fibrous connective tissue
elastin	elastic connective tissue
glycoproteins	cell walls
α-keratin	skin, nails, feathers, hoofs
contractile	
actin	thin filaments in muscles
myosin	thick filaments in muscles
dynein	involved in movement of cilia and flagella
transport	
hemoglobin	transports oxygen in blood
myoglobin	transports oxygen in muscle
serum albumin	transports fatty acids in blood
transferrin	transports iron in blood
hormones	
adrenocorticotropic hormone	regulates corticosteroid synthesis
growth hormone	stimulates growth of skeletal structure
insulin	regulates metabolism of glucose
storage	
casein	milk protein
gliadin	stores amino acids in wheat
ovalbumin	protein in egg white
zein	stores amino acids in corn
protection	
antibodies	react with foreign proteins
fibrinogen	involved in blood clotting
thrombin	involved in blood clotting

proteins involved in the movement of chromosomes in cell division and the swimming motion of sperm.

3. *Transport proteins* circulate and carry other molecules. For example, hemoglobin carries the small molecule oxygen. Serum albumin, which carries fatty acids between adipose tissue and other tissues, is another transport protein.

4. *Enzymes* are usually globular-shaped proteins and represent the largest variety of proteins. Without enzymes and their enormous catalytic power, chemical reactions would not occur fast enough to sustain life.

5. *Protein hormones*, which regulate cell processes, are chemical messengers produced in endocrine glands. One such example is insulin, which along with the lipid cortisone, regulates the blood glucose level.

6. *Protection proteins* include antibodies and blood-clotting proteins. Antibodies are proteins that recognize and combine with foreign substances, such as viruses, that have entered the blood or tissues. Antibodies provide immune protection and defend animals from outside invaders.

7. *Storage proteins* store small molecules or ions. Ovalbumins store amino

acids used as nutrients by chicken embryos in eggs. Ferritin, a liver protein, stores iron in humans and other animals.

23.2
Amino Acids

As the name indicates, **amino acids** *contain both an amino group and a carboxylic acid group.* Although there could be millions of possible amino acids, only about 250 have been found in natural sources. Furthermore, only about 20 of them occur in large amounts, especially in protein.

The amino acids of proteins in plants and animals have an amino group attached to the carbon atom adjacent to the carboxylic acid group. Such amino acids are α-amino acids. The general formula for an α-amino acid is

$$H_2N-\underset{\underset{\displaystyle R}{|}}{\overset{\overset{\displaystyle CO_2H}{|}}{C}}-H$$

α-amino acid

The R of this general formula may be hydrogen, an alkyl group, an aromatic ring, or a heterocyclic ring.

Except for glycine, in which R is hydrogen, naturally occurring amino acids contain at least one center of asymmetry and are of the L configuration. The human body, which consists of proteins composed of L-amino acids, either synthesizes the L-amino acids it needs or uses only the L-amino acids it obtains from nutritional sources.

Classification of Amino Acids

The amino acids contained in proteins and their common names are given in Figure 23.1. Three-letter abbreviations are used to help describe protein composition and the sequence of amino acids in proteins. The table is divided into neutral, basic, and acidic amino acids.

Neutral amino acids *contain only one amino group and one carboxyl group.* The simplest neutral amino acids have an alkyl group at the α-carbon. Proline differs from these compounds since it contains a secondary rather than a primary amino group. Two of the neutral amino acids—serine and threonine—are also alcohols. Three of the neutral amino acids—phenylalanine, tyrosine, and tryptophan—contain aromatic rings. Two of the neutral amino acids—cysteine and methionine—contain a sulfur atom.

Amino acids that have an additional amino or carboxyl group are classified as **basic** *or* **acidic amino acids,** respectively. The three basic amino acids are lysine, arginine, and histidine. The acidic amino acids are aspartic acid and glutamic acid; conversion of the second carboxylic acid group of these two amino acids into an amide yields asparagine and glutamine, which are neutral amino acids.

In proteins, the amino and carboxylic acid groups are combined in peptide linkages. The differences in the properties of proteins then depend on the nature of the R group attached to the α-carbon. These R groups are classed as hydrophilic or polar and hydrophobic or nonpolar. The hydrophilic groups may have polar functional groups such as the amino and carboxylic acid groups in basic and acidic amino acids. In addition, the hydroxyl group of serine, threonine, and tyrosine causes these amino acids to be hydrophilic. Hydrophobic amino acids have alkyl or aromatic R groups, which do not hydrogen-bond to water.

Figure 23.1
Structures of
Amino Acids.

Neutral Amino Acids

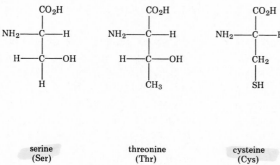

glycine
(Gly)

alanine
(Ala)

valine
(Val)

leucine
(Leu)

isoleucine
(Ile)

proline
(Pro)

phenylalanine
(Phe)

tyrosine
(Tyr)

tryptophan
(Trp)

serine
(Ser)

threonine
(Thr)

cysteine
(Cys)

methionine
(Met)

asparagine
(Asn)

glutamine
(Gln)

Basic Amino Acids

Acidic Amino Acids

aspartic acid
(Asp)

glutamic acid
(Glu)

lysine
(Lys)

arginine
(Arg)

histidine
(His)

527

Table 23.2

Essential Amino Acids and Estimated Daily Requirements

	Minimum daily requirements (g)	
Amino acids	Women	Men
isoleucine	0.45	0.7
leucine	0.6	1.1
lysine	0.5	0.8
methionine	0.55	1.0
phenylalanine	1.1	1.4
threonine	0.3	0.5
tryptophan	0.15	0.25
valine	0.65	0.8

Essential Amino Acids

Adequate amounts of about 10 of the 20 amino acids can be synthesized from carbohydrates and lipids in the body providing a source of nitrogen is also available. The remaining amino acids must be obtained from food because there is no biochemical pathway to produce them or the pathway produces inadequate amounts for proper nutrition. *Amino acids that must be obtained from food are called* **essential amino acids.** Of course, all amino acids are necessary for normal development, but the term essential is reserved for those amino acids that must be obtained in the diet.

The essential amino acids and the estimated minimum daily requirements for men and women are listed in Table 23.2. Phenylalanine is converted into tyrosine by the body. Thus, without phenylalanine in the diet we would have to classify tyrosine as an essential amino acid. Histidine is essential for growth in infants and may be essential for adults as well. The rate of synthesis of histidine cannot meet the needs of a growing body, and thus histidine may be considered an essential amino acid depending on the age and state of health of the individual.

23.3
Acid–Base Properties of Amino Acids

The structures of the amino acids have been drawn as uncharged electrically neutral molecules. However, the properties of amino acids resemble those of salts rather than uncharged molecules. Consider ethylamine, acetic acid, and glycine.

$$CH_3CH_2NH_2 \qquad CH_3CO_2H \qquad NH_2CH_2CO_2H$$

melting point: $\quad$ −84°C $\qquad\qquad$ 16°C $\qquad\qquad$ 232°C

Ethylamine is a gas and acetic acid is a liquid at room temperature, whereas glycine is a solid.

The solubility of amino acids in organic solvents is quite low, but they are moderately soluble in water. This physical property is quite unlike the majority of organic compounds. The properties of amino acids more closely resemble those of charged ionic compounds.

The presence of the acidic carboxylic acid group and the basic amino group in the same molecule makes possible an acid–base reaction. When the amino group accepts a proton, a positive ammonium group is formed at one end of the molecule. The loss of a proton from the carboxylic acid forms a negative carboxylate group.

AN ASIDE

The composition of dietary protein must match the requirements for the formation of body protein. If one or more amino acids are not available at the time of necessary synthesis of a vital protein, then the protein is not made. Protein synthesis is an "all or nothing" event.

Dietary proteins are rated in terms of biological value on a percentage scale. **Complete proteins** *have a high biological value and supply all the required amino acids in the amounts required for normal growth.* A list of protein sources and their biological value is given in Table 23.3. Note that hen's egg, cow's milk, and fish provide high-biological-value proteins. Plant proteins vary more in biological value than animal proteins. However, not all plant proteins are deficient in the same amino acids. Gliadin, a wheat protein, is low in lysine; zein, a corn protein, is low in both lysine and tryptophan. Societies that eat large amounts of corn or wheat products must have other sources of lysine.

Vegetarians must carefully choose their food so that all of the essential amino acids are available. For example, beans are high in lysine, whereas wheat is low in lysine. Similarly, wheat is high in cysteine and methionine, whereas beans are low in these two amino acids. By eating both beans and wheat, the vegetarian increases the percentage of the usable proteins.

The diets in some areas of the world today fall below the minimum daily requirement of protein owing to economic conditions and in some cases to social and religious customs. As income decreases, the more costly animal protein is replaced by cereal grains and other incomplete protein sources. If a variety of plant proteins is not available, a number of diseases in young children result. Kwashiorkor is a disease in young children after weaning, when their diet is changed to starches. The disease is characterized by bloated bellies and patchy skin. After a certain point, death is inevitable. Some forms of mental retardation also result from incomplete nutrition.

One solution to the problem of poor nutrition is education of people to take advantage of locally available foods to balance their diet. However, dietary customs are slow to change. Another solution is the development of plant mutants that will have a higher protein content or biological value.

Table 23.3

Biological Value of
Dietary Protein

Food	Biological value (%)
whole hen's egg	94
whole cow's milk	84
fish	83
beef	73
soybeans	73
white potato	67
whole grain wheat	65
whole grain corn	59
dry beans	58

This double ion which results from the transfer of a proton from one functional group to another within a molecule is called a **zwitterion.** Amino acids tend to behave as ionic substances since they have sites of charge even though they are electrically neutral. In addition, the ammonium and carboxylate sites within the zwitterion can act as an acid and base, respectively.

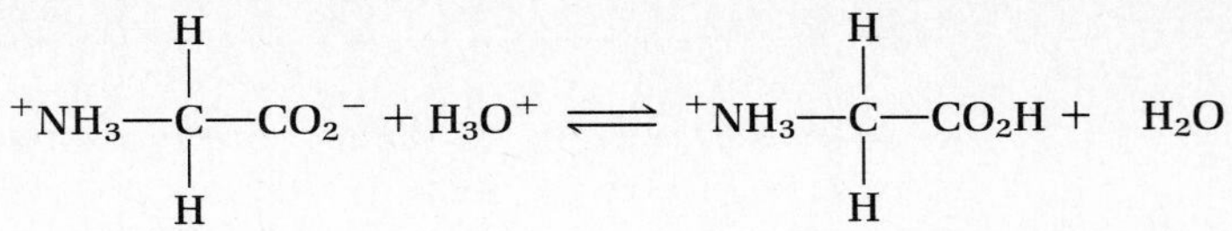

$$^+NH_3-\underset{\underset{\displaystyle H}{|}}{\overset{\overset{\displaystyle H}{|}}{C}}-CO_2^- + H_3O^+ \rightleftharpoons {}^+NH_3-\underset{\underset{\displaystyle H}{|}}{\overset{\overset{\displaystyle H}{|}}{C}}-CO_2H + H_2O$$

zwitterion   acid   conjugate   conjugate
                                          acid   base

In strongly acidic solution the positively charged ammonium salt of the amino acid is formed. The negatively charged carboxylate ion is formed in strongly basic solution. At intermediate acidity the electrically neutral zwitterion predominates.

Example 23.1

Draw the zwitterion, conjugate base, and conjugate acid of alanine.

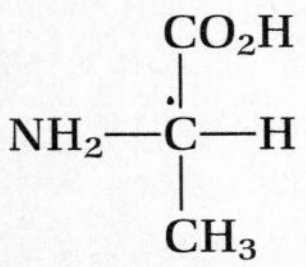

$$NH_2-\underset{\underset{\displaystyle CH_3}{|}}{\overset{\overset{\displaystyle CO_2H}{|}}{C}}-H$$

alanine

Solution. The zwitterion can be written by removing a proton from the carboxylate group and transferring it to the nitrogen atom.

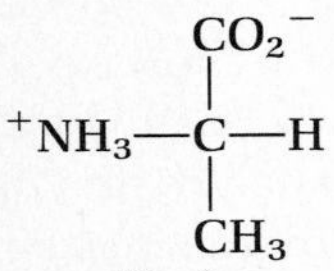

$$^+NH_3-\underset{\underset{\displaystyle CH_3}{|}}{\overset{\overset{\displaystyle CO_2^-}{|}}{C}}-H$$

zwitterion

The conjugate base is written by removing a proton from the ammonium ion site of the zwitterion. The nitrogen atom becomes neutral, while the carboxylate ion retains a negative charge.

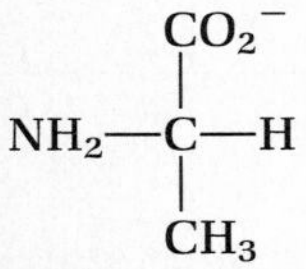

$$NH_2-\underset{\underset{\displaystyle CH_3}{|}}{\overset{\overset{\displaystyle CO_2^-}{|}}{C}}-H$$

conjugate base

The conjugate acid is written by adding a proton to the carboxylate site of the zwitterion. The resulting carboxyl group is neutral, while the ammonium ion site retains a positive charge.

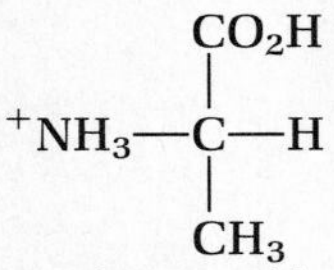

$$^+NH_3-\underset{\underset{\displaystyle CH_3}{|}}{\overset{\overset{\displaystyle CO_2H}{|}}{C}}-H$$

conjugate acid

[Additional examples may be found in **23.31–23.34** at the end of the chapter.]

Table 23.4

pK_a Values for Some Amino Acids at 25°C

Amino acid	pK_a (COOH)	pK_a (NH$_3^+$)	pK_a (side chain)
glycine	2.3	9.6	
alanine	2.3	9.7	
serine	2.2	9.2	
aspartic acid	2.1	9.8	3.9
glutamic acid	2.2	9.7	4.3
histidine	1.8	9.2	6.0
lysine	2.2	9.0	10.5
arginine	2.2	9.0	12.5

Amino acids have two pK_a (Sec. 18.4) values since ionization can occur at the ammonium group (—NH$_3^+$) or at the carboxylic acid group (—CO$_2$H).

$$R—NH_3^+ + H_2O \rightleftharpoons R—NH_2 + H_3O^+$$

$$R—CO_2H + H_2O \rightleftharpoons R—CO_2^- + H_3O^+$$

The acidic and basic amino acids have a third pK_a because of the additional ionizable functional group in the chain attached to the α-carbon atom.

For glycine the pK_as of the —CO$_2$H and —NH$_3^+$ groups are 2.3 and 9.6, respectively. The pK_a values of other amino acids are given in Table 23.4.

$$^+NH_3—\overset{\overset{\displaystyle H}{|}}{\underset{\underset{\displaystyle H}{|}}{C}}—CO_2H + H_2O \rightleftharpoons {}^+NH_3—\overset{\overset{\displaystyle H}{|}}{\underset{\underset{\displaystyle H}{|}}{C}}—CO_2^- + H_3O^+ \qquad K_a = 5 \times 10^{-3}$$
$$pK_a = 2.3$$

$$^+NH_3—\overset{\overset{\displaystyle H}{|}}{\underset{\underset{\displaystyle H}{|}}{C}}—CO_2^- + H_2O \rightleftharpoons NH_2—\overset{\overset{\displaystyle H}{|}}{\underset{\underset{\displaystyle H}{|}}{C}}—CO_2^- + H_3O^+ \qquad K_a = 1.6 \times 10^{-10}$$
$$pK_a = 9.6$$

Recall that when the pH of a solution is equal to the pK_a of an acid, the concentrations of the acid and its conjugate base are equal (Section 18.4). Under these conditions, the solution is an effective buffer. At pH = 2.3, the concentrations of the ammonium ion of glycine and its zwitterion are equal, and the mixture is an effective buffer. At pH = 9.6, the concentrations of the carboxylate ion of glycine and its zwitterion are equal, and that solution is also an effective buffer. At pH values between 2.3 and 9.6, the zwitterion is the major form of the amino acid.

Example 23.2

In what form will serine exist in a solution of 0.1 M HCl?

Solution. A 0.1 M solution of HCl has a pH = 1.0.

$$\begin{aligned} pH &= -\log[H_3O^+] \\ &= -\log(1 \times 10^{-1}) \\ &= -\log(1) - \log(10^{-1}) \\ &= -0 - (-1) = +1 \end{aligned}$$

The pK_a values of serine are 2.2 and 9.2 (Table 23.4). Thus, at pH = 1, serine will exist as the conjugate acid.

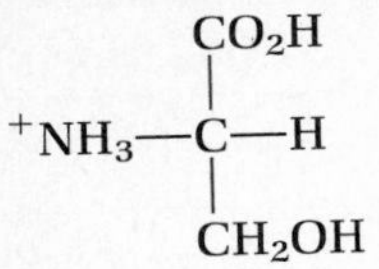

$$
\begin{array}{c}
CO_2H \\
| \\
^{+}NH_3-C-H \\
| \\
CH_2OH
\end{array}
$$

[Additional examples may be found in **23.31–23.34** at the end of the chapter.]

The R groups of acidic and basic amino acids have a third pK_a. The third pK_a of glutamic acid (4.3) is between the other two pK_a values. Thus, in the pH range near the third pK_a, the amino acid exists as the zwitterion. The third pK_a is for the following equilibrium.

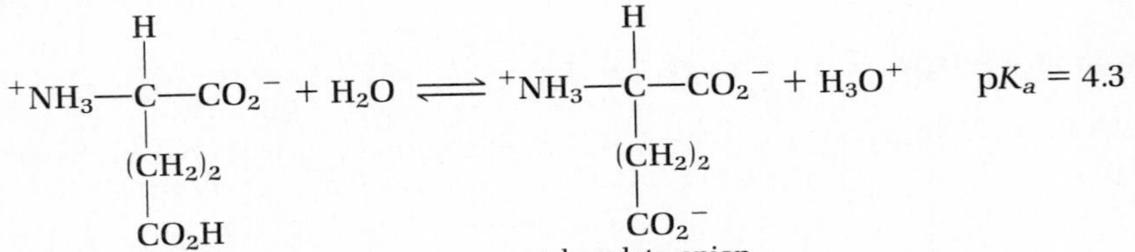

$$
\begin{array}{c}
H \\
| \\
^{+}NH_3-C-CO_2^{-} + H_2O \\
| \\
(CH_2)_2 \\
| \\
CO_2H \\
\text{zwitterion}
\end{array}
\rightleftharpoons
\begin{array}{c}
H \\
| \\
^{+}NH_3-C-CO_2^{-} + H_3O^{+} \qquad pK_a = 4.3 \\
| \\
(CH_2)_2 \\
| \\
CO_2^{-} \\
\text{carboxylate anion}
\end{array}
$$

At pH = 4.3, a solution of glutamic acid contains equal amounts of a zwitterion and a carboxylate anion derived from the second carboxyl group. At lower pH values, the zwitterion concentration increases, whereas at higher pH values the concentration of the carboxylate anion increases.

Example 23.3 ─────────────────────────────

Describe how aspartic acid exists in a solution at pH = 3.9.

Solution. The third pK_a of aspartic acid is 3.9 (Table 23.4). Thus, at pH = 3.9, a solution will contain equal amounts of a zwitterion and the carboxylate anion derived from the second carboxyl group.

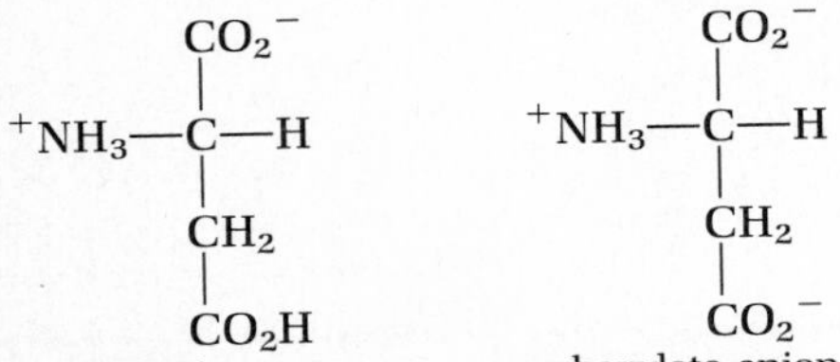

$$
\begin{array}{c}
CO_2^{-} \\
| \\
^{+}NH_3-C-H \\
| \\
CH_2 \\
| \\
CO_2H \\
\text{zwitterion}
\end{array}
\qquad
\begin{array}{c}
CO_2^{-} \\
| \\
^{+}NH_3-C-H \\
| \\
CH_2 \\
| \\
CO_2^{-} \\
\text{carboxylate anion}
\end{array}
$$

[Additional examples may be found in **23.31–23.34** at the end of the chapter.]

For lysine the third pK_a (10.5) is larger than the pK_a of the α-ammonium ion (9.0) of the amino acid. Thus, in basic solutions of pH greater than 9.0, the predominant species are given by the following equilibrium.

$$NH_2\!-\!\underset{\underset{\overset{|}{{}^+NH_3}}{\overset{|}{(CH_2)_4}}}{\overset{|}{\underset{|}{C}}}\!-\!CO_2^- + H_2O \rightleftharpoons NH_2\!-\!\underset{\underset{\overset{|}{NH_2}}{\overset{|}{(CH_2)_4}}}{\overset{|}{\underset{|}{C}}}\!-\!CO_2^- + H_3O^+ \qquad pK_a = 10.5$$

At pH = 10.5, a solution of this amino acid contains equal concentrations of the conjugate base of lysine and the zwitterion involving the amine of the side chain and the carboxylate ion. At lower pH values, the zwitterion concentration is larger than that of the conjugate base, whereas at higher pH values, the opposite is true.

The third pK_a value of histidine is close to the pH of a neutral solution. Histidine is the only amino acid whose side chain allows the formation of a buffer in the pH region near 7. For this reason histidine is a good buffer to maintain the pH of living organisms. Hemoglobin contains histidine, which contributes to maintaining the pH of blood.

By changing the pH, the structure of an amino acid is altered. At pH values smaller than the pK_a of the carboxylic acid, the amino acid has a net positive charge. At pH values higher than the pK_a of the ammonium group, the amino acid has a net negative charge. At pH values between the two pK_a values, the zwitterion predominates.

The pH at which the concentration of the zwitterion is at a maximum is called the **isoelectric point.** Above the isoelectric point, the collection of species tends to have a net negative charge. At a pH below the isoelectric point, the cationic forms of amino acids tend to exist. The isoelectric point of some amino acids is given in Table 23.5. For alanine, a neutral amino acid, the isoelectric point is 6.0. This value is the average of the two pK_a values of alanine.

The calculation of the isoelectric point of acidic and basic amino acids is somewhat complicated since there are so many equilibria to consider. However, note that for an acidic amino acid the isoelectric point is in the acid range, whereas for a basic amino acid the isoelectric point is in the basic range.

The isoelectric point of an amino acid is easily measured. If a charged molecule is placed in an electric field, it will migrate to the electrode of opposite charge. At low pH an amino acid has a net positive charge and will migrate toward the negative electrode (Figure 23.2). At high pH an amino acid has a net negative charge and will migrate toward the positive electrode. At the isoelectric

23.4
Isoelectric Point

Table 23.5

Isoelectric Points of Some Amino Acids and Proteins

Amino Acid	Isoelectric point	Protein	Isoelectric point
aspartic acid	2.8	pepsin (enzyme)	1.1
glutamic acid	3.2	casein (milk protein)	4.6
serine	5.7	egg albumin	4.7
valine	6.0	urease (enzyme)	5.0
alanine	6.1	insulin (hormone)	5.3
lysine	9.7	hemoglobin	6.8
arginine	10.8	ribonuclease (enzyme)	9.5
		chymotrypsin (enzyme)	9.5

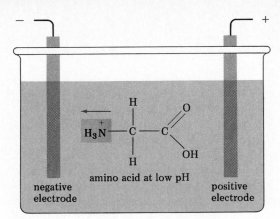

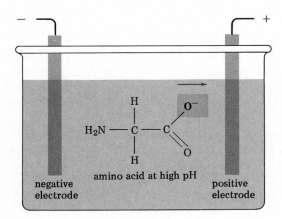

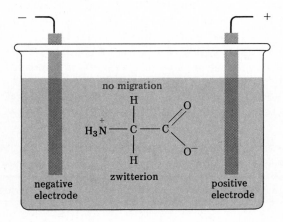

Figure 23.2
Determination of the Isoelectric
Point.

point the zwitterion, which will not migrate toward either electrode, predominates. The isoelectric point is then a characteristic property of each amino acid.

Mixtures of amino acids can be separated and identified by a process called **electrophoresis,** *which is based on the charge of an amino acid at a given pH* (Figure 23.3). A paper strip saturated with a buffer solution at a selected pH bridges two vessels containing the buffer. A sample of the amino acid mixture is

Figure 23.3

Electrophoresis of Amino Acids.
The protein mixture has been separated into
three components. Amino acid B has not migrated from the origin of the sample and must
exist as the zwitterion at the pH of the buffer.
Amino acid A has migrated toward the negative
electrode and must have a net positive charge.
Amino acid C has migrated toward the positive
electrode and must have a net negative charge.

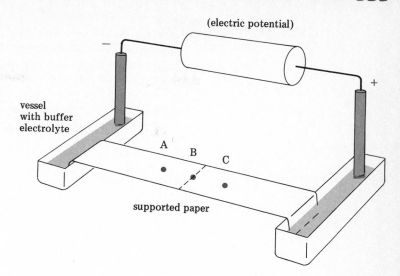

placed at the center of the paper, and an electric potential is applied between the
two vessels. If the buffer pH equals the isoelectric point of an amino acid, the
zwitterion predominates, and it will not migrate. An amino acid existing as the
conjugate base has a negative charge and will migrate toward the positive electrode. An amino acid existing as the conjugate acid has a positive charge and will
migrate toward the negative electrode. After a period of time the original "spot" of
the amino acid sample is separated into spots corresponding to each of the
amino acids present.

Proteins have isoelectric points that are characteristic of their component
amino acids. Some of these values are listed in Table 23.5. Proteins have no net
charge at their isoelectric point and are then the least soluble. As a consequence
proteins tend to precipitate from solution at the isoelectric point. For example,
casein will be least soluble at pH 4.7. Cow's milk has pH 6.3 and casein has a net
negative charge. If the milk is made more acidic, the casein is protonated and
precipitates. Casein, which is used in making cheese, can be obtained by adding
an acid to milk or by bacterial action that produces lactic acid and causes a
decrease in pH of milk.

Electrophoretic separation of proteins is an important tool in clinical laboratories. Because proteins have a number of different charges and molecular
weights, they move at different rates in the electrophoretic apparatus. The most
common use of electrophoresis is the analysis of blood serum. The identification
of certain enzymes (which are proteins) in the blood is diagnostic for a myocardial infarction.

Proteins consist of hundreds of amino acids bonded by peptide bonds. The components of the protein are called amino acid residues. Hydrolysis of a protein
yields a mixture of its constituent amino acids.

23.5
Peptides

$$-\underset{\underset{H}{|}}{N}-\underset{\underset{R}{|}}{CH}-\overset{\overset{O}{||}}{C}-\underset{\underset{H}{|}}{N}-\underset{\underset{R}{|}}{CH}-\overset{\overset{O}{||}}{C}- + H_2O \longrightarrow -\underset{\underset{H}{|}}{N}-\underset{\underset{R}{|}}{CH}-\overset{\overset{O}{||}}{C}-OH + H-\underset{\underset{H}{|}}{N}-\underset{\underset{R}{|}}{CH}-\overset{\overset{O}{||}}{C}-$$

Peptides are lower-molecular-weight materials containing fewer amino acids than proteins. A peptide containing two amino acid units is called a dipeptide; one containing three amino acids, a tripeptide; and one containing a large number, but less than 50, a polypeptide. Peptides are named by combining the names or abbreviations of the individual amino acids. The name starts with the amino acid whose amino group is free (the N-terminal amino acid) and ends with that whose carboxyl group is free (the C-terminal amino acid). Two examples of this nomenclature are

$$
\underbrace{NH_2CH_2\overset{\displaystyle O}{\overset{\|}{C}}}_{\substack{\text{glycine}\\\text{residue}}}-NH-\underbrace{\overset{\displaystyle CH_3}{\overset{|}{C}}HCO_2H}_{\substack{\text{alanine}\\\text{residue}}}
\qquad
\underbrace{NH_2\overset{\displaystyle CH_3}{\overset{|}{C}}H-\overset{\displaystyle O}{\overset{\|}{C}}}_{\substack{\text{alanine}\\\text{residue}}}-\underbrace{NHCH_2CO_2H}_{\substack{\text{glycine}\\\text{residue}}}
$$

glycylalanine (Gly-Ala)   alanylglycine (Ala-Gly)

Although there are only two dipeptides possible from a given pair of amino acids (see above), the number of possible isomers increases rapidly with more amino acids. With three different amino acids in a tripeptide there are six possible isomers. The amino acids glycine, alanine, and valine can be combined to give Gly-Ala-Val, Gly-Val-Ala, Val-Gly-Ala, Val-Ala-Gly, Ala-Gly-Val, and Ala-Val-Gly.

The number of isomeric peptides containing one molecule each of n different amino acids is equal to $n!$, where $n! = 1 \times 2 \times 3 \times \cdots \times (n-1) \times n$. For peptides containing one molecule each of 20 different amino acids the number of isomers possible is 2,432,902,008,176,640,000. The higher peptides and proteins require duplication of amino acids, and the $n!$ formula no longer applies. However, the number of isomers continues to rise astronomically. Nevertheless, when proteins are synthesized by living organisms, the amino acids are not randomly joined together and isomeric compounds are not formed. Each protein has a specific sequence of amino acids, which are assembled under the direction and control of nucleic acids.

Example 23.4

Identify the components of the following dipeptide and assign its name.

$$
NH_2-\underset{\underset{\displaystyle CH_2OH}{|}}{C}H-\overset{\displaystyle O}{\overset{\|}{C}}-NH-CH-CO_2H
$$

Solution. The residue on the left is the N-terminal amino acid and is serine. The residue on the right is the C-terminal amino acid and is phenylalanine. Thus the name is serylphenylalanine. The abbreviated name is Ser-Phe.

[Additional examples may be found in **23.45–23.53** at the end of the chapter.]

23.6 Structure of Peptides

The structure and function of proteins have been studied for many years, but only recently have peptides of physiological importance been examined. Although it might seem that peptides should have been studied first, there are experimental difficulties in working with peptides. The peptides are produced and released in small amounts and are rapidly metabolized since their physiological action may be necessary for only a short time. For example, the physiological action of peptides includes pain relief and control of blood pressure. Examples of some peptides with simple structures are given in Figure 23.4.

Enkephalins *are a class of peptides that bind at receptor sites in the brain to reduce pain.* The action of some drugs such as morphine is based on binding at

Figure 23.4
Structures of Some Peptide Hormones.

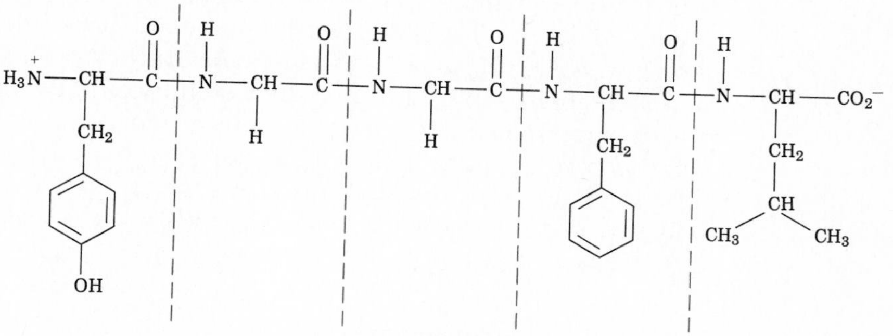

Tyr-Gly-Gly-Phe-Leu
(an enkephalin)

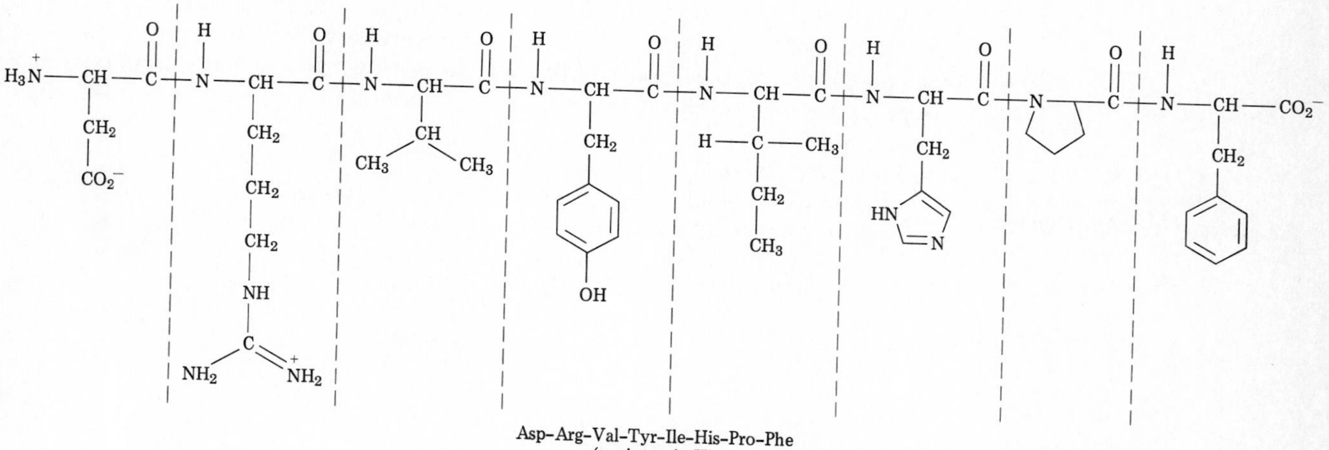

Asp-Arg-Val-Tyr-Ile-His-Pro-Phe
(angiotensin II)

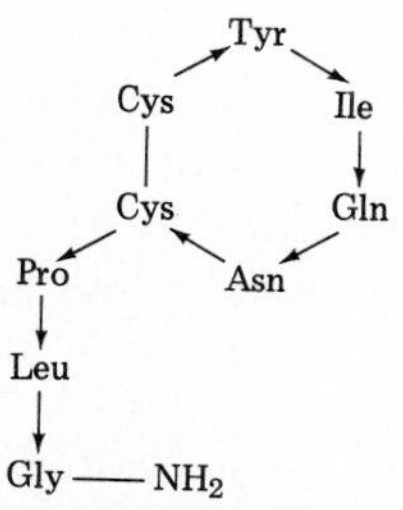

oxytocin

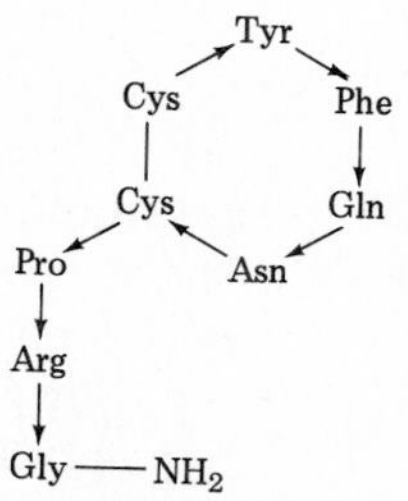

vasopressin

these same receptor sites. The receptor sites are called opiate receptors based on the effect of opiates on them. It was suggested that opiates mimic in part some compounds normally present in the body to mitigate pain. The enkephalins were finally isolated in 1975. The enkaphalins, then, are nature's own painkillers.

Angiotensin II (Figure 23.4) is an octapeptide that is produced in the kidneys. It causes constriction of the blood vessels and thus increases blood pressure. Angiotensin II is the most potent vasoconstrictor known and is involved in some cases of hypertension.

The structures of two cyclic peptides that are important hormones are given in Figure 23.4. The arrows indicate the direction of attachment from a carboxyl group to an amino group. Note the presence of the neutral amino acids asparagine (Asn) and glutamine (Gln). These amino acids are derived from aspartic acid and glutamic acid by reaction with ammonia to form an amide. Oxytocin is formed by the pituitary gland and causes the contraction of smooth muscle, such as that of the uterus. It is used to induce delivery or to increase the effectiveness of uterine contractions. Vasopressin is also produced in the pituitary gland. It is one of the hormones that regulate the excretion of water by the kidneys, and it affects blood pressure. The structure of vasopressin differs from that of oxytocin by only two amino acids. It is remarkable how these changes cause the difference in physiological action.

In order to understand the properties and function of proteins it is necessary to study their bonding and structure. In this section four types of bonds—the peptide bond, the disulfide bond, the hydrogen bond, and the ionic bond—will be first discussed. In addition, hydrophobic interactions will be considered. The characteristics of these bonds will then be used to describe the structure and conformation of proteins in Section 23.8.

23.7
Bonding in Proteins

The Peptide Bond

The peptide bond is the strongest and most important bond in a protein. Recall from Section 19.7 that there is restricted rotation about the carbon–nitrogen bond of amides. This restricted rotation is the result of some double bond character which is pictured in resonance forms.

Peptides tend to exist in *trans* conformations about the carbon–nitrogen bond.

The peptide unit is rigid and planar, a feature that is important in the overall structure of proteins. In contrast, the bond between the α-carbon atom and the carbonyl carbon atom is a single bond that is rotationally free. Similarly, the single bond between the nitrogen atom and the α-carbon atom is also rotationally free. Thus a protein chain consists of rigid peptide units connected to one another by freely rotating single bonds. There is, in addition, free rotation about the bond between the α-carbon atom and the R group. Both the rigidity and flexibility of portions of the protein chain are important in determining its conformation.

The Disulfide Bond

The disulfide bond (Sec. 16.11) is a covalent bond between two sulfur atoms. A disulfide bond results from the oxidation of the —SH (sulfhydryl) groups of two cysteine molecules to form cystine.

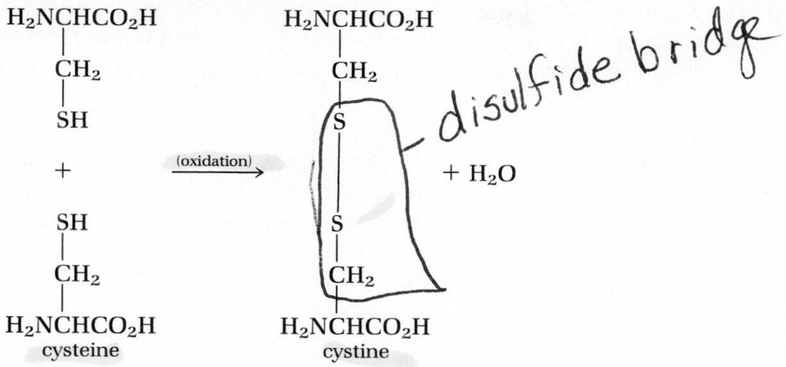

$$
\begin{array}{ccc}
H_2NCHCO_2H & & H_2NCHCO_2H \\
| & & | \\
CH_2 & & CH_2 \\
| & & | \\
SH & & S \\
 & & | \\
+ & \xrightarrow{\text{(oxidation)}} & | \quad + H_2O \\
 & & S \\
SH & & | \\
| & & CH_2 \\
CH_2 & & | \\
| & & H_2NCHCO_2H \\
H_2NCHCO_2H & & \\
\text{cysteine} & & \text{cystine}
\end{array}
$$

disulfide bridge

The disulfide bond significantly affects the shape of protein molecules as it both determines conformation and restricts the flexibility of the polypeptide chain. Interchain disulfide bonds occur to link cysteine in one polypeptide chain with cysteine in another polypeptide chain. Intrachain disulfide bonds also occur to form cyclic structures.

Hydrogen Bonds

In general, hydrogen bonds are much weaker than covalent bonds, such as peptide and disulfide bonds. However, there are numerous ways in which hydrogen bonds may occur in proteins. The peptide bond itself can form hydrogen bonds. The electron pairs of oxygen can form a hydrogen bond with a positive hydrogen atom. The hydrogen atom of the nitrogen–hydrogen bond can hydrogen bond with an electron pair on an electronegative atom such as oxygen or nitrogen.

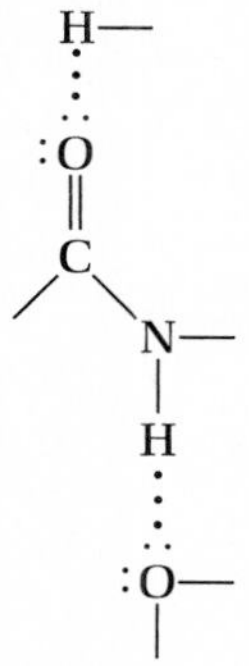

$$
\begin{array}{l}
H— \\
\vdots \\
:O \\
\| \\
C \\
\diagup \quad \diagdown \\
\qquad N— \\
\qquad | \\
\qquad H \\
\qquad \vdots \\
\qquad :O—
\end{array}
$$

Other sites for hydrogen bonding include the —OH groups of serine, threonine, and tyrosine as well as the —NH in the ring of tryptophan. Hydrogen bonding between peptide chains and within peptide chains is known.

Ionic Bonds

At physiological pH some of the R groups attached to the polypeptide chain are charged. Ionic attractive forces between the carboxylate group and the ammonium group pull portions of chains together. *The intrachain ionic bond is called a* **salt bridge.** Ionic bonds are important in acidic and basic amino acids.

Hydrophobic Interactions

Hydrophobic interactions are similar to those discussed in the micelle of a soap or the bilayer of lipids in membranes. The nonpolar R groups of the polypeptides tend to cluster together as a result of London forces. Polar portions of the polypeptide chain as well as water molecules are excluded from the region of hydrophobic interactions. Nonpolar amino acids such as leucine and phenylalanine tend to associate by means of hydrophobic interactions.

23.8
Structure of Proteins

The compositions of selected proteins are listed in Table 23.6. Note that frequently a particular amino acid occurs several times within a protein. A list of the molecular weights of some proteins is given in Table 23.7. However, as indicated several times in this book, composition and structure are quite different considerations.

The unique biological activity of proteins is dependent on the three-dimen-

Table 23.6

Numbers of Component Amino Acids in Proteins

Amino acid	Human insulin	Horse myoglobin	Pepsin	Egg albumin
glycine	4	13	29	19
alanine	1	15		12
serine	3	6	40	36
threonine	3	7	28	16
valine	4	6	21	28
leucine	6	22	27	32
isoleucine	2		28	25
proline	1	5	16	14
cysteine			2	5
cystine	3		2	1
methionine		2	4	16
phenylalanine	3	5	13	21
tyrosine	4	2	16	9
tryptophan		2	4	3
aspartic acid	3	10	41	32
glutamic acid	7	19	28	52
arginine	1	2	2	15
lysine	1	18	2	20
histidine	2	7	9	2

Table 23.7

Protein Molecular Weights

Protein	Molecular weight
insulin (a hormone)	5,700
ribonuclease (an enzyme)	12,700
lactalbumin (in milk)	15,500
trypsin (an enzyme)	24,000
ovalbumin (in egg white)	40,000
zein (in wheat)	40,000
catalase (an enzyme)	250,000
urease (an enzyme)	480,000

sional shape of the molecule. Any alteration of structure by breaking any of the types of bonds usually completely eliminates the physiological function of the protein. *The overall shape and structure of a protein constitute its* **native state** *or* **native conformation.**

Protein structure may be described at four levels called primary, secondary, tertiary, and quaternary. Each of these divisions is somewhat arbitrary since it is the total structure of the protein that controls function. Nevertheless, it is useful to consider the levels of structure one by one.

Primary Structure

The sequence of amino acids in a protein and the location of disulfide bonds is called its **primary structure.** The determination of amino acid sequences in proteins has involved years of work because for a given composition there are many isomers possible. There are approximately 10^{64} isomeric proteins consisting of 50 constituent amino acids!

Insulin consists of two peptide chains linked by two disulfide bonds. The A chain has 21 amino acids, and the B chain has 30 amino acids (Figure 23.5). An intrachain disulfide bond is present in the A chain. The Nobel prize was awarded to the English biochemist F. Sanger in 1958 for his structure determination of insulin, the first protein structure determined.

The insulin from different animals has slightly different sequences (Figure 23.5). Because the sequence within the cyclic portion of the shorter chain does not affect the physiological function of the insulin, diabetics who become allergic to one type of insulin can use insulin from another animal source.

Human hemoglobin consists of heme and globin. Globin is composed of four protein chains: two α-chains and two β-chains. The numbers of amino acids in the α- and β-chains are 141 and 146, respectively. The amino acid sequence of the β-chain is given in Figure 23.6. In some people the sixth amino acid from the N-terminal position of their hemoglobin β-chains is valine rather than glutamic

Figure 23.5
Primary Structure of Insulin.

	Positions		
Animal	**8**	**9**	**10**
Sheep	Ala	Gly	Val
Cow	Ala	Ser	Val
Pig	Thr	Ser	Ile
Horse	Thr	Gly	Ile

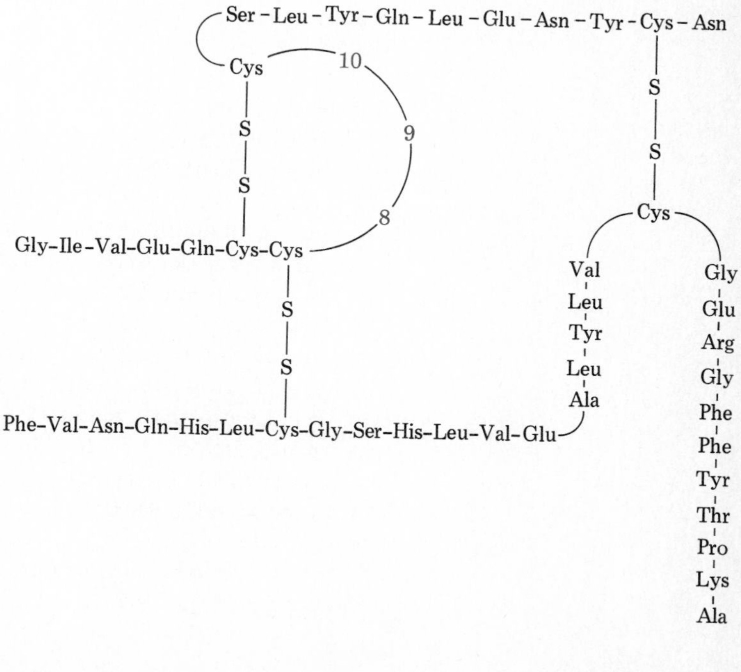

Val-His-Leu-Thr-Pro-Glu-Glu-Lys-Ser-Ala-Val-Thr-Ala-Leu-Trp-Gly-Lys-Val-Asp-Val-Asp-Glu-Val-Gly-Gly-Glu-Ala-Leu-Gly-Arg-
1 2 3 4 5 ⑥ 7 8 9 10 11 12 13 14 15 16 17 18 19 20 21 22 23 24 25 26 27 28 29 30

Leu-Leu-Val-Val-Tyr-Pro-Trp-Thr-Glu-Arg-Phe-Phe-Glu-Ser-Phe-Gly-Asp-Leu-Ser-Thr-Pro-Asp-Ala-Val-Met-Gly-Asp-Pro-Lys-Val-
31 32 33 34 35 36 37 38 39 40 41 42 43 44 45 46 47 48 49 50 51 52 53 54 55 56 57 58 59 60

Lys-Ala-His-Gly-Lys-Lys-Val-Leu-Gly-Ala-Phe-Ser-Asp-Gly-Leu-Ala-His-Leu-Asp-Asp-Leu-Lys-Gly-Thr-Phe-Ala-Thr-Leu-Ser-Glu-
61 62 63 64 65 66 67 68 69 70 71 72 73 74 75 76 77 78 79 80 81 82 83 84 85 86 87 88 89 90

Leu-His-Cys-Asp-Lys-Leu-His-Val-Asp-Pro-Glu-Asp-Phe-Arg-Leu-Leu-Gly-Asp-Val-Leu-Val-Cys-Val-Leu-Ala-His-His-Phe-Gly-Lys-
91 92 93 94 95 96 97 98 99 100 101 102 103 104 105 106 107 108 109 110 111 112 113 114 115 116 117 118 119 120

Glu-Phe-Thr-Pro-Pro-Val-Glu-Ala-Ala-Tyr-Glu-Lys-Val-Val-Ala-Gly-Val-Ala-Asp-Ala-Leu-Ala-His-Lys-Tyr-His-
121 122 123 124 125 126 127 128 129 130 131 132 133 134 135 136 137 138 139 140 141 142 143 144 145 146

Figure 23.6
Primary Structure of the Beta Chain of Hemoglobin.
The N-terminal amino acid is valine. The C-terminal amino acid is histidine. Substitution of glutamic acid by valine at position 6 causes the red blood cell to sickle.

acid. This difference of a single amino acid out of 146 in the chain causes changes in the shape of red blood cells. The cells tend to sickle, and as a result their passage through the blood vessels is restricted. The circulatory problems are known as sickle cell anemia.

Differences in hemoglobin do not affect its oxygen-carrying capacity in other animals. The β-chains of the gorilla and human hemoglobins are identical except for the position 104, in which lysine replaces another basic amino acid, arginine, of humans. The pig β-chain differs at 17 sites and that of the horse, at 26 sites, compared to human hemoglobin. In spite of the variability of composition of hemoglobin in different animals, there are nine positions that contain the same amino acids in all hemoglobin molecules. These positions are important to the oxygen-binding function of hemoglobin.

Secondary Structure

The specific spatial arrangement of the amino acid residues that are close to one another in the polypeptide chain is called the **secondary structure** (Figure 23.7). Recall that there is free rotation of the bonds separating the planar and rigid peptide units (Section 23.7). These peptide units are oriented at angles to each other and can exist in a variety of conformations. Many proteins consist of chains coiled into a spiral known as a helix. Such a helix may be either right- or left-handed, as in the case of screws. For proteins consisting of L-amino acids the right-handed (or α) helix is more stable than the left-handed helix. The spiral is held together by hydrogen bonds between the proton of the N—H group of one amino acid and the oxygen of the C=O group of another amino acid in the next turn of the helix.

In contrast to the intrachain hydrogen bonding that occurs in the α-helix, there are proteins in which interchain hydrogen bonding is important. A general representation of interchain bonding appears in Figure 23.7. Proteins in which interchain hydrogen bonding occurs include fibrin (the blood-clotting protein of blood), myosin (a protein of muscle), and keratin (the protein of hair).

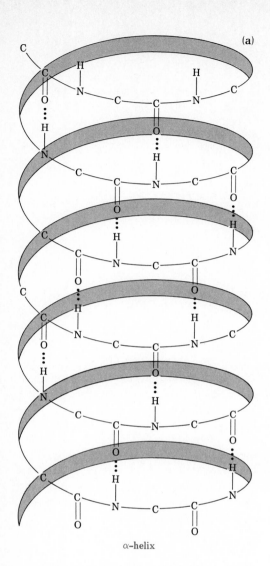

(a)

α–helix

Figure 23.7
Examples of Intramolecular and Intermolecular Hydrogen Bonding in Proteins.
(a) The intramolecular hydrogen bonds between coils of the helix are shown only on the "front." This type of protein is found in both fibrous and globular proteins. (b) The intermolecular hydrogen bonds between the chains of proteins cause a regular pleated or partially folded structure. This type of structure is found in the protein in silk.

(b)

β–pleated sheet

Tertiary Structure

The **tertiary structure** *refers to the spatial arrangement of amino acids residues that are far apart in the polypeptide chain.* The spatial arrangement is the result of the three-dimensional shape of the protein. Determination of the tertiary structure provides information about the spatial separation between amino acids. Two amino acids that are separated by many intervening amino acids may actually be close together in the folded structure. The proximity of amino acids in the tertiary structure is responsible for the activity of many enzymes (Section 24.4).

The three-dimensional folded shape of a globular protein (Figure 23.8) is the result of its primary and secondary structures, which together make possible the long-range interaction between amino acids. The forces of attraction between amino acids include ionic bonds, hydrogen bonds, and hydrophobic interactions.

The hydrophobic portions of the protein tend to associate within the interior of the folded structure. The polar or charged groups tend to be located at the

carboxyl end

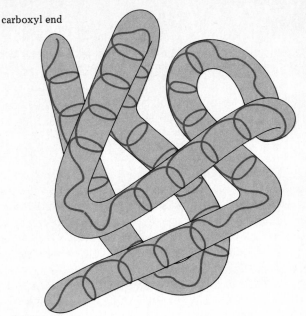

amino end

Figure 23.8
Tertiary Structure of a Protein.
The helix is shown within the volume occupied by the protein.

surface near water molecules. Thus there is no single reason for a specific shape of a globular protein. Each shape is unique and is the result of both the composition and sequence of the polypeptide chain.

Quaternary Structure

The **quaternary structure** *of a protein is the organization or the association of several protein chains or subunits into a closely packed arrangement.* Each of the subunits has its own primary, secondary, and tertiary structure. The subunits fit together because of their shape and are held together by noncovalent interactions. Single-chain proteins have no quaternary structure.

The subunits in a quaternary structure are specifically arranged in order for the entire protein to function properly. Any alteration in the structure of the subunits or the way in which the subunits are associated results in marked changes in biological activity. A list of some proteins having quaternary structure is given in Table 23.8.

Table 23.8
Quaternary-Structured Proteins

Protein	Molecular weight	Number of subunits	Biological function
alcohol dehydrogenase	80,000	4	enzyme for alcohol fermentation
aldolase	150,000	4	enzyme for glycolysis
fumarase	194,000	4	enzyme in the citric acid cycle
hemoglobin	65,000	4	oxygen transport in blood
insulin	11,500	2	hormone regulating metabolism of glucose

Figure 23.9
Quaternary Structure of Hemoglobin.
The iron of the heme is shown as spheres
within the folds of the four protein chains.
The large organic molecule (heme) surrounding the iron is represented by the
plane around the iron.

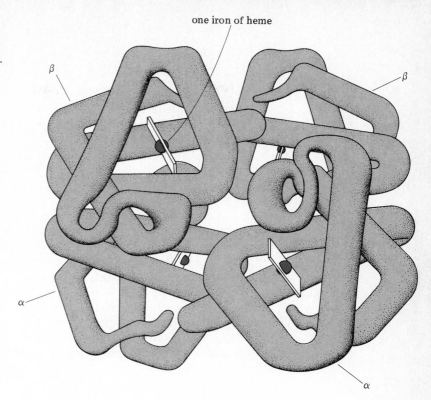

one iron of heme

β

β

α

α

Hemoglobin consists of two pairs of different proteins, each protein enfold-
ing a molecule of heme. The two identical α-chains and two identical β-chains
are arranged tetrahedrally in a three-dimensional structure (Figure 23.9). All of
these units are held together by forces other than covalent bonds. Hydrophobic
interactions, hydrogen bonding, and salt bridges may all contribute. The result of
fitting together the proteins is an almost spherical shape. The four protein subunits of hemoglobin do not behave independently. When one heme molecule
binds O_2, the conformation of the surrounding protein chain is slightly altered.
*Changes in conformation at one site caused by a change in a spatially separated
site of a protein molecule are called* **allosteric effects.** These interactions commonly occur in proteins containing subunits and are critical in the regulation of
metabolism by enzymes. As a result of allosteric effects, each heme in the other
subunits then can bind more easily to additional oxygen molecules. As each
oxygen binds, there are further conformational charges in the remaining protein
chains that enhance their binding capability. As a consequence, once oxygena-
tion occurs at one heme, there is cooperation at all other sites in hemoglobin,
which then can carry its full "load" of four oxygen molecules.

$$Hb + O_2 \rightleftharpoons HbO_2 \qquad K_1 = \frac{[HbO_2]}{[Hb][O_2]}$$

$$HbO_2 + O_2 \rightleftharpoons Hb(O_2)_2 \qquad K_2 = \frac{[Hb(O_2)_2]}{[HbO_2][O_2]}$$

$$Hb(O_2)_2 + O_2 \rightleftharpoons Hb(O_2)_3 \qquad K_3 = \frac{[Hb(O_2)_3]}{[Hb(O_2)_2][O_2]}$$

$$Hb(O_2)_3 + O_2 \rightleftharpoons Hb(O_2)_4 \qquad K_4 = \frac{[Hb(O_2)_4]}{[Hb(O_2)_3][O_2]}$$

$$K_4 > K_3 > K_2 > K_1$$

The cooperative effect in hemoglobin also works in reverse. When one oxygen molecule is "unloaded" from one subunit, the other subunits lose their oxygen molecules more readily. As a consequence, hemoglobin readily loses all four oxygen molecules. In summary, hemoglobin is a very efficient transporter of oxygen. It takes on a "full load" and "unloads" its entire supply as a cooperative unit rather than each site operating independently.

23.9
Hydrolysis of Proteins

Recall from Section 19.9 that amide bonds are quite unreactive toward hydrolysis. An amide can be hydrolyzed by heating with concentrated solutions of either strong acid or base. Similarly, the hydrolysis of peptides or proteins in 6 N HCl at 100°C requires hours. Complete hydrolysis eventually produces the constituent amino acids of the peptide or protein. Digestion of proteins in the body must occur at a faster rate and at much lower acid concentration.

Enzymes hydrolyze proteins in the body and can be used to hydrolyze large peptides and proteins in the laboratory. Enzymes have been used to determine the structure of peptides and proteins. *Enzymes that catalyze the hydrolysis of proteins are called* **proteases.** The proteases pepsin, chymotrypsin, and trypsin have preferences for certain types of amide groups in a protein. Pepsin will hydrolyze the peptide bond at the nitrogen of phenylalanine, tyrosine, or tryptophan. All three amino acids contain an aromatic ring. Chymotrypsin hydrolyzes the peptide bond of the same three amino acids but at the carboxyl end of the amino acid. Trypsin hydrolyzes peptide bonds of the basic amino acids lysine and arginine at the carboxyl end of the molecule.

$$\text{Lys-Glu-Tyr-Leu} \xrightarrow{\text{pepsin}} \text{Lys-Glu + Tyr-Leu}$$

$$\text{Lys-Glu-Tyr-Leu} \xrightarrow{\text{chymotrypsin}} \text{Lys-Glu-Tyr + Leu}$$

$$\text{Lys-Glu-Tyr-Leu} \xrightarrow{\text{trypsin}} \text{Lys + Glu-Tyr-Leu}$$

Other enzymes hydrolyze peptide bonds from one end of the molecule. **Carboxypeptidases** *sequentially hydrolyze peptides only from the end that has the free carboxyl group.* **Aminopeptidases** *sequentially hydrolyze peptides only from the end containing the free amino group.* By determining the amino acids produced by a carboxypeptidase or an aminopeptidase at various time intervals, the sequence of amino acids can be determined. The tetrapeptide Arg-Glu-Lys-Trp, when reacted with a carboxypeptidase, first liberates tryptophan. The remaining tripeptide, Arg-Glu-Lys, will eventually yield lysine, and subsequently both glutamic acid and arginine will be formed. When an aminopeptidase is used, the sequential formation of amino acids will be arginine followed by glutamic acid and then both lysine and tryptophan.

Example 23.5 ─────────────────────────────

Predict the products of the chymotrypsin-catalyzed hydrolysis of the enkephalin given in Figure 23.4.

Solution. Chymotrypsin catalyzes hydrolysis at the carboxyl end of aromatic amino acids. The enkephalin in Figure 23.4 contains both phenylalanine and tyrosine.

Tyr-Gly-Ala-Phe-Leu

The tyrosine is the N-terminal amino acid, and hydrolysis at the carboxyl end results in free tyrosine.

Tyr-Gly-Ala-Phe-Leu $\longrightarrow$ Tyr + Gly-Ala-Phe-Leu

The phenylalanine is bonded to the C-terminal amino acid leucine. Hydrolysis at the carboxyl group of phenylalanine frees leucine. A tripeptide results.

Gly-Ala-Phe-Leu $\longrightarrow$ Gly-Ala-Phe + Leu

[Additional examples may be found in **23.75–23.83** at the end of the chapter.]

23.10 Denaturation of Proteins

Denaturation *is the loss or destruction of the native conformation of a protein.* As a result, denaturation of a protein destroys the biochemical function of the molecule. In denaturation, the peptide bond is not affected. The hydrogen bonds, disulfide bonds, ionic bonds, and hydrophobic interactions responsible for secondary, tertiary, and quaternary structures are disrupted.

When there is a quaternary structure, denaturation first affects the forces holding the subunits together. Next, the tertiary structure is disrupted and the protein unfolds. Finally, the hydrogen bonds are broken and a random coil protein results.

Denaturation may be reversible or irreversible. *The re-formation of a native protein from a denatured protein is called* **renaturation** (Figure 23.10). If the enzyme aldolase is treated with $4\,M$ urea, the tetramer separates into its four subunits and these protein chains unfold to some degree. However, when the protein is separated from the urea, the chains refold and the subunits rejoin. The resulting material has about 75% of the original aldolase activity, and the denaturation process is reversible. On the other hand, if the disulfide bonds in insulin are cleaved by reduction and then reoxidized, the resulting molecule has none of its former biological activity. Denaturation is irreversible in this case.

Denaturation may be caused by heat, radiation, acids or bases, oxidizing or reducing agents, hydrogen-bonding solvents, and heavy metal salts. The action of each of these denaturation agents will now be considered.

Heat or Radiation

If energy from heat or radiation is added to a protein, vibration and motion within the molecule increase. Hydrophobic interactions and hydrogen bonds are

Figure 23.10
Denaturation and Renaturation of a Protein.

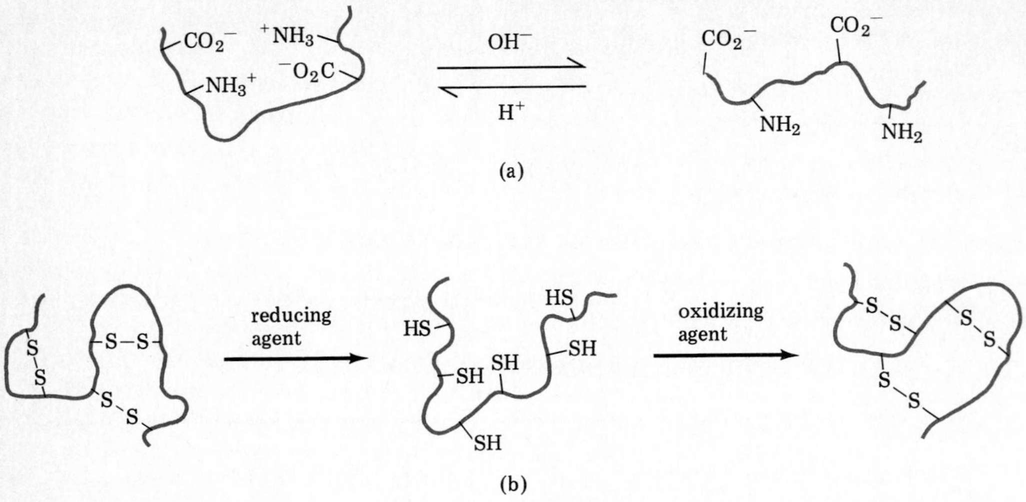

Figure 23.11
Types of Denaturation Processes.
The denaturation of a protein by acid or base may be reversible if the pH changes are small and the degree of unfolding of the chain is small. Reduction of disulfide bonds followed by oxidation may not regenerate the native protein.

disrupted, and the protein is denatured. Since both types of bonds are weak, very little heat is required for denaturation. Most proteins are denatured when heated above 50°C, a temperature above the normal body temperature of 37°C. The most common example of denaturation by heat is frying an egg.

Burns, as well as sunburn, cause denaturation of protein in the skin. In surgery, cauterization or heating of body protein is used to seal small blood vessels.

Acids and Bases

When the pH of a solution of a protein is changed, the carboxyl group may exist as such or as the carboxylate ion. Similarly, the amino group may exist as such or as an ammonium ion. As a consequence, the ionic bonds in the protein are disrupted. The degree of destruction of the protein structure depends on how much the pH is changed. If the change is slight, reversal of the pH could result in complete regeneration of structure (Figure 23.11).

Some burn ointments contain acids such as tannic acid. The acids denature and coagulate the proteins in burned skin. A protective coating is formed that covers the burn area and prevents fluid loss.

Oxidizing and Reducing Agents

Reducing agents convert disulfide bonds to free sulfhydryl groups. Since the disulfide bond is part of the primary structure, reducing agents seriously disrupt the structure of proteins (Figure 23.11). Oxidizing agents can convert sulfhydryl groups back into disulfide bonds, but the structure of the protein may be significantly different than the native protein.

Hydrogen-Bonding Solvents

Hydrogen-bonding solvents denature proteins by disrupting the hydrogen bonds within the protein molecule. Ethanol is one such solvent. The —OH group effectively competes with hydrogen bonds within the protein. As a result, ethanol is a good antiseptic. A 70% solution of ethanol is able to pass across the cell wall of bacteria. Once inside the cell, it denatures the bacterial proteins.

Heavy Metal Salts

Heavy metal ions such as Ag^+, Hg^{2+}, and Pb^{2+} denature proteins by reacting with sulfhydryl bonds to form stable metal–sulfur bonds. The metal ions also combine with the carboxylate ions in the side chains of acidic amino acids.

Proteins are precipitated by heavy metal ions. A 1% solution of Ag^+ formerly was used to treat the eyes of newborn babies. This method kills the bacteria causing gonorrhea.

Summary

Proteins, which are polymers of 20 α-amino acids, are classified into two broad categories, globular and fibrous. Hydrolysis of proteins yields amino acids, but other materials are also formed from conjugated proteins.

Proteins play crucial roles in many biological processes, and the functions of proteins are used to classify them. Proteins provide structural support, transport material, act as chemical messengers, cause motion, serve as storage, and provide protection.

The 20 amino acids all have an α-amino group and, with the exception of the optically inactive glycine, are of the L configuration. Based on the remainder of the molecule, the amino acids are classified as neutral, acidic, or basic. At physiological pH most amino acids exist as neutral zwitterions.

The pH at which the molecule of an amino acid has a zero net charge is the isolectric point. The isoelectric points of neutral amino acids are in the 5.5–6.5 range. Basic amino acids have higher isoelectric points, and acidic amino acids have lower isoelectric points. Electrophoresis is a process used to separate amino acids and proteins based on the electric charge in a chosen buffer solution.

The structure of proteins is described in four parts. The primary structure is the sequence of the amino acids in the protein. Secondary structure is the arrangement or conformation about the peptide backbone. Hydrogen bonding is the principal contributor to secondary structure. Tertiary structure refers to the overall folding and bending of the peptide backbone. Quaternary structure is the arrangement of subunits in those proteins that have several protein chains held together by hydrophobic interactions.

Denaturation is the loss of biological activity of a protein caused by changes in the secondary, tertiary, or quaternary structure. Denaturation is caused by heat, changes in pH, oxidation or reducing agents, certain metal ions, salts, and some organic solvents.

Learning Objectives

As a result of studying Chapter 23 you should be able to

- Describe the ways in which proteins are classified.
- List some of the functions of proteins.
- Classify amino acids as neutral, acidic, or basic from the structural formula.
- Explain why some amino acids are designated as essential.
- Write the structures of amino acids as they occur in solutions of different pH.
- Define zwitterion and isoelectric point.
- Name a peptide, given a structural formula.
- Describe the biological function of the low-molecular-weight proteins that are hormones.
- List the types of bonds present in proteins.
- Define primary, secondary, tertiary, and quaternary structures of proteins.
- Write the products of enzyme-catalyzed hydrolysis of peptides.
- List the ways to denature a protein and give an example of each.

New Terms

Acidic amino acids have an additional carboxyl group.

An **allosteric effect** is a change in conformation at one site caused by change in conformation at a second, spatially separated site.

An **amino acid** is an amino-substituted carboxylic acid. In proteins the amino group is at the α position.

Aminopeptidases are enzymes that sequentially hydrolyze peptides only from the end that has the free amino group.

Antibodies are proteins that recognize and combine with foreign substances.

Basic amino acids have an additional amino group.

Carboxypeptidases are enzymes that sequentially hydrolyze peptides only from the end that has the free carboxyl group.

Conjugated proteins consist of a protein combined with a nonprotein material.

Denaturation is the loss or destruction of the native conformation of a protein.

Electrophoresis is a method of separating amino acids or proteins based on their net charge at a selected pH.

Enkephalins are peptides that bind at receptor sites in the brain to reduce pain.

Essential amino acids must be obtained from food.

Fibrous proteins are proteins arranged in sheet-like structures.

Globular proteins are tightly folded proteins that are roughly spherical in shape.

HDL refers to **high density lipoproteins,** which transport cholesterol via the blood away from body tissues to the liver.

Hormones are chemical messengers that are produced by endocrine glands. Some hormones are proteins.

The **isoelectric point** is the pH at which there is no net charge on an amino acid or protein.

LDL refers to **low density lipoproteins,** which transport cholesterol to tissues via the blood.

Native state or **conformation** of a protein is its overall shape and structure.

Neutral amino acids contain only one amino group and one carboxyl group.

A **peptide bond** is the amide bond in a polypeptide or protein.

The **primary structure** of a protein is the sequence of amino acids.

A **prosthetic group** is a nonprotein material combined in a conjugated protein.

Proteases are enzymes that catalyze the hydrolysis of proteins.

A **protein** is a polymer of amino acids joined by peptide bonds.

The **quaternary structure** of a protein is the manner in which protein subunits (chains) are assembled to give the whole protein.

Renaturation is the re-formation of a native protein from a denatured protein.

A **salt bridge** is an ionic bond between parts of a protein chain.

The **secondary structure** of a protein is the spatial arrangement of amino acid residues that are close to each other in the polypeptide chain.

The **tertiary structure** refers to the spatial arrangement of amino acid residues that are far apart in the polypeptide chain.

VLDL refers to **very low density lipoproteins.**

A **zwitterion** is an electrically neutral ion resulting from transfer of a proton from an acidic to a basic site in a molecule.

Questions and Problems

Terminology

23.1 What distinguishes acidic from basic amino acids?

23.2 What do the terms primary, secondary, tertiary, and quaternary structures mean?

23.3 What are zwitterions, and under what conditions do they exist in solutions of amino acids?

23.4 What is the difference between a peptide and a protein?

Classification of Proteins

23.5 What is the basis of the distinction between globular and fibrous proteins?

23.6 What is a conjugated protein?

23.7 Name the classes of conjugated proteins.

23.8 Give the class of each of the following conjugated proteins.
(a) hemoglobin (b) heparin (c) casein
(d) LDL

23.9 Explain the role of lipoproteins in controlling blood cholesterol level.

23.10 How is exercise apparently related to blood cholesterol level?

23.11 List the six principal functions of proteins.

23.12 Give one example of a protein for each function.
(a) hormonal action (b) storage
(c) transport (d) structure
(e) protection (f) motion

Amino Acids

23.13 Amino acids in proteins are designated as α-amino acids. What does the term α mean?

23.14 Amino acids in proteins are optically active and have the L configuration. What does the term L mean?

23.15 Draw the structures of two amino acids that contain more than one asymmetric carbon atom.

23.16 The following compound is an unusual amino acid found in collagen. From what amino acids could this compound be derived?

$$NH_2-CH_2-\underset{\underset{OH}{|}}{CH}-CH_2-CH_2-\underset{\underset{NH_2}{|}}{CH}-CO_2H$$

23.17 The following compound is an unusual amino acid that is a neurotransmitter. Classify this amino acid, and assign its IUPAC name.

$$NH_2-CH_2-CH_2-CH_2-CO_2H$$

23.18 A number of D-amino acids are found in bacteria. D-Glutamic acid is found in bacterial cell walls. Draw a projection formula for the amino acid.

23.19 Earthworms have some D-serine. Draw the projection formula of this amino acid.

23.20 Give the name and write the structure for each of the following acids.
(a) Gly (b) Ala (c) Phe (d) Pro (e) Ser
(f) Cys

23.21 Name the three basic amino acids. Name the two acidic amino acids.

23.22 Why is asparagine a neutral amino acid, while aspartic acid is an acidic amino acid?

Essential Amino Acids

23.23 Why are some amino acids called essential?

23.24 Name the essential amino acids for humans.

23.25 Tyrosine can be synthesized by the body. Why, then, is tyrosine considered an essential amino acid?

23.26 Why is histidine not listed as an essential amino acid in Table 23.2?

23.27 How do the amino acids in fish, meat, and cereal grains compare in terms of biological value?

23.28 Why is the soybean useful as a substitute for beef or as an extender of beef products?

23.29 What precautions must vegetarians take in their diets?

23.30 Ovo-lacto vegetarians eat eggs and drink milk. Do these individuals have as many dietary concerns as total vegetarians?

Zwitterions

23.31 Draw the structures of the following amino acids at pH = 1, at pH = 7, and pH = 12.
(a) alanine   (b) glutamic acid   (c) lysine

23.32 Write the structure for the zwitterion of alanine.

23.33 How could you distinguish between a solution of serine and a solution of lysine?

23.34 Would you expect a solution of lysine to be neutral, acidic, or basic?

Isoelectric Points

23.35 What structure of alanine predominates at the isoelectric point? What structure becomes more important at pH values larger than the isoelectric point?

23.36 Describe the values for the isoelectric point of neutral, acidic, and basic amino acids.

23.37 The isoelectric points of glutamic acid and glutamine are 3.2 and 5.7, respectively. Explain the reason for this difference of the two related compounds.

23.38 Examine the structures of oxytocin and vasopressin in Figure 23.4. Which should have the higher isoelectric point?

23.39 Examine the structure of the enkephalin given in Figure 23.4 and estimate its isoelectric point.

23.40 Estimate the isoelectric points of the following tripeptides.
(a) Ala-Val-Gly   (b) Ser-Val-Asp
(c) Lys-Ala-Val

23.41 List the number of neutral, acidic, and basic amino acids in sheep insulin. On this basis, predict the isoelectric point of insulin.

23.42 The isoelectric point of chymotrypsin is 9.5. What does this value indicate about the composition of chymotrypsin?

23.43 The isoelectric point of pepsin is 1.1. What does this value indicate about the composition of pepsin?

23.44 The isoelectric point of hemoglobin is 6.8. What does this value indicate about the composition of hemoglobin?

Peptides

23.45 Write the complete formula and the condensed formula for alanylserine.

23.46 How does glycylserine differ from serylglycine?

23.47 Which amino acids can form peptides with carboxyl groups or carboxylate groups at internal positions in the peptide chain?

23.48 Write the condensed formulas for the 24 isomeric tetrapeptides containing glycine, alanine, leucine, and isoleucine.

23.49 How many isomeric tetrapeptides containing two molecules each of alanine and leucine are there? Write their condensed formulas.

23.50 In what form would glycylalanine exist in a neutral solution?

23.51 How many isomers of the enkephalin Tyr-Gly-Gly-Phe-Leu are there?

23.52 Identify the amino acids contained in each of the following tripeptides.

23.53 Name each of the tripeptides in **23.52.**

Proteins

23.54 What types of bonds occur in proteins? Which bonds are covalent?

23.55 Which amino acids can form salt bridges in proteins?

23.56 Which amino acids have R groups that form hydrogen bonds?

23.57 What class of amino acids is responsible for hydrophobic interactions in proteins?

23.58 Describe the primary structure of insulin.

23.59 What peptides would result if the chain of insulin containing 21 amino acids were treated with pepsin?

23.60 Noting that proline is a secondary amine, explain how proline can disrupt the α-helix of a protein.

23.61 Examine the structure of valine and glutamic acid and suggest a reason why human hemoglobin is affected by the substitution of valine for glutamic acid at position 6 in the chain.

23.62 Which of the following amino acids are likely to exist in the interior of a globular protein dissolved in an aqueous solution?

(a) glycine (b) phenylalanine
(c) glutamic acid (d) arginine
(e) proline (f) cysteine
(g) glutamine (h) aspartic acid

23.63 If a globular protein is embedded in a lipid bilayer, which of the amino acids listed in **23.62** would be in contact with the interior of the bilayer?

23.64 Myoglobin and the α- and β-chains of hemoglobin are globular proteins having similar shapes. Myoglobin exists as a monomer in aqueous solution. However, the protein chains of hemoglobin assemble to form a tetramer. Which protein chains have the higher percentage of nonpolar amino acids?

23.65 Indicate how each of the following pairs of amino acids in a protein could interact if the tertiary structure of the protein brought them close together.

(a) aspartic acid and arginine
(b) serine and glutamic acid
(c) phenylalanine and valine
(d) leucine and isoleucine

23.66 Describe the cooperative effect of hemoglobin that affects the transport of oxygen.

23.67 What is an allosteric effect?

Denaturation of Proteins

23.68 What is meant by the term denaturation?

23.69 Explain how pH changes could change the tertiary structure of a protein that has a large number of salt bridges as part of the tertiary structure.

23.70 What would occur to the structure of insulin if treated with a reducing agent?

23.71 What would be the effect of increasing the pH in a solution of a protein containing large amounts of lysine?

23.72 Why does cooking an egg denature egg proteins?

23.73 Explain how the addition of a small amount of an acid followed by the addition of a small amount of a base may not cause denaturation of a protein.

23.74 Explain how egg whites can be used to treat lead poisoning.

Enzymatic Hydrolysis

23.75 How does the action of pepsin differ from that of chymotrypsin?

23.76 Explain the enzymatic action of an aminopeptidase.

23.77 What would be the result of treating samples of Lys-Gly-Tyr-Leu with each of the following enzymes?

(a) pepsin (b) chymotrypsin (c) trypsin

23.78 Indicate which of the following tripeptides will be cleaved by trypsin and name the products.

(a) Arg-Gly-Tyr (b) Glu-Asp-Gly
(c) Phe-Trp-Ser (d) Ser-Phe-Asp
(e) Asp-Lys-Ser (f) Lys-Tyr-Cys
(g) Asp-Gly-Lys (h) Arg-Glu-Ser

23.79 Indicate which of the tripeptides in **23.78** will be cleaved by pepsin and name the products.

23.80 Indicate which of the tripeptides in **23.78** will be cleaved by chymotrypsin and name the products.

23.81 What amino acid would be formed first when Gly-Ala-Phe is treated with an aminopeptidase?

23.82 What amino acid would be formed first when Gly-Ala-Phe is treated with a carboxypeptidase.

23.83 List the products formed from the peptide Trp-Cys-Ala-Lys-Leu-Gly-Phe-Pro-Cys when treated in separate experiments with each enzyme.

(a) pepsin (b) trypsin (c) chymotrypsin

Enzymes

This bread was raised by the action of enzymes in yeast prior to being baked.

Many of the chemical reactions discussed in the preceding chapters also occur in living systems. For example, esters, acetals, and amides can be formed or hydrolyzed. However, unlike laboratory processes, life processes must occur in aqueous solutions and at relatively low temperatures. Moreover, the reactions must occur rapidly enough to provide the response necessary to support a living system.

In this chapter the largest group of proteins, known as enzymes, will be discussed. Enzymes are extraordinary catalysts since they may increase the rate of a reaction by a factor of a million or more. Reactions that otherwise would require years at neutral pH and at 37°C can occur in seconds in the presence of an enzyme. For example, carbonic anhydrase catalyzes the hydration of carbon dioxide.

$$H_2O + CO_2 \xrightarrow{\text{carbonic anhydrase}} H_2CO_3$$

Each molecule of enzyme catalyzes the hydration of 10^5 molecules of carbon dioxide in one second. The catalyzed reaction occurs 10^7 times faster than the uncatalyzed reaction. Without the catalyst, the transfer of CO_2 from the tissues into the blood would occur too slowly to maintain life processes.

Substrate and Enzyme Specificity

A **substrate** *is a reactant in a biological reaction catalyzed by an enzyme.* Enzymes catalyze specific reactions of certain substrates. *The limitation of the type of reaction catalyzed and the substrate is called* **specificity.** Some enzymes catalyze a single reaction for a variety of substrates such as the hydrolysis of esters. Other enzymes have a high degree of specificity and catalyze the reaction of only a few substrates. Because of the specificity of enzymes, the cell needs many enzymes to maintain its functions. Several thousand enzymes may be present in the cell.

Enzymes that act on one substrate and no others have **absolute specificity.** Succinate dehydrogenase is an enzyme of absolute specificity in the citric acid cycle. It catalyzes the oxidation of succinic acid to fumaric acid by the oxidizing agent FAD.

$$FAD + HO_2CCH_2CH_2CO_2H \xrightarrow{\text{succinate dehydrogenase}} \underset{\text{fumaric acid}}{\overset{H}{\underset{HO_2C}{}}C=C\overset{CO_2H}{\underset{H}{}}} + FADH_2$$

succinic acid   fumaric acid

No other dicarboxylic acid is oxidized by succinate dehydrogenase. Moreover, note that the product is exclusively fumaric acid, not maleic acid, its *cis* isomer.

Aspartase is another example of an enzyme with absolute specificity. It catalyzes the interconversion of fumaric acid and L-aspartic acid.

$$\underset{\text{fumaric acid}}{\overset{H}{\underset{HO_2C}{}}C=C\overset{CO_2H}{\underset{H}{}}} + NH_4^+ \xrightarrow{\text{aspartase}} \underset{\text{L-aspartic acid}}{NH_2-\overset{CO_2H}{\underset{CH_2CO_2H}{|}}{C}-H} + H^+$$

In this case not only is the enzyme absolutely specific for fumaric acid but only the L isomer of aspartic acid is formed.

Example 24.1

Will aspartase catalyze the following reaction?

$$
\begin{array}{c}
\underset{CH_3O_2C}{\overset{H}{\diagdown}}C=C\underset{H}{\overset{CO_2CH_3}{\diagup}} + NH_4^+ \longrightarrow NH_2\underset{CH_2CO_2CH_3}{\overset{CO_2CH_3}{\underset{|}{\overset{|}{-}}}}C-H + H^+
\end{array}
$$

Solution. The enzyme aspartase has absolute specificity for fumaric acid. This means that fumaric acid and only fumaric acid will react. The enzyme cannot react with compromise molecules no matter how similar they are to the correct substrate. In this case the dimethyl fumarate does not react as indicated in an enzyme-catalyzed reaction.

[Additional examples may be found in **24.6–24.11** at the end of the chapter.]

Energy and Biochemical Reactions

Biochemical reactions catalyzed by enzymes follow the same laws of conservation of mass and energy as do ordinary chemical reactions. The reactants and products of the enzyme-catalyzed reactions are no different from those of an ordinary chemical reaction. However, an enzyme provides a reaction pathway with a lower activation energy than is involved in an uncatalyzed reaction (Figure 24.1). With a lower activation energy, the substrate molecules react at a faster rate (Section 10.5).

Enzymes do not change the normal position of a chemical equilibrium. The product is only formed faster, not in larger quantity. In addition, the energy released in the reaction is not changed by an enzyme. Enzymes do not promote energetically unfavorable reactions, but only accelerate energetically favorable reactions.

Figure 24.1
Energy Diagram for the Progress of a Chemical Reaction and the Effect of an Enzyme.
The enzyme provides a pathway of lower activation energy that allows the conversion of reactant to product to occur at a faster rate.

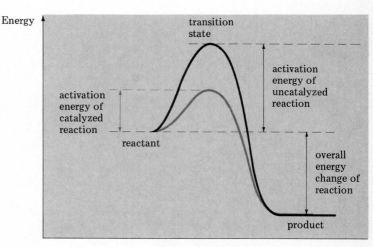

AN ASIDE

Enzymes have long been recognized as useful catalytic compounds. They have been used for a number of chemical reactions in foods for many years. For centuries yeast has been used in the brewing of beer and other alcoholic beverages. The leavening of bread and formation of vinegar are other processes that are enzyme catalyzed. Because the earliest fermentation processes are catalyzed by yeast, the words *en zume* from the Greek meaning "in yeast" were chosen about 1880 as a name for such catalytic material.

The chemical identity of enzymes was suggested by an American biochemist, James B. Sumner, in 1926. Sumner obtained the enzyme urease as a crystalline compound and demonstrated that it is a protein. Although some other biochemists did not accept his suggestion that enzymes are proteins, scores of other enzymes have been isolated and crystallized since that time. In every case the enzyme has been shown to be a protein. Sumner was finally recognized for his contribution to chemistry in 1946 when he was awarded the Nobel prize.

24.2
Classification of Enzymes

As in the case of the development of organic chemistry, the biochemists first assigned common names to enzymes. In the last chapter you encountered three such examples—pepsin, trypsin, and chymotrypsin, which catalyze the hydrolysis of proteins. A biochemist has to learn to associate the common name of each enzyme with the specific reaction that it catalyzes. There is nothing in the name to tell what type of substrate the enzyme acts on or what type of reaction will be catalyzed.

Some improvement in nomenclature developed when the suffix **-ase** was added to the name of the substrate on which the enzyme acts. For example, urease catalyzes the hydrolysis of urea.

$$H_2O + NH_2-\overset{\overset{\textstyle O}{\|}}{C}-NH_2 \xrightarrow{\text{urease}} 2\,NH_3 + CO_2$$

However, a substrate such as urea can undergo several possible reactions, and an enzyme such as urease catalyzes only one of the reactions. As a consequence, it was determined to indicate the name of the substrate and the type of reaction catalyzed when naming an enzyme. Urease catalyzes the hydrolysis of amide bonds in urea. The name urea amidohydrolase conveys this information.

In 1961, the International Union of Biochemistry decided on a system of

Table 24.1
Classification of Enzymes

Enzyme	Type of reaction catalyzed
oxidoreductases	oxidation–reduction
transferases	transfer of a group from one compound to another
hydrolases	hydrolysis
lyases	nonhydrolytic addition or removal of groups
isomerases	conversion of a substance into one of its isomers
ligases	synthesis of two small molecules into one larger molecule

classifying and cataloging enzymes. Each enzyme is assigned to one of six groups listed in Table 24.1.

Within each class of enzymes there are subclasses. For example, many substrates can undergo hydrolysis reactions catalyzed by hydrolases. The hydrolases that catalyze the hydrolysis of peptides are known as peptidases, whereas those that catalyze the hydrolysis of esters are esterases.

Oxidoreductases

Oxidoreductases catalyze oxidation–reduction reactions involving the transfer of protons and electrons to or from a substrate. Oxidoreductases are often given other names such as oxidase, dehydrogenase, or reductase. The oxidation of lactic acid to pyruvic acid involves the removal of two hydrogen atoms and is catalyzed by lactic acid dehydrogenase.

$$CH_3 \overset{\overset{\displaystyle OH}{|}}{\underset{\underset{\displaystyle H}{|}}{C}} CO_2H \xrightarrow[\text{dehydrogenase}]{\text{lactic acid}} CH_3 \overset{\overset{\displaystyle O}{\|}}{C} CO_2H + 2\,H^+ + 2e^-$$

lactic acid   pyruvic acid

Recall from Chapter 10 that oxidation, which involves loss of electrons, must be accompanied by reduction. Oxidation is accomplished by an oxidizing agent, which then is reduced. The enzyme that catalyzes an oxidation must accept electrons. Such enzymes are combined with coenzymes such as NAD^+ and FAD that accept electrons and protons. These processes will be discussed in the next chapter.

Transferases

Transferases catalyze the transfer of groups of atoms from one compound to another. There are numerous subclasses of transferases, which are named according to the groups that are transferred. An important group of enzymes, which catalyzes the transfer of amino groups from amino acids to keto acids, is called aminotransferase or transaminase

$$CH_3 \overset{\overset{\displaystyle NH_2}{|}}{\underset{\underset{\displaystyle H}{|}}{C}} CO_2H + \overset{\overset{\displaystyle O}{\|}}{C} \underset{\underset{\underset{\displaystyle CO_2H}{|}}{(CH_2)_2}}{} CO_2H \xrightarrow[\text{transaminase}]{\text{alanine–glutamate}} CH_3 \overset{\overset{\displaystyle O}{\|}}{C} CO_2H + H \overset{\overset{\displaystyle NH_2}{|}}{\underset{\underset{\underset{\displaystyle CO_2H}{|}}{(CH_2)_2}}{C}} CO_2H$$

alanine   α-ketoglutaric acid   pyruvic acid   glutamic acid

Hydrolases

Various hydrolases catalyze the hydrolysis of esters, glycosidic bonds in carbohydrates, and peptide bonds in proteins. The hydrolases are important in digestive processes. The digestive enzymes pepsin, trypsin, and chymotrypsin (Section 23.9) are all hydrolases.

Lys-Glu-Tyr-Leu $\xrightarrow{\text{pepsin}}$ Lys-Glu + Tyr-Leu

Lyases

Lyases catalyze the addition reaction of water, ammonia, or carbon dioxide to double bonds such as C=C and C=N, as well as the reverse reactions, which are known as elimination reactions. The addition of water to fumaric acid, one of the steps in the citric acid cycle, is catalyzed by fumarase.

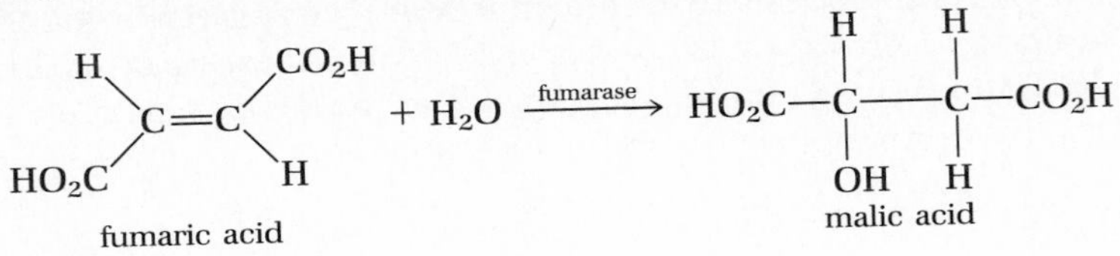

fumaric acid malic acid

Isomerases

Isomerases catalyze isomerization reactions such as the conversion of glucose 6-phosphate to fructose 6-phosphate in glycolysis reactions.

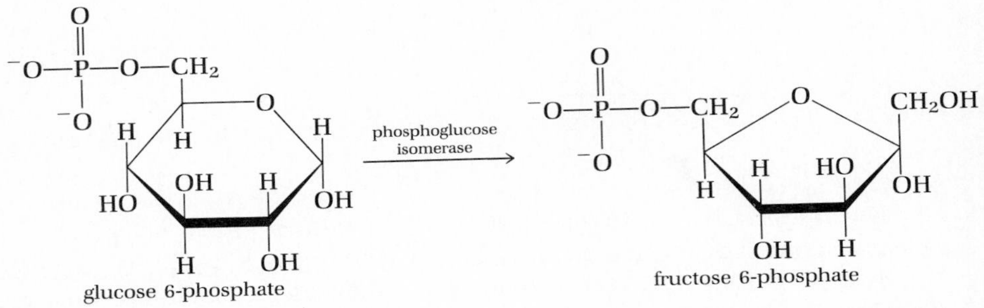

glucose 6-phosphate fructose 6-phosphate

Ligases

Ligases catalyze the formation of bonds by joining two molecules to form a larger one. These reactions are not spontaneous and require a source of energy. The necessary energy is provided by ATP. Because ligases catalyze synthesis reactions, these enzymes are also called synthetases. The conversion of the acidic amino acid glutamic acid into the neutral amino acid glutamine is an example. Ligases are also involved in the synthesis of proteins, lipids, carbohydrates, and nucleic acids.

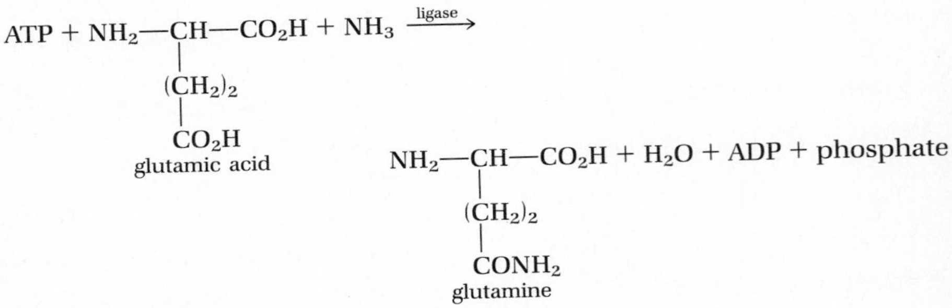

glutamic acid glutamine

Example 24.2

Classify the reaction given in Example 24.1.

Solution. The forward reaction is an addition reaction with ammonia, whereas the reverse is an elimination reaction. Such reactions are catalyzed by lyases. The forward reaction is known as an amination reaction, and the reverse is a deamination reaction.

AN ASIDE

The detection of enzymes in medical laboratories is an important clinical tool in the diagnosis of disease. Some enzymes are located in high levels in certain tissues and occur in only low concentration in blood. Thus, an abnormally high level of an enzyme in the blood is indicative of tissue damage.

Lactate dehydrogenase (LDH), aspartate transferase (AST), and creatinine phosphokinase (CPK) are all contained in heart tissue. When a myocardial infarction occurs, these enzymes are released into the blood (Figure 24.2). In the blood, AST is usually called serum glutamate–oxaloacetate transaminase (SGOT). The amount of the three enzymes in the blood indicates the extent of heart damage.

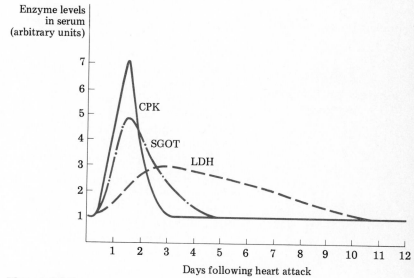

Figure 24.2

Changes in Enzyme Concentration in the Blood After a Heart Attack.

Example 24.3

What type of enzyme is required for the following reaction? Suggest a name for the enzyme that catalyzes the reaction.

$$\text{sucrose} + H_2O \longrightarrow \text{glucose} + \text{fructose}$$

Solution. The reaction involves hydrolysis of an acetal of glucose and a ketal of fructose (Chapter 21). The necessary enzyme is a hydrolase. A reasonable name to suggest is sucrase to indicate that sucrose is the substrate. However, this name does not indicate specifically which reaction of sucrose is being catalyzed.

Example 24.4

Phosphate esters play important biological roles. These esters are formed by passing a phosphate group from a higher-energy phosphate ester to another alcohol to yield a lower-energy phosphate ester. What type of enzyme catalyzes this reaction?

$$R'{-}O{-}\overset{\overset{\displaystyle O}{\|}}{\underset{\underset{\displaystyle O^-}{|}}{P}}{-}O^- + R{-}OH \longrightarrow R'{-}OH + R{-}O{-}\overset{\overset{\displaystyle O}{\|}}{\underset{\underset{\displaystyle O^-}{|}}{P}}{-}O^-$$

Solution. A phosphate group is transferred from one molecule to another. Transferases catalyze such reactions. The name of the type of enzyme required for this reaction is phosphotransferase.

[Additional examples may be found in **24.12–24.24** at the end of the chapter.]

24.3
Models of Enzyme Action

Recall from Chapter 10 that chemical reactions involve collisions between reacting molecules. Furthermore, catalysts must be involved in some contact with the reactants as well. In an enzyme-catalyzed reaction, the enzyme and substrate are pictured as binding to form a complex that subsequently reacts. The binding may involve an ionic bond, hydrogen bonding, or hydrophobic interactions. The specificity of the enzyme depends, then, on the ability of an enzyme to bind selectively to a substrate by a combination of unique interactions. For example, an ionic bond may be formed with an ammonium ion of the lysine residue in the enzyme and a negative site in the substrate. One or more types of interactions may be involved in the binding of the substrate to the enzyme. The binding energy for such complexes is usually in the 3–12 kcal/mole range.

The binding of an enzyme and a substrate is usually shown by the following equations.

$$\underset{\text{(enzyme)}}{E} + \underset{\text{(substrate)}}{S} \rightleftharpoons \underset{\substack{\text{(enzyme–substrate} \\ \text{complex)}}}{ES}$$

$$ES \longrightarrow E + \underset{\text{(product)}}{P}$$

The enzyme–substrate complex may or may not react with additional molecules to form a product. If the enzyme is a hydrolase, water is required. In the case of a dehydrogenase, an oxidizing agent is required.

According to the current model of the action of enzymes, the binding occurs at a cleft or crevice formed by the folded protein chain. In order for the substrate and the enzyme to bind, there must be an attraction between a portion of the enzyme and the substrate. However, it is also necessary that the two molecular surfaces have complementary shapes to achieve a fit (Figure 24.3).

Two models for obtaining enzyme–substrate complexes have been suggested: the lock-and-key theory and the induced-fit hypothesis.

Lock-and-Key Theory

The first model for enzyme action was put forward by the German chemist Emil Fischer in 1890. He suggested that enzymes reacted with substrates because of complementary molecular geometry. His theory has been developed based on the idea that molecules have rigid geometry. One analogy for the complementary

Figure 24.3

Lock and Key Model.
In order for the enzyme and substrate to combine and form a complex, it is postulated that the two substances must have a complementary shape.

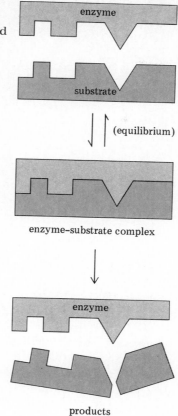

enzyme

substrate

(equilibrium)

enzyme–substrate complex

enzyme

products

shape of rigid materials is the relationship between a lock and a key. Only the proper key will fit a particular lock. This analogy is called the lock-and-key model for enzyme-catalyzed reactions. *The* **lock-and-key theory** *pictures an enzyme as conformationally rigid and having a shape suitable for binding a substrate.* In actuality, the fit that is required between an enzyme and a substrate is much more precise than any lock and key. A slightly worn key may still open a lock. However, small changes in the substrate and especially in the enzyme can totally eliminate chemical reactivity.

Induced-Fit Hypothesis

As the structures of some enzymes were established, it became clear that not all enzymes have the specific shapes suggested by the lock-and-key theory. Moreover, some enzymes, such as subtilisin found in certain bacteria, are peptidases with very low specificity. In order to account for the action of such enzymes, Daniel F. Koshland, Jr., an American biochemist, suggested an alternative model, the induced-fit hypothesis.

The **induced-fit hypothesis** *pictures an enzyme as conformationally flexible and subject to change as the substrate approaches and starts to bind.* The substrate induces a change in the shape of the enzyme so that the enzyme–substrate complex achieves the proper complementary shape. The specificity of the fit still requires complementarity between the enzyme and the substrate. Both carboxypeptidase A, a protein-digesting enzyme, and hexokinase, an enzyme that

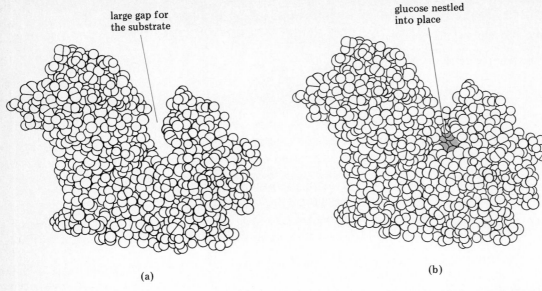

large gap for
the substrate

(a)

glucose nestled
into place

(b)

Figure 24.4
Induced-Fit Hypothesis for Enzyme Action.
The enzyme hexokinase (a) has a gap or cleft into which glucose can enter. In (b) notice how the gap has closed to achieve an appropriate complementary fit between substrate and the enzyme.

catalyzes the conversion of glucose into glucose 6-phosphate, operate by induced-fit pathways (Figure 24.4).

Asymmetry of Substrates

Enzymes can distinguish between enantiomers and bind only one of the two isomers. Furthermore, asymmetric compounds formed by enzymes are of a single configuration. A simple model explains these observations. Enzymes themselves are also optically active since they contain many asymmetric centers. Since

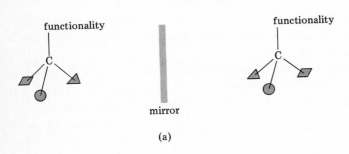

functionality

mirror

(a)

functionality

Figure 24.5
Specificity of Enzymes for an Enantiomer.
(a) Enantiomers are mirror images. (b) One enantiomer fits the surface, but the other does not.

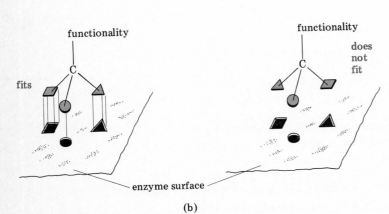

functionality

fits

functionality

does
not
fit

enzyme surface

(b)

the enzymes have a "handedness," they will fit and bind only with the proper "handed" molecule (Figure 24.5). The enantiomer will not fit. For example, the formation of L-tryptophan from L-serine and indole is catalyzed by tryptophan synthetase. The enantiomer, D-serine, does not bind to the enzyme and is not converted into tryptophan.

indole   L-serine   L-tryptophan

24.4
Active Sites

Let's now examine what is responsible for the high rate of reactions catalyzed by enzymes. In an enzyme-catalyzed reaction, the reactants are brought close together as they bind to the enzyme. Binding increases the effective local concentration, which results in an increased reaction rate. Even in a very dilute solution, the necessary reactants are brought together in the proper amounts at the enzyme.

Precisely defined alignment of reactants is another reason for the fast reaction of enzyme-catalyzed reactions. In solution a reaction occurs between reactants only if the molecules are properly aligned when they collide. Only then can bonds be ruptured and formed. In an enzyme-catalyzed reaction the substrate is already precisely positioned for reaction.

Although high effective concentration and specific reactant alignment are important in enzyme-catalyzed reactions, there is a third, even more important, feature of enzymes called the active site. *Functional groups within the enzyme form the* **active site,** which is actually responsible for the catalysis.

An active site is usually only a small portion of the enzyme molecule. One or two amino acids at precisely placed points in a chain of hundreds of amino acids may be the active site. These amino acids actually participate in the reaction and alter the reaction pathway so that there is a lower activation energy compared to the uncatalyzed pathway. The remainder of the protein provides the cleft for selecting and binding the substrate in place. In the simple model of Figure 24.6, the active site would be at the point where reaction of the substrate occurs.

The active site in lysozyme, an enzyme that cleaves polysaccharides in the cell walls of bacteria, has been well defined. The amino acid residues at positions 35, 52, 62, 63, and 101 in the sequence of 129 amino acids are involved in creating the active site. Although these amino acids are in different parts of the primary structure, they are close in space as a result of the tertiary structure of the enzyme.

The active sites in chymotrypsin and trypsin are identical. In both substances there are two histidine molecules that, although separated by 16 amino acids, are closely situated in the three-dimensional structure of the molecules. The related placements are due to disulfide bonds between cysteines 31 and 47 in trypsin and between cysteines 42 and 58 in chymotrypsin (Figure 24.7). Chymotrypsin and trypsin also contain serine molecules that are bound in similar sequences between two cysteine molecules. The tertiary structures of the

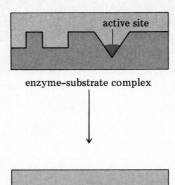

active site

enzyme–substrate complex

products

Figure 24.6
Lock and Key Model for the Active Site.
The active site is only a small portion of the enzyme
molecule at which the enzyme-catalyzed reaction
occurs.

molecules are such that both cyclic portions created by the disulfide bonds are
located near each other, and as a result the two histidines and one serine constitute the active site for hydrolysis reactions.

The difference in the total number of amino acids in the chains of chymotrypsin and trypsin does not alter the effectiveness of the active site, nor do the
small variations in amino acid content within the two cyclic portions containing
the active site.

24.5
Cofactors

Some enzymes are comprised only of protein. However, other enzymes require
one or more nonprotein components called cofactors to be biologically active. *A*
cofactor *may be a metal ion, or it may be an organic molecule called a* **co-**

histidine region

chymotrypsin

-His-Phe-Cys-Gly-Gly-Ser-Leu- Ile -Asn-Glu-Asn-Trp-Val-Val-Thr-Ala-Ala-His-Cys-
40   42   57 58

trypsin

-His-Phe-Cys-Gly-Gly-Ser-Leu- Ile -Asn-Ser-Gln-Trp-Val-Val-Ser-Ala-Ala-His-Cys-
29   31   46 47

serine region

chymotrypsin

-Cys-Met-Gly-Asp-Ser-Gly-Gly-Pro-Leu-Val-Cys-
191   195   201

trypsin

-Cys-Gln-Gly-Asp-Ser-Gly-Gly-Pro-Val-Val-Cys-
179   183   189

Figure 24.7
The Amino Acids in the Region of the
Active Site.

enzyme. *The enzyme–cofactor complex is called a* **holoenzyme;** *the inactive enzyme alone is called an* **apoenzyme.**

apoenzyme + metal ion $\longrightarrow$ holoenzyme

apoenzyme + coenzyme $\longrightarrow$ holoenzyme

Metal Ion Cofactors

Usually only one metal ion per molecule of enzyme is required to form the active holoenzyme, which in this case is sometimes called a metalloenzyme. Because of the high molecular weight of apoenzymes, the weight percent of the metal in the holoenzyme is in the range of tenths of a percent. Thus we can see why only small amounts of certain metal ions are required in our diets.

Aminopeptidases contain manganese or magnesium, whereas carboxypeptidases contain zinc. In the case of carboxypeptidases the removal of zinc results in the loss of the ability of the enzyme to hydrolyze amides. Replacement of cadmium for zinc alters the selectivity of carboxypeptidase so that it hydrolyzes esters but does not hydrolyze amides. Replacement of zinc by copper results in complete loss of biological activity. The zinc is known to be located on the surface of the enzyme. It is at the center of a tetrahedral site surrounded by two histidine residues, a glutamic acid residue, and a water molecule.

Coenzymes

Coenzymes are of lower molecular weight than apoenzymes. Although models have not been established for the action of all coenzymes, it appears that coenzymes join with the apoenzyme to form the special conformation required for the reaction that the enzyme catalyzes. The coenzyme itself may contain the active site (Figure 24.8), but may need also to be attached to the larger apoenzyme in order to establish the proper conformation of the final active enzyme.

Coenzymes often serve as shuttles for exchanges of groups of atoms among the many metabolic reactions. If an enzyme oxidizes a substrate, the coenzyme is reduced, and in another reaction the reduced coenzyme acts as a reducing agent with a different substrate. If the substrate releases a phosphate unit, the coenzyme accepts it and later transfers a phosphate unit to another substrate.

Many coenzymes incorporate vitamins in their structure. In fact, the principal biochemical role of the water-soluble vitamins appears to be their activity as coenzymes (Figure 24.9). The coenzymes are usually phosphate esters of the vitamins. The vitamins in these coenzymes can be obtained from a properly balanced diet. Further discussions of coenzymes will be given in Chapter 25.

Figure 24.8
Models of Combination of Apoenzyme and Coenzyme.

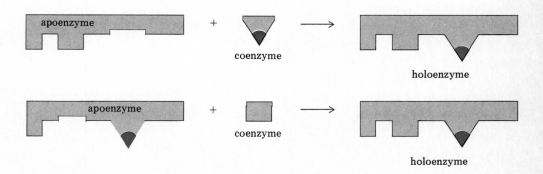

Figure 24.9
Structures of Some Coenzymes.

Coenzyme

Related vitamin

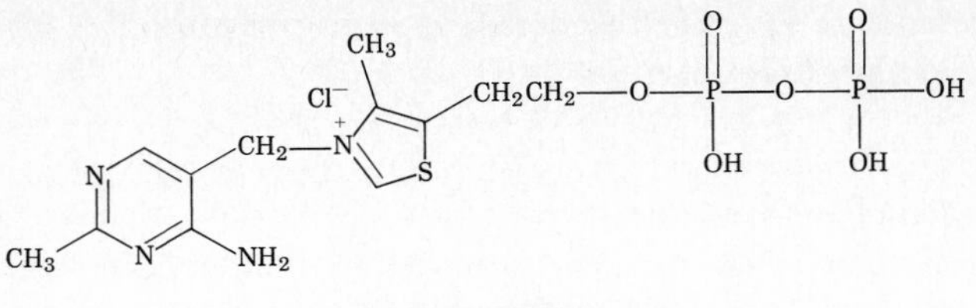

thiamine pyrophosphate — B₁

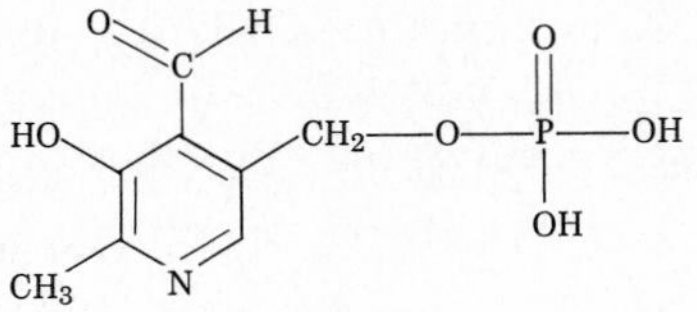

pyridoxal phosphate — B₆

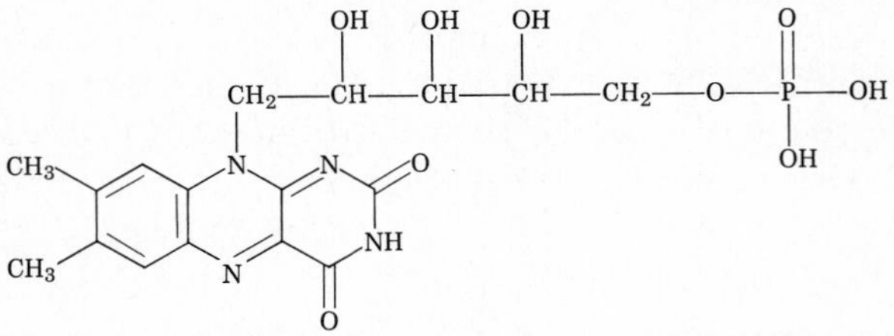

riboflavin phosphate — B₂

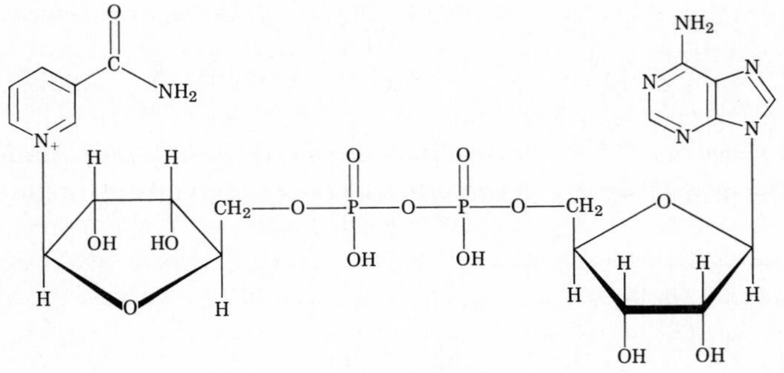

nicotinamide adenine dinucleotide — niacin

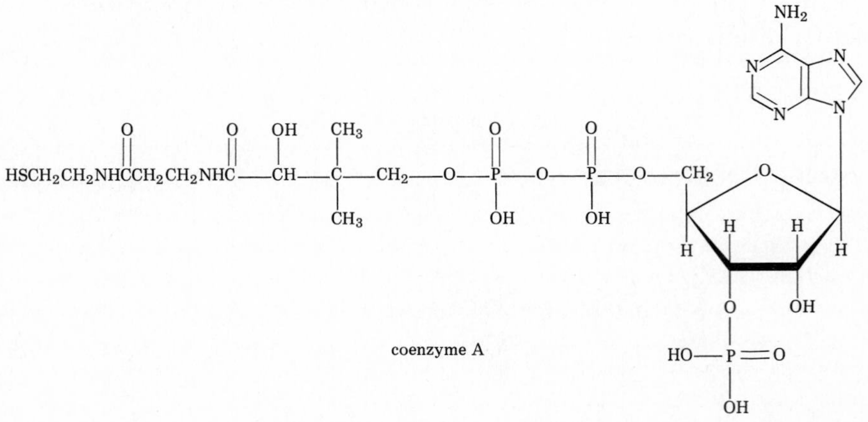

coenzyme A — B₃

566

The rate of an enzyme reaction depends on the relative amounts of substrate and enzyme as well as the temperature and pH.

For an ordinary chemical reaction, the rate of formation of product usually increases with increasing concentration of the reactants. Enzyme-catalyzed reactions behave differently. The rate increases with increasing substrate concentration but only up to a point. A maximum value is attained no matter how large the substrate concentration (Figure 24.10). The reason for this maximum rate or velocity (V_{max}) is that most of the enzyme has been converted to the reactive enzyme–substrate complex. The enzyme is then "saturated."

$$E + S \rightleftharpoons ES$$

The saturation of an enzyme is expected from Le Châtelier's principle. Enzymes are present in low concentrations. Thus increasing the substrate concentration shifts the reaction toward the product side of the enzyme–substrate equilibrium. Once all enzyme molecules are bound to substrate, the reaction depends only on how fast the enzyme–substrate complex may be converted into product. In the human body, enzymes do not work at maximum velocity because substrate concentrations are usually below the saturation level.

The V_{max} for an enzyme-catalyzed reaction is directly proportional to the concentration of the enzyme present. Larger amounts of an enzyme under

24.6
Kinetics of Enzyme Reactions

Figure 24.10
Comparison of an Enzyme-Catalyzed Reaction and an Ordinary Chemical Reaction.

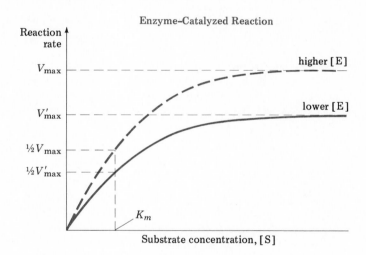

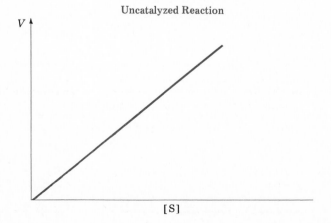

Table 24.2

Turnover Numbers of Some Enzymes

Enzyme	Turnover number (min^{-1})
carbonic anhydrase	36,000,000
ketosteroid isomerase	17,100,000
fumarase	1,200,000
β-amylase	1,100,000
β-galactosidase	12,500
phosphoglucomutase	1,240
succinate dehydrogenase	1,150

saturation conditions result in higher concentrations of the enzyme–substrate complex. Each enzyme has an intrinsic activity, which is the ratio of V_{max} to [E] and is known as the turnover number.

$$\text{turnover number} = \frac{V_{max}}{[E]}$$

The **turnover number** *is the number of substrate molecules converted into product per minute for each molecule of enzyme under saturation conditions.* The turnover number of enzymes can be very high (Table 24.2), as in the case of carbonic anhydrase. The different values reflect the biological requirements of the living organism.

Stability of Enzyme–Substrate Complexes

The equilibrium constant for formation of the enzyme–substrate complex has been determined for many enzymes. However, the quantity K_m is used to express the reverse reaction, that is, the dissociation of the complex.

$$\text{ES} \rightleftharpoons \text{E} + \text{S}$$

$$K_m = \frac{[E][S]}{[ES]}$$

A large value of K_m, or large dissociation of the enzyme–substrate complex, indicates weak binding. A small K_m means that strong binding occurs in the enzyme–substrate complex. A list of K_m values is given in Table 24.3.

The smaller the K_m, the less readily ES dissociates. As a result the conversion of ES into product can occur at low [S]. A large K_m means that ES tends to dissociate into E and S, and product formation can only occur when the [S] is large and forces the above equilibrium to the left.

Table 24.3

Stability of Enzyme–Substrate Complexes

Enzyme	Substrate	K_m (M)
chymotrypsin	acetyl-L-tryptophanamide	5×10^{-3}
lysozyme	hexa-N-acetylglucosamine	6×10^{-6}
β-galactosidase	lactose	4×10^{-3}
threonine deaminase	threonine	5×10^{-3}
carbonic anhydrase	CO_2	8×10^{-3}
pyruvate carboxylase	pyruvate	4×10^{-4}

Figure 24.11
Effect of Temperature on an Enzyme-Catalyzed Reaction.

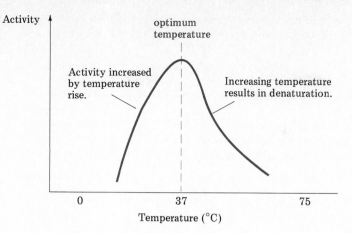

Effect of Temperature on Reaction Rate

Ordinary chemical reactions occur at faster rates as the temperature is increased. Recall that an increase of 10°C approximately doubles or triples the rate. Enzyme-catalyzed reactions also increase as the temperature is increased. *The temperature at which the rate is a maximum is the* **optimum temperature** (Figure 24.11). For humans, this optimum temperature is close to 37°C, the normal body temperature. Above the optimum temperature, the rate of reaction decreases sharply.

Up to the optimum temperature an increase in heat energy more rapidly converts the enzyme–substrate complex into product. However, enzymes, being proteins, are denatured by excessive heat. Denaturation inactivates the enzyme by altering its structure so that it does not form the enzyme–substrate complex necessary for reaction.

Effect of pH on Reaction Rate

Most enzymes have a maximum activity at a specific or **optimum pH.** Trypsin is most active at pH = 8, whereas pepsin is most active at pH = 2 (Figure 24.12). There are two reasons for the observed effect of pH on rate. First, the tertiary

Figure 24.12
Effect of pH on Enzyme-Catalyzed Reactions.

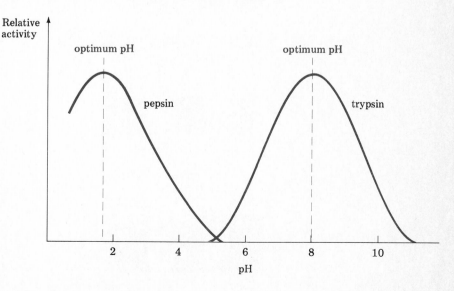

structure depends in part on ionic bonds, which are affected by pH. Changes in tertiary structure alter the binding efficiency of the enzyme. Second, the acidic and basic amino acid side chains that may be part of the active site can lose or gain protons. A change in the active site alters the reactivity of the enzyme.

Example 24.5 ——————————————————————————————

The catalytic activity of lysozyme is maximum at pH = 5.0. The active site involves the side chains of aspartic acid and glutamic acid whose pK_a values are 3.9 and 4.3, respectively. In what form does each of these side chains exist at maximum catalytic activity?

Solution. At the pH equal to the pK_a of the side chain, there is an equal concentration of the carboxylic acid and the carboxylate ion. Since the pH of maximum activity is higher than 3.9 and 4.3, the solution is less acidic (more basic). As a consequence, a higher fraction of each side chain will exist as the carboxylate ion. Since the side chain of aspartic acid is more acidic than that of glutamic acid, there will be a larger amount of the conjugate base of the aspartic acid side chain than of the glutamic acid side chain.

————————————————————————————————————

[Additional examples are found in **24.33**, **24.35**, and **24.43–24.53** at the end of the chapter.]

24.7
Enzyme Inhibition

Enzyme inhibitors *are compounds that destroy or deactivate enzymes*. There are two types of inhibitors, competitive and noncompetitive enzyme inhibitors. In addition, some products of biological reactions inhibit the enzymes that are required for their formation. This process is called feedback inhibition (Section 24.8).

Competitive Inhibition

In **competitive inhibition** *the enzyme's active site is occupied by an inhibitor that has a shape similar to the substrate* (Figure 24.13). The strength of the binding of the inhibitor is characterized by the terms irreversible and reversible. In irre-

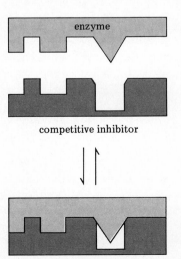

Figure 24.13
Model for Competitive Inhibition of Enzyme Activity.

versible competitive inhibition, the inhibitor forms a strong covalent bond to the enzyme at the active site and enzymatic activity is eliminated. Many toxins, as well as nerve gas poisons, are irreversible competitive inhibitors.

Reversible competitive inhibitors form a complex in equilibrium with the enzyme. Malonic acid is an example of a reversible competitive inhibitor. This compound structurally resembles succinic acid, the substrate for succinate dehydrogenase. If malonic acid is present, it binds with the enzyme to form an enzyme–substrate complex that cannot lead to product. The succinic acid that is in competition for active sites with the malonic acid cannot be as rapidly converted to product.

succinic acid

malonic acid

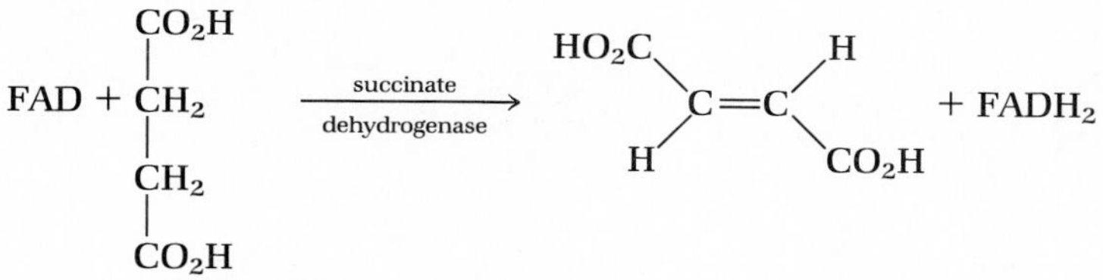

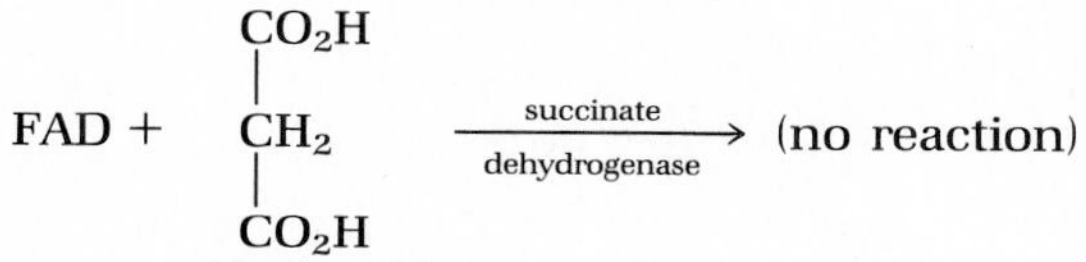

It is quite common for products of an enzyme-catalyzed reaction to be reversible competitive inhibitors. This is especially true for products that closely resemble the reactant. For example, 2,3-diphosphoglycerate formed from 1,3-diphosphoglycerate is a reversible competitive inhibitor for diphosphoglycerate mutase. Thus, as the product is formed, the enzyme activity is decreased and the enzymatic activity is regulated. Additional details of enzyme regulation will be given in Section 24.8.

1,3-diphosphoglycerate 2,3-diphosphoglycerate

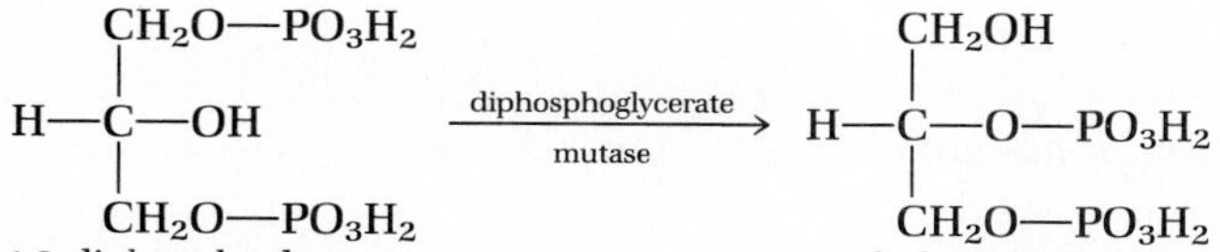

Competitive enzyme inhibitors are useful in treating bacterial infection. For example, some bacteria may require *p*-aminobenzoic acid for the synthesis of the vitamin folic acid. The sulfa drug sulfanilamide has a similar structure.

p-aminobenzoic acid

sulfanilamide

folic acid

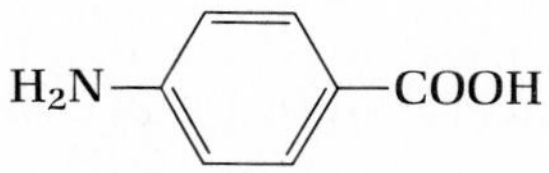

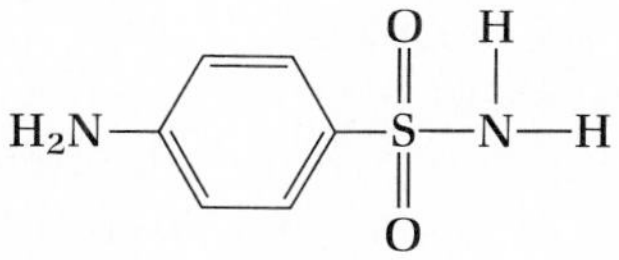

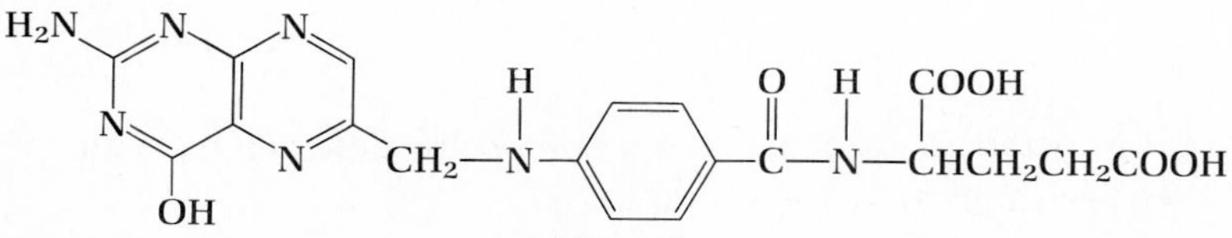

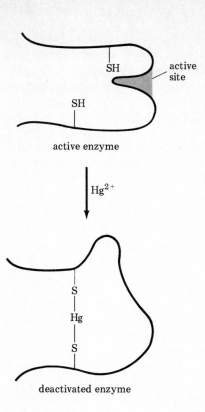

Figure 24.14
Model for Noncompetitive Inhibition of an Enzyme by Hg^{2+}.

A bacterial enzyme mistakes sulfanilamide for *p*-aminobenzoic acid and incorporates that molecule instead. The altered folic acid that results cannot function as a coenzyme, and the bacteria then die. Sulfa drugs do not interfere with the metabolism of humans because humans do not synthesize their own folic acid but rather obtain it from foodstuffs.

Noncompetitive Inhibition

Noncompetitive inhibition *occurs when an inhibitor binds with some site other than the active site of the enzyme.* The enzyme is then changed so that the catalytic properties of the active site are altered and the turnover number is decreased. Noncompetitive inhibitors may bind at some point far removed from the active site, but the resulting conformational change is drastic enough to alter the catalytic activity at the binding site. These changes are called allosteric effects. Recall that allosteric effects are important in the binding of oxygen to hemoglobin, an example of an enhancement of biological activity. In many cases noncompetitive inhibition is reversible.

Heavy metal salts are irreversible noncompetitive inhibitors of many enzymes. Metal ions such as mercury combine with sulfhydryl groups. Links within the protein chain can result (Figure 24.14) that alter the shape of the binding site. The toxic properties of mercury are a concern when mercury from industrial wastes gets into rivers.

Detection of Inhibition

Competitive and noncompetitive inhibition can be detected by kinetic methods. The effect of inhibitors on V_{max} is different.

Figure 24.15
Relation Between Rate of
Reaction and Substrate
Concentration in Inhib-
ited Reactions.

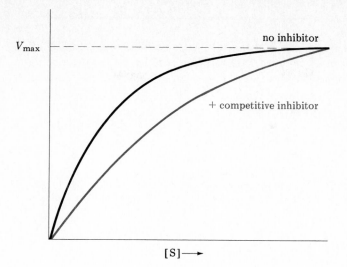

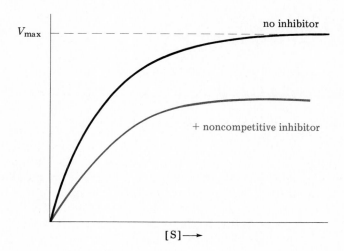

Competitive inhibition means that both the inhibitor and the substrate compete for binding to the enzyme. When the inhibitor occupies binding sites, some
of the substrate molecules cannot react. However, for a given enzyme concentration, the substrate can be made more competitive for binding sites by increasing
its concentration. As the substrate concentration is increased, the enzyme sites
will be increasingly occupied by substrate. The V_{max} then will be the same as in
the absence of the competitive inhibitor (Figure 24.15).

The decrease in rate caused by a noncompetitive inhibitor cannot be overcome by increasing the concentration of substrate, since the binding site is not
occupied by the inhibitor. Rather, the enzyme molecules combined with the
noncompetitive inhibitor have lower turnover numbers or are no longer active.
The effect then is a decreased fraction of enzyme molecules capable of catalyzing
the normal reaction. As a result the reaction velocity is decreased in the same way
as it would be if the concentration of the enzyme were decreased, and the V_{max} is
decreased (Figure 24.15).

Example 24.6 ───────────────────────────────

Elastase, a protease, has valine and threonine at its binding site. The following compound is an inhibitor of elastase. What amino acid of the binding site is likely to be involved in forming a covalent bond with the aldehydic carbon?

$$CH_3\overset{\overset{\displaystyle O}{\|}}{C}\text{—Pro—Ala—Pro—}\overset{}{\underset{\underset{\displaystyle H}{|}}{N}}\text{—}\overset{\overset{\displaystyle CH_3}{|}}{\underset{\underset{\displaystyle H}{|}}{C}}\text{—}\overset{\overset{\displaystyle O}{\|}}{C}\text{—H}$$

Solution. Valine has no functional groups, whereas serine has a hydroxyl group. An alcohol and an aldehyde react to form a hemiacetal. The covalent bond is depicted as

$$CH_3\overset{\overset{\displaystyle O}{\|}}{—C}\text{—Pro-Ala-Pro—}\overset{}{\underset{\underset{\displaystyle H}{|}}{N}}\text{—}\overset{\overset{\displaystyle CH_3}{|}}{\underset{\underset{\displaystyle H}{|}}{C}}\text{—}\overset{\overset{\displaystyle OH}{|}}{\underset{\underset{\displaystyle H}{|}}{C}}\text{—O—}CH_2\text{—}\overset{\overset{\displaystyle N—H}{\overset{|}{\overset{\displaystyle Protein}{}}}}{\underset{\underset{\displaystyle C=O}{|}}{CH}}$$
Protein

[Additional examples may be found in **24.54–24.64** at the end of the chapter.]

──

24.8
Regulatory Enzymes

The activity of enzymes within cells must be initiated or shut off according to the demands of life processes. When a compound is absent or in short supply, an enzyme must catalyze the formation of the vital compound. Once the concentration of the necessary compound is sufficient, its production must be terminated and the enzyme activity must be switched off.

Concentration or availability of cofactors as well as temperature and pH affect enzyme reactivity, but more complex control mechanisms are required for an organism to make the most effective use of its enzymes. Two of the most important control mechanisms are the operation of regulatory enzymes in feedback inhibition and the formation of enzymes from zymogens. The chemistry of zymogens is the subject of the next section.

Feedback Inhibition

One of the ways of controlling the formation of compounds is with a regulatory enzyme. *A* **regulatory enzyme** *catalyzes the first reaction of a sequence of reactions and is inhibited by the end product of the sequence. A* **negative modulator** *is a product at the end of a series of reactions that is an inhibitor for some earlier step in the same reaction series.* The negative modulator "feeds back" to inhibit some preceding step. Consider the sequence of reactions for A to B to C and eventually to D. Each step may be catalyzed by enzymes labeled E(a), E(b), and E(c).

$$A \xrightarrow{\text{E(a)}} B \xrightarrow{\text{E(b)}} C \xrightarrow{\text{E(c)}} D$$

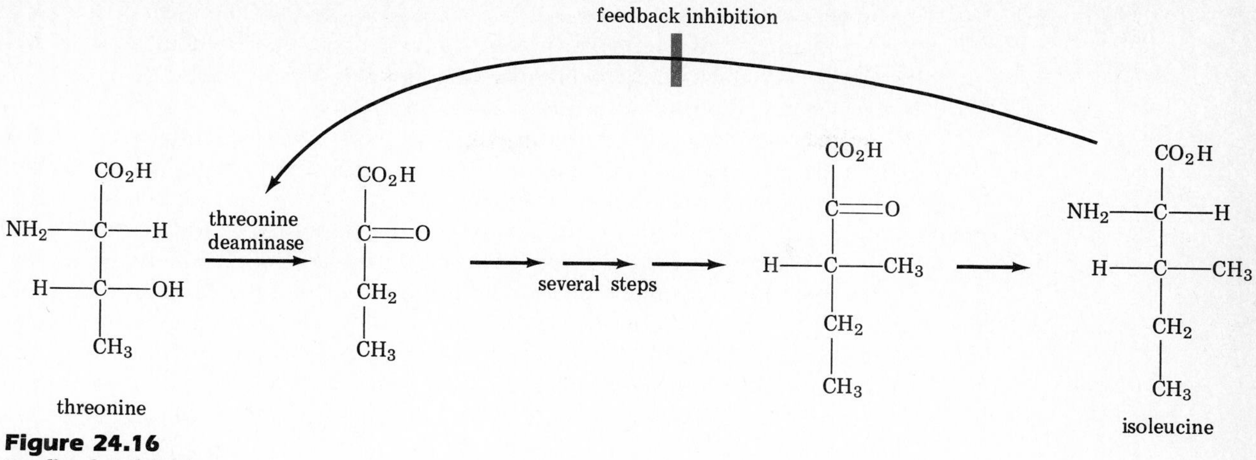

feedback inhibition

Figure 24.16
Feedback Inhibition.
In the biosynthesis of isoleucine from threonine, the enzyme threonine deaminase is inhibited by isoleucine.

In this sequence of steps the final product, D, cannot be formed unless the product C and, in turn, the product B are first formed, and these products cannot be formed without the appropriate enzymes.

If D is a competitive inhibitor of enzyme E(a), the rate of formation of B is decreased and as a consequence the rate of formation of D is decreased. The whole sequence of reactions may be shut off as the concentration of D reaches a certain level. *The process of controlling a reaction sequence by a negative modulator and a regulatory enzyme is* **feedback inhibition.**

$$E(a) + D \rightleftharpoons E(a)-D$$

As long as D is present in sufficient quantity, the concentration of E(a) is diminished by complex formation. As D is used up by the organism, the equilibrium shifts and E(a) is released and the entire sequence of conversion of A to D commences. In this example D is the negative modulator and E(a) is called the regulatory enzyme.

The process by which a product regulates or controls an enzyme in a sequence of reactions is understood for a few examples. The conversion of threonine into isoleucine involves several steps, the first of which is catalyzed by threonine deaminase (Figure 24.16). Isoleucine occupies a site on threonine deaminase different from the one that binds to threonine. This second site is called the allosteric site because, when isoleucine binds, a change in the conformation of the enzyme occurs. The binding characteristics are altered so that threonine now binds poorly. *Regulation of enzyme activity by noncompetitive inhibition involving allosteric changes is called* **allosteric regulation.**

Allosteric enzymes are usually composed of two or more protein subunits. Frequently one subunit binds the substrate, whereas the other subunit binds the negative modulator. Binding of the negative modulator in one subunit changes the conformation of the other subunit so its catalytic activity decreases. The action of negative modulators in the feedback inhibition of allosteric enzymes is important to the economical operation of a cell. In this way organisms do not expend energy and compounds in a complex series of reactions to make compounds already present in sufficient quantities.

24.9
Zymogens

Some enzymes are synthesized and stored in an inactive form. They are transformed into active enzymes when the organism needs them as catalysts. *The inactive form of an enzyme is called a* **zymogen** *or proenzyme.* Zymogens provide a source of potential enzymatic activity.

The peptidases trypsin and chymotrypsin are both produced and stored in the pancreas as the zymogens trypsinogen and chymotrypsinogen, respectively. This is a necessary protection for the pancreas, which produces them, for if the peptidases were produced in active form the pancreas would be destroyed by its own protein-digesting enzymes. There are two additional safety features for the pancreas. The zymogens produced in the pancreas travel from the endoplasmic reticulum to the Golgi apparatus where they are surrounded by membranes made of protein and lipid. These zymogen granules are secreted into a duct to the duodenum when stimulated by a hormone. The pancreas also contains an inhibitor for the enzyme required for the conversion of the zymogen into an active enzyme.

The zymogen chymotrypsinogen, which contains 245 amino acids and has five disulfide bonds (Figure 24.17), reacts in the small intestine by cleaving the polypeptide between arginine at position 15 and isoleucine at position 16. The resulting π-chymotrypsin is an enzyme, but further reactions occur to form α-chymotrypsin. A dipeptide from positions 14 and 15 and a dipeptide from positions 147 and 148 are released. These additional cleavages are superfluous because the initial cleavage between amino acids 15 and 16 gives an active enzyme. The striking aspect for the activation of the zymogen chymotrypsinogen is that the cleavage of a single peptide bond out of the 244 peptide bonds converts an inactive molecule into an enzyme. However, the resulting enzyme undergoes a marked conformational change that results in the enzymatic activity.

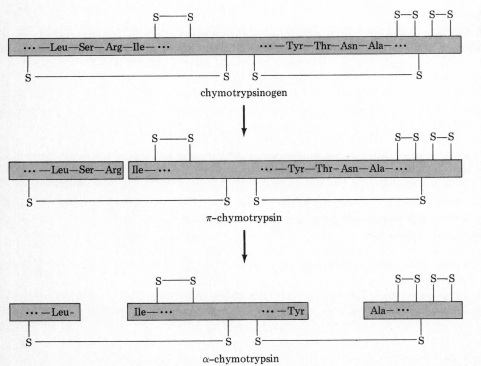

Figure 24.17
Activation of Chymotrypsinogen.

Summary

Enzymes are proteins that are biological catalysts. They increase the rates of reactions enormously and are highly specific in the reactions they catalyze. Reactant molecules in enzyme-catalyzed reactions are substrates. Names of enzymes are based on the names of the substrate and the type of reaction catalyzed. The classes of enzymes are oxidoreductases, transferases, hydrolases, lyases, isomerases, and ligases.

Enzymes change reaction rates by providing a reaction pathway with a lower activation energy. However, the concentrations at equilibrium are not affected by an enzyme. The reaction pathway with an enzyme involves formation of a complex in which the substrate and enzyme bind because of complementary shapes. Two models account for the formation of the complex: the lock-and-key theory and the induced-fit hypothesis. The lock-and-key theory regards enzymes and substrates as rigid molecules, whereas the induced-fit hypothesis recognizes the flexibility of an enzyme and its ability to assume a shape induced by the substrate. Enzymatic activity occurs at a small portion of the enzyme called the active site, which contains a unique arrangement of amino acid side chains as a consequence of the tertiary structure.

Many enzymes require a cofactor for catalytic activity. The complete catalyst, or holoenzyme, consists of the protein part, or apoenzyme, and the cofactor. Some cofactors are metal ions. Organic cofactors are coenzymes. Most water-soluble vitamins or their derivatives are coenzymes.

The kinetics of enzyme reactions are characterized by a turnover number and an equilibrium constant. The turnover number is the number of substrate molecules converted into product per minute for each molecule of enzyme under saturation conditions. A K_m value indicates the stability of the enzyme–substrate complex.

Enzymatic reactions have a maximum velocity at a characteristic optimum temperature and pH. The optimum temperature for most enzymes in the human body is 37°C. Higher temperatures decrease the rate of reaction as a result of alteration of the enzyme structure and eventual denaturation. Changes in pH alter the tertiary structure of the enzyme and affect its binding efficiency.

The activity of enzymes is inhibited by molecules called inhibitors. Competitive inhibitors compete with the substrate for binding at the active site and are usually structurally similar to the natural substrate. Non-competitive inhibitors bind with the enzyme at some site other than the active site. Generally, noncompetitive inhibitors have no structural relationship to the substrate.

Regulation of enzyme activity is required for economic use of the energy and materials of the cell. Allosteric enzymes provide feedback inhibition by means of a negative modulator. The negative modulator produced at the end of a series of steps is a reversible inhibitor for an earlier reaction. A zymogen is an inactive form of an enzyme that is activated when required by a cell.

Learning Objectives

As a result of studying Chapter 24 you should be able to:

- Explain why enzymes are important in living cells.
- List six classes of enzymes and give an equation for a typical reaction of each class.
- Explain the contribution of binding of an enzyme to a substrate on enzyme activity.
- Describe the types of coenzymes and their functions.
- Outline the structural features of active sites.
- Interpret the effect of pH and temperature on enzyme activity.
- Compare competitive and noncompetitive inhibition of enzyme reactions and give one example of each inhibitor.
- Outline the steps that occur in the regulation of enzyme activity.
- Explain the allosteric effect in the regulation of enzyme activity.
- Describe how zymogens differ from enzymes.

New Terms

Absolute specificity in an enzyme indicates catalysis of the reaction of a single substrate.

An **active site** is a region in an enzyme that has a unique arrangement of amino acid side chains required for catalytic activity.

Allosteric regulation involves noncompetitive inhibition causing conformation changes.

An **apoenzyme** is an inactive enzyme without the cofactor required to form the holoenzyme.

A **coenzyme** is a cofactor that is an organic molecule.

A **cofactor** is a metal ion or coenzyme required to combine with an apoenzyme.

Competitive inhibition involves an equilibrium of an inhibitor having the same shape as the substrate with the active site of the enzyme.

Complementary shapes refer to two structures that fit one in the other to form a unit.

An **enzyme** is a catalyst that is predominately protein.

An **enzyme–substrate complex** is an intermediate required for enzyme-catalyzed reactions.

Feedback inhibition is a method of control of enzyme level in which a product of a series of reactions is an inhibitor of an enzyme required for an earlier step.

A **holoenzyme** is an active enzyme consisting of an apoenzyme and a cofactor.

The **induced-fit hypothesis** pictures an enzyme as conformationally flexible.

Inhibitors are compounds that destroy or deactivate enzymes.

The **lock-and-key theory** pictures an enzyme as conformationally rigid.

A **negative modulator** is a product at the end of a series of reactions that is an inhibitor for an earlier step.

Noncompetitive inhibition involves a combination of an inhibitor with an enzyme at a point other than the active site.

The **optimum pH** is the pH at which an enzyme-catalyzed reaction occurs at the maximum rate.

The **optimum temperature** for an enzyme is the temperature at which the reaction rate is a maximum.

A **regulatory enzyme** catalyzes the first reaction of a sequence of reactions and is inhibited by the end product of the sequence.

Specificity indicates the selectivity of enzymes for the individual reactions that they catalyze.

A **substrate** is a reactant in an enzyme-catalyzed reaction.

The **turnover number** is the number of substrate molecules converted into product per minute per enzyme molecule under saturated conditions.

A **zymogen** is an inactive storage form of an enzyme.

Questions and Problems

Terminology
24.1 What is the relationship between an apoenzyme and a coenzyme?

24.2 How do competitive and noncompetitive inhibition differ?

24.3 How does a zymogen differ from an enzyme in structure and in activity?

24.4 What is a regulatory enzyme?

24.5 What is a negative modulator?

Enzymes
24.6 How do enzymes differ from ordinary catalysts? In what ways are they similar?

24.7 What is meant by the term specificity?

24.8 How is enzyme specificity explained?

24.9 What is absolute specificity?

24.10 Why won't D-aspartic acid form fumaric acid in a reaction catalyzed by aspartase?

24.11 Lactate dehydrogenase will oxidize D-lactic acid to pyruvic acid. Explain what will result when lactate dehydrogenase is placed in solution with racemic lactic acid.

Classification of Enzymes
24.12 What is the ending of the name of enzymes?

24.13 Suggest a name for an enzyme that hydrolyzes lactose.

24.14 What type of enzyme would catalyze each of the following reactions?
(a) sucrose $\longrightarrow$ glucose + fructose
(b) glucose $\longrightarrow$ fructose
(c) aspartic acid + NH_3 $\longrightarrow$ asparagine
(d) pyruvic acid $\longrightarrow$ lactic acid
(e) alanylserine $\longrightarrow$ alanine + serine

24.15 What is the difference between lactose and lactase?

24.16 What substrates should react, catalyzed by each of the following enzymes?
(a) esterase   (b) phosphatase   (c) peptidase

24.17 What type of enzyme is required for the following reaction?

$$\begin{array}{ccc} CO_2^- & & CO_2^- \\ | & & | \\ H-C-OH & \longrightarrow & H-C-OPO_3^{2-} \\ | & & | \\ CH_2OPO_3^{2-} & & CH_2OH \end{array}$$

24.18 What type of enzyme is required for the following reaction?

$$\begin{array}{c} CO_2H \\ | \\ C{=}O \longrightarrow CH_3CHO + CO_2 \\ | \\ CH_3 \end{array}$$

24.19 What type of enzyme is required for the following reaction, which occurs in the citric acid cycle?

$$\begin{array}{ccc} CH_2-CO_2H & & CH_2-CO_2H \\ | & & | \\ HO-C-CO_2H & \longrightarrow & C-CO_2H \quad + H_2O \\ | & & || \\ CH_2-CO_2H & & CH-CO_2H \end{array}$$

24.20 What type of enzyme is required for the following reaction of a fat?

$$\begin{array}{c} \quad\quad O \\ \quad\quad || \\ CH_2-O-C-R \\ \\ \quad\quad O \quad\quad\quad\quad CH_2OH \\ \quad\quad || \quad\quad\quad\quad | \\ CH-O-C-R \xrightarrow{H_2O} CHOH \ + 3\ RCO_2H \\ \quad\quad\quad\quad\quad\quad\quad\quad | \\ \quad\quad\quad\quad\quad\quad\quad\quad CH_2OH \\ \\ \quad\quad O \\ \quad\quad || \\ CH_2-O-C-R \end{array}$$

24.21 What type of enzyme is required to convert a protein into peptides?

24.22 What type of enzyme is required for the following reaction, which is part of the pentose phosphate pathway?

$$
\begin{array}{ccc}
\text{CH}_2\text{OH} & & \text{CHO} \\
| & & | \\
\text{C}=\text{O} & & \text{H}-\text{C}-\text{OH} \\
| & & | \\
\text{H}-\text{C}-\text{OH} & \longrightarrow & \text{H}-\text{C}-\text{OH} \\
| & & | \\
\text{H}-\text{C}-\text{OH} & & \text{H}-\text{C}-\text{OH} \\
| & & | \\
\text{CH}_2\text{OPO}_3{}^{2-} & & \text{CH}_2\text{OPO}_3{}^{2-}
\end{array}
$$

24.23 What type of enzyme is required for the following reaction, which occurs in the citric acid cycle?

$$
\begin{array}{ccc}
\text{CO}_2{}^- & & \text{CO}_2{}^- \\
| & & | \\
\text{H}-\text{C}-\text{OH} + \text{NAD}^+ & \longrightarrow & \text{C}=\text{O} + \text{NADH} + \text{H}^+ \\
| & & | \\
\text{CH}_2 & & \text{CH}_2 \\
| & & | \\
\text{CO}_2{}^- & & \text{CO}_2{}^-
\end{array}
$$

24.24 What type of enzyme is required for the following reaction, which is part of the urea cycle that eliminates excess nitrogen from the body?

$$
\text{CO}_2 + \text{NH}_3 + 2\,\text{ATP} + \text{H}_2\text{O} \longrightarrow
$$

$$
\begin{array}{cc}
\quad\ \text{O} & \ \ \text{O} \\
\quad\ \| & \ \ \| \\
\text{NH}_2-\text{C}-\text{O}-\text{P}-\text{O}^- + 2\,\text{ADP} + \text{HPO}_4{}^{2-} + \text{H}^+ \\
\quad\quad\quad\quad | \\
\quad\quad\quad\quad \text{O}^-
\end{array}
$$

Theory of Enzyme Action

24.25 What two models are used to describe the specificity of enzymes?

24.26 What is meant by complementary shapes?

24.27 What is an enzyme–substrate complex?

24.28 How does the lock-and-key theory explain enzyme specificity?

24.29 Why are enzymes specific only for one enantiomer?

24.30 What type of side chains in a protein might bind a substrate that is a carboxylate ion?

24.31 What type of side chains in a protein might bind a substrate that is a protonated amine?

24.32 How does the induced-fit hypothesis explain the action of enzymes with broad specificities?

Active Site

24.33 Explain how only a small portion of a protein can serve as an active site.

24.34 Histidine is part of the active site in many enzymes that catalyze reactions involving protonation and deprotonation. Explain the suitability of histidine using one of its pK_a values.

24.35 How can two amino acids separated by many other amino acids in a chain still combine in an active site?

Cofactors and Coenzymes

24.36 Name two oxidoreductases that contain vitamins as coenzymes.

24.37 Why are higher concentrations of coenzymes required as compared to the enzyme concentration?

24.38 Why are only low concentrations of metals required nutritionally even though the ions are cofactors?

24.39 What does each of the following terms mean?
(a) apoenzyme   (b) holoenzyme
(c) metalloenzyme

Enzyme Activity

24.40 How does each of the following affect the rate of formation of product in an enzyme-catalyzed reaction?
(a) increased substrate concentration
(b) increased enzyme concentration
(c) increase in temperature
(d) increase in the pH

24.41 What is meant by each of the following terms?
(a) optimum temperature   (b) optimum pH

24.42 How is V_{max} determined?

24.43 The enzyme acetylcholinesterase contains serine, histidine, and aspartic acid at the active site. The enzyme is inactive in acidic solution, but becomes active as the pH is increased. Explain this result.

24.44 The enzyme pepsin catalyzes the hydrolysis of dietary protein in the stomach but not in the intestine where the pH of intestinal juices is slightly basic. Explain the difference in the activity of pepsin.

24.45 What is meant by saturation in enzyme reactions?

24.46 What is the significance of the turnover number of enzymes?

24.47 What do the K_m values of enzymes indicate about the rate of enzyme-catalyzed reactions?

24.48 Some leukemia requires asparagine, and the cancer may be suppressed by intravenous administration of asparaginase, which catalyzes the hydrolysis of asparagine to aspartic acid. The K_ms of asparaginases from various sources differ. What asparaginase is the most effective in fighting leukemia, one with a low or high K_m?

24.49 Sketch the shape of a curve showing the rate of an enzyme-catalyzed reaction versus concentration of the substrate at constant enzyme concentration.

24.50 Most enzymes have a maximum reaction velocity in a narrow pH range. Why does the velocity decrease at both higher and lower pH?

24.51 Why do most enzymes lose their catalytic activity when heated to 50–60°C?

24.52 Sketch the shape of a curve showing the rate of an enzyme-catalyzed reaction as a function of temperature.

24.53 A receptor protein for progesterone is active at pH 7.0. The side chains of the protein form hydrogen bonds with the two carbonyl groups of progesterone. At this pH what amino acids might be involved?

Enzyme Inhibition

24.54 Malonic acid, $HO_2CCH_2CO_2H$, is a competitive inhibitor for the enzyme succinate dehydrogenase. Explain how malonic acid acts as an inhibitor.

24.55 Explain how a competitive inhibitor can be experimentally distinguished from a noncompetitive inhibitor.

24.56 Why is Pb^{2+} a poison even at low concentrations in the body?

24.57 Arsenate, AsO_4^{3-}, is toxic. Explain how arsenate might act as a competitive inhibitor.

24.58 How does a sulfa drug act as an antiobiotic?

24.59 Methanol is poisonous and will cause blindess due to its oxidation to formaldehyde in a reaction catalyzed by a dehydrogenase. Why does immediately drinking ethanol after methanol poisoning allow the methanol to be harmlessly excreted from the system?

24.60 Explain why lead in paint is poisonous.

24.61 Iodoacetamide reacts with an enzyme sulfhydryl group by the following reaction. Is this an example of reversible or irreversible inhibition?

$$E-S-H + I-CH_2-\overset{\overset{\displaystyle O}{\|}}{C}-NH_2 \longrightarrow$$

$$E-S-CH_2-\overset{\overset{\displaystyle O}{\|}}{C}-NH_2 + HI$$

24.62 Diisopropylphosphofluoridate is a nerve gas poison that reacts with the hydroxyl group of serine, which is part of the active site of acetylcholinesterase.

$$(CH_3)_2CH-O-\overset{\overset{\displaystyle F}{|}}{\underset{\underset{\displaystyle O}{\|}}{P}}-O-CH(CH_3)_2$$

diisopropylphosphofluoridate

A mole of HF is produced in the reaction. Represent the structure of the product.

24.63 Consider the following rates of an enzyme-catalyzed reaction in the presence and absence of an inhibitor. Is the inhibitor competitive or noncompetitive?

	Rate (μmole/min)	
[S] (μmole/L)	No inhibitor	Inhibitor, 2×10^{-3} M
3	10.0	4.0
5	14.5	6.5
10	22.5	11.5
30	34.0	22.5
90	40.5	34.0

24.64 Consider the following rates of an enzyme-catalyzed reaction in the presence and absence of an inhibitor. Is the inhibitor competitive or noncompetitive?

	Rate (μmole/min)	
[S] (μmole/L)	No inhibitor	Inhibitor, 1×10^{-4} M
3	10.5	2.0
5	14.5	3.0
10	22.5	4.5
30	34.0	7.0
90	40.5	8.0

Regulatory Enzymes

24.65 Explain the action of a modulator?

24.66 Explain how allosteric enzymes function.

24.67 Explain why biological systems require regulatory enzymes.

Zymogens

24.68 Trypsinogen is formed in the pancreas. Explain why the pancreas contains a trypsin inhibitor.

24.69 How does a peptidase convert a zymogen into an active enzyme?

24.70 Explain how a small change in the primary structure of a zymogen results in the formation of an active enzyme.

24.71 Acute pancreatitis is a severe and usually lethal disease in which the zymogens produced by the pancreas are activated while still inside the pancreas. Explain why this condition may be lethal.

24.72 Clotting of blood occurs by a series of zymogen activations. Explain why the enzymes for blood clotting are produced as zymogens.

Biochemical Energy

A runner training for a marathon depends on biochemical energy derived from the metabolism of carbohydrates and lipids.

25.1
Chemical Energy

In Chapter 2 you learned that according to the law of conservation of energy, energy may be converted from one form to another, but may be neither created nor destroyed. In Chapter 10 you learned that chemical reactions occur in the direction in which energy is released. The more energy released, the more complete is the reaction in that direction and the larger is the equilibrium constant. These important concepts apply to biochemical reactions as well as to ordinary chemical reactions. Cells extract energy from food and also convert foodstuffs into cell components by a highly integrated series of chemical reactions called metabolism. The energetics of metabolic reactions can be understood in terms of the concept of free energy.

Free Energy

In Section 10.6 you learned that a free energy change, symbolized by $\Delta G°$, expresses the energy difference between compounds and hence the energy change in a chemical reaction. Free energy is contained in the chemical bonds of both reactants and products. In a chemical reaction, bonds are broken and formed and there is a change in the free energy. If the free energy of the product is less than the free energy of the reactant, some free energy is released and the reaction is exergonic. If the free energy of the product is greater than the free energy of the reactant, free energy must be absorbed and the reaction is endergonic. These two examples are shown in Figure 25.1.

An understanding of free energy in chemistry is important since $\Delta G°$ indicates the direction of a chemical reaction. An exergonic reaction has a negative $\Delta G°$ and is spontaneous. The term spontaneous indicates that there is a natural tendency or potential for a reaction to occur.

Consider the analogy of water at the top of a hill. We know that the water will tend to go downhill spontaneously. The potential energy is released as the water moves toward the bottom of the hill. The energy released does not depend on the actual path that the water takes but only on the vertical difference between the top and bottom of the hill.

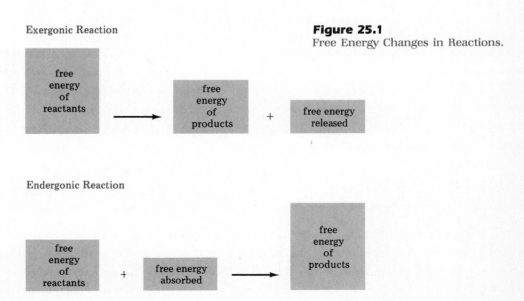

Exergonic Reaction

Endergonic Reaction

Figure 25.1
Free Energy Changes in Reactions.

Additivity of Free Energy

For any particular reaction the $\Delta G°$ is always the same, whether the end product is obtained in one or in several steps. Thus $\Delta G°$ for a chemical reaction can be calculated from the $\Delta G°$ of several other reactions. Consider the combustion (oxidation) of methane in a single step.

$$CH_4 + 2\,O_2 \longrightarrow CO_2 + 2\,H_2O \qquad \Delta G°_{combustion} = -201.4\,kcal/mole$$

The same overall reaction can be obtained by adding the following three reactions.

$$
\begin{array}{ll}
CH_4 \longrightarrow C + 2\,H_2 & \Delta G°_1 = \ \ +12.2\,kcal \\
C + O_2 \longrightarrow CO_2 & \Delta G°_2 = \ \ -94.3\,kcal \\
\underline{2\,H_2 + O_2 \longrightarrow 2\,H_2O} & \underline{\Delta G°_3 = -109.3\,kcal} \\
CH_4 + 2\,O_2 \longrightarrow CO_2 + 2\,H_2O & \Delta G°_1 + \Delta G°_2 + \Delta G°_3 = -201.4\,kcal/mole
\end{array}
$$

The sum of the individual $\Delta G°$ values equals $\Delta G°_{combustion}$ since the same net reaction is involved. The individual reactions in the sequence of three listed need not occur in order to oxidize methane. They are only chosen to illustrate a way to calculate the desired $\Delta G°$ for the oxidation of methane.

The additivity of $\Delta G°$ values is useful in discussing series of biochemical reactions. Some individual reactions may be endergonic, whereas others are exergonic. In total the net $\Delta G°$ for the series of reactions is the sum of the individual $\Delta G°$ values.

Example 25.1

The $\Delta G°$ for the oxidation of ethanol to form acetic acid is $-106.7\,kcal/mole$. The $\Delta G°$ for the oxidation of acetic acid to carbon dioxide is $-204.0\,kcal/mole$. What is the $\Delta G°$ for the oxidation of ethanol to carbon dioxide?

Solution. First write the balanced equations for each reaction.

$$C_2H_5OH + O_2 \longrightarrow CH_3CO_2H + H_2O$$

$$CH_3CO_2H + 2\,O_2 \longrightarrow 2\,CO_2 + 2\,H_2O$$

$$C_2H_5OH + 3\,O_2 \longrightarrow 2\,CO_2 + 3\,H_2O$$

The third equation for the oxidation of ethanol to carbon dioxide and water is the sum of the other two equations. Thus the $\Delta G°$ for the third equation is calculated as $-106.7 + (-204.0) = -310.7\,kcal/mole$.

As indicated in Chapter 20, the sun provides energy to produce plants, which then have stored chemical energy. In terms of free energy, plant molecules have higher free energy than the compounds from which they were formed, and photosynthesis is an endergonic reaction (Figure 25.2).

$$6\,CO_2 + 6\,H_2O \longrightarrow \underset{\text{(glucose)}}{C_6H_{12}O_6} + 6\,O_2 \qquad \Delta G° = 686\,kcal/mole$$

It should be pointed out that photosynthesis is an extremely complex process involving more than 100 steps, each catalyzed by enzymes. However, the net

25.2
Energy and Oxidation State of Carbon Compounds

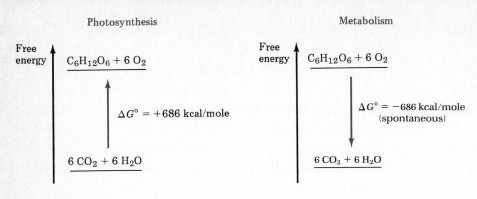

Photosynthesis

Free
energy $\quad$ $C_6H_{12}O_6 + 6\,O_2$

$\Delta G^\circ = +686\ \text{kcal/mole}$

$6\,CO_2 + 6\,H_2O$

Metabolism

Free
energy $\quad$ $C_6H_{12}O_6 + 6\,O_2$

$\Delta G^\circ = -686\ \text{kcal/mole}$
(spontaneous)

$6\,CO_2 + 6\,H_2O$

Figure 25.2
Energy Cycle of Photosynthesis
and Metabolism.
The energy stored in plants by
photosynthesis is released in
metabolism (or combustion) in
exactly the same amount.

free energy change of all of the steps is equal to the difference in the free energies
of the initial reactants and the final products.

When glucose is oxidized, either by burning in air or in metabolic reactions
in animals, the 686 kcal/mole of free energy is released (Figure 25.2).

$$C_6H_{12}O_6 + 6\,O_2 \longrightarrow 6\,CO_2 + 6\,H_2O \qquad \Delta G^\circ = -686\ \text{kcal/mole}$$

Life then depends on a cycle of carbon compounds and the energy associated
with the reactions of carbon compounds. Plants synthesize molecules with a
high free energy content. Animals eat plants and release the free energy stored in
the bonds of plant molecules and the carbon dioxide formed is returned to be
recycled by plants in photosynthetic reactions (Figure 25.3).

Oxidation State of Carbon

The concept of oxidation state (Section 6.7) provides another way of looking
at biochemicals and the reactions of life processes. The oxidation number of
carbon combined with oxygen in carbon dioxide is +4, its maximum oxidation
number. When carbon combines with hydrogen, the oxidation state of carbon is
decreased; the lowest oxidation state of carbon is −4 in methane. The oxidation
states of some simple compounds of carbon with oxygen and hydrogen illustrate
this trend (Figure 25.4).

Photosynthesis yields carbon compounds in a reduced state compared to
carbon dioxide. A molecule such as glucose is a "reduced molecule" compared to

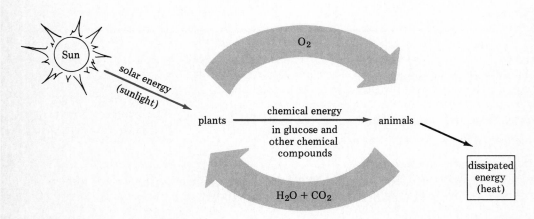

Figure 25.3
The Carbon Cycle.
Living organisms depend
on each other for food.
The carbon dioxide and
oxygen in the atmos-
phere, as well as water,
are cycled by plants and
animals.

Figure 25.4
Oxidation Number and En-
ergy Content of Carbon
Compounds.

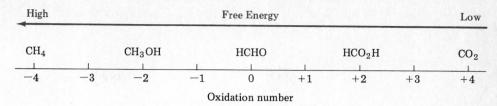

carbon dioxide. Reduced molecules are "energy-rich" molecules, and they can
react to release energy in reactions producing oxidized molecules (Figure 25.4).

Catabolic reactions in animals break down reactants in plant foodstuffs into
products of a higher oxidation state by oxidative degradation reactions. Anabolic
reactions, which synthesize products in a lower oxidation state, hence a more
reduced state, are known as reductive biosynthetic reactions.

Example 25.2

Fats have more stored energy than carbohydrates on a per gram basis. Why?

Solution. Fats consist of esters of glycerol and long-chain fatty acids. The majority of the carbon atoms in a fat have two bonded hydrogen atoms.

$$
\begin{array}{c}
\text{O} \\
\| \\
\mathrm{CH_2{-}O{-}C{-}(CH_2)_{16}CH_3} \\
| \\
\text{O} \\
\| \\
\mathrm{CH{-}O{-}C{-}(CH_2)_{16}CH_3} \\
| \\
\text{O} \\
\| \\
\mathrm{CH_2{-}O{-}C{-}(CH_2)_{16}CH_3} \\
\text{a fat}
\end{array}
$$

In carbohydrates such as glucose, each carbon atom has an oxygen atom
bonded to it.

$$
\begin{array}{c}
\mathrm{CHO} \\
| \\
\mathrm{H{-}C{-}OH} \\
| \\
\mathrm{HO{-}C{-}H} \\
| \\
\mathrm{H{-}C{-}OH} \\
| \\
\mathrm{H{-}C{-}OH} \\
| \\
\mathrm{CH_2OH} \\
\text{glucose}
\end{array}
$$

Since fats are more reduced than carbohydrates, they have more stored energy.

[Additional examples may be found in **25.7–25.10** at the end of the chapter.]

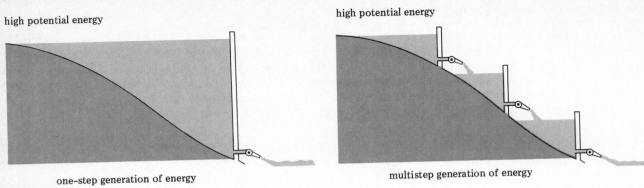

high potential energy

one–step generation of energy

high potential energy

multistep generation of energy

Figure 25.5
An Analogy for Direct Combustion and Metabolic Oxidation.

Metabolic Reactions

Although the same equation represents the net change that occurs in direct combustion and in oxidative degradation of glucose, there are differences in the individual steps. In direct combustion the energy is produced as heat energy. If the identical process occurred in a living organism, the cells would be destroyed by the high temperature. The energy liberated would be more than the organism would need at the time of combustion. The human body needs steadily produced heat in order to maintain itself within a living temperature range. However, the body also needs energy for processes such as muscle movement, active transport, and transmission of nerve impulses, as well as for anabolic reactions that produce polysaccharides, lipids, proteins, and nucleic acids. These classes of compounds will be discussed in the following four chapters.

Oxidative degradations occur in a series of steps and produce energy in small usable packets. *The intermediate compounds of metabolism are called* **metabolites.** Heat energy is produced, but in addition some steps produce adenosine triphosphate (ATP). ATP is the body's storehouse of energy. It is a high-energy compound that provides energy for the chemistry of the body when it is needed. In Figure 25.5 a comparison between direct combustion and metabolic reactions is given by analogy with the potential energy of water.

If a single dam is used to generate electric power, the energy produced comes from a single step. This is analogous to the direct-combustion process. A series of dams, each with its individual turbines, is analogous to metabolic reactions. Ultimately the same amount of energy is released from the water as it completes its course to the bottom of the hill, but is produced in smaller units. In the dam analogy the released energy is converted into electricity. In direct combustion the energy is released as heat. In oxidative degradation some of the energy is released as heat, but a substantial fraction is used to produce ATP from ADP.

$$\text{ADP} + \text{H}_3\text{PO}_4 \longrightarrow \text{ATP} + \text{H}_2\text{O} \qquad \Delta G^\circ = +7.3\ \text{kcal/mole}$$

Thus ATP has stored chemical energy of 7.3 kcal/mole in one of its phosphoric acid anhydride bonds. This energy is released during hydrolysis in the organism at a time of need.

$$\text{ATP} + \text{H}_2\text{O} \longrightarrow \text{ADP} + \text{H}_3\text{PO}_4 \qquad \Delta G^\circ = -7.3\ \text{kcal/mole}$$

Example 25.3 _____

The $\Delta G°$ for the formation of glucose 6-phosphate from glucose and a source of phosphate is $+3.3$ kcal/mole.

$$\text{glucose} + H_3PO_4 \longrightarrow \text{glucose 6-phosphate} + H_2O \quad \Delta G° = +3.3 \text{ kcal/mole}$$

What is the $\Delta G°$ for the hydrolysis of glucose 6-phosphate?

Solution. The hydrolysis of glucose 6-phosphate is the reverse of the reaction forming it. Thus the $\Delta G°$ is -3.3 kcal/mole.

$$\text{glucose 6-phosphate} + H_2O \longrightarrow \text{glucose} + H_3PO_4 \quad \Delta G° = -3.3 \text{ kcal/mole}$$

Living organisms require a source of free energy for mechanical work, active transport of materials, and synthesis of molecules. The free energy used for these processes is derived from oxidation of food by animals and from the sun by plants.

The free energy released in catabolic reactions in animals is partially converted into molecules that can serve as carriers and suppliers of energy. The most common and widely distributed carrier of biochemical energy is ATP. ATP has stored free energy that becomes available when needed by the cell (Figure 25.6).

The structure of ATP (Figure 25.7) consists of an organic base, adenine, bonded to ribose, which in turn is bonded to a triphosphate group. Adenine bonded to ribose is called adenosine. At physiological pH, ATP has a -4 charge, but for convenience, we use either the protonated molecule or shorthand structures.

The hydrolysis of ATP to ADP is shown in Figure 25.8. The reaction involves the hydrolysis of an anhydride bond (Section 18.9) of the triphosphate part of the molecule. The shorthand equation uses P_i to represent the ions of phosphoric acid that exist at physiological pH.

$$\text{ATP} + H_2O \longrightarrow \text{ADP} + P_i \quad \Delta G° = -7.3 \text{ kcal/mole}$$

The anhydride bond in ADP can also hydrolyze to produce adenosine monophosphate (AMP).

$$\text{ADP} + H_2O \longrightarrow \text{AMP} + P_i \quad \Delta G° = -6.5 \text{ kcal/mole}$$

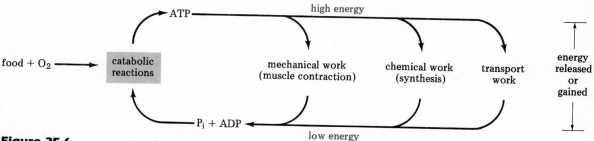

Figure 25.6

ATP—The Energy Currency.
The energy stored in ATP by catabolic reactions is released in subsequent biological processes, and ADP is formed.

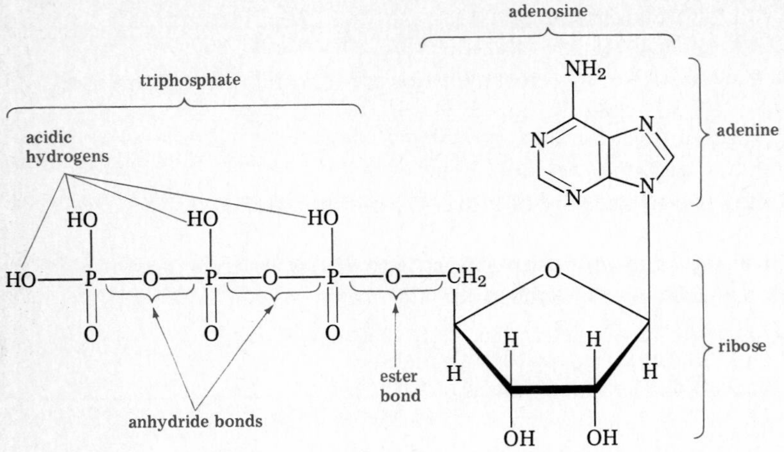

Figure 25.7
Structure of Adenosine Triphosphate.

In the metabolism of fatty acids (Chapter 27), one reaction involves the hydrolysis of ATP to AMP and pyrophosphoric acid (Figure 25.8). In biological systems, pyrophosphoric acid exists as the pyrophosphate ion. The shorthand equation for this hydrolysis uses PP_i to represent the ions of pyrophosphoric acid that exist at physiological pH.

$$ATP + H_2O \longrightarrow AMP + PP_i \qquad \Delta G° = -8.6 \text{ kcal/mole}$$

An important feature of these hydrolysis reactions is that they may be reversed with a source of energy. The reverse process is phosphorylation. Without

Figure 25.8
Hydrolysis Reactions
of ATP.

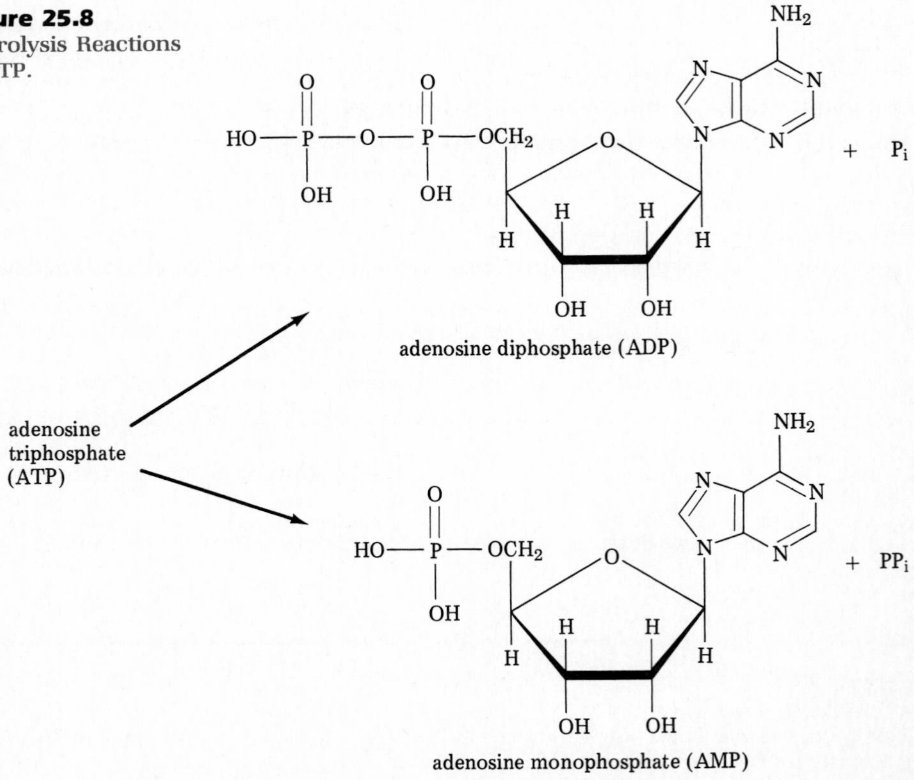

adenosine diphosphate (ADP)

adenosine monophosphate (AMP)

phosphorylation all ATP would be eventually converted to ADP or AMP. However, free energy stored in food molecules is used to convert ADP or AMP back to ATP, and cells then maintain a reservoir of ATP. With sufficient food sources, enough ATP is produced to maintain the cell. Any excess ATP is used in anabolic reactions to synthesize compounds such as glycogen and fats, which store free energy. When food is not available, the ATP requirements for mechanical work and transport are obtained from the stored compounds, which undergo catabolism and produce ATP.

In a typical cell, an ATP molecule is consumed within minutes of its formation. Thus there is a high turnover of ATP. In fact, the human body in a resting state consumes about 40 kg of ATP per day. However, the amount of ATP present at any time is only a small fraction of this value.

Before considering further the role of ATP in biochemical reactions it is useful to consider ATP as a currency in the world of energy. ATP as an energy currency comes in multiples of 7.3 kcal/mole. Neither this amount nor its multiples fit perfectly with the energy requirements of molecules in chemical reactions.

Let's consider an analogy to ATP by using money. What would happen to you if you were restricted to carrying only $10 bills and to spending and saving only $10 bills? If you wanted to buy a $7 item you would have to use a $10 bill and lose your "change." What would happen if you worked to make $15? Since you can only carry $10 bills, you would lose $5 and could only accept the $10 bill. Some money is left over in each example. A very similar situation is involved in ATP–ADP conversions.

The conversion of fructose and glucose into sucrose requires energy.

$$\text{glucose} + \text{fructose} \longrightarrow \text{sucrose} + H_2O \qquad \Delta G° = 5.5 \text{ kcal/mole}$$

If ATP provided the energy for the reaction, there will be 1.8 kcal/mole left over as "change" and the energy would be used for temperature maintenance. If a reaction releases energy in excess of 7.3 kcal/mole, only a portion of the energy may be saved as ATP. The hydrolysis of creatine phosphate releases 10.3 kcal/mole.

$$\text{creatine phosphate} \longrightarrow \text{creatine} + P_i \qquad \Delta G° = -10.3 \text{ kcal/mole}$$

If this energy is saved by conversion of ADP into ATP, then 3.0 kcal/mole will be left over and released as heat.

AN ASIDE

In order for muscle to contract, chemical energy must be converted into mechanical energy. Muscles contract by shortening many individual muscle cells. The shortening of muscle cells occurs as a result of many enzyme-catalyzed reactions forming bonds that require a supply of energy from ATP.

Muscle cells contain many myofibrils, which in turn are made of thick and thin filaments. The thick filament is made up of the protein myosin. The thin filament consists of three proteins, actin, troponin and tropomyosin. The arrangement of the filaments is in a striated pattern called a sarcomere. In a relaxed myofibril the sarcomeres have a characteristic arrangement that changes in the contracted state. The change results from bonds formed between the thick and thin filaments. It is this bonding that requires ATP as a source of energy.

There is no guarantee that any energy liberated in a chemical reaction will be saved as ATP. In order for energy to be saved there must be a harnessing or coupling of two reactions. Similarly, ATP is useful only if its hydrolysis is coupled to a second reaction.

25.4
Coupling of Reactions

In order for a reaction to occur spontaneously, the free energy change must be negative. A reaction with a positive $\Delta G°$ can occur only if energy is supplied from another source. Let's now see how these principles can be used to explain biochemical reactions.

Consideration of a reaction with a positive $\Delta G°$ along with another reaction with a negative $\Delta G°$ is important is biochemistry. Providing the increase in $\Delta G°$ of one reaction is smaller than the decrease in $\Delta G°$ of another reaction, the net $\Delta G°$ for the two reactions is thermodynamically favorable. *When an energy-releasing reaction and an energy-consuming reaction occur together, they are* **coupled reactions.**

One step in the citric acid cycle is the oxidation of malic acid by the coenzyme nicotinamide adenine dinucleotide (NAD^+).

$$\text{malic acid} + NAD^+ \longrightarrow \text{oxaloacetic acid} + NADH + H^+$$
$$\Delta G° = +6.1 \text{ kcal/mole}$$

The positive $\Delta G°$ means that this single reaction is not spontaneous and at equilibrium there would be little oxaloacetic acid. The subsequent step in the citric acid cycle is the conversion of oxaloacetic acid into citric acid.

$$\text{oxaloacetic acid} + \text{acetyl coenzyme A} \longrightarrow \text{citric acid} + \text{coenzyme A}$$
$$\Delta G° = -7.7 \text{ kcal/mole}$$

This reaction is favorable and uses oxaloacetic acid formed in the preceding step. As the oxaloacetic acid is used in the second reaction, more malic acid must be converted into oxaloacetic acid in the first reaction. This effect is another of the many examples of Le Châtelier's principle. The second reaction "pulls" the first reaction by removing one of its products. In terms of free energy the net result of the two coupled reactions is the release of 1.6 kcal/mole.

As part of the metabolism of glucose (Chapter 26), the monosaccharide is converted to a phosphate ester. Consider the possible phosphorylation reaction with inorganic phosphate as a reagent.

This reaction will not proceed spontaneously. In fact, the reverse of the reaction, the hydrolysis of glucose 6-phosphate, is spontaneous with $\Delta G° = -3.3 \text{ kcal/mole}$.

Now let's consider the possibility of coupling the endergonic phosphylation of glucose with the exergonic hydrolysis of ATP.

$$ATP + H_2O \longrightarrow ADP + P_i \qquad \Delta G° = -7.3 \text{ kcal/mole}$$

$$\text{glucose} + P_i \longrightarrow \text{glucose 6-phosphate} + H_2O \qquad \Delta G° = +3.3 \text{ kcal/mole}$$

For these two coupled reactions the net $\Delta G°$ is negative and 4.0 kcal/mole is released.

The two separate reactions do not have to occur as written. In other words, ATP does not have to hydrolyze and form P_i in order to phosphorylate glucose. The same net result of adding together the two equations can occur directly in a cell.

$$\text{ATP} + \text{glucose} \longrightarrow \text{ADP} + \text{glucose 6-phosphate} \qquad \Delta G° = -4.0 \text{ kcal/mole}$$

The phosphate is transfered from ATP to glucose at the active site of the enzyme hexokinase. The spontaneous reaction with $\Delta G° = -4.0$ kcal/mole occurs because the sum of the free energies of glucose and ATP is larger than the sum of the free energies of glucose 6-phosphate and ADP.

In biochemical equations it is often convenient to use the following representation to save space.

$$\text{glucose} \xrightarrow[\text{ATP} \quad \text{ADP}]{} \text{glucose 6-phosphate}$$

The curved arrow is used for some simultaneous chemical changes that occur with a number of central reactions in biochemistry. Thus ATP and ADP are usually not written on the same line as the other reactants and products.

Example 25.4 _____

The phosphorylation of fructose to give fructose 6-phosphate has $\Delta G° = +3.8$ kcal/mole and is not spontaneous. Can a coupled reaction of fructose with ATP be spontaneous?

Solution. The phosphorylation of fructose could occur with the transfer of a phosphate group from ATP. Since the hydrolysis of ATP to ADP releases more free energy than that required to phosphorylate fructose, a coupled reaction is feasible.

fructose + P_i $\longrightarrow$ fructose 6-phosphate + H_2O	$\Delta G° = +3.8$ kcal/mole
ATP + H_2O $\longrightarrow$ ADP + P_i	$\Delta G° = -7.3$ kcal/mole
fructose + ATP $\longrightarrow$ fructose 6-phosphate + ADP	$\Delta G° = -3.5$ kcal/mole

[Additional examples may be found in **25.17–25.20** at the end of the chapter.]

25.5
Substrate Level Phosphorylation

Substrate level phosphorylation *involves a transfer of a phosphate group from ATP or some other phosphorylating agent to another compound.* The direction of the transfer is governed by $\Delta G°$. Therefore, a compound with a high free energy will transfer its phosphate to a compound of lower free energy. Several phosphorylated compounds that are found in cells are listed in Table 25.1 with the free energy change for their hydrolysis. Compounds toward the top of the table release a large amount of energy in hydrolysis. These compounds can transfer phosphate to ADP and convert it to ATP.

Table 25.1

Free Energy of Hydrolysis of Phosphorylating
Agents

Compound	$\Delta \Gamma$ at pH 7.0 (cal)
phosphoenolpyruvate (PEP)	−14,800
1,3-diphosphoglycerate	−11,800
phosphocreatine	−10,300
acetyl phosphate	−10,100
pyrophosphate	−8,000
acetyl CoA	−7,500
ATP to ADP and P_i	−7,300
ATP to AMP and pyrophosphate	−8,600
ADP	−6,500
glucose 1-phosphate	−5,000
fructose 6-phosphate	−3,800
glucose 6-phosphate	−3,300
L-glycerol 3-phosphate	−2,200

$$
\begin{array}{ll}
PEP + H_2O \longrightarrow \text{pyruvic acid} + P_i & \Delta G^\circ = -14.8 \text{ kcal/mole} \\
ADP + P_i \longrightarrow ATP + H_2O & \Delta G^\circ = +7.3 \text{ kcal/mole} \\
\hline
PEP + ADP \longrightarrow \text{pyruvic acid} + ATP & \Delta G^\circ = -7.5 \text{ kcal/mole}
\end{array}
$$

The phosphates below ATP in the table have a smaller tendency to hydrolyze or transfer phosphate. ATP is a better phosphorylating agent than those substances below it.

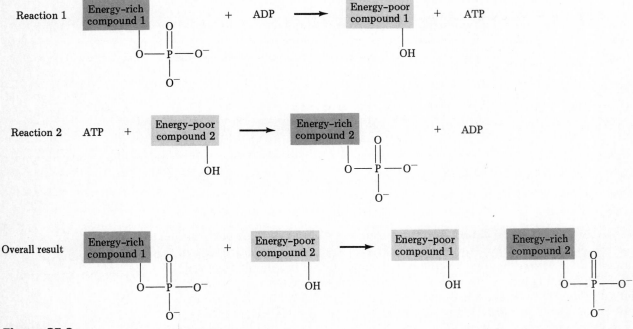

Figure 25.9
Transfer of Phosphate Groups by ATP and ADP.

$$ATP + H_2O \longrightarrow ADP + P_i \qquad\qquad \Delta G° = -7.3 \text{ kcal/mole}$$
$$\underline{glycerol + P_i \longrightarrow glycerol\ 3\text{-phosphate} + H_2O \qquad \Delta G° = +2.2 \text{ kcal/mole}}$$
$$ATP + glycerol \longrightarrow ADP + glycerol\ 3\text{-phosphate} \qquad \Delta G° = -5.1 \text{ kcal/mole}$$

The intermediate position of ATP in the table indicates it is an effective carrier of phosphate groups. ADP can accept phosphate from substances of high free energy, and then ATP can transfer phosphate to substances of low free energy (Figure 25.9).

25.6

Oxidation–Reduction

Recall from Section 10.4 that oxidation was defined as loss of electrons. This definition is useful in considering the oxidation of Fe^{2+} to Fe^{3+}. However, in the reaction of organic compounds oxidation is more readily seen as a gain of oxygen or a loss of hydrogen. Each example given in Section 10.5 used an oxidizing agent represented as AO. In the oxidation of acetaldehyde to form acetic acid there is a gain of oxygen. In the oxidation of ethanol to form acetaldehyde there is a loss of hydrogen. In each case a carbon atom has been oxidized, and it has "lost" electrons to oxygen.

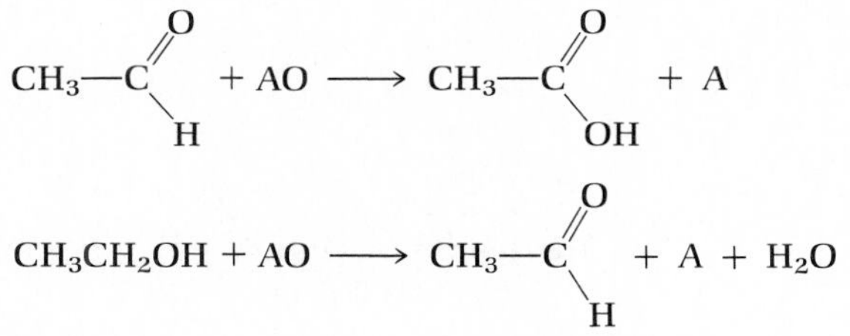

These reactions also occur in cells. Oxidation in cells often involves the removal of hydrogen atoms as protons as well as the electrons originally present in the carbon–hydrogen bonds.

Oxygen is an oxidizing agent and is brought into tissues as a result of respiration. Therefore you might think that oxygen oxidizes the foods in cells to produce CO_2 and H_2O. However, the process is not that straightforward. The oxygen molecule does not react directly with metabolites. Oxygen is the ultimate oxidizing agent, but many steps are involved in the transfer of electrons.

A small group of oxidizing agents are the first used in all biochemical reactions to remove hydrogen. The structures of the coenzymes NAD^+ and flavin adenine dinucleotide (FAD), which are oxidizing agents, are shown in Figure 25.10.

NAD$^+$

The portion of NAD^+ that serves as an oxidizing agent is nicotinamide (niacin), which is vitamin B_3. Only this portion of the molecule is shown when it serves as an oxidizing agent.

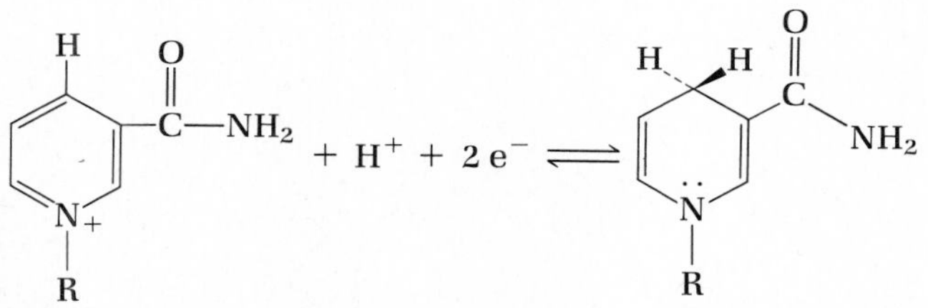

Figure 25.10
Structure of NAD and FAD.

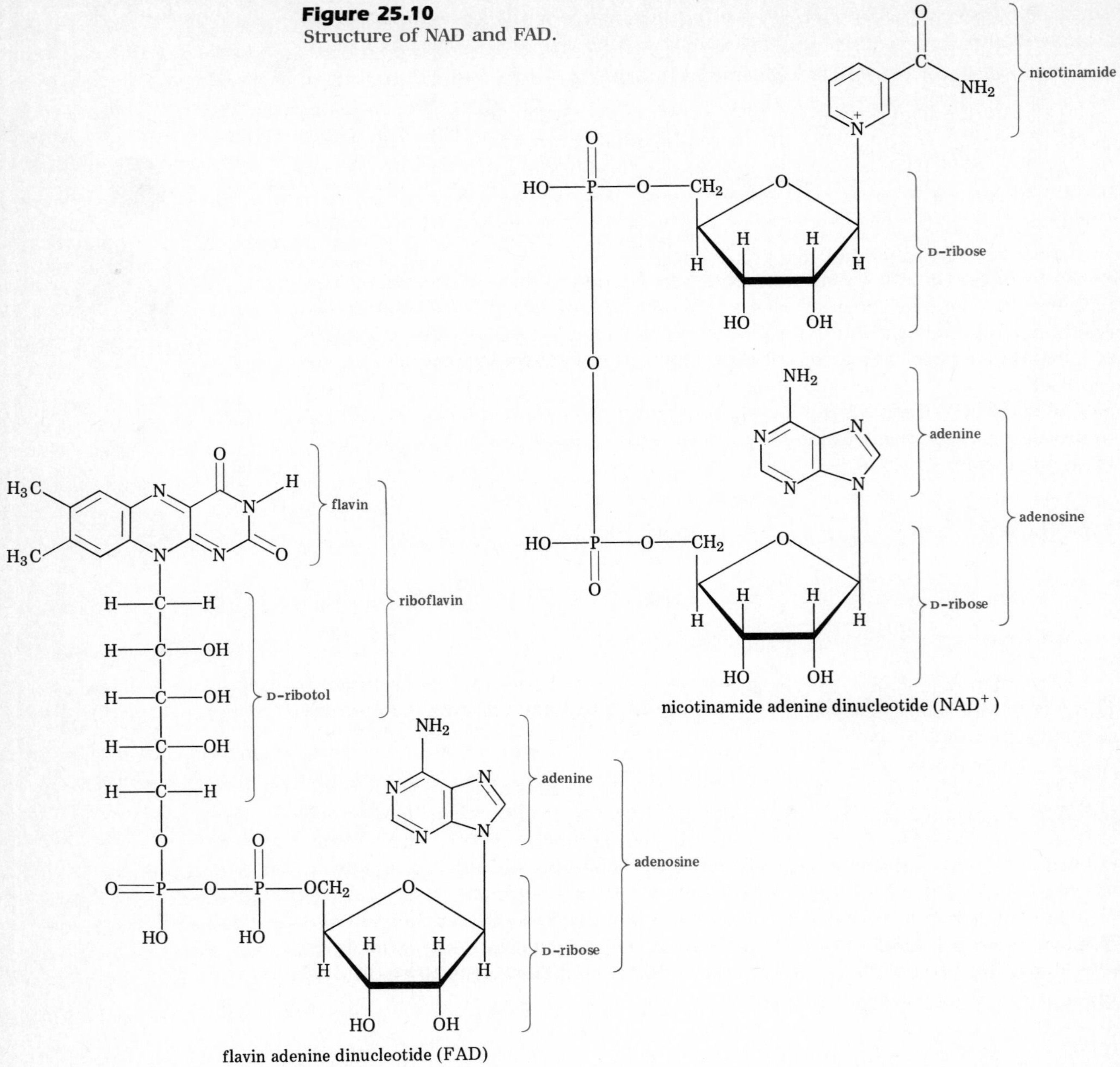
nicotinamide
D-ribose
adenine
adenosine
D-ribose
nicotinamide adenine dinucleotide (NAD$^+$)
flavin
riboflavin
D-ribotol
adenine
adenosine
D-ribose
flavin adenine dinucleotide (FAD)

NAD$^+$ gains two electrons and is reduced. A substrate that NAD$^+$ oxidizes has to lose two electrons. The substrate may also lose two protons, but only one proton combines with NAD$^+$. The overall reaction for oxidation of ethanol to acetaldehyde is

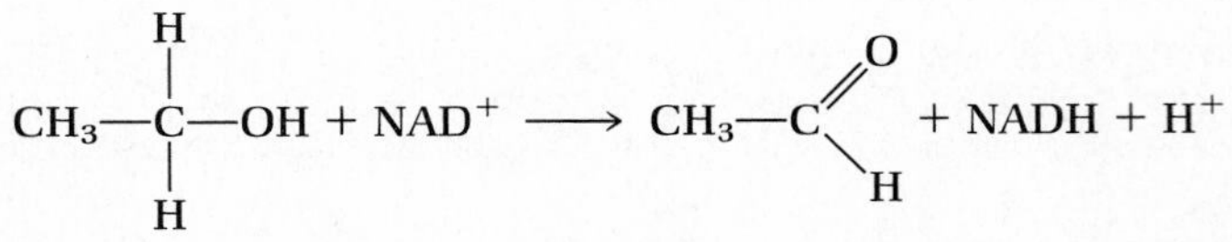
CH3—C—OH + NAD+ ⟶ CH3—C + NADH + H+

NAD^+ can also oxidize aldehydes to carboxylic acids by adding an oxygen atom from water.

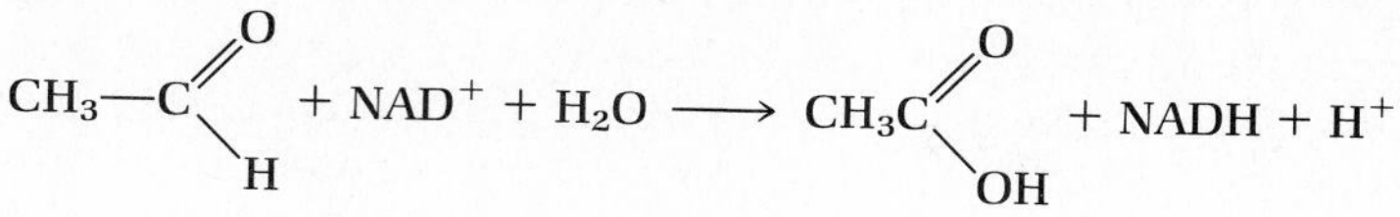

$$CH_3-C{\overset{O}{\underset{H}{}}} + NAD^+ + H_2O \longrightarrow CH_3C{\overset{O}{\underset{OH}{}}} + NADH + H^+$$

As substrates become oxidized, their free energy is decreased and some of this energy is conserved by the reduction of NAD^+. The reduced form, NADH, then can act in two ways. It may reduce other substrates, or it can reduce oxygen to water in cellular respiration via an electron transport chain (Section 25.7).

FAD

The structure of FAD given in Figure 25.10 contains riboflavin, vitamin B_2. The flavin portion of the molecule is shown in an equation in which it is reduced by accepting two electrons and two protons.

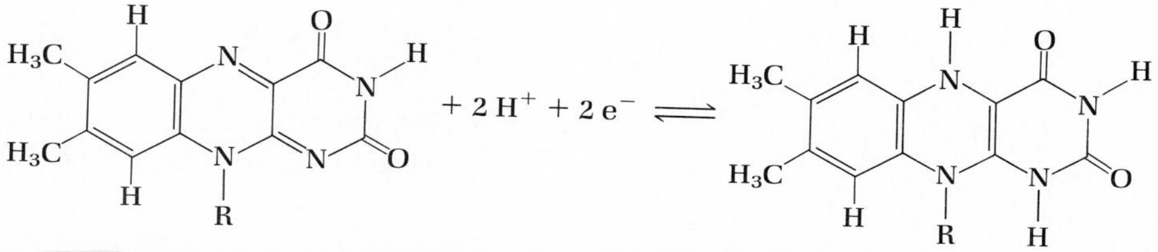

$$+ 2 H^+ + 2 e^- \rightleftharpoons$$

FAD oxidizes different substances than does NAD^+. It is involved in dehydrogenating $—CH_2CH_2—$ units to form a double bond. For example, in the metabolism of fatty acids, one step is the removal of hydrogen atoms from the α and β positions of a coenzyme A–activated compound. The *trans* isomer is formed exclusively.

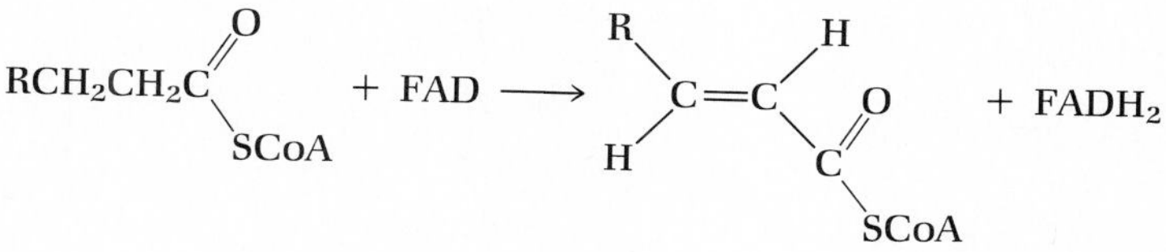

$$RCH_2CH_2C{\overset{O}{\underset{SCoA}{}}} + FAD \longrightarrow {\underset{H}{\overset{R}{}}}C=C{\overset{H}{\underset{C\ SCoA}{}}} + FADH_2$$

Reoxidation of Coenzymes

Reduction of the coenzymes NAD^+ and FAD cannot continue indefinitely, since they are available in limited amounts. In order for the coenzymes to continue as oxidizing agents, their reduced forms must be reoxidized.

Oxidation in biological systems eventually involves oxygen. Let's consider the possibility of the oxidation of NADH by O_2.

$$NADH + H^+ + \tfrac{1}{2}O_2 \longrightarrow NAD^+ + H_2O \qquad \Delta G° = -53 \text{ kcal/mole}$$

If this reaction were to occur in a single step, the energy released would be in a larger "packet" than the cell could use. Furthermore, if used only as heat energy, much of the value of the oxidation of the food would be lost. Oxidation of NADH or $FADH_2$ occurs in several steps so that the energy can be stored in other compounds. In the next section we will see how this is done. As you might expect, the series of coupled reactions involves the formation of ATP since it is the energy currency in the cell.

25.7
Cellular Respiration

In cellular respiration, NADH and $FADH_2$ are oxidized to NAD^+ and FAD, respectively. In an indirect way, oxygen is the oxidizing agent, but the *actual process of oxidation of the reduced forms of the coenzymes involves many steps called the* **electron transport chain.** In the case of NADH there are six steps in which the electrons removed are eventually passed along to oxygen. Each step involves some type of electron carrier; together the steps comprise an electron transport chain and are arranged in order of their reaction in Figure 25.11.

The electron transport process is very much like an old-time bucket brigade. Water obtained from the water source is handed in a bucket from one person to another until at the end of the line the water is thrown on the fire. Each individual goes through a sequence of having a bucket of water and then not having one as the bucket of water is passed on. In the electron transport chain, each substance receives electrons from the preceding substance. Then the reduced material is oxidized back by passing electrons along to the next substance in the chain.

The first step in the electron transport chain is the oxidation of NADH by a flavoprotein, FP_1, which is similar to FAD.

$$NADH + H^+ + FP_1 \xrightarrow[\text{ADP} \quad \text{ATP}]{} NAD^+ + FP_1H_2$$

In this process a mole of ATP is produced in a coupled reaction, and some of the energy stored in NADH is saved. In the second step FP_1H_2, produced from NADH, is oxidized by coenzyme Q (CoQ).

$$FP_1H_2 + CoQ \longrightarrow FP_1 + CoQH_2$$

Coenzyme Q is a lipid-soluble molecule that can shuttle between the flavoprotein and the cytochromes, which are the next carriers in the chain.

Cytochromes are proteins that contain heme as a prosthetic group. The differences between individual cytochromes are in the proteins surrounding the heme. Since heme contains iron, the oxidation–reduction reactions are

$$Fe^{2+} \underset{\text{reduction}}{\overset{\text{oxidation}}{\rightleftharpoons}} Fe^{3+} + 1\,e^-$$

Each step of the series of reactions of cytochromes passes along the electrons, which eventually reach oxygen to form water. The important fact about this process is that formation of ATP occurs in two of the steps. Recall that 1 mole of ATP is formed in the reduction of FP_1, and therefore the oxidation of 1 mole of NADH produces 3 moles of ATP. Thus, some of the energy stored in NADH from the oxidation of substrate is saved in coupled reactions in which ATP is formed. In the process all members of the chain are reduced and then oxidized back. The electrons are passed between the members of the chain, but the protons are not passed to the cytochromes. In terms of the stoichiometry of the reaction, the protons are required to combine with oxygen to form water. There are, however, some additional consequences to the release of protons from a substrate. This subject will be discussed in Section 25.10.

$FADH_2$ also is oxidized and transfers two electrons in the electron transport chain, but $FADH_2$ enters the chain in a different way than NADH. $FADH_2$ cannot reduce FP_1, but reduces a different flavoprotein, FP_2, to FP_2H_2. There is another significant difference in this step. ATP is not produced in a coupled reaction since the reaction is not sufficiently exergonic. FP_2H_2 then reduces CoQ to $CoQH_2$, and the remaining steps are identical to the series described for NADH (Figure 25.12).

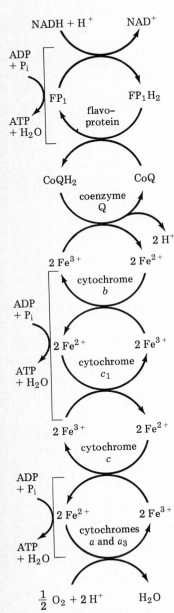

Figure 25.11
Coupled Reactions of NADH in the Electron Transport Chain.

In summary, $FADH_2$ passes two electrons along a chain to oxygen and releases two protons. There are only 2 moles of ATP formed per mole of $FADH_2$.

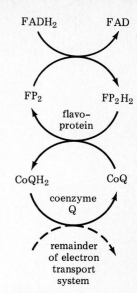

Figure 25.12
Coupled Reactions of $FADH_2$ in the Electron Transport Chain.

Oxidative Phosphorylation

Energy is saved in cellular respiration by the phosphorylation of ADP as the electrons flow through the electron transport chain. Only three of the steps in the oxidation of NADH are exergonic enough to transfer energy to ADP and form ATP. In the oxidation of $FADH_2$ two steps produce ATP. *The formation of ATP in these coupled reactions of the electron transport chain is called* **oxidative phosphorylation.** The ATP is produced under aerobic conditions. Aerobic means with or requiring oxygen; in the case of the electron transport chain it applies to the last step of the chain.

The exact mechanism by which free energy is transferred into ATP in the electron transport chain is still a subject of research. However, each component of the electron transport chain is anchored in the inner membrane of the mitochondrion (Section 25.10). Somehow the individual components are arranged for efficient transfer of electrons.

The free energy change for the direct oxidation of NADH by oxygen is -53 kcal/mole.

$$NADH + H^+ + \tfrac{1}{2}O_2 \longrightarrow NAD^+ + H_2O \qquad \Delta G^\circ = -53 \text{ kcal/mole}$$

The energy stored in ATP by the electron transport chain is $+21.9$ kcal/mole, since three moles of ATP are produced. The fraction of the energy saved by ATP for each mole of NADH oxidized is about 42%.

$$\frac{3 \times 7.3 \text{ kcal/mole ATP}}{53.0 \text{ kcal/mole NADH}} \times 100 = 42\% \text{ conserved}$$

The remaining free energy is released primarily as heat energy.

Example 25.5

The free energy change for the oxidation of $FADH_2$ by oxygen is -40.5 kcal/mole. Calculate the percentage of the free energy stored as ATP.

$$FADH_2 + \tfrac{1}{2}O_2 \longrightarrow FAD + H_2O \qquad \Delta G^\circ = -40.5 \text{ kcal/mole}$$

Solution. When one mole of $FADH_2$ is oxidized in the electron transport chain, there are two moles of ATP produced.

$$\frac{2 \times 7.3 \text{ kcal/mole ATP}}{40.5 \text{ kcal/mole FAD}} \times 100 = 36\% \text{ conserved}$$

25.8
Citric Acid Cycle

In the **citric acid cycle** *a two-carbon-atom group called the acetyl group is processed and eventually oxidized to* CO_2 *and* H_2O. [The citric acid cycle is also called the Krebs cycle or tricarboxylic acid cycle (TCA).] Although the details of the sources of the acetyl group will be described in Chapter 26, an overview of the sources of the acetyl group and its relationship to the citric acid cycle and cellular respiration is shown in Figure 25.13. Entry into the cycle occurs via the complex compound acetyl coenzyme A.

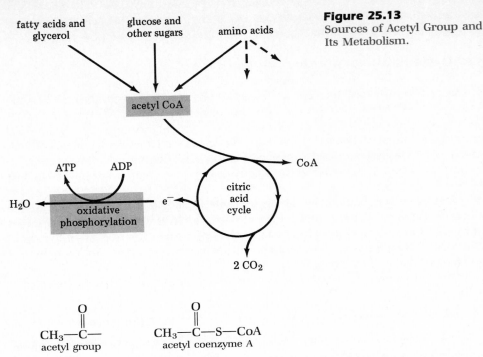

Figure 25.13
Sources of Acetyl Group and Its Metabolism.

Some of the energy released in the oxidation of the acetyl group is saved by forming NADH and FADH$_2$, both of which yield ATP in the electron transport chain where the reduced coenzymes are oxidized.

The reactions of the Krebs cycle are something like a chemical merry-go-round. The product of the last of a series of reactions becomes a reactant for the first reaction in another trip around the cycle (Figure 25.14). For convenient reference the steps of the cycle are numbered. The acids actually exist as anions at physiological pH.

The overall cycle starts at Step 1 in which a six-carbon citric acid is formed from the four-carbon oxaloacetic acid and acetyl coenzyme A. The reaction is an aldol type condensation (Section 17.7) in which the α-carbon atom of the acetyl coenzyme A adds to the carbonyl carbon atom of oxaloacetic acid.

Although each subsequent step could be explained using principles of organic reactions, the objective of this section is to summarize what occurs in the citric acid cycle.

During the cycle the citric acid is converted to a five-carbon compound in Step 3 and to a four-carbon compound in Step 4. The carbon atoms are released as carbon dioxide, the most oxidized form of carbon.

Oxidations that occur in the cycle are dehydrogenations, but no oxygen is required. The hydrogen atoms and electrons are donated to NAD$^+$ and FAD. The reduced coenzymes can then be reconverted to oxidized forms in the electron transport chain. In the citric acid cycle, NADH is produced in coupled reactions 3, 4, and 7, and FADH$_2$ is produced in reaction 5. In Step 4 a mole of guanosine triphosphate (GTP) is formed from guanosine diphosphate (GDP). The GTP has a somewhat higher free energy than ATP. As a consequence, GTP can transfer a phosphate group to ADP by substrate level phosphorylation.

$$GTP + ADP \longrightarrow GDP + ATP$$

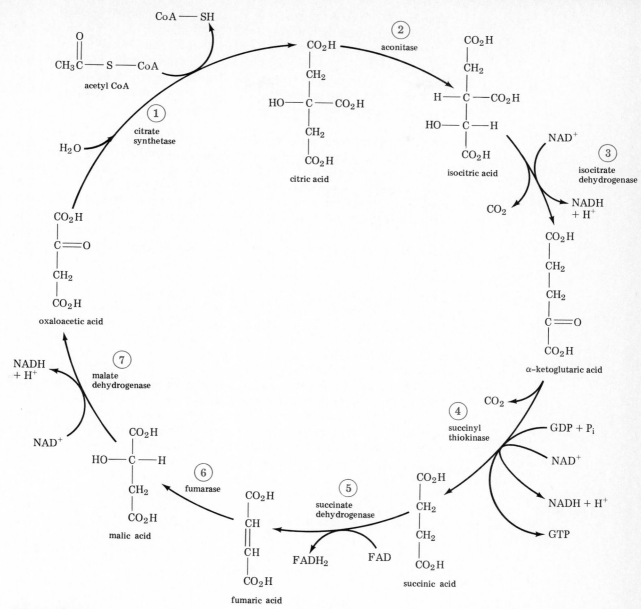

Figure 25.14
Reactions of the Krebs Cycle.

Now let's summarize how some of the free energy of one mole of an acetyl group is conserved in the citric acid cycle. The net reaction of the citric acid cycle is

$$\text{acetyl CoA} + 3 \text{ NAD}^+ + \text{FAD} + \text{GDP} + P_i + 2 H_2O \longrightarrow$$
$$2 CO_2 + 3 \text{ NADH} + \text{FADH}_2 + \text{GTP} + 2 H^+ + \text{CoA}$$

The three moles of NADH produce nine moles of ATP in the electron transport chain, and the one mole of $FADH_2$ produces two moles of ATP. Since one mole of

ATP is formed from GTP in the cycle, there is a total of twelve moles of ATP ultimately produced by the oxidation of one mole of an acetyl group.

The net reaction of the citric acid cycle does not directly involve molecular oxygen. However, the cycle can occur only under aerobic conditions because NAD^+ and FAD are required. These two coenzymes are available only if they are generated in the electron transport process in which electrons are transferred ultimately to molecular oxygen as the oxidizing agent.

Control of the Citric Acid Cycle

The rate at which the citric acid cycle runs and oxidizes acetyl groups is adjustable to meet the ATP needs of the cell. The citric acid cycle requires a supply of NAD^+ and FAD, which are regenerated from NADH and $FADH_2$ by oxidation in the electron transport chain. Along with the oxidation of NADH and $FADH_2$ there is an accompanying formation of ATP. If ATP is needed and is formed, then NADH and $FADH_2$ are reconverted into NAD^+ and FAD, which, in turn, are used in the citric acid cycle. Thus ATP is responsible for the control of the citric acid cycle. This control is exerted at three points in the cycle (Figure 25.15).

The synthesis of citric acid from oxaloacetic acid and acetyl coenzyme A in Step 1 is an important control point. Unless the acetyl unit enters the cycle, there will be no subsequent reactions. ATP is an allosteric inhibitor of the enzyme citrate synthetase required in this step. As the concentration of ATP increases, the enzyme is less effective in binding acetyl coenzyme A and so the rate of the citric acid cycle is decreased.

A second control point of the citric acid cycle occurs at step 3. Both ADP and NAD^+ bind at the enzyme isocitrate dehydrogenase and allosterically enhance the activity of the enzyme for isocitric acid. Thus, as the ATP concentration decreases and the ADP concentration increases, the citric acid cycle is stimulated. In addition, NADH is a competitive inhibitor of the enzyme by displacing the required NAD^+. Thus, NADH generated in the citric acid cycle inhibits the cycle

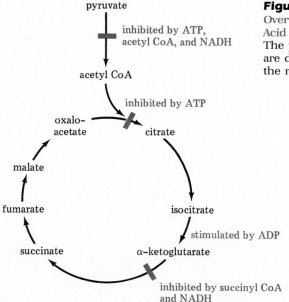

Figure 25.15
Overview of Control Points in the Citric Acid Cycle.
The points at which inhibition occurs are designated by colored lines across the reaction arrows.

itself. If ATP is required by the cell, oxidation of NADH occurs in the electron transport chain. The citric acid cycle then proceeds since the concentration of the competitive inhibitor is reduced.

Control of the citric acid cycle also occurs at Step 4 which is inhibited by NADH. The product of step 4, succinyl coenzyme A, is also a competitive inhibitor of the reaction. However, its concentration is decreased by the subsequent reactions.

In summary, the citric acid cycle is controlled through a number of complementary chemical reactions. The introduction of the acetyl unit into the cycle, as well as its rate of conversion, is reduced when the ATP level of the cell is high.

stop

25.9
Food and Energy

skip rest of chapter go to 26 1-5

The generation of energy via catabolic reactions of foodstuffs can be described in three stages (Figure 25.16). The first stage, digestion, is a process that occurs after eating food. Digestion is essentially a hydrolytic process in which large molecules are converted into smaller molecules to be absorbed by the body. Glycosidic bonds in disaccharides and polysaccharides are hydrolyzed to produce monosaccharides. Ester bonds of fats and oils are hydrolyzed to produce glycerol and fatty acids. Amide bonds of proteins are hydrolyzed to produce amino acids. The hydrolysis of each class of nutrients is catalyzed by specific enzymes. However, these hydrolysis reactions are not coupled to reactions to store energy.

In the second stage of catabolic reactions, the hydrolysis products of digestion are degraded to the few simple compounds that are central to metabolism. Each of these compounds is converted into the acetyl group of acetyl coenzyme A. In this stage of metabolism of foodstuffs, some ATP is generated. However, the amount of energy stored in ATP in this stage is only a small fraction of the amount potentially available by further oxidation in stage III.

In the third stage of metabolism, cells conserve and store some of the energy

Figure 25.16
Stages of Metabolism.

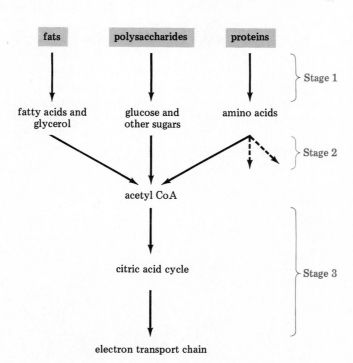

contained in foodstuffs. Acetyl coenzyme A is oxidized in the citric acid cycle to yield carbon dioxide. Electrons are transferred to NAD^+ and FAD to form NADH and $FADH_2$. The electrons are then transferred to oxygen via the electron transport chain. It is this latter process that generates the largest quantity of ATP.

It will be necessary in subsequent chapters to keep an energy "score card" to understand how much energy from food is converted into ATP. The fraction of the energy stored then will allow us to calculate the efficiency of the metabolic reactions. The flow of energy in metabolic reactions is not always in one direction. Usually there is an "investment" phase in which some of the energy "currency" is used. ATP may be consumed in a reaction required to activate a substrate for subsequent reactions. In other reactions an energy "recovery" is involved and ATP is formed. Finally, the metabolic reactions enter a "profit" phase to produce additional ATP. The oxidation of an acetyl group in the citric acid cycle followed by cellular respiration is an example of a profitable sequence of reactions.

The metabolic reactions of stages II and III and their interrelationships are affected by where the reactions occur. Glycolysis and the pentose phosphate pathway occur in the cytoplasm. The citric acid cycle, oxidative phosphorylation, and oxidation of fatty acids occur in the mitochondria. Finally, gluconeogenesis and the urea cycle (Chapter 28) depend on both the cytoplasm and the mitochondria. Thus, the interrelations between the various metabolic reactions depend on flow of compounds across the mitochondrial membranes.

25.10
The Mitochondria

The mitochondria in the cell are called the power plants because they obtain energy from oxidation and produce ATP. A mitochondrion consists of an outer membrane with a convoluted inner membrane (Figure 25.17). *The inner membrane creates a structure called a* **crista.** *The space between the outer membrane and the inner membrane is the* **intermembrane space.** *The space contained within the crista is the* **matrix.**

The reactions of the electron transport chain occur on the inner membrane. The citric acid cycle occurs in the matrix where necessary enzymes are located. The reduced coenzymes NADH and $FADH_2$ from the citric acid cycle in the matrix can be oxidized in electron transport on the membrane. Thus mitochondria are efficient factories for the conversion of chemicals into properly stored energy.

Each of the membranes of a mitochondrion contains lipid and protein molecules. However, the membranes are quite different in their permeability. The outer membrane is permeable to most small molecules. The inner membrane is

Figure 25.17
Structure of the Mitochondrion.

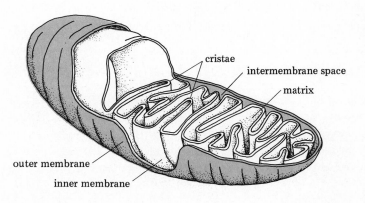

cristae

intermembrane space

matrix

outer membrane

inner membrane

impermeable to nearly all ions and most molecules. However, oxygen, carbon dioxide, and water can freely cross the inner membrane. Large molecules such as fatty acids and charged materials such as ADP and ATP must be actively transported across the inner membrane, and energy is required.

The flow of ADP and ATP across the inner mitochondrial membrane is controlled. As ADP enters the matrix via a carrier, ATP exits via a carrier. *The controlled flow in both directions across a membrane is called* **facilitated exchange diffusion.** The carrier is ATP-ADP translocase, a protein that constitutes about 6% of the membrane.

There are a variety of other carriers for ions and ionic metabolites. For example, there is a dicarboxylate carrier for malic acid, succinic acid, and fumaric acid. Further discussion of these is beyond the scope of this text.

How is the oxidation of NADH coupled to the phosphorylation of ADP? In 1961 it was proposed that these two processes are coupled as a result of a difference in proton (hydronium ion) concentration across the inner mitochondrial membrane. According to this model, protons generated in the electron transport chain are pumped from the matrix to the intermembrane space by carriers of the inner mitochondrial membrane.

It has been shown experimentally that the proton concentration of the intermembrane space is higher since its pH is 1.4 units lower than the matrix. Therefore, not only are there "pumps" in the inner membrane but also the inner membrane must be quite impermeable to proton flow in the opposite direction.

The difference in proton concentration causes a voltage difference of 0.22 volt between the matrix and the intermembrane space. It is the voltage difference that drives the synthesis of ATP in an enzyme-catalyzed reaction, by a mechanism that even today is poorly understood.

25.11 Biosynthesis and NADPH

In biosynthesis, energy provided by ATP is required to convert low-energy compounds into high-energy molecules. Furthermore, since the products are more reduced than the reactants, a reducing agent is required. Although NADH and $FADH_2$ are available in cells from oxidative degradation, these compounds are used primarily for the generation of ATP in the electron transport chain.

Nicotinamide adenine dinucleotide phosphate (NADPH) is the reducing agent of reductive biosynthesis. It is structurally related to NADH, having a hydroxyl group of adenosine esterified with phosphate (Figure 25.18). The extra phosphate group can be discerned by enzymes, and NADPH is the currency of reducing power for reductive biosynthetic reactions. NADPH is generated as glucose 6-phosphate is oxidized to ribose 5-phosphate in the pentose phosphate pathway (Section 26.7). The activity of this process is low in muscle but is very high in adipose (fat storage) tissue where NADPH is used to synthesize fatty acids, relatively reduced compounds.

25.12 Regulation of Metabolic Reactions

It should be evident from observing life forms that oxidative degradation and reductive biosynthesis must be carefully regulated. In addition, organisms must be able to respond to changes in their environment.

Enzymes play an important role in biochemical reactions, as outlined in the previous chapter. Enzymatic activity is controlled by the availability of cofactors and by the rate of synthesis of enzymes as well as their rate of degradation. In

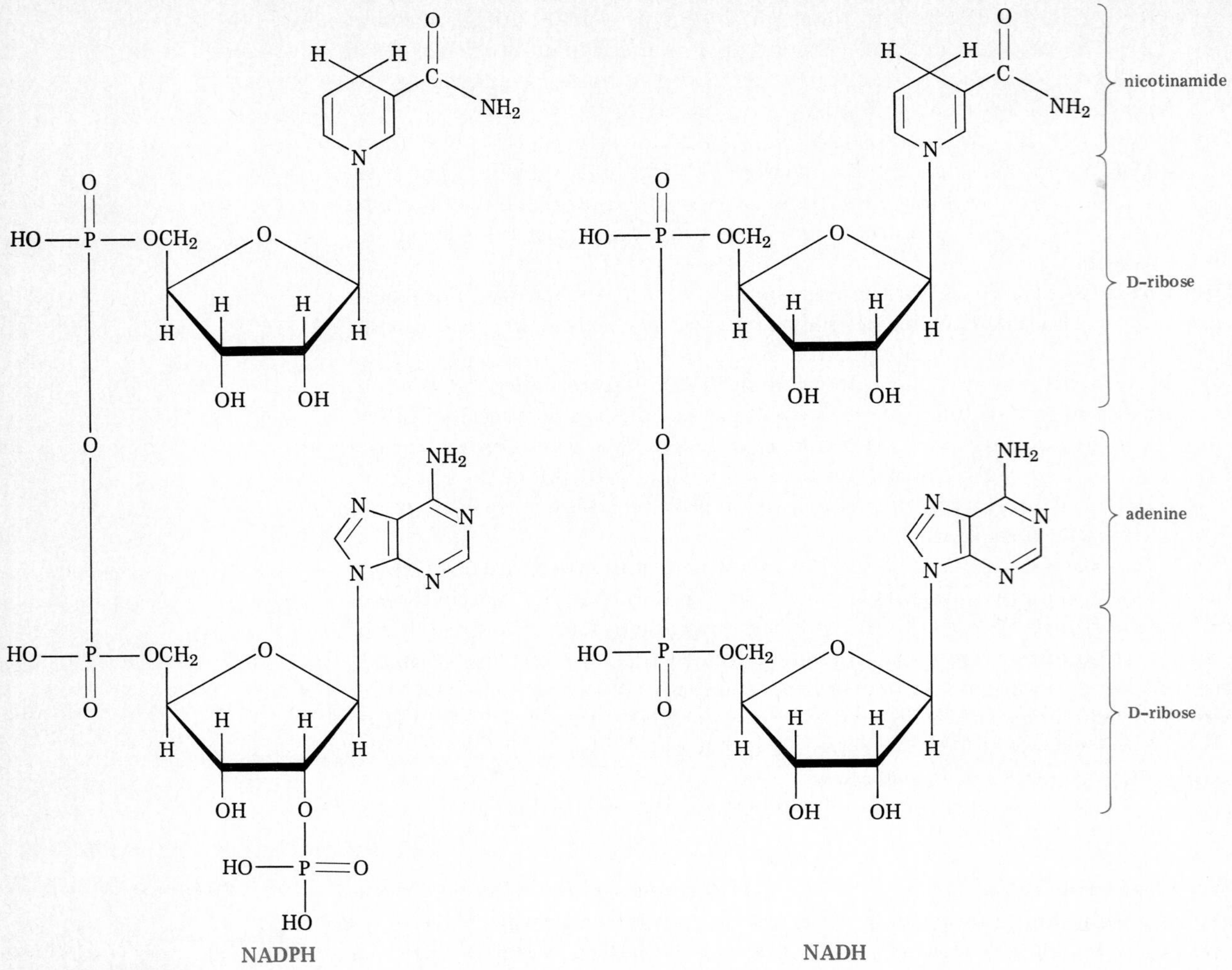

Figure 25.18
Comparison of the Structures of NADPH and NADH.

addition, allosteric control in feedback inhibition occurs in both oxidative degradation and reductive biosynthesis. Given the availability of the proper enzymes,
how do these pathways stay separate from each other? Furthermore, what features of the cells determine which reactions are needed?

Degradative and biosynthetic pathways are separately maintained by com-
partmentation in organelles. Thus, oxidation of fatty acids occurs in mitochondria, but fatty acid biosynthesis occurs in the cytosol (soluble portion of the
cytoplasm).

Metabolic reactions are governed by the energy charge of the cell. This energy charge is somewhat analogous to the charge of a car battery. If the charge is
low, oxidation of gasoline is used to recharge the battery. When the battery is fully
charged, it accepts no further charge and it may be used to supply energy to the
car. Under balanced conditions the battery charge is maintained and the car
continues to run. *The **energy charge** is proportional to the ATP level of the cell.*

$$\text{energy charge} = \frac{[\text{ATP}] + \frac{1}{2}[\text{ADP}]}{[\text{ATP}] + [\text{ADP}] + [\text{AMP}]}$$

Figure 25.19
Energy Change of a Cell and ATP Pathways.

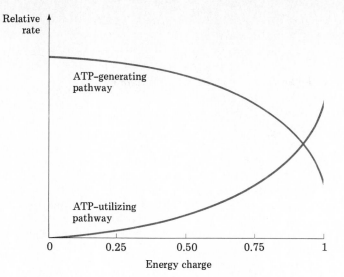

The factor of $\frac{1}{2}$ in the numerator indicates that there is only one phosphate anhydride bond in ADP compared to two in ATP. According to this relationship the energy charge would be 0 if only AMP were present and 1 if only ATP were present.

Pathways generating ATP are inhibited by a high energy charge of the cell, whereas pathways using ATP are enhanced by high energy charge (Figure 25.19). Thus ATP level responds to changes as predicted by Le Châtelier's principle. Note that the plots of the rates for formation and use of ATP are steep at relatively high energy charges and intersect at 0.9. Thus control of these two pathways can be maintained in the cell. In a sense this control is similar to that of a buffer. For most cells the energy charge is in the 0.85–0.95 range.

Summary

Free energy change, symbolized by $\Delta G°$, represents the energy from a chemical reaction that can be used to do work. An endergonic reaction occurs with an increase in free energy, whereas an exergonic reaction occurs with a decrease in free energy. Free energies are additive and may be used to calculate the $\Delta G°$ of a chemical reaction.

The oxidation state of carbon determines the energy content of biological molecules. Photosynthesis produces relatively reduced or energy-rich compounds. Metabolism results in oxidized or energy-poor compounds.

ATP is the most widely distributed carrier and supplier of biochemical energy. ATP is produced predominantly by the electron transport chain in reactions coupled with the oxidation of NADH. ATP is the source of free energy for mechanical work, active transport, and biosynthesis.

A coupled reaction involves an energy-releasing reaction and an energy-consuming reaction. The two reactions do not have to occur separately. Both reactions may be joined as a single reaction at an active site.

Substrate level phosphorylation occurs between a phosphate donor and acceptor. The reaction occurs to transfer phosphate from a high-energy to a low-energy compound. ATP is an intermediate energy carrier. Compounds whose free energy of hydrolysis is more negative than ATP's are high-energy compounds.

Oxidation of biochemicals occurs with NAD^+ to give NADH and H^+ or with FAD to give $FADH_2$. The reoxidation of the reduced coenzymes occurs in the electron transport chain.

Cellular respiration results in the oxidation of NADH and $FADH_2$ in steps of the electron transport chain. Electrons are passed along the components of the chain and eventually accepted by oxygen. Coupled reactions result in the formation of 3 moles of ATP from NADH and 2 moles of ATP from FADH$_2$.

An acetyl group, one of the central compounds of metabolism, is oxidized in the citric acid cycle to produce carbon dioxide. Three moles of NADH, 1 mole of FADH$_2$, and 1 mole of GTP are produced. Thus, the equivalent of 12 moles of ATP is produced in the oxidation of 1 mole of an acetyl group.

The membranes of the mitochondria have selective permeability to control the production of energy from oxidation in the form of ATP. The electron transport chain occurs on the inner membranes. The citric acid cycle occurs in the matrix. The flow of ADP and ATP is controlled by a protein carrier called ATP-ADP translocase.

Reductive biosynthesis requires ATP as well as the reducing agent NADPH. The extra phosphate of NADPH compared to NADH serves as a cellular distinction for their proper use as reducing agents.

Metabolic reactions are catalyzed by enzymes but controlled by the energy charge of the cell. Pathways generating ATP are inhibited by a high energy charge of a cell.

Learning Objectives

As a result of studying Chapter 25 you should be able to

- Compare free energy changes in ordinary chemical reactions and biochemical reactions.
- Describe the changes in energy and in oxidation state in the carbon cycle.
- Write equations for the hydrolysis of ATP at the anhydride bonds.
- Explain how ATP fulfills a role in cellular energy changes.
- Describe how biochemical reactions may be coupled.
- Distinguish between substrate level phosphorylation and oxidative phosphorylation.
- Classify NAD, NADH, FAD, and FADH$_2$ as oxidizing agents or reducing agents.
- Write net equations for the oxidation of NADH and FADH$_2$ in cellular respiration.
- Explain what changes occur in the electron transport chain.
- Indicate the number of moles of ATP formed per mole of NADH or FADH$_2$.
- Describe the function of the citric acid cycle and calculate the moles of ATP formed from a mole of an acetyl group.
- Describe the mitochondrion and the reactions that occur in it.
- Contrast the uses of NADH and NADPH in the cell.
- Explain how ATP level or energy charge of the cell regulates metabolism.

New Terms

The **citric acid cycle** oxidizes an acetyl group to carbon dioxide and water and saves stored energy in NADH, FADH$_2$, and ATP.

Coupled reactions are reactions that occur together. One is energy-releasing, and the other is energy-consuming.

The **crista** is a structure created by the inner membrane of the mitochondrion.

The **electron transport chain** is a series of electron carriers that remove electrons from coenzymes and transfers them to oxygen.

The **energy charge** of a cell is proportional to the ATP level.

Facilitated exchange diffusion is the controlled flow of material across a membrane via a carrier.

The **intermembrane space** is the space between the outer and inner membranes in the mitochondrion.

The **Krebs cycle** is the citric acid cycle.

Metabolites are intermediate compounds in metabolism.

The **matrix** is a space within the crista.

Nicotinamide adenine dinucleotide phosphate (**NADPH**) is a reduced coenzyme used in reductive biosynthesis.

Oxidative phosphorylation is the process of coupled formation of ATP in the electron transport chain.

Substrate level phosphorylation is the phosphorylation of a low-energy compound by a high-energy phosphate.

The **tricarboxylic acid cycle** (**TCA**) is the citric acid cycle.

Questions and Problems

Terminology

25.1 What are coupled reactions?

25.2 What is the difference between substrate level phosphorylation and oxidative phosphorylation?

25.3 What is the citric acid cycle?

Free Energy

25.4 What is the sign of the free energy change for a spontaneous reaction?

25.5 What is meant by the additivity of free energy changes?

25.6 What free energy changes are associated with photosynthesis and with the metabolism of glucose?

Oxidation

25.7 What changes in oxidation state occur in photosynthesis and the metabolism of glucose?

25.8 Why might carbon dioxide be called an "energy-poor" molecule?

25.9 Which of the following conversions represent oxidation or reduction reactions?
(a) ethanol $\longrightarrow$ acetaldehyde
(b) acetic acid + ethanol $\longrightarrow$ ethyl acetate
(c) acetic acid $\longrightarrow$ acetaldehyde

25.10 An enzyme catalyzes the following change of a functional group. Is this an oxidation or reduction reaction?

$$\begin{array}{c} \text{H} \\ | \\ -\text{C}-\text{NH}_2 \\ | \end{array} \longrightarrow \begin{array}{c} \\ \diagdown \\ \diagup \end{array}\text{C}=\text{NH}$$

ATP

25.11 What are the structural features of ATP?

25.12 What is the charge on the ion after the removal of all acidic protons from each of the following substances?
(a) ATP (b) ADP (c) AMP (d) PP_i (e) P_i

25.13 Using symbolic formulas write an equation for the hydrolysis of ATP to ADP. Do the same for the hydrolysis of ATP to AMP.

25.14 Is the hydrolysis of pyrophosphate to phosphate expected to be exergonic or endergonic?

25.15 Explain why 100% of biochemical energy for a reaction cannot be stored in ATP.

25.16 Why is ATP called the energy currency in biological systems?

Coupled Reactions

25.17 Can two endergonic reactions be coupled?

25.18 Explain how a series of endergonic reactions can occur providing a strongly exergonic reaction occurs at the end of the series.

25.19 What convention is used in equations to represent common biochemical conversions as ATP → ADP?

25.20 Explain how the hydrolysis of ATP may be part of a coupled reaction.

Substrate Level Phosphorylation

25.21 Which of the following has a higher free energy for hydrolysis than ATP?
(a) glucose 6-phosphate
(b) PEP
(c) adenosine diphosphate
(d) phosphocreatine

25.22 Using data in Table 25.1 determine if phosphoenol pyruvate should phosphorylate ADP.

25.23 Using data in Table 25.1 determine the $\Delta G°$ of each of these isomerization reactions.
(a) glucose 6-phosphate $\longrightarrow$ glucose 1-phosphate
(b) glucose 6-phosphate $\longrightarrow$ fructose 6-phosphate

25.24 Why is phosphoenolpyruvate called a high-energy phosphate but glycerol 3-phosphate is not?

25.25 The following reaction of guanosine triphosphate is spontaneous. What statement can be made about the free energy of hydrolysis of GTP to GDP?

$$\text{GTP} + \text{ADP} \longrightarrow \text{GDP} + \text{ATP}$$

Cellular Respiration

25.26 What is the first step in the electron transport chain starting with NADH?

25.27 What is the general name for the heme-containing proteins in the electron transport chain? What metal and what oxidation states are involved?

25.28 What is the first step in the electron transport chain starting with $FADH_2$?

25.29 List the following in the order in which they accept electrons.
(a) coenzyme Q (b) FP_1 (c) cytochrome b

25.30 How many moles of ATP are produced in coupled reactions of one mole of NADH in the electron transport chain?

25.31 Why does $FADH_2$ yield less ATP than does NADH in the electron transport chain?

25.32 What substances cause the electron transport chain to produce ATP?

25.33 Are the protons from NADH and $FADH_2$ passed along the electron transport chain?

25.34 How is water produced in the electron transport chain?

25.35 Why is only 42% of the energy of oxidation of NADH saved in the form of ATP?

25.36 What happens to a cytochrome when it accepts an electron?

25.37 How is FAD involved in the electron transport chain?

25.38 How many points in the electron transport chain produce ATP?

25.39 Cyanide forms a covalent bond with cytochromes. Explain why cyanide is poisonous.

Citric Acid Cycle

25.40 What group is oxidized in the citric acid cycle? What are the carbon products?

25.41 What is the six-carbon compound formed from acetyl coenzyme A?

25.42 How many steps in the citric acid cycle involve oxidation? What reduced forms of coenzymes are formed?

25.43 How much ATP is produced from one mole of the acetyl group?

25.44 Why are the citric acid cycle and the electron transport chain interrelated processes?

25.45 Explain how ATP, NAD^+, and NADH control the rate of conversion of acetyl groups into CO_2 in the citric acid cycle.

25.46 Malonic acid is a competitive inhibitor in the citric acid cycle. What enzyme does it inhibit?

25.47 Why is succinyl coenzyme A a competitive inhibitor of the citric acid cycle?

25.48 How does ATP control the citric acid cycle?

Mitochondria

25.49 The number of mitochondria in a cell varies according to the function of the cell. Explain why liver cells have a large number of mitochondria.

25.50 The mitochondria have transport carriers for substances such as citric acid and α-ketoglutaric acid. Explain why these carriers are needed.

25.51 How is the flow of ADP and ATP across the mitochondrial membrane controlled?

25.52 Across which cellular membrane are the protons of the respiratory chain pumped?

25.53 If the inner mitochondrial membrane is broken, the oxidation of NADH can still occur, but the formation of ATP stops. Explain why.

Metabolism of Carbohydrates

A cow can metabolize the polysaccharide cellulose in grass and produce the disaccharide lactose in milk.

26.1
Digestion of Carbohydrates

Carbohydrates provide about 50% of the daily energy needs of a typical American. The digestion of starch, the principal carbohydrate in diets, starts in the mouth where the enzyme ptyalin catalyzes hydrolysis of starch to give a mixture of lower-molecular-weight carbohydrates called dextrins and maltose. Hydrolysis of starch and dextrin continues in the esophagus, but is slowed in the acid fluid of the stomach where the enzyme activity is low. Digestion then continues in the intestine where the enzyme maltase catalyzes the hydrolysis of maltose to produce glucose.

Other carbohydrates such as the disaccharides sucrose and lactose are also hydrolyzed in the digestive process. Sucrose yields glucose and fructose, whereas lactose yields glucose and galactose. The three monosaccharides glucose, fructose, and galactose are absorbed through the wall of the small intestine into the blood.

Fructose is converted in the liver into fructose 1-phosphate, which is then split into dihydroxyacetone phosphate and glyceraldehyde. Phosphorylation of glyceraldehyde yields glyceraldehyde 3-phosphate, which can enter the same glycolysis pathway that is used to metabolize glucose.

Galactose is converted into galactose 6-phosphate and eventually into glucose 1-phosphate by a series of four steps of a galactose–glucose interconversion pathway. One of these steps involves epimerization at the number 4 carbon atom. Glucose 1-phosphate is converted into glycogen, which then can produce glucose at a later time.

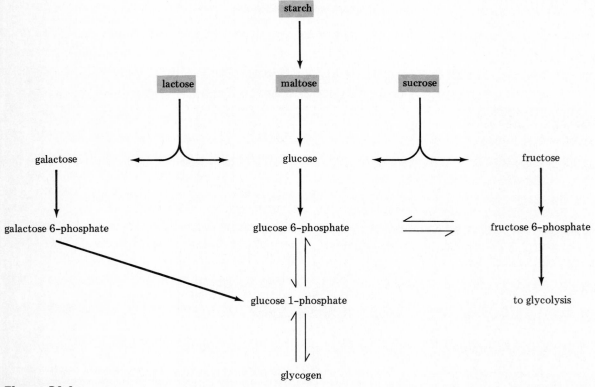

Figure 26.1
Common Metabolism of Dietary Carbohydrates.

In summary, the three principal aldoses are transformed by reactions to glucose derivatives or the metabolic intermediates derived from glucose (Figure 26.1). It thus is appropriate to consider what happens to glucose in the body.

The levels of glucose in the blood and of glycogen, the storage form of glucose, are regulated by many interrelated metabolic pathways. Each of these carefully balanced sequences of reactions is designed to produce energy, store chemical energy in other forms, or produce necessary intermediates for biosynthesis (Figure 26.2).

26.2
Metabolic Overview of Glucose

Glycogenesis and Glycogenolysis

If there is too much glucose in the blood, glycogenesis (Section 21.4) occurs to store the glucose as glycogen. When the glucose level in the blood drops, glycogen is hydrolyzed in a series of steps called glycogenolysis (Section 21.4). Thus glucose, which is virtually the sole fuel for the brain, must be obtained from the diet or from stored glycogen. Only under conditions of starvation do other materials substitute for glucose in maintaining the brain. Note that only glycerol from fats and pyruvic acid derived from some amino acids can be converted into glucose (Figure 26.2).

Glycolysis

Glucose is oxidized in all cells to provide energy as well as pyruvic acid and acetyl coenzyme A both of which can be used to produce fats and some amino acids. The biosynthetic pathways of fats and amino acids will be discussed in subsequent chapters.

The catabolism of glucose can be considered in two parts. The first part is a series of reactions known as glycolysis (Section 21.4) or the Embden–Meyerhof pathway. *In* **glycolysis** *a mole of glucose is converted into 2 moles of pyruvic acid, 2 moles of NADH, and 2 moles of ATP.* The fate of pyruvic acid and NADH in the second part of the catabolism of glucose depends on the availability of oxygen in the cell. If the supply of oxygen is insufficient, pyruvic acid is reduced to lactic acid by NADH. In the liver the lactic acid is converted back into pyruvic acid and then into glucose via gluconeogenesis. With sufficient oxygen, the NADH is oxidized in the electron transport chain in the mitochondria and the pyruvic acid is converted into acetyl coenzyme A which is then oxidized in the citric acid cycle.

In terms of ATP glycolysis stores only a small amount of energy, since only 2

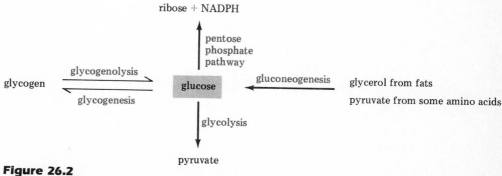

Figure 26.2
Metabolic Pathways of Glucose.

moles of ATP are formed per mole of glucose. However, the steps in the pathway include an investment, recovery, and small-profit part of the metabolism of glucose. The citric acid cycle and the electron transport chain constitute a major energy-profit part of the metabolic reactions. The majority of this chapter will be devoted to a detailed analysis of the catabolic reactions of glucose.

Pentose Phosphate Pathway

Glucose in the form of glucose 6-phosphate is involved in the pentose phosphate pathway (Figure 26.2), one of the important anabolic reactions of carbohydrates. The relationship of the amount of ribose 5-phosphate and NADPH generated in this pathway will be discussed later in this chapter.

26.3
Glycolysis

The reactions of glycolysis are summarized in Figure 26.3. The subsequent discussion gives the structures of the intermediates, for each of the individual steps. Each structure is given in the ionized form.

Step 1: Glucose undergoes substrate level phosphorylation by ATP as part of an energy-investment phase of glycolysis. The reaction is catalyzed by hexokinase.

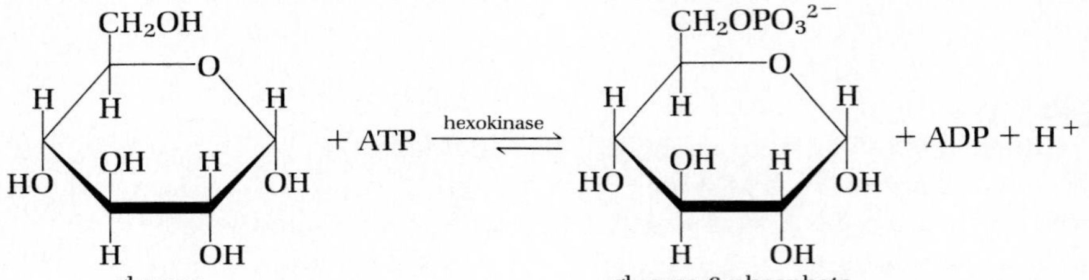

glucose $\xrightarrow{\text{hexokinase}}$ glucose 6-phosphate $+$ ADP $+$ H$^+$ ($+$ ATP)

Step 2: Glucose 6-phosphate is isomerized into fructose 6-phosphate in a reaction catalyzed by phosphoglucose isomerase.

glucose 6-phosphate $\xrightleftharpoons{\text{phosphoglucose isomerase}}$ fructose 6-phosphate

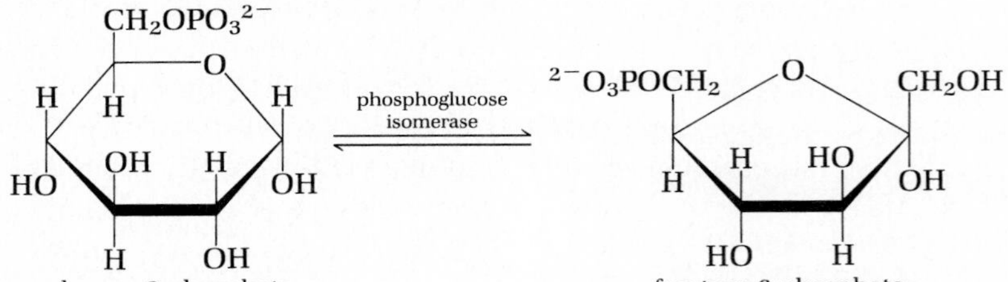

Step 3: Fructose 6-phosphate is phosphorylated by ATP. This reaction is another energy investment.

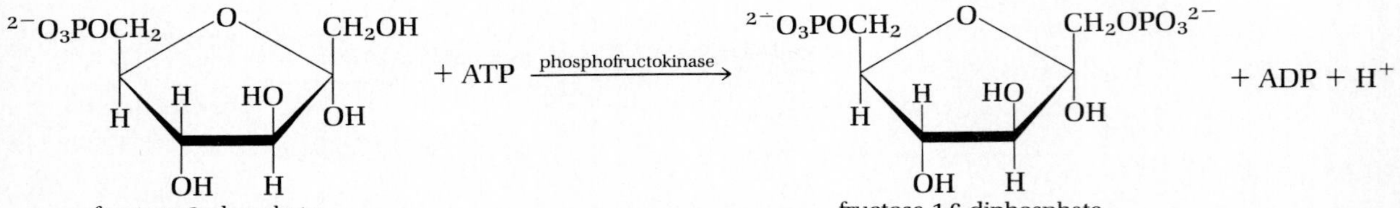

fructose 6-phosphate $+$ ATP $\xrightarrow{\text{phosphofructokinase}}$ fructose 1,6-diphosphate $+$ ADP $+$ H$^+$

Step 4: Fructose 1,6-diphosphate is cleaved into glyceraldehyde 3-phosphate and dihydroxyacetone phosphate. (Fructose 1,6-diphosphate is now written in the

Figure 26.3
Glycolysis Reactions.

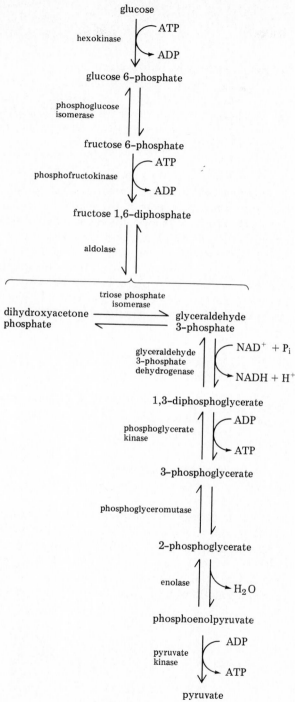

open-chain form because it makes this step easier to see.) This step is a reverse of an aldol reaction (Section 17.7).

$$
\underset{\substack{\text{fructose}\\\text{1,6-diphosphate}}}{
\begin{array}{c}
CH_2OPO_3{}^{2-}\\
|\\
C=O\\
|\\
HO-C-H\\
|\\
H-C-OH\\
|\\
H-C-OH\\
|\\
CH_2OPO_3{}^{2-}
\end{array}}
\;\xrightleftharpoons{\text{aldolase}}\;
\underset{\substack{\text{dihydroxyacetone}\\\text{phosphate}}}{
\begin{array}{c}
CH_2OPO_3{}^{2-}\\
|\\
C=O\\
|\\
HO-C-H\\
|\\
H
\end{array}}
\;+\;
\underset{\substack{\text{glyceraldehyde}\\\text{3-phosphate}}}{
\begin{array}{c}
H\quad O\\
\diagdown\;//\\
C\\
|\\
H-C-OH\\
|\\
CH_2OPO_3{}^{2-}
\end{array}}
$$

Fructose 1-phosphate, obtained by phosphorylation of fructose in the liver, also undergoes a reverse aldol reaction. The products are dihydroxyacetone phosphate and glyceraldehyde, which is subsequently phosphorylated. Thus the metabolic paths of glucose and fructose converge at this point.

Step 5: Dihydroxyacetone phosphate and glyceraldehyde 3-phosphate are isomers. They may be interconverted via an enediol intermediate (Section 21.9). Dihydroxyacetone phosphate cannot be used directly by cells. Therefore, dihydroxyacetone phosphate is isomerized into glyceraldehyde 3-phosphate, which can continue in the subsequent metabolic reactions.

$$
\underset{\substack{\text{dihydroxyacetone phosphate}\\\text{(a ketose)}}}{
\begin{array}{c}
CH_2OH\\
|\\
C=O\\
|\\
CH_2OPO_3{}^{2-}
\end{array}}
\;\xrightleftharpoons[\text{isomerase}]{\substack{\text{triose phosphate}}}\;
\underset{\substack{\text{glyceraldehyde 3-phosphate}\\\text{(an aldose)}}}{
\begin{array}{c}
O\quad H\\
\diagdown\;//\\
C\\
|\\
H-C-OH\\
|\\
CH_2OPO_3{}^{2-}
\end{array}}
$$

After this point you will have to remember that one glucose molecule produces two glyceraldehyde 3-phosphate molecules. In all subsequent reactions the energy produced per mole of substrate must be multiplied by 2 to obtain energy per mole of glucose.

Step 6: Glyceraldehyde 3-phosphate is oxidized and phosphorylated to form 1,3-diphosphoglycerate (DPG). NAD$^+$ is the oxidizing agent and 1 mole of NADH is produced. Remember that this quantity must be multiplied by 2 since 1 mole of glucose gave 2 moles of substrate for this reaction. The fate of NADH depends on whether oxygen is present or absent. We will return to this question when the possible reactions of pyruvic acid are considered.

$$
NAD^+ + P_i +
\underset{\substack{\text{glyceraldehyde}\\\text{3-phosphate}}}{
\begin{array}{c}
O\quad H\\
\diagdown\;//\\
C\\
|\\
H-C-OH\\
|\\
CH_2OPO_3{}^{2-}
\end{array}}
\;\xrightleftharpoons[\text{dehydrogenase}]{\substack{\text{glyceraldehyde}\\\text{3-phosphate}}}\;
\underset{\text{1,3-diphosphoglycerate}}{
\begin{array}{c}
O\quad OPO_3{}^{2-}\\
\diagdown\;//\\
C\\
|\\
H-C-OH\\
|\\
CH_2OPO_3{}^{2-}
\end{array}}
\;+\; NADH
$$

Step 7: The high-energy 1,3-diphosphoglycerate phosphorylates ADP. This reaction results in the first ATP formed in glycolysis. Multiplying by 2, there are two

ATP molecules formed from the original glucose molecule. However, two ATP molecules were required earlier in the sequence and as a result Step 7 only results in a recovery of the initial investment of ATP.

Step 8: 3-Phosphoglycerate is isomerized to 2-phosphoglycerate.

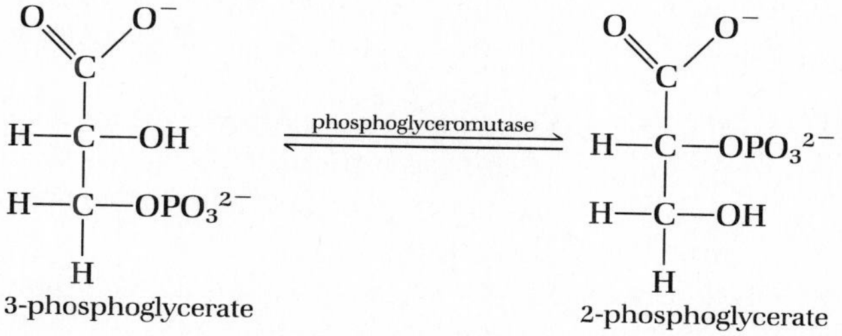

Step 9: Dehydration of 2-phosphoglycerate yields phosphoenolpyruvate, a high-energy phosphorylating agent.

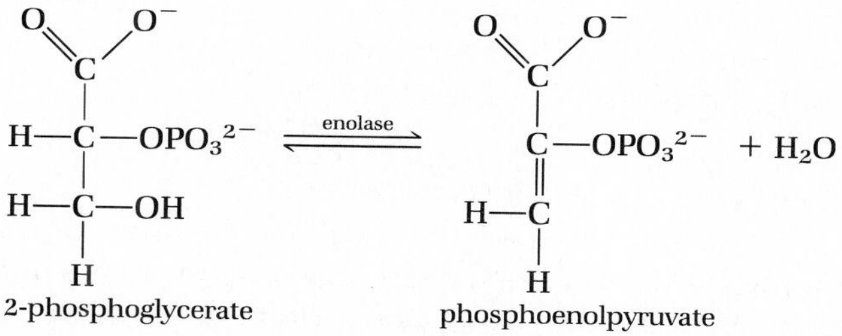

Step 10: The high-energy phosphoenolpyruvate converts ADP into ATP, and pyruvic acid is formed. Remembering to multiply by 2, we now have the first "profit" of 2 moles of ATP from the original 1 mole of glucose.

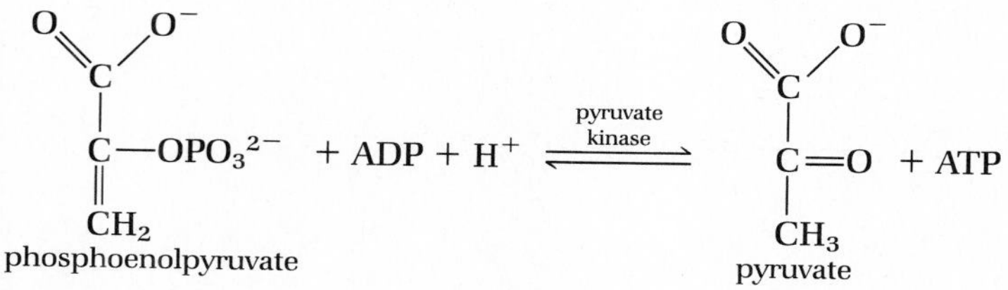

The net reaction in the transformation of glucose into pyruvic acid is

glucose + 2 P_i + 2 ADP + 2 NAD$^+$ $\longrightarrow$
$$2 \text{ pyruvic acid} + 2 \text{ ATP} + 2 \text{ NADH} + 2 \text{ H}^+ + 2 \text{ H}_2\text{O}$$

The four steps of glycolysis in which ATP is generated or consumed are summarized in Table 26.1.

Table 26.1

Reactions of Glycolysis That Use and Produce ATP

Reaction	ATP change per glucose
glucose $\longrightarrow$ glucose 6-phosphate	-1
fructose 6-phosphate $\longrightarrow$ fructose 1,6-diphosphate	-1
2 1,3-diphosphoglycerate $\longrightarrow$ 2 3-phosphoglycerate	$+2$
2 phosphoenolpyruvate $\longrightarrow$ 2 pyruvate	$+2$
Net	$+2$

26.4
Pyruvic Acid—A Key Compound

In Chapter 20, it was stated that pyruvic acid, glucose, and acetyl coenzyme A are three key compounds in metabolic reactions. The centrality of pyruvic acid is shown in Figure 26.4.

Lactic acid and pyruvic acid are interrelated by a simple oxidation–reduction reaction. Under conditions such as exist in strenuous exercise, pyruvic acid is reduced to lactic acid, which then is reoxidized to pyruvic acid in the liver. This reaction will be further discussed in this section.

Another related set of reactions is the interconversion of pyruvic acid and alanine by a process called transamination. By this process a link is established between the metabolism of amino acids and carbohydrates.

Pyruvic acid can be converted into oxaloacetic acid inside the mitochondria. This process is a source to replenish the citric acid cycle intermediates. In addition, the oxaloacetic acid is used in gluconeogenesis.

Finally, pyruvic acid can be converted into acetyl coenzyme A if ATP is needed. This process will be discussed further in this section.

Fate of Pyruvic Acid Under Anaerobic Conditions

Anaerobic *means in the absence of or not requiring oxygen.* Under anaerobic conditions, NADH cannot be oxidized via the electron transport chain, and the supply of NAD^+ needed to serve as an oxidizing agent would not be replenished.

$$2\,NADH + 2\,H^+ + O_2 \longrightarrow 2\,NAD^+ + 2\,H_2O$$

Since NAD^+ is required for glycolysis, this important pathway would stop if there were not an alternative way of producing NAD^+. During strenuous exercise, the alternative pathway involves the reduction of pyruvic acid to lactic acid.

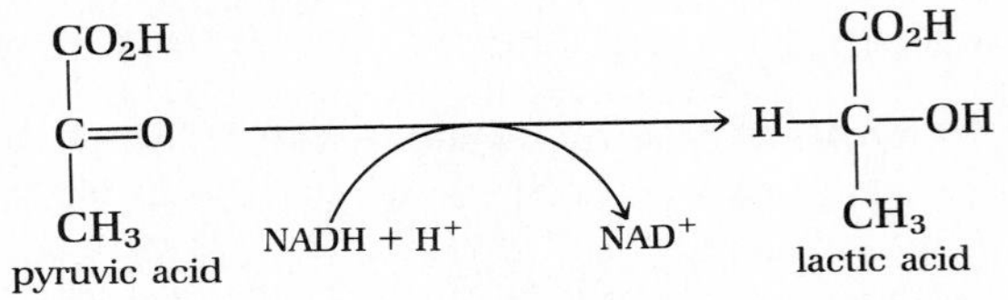

Adding this reaction to the net reaction of glycolysis gives the net reaction for anaerobic glycolysis.

$$glucose + 2\,ADP + 2\,P_i \longrightarrow lactic\ acid + 2\,ATP$$

Note that this reaction does not involve oxidation. The oxidation in glycolysis is compensated by the reduction of pyruvic acid.

The anaerobic process generates some ATP to be used by the muscles, but

Figure 26.4
Metabolic Pathways of
Pyruvic Acid.

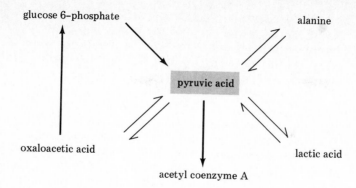

very little of the chemical energy of glucose has been released. There is another penalty to the body under anaerobic conditions. If the muscle tissues cannot obtain enough oxygen, the lactic acid concentration increases in the muscles and blood and causes fatigue. *The accumulated lactic acid constitutes an* **oxygen debt.**

Lactic acid must be oxidized back to pyruvic acid in order to be metabolized. Lactic acid diffuses out of the muscle into the blood and is carried to the liver where the Cori cycle occurs. In the **Cori cycle** *lactic acid is oxidized to pyruvic acid, which is then converted into glucose via gluconeogenic processes.* The net result of the Cori cycle is to shift the metabolic burden from the muscles to the liver (Figure 26.5).

Fate of Pyruvic Acid Under Aerobic Conditions

Aerobic *means in the presence of oxygen.* Under aerobic conditions, pyruvic acid reacts with coenzyme A to produce acetyl coenzyme A in a process that involves many steps. However, we will refer to it as Step 11 (aerobic). This step, which is the link between glycolysis and the citric acid cycle, occurs in the matrix of the mitochondria. On the basis of the original mole of glucose, 2 moles of NADH are produced.

$$CH_3-\overset{O}{\overset{\|}{C}}-CO_2H + CoA \xrightarrow[\text{NAD}^+ \quad \text{NADH} + \text{H}^+]{} CH_3-\overset{O}{\overset{\|}{C}}-CoA + CO_2$$

Figure 26.5
The Cori Cycle.
Lactic acid formed in the muscle under anaerobic conditions is carried to the liver to re-form glucose. Since gluconeogenesis requires energy, the metabolic burden is shifted to the liver.

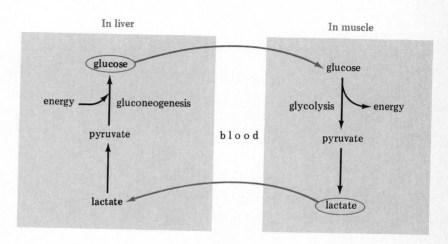

At this point we can see how the biological system can really "profit" from the energy of glucose. In total, the 1 mole of glucose has produced 2 moles of ATP, 4 moles of NADH, and 2 moles of acetyl coenzyme A. Acetyl coenzyme A then produces NADH, FADH$_2$, and ATP in the citric acid cycle. The NADH and FADH$_2$ cause the formation of ATP in the electron transport chain.

26.5
The ATP Profit from Glucose

There is one fine point to be considered before the final "profit" from the metabolism of glucose can be calculated. Glycolysis enzymes are in the cytoplasm of the cell and 2 moles of NADH per mole of glucose (Step 6) are released in the cytoplasm. However, the 2 moles of NADH per mole of glucose produced in converting pyruvic acid into acetyl coenzyme A are released in the mitochondria.

Since the reactions of the citric acid cycle, as well as those of the electron transport chain, occur in the mitochondria, it is necessary to have a mechanism to transport cytoplasmic NADH into the mitochondria. However, mitochondria are impermeable to NADH and NAD$^+$! The solution is that the electrons from NADH are shuttled across the mitochondrial membrane, but not NADH itself. The two shuttles for electrons are called the glycerol 3-phosphate shuttle and the malate–aspartate shuttle.

In the glycerol 3-phosphate shuttle, the reduced carrier diffuses across the mitochondrial membrane and is oxidized by FAD (Figure 26.6). The dihydroxyacetone phosphate then diffuses out of the mitochondria into the cytoplasm. The net reaction is

$$\underset{\text{(cytoplasmic)}}{\text{NADH}} + \text{H}^+ + \underset{\text{(mitochondrial)}}{\text{FAD}} \longrightarrow \underset{\text{(cytoplasmic)}}{\text{NAD}^+} + \underset{\text{(mitochondrial)}}{\text{FADH}_2}$$

As a consequence, only 2, rather than 3, ATP are formed for each cytoplasmic NADH. The price that is paid for this transport is not wasted. Without such a shuttle, the potential energy stored in cytoplasmic NADH would not be available.

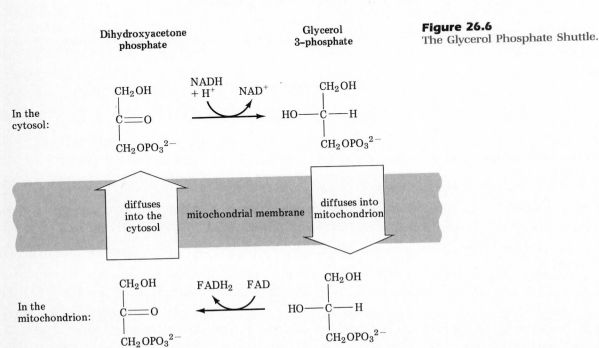

Figure 26.6
The Glycerol Phosphate Shuttle.

Table 26.2

Yield of NADH, FADH$_2$, and ATP from the Metabolism of Glucose

Reaction	ATP yield per glucose
Glycolysis: glucose $\longrightarrow$ pyruvate (in the cytosol)	
Phosphorylation of glucose	−1
Phosphorylation of fructose 6-phosphate	−1
Dephosphorylation of 2 molecules of 1,3-DPG	+2
Dephosphorylation of 2 molecules of phosphoenolpyruvate	+2
2 NADH formed in the oxidation of 2 molecules of glyceraldehyde 3-phosphate	
Pyruvate $\longrightarrow$ acetyl CoA (in mitochondria)	
2 NADH formed	
Citric acid cycle (in mitochondria)	
2 molecules of guanosine triphosphate are formed from 2 molecules of succinyl CoA	
6 NADH formed	+2
2 FADH$_2$ formed	
Oxidative phosphorylation (in mitochondria)	
2 NADH formed in glycolysis; each yields 2 ATP (assuming transport of NADH by the glycerol phosphate shuttle)	+4
2 NADH formed in the oxidative decarboxylation of pyruvate; each yields 3 ATP	+6
2 FADH$_2$ formed in the citric acid cycle; each yields 2 ATP	+4
6 NADH formed in the citric acid cycle; each yields 3 ATP	+18
Net	+36

In the malate–aspartate shuttle, electrons from NADH in the cytoplasm are transferred to malate, which can cross the inner mitochondrial membrane. Then this carrier of electrons is oxidized by NAD$^+$ within the mitochondrial matrix to form NADH. The net reaction of the shuttle is

$$\underset{\text{(cytoplasmic)}}{\text{NADH}} \; + \; \underset{\text{(mitochondrial)}}{\text{NAD}^+} \; \longrightarrow \; \underset{\text{(cytoplasmic)}}{\text{NAD}^+} \; + \; \underset{\text{(mitochondrial)}}{\text{NADH}}$$

This process is reversible, and no energy is consumed. Thus, the NADH produced in the cytoplasm will yield 3 moles of ATP in the electron transport chain in the mitochondria. However, in spite of the energy-conserving feature of this shuttle, there is a limitation. The shuttle cannot occur against a high NADH concentration in the mitochondria.

Now let's summarize the numerous steps in the metabolism of 1 mole of glucose in terms of the ATP produced (Table 26.2). In this summary the glycerol 3-phosphate shuttle is used. If the malate-aspartate shuttle were used, a total of 38 ATP rather than 36 ATP would be formed.

When 1 mole of glucose is burned to produce carbon dioxide and water, 686 kcal of energy is liberated. The same oxidation in cells produces 36 moles of ATP, which accounts for $36 \times 7.3 = 263$ kcal of energy stored. Thus the metabolic process is 38% efficient.

$$\frac{263}{686} \times 100 = 38\% \text{ efficient}$$

While this efficiency may not seem high, the internal combustion engines of cars are far less efficient. Furthermore, the human body does not really need a 100% efficient energy storage system, since some energy is used to maintain the temperature of the body.

26.6 Gluconeogenesis

Synthesis of glucose from noncarbohydrate precursors is gluconeogenesis. Without this important pathway, the brain, which is dependent on glucose as a fuel, would be starved if food intake were restricted even for one day. The sources of material to produce glucose are lactic acid, amino acids, and glycerol.

Lactic acid is formed in the muscles when the rate of glycolysis exceeds the rate of the citric acid cycle and the electron transport chain. This anaerobic state occurs in active exercise. Amino acids are obtained from proteins in the diet. However, during starvation, proteins in muscle are hydrolyzed and oxidized. The hydrolysis of triglycerides yields glycerol and fatty acids. Glycerol can be converted into pyruvic acid for gluconeogenesis, but fatty acids cannot.

The steps in gluconeogenesis are listed in Figure 26.7. At first glance, these reactions look like the reverse of the glycolysis reactions. However, a closer inspection reveals that gluconeogenesis is not the reverse of glycolysis.

The free energy released in the formation of pyruvic acid from glucose is about −20 kcal/mole. This large negative $\Delta G°$ effectively makes glycolysis irreversible. The three reactions that cannot be reversed are steps 1, 3, and 10.

$$\text{glucose} + \text{ATP} \xrightarrow{\text{hexokinase}} \text{glucose 6-phosphate} + \text{ADP}$$

$$\text{fructose 6-phosphate} + \text{ATP} \xrightarrow{\text{phosphofructokinase}} \text{fructose 1,6-diphosphate} + \text{ADP}$$

$$\text{phosphoenolpyruvate} + \text{ADP} \xrightarrow{\text{pyruvate kinase}} \text{pyruvate} + \text{ATP}$$

These steps must be bypassed by different reactions in gluconeogenesis.

Pyruvic acid is converted to phosphoenolpyruvate by way of oxaloacetic acid, which is not involved in the glycolysis pathway. The two reactions are fueled by ATP and GTP in order to overcome the unfavorable free energy difference between pyruvic acid and phosphoenolpyruvate.

pyruvic acid　　　　oxaloacetic acid

oxaloacetic acid　　　　phosphoenolpyruvic acid

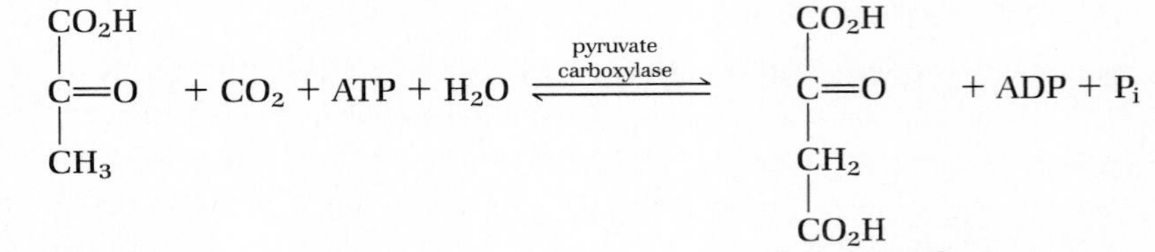

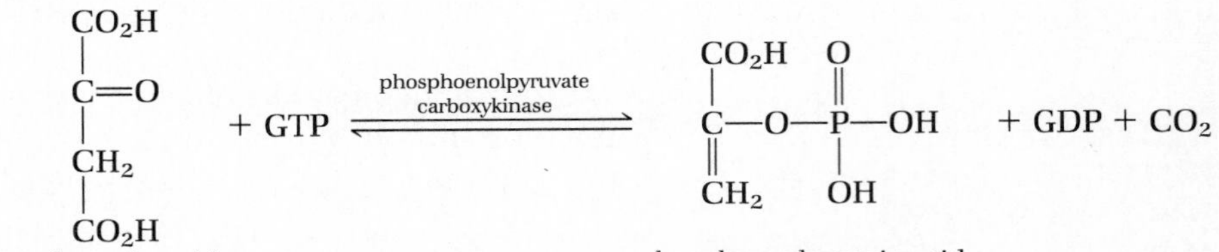

Figure 26.7
Gluconeogenesis Pathway.
The distinctive reactions of gluconeogenesis are indicated by
colored arrows. The other reactions are common to glycolysis.
Entry points of other materials
are indicated.

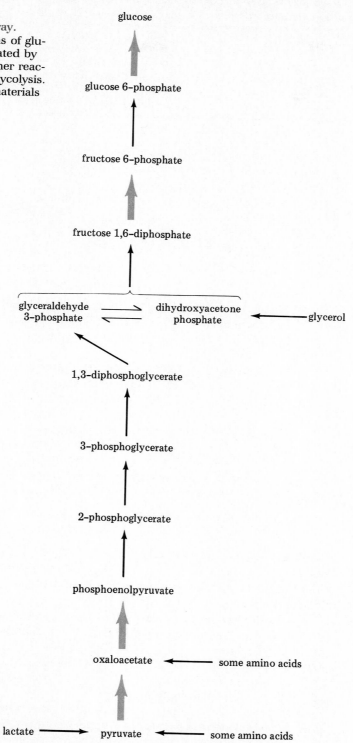

Table 26.3 ─────────────────────

Enzymatic Differences Between Glycolysis
and Gluconeogenesis

Glycolysis	Gluconeogenesis
hexokinase	glucose 6-phosphatase
phosphofructokinase	fructose 1,6-diphosphatase
pyruvate kinase	pyruvate carboxylase
	phosphoenolpyruvate carboxykinase

Conversion of fructose 1,6-diphosphate into fructose 6-phosphate is catalyzed by fructose 1,6-diphosphatase, a different enzyme than required for the reverse glycolysis reaction (Table 26.3).

Hydrolysis of glucose 6-phosphate into glucose is catalyzed by glucose 6-phosphatase, also a different enzyme than required for the reverse glycolysis reaction (Table 26.3).

The net reaction for gluconeogenesis is

$$2 \text{ pyruvic acid} + 4\,\text{ATP} + 2\,\text{GTP} + 2\,\text{NADH} + 2\,H_2O \longrightarrow$$
$$\text{glucose} + 4\,\text{ADP} + 2\,\text{GDP} + 6\,P_i + 2\,\text{NAD}^+$$

The free energy change for the reaction is -9 kcal/mole. Since glycolysis has $\Delta G° = -20$ kcal/mole, a direct reverse of the glycolysis pathway would require a free energy input of 20 kcal/mole. It is for this reason that gluconeogenesis is not the reverse of glycolysis.

Glycolysis to produce pyruvic acid generates 2 ATP. In order to reconvert pyruvic acid to glucose in gluconeogenesis, six high-energy phosphates (4 ATP + 2 GTP) are required. Thus four high-energy phosphates are lost to return pyruvate to glucose. The gluconeogenesis pathway then can occur only at the expense of other ATP-producing metabolic processes.

26.7
The Pentose Phosphate Pathway

In the **pentose phosphate pathway,** *glucose 6-phosphate is oxidized to ribose 5-phosphate and NADPH is produced.* The net reaction is

$$\text{glucose 6-phosphate} + 2\,\text{NADP}^+ + H_2O \longrightarrow$$
$$\text{ribose 5-phosphate} + 2\,\text{NADPH} + 2\,H^+ + CO_2$$

The ribose 5-phosphate is used to form ATP, NAD$^+$, FAD, coenzyme A, RNA, and DNA. The NADPH is used in reductive biosynthetic reactions. The pathway is also called the pentose shunt, the hexose monophosphate pathway, or phosphogluconate oxidative pathway.

The pentose phosphate pathway has two sections to be considered: an oxidative section and a nonoxidative section. The reactions of the oxidative section (steps 1–3) are given in Figure 26.8. The enzyme for step 1 is very specific for NADP$^+$ since the K_m value is about 1000 times greater than the K_m value for NAD$^+$. Thus these two coenzymes are not confused, and their separate identities and chemistry are maintained.

The nonoxidative section of the pentose phosphate pathway is shown in Figure 26.9. Although these reactions appear complex, you have already learned the basis of reactions 4 and 5, which occur via enediol intermediates (Section 21.9).

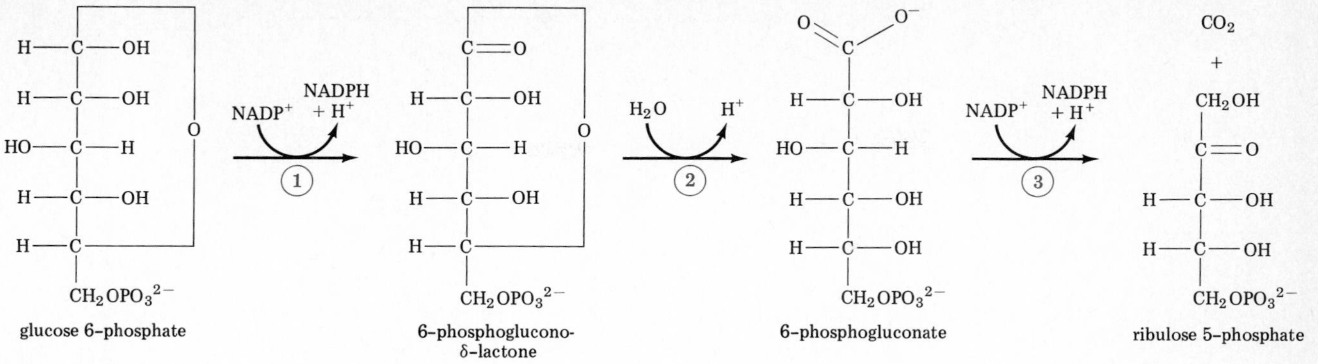

Figure 26.8
Oxidative Section of Pentose Phosphate Pathway.

Reactions 6, 7, and 8 all involve either transfer of a two-carbon unit, catalyzed by a transketolase, or transfer of a three-carbon unit, catalyzed by a transaldolase. The donor of the unit is a ketose, whereas the acceptor is an aldose.

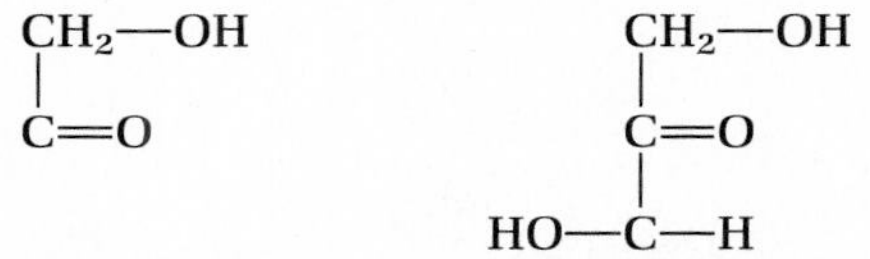

(transferred by transketolase   (transferred by transaldolase)

Reactions 6, 7, and 8 can be viewed more easily if only the numbers of carbon atoms in the compounds are shown.

$$C_5 + C_5 \xrightleftharpoons{\text{transketolase}} C_3 + C_7$$

$$C_3 + C_7 \xrightleftharpoons{\text{transaldolase}} C_4 + C_6$$

$$C_5 + C_4 \xrightleftharpoons{\text{transketolase}} C_3 + C_6$$

The sum of these three reactions is

$$3\,C_5 \rightleftharpoons 2\,C_6 + C_3$$

Thus, three pentoses give two hexoses and a triose.

Note that these reactions are reversible. Since C_6 and C_3 compounds are part of the glycolysis pathway, the nonoxidative section of the pentose phosphate pathway and glycolysis are linked by common intermediates.

In the balanced equation given for the oxidative section of the pentose phosphate pathway, 2 moles of NADPH are produced per mole of ribose 5-phosphate. However, there are times when the biological needs for these two substances are not in a 2:1 ratio. If more ribose 5-phosphate is required, it is formed by the nonoxidative branch from glycolysis intermediates. If more NADPH is required, then the ribose 5-phosphate is converted into glycolysis intermediates and is oxidized.

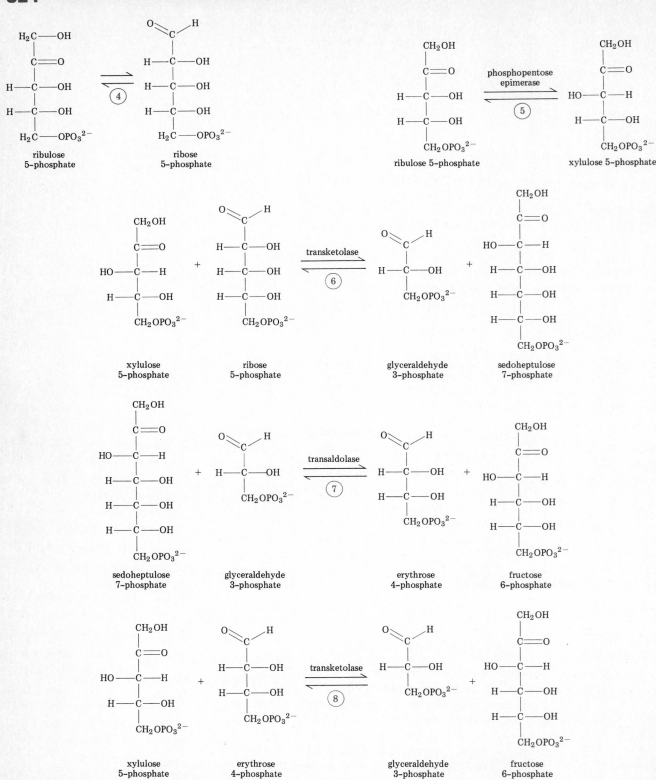

Figure 26.9
Nonoxidative Section of Pentose Phosphate Pathway.

Glycogenolysis and glycogenesis are not the simple reverse reactions of each other. The reactions of glycogenolysis are

$$\text{glycogen} + P_i \xrightarrow{\text{glycogen phosphorylase}} \text{glucose 1-phosphate} + \text{glycogen}$$
$$\text{(n residues)} \hspace{6cm} \text{($n-1$ residues)}$$

$$\text{glucose 1-phosphate} \xrightarrow{\text{phosphoglucomutase}} \text{glucose 6-phosphate}$$

$$\text{glucose 6-phosphate} + H_2O \xrightarrow{\text{glucose phosphatase}} \text{glucose} + P_i$$

Phosphorylated glucose cannot leave a cell, whereas glucose can. The enzyme glucose phosphatase is present only in the kidneys, intestine, and liver. Other organs, such as muscles and the brain, lack this enzyme and cannot release glucose to the blood. Thus, the glucose is maintained in areas requiring fuel for the formation of ATP.

The reactions of glycogenesis involve uridine diphosphate glucose as a donor of glucose (Figure 26.10). The reactions are

$$\text{glucose 1-phosphate} + \text{UTP} \xrightarrow{\text{UDP-glucose pyrophosphorylase}} \text{UDP-glucose} + PP_i$$

$$PP_i + H_2O \longrightarrow 2 P_i$$

$$\text{UDP-glucose} + \text{glycogen} \xrightarrow{\text{glycogen synthetase}} \text{glycogen} + \text{UDP}$$
$$\text{(n residues)} \hspace{6cm} \text{($n+1$ residues)}$$

The hydrolysis of pyrophosphoric acid (PP_i) is essentially irreversible and is required for the glycogenesis pathway.

Figure 26.10
Structure of Uridine Diphosphate Glucose.

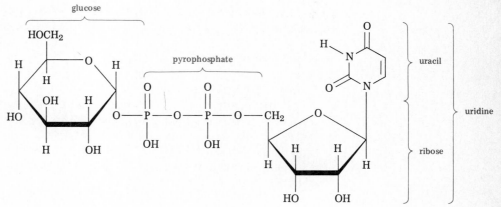

There are several control points for regulation of the metabolic reactions of carbohydrates. Each of the major pathways has such regulatory mechanisms.

Glycolysis

The phosphofructokinase-catalyzed step is controlled by the ATP/AMP ratio and is inhibited by a high energy charge on the cell. AMP allosterically stimulates the enzyme, and ATP inhibits it. Thus, when the energy charge of the cell is high,

AN ASIDE

There are a number of genetically inherited diseases that involve the storage of glycogen. The first of these, Von Gierke's disease, was described by Edgar Von Gierke in 1929. The liver lacks the enzyme glucose 6-phosphatase, which is required to hydrolyze glucose 6-phosphate so that it may leave the liver. Thus glycogen continues to be stored and the liver becomes very large. As a consequence, the rate of glycolysis in the liver is increased, leading to the release of pyruvic acid and lactic acid in the blood. Patients with Von Gierke's disease rely more on fat metabolism.

In McArdle's disease the muscles lack the enzyme phosphorylase, which is required for formation of glucose 1-phosphate from glycogen. Without a ready supply of glucose from glycogen the patient cannot perform strenuous exercise.

Glycogen contains $\alpha(1{\rightarrow}6)$ glucose branches. An enzyme is required to debranch glycogen in glycogenolysis and to branch glycogen in glycogenesis. In Cori's disease the liver lacks the debranching enzyme and glucose is only partially released from glycogen. In Andersen's disease both the liver and the spleen lack the enzyme to form branches in glycogen. As a consequence, progressive cirrhosis of the liver occurs. With liver failure by about the age of 2, the patient dies.

---

glycolysis does not occur, and as a consequence glycogenesis occurs resulting in storage of glucose as glycogen.

Gluconeogenesis

The major control point is the conversion of fructose 1,6-diphosphate into fructose 6-phosphate, which is catalyzed by fructose 1,6-diphosphatase. The enzyme is activated by citric acid and inhibited by AMP, the reverse of the effect of the glycolysis pathway on phosphofructokinase. When the energy charge of the cell is low, it is not energetically economical to produce glucose by a pathway that requires energy.

Pentose Phosphate Pathway

If the needs for NADPH and ribose 5-phosphate are balanced, no ATP is required in the pentose phosphate pathway. The reaction is controlled by $NADP^+$, which is distinctly different from NAD^+. Thus it is possible to have both a high $NADPH/NADP^+$ ratio and a high $NAD^+/NADH$ ratio in a cell. Reductive biosynthesis and glycolysis reactions can then both occur at a high rate when required by the cell.

Hormones

The pancreatic hormone glucagon stimulates the breakdown of glycogen in the muscles by activating glycogen phosphorylase. The hormone also inactivates glycogen synthetase, the enzyme that controls glycogenesis. Epinephine, another hormone, performs functions similar to those of glucagon.

Insulin, another pancreatic hormone, binds to protein receptors in cell membranes and facilitates entry of glucose into the cell. The release of insulin is triggered by high blood glucose levels. Insulin also increases the rate of synthesis of glycogen and stimulates glycolysis.

Summary

Dietary carbohydrates are digested to the monosaccharides glucose, fructose, and galactose. The latter two compounds are converted into intermediates that are metabolized by the pathways of glucose.

Because of the need of cells for glucose, the body has a series of interrelated metabolic pathways to maintain the requisite glucose level. Formation of glycogen and its subsequent hydrolysis provide the source of glucose during periods of low carbohydrate intake or when glucose is being rapidly used.

Glycolysis is a series of enzyme-catalyzed reactions producing 2 moles each of pyruvic acid, ATP, and NADH. Under anaerobic conditions NADH reduces pyruvic acid to lactic acid. Lactic acid is oxidized back to pyruvic acid in the liver. Under aerobic conditions pyruvic acid is converted to acetyl coenzyme A, which is oxidized in the citric acid cycle.

Gluconeogenesis, which occurs in the liver, converts pyruvic acid into glucose, which then is released into the blood to be transported to other organs. Transport of lactic acid to the liver, where it is converted to pyruvic acid and then glucose, is called the Cori cycle.

The pentose phosphate pathway results in the formation of ribose 5-phosphate and NADPH. The pentose can be interconverted with glycolysis intermediates. NADPH is used in reductive biosynthesis.

Carbohydrate metabolism is regulated by a combination of enzymes and hormones. The principal enzyme controls are phosphofructokinase in glycolysis and pyruvate carboxylase in gluconeogenesis. The hormones epinephrine and glucagon inactivate glycogen synthetase and activate glycogen phosphorylase. Insulin facilitates the entry of glucose into cells.

Learning Objectives

As a result of studying Chapter 26 you should be able to
- Give a metabolic overview of glucose.
- Outline the glycolysis of glucose.
- Describe the metabolic pathways of pyruvic acid.
- Give an accounting of ATP produced in the metabolism of glucose.
- Show why gluconeogenesis and glycolysis are not simple reverse reactions of each other.
- Describe the products of the oxidative section of the pentose phosphate pathway.
- Show how some glycolysis products and the compounds of the nonoxidative section of the pentose phosphate pathway are related.
- Show why glycogenolysis and glycogenesis are not simple reverse reactions of each other.
- Outline the control points of the metabolism of glucose.

New Terms

Aerobic processes occur in the presence of oxygen.
Anaerobic processes do not require oxygen.
In the **Cori cycle,** lactic acid is oxidized to pyruvic acid in the liver and then converted into glucose.
The **Embden–Meyerhof pathway** is another name for glycolysis.
Oxygen debt is a condition of high lactic acid concentration.
The **pentose phosphate pathway** produces ribose 5-phosphate and NADPH.

Questions and Problems

Terminology
26.1 Distinguish between glycogenolysis and glycogenesis.
26.2 Distinguish between glycolysis and gluconeogenesis.
26.3 Distinguish between aerobic and anaerobic.

Digestion of Carbohydrates
26.4 Where does the enzymatic hydrolysis of starch first occur?
26.5 Why is the enzymatic hydrolysis of starch slow in the stomach?
26.6 How are the metabolisms of fructose and galactose related to the metabolism of glucose?
26.7 Suggest a way in which mannose may be converted to glucose in a metabolic reaction.
26.8 What are the monosaccharides produced by hydrolysis of higher saccharides?

Glucose Metabolism
26.9 What organ uses glucose as its sole fuel?
26.10 What is the role of glycogen in controlling blood glucose level? What two processes involve glycogen formation and hydrolysis?
26.11 What are the fates of pyruvic acid under anaerobic and aerobic conditions?
26.12 What substances are produced in the pentose phosphate pathway?

Glycolysis
26.13 What is the end product of glycolysis?
26.14 Which coenzyme functions as the oxidizing agent in glycolysis?

26.15 How is NADH oxidized under aerobic conditions?

26.16 How is NADH used in glycolysis under anaerobic conditions?

26.17 How many moles of pyruvic acid are produced from glucose by glycolysis?

26.18 How many moles of ATP are produced from 1 mole of glucose in glycolysis?

26.19 How many moles of NADH are produced in the glycolysis of glucose?

26.20 What is the first step in glycolysis, and what enzyme catalyzes the reaction?

26.21 How many steps are there in glycolysis?

26.22 What type of enzyme catalyzes each of the following reactions?
   (a) glucose 6-phosphate $\longrightarrow$ fructose 6-phosphate
   (b) fructose 6-phosphate $\longrightarrow$
   fructose 1,6-diphosphate
   (c) 3-phosphoglycerate $\longrightarrow$ 2-phosphoglycerate

26.23 How many steps in glycolysis require ATP? How many steps produce ATP?

Pyruvic Acid

26.24 What reactions occur with pyruvic acid under anaerobic conditions?

26.25 Where is lactic acid oxidized to pyruvic acid?

26.26 Outline the Cori cycle.

26.27 What reaction occurs with pyruvic acid under aerobic conditions?

Metabolism of Glucose

26.28 How many molecules of acetyl coenzyme A are formed from one molecule of glucose? How many molecules of CO_2 are formed in this reaction?

26.29 How does the NADH formed in glycolysis differ from the NADH formed in the citric acid cycle?

26.30 Why are shuttles required to make use of the electrons of NADH produced in the cytoplasm?

26.31 What two shuttles operate in the mitochondria to transport reducing power across the membrane? How do the shuttles differ?

26.32 Consider the metabolism of fructose. Does the amount of ATP produced differ from that obtained from glucose?

Gluconeogenesis

26.33 What are the sources of compounds for gluconeogenesis?

26.34 What portion of a fat can be used for gluconeogenesis? What portion cannot be used?

26.35 How many steps of glycolysis must be bypassed in gluconeogenesis?

26.36 What intermediate in gluconeogenesis is common with the citric acid cycle?

26.37 What high-energy phosphates are used in gluconeogenesis?

26.38 What is the net energy cost to convert glucose to pyruvic acid and back to glucose?

Pentose Phosphate Pathway

26.39 How does NADPH differ from NADH in biological function?

26.40 What compound contains the carbon atom lost from glucose in its conversion to ribose?

26.41 Why does the first enzyme in the pentose phosphate pathway use $NADP^+$ rather than NAD^+.

26.42 What unit does a transaldolase transfer? What is the acceptor?

26.43 What unit does a transketolase transfer? What is the donor? What is the acceptor?

26.44 How many moles of NADPH are produced per mole of ribose?

26.45 How is ribose produced when NADPH is not required?

26.46 What occurs to the ribose that is produced because of a high NADPH requirement but is not itself required by the organism?

Glycogenolysis and Glycogenesis

26.47 What reaction determines whether glucose formed by glycogenolysis can leave a cell?

26.48 Which organs contain the enzyme glucose 6-phosphatase?

26.49 What reaction makes glycogenesis essentially irreversible?

26.50 What triphosphate is involved in the first step of glycogenesis?

Glycogen Storage Diseases

26.51 How do Von Gierke's and McArdle's diseases differ?

26.52 How do Cori's and Andersen's diseases differ?

Control of Carbohydrate Metabolism

26.53 Which reaction in glycolysis is a control point?

26.54 Which reaction in gluconeogenesis is a control point?

26.55 What hormones are involved in the control of glucose metabolism?

Metabolism of Lipids

27

A camel produces and stores fat that can be metabolized later for energy.

27.1
Fats as Fuel

Carbohydrates, which are a source of quick energy, are digested and absorbed very rapidly. The body stores glucose as glycogen, which can be broken down in seconds. It is glycogen that provides the body with an immediate source of energy.

Fats are digested very much more slowly than carbohydrates and stay in the digestive system to provide a source of energy well past your last meal. Fats are also stored in the body in large amounts, much to the chagrin of many individuals. However, it is these fat deposits that can be made available to keep the body running after the glycogen stores are used. Thus fats are a long-term energy storage, whereas carbohydrates are a short-term energy storage.

About 35% of the American diet consists of fats. As indicated in the previous chapter, fats cannot be used to provide the glucose necessary for the body. Thus fats in the diet are converted into other fats in the body. In addition, once the glycogen stores are full from carbohydrate sources, any additional carbohydrates that an animal takes in are converted into fat.

The fatty acids in triglycerides are a very compact storage form of energy. About 9 kcal is available from 1 g of fat, whereas 4 kcal is available from 1 g of anhydrous glycogen. Since about 2 g of water are actually stored with the very polar glycogen, the efficiency of fats as a store of energy is even greater on a weight basis; 1 g of fat has six times the energy content of 1 g of hydrated glycogen.

27.2
Digestion of Triglycerides

The digestion of triglycerides does not begin until they reach the small intestine. However, the hydrolysis is not complete, and a mixture of monoglycerides and diglycerides results in addition to glycerol and fatty acids. After passage through the intestinal wall these fragments are reassembled into triglycerides, which then are absorbed into the lymph system. These droplets of triglycerides give the usually clear lymph fluid a milky appearance.

Triglycerides are not soluble in blood, but are maintained in micelles with plasma proteins and lipoproteins (Section 23.1). Triglycerides that enter the blood are distributed as tissue lipid or as storage fat called depot fat. **Tissue lipids** *are contained in cell membranes; they consist of phosphoglycerides, phosphosphingolipids, and glycosphingolipids.* These lipids are used to maintain cell membranes and the membranes of cell organelles like the mitochondria. Except in extreme cases of starvation, the tissue lipids are not used to provide energy.

Depot fat, *also called* **adipose tissue,** *is connective tissue in the body that contains cellular fat.* Adipose tissue often occurs as a subcutaneous layer, particularly in the abdominal area and also the areas surrounding vital organs such as the kidneys and the spleen. Protection against physical shock and insulation for the organ are provided by adipose tissue, but the primary function of adipose tissue is to serve as a reserve of stored chemical energy. Given sufficient water the average adult can sustain life for approximately 40 days from adipose tissue.

When food is continually ingested and exercise is frequent, the human adipose tissue is constantly undergoing change. Newly absorbed fats are used to form new adipose tissue, and old adipose tissue is metabolized. A state of equilibrium can be achieved by a proper diet and regular exercise. There is no way to excrete excess lipid material from the body; lipids are only removed by oxidation. As a consequence, an intake of foodstuffs with little or no exercise will lead to obesity. In most cases excessive fat arises not from ingested fat but from carbohydrates that are converted to fat.

The release of fatty acids from adipose tissue is called **fatty acid mobilization.** The reaction involves hydrolysis of the ester bonds by enzymes called lipases. The lipases differ depending on their location in the body and the way in which they are regulated.

$$CH_2-O-\overset{\overset{O}{\|}}{C}-R$$
$$CH-O-\overset{\overset{O}{\|}}{C}-R \ +\ 3\,H_2O \xrightarrow{\text{lipase}} \begin{array}{l} CH_2OH \\ | \\ CHOH \\ | \\ CH_2OH \end{array} +\ 3\,RCO_2H$$
$$CH_2-O-\overset{\overset{O}{\|}}{C}-R$$

The free fatty acids are transported by serum albumin to various parts of the body where they may enter cells. Unlike glucose, which requires insulin and a membrane transport system to enter the cells, fatty acids easily cross the lipid bilayer of cell membranes. Thus, given a choice, most cells will oxidize fatty acids in preference to glucose. As a consequence, the body's glycogen storage is conserved so that the brain may use glucose. Brain cells cannot use fatty acids because they obtain their nutrients from cerebrospinal fluid rather than directly from the blood. This so-called blood–brain barrier is permeable for glucose, but not albumin-bound fatty acids.

The two components of triglycerides that are obtained by the action of lipases are glycerol and fatty acids. However, the glycerol contains only a small fraction of the carbon atoms of a triglyceride. The major source of energy from triglycerides, then, is from the fatty acids. Glycerol is phosphorylated in the liver to form glycerol 3-phosphate, which is then oxidized to dihydroxyacetone phosphate.

$$\text{glycerol} \xrightarrow[\underset{\text{ATP}\quad\text{ADP}}{}]{\text{glycerol kinase}} \text{glycerol 3-phosphate}$$

$$\text{glycerol 3-phosphate} \xrightarrow[\underset{\text{NAD}^+\quad\text{NADH + H}^+}{}]{\text{dehydrogenase}} \text{dihydroxyacetone phosphate}$$

These two steps provide a net of 2 moles of ATP. The dihydroxyacetone phosphate is then isomerized to glyceraldehyde 3-phosphate and enters the glycolysis pathway described for glucose. As a source of energy, the metabolism of glycerol is not as important as the metabolism of the fatty acids because glycerol is more highly oxidized. In glyceryl tristearate, fatty acids are highly reduced —CH_2— units. As a consequence, there is a large quantity of energy stored in fatty acids.

The fatty acids are metabolized by forming two-carbon units of acetyl coenzyme A, which then is oxidized in the citric acid cycle. The catabolic reactions of fatty acids occur in the mitochondria and yield acetyl coenzyme A in a series of five steps: activation, dehydrogenation, hydration, oxidation, and cleavage. These steps are summarized in Figure 27.1.

Step 1: Activation of the acid occurs by conversion to an acyl coenzyme A, which is a thioester. This process requires energy derived from ATP and is catalyzed by

Figure 27.1
Metabolism of
Fatty Acids.

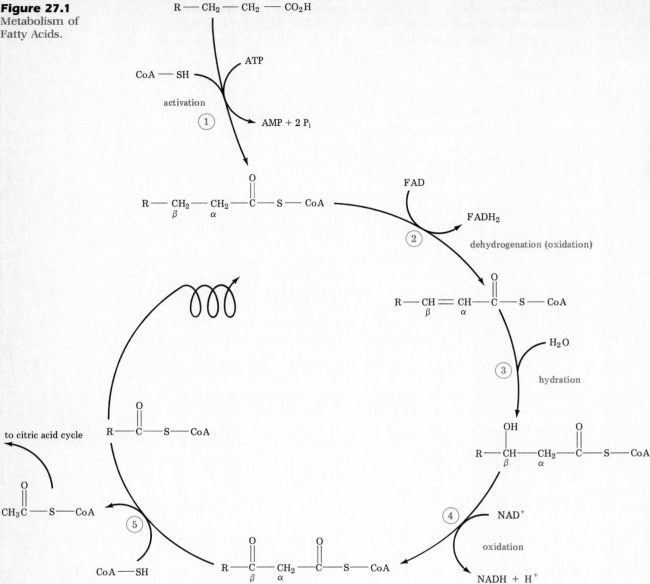

thiokinase. There are actually several thiokinase enzymes, each of which has a specificity for certain lengths of fatty acid chains.

$$R-CH_2-CH_2-\overset{\overset{\displaystyle O}{\|}}{C}-OH + CoA-SH \xrightarrow[\text{ATP}\quad\text{AMP}+\text{PP}_i]{\text{thiokinase}} R-CH_2-CH_2-\overset{\overset{\displaystyle O}{\|}}{C}-S-CoA + H_2O$$
a fatty acid   an acyl SCoA

Note that ATP is converted to AMP in this step because the energy required to form the activated fatty acid is about 8 kcal/mole.
Step 2: The enzyme acyl coenzyme A dehydrogenase catalyzes the dehydrogenation of the activated acid to form a *trans* double bond between the α- and β- carbon atoms of the acid.

$$\text{R}-\underset{\beta}{\text{CH}}-\underset{\alpha}{\text{CH}}-\overset{\text{O}}{\overset{\|}{\text{C}}}-\text{S}-\text{CoA} \quad \xrightarrow[\text{FAD} \quad \text{FADH}_2]{\text{dehydrogenase}} \quad \text{R}-\text{C}=\text{C}-\overset{\text{O}}{\overset{\|}{\text{C}}}-\text{S}-\text{CoA}$$

The four different enzymes that can catalyze this reaction require FAD as an oxidizing agent. Each specific enzyme catalyzes the dehydrogenation of acids most efficiently for a given range of chain lengths. Since 1 mole of $FADH_2$ yields 2 moles of ATP in the electron transport chain, this step provides both the "recovery" and "profit" part of the metabolism of fatty acids. Recall that the equivalent of 2 moles of ATP was required in step 1 in which ATP is converted to AMP.

Step 3: This step is a hydration of the double bond catalyzed by enolhydrase. The addition of water occurs so that the hydroxyl group is formed on the β-carbon atom and yields the L isomer exclusively.

$$\text{H}_2\text{O} + \text{R}-\text{C}=\text{C}-\overset{\text{O}}{\overset{\|}{\text{C}}}-\text{S}-\text{CoA} \longrightarrow \text{R}-\text{C}-\text{C}-\overset{\text{O}}{\overset{\|}{\text{C}}}-\text{S}-\text{CoA}$$

Step 4: The 3-hydroxyacyl CoA is oxidized by NAD^+ in a reaction catalyzed by the enzyme L-3-hydroxyacyl CoA dehydrogenase. The enzyme is specific for the L stereoisomer, but works for any chain length. With the formation of NADH, 3 moles of ATP are ultimately generated in the electron transport chain, and this series of metabolic reactions continues in the "profit" phase.

$$\text{R}-\underset{\underset{\text{H}}{|}}{\overset{\overset{\text{OH}}{|}}{\text{C}}}-\text{CH}_2-\overset{\text{O}}{\overset{\|}{\text{C}}}-\text{S}-\text{CoA} \quad \xrightarrow[\text{NAD}^+ \quad \text{NADH} + \text{H}]{\text{dehydrogenase}} \quad \text{R}-\overset{\text{O}}{\overset{\|}{\text{C}}}-\text{CH}_2-\overset{\text{O}}{\overset{\|}{\text{C}}}-\text{S}-\text{CoA}$$

Step 5: This step is a thiolytic cleavage in which a molecule of CoA reacts with acyl CoA to cleave a two-carbon fragment from the carboxyl end of the acid. The two-carbon fragment is released as acetyl coenzyme A. As a result a new acid shortened by two carbon atoms is still present as a coenzyme A ester which then proceeds through a repeat of the steps described.

$$\text{R}-\underset{\beta}{\overset{\text{O}}{\overset{\|}{\text{C}}}}-\underset{\alpha}{\text{CH}_2}-\overset{\text{O}}{\overset{\|}{\text{C}}}-\text{S}-\text{CoA} + \text{CoA}-\text{SH} \quad \xrightarrow{\text{thiolase}}$$

$$\underset{\text{an acyl CoA}}{\text{R}-\overset{\text{O}}{\overset{\|}{\text{C}}}-\text{S}-\text{CoA}} + \underset{\text{acetyl CoA}}{\text{CH}_3-\overset{\text{O}}{\overset{\|}{\text{C}}}-\text{S}-\text{CoA}}$$

Spiral Reactions

Fatty acids are metabolized in a **fatty acid spiral,** *which degrades the acid into two-carbon-atom chunks in a series of five steps.* These reactions resemble the set of reactions of the citric acid cycle except that in each cycle the substance being metabolized becomes degraded by an additional two carbon atoms. The spiral reaction sequence is shown in Figure 27.1. Note that step 1 does not occur again; the product of step 5 becomes the reactant of step 2. Each time the spiral reaction occurs the products are NADH, $FADH_2$, acetyl coenzyme A, and a shortened acid molecule combined with coenzyme A.

The Energy Scorecard

Now let's consider the reaction of palmitic acid. In order to cleave this 16-carbon acid into acetyl coenzyme A groups, the spiral reaction sequence must occur seven times. In each of the first six times one acetyl coenzyme A is formed but in the seventh cycle the four-carbon acid is cleaved into two acetyl coenzyme A groups.

Each cycle produces 1 mole of NADH and 1 mole of $FADH_2$. The two reduced coenzymes yield 5 moles of ATP in the electron transport chain. For palmitic acid there is the equivalent of $7 \times 5 = 35$ moles of ATP formed as a result of the seven spiral cycles. In addition, 8 moles of acetyl coenzyme A is formed, which gives $8 \times 12 = 96$ moles of ATP in the citric acid cycle coupled with the electron transport chain. The only energy cost to initiate the oxidation of the fatty acid is the hydrolysis of 1 mole of ATP to AMP, which is the equivalent of hydrolyzing two moles of ATP to ADP. The net number of moles of ATP produced is $35 + 96 - 2 = 129$. This quantity of ATP is equal to $129 \times 7.3 = 942$ kcal. Complete oxidation of palmitic acid in a combustion process liberates 2340 kcal/mole. The biological energy storage process, then, is 40% efficient.

$$\frac{942}{2340} \times 100 = 40\%$$

Example 27.1

How many moles of acetyl coenzyme A are produced from the metabolism of $CH_3(CH_2)_{12}CO_2H$? How many times must the fatty acid cycle occur?

Solution. The compound contains 14 carbon atoms and 7 acetyl coenzyme A units will be produced. Since the last cycle of metabolism produces 2 acetyl groups from a four-carbon acid, the cycle occurs only six times.

Example 27.2

What quantity of ATP is produced from the metabolism of $CH_3(CH_2)_{12}CO_2H$?

Solution. The compound, which contains 14 carbon atoms, must pass through the cycle six times to produce 7 acetyl coenzyme A units. The six cycles result in formation of six units of $FADH_2$ and six units of NADH. These reduced coenzymes yield a total of 30 ATP by oxidative phosphorylation.

$$6 \text{ NADH} \times \frac{3 \text{ ATP}}{1 \text{ NADH}} + \left(6 \text{ FADH}_2 \times \frac{2 \text{ ATP}}{1 \text{ FADH}_2} \right) = 30 \text{ ATP}$$

The 7 acetyl coenzyme A yield 84 ATP in the citric acid cycle.

$$7 \text{ acetyl CoA} \times \frac{12 \text{ ATP}}{1 \text{ acetyl CoA}} = 84 \text{ ATP}$$

Two ATP must be deducted for the activation of the acid. The total moles of ATP produced per mole of the acid is $30 + 84 - 2 = 112$.

[Additional examples may be found in **27.20–27.22** at the end of the chapter.]

27.4
Ketone Bodies

Excessive fat metabolism occurs in starvation, diabetes, and in high fat diets. In starvation, stored adipose tissue becomes the sole energy source. In diabetes, a condition in which glucose is not used efficiently, fat becomes the alternative energy source. In each of the conditions, there is a decrease in the level of oxaloacetic acid, which comes from carboxylation of pyruvic acid. Therefore, there is insufficient oxaloacetic acid to react with acetyl coenzyme A produced from fatty acid oxidation.

When acetyl coenzyme A cannot enter the citric acid cycle, it combines with itself in the liver to form acetoacetyl coenzyme A and liberates coenzyme A.

$$CH_3-\overset{O}{\overset{\|}{C}}-S-CoA + CH_3-\overset{O}{\overset{\|}{C}}-S-CoA \longrightarrow$$

$$CH_3\overset{O}{\overset{\|}{C}}-CH_2-\overset{O}{\overset{\|}{C}}-S-CoA + CoA-SH$$

This reaction is essentially a Claisen condensation (Section 18.7). Acetoacetyl coenzyme A forms *acetoacetic acid, β-hydroxybutyric acid, and acetone, collectively known as* **ketone bodies.**

$$\underset{\text{acetoacetic acid}}{CH_3\overset{O}{\overset{\|}{C}}CH_2CO_2H} \qquad \underset{\beta\text{-hydroxybutyric acid}}{CH_3\overset{OH}{\overset{|}{C}}HCH_2CO_2H} \qquad \underset{\text{acetone}}{CH_3\overset{O}{\overset{\|}{C}}CH_3}$$

In normal metabolism the blood level of ketone bodies is usually less than 3 mg/dL. In diabetes there is an increase in ketone bodies in the blood. *A level of ketone bodies in the blood in excess of 20 mg/dL is called* **ketonemia.** Continued production of ketone bodies may exceed the amount that tissue cells can use. When the blood level reaches 70 mg/dL, the renal threshold for ketone bodies, the excess ketone bodies are excreted in the urine. *The condition of "ketones in the urine" is called* **ketonuria.** At still higher ketone bodies concentration, **acetone breath** *results from excreting acetone via the lungs.*

The conditions of ketonemia, ketonuria, and acetone breath are collectively called **ketosis** *or* ketoacidosis. The excess acids (acetoacetic acid and β-hydroxybutyric acid) in the blood lower the blood pH. The pH of blood is normally maintained by bicarbonate ion controlled by the kidneys. However, when ketosis occurs, there is insufficient bicarbonate ion to neutralize the acid.

$$H_3O^+ + HCO_3^- \rightleftharpoons H_2CO_3 + H_2O$$

A low blood pH causes hemoglobin to become less efficient in transporting oxygen. Thus, brain cells can become oxygen starved. Coma and death may result from severe ketosis.

27.5
Biosynthesis of Fatty Acids

Fats are synthesized when more nutrients are digested than required for energy and the biosynthetic reactions that maintain the organism. *The* **biosynthesis of fatty acids** *from acetyl coenzyme A occurs in the cytoplasm of cells in the liver and in adipose tissue.* The reaction is not the reverse of fatty acid oxidation. As was cited in the case of glycolysis and gluconeogenesis, apparent reverse reactions do not occur by the same set of reactions. Furthermore, different enzymes are usually involved. In the case of fatty acid degradation and synthesis, the two pathways even occur in different cellular compartments. Degradation occurs in the mitochondria and biosynthesis occurs in the cytoplasm. Thus, the two sets of reactions can occur at the same time, each controlled separately.

The following is a useful overview of the differences between oxidative degradation and biosynthesis of fatty acids.

1. *The intermediates of fatty acid synthesis are bonded to an* **acyl carrier protein (ACP)**, whereas, in oxidative degradation, coenzyme A is bonded to the intermediates.
2. *The enzymes in reductive biosynthesis are organized in a multienzyme complex called a* **fatty acid synthetase system.** All of the intermediates remain with the system until all reactions are completed. Degradative enzymes in the cytoplasm are not associated, and the steps occur independently.
3. The reduction occurs with the coenzyme NADPH, rather than the NAD^+ used in oxidation.
4. The growing fatty acid chain receives its two-carbon units from malonyl ACP. Acetyl coenzyme A is used to form malonyl ACP, but it is not involved in subsequent steps. Coenzyme A derivatives are involved in all steps in oxidative degradation.

A preliminary step before fatty acid synthesis begins converts acetyl coenzyme A into malonyl coenzyme A by reaction with bicarbonate.

$$CH_3\overset{O}{\underset{\|}{C}}-S-CoA + HCO_3^- + ATP \longrightarrow HO_2CCH_2\overset{O}{\underset{\|}{C}}-S-CoA + ADP + P_i$$

The reaction can proceed only if the ATP level is high. This situation occurs when the energy charge of the cell is high and the level of acetyl coenzyme A produced from nutrients such as carbohydrates is high. The reaction is catalyzed by an allosteric enzyme, acetyl CoA carboxylase, which requires the vitamin biotin as a cofactor.

A series of four steps occurs in the multienzyme fatty acid synthetase system. There are six enzymes in the system and ACP, which bonds to all intermediates. All acetyl coenzyme A units are converted into acetyl ACP units.

$$CH_2\overset{O}{\underset{\|}{C}}-S-CoA + ACP-SH \longrightarrow CH_3\overset{O}{\underset{\|}{C}}-S-ACP + CoA-SH$$

The four steps in fatty acid synthesis that are analogous to those of oxidative degradation are

1. Carbon–carbon bond formation.
2. Reduction of a ketone.
3. Dehydration of an alcohol.
4. Reduction of an alkene.

The steps are shown in Figure 27.2. The product of Step 4 in the first series of four steps becomes the reactant in the next series of four steps. Thus, initially acetyl ACP reacts with malonyl ACP to produce a butyryl ACP. Subsequent series produce compounds having six, eight, ten, etc., carbon atoms.

In order to produce palmitic acid from eight acetyl coenzyme A precursors, the series of steps must occur seven times.

$$8 \; CH_3\overset{\overset{\textstyle O}{\|}}{C}-S-CoA + 14 \; NADPH + 7 \; ATP \longrightarrow$$

$$CH_3(CH_2)_{14}CO_2H + 8 \; CoA-SH + 14 \; NADP^+ + 7 \; ADP + 7 \; P_i + 6 \; H_2O$$

Each NADPH is equivalent to 3 ATP and the total energy requirements for the biosynthesis are $7 + 3(14) = 49$ ATP.

Figure 27.2
Biosynthesis of Fatty
Acids.

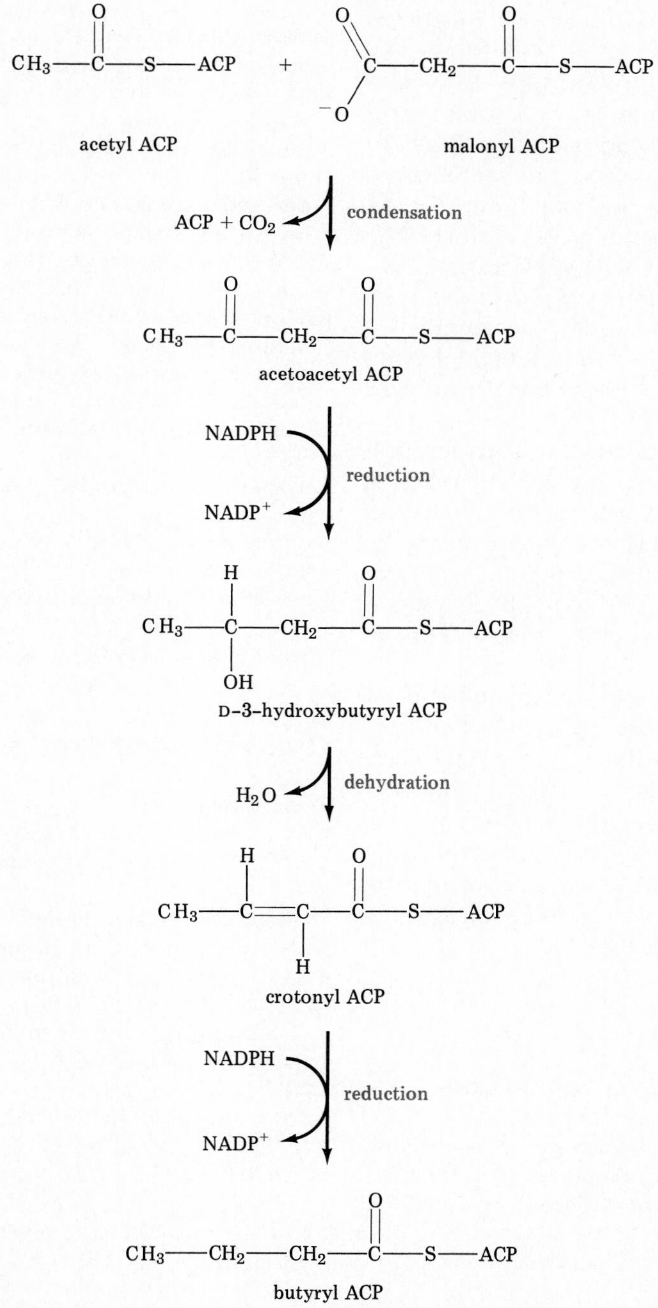

acetyl ACP

malonyl ACP

ACP + CO₂ ← condensation

acetoacetyl ACP

NADPH → reduction → NADP⁺

D-3-hydroxybutyryl ACP

H₂O ← dehydration

crotonyl ACP

NADPH → reduction → NADP⁺

butyryl ACP

Summary

About 35% of the American diet consists of fats. Fatty acids in fats cannot be converted into glucose. Fats have a higher caloric content than carbohydrates, which are more oxidized and are hydrated.

Fats are digested in the intestine and then reassembled after passage through the intestinal wall. The fatty acids are used to form tissue lipids and storage or depot lipids. The depot lipids serve as a long-term source of chemical energy.

Fatty acids are degraded two carbon atoms at a time and form acetyl coenzyme A. There are four repeating reactions in the fatty acid oxidation spiral. In each turn of the spiral one molecule of acetyl coenzyme A and a fatty acid having two fewer carbon atoms are produced. A turn of the cycle produces one molecule each of NADH and $FADH_2$. The reduced coenzymes are oxidized in the electron transport chain, and energy is stored as ATP. The acetyl coenzyme A is oxidized in the citric acid cycle, which, when followed by the oxidation of reduced coenzymes, produces more ATP.

Acetoacetic acid, β-hydroxybutyric acid, and acetone are ketone bodies that are produced as a result of the excessive oxidation of fatty acids. Accumulation of the ketone bodies results in ketosis, which affects the pH of blood and therefore the oxygen-carrying efficiency of hemoglobin. Ketosis occurs in starvation and may occur in high fat diets or as a result of diabetes mellitus.

The biosynthesis of fatty acids occurs in the cytoplasm of the cell and is catalyzed by a special enzyme complex called fatty acid synthetase. The intermediates in the reaction are bonded to an acyl carrier protein. In the first step a bond is formed between acetyl ACP and malonyl ACP. Subsequent steps involve reduction of a β-carbonyl group, dehydration of an alcohol, and reduction of an alkene. The reducing agent in fatty acid biosynthesis is NADPH.

Learning Objectives

As a result of studying Chapter 27 you should be able to

- Describe the digestive processes for triglycerides and how these lipids are distributed in the body.
- Compare fatty acid mobilization to glycogen hydrolysis as a source of chemical energy.
- Outline the reactions of the fatty acid spiral.
- Calculate the energy stored in the oxidation of a fatty acid.
- Write the equation responsible for the formation of ketone bodies.
- Define the terms ketosis, ketonemia, ketonuria, and acetone breath.
- Outline the steps of fatty acid biosynthesis.

New Terms

Acetone breath results from release of acetone from the lungs from a ketosis condition.

Acyl carrier protein is the carrier of acyl groups in fatty acid biosynthesis.

Adipose tissue is another name for depot fat.

Depot fat is connective tissue that contains cellular fat.

Fatty acid mobilization is the release, by hydrolysis, of fatty acids from adipose tissue.

The **fatty acid spiral** is a sequence of reactions which degrade a fatty acid chain by a two-carbon-atom acetyl unit.

Fatty acid synthetase is a complex of enzymes used in fatty acid biosynthesis.

Fatty acid biosynthesis occurs in two-carbon-atom acetyl units in the cytoplasm of the cell.

Ketone bodies are acetoacetic acid, β-hydroxybutyric acid, and acetone.

Ketonemia is a condition of high ketone bodies concentration in the blood.

Ketonuria is a condition of high ketone bodies concentration in the urine.

Ketosis is a condition of high concentrations of ketone bodies.

Tissue lipids are lipid materials in cell membranes.

Questions and Problems

Terminology

27.1   What are depot lipids?

27.2   What is fatty acid mobilization?

Digestion of Triglycerides

27.3   Where does the digestion of triglycerides begin?

27.4   What are the products of digestion of triglycerides?

27.5   What happens to the products of lipid digestion after they pass through the intestinal wall?

27.6   How are lipids carried in the blood?

27.7   In what forms are lipids distributed in the body?

27.8   How do tissue lipids differ from adipose tissue?

27.9   How can carbohydrates be converted into fat?

27.10 What process releases fatty acids from adipose tissue?

27.11 What enzymes catalyze the hydrolysis of triglycerides?

27.12 Compare glucose and fatty acids in terms of ease of transport into cells.

27.13 Explain why fatty acids cannot serve as fuel for the brain.

Catabolic Reactions of Triglycerides

27.14 Why is the metabolic pathway of fatty acids called a spiral rather than a cycle?

27.15 Why is ATP hydrolyzed to AMP rather than ADP in the activation of a fatty acid?

27.16 What coenzyme dehydrogenates fatty acids?

27.17 What is the configuration of the unsaturated fatty acid formed by dehydrogenation?

27.18 What is the configuration of the β-hydroxy fatty acid?

27.19 Explain why the metabolism of oleic acid would produce less energy than the metabolism of stearic acid.

27.20 How many moles of acetyl coenzyme A are produced in the metabolism of stearic acid? How many times does the spiral reaction sequence occur?

27.21 How many moles of acetyl coenzyme A would be produced in the oxidation of decanoic acid?

27.22 How many times does the spiral reaction sequence occur for palmitic acid?

27.23 What reactions occur in the metabolism of glycerol?

27.24 How is FAD recovered from the $FADH_2$ produced in the fatty acid cycle?

27.25 How is NAD^+ re-formed from the NADH produced in the fatty acid cycle?

27.26 The fatty acid cycle is also called beta-oxidation. Explain why.

27.27 Why can't fatty acids be converted into glycogen?

27.28 Fats can provide a small quantity of glucose. What part of the fats can undergo gluconeogenic reactions.

27.29 What fraction of the carbon atoms in glyceryl tristearate can undergo gluconeogenic reactions.

Ketone Bodies

27.30 What are ketone bodies?

27.31 What reaction produces ketone bodies?

27.32 What conditions cause the formation of ketone bodies?

27.33 Define each of the following terms.
(a) ketonemia (b) ketonuria (c) acetone breath

27.34 What is ketosis?

27.35 Why can ketosis result in death?

Biosynthesis of Fatty Acids

27.36 What binds the intermediates in fatty acid biosynthesis?

27.37 What special features serve to catalyze the reactions of fatty acid biosynthesis?

27.38 Where does fatty acid biosynthesis occur?

27.39 How is acetyl coenzyme A converted into malonyl coenzyme A?

27.40 What types of reactions occur in each of the four steps of fatty acid biosynthesis?

27.41 How much NADPH and ATP is required for each two carbon unit placed in a fatty acid chain?

27.42 How many kcal/mole are required to place each two-carbon unit in a fatty acid chain?

27.43 What quantities of NADPH and ATP are required to synthesize myristic acid?

28

Metabolism of Amino Acids

Turkey and chicken are becoming an important source of protein in the American diet.

Amino acids are used by living organisms in three ways.

1. For synthesis of proteins.
2. To provide carbon and nitrogen atoms for biosynthetic reactions.
3. To produce energy for the organism.

About 75% of amino acids that enter the amino acid pool of the body are used to build proteins (Figure 28.1). The **amino acid pool** *is a reserve of free amino acids in a state of equilibrium.* These amino acids circulate in blood and other fluids throughout the body, and the composition of the pool changes as some amino acids leave, while others enter.

Protein Synthesis

Each protein in the body is degraded and rebuilt constantly. *The time that it takes for a protein to be degraded and replaced is the* **protein turnover rate.** Peptide hormones and enzymes have turnover rates of a few minutes. For liver and blood protein the turnover rate is about six days. Muscle proteins have a turnover rate of about a half year. Collagens, the structural proteins, are replaced in about three years. Protein synthesis will be discussed in Chapter 29.

As indicated in Chapter 23, humans can only synthesize about half of the 20 amino acids. The other amino acids, which must be obtained in the diet, are essential amino acids (see Table 23.2). There is no way to store amino acids for later use. Thus, a balanced diet of essential amino acids is needed to synthesize proteins when required by the organism.

Nitrogen Balance

A **nitrogen balance** *relates the daily intake of nitrogen compounds and the loss of nitrogen compounds by excretion.* During periods of growth the nitrogen balance must be positive so that vital proteins and other nitrogen-containing compounds can be synthesized. During fasting, when the nitrogen intake is less than the loss of nitrogen, a negative balance results.

Amino acids also supply the nitrogen atoms used to form nonprotein nitrogen-containing compounds. Examples include the heme of hemoglobin, the choline and ethanolamine of phosphoglycerides, and the purines and pyrimidines of nucleic acids.

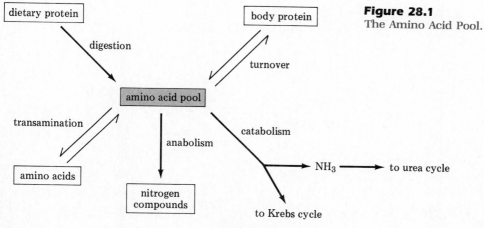

Figure 28.1
The Amino Acid Pool.

An organism with a positive nitrogen balance cannot synthesize nitrogen-containing storage forms. Thus, amino acids differ from carbohydrates and fats, which are stored for later use as an energy source by the organism. The carbon skeleton of an unneeded amino acid is degraded into intermediates of the citric acid cycle. The compounds then are oxidized or are stored as glycogen or fat. However, amino acids provide only about 15% of the total energy needs of an adult. The oxidation of amino acids is presented in Section 28.3.

Although the carbon atoms of amino acids can be stored in alternative forms, there is no permanent storage of nitrogen atoms. Most of the excess nitrogen atoms are converted to urea and excreted in the urine. The reactions to form urea are presented in Section 28.4.

28.2
Digestion of Proteins

About 10% of the peptide bonds in proteins are hydrolyzed in the stomach. The gastric juices denature the protein and make the peptide bonds more susceptible to enzymatic hydrolysis. The major enzyme in the stomach is pepsin, which enters the stomach as the zymogen pepsinogen. Pepsin catalyzes the hydrolysis of only certain peptide bonds, and as a result peptides, not amino acids, are formed.

Protein digestion in the small intestine is catalyzed by the digestive juices from the pancreas. These juices contain the zymogens chymotrypsinogen and trypsinogen as well as procarboxypeptidases. Once released in active form, the three enzymes have a broad range of hydrolytic capabilities. The resulting amino acids, once formed, are absorbed into the blood and are carried to the liver for metabolism.

Each amino acid has a different oxidative degradative pathway. However, we can separate the common features of amino acid degradation into two parts (Figure 28.2). One part is the carbon skeleton, which is degraded into carbon dioxide and water (Section 28.3). The second part is the α-amino groups, which are converted into urea (Section 28.4).

28.3
The Carbon Skeleton

There are as many sequences of oxidation as there are amino acids. Each amino acid has a multistep sequence with numerous specialized enzymes. Although the details of these steps are beyond the scope of this book, they all eventually con-

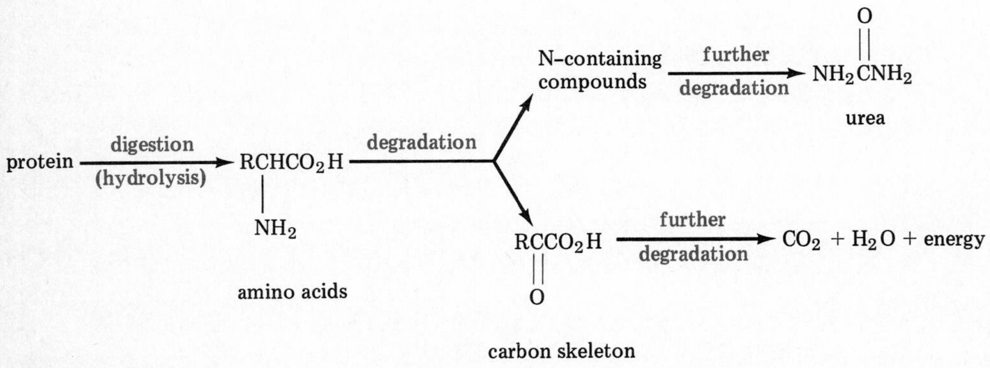

Figure 28.2
Metabolism of Proteins.

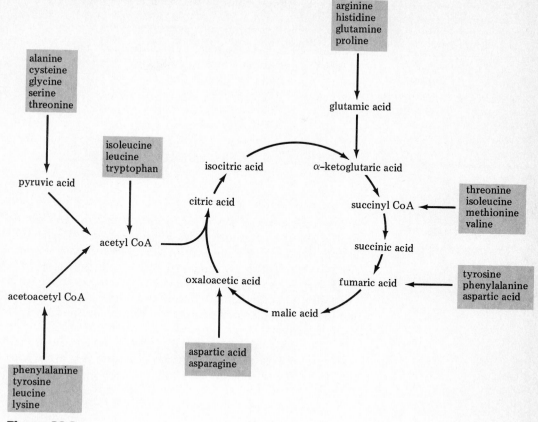

Figure 28.3
Metabolism of Amino Acids.

verge in steps that lead to the citric acid cycle (Figure 28.3). There are four points of entry to compounds directly involved in the citric acid cycle. In addition, some amino acids can form acetyl coenzyme A, and others can form pyruvic acid or acetoacetyl coenzyme A, both of which can form acetyl coenzyme A.

Amino acids that are degraded to pyruvic acid, α-ketoglutaric acid, succinyl coenzyme A, and oxaloacetic acid are classified as **glucogenic.** Glucogenic amino acids produce intermediates that can synthesize glucose via gluconeogenesis. *Amino acids that are degraded to acetyl coenzyme A or acetoacetyl coenzyme A are called* **ketogenic.** Ketogenic amino acids cannot be transformed into glucose. This fact illustrates one of the reasons that a balanced protein diet is important for good health. In addition, if one eats too much protein with little carbohydrates, ketosis may result.

Fourteen of the amino acids are glucogenic; only one amino acid, leucine, is exclusively ketogenic. Five amino acids can be converted by multiple pathways and are both glucogenic and ketogenic (Table 28.1).

There are two types of reactions that remove amino groups from amino acids. These are transamination and oxidative deamination. As a result of these processes ammonia, which is toxic, is formed and is eliminated from the body as urea.

Table 28.1 _____

Glucogenic and Ketogenic Amino Acids

Glucogenic		Glucogenic and ketogenic	Ketogenic
			leucine
alanine	glycine	isoleucine	
arginine	histidine	lysine	
asparagine	methionine	phenylalanine	
aspartic acid	proline	tryptophan	
cysteine	serine	tyrosine	
glutamic acid	threonine		
glutamine	valine		

28.4
Transamination

Transamination *is an interconversion of keto and amino compounds.* The α-amino group in alanine, arginine, asparagine, aspartic acid, cysteine, isoleucine, lysine, phenylalanine, tryptophan, tyrosine, or valine is removed by transamination. In these reactions the α-amino group is transferred to the α-carbon of one of the three α-keto acids—pyruvic acid, α-ketoglutaric acid, and oxaloacetic acid—the latter two of which are part of the citric acid cycle. The keto acid corresponding to the amino acid is produced, and one of the three α-keto acids, is converted to its analogous α-amino acid.

The three aminotransferases that catalyze the transamination are alanine transaminase, glutamate transaminase, and aspartate transaminase.

$$\alpha\text{-amino acid} + \text{pyruvic acid} \xrightarrow{\text{alanine transaminase}} \alpha\text{-keto acid} + \text{alanine}$$

$$\alpha\text{-amino acid} + \alpha\text{-ketoglutaric acid} \xrightarrow{\text{glutamate transaminase}} \alpha\text{-keto acid} + \text{glutamic acid}$$

$$\alpha\text{-amino acid} + \text{oxaloacetic acid} \xrightarrow{\text{aspartate transaminase}} \alpha\text{-keto acid} + \text{aspartic acid}$$

These three aminotransferases allow the collection and interconversion of the three α-keto acids into three amino acids—alanine, glutamic acid, and aspartic acid. These three amino acids are further collected in the form of only one amino acid, glutamic acid. These conversions are catalyzed by glutamic-pyruvic transaminase (also named alanine-glutamate transaminase) and glutamic-oxaloacetic transaminase (also named aspartate-glutamate transaminase).

$$\text{alanine} + \alpha\text{-ketoglutaric acid} \xrightarrow{\text{glutamic-pyruvic transaminase}} \text{pyruvic acid} + \text{glutamic acid}$$

$$\text{aspartic acid} + \alpha\text{-ketoglutaric acid} \xrightarrow{\text{glutamic-oxaloacetic transaminase}}$$
$$\text{oxaloacetic acid} + \text{glutamic acid}$$

Whatever the pathways of transamination, the final acceptor of an amino group for many amino acids is α-ketoglutaric acid, which produces glutamic acid.

Figure 28.4
Glutamic Acid as a
Universal Donor.

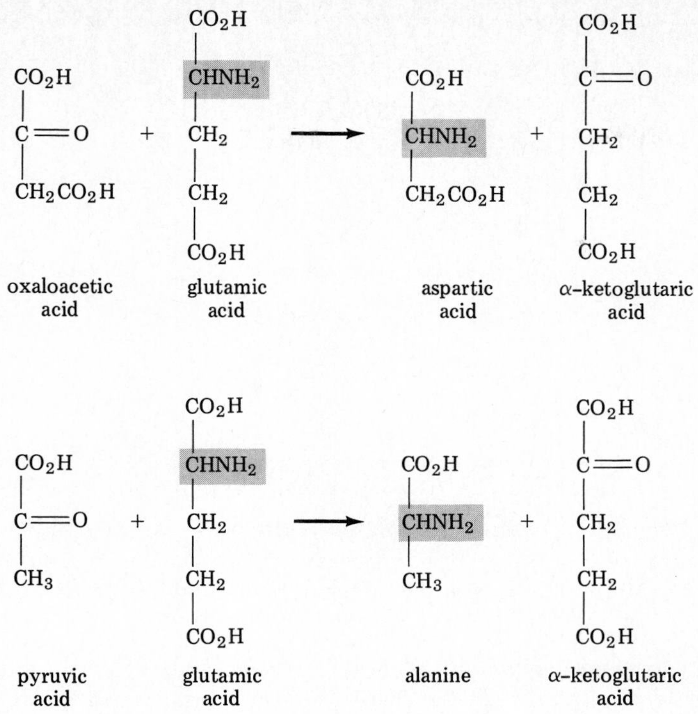

oxaloacetic acid + glutamic acid ⟶ aspartic acid + α-ketoglutaric acid

pyruvic acid + glutamic acid ⟶ alanine + α-ketoglutaric acid

The reverse of the collection reactions can also occur. Glutamic acid acts as a "universal" donor of an amino group to keto acids (Figure 28.4). The enzyme glutamic-pyruvic transaminase (GPT) is released from the liver into the blood in cases of severe liver damage. The increase in serum glutamic-pyruvic transaminase (SGPT) is clinical indication of liver damage.

The enzyme glutamic-oxaloacetic transaminase (GOT) is abundant in heart muscle. In myocardial infarction, the damaged heart muscle releases its transaminase into the blood. The increase in serum glutamic-oxaloacetic transaminase (SGOT) is a clinical indication of myocardial infarction.

28.5 Oxidative Deamination

Oxidative deamination *is the conversion of an amino acid into a keto acid with the release of an ammonium ion.* Oxidative deamination of the glutamic acid formed by the action of the various transaminations unloads the nitrogen of amino acids in the form of ammonium ions, which are then converted to urea in the urea cycle. The oxidative deamination is catalyzed by glutamic dehydrogenase; NAD^+ is the coenzyme for the reaction,

$$\text{glutamic acid} + NAD^+ + H_2O \longrightarrow \alpha\text{-ketoglutaric acid} + NADH + NH_4^+$$

The α-ketoglutaric acid formed can enter the citric acid cycle, and ammonia is incorporated into the urea cycle.

The initial reaction in oxidative deamination involves dehydrogenation to form an imine, which is then hydrolyzed to a keto acid.

$$\text{H}_2\text{N}-\underset{\underset{\text{CH}_2}{|}}{\overset{\overset{\text{CO}_2\text{H}}{|}}{\text{C}}}-\text{H} \xrightarrow[\text{NAD}^+ \quad \text{NADH} + \text{H}^+]{\text{glutamate dehydrogenase}} \text{HN}=\text{C}$$

(reaction scheme: glutamic acid, with CO₂H, CH₂, CH₂, CO₂H, converted via glutamate dehydrogenase with NAD⁺ → NADH + H⁺ to the imine HN=C with CO₂H, CH₂, CH₂, CO₂H)

$$\text{H}_2\text{O} + \text{HN}=\text{C} \longrightarrow \text{C}=\text{O} + \text{NH}_3$$

(second reaction: the imino intermediate HN=C (with CO₂H, CH₂, CH₂, CO₂H) plus H₂O yields the keto acid C=O (with CO₂H, CH₂, CH₂, CO₂H) plus NH₃)

The activity of glutamate dehydrogenase is allosterically regulated. Both ATP and GTP inhibit the enzyme, whereas ADP and GDP activate it. Thus, a decrease in the energy charge of the cell accelerates the oxidative deamination of amino acids.

28.6
The Urea Cycle

The **urea cycle** *is a metabolic pathway that converts ammonia into urea.* Urea is then excreted in urine. The synthesis of urea, which occurs primarily in the liver, involves intermediates in the citric acid cycle and two amino acids that do not occur in proteins, ornithine and citrulline.

Elevated levels of NH_4^+ *in the blood is called* **hyperammonemia.** A deficiency of any of the enzymes of the urea cycle will cause this condition. High levels of NH_4^+ are toxic, since a shift that occurs in the equilibrium catalyzed by glutamate dehydrogenase converts α-ketoglutaric acid to glutamic acid. This process depletes α-ketoglutaric acid, which is a vital intermediate of the citric acid cycle. A depletion of α-ketoglutaric acid decreases the rate of formation of ATP. Therefore, ammonium ions produced in the oxidative deamination of amino acids must be removed. In addition to urea, some of the ammonium ions are excreted in the urine as uric acid and purines.

There are five steps in the cycle (Figure 28.5). In order to keep track of the carbon and nitrogen involved in the net reaction, these atoms appear in color for each substance. Hydrogen and oxygen are exchanged with other substances during the reactions. The stoichiometry of the urea cycle is

$$CO_2 + NH_4^+ + 2\,H_2O + \text{aspartic acid} + 3\,ATP \longrightarrow$$
$$\text{urea} + \text{fumaric acid} + 2\,ADP + 2\,P_i + AMP + PP_i$$

Carbon dioxide and one of the nitrogen atoms are brought into the cycle in the first step. The other nitrogen atom comes from aspartic acid, which is brought into the cycle in step 3.

Ornithine is the carrier of the carbon and nitrogen atoms to form urea. This modified amino acid is regenerated in step 5 as urea is released. The ornithine then combines with the carbamyl phosphate formed in step 1.

Fumaric acid is released in step 4 and is hydrated to malic acid, which in turn is oxidized to oxaloacetic acid. The oxaloacetic acid can be transaminated to re-form aspartic acid necessary for continued operation of the urea cycle. Alternatively, the oxaloacetic acid can be converted into glucose by gluconeogenesis

Figure 28.5
The Urea Cycle.

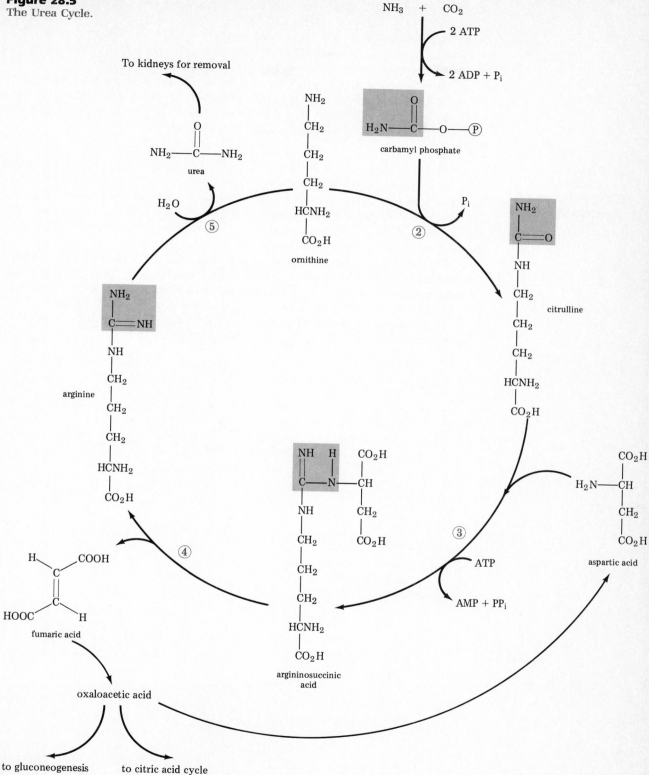

or into citric acid by reaction with acetyl coenzyme A as part of the citric acid cycle.

The formation of NH_4^+ from amino acids, conversion into carbamyl phosphate (step 1), and synthesis of citrulline (step 2) occur in the mitochondrial matrix. However, steps 3–5 occur in the cytoplasm.

The urea cycle requires three molecules of ATP per molecule of urea formed. Two of the ATP are used in step 1, and one ATP is used in step 3. However, since ATP is converted to AMP in step 3 and the PP_i is rapidly hydrolyzed, the energy equivalent of four ATP molecules has been expended.

28.7
Biosynthesis of Amino Acids

The ability of animals to synthesize amino acids varies with the species. Recall that humans cannot synthesize about one-half of the amino acids required for protein formation. The synthetic pathways of the nonessential amino acids are reversals of transamination and oxidative deamination in many cases. Thus, glutamic acid, alanine, and aspartic acid can all be formed by transamination reactions from a corresponding α-keto acid.

Proline is formed in several steps from glutamic acid. These steps include a reduction of the side chain acid to an aldehyde followed by a cyclic addition reaction to produce an imine (Figure 28.6). Reduction of the imine gives proline. Note that NADH and NADPH are involved in the two reduction steps.

Tyrosine is formed from the essential amino acid phenylalanine. The oxidative reaction involves hydroxylation at the para position.

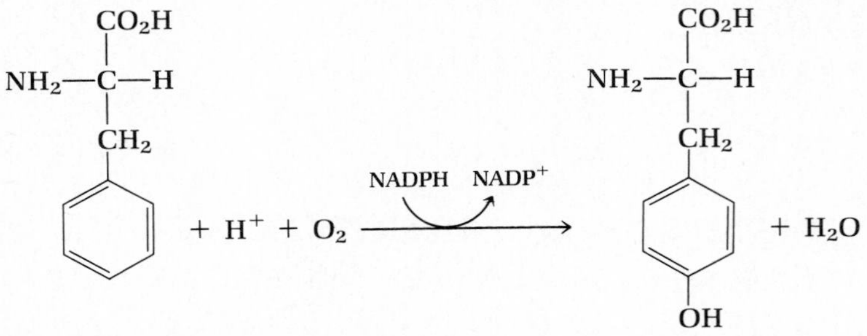

Figure 28.6
Biosynthesis of Proline.

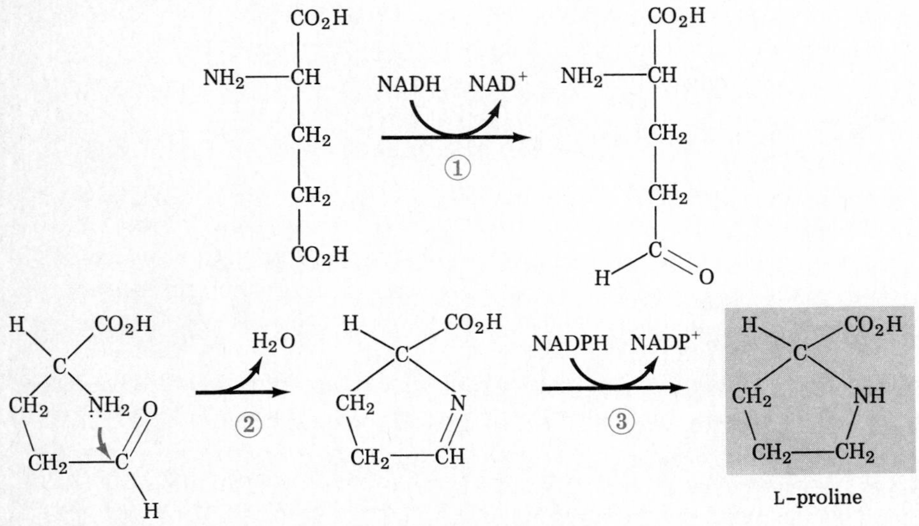

Figure 28.7
Biosynthesis of Serine.

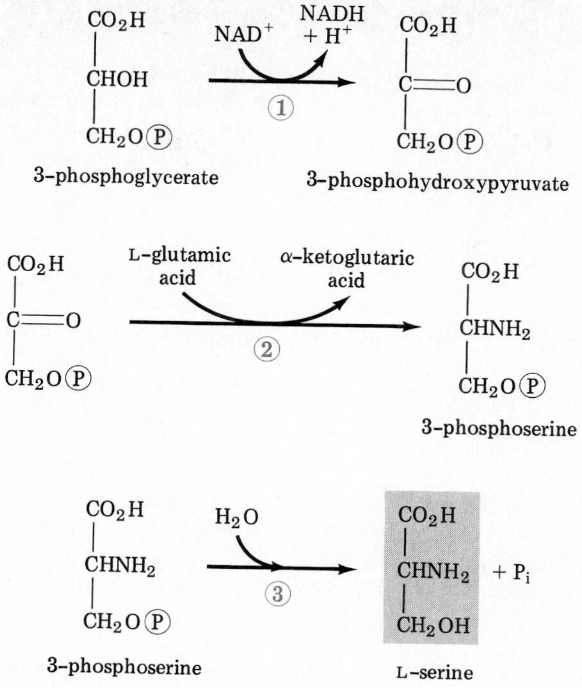

$$
\begin{array}{c}
\text{CO}_2\text{H} \\
| \\
\text{CHOH} \\
| \\
\text{CH}_2\text{O}\,\text{(P)}
\end{array}
\xrightarrow[\substack{(1)}]{\text{NAD}^+ \quad \substack{\text{NADH} \\ +\ \text{H}^+}}
\begin{array}{c}
\text{CO}_2\text{H} \\
| \\
\text{C}=\text{O} \\
| \\
\text{CH}_2\text{O}\,\text{(P)}
\end{array}
$$

3-phosphoglycerate   3-phosphohydroxypyruvate

$$
\begin{array}{c}
\text{CO}_2\text{H} \\
| \\
\text{C}=\text{O} \\
| \\
\text{CH}_2\text{O}\,\text{(P)}
\end{array}
\xrightarrow[\substack{(2)}]{\substack{\text{L-glutamic} \\ \text{acid}} \quad \substack{\alpha\text{-ketoglutaric} \\ \text{acid}}}
\begin{array}{c}
\text{CO}_2\text{H} \\
| \\
\text{CHNH}_2 \\
| \\
\text{CH}_2\text{O}\,\text{(P)}
\end{array}
$$

3-phosphoserine

$$
\begin{array}{c}
\text{CO}_2\text{H} \\
| \\
\text{CHNH}_2 \\
| \\
\text{CH}_2\text{O}\,\text{(P)}
\end{array}
\xrightarrow[\substack{(3)}]{\text{H}_2\text{O}}
\begin{array}{c}
\text{CO}_2\text{H} \\
| \\
\text{CHNH}_2 \\
| \\
\text{CH}_2\text{OH}
\end{array} + \text{P}_i
$$

3-phosphoserine   L-serine

Serine is formed from 3-phosphoglycerate as shown in Figure 28.7. The steps are

1. Oxidation of a secondary alcohol.
2. Transamination of a ketone.
3. Hydrolysis of a phosphate ester.

In each of the syntheses shown in this section, the amino acid is produced from a substance having the correct carbon skeleton. The essential amino acids, which must be obtained from the diet, cannot be synthesized because our bodies do not have the correct carbon skeletons available. Furthermore, we do not have enzyme-catalyzed reactions to produce these skeletons.

Summary

The enzymes pepsin, chymotrypsin, trypsin, and carboxypeptidases catalyze the hydrolysis of proteins into amino acids. The amino acids are absorbed into the blood to be used for the formation of proteins and other nitrogen-containing biological molecules. Unneeded amino acids are degraded in the liver to produce either carbohydrates or fats or are oxidized directly to carbon dioxide. Nitrogen atoms in excess of immediate biosynthetic needs are converted to ammonia and urea and are excreted in the urine.

The carbon skeletons of all 20 amino acids are eventually converted into five compounds: acetyl coenzyme A, α-ketoglutaric acid, succinyl coenzyme A, fumaric acid, and oxaloacetic acid. Several amino acids produce either pyruvic acid or acetoacetyl coenzyme A, which can be transformed into acetyl coenzyme A.

The amino acids that can be converted into pyruvic acid, α-ketoglutaric acid, succinyl coenzyme A, and oxaloacetic acid can finally produce glucose. These amino acids are glucogenic. Amino acids that are degraded to acetyl coenzyme A or acetoacetyl coenzyme A are ketogenic. Fourteen amino acids are glucogenic and only leucine is exclusively ketogenic. Five amino acids are both glucogenic and ketogenic.

The α-amino groups of many amino acids are transferred to α-ketoglutaric acid in transamination reactions. Oxidative deamination of glutamic acid yields ammonia, which then is converted to urea in the urea cycle.

Glutamic acid provides the α-amino group required for the synthesis of most of the nonessential amino acids by transamination of keto acids. Independent pathways exist for the syntheses of serine, tyrosine, and proline.

Learning Objectives

As a result of studying Chapter 28 you should be able to

- Describe the three ways that living organisms use amino acids obtained in the diet.
- Explain the significance of the nitrogen balance in the body.
- List the enzymes that catalyze the hydrolysis of proteins.
- Describe what happens to the carbon atoms of amino acids that are not used in forming body protein.
- Explain the link between amino acid metabolism and the citric acid cycle.
- Summarize the steps of the urea cycle.
- Show how transamination reactions produce nonessential amino acids.
- Give examples of specialized reactions to produce amino acids.

New Terms

The **amino acid pool** is a reserve of free amino acids in the body.

Glucogenic amino acids are amino acids that can be converted to pyruvic acid, oxaloacetic acid, succinyl coenzyme A, or α-ketoglutaric acid.

Glutamic-oxaloacetic transaminase (GOT) is a transaminase found in the heart muscle.

Glutamic-pyruvic transaminase (GPT) is a transaminase found in the liver.

Hyperammonemia is an elevated level of NH_4^+ in the blood.

Ketogenic amino acids are amino acids that can be converted into acetoacetyl coenzyme A or acetyl coenzyme A.

The **nitrogen balance** relates the daily intake of nitrogen compounds and the quantity of nitrogen compounds excreted.

Oxidative deamination is the conversion of an amino acid into a keto acid with the release of an ammonium ion.

Protein turnover rate refers to the time required for the degradation and replacement of proteins in the cells of the body.

Transamination is an interconversion process between keto compounds and amino compounds.

The **urea cycle** is a metabolic pathway that converts ammonia into urea.

Questions and Problems

Terminology

28.1 What is the amino acid pool?

28.2 What is meant by protein turnover?

28.3 What occurs in a transamination reaction?

Uses of Amino Acids

28.4 List the three functions of amino acids in the body.

28.5 What percent of digested amino acids is used to build proteins?

28.6 What percent of human energy needs is provided by amino acids?

28.7 How does the turnover rate for enzymes differ from that of muscle protein?

28.8 Why must there be a positive nitrogen balance in a period of active growth?

28.9 What happens to the nitrogen balance during a fast?

28.10 What happens to the carbon atoms of unneeded amino acids?

28.11 What happens to excess nitrogen atoms of amino acids?

Digestion of Proteins

28.12 What percent of peptide bonds in proteins is hydrolyzed in the stomach?

28.13 What is the major enzyme for protein digestion in the stomach?

28.14 What digestive enzymes are provided by the pancreas?

Degradation of Amino Acids

28.15 How many points of entry to the citric acid cycle are involved in amino acid metabolism?

28.16 What compounds are produced from glucogenic amino acids?

28.17 What compounds are produced from ketogenic acids?

28.18 What amino acid is exclusively ketogenic?

28.19 Name the five amino acids that are both glucogenic and ketogenic.

Transamination

28.20 Explain the function of the transamination reaction in the metabolism of amino acids.

28.21 Write an equation for the transamination of alanine and α-ketoglutaric acid.

28.22 Write an equation for a reaction catalyzed by GOT.

28.23 Write an equation for a reaction catalyzed by GPT.

28.24 What role does α-ketoglutaric acid play in transamination reactions?

28.25 What does the presence of SGPT in the blood indicate?

28.26 What does the presence of SGOT in the blood indicate?

28.27 The following α-keto acid can substitute for an essential amino acid in a diet. What is the amino acid? How can the keto acid substitute for an essential amino acid?

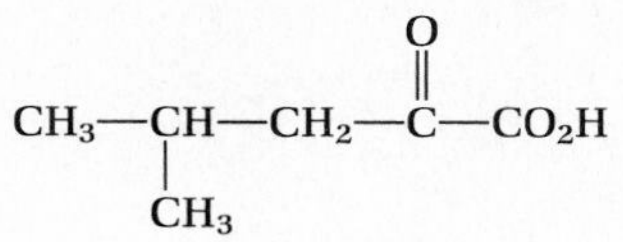

$$CH_3-\underset{\underset{CH_3}{|}}{CH}-CH_2-\overset{\overset{O}{\|}}{C}-CO_2H$$

Oxidative Deamination

28.28 What is oxidative deamination?

28.29 Write the product of oxidative deamination of glutamic acid.

28.30 What coenzyme is required for oxidative deamination?

28.31 What is the fate of α-ketoglutaric acid formed in oxidative deamination?

Urea Cycle

28.32 What two amino acids are involved in the urea cycle?

28.33 Where does the synthesis of urea occur?

28.34 What is hyperammonemia?

28.35 Why are high levels of ammonium ion toxic?

28.36 What substance reacts and is formed in the urea cycle?

28.37 What is the fate of fumaric acid formed in the urea cycle?

28.38 What are the energy requirements of the urea cycle?

28.39 What steps in the urea cycle occur in the mitochondrial matrix?

28.40 What steps in the urea cycle occur in the cytoplasm?

Biosynthesis of Amino Acids

28.41 What is the major pathway for the synthesis of non-essential amino acids?

28.42 Outline how proline is formed from glutamic acid.

28.43 What substance is used to synthesize tyrosine?

28.44 What substance is used to synthesize serine?

29

Nucleic Acids

This child is learning from her mother. However, genetic information, contained in DNA, has already been transferred to the child.

Most people know that genes control heredity from parents to their children. However, they do not understand that there is more to genes than physical features and individual talents. Genes are present in all life forms and are responsible for the reproduction of cells and all of their day-to-day functions. Genes determine what structural proteins will be synthesized to help form a cell. They are also responsible for the synthesis of substances such as the enzymes that control so many of the biochemical reactions.

There are many thousands of genes in the chromosomes of the nucleus of the many types of cells. **Chromosomes** *are contained in somatic cells and possess genetic information.* In humans each cell has 46 chromosomes arranged in 23 pairs. It has been estimated that there are well over a million different genes in the nucleus of human cells. *These genes contain the* **genetic code,** *which is all of the information to maintain the individual's identity.* The gene is comprised of a nucleic acid called deoxyribonucleic acid (DNA). DNA controls the formation of various ribonucleic acids (RNA), which in turn control the formation of all proteins.

Nucleic acids were first isolated in crude form from the nucleus by Johannes Miescher in 1869. The material was shown to contain not only carbon, hydrogen, and nitrogen but phosphorus as well. The proportion of nitrogen and the presence of phosphorus showed that nucleic acids are quite different from carbohydrates, fats, and proteins.

In order to understand chemistry it is necessary somehow to translate composition into structure. In the case of nucleic acids, structural information became available only in the 1950s!

By the 1950s biochemists established that both DNA and RNA are polymers of nucleotides or polynucleotides. *A* **nucleotide** *consists of a heterocyclic nitrogen base bonded to a monosaccharide, which in turn is bonded to a phosphate. The combination of the nitrogen base with a monosaccharide is a* **nucleoside.** Phosphorylation of a nucleoside forms a nucleotide.

The monosaccharide in RNA is the β anomer of D-ribose. The name RNA is based on this sugar unit. The monosaccharide in DNA is the β anomer of D-2-deoxyribose, hence the name deoxyribonucleic acid.

β-D-ribose β-D-2-deoxyribose

Four heterocyclic nitrogen bases are found in DNA (Figure 29.1). Two of the bases, *cytosine* (C) and *thymine* (T), are **pyrimidines;** the other two bases, *adenine* (A) and *guanine* (G), are **purines.** The capital letters are used in shorthand representations of the nucleic acids.

Four heterocyclic nitrogen bases are also found in RNA. Adenine (A), guanine (G), and cytosine (C) are present, but the pyrimidine *uracil* (U) is found instead of thymine. Note the similarity of structure of thymine and uracil (Figure 29.1).

The nitrogen base in nucleosides is attached to the carbohydrate by a nitrogen glycosidic bond (Figure 29.2). The ring atoms in the base are numbered, and

Figure 29.1
Nitrogen Bases in Nucleic Acids.

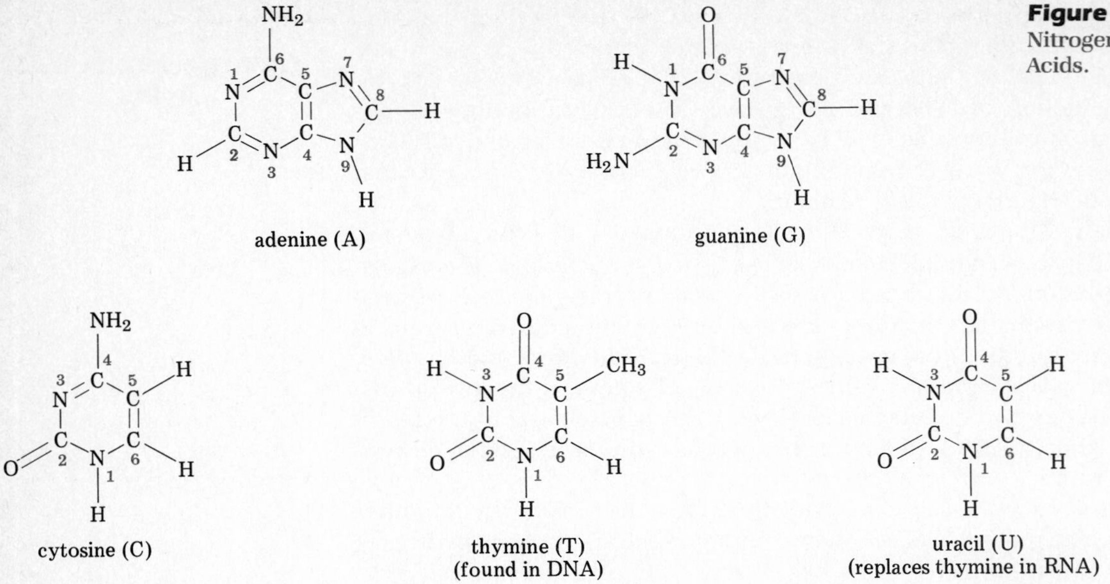
adenine (A)
guanine (G)
cytosine (C)
thymine (T)
(found in DNA)
uracil (U)
(replaces thymine in RNA)

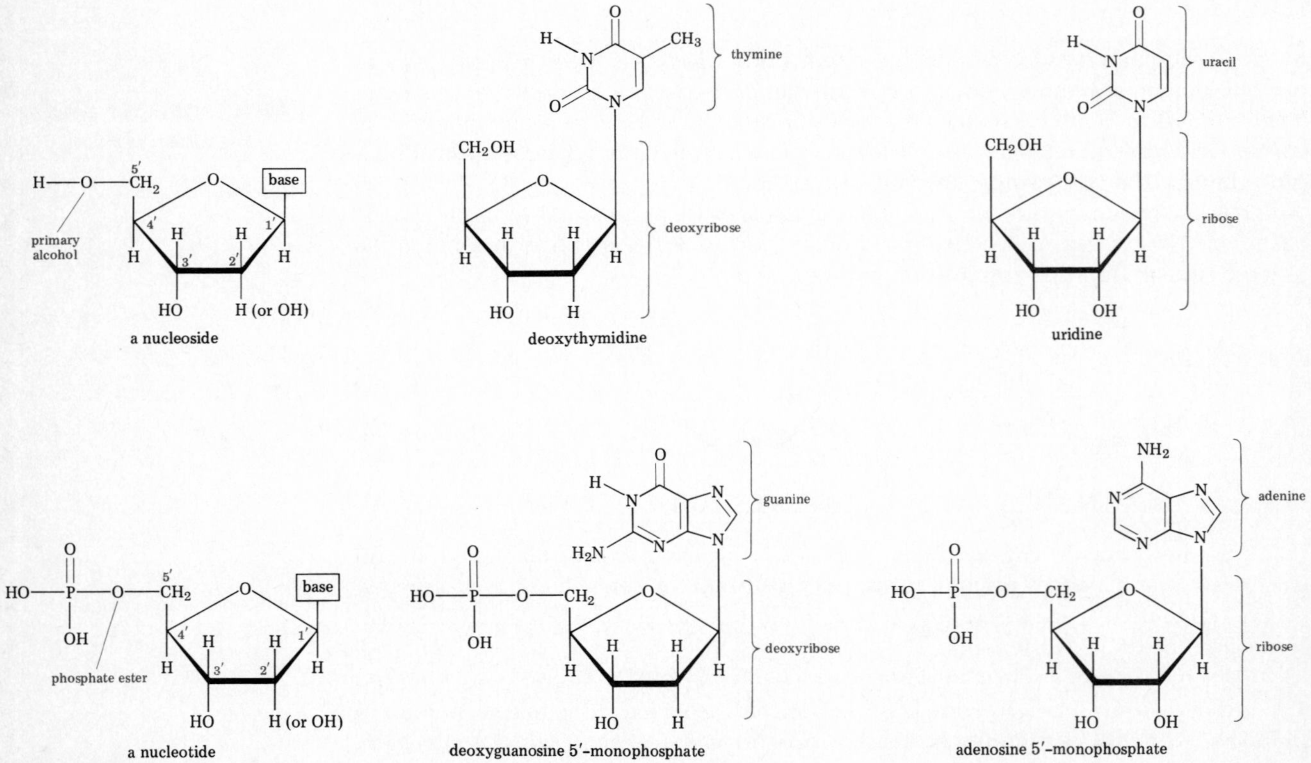
primary alcohol
base
a nucleoside
thymine
deoxyribose
deoxythymidine
uracil
ribose
uridine
phosphate ester
a nucleotide
guanine
deoxyguanosine 5′-monophosphate
adenine
adenosine 5′-monophosphate

Figure 29.2
Structure of Nucleosides and Nucleotides.

Table 29.1 ———————————————————————————

Names of Bases, Nucleosides, and Nucleotides Found in DNA and RNA

Base	Nucleoside	Nucleotide	
		DNA	
adenine (A)	deoxyadenosine	deoxyadenosine 5'-monophosphate (dAMP)	(deoxyadenylic acid)
guanine (G)	deoxyguanosine	deoxyguanosine 5'-monophosphate (dGMP)	(deoxyguanylic acid)
thymine (T)	deoxythymidine	deoxythymidine 5'-monophosphate (dTMP)	(thymidylic acid)
cytosine (C)	deoxycytidine	deoxycytidine 5'-monophosphate (dCMP)	(deoxycytidylic acid)
		RNA	
adenine (A)	adenosine	adenosine 5'-monophosphate (AMP)	(adenylic acid)
guanine (G)	guanosine	guanosine 5'-monophosphate (GMP)	(guanylic acid)
uracil (U)	uridine	uridine 5'-monophosphate (UMP)	(uridylic acid)
cytosine (C)	cytidine	cytidine 5'-monophosphate (CMP)	(cytidylic acid)

in order to avoid confusion primed numbers are used for the atoms of the carbohydrate. The nitrogen base is always attached to the 1' carbon atom of the carbohydrate, and there is a primary hydroxyl group located at the 5' carbon atom.
Ribose has secondary hydroxyl groups at the 2' and 3' carbon atoms, whereas
deoxyribose has a secondary hydroxyl group only at the 3' carbon atom. The
names of the nucleosides are listed in Table 29.1.

The phosphate group forms an ester with the hydroxyl group at the 5' carbon atom to form a nucleotide (Figure 29.2). The nucleotides are usually ionized
at physiological pH. The names of the nucleotides are also listed in Table 29.1.

Example 29.1 ——————————————————————————

Classify the following structure, identify its components, and assign its
name.

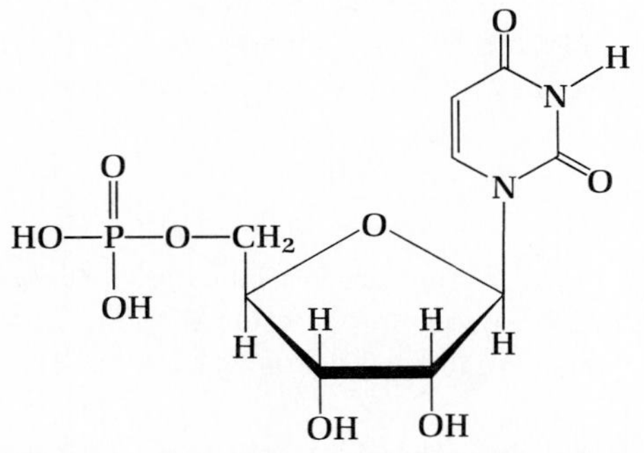

Solution. The compound contains a phosphate group and is a nucleotide. The
sugar is ribose, which means that the compound is a ribonucleotide. The base is
uracil, which can only be found in ribonucleotides. From Table 29.1 we find that
the name of the compound is uridine 5'-monophosphate (UMP) or uridylic acid.

———————————————————————————————————————

[Additional examples may be found in **29.13–29.24** at the end of the chapter.]

Table 29.2

Composition of DNA of Various Species

Species	Mole %			
	A	**T**	**G**	**C**
Escherichia coli	25	25	25	25
brewer's yeast	32	32	18	18
wheat	27	27	23	23
bovine	28	28	22	22
human	30	30	20	20

29.3
Polynucleotides

Composition of Polynucleotides

Measurement of the amounts of A, T, G, and C in the DNA of many animals, plants, and microorganisms reveals an interesting pattern (Table 29.2). The amount of C is equal to the amount of G; the amount of A is equal to the amount of T. The amounts of each of the pairs is different and varies from one life form to another. However, the sum of the percentages of A and G, the two purine bases, equals the sum of the percentages of C and T, the two pyrimidine bases. The relationships are

$$\% \, C = \% \, G \qquad \% \, A = \% \, T \qquad \% \, A + \% \, G = \% \, C + \% \, T$$

We will return to these interesting relationships of the bases in DNA in Section 29.5. There are no such simple relationships in the bases of RNA.

Structure of Polynucleotides

Nucleotides, bonded through phosphodiester linkages, make up the primary structure of RNA and DNA molecules (Figure 29.3). **Phosphodiesters** *are formed between the 3′ hydroxyl group of one nucleotide and the 5′ phosphate ester of another nucleotide.* Note that each phosphodiester has one acidic hydrogen, hence the name nucleic acids.

To discuss nucleic acids it is necessary to select a direction for listing the sequence of nucleotides. The direction selected is to start from the 5′ end, which exists as a phosphate ester, and proceed to the 3′ end, which is the hydroxyl group.

Shorthand Structures

DNA and RNA consist of a backbone of the appropriate units linked as phosphodiesters. The variation in structure is due to the order of the bases, and thus biochemists write the sequence of bases to describe a particular polynucleotide.

The letters represent the individual bases and are listed starting from the base on the nucleotide at the 5′ end. The order of bases is separated by dashes in the direction toward the 3′ end. For a ribonucleic acid the shorthand is

A—C—G——···—U

5′ ⟶ 3′

RNA

The same process is used for DNA. A lowercase d can be placed at the left of the base to indicate the deoxyribose in the backbone. However, the presence of

Figure 29.3
Structure of Polynucleotides.

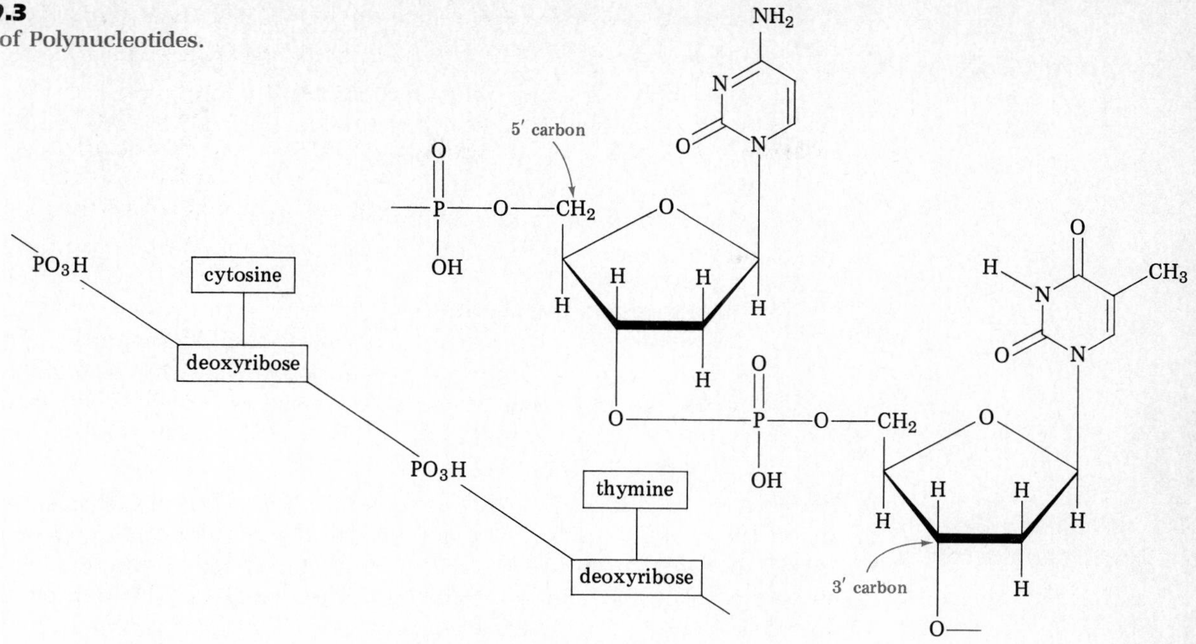

a portion of deoxyribonucleic acid

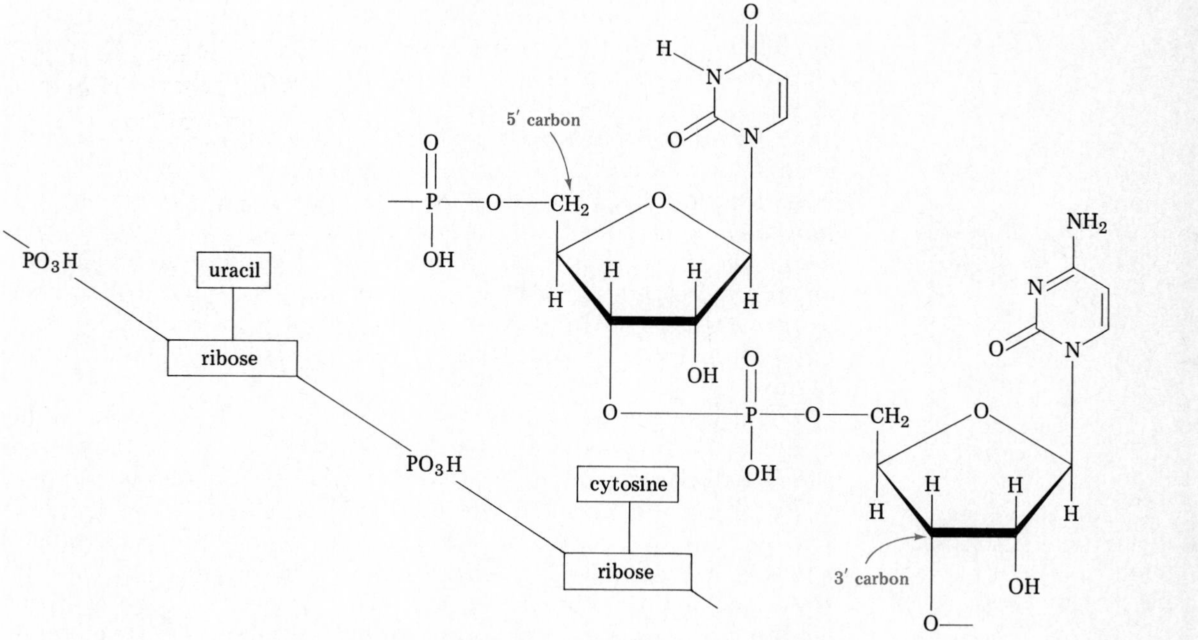

a portion of ribonucleic acid

thymine rather than uracil also indicates that the nucleic acid is deoxyribonucleic acid.

$$dT—C—C—\cdots—A$$

$5'$ $\longrightarrow$ $3'$

DNA

29.4
The Central Dogma

The **central dogma** *is a series of theories of the transmission of hereditary information and protein synthesis.* The tenets of this dogma are

1. DNA stores and transmits all hereditary information.
2. DNA is replicated in the cell nucleus when cells are ready to divide. **Replication** *is a copying process by which DNA is supplied to the new cells formed by cell division.*
3. Information for protein synthesis is passed from DNA to a form of RNA called messenger RNA (mRNA). *The transmission of information is called* **transcription.**
4. mRNA directs the synthesis of proteins by a process called translation. **Translation** *is a conversion of information in the sequence of nucleotides into a sequence of amino acids in a protein chain.* Amino acids are brought to the site of protein synthesis attached to transfer RNA (tRNA).

29.5
The Double Helix

The % C = % G and % A = % T relationship was used by J. D. Watson and F. Crick in 1952 to postulate a structure of DNA. They suggested that the pairing of A and T and also of C and G in DNA is the result of hydrogen bonding. Cytosine forms three hydrogen bonds to guanine, and adenine forms two hydrogen bonds to thymine (Figure 29.4). *The matching of bases through hydrogen bonds is called* **complementary base pairing.** For each C molecule in DNA there has to be a molecule of the complementary base G; for each A molecule there must be a complementary base T. Note that this requirement based on complementarity also means that for each purine base there is a pyrimidine base. Thus the Watson–Crick model is consistent with the experimentally observed relationship % A + % G = % C + % T.

Watson and Crick also proposed that DNA exists as two polynucleotide chains coiled around each other in a double helix (Figure 29.4). *In the* **double helix** *the individual chains of polydeoxyribonucleotides run in the opposite directions and are said to be antiparallel.* The bases attached to the sugar–phosphate backbone of each chain extend inward toward the other chain. One base in a chain is hydrogen-bonded to a complementary base in the other chain. Cytosine is opposite guanine, and these two bases are held together by three hydrogen bonds. Adenine is opposite thymine, and they are held together by two hydrogen bonds.

The arrangement of the bases in A–T and C–G pairs results in the maximum number of hydrogen bonds to stabilize the DNA molecule. However, spatial considerations are also important. Two pyrimidine molecules would be too small to bridge across the gap in the interior of DNA, and thus C–C, T–T, and C–T pairs do not occur. Similarly, two purine molecules across from each other exceed the space of the interior of DNA, and thus A–A, G–G, and A–G pairs do not occur. Only a small pyrimidine paired with a large purine results in the proper fit. This feature also accounts for the observation that the percentage of the purine bases equals the percentage of the pyrimidine bases.

29.6
DNA Replication

Once we know the structure of DNA, we can begin to consider how its structure determines the transmission of hereditary information. As new cells are formed, this information is passed on to new DNA molecules by a duplication process known as replication, which is controlled by the hydrogen-bonding properties of

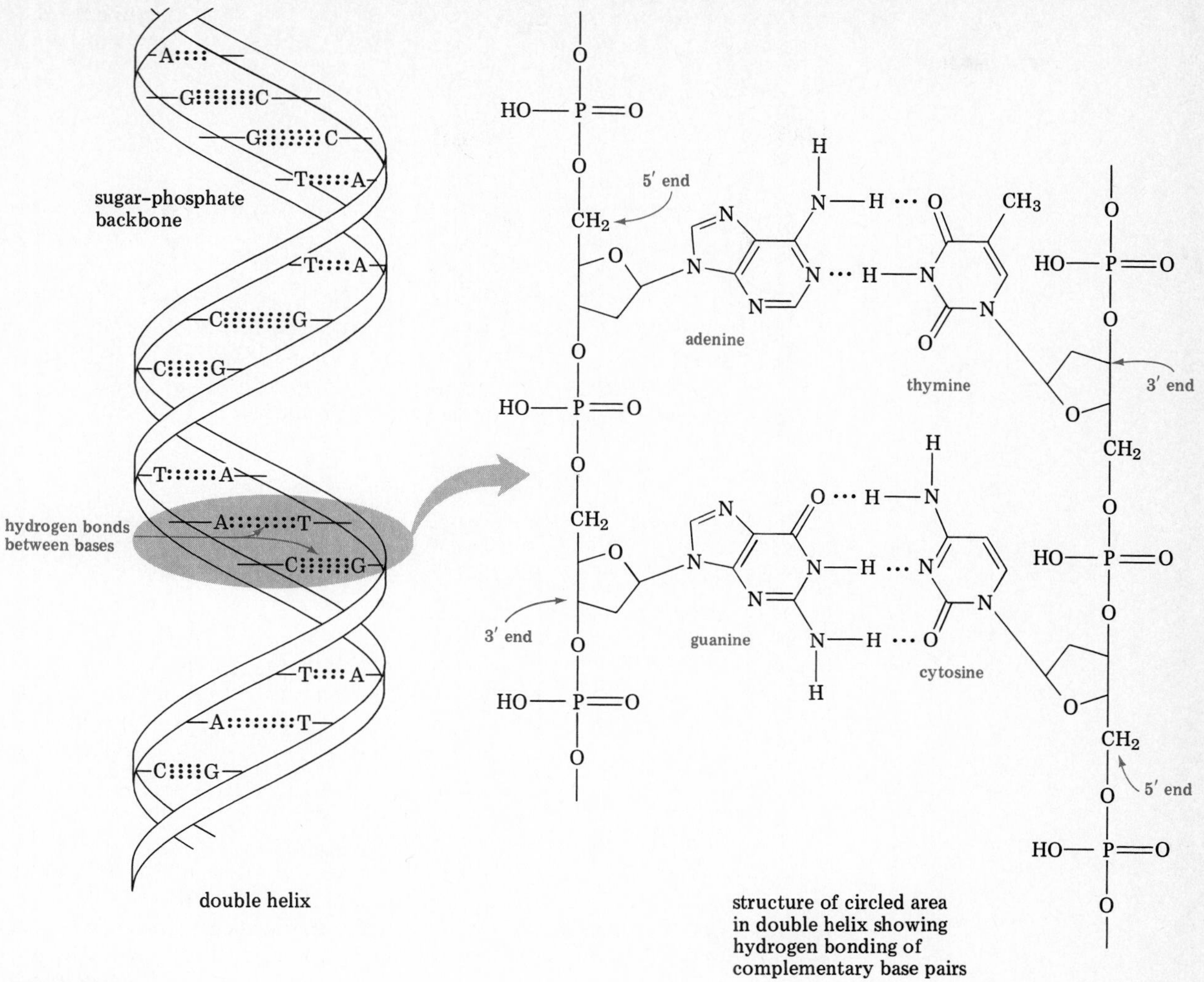

Figure 29.4
Hydrogen Bonding in the Double Helix.

DNA. *Each strand of DNA serves as a pattern or* **template** *to produce a new complementary strand.* This newly formed complementary strand is identical to the original complementary strand. Since DNA is a double-stranded molecule, each strand replicates its own complementary strand.

Replication of DNA occurs by a **semiconservative process;** *that is, each new DNA molecule contains one strand of the parent DNA molecule.* Thus each new generation of DNA molecules has one-half of the DNA or one strand of the previous generation. The remaining necessary strands are formed by the replication process (Figure 29.5). A shorthand representation of semiconservative replication indicating the base sequences in each chain is also shown in Figure 29.5.

In replication it is necessary for DNA to unwind so that the new complementary chain can be formed. *The enzyme* **DNA polymerase** *catalyzes the replication.* Although the entire process is quite complex, a simple model for illustrating the process of replication (Figure 29.6) involves a replicating fork. *The* **replicating fork** *occurs as DNA unravels and creates a place of DNA synthesis.* The process of

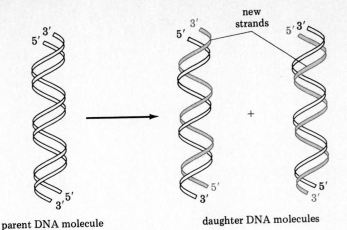

Figure 29.5
Semiconservative Replication of DNA.

parent DNA molecule   daughter DNA molecules

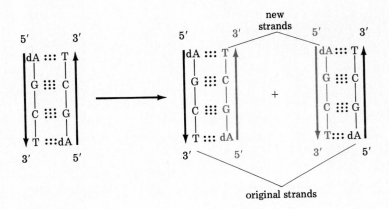

original strands

replication occurs with bonding of the proper nucleosides from the 5' position of one unit to the 3' position of another unit to form the backbone. This process is known as 5'→3' replication. Since the strands of DNA are antiparallel, one of the new strands is synthesized toward the replicating fork and the other strand away from the fork.

Example 29.2

Part of the nucleotide sequence in one chain of DNA is given below. Write a representation of the complementary DNA chain.

Solution. The complementary base pairs in DNA are A with T and G with C. For each A in one chain there is a T in the other. For each C in one chain there is G in the other. The complementary chains are

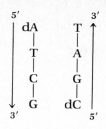

```
5'              3'
|   dA      T   ↑
|   |       |   |
|   T       A   |
|   |       |   |
|   C       G   |
↓   |       |   |
    G      dC   |
3'              5'
```

[Additional examples may be found in **29.25–29.33** at the end of the chapter.]

Figure 29.6
Replication of DNA.

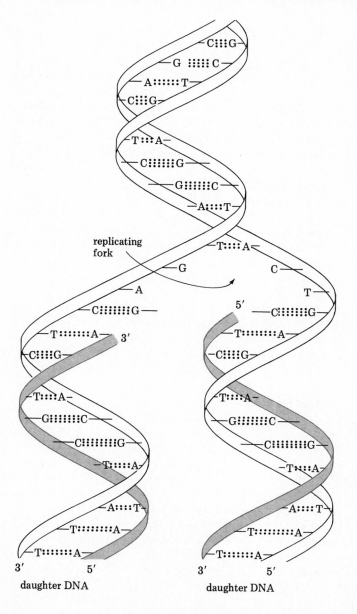

daughter DNA   daughter DNA

29.7
RNA and Transcription in Protein Synthesis

DNA forms RNA by a transcription process in which DNA serves as a template (Figure 29.7) for base pairing. The RNA is then a complementary strand. However, transcription is distinguished from replication because

1. RNA polymerase rather than DNA polymerase is the catalyst.
2. Uracil rather than thymine is incorporated in RNA.
3. The RNA molecules formed are single strands rather than double strands.

A shorthand representation of transcription is given below.

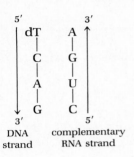

DNA actually produces three types of RNA—ribosomal RNA (rRNA), messenger RNA (mRNA), and transfer RNA (tRNA). All are formed as complementary strands against a portion of DNA as a template. Each type of RNA will now be described; their functions will be considered further in Section 29.9.

Ribosomal RNA

Ribosomal RNA *combines with over fifty different proteins to form complex structures called ribosomes* whose molecular weight is about 3 million. **Ribosomes** *are the site of protein synthesis*, but the function of ribosomal RNA is not informational. The ribosomes are like factories that of themselves cannot manufacture products since they need directions and workers.

Messenger RNA

Messenger RNA *carries instructions or directions for protein synthesis in a code or codon composed of a series of nitrogen bases.* The molecular weight of mRNA depends on the length of the protein whose synthesis it must direct. There must be three nucleotides for every amino acid in the protein to be formed. The reason for this relationship will be given in Section 29.8.

Transfer RNA

The function of **transfer RNA** *is to deliver individual amino acids from the amino acid pool to the site of protein synthesis.* Most tRNA molecules contain fewer than 100 nucleotides. A specific tRNA will only carry one type of amino acid. However, as will be shown in Section 29.8, several different tRNA molecules can transport the same amino acid. Considerable structural information has been obtained for several tRNA molecules. Each tRNA has an amino acid attachment site and a template recognition site. Each 3′ end has a CCA sequence at the amino acid attachment site. The amino acid is bonded to adenylic acid. The base sequence of the template recognition site will be discussed in Section 29.3.

In 1965 R. Holley of Cornell University determined the entire sequence of a tRNA for alanine. He suggested that the single strand molecule could form

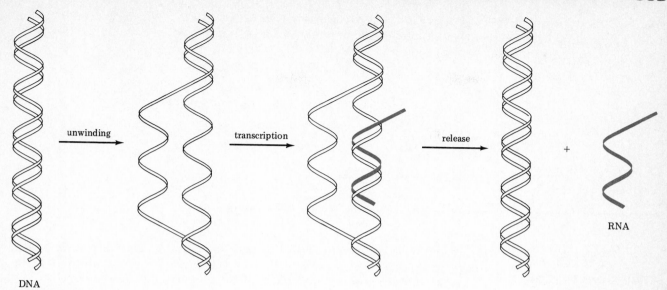

unwinding → transcription → release +

DNA

RNA

Figure 29.7
Transcription by DNA to Form RNA.

intrachain hydrogen bonds between some complementary base pairs. By constructing the maximum number of hydrogen bonds, he postulated a "cloverleaf" model for tRNA of alanine (Figure 29.8). The base pairings have been confirmed experimentally, but the actual shape of the three-dimensional molecule is not a planar cloverleaf. Nevertheless, the model is useful in discussing the chemistry of tRNA.

As shown in Figure 29.8, there are several bases other than A, U, C, and G in tRNA that occur in each of the leaves. One of the leaves critically involved in protein synthesis has an important three-nucleotide sequence called an anti-codon. *An* **anticodon** *is the complement of a trinucleotide sequence called a codon, which is located in mRNA* (Section 29.8). The interaction of the codon of mRNA and the anticodon of tRNA will be described in Section 29.9.

Example 29.3 ————————————————————————

What portion of a messenger RNA chain will be produced from the following portion of a DNA strand?

```
 5'
 |   dG
 |    |
 |    C
 |    |
 |    A
 ↓    |
 3'   T
```

Solution. Base pairing in RNA will result in C in RNA opposite G in DNA and G in RNA opposite C in DNA. The base T in DNA will pair with A in RNA, but A in DNA will pair with U in RNA.

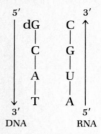

29.8
The Genetic Code

How can only four different bases in a base sequence of messenger RNA control the synthesis of a protein molecule? George Gamow, a theoretical physicist at the University of Colorado, pointed out that the four bases could be considered a four-letter alphabet that combine to form code words or codons. In order to provide a minimum of 20 codons sufficient for 20 different amino acids, it is necessary to use three-letter words. With one-letter words there are only four codons. With two-letter words $4 \times 4 = 16$ codons could be formed. However, with three-letter words $4 \times 4 \times 4 = 64$ codons are possible (Table 29.3), and thus the number of three-letter words exceeds the number of amino acids.

Codons *are the three-base sequences present in messenger RNA.* The portion of the DNA molecule that synthesizes messenger RNA contains the base

Table 29.3
Codons and Amino Acids

First base		Second base								Third base
		U		**C**		**A**		**G**		
U		UUU	Phe	UCU	Ser	UAU	Tyr	UGU	Cys	**U**
		UUC	Phe	UCC	Ser	UAC	Tyr	UGC	Cys	**C**
		UUA	Leu	UCA	Ser	UAA		UGA		**A**
		UUG	Leu	UCG	Ser	UAG		UGG	Trp	**G**
C		CUU	Leu	CCU	Pro	CAU	His	CGU	Arg	**U**
		CUC	Leu	CCC	Pro	CAC	His	CGC	Arg	**C**
		CUA	Leu	CCA	Pro	CAA	Gln	CGA	Arg	**A**
		CUG	Leu	CCG	Pro	CAG	Gln	CGG	Arg	**G**
A		AUU	Ile	ACU	Thr	AAU	Asn	AGU	Ser	**U**
		AUC	Ile	ACC	Thr	AAC	Asn	AGC	Ser	**C**
		AUA	Ile	ACA	Thr	AAA	Lys	AGA	Arg	**A**
		AUG	Met	ACG	Thr	AAG	Lys	AGG	Arg	**G**
G		GUU	Val	GCU	Ala	GAU	Asp	GGU	Gly	**U**
		GUC	Val	GCC	Ala	GAC	Asp	GGC	Gly	**C**
		GUA	Val	GCA	Ala	GAA	Glu	GGA	Gly	**A**
		GUG	Val	GCG	Ala	GAG	Glu	GGG	Gly	**G**

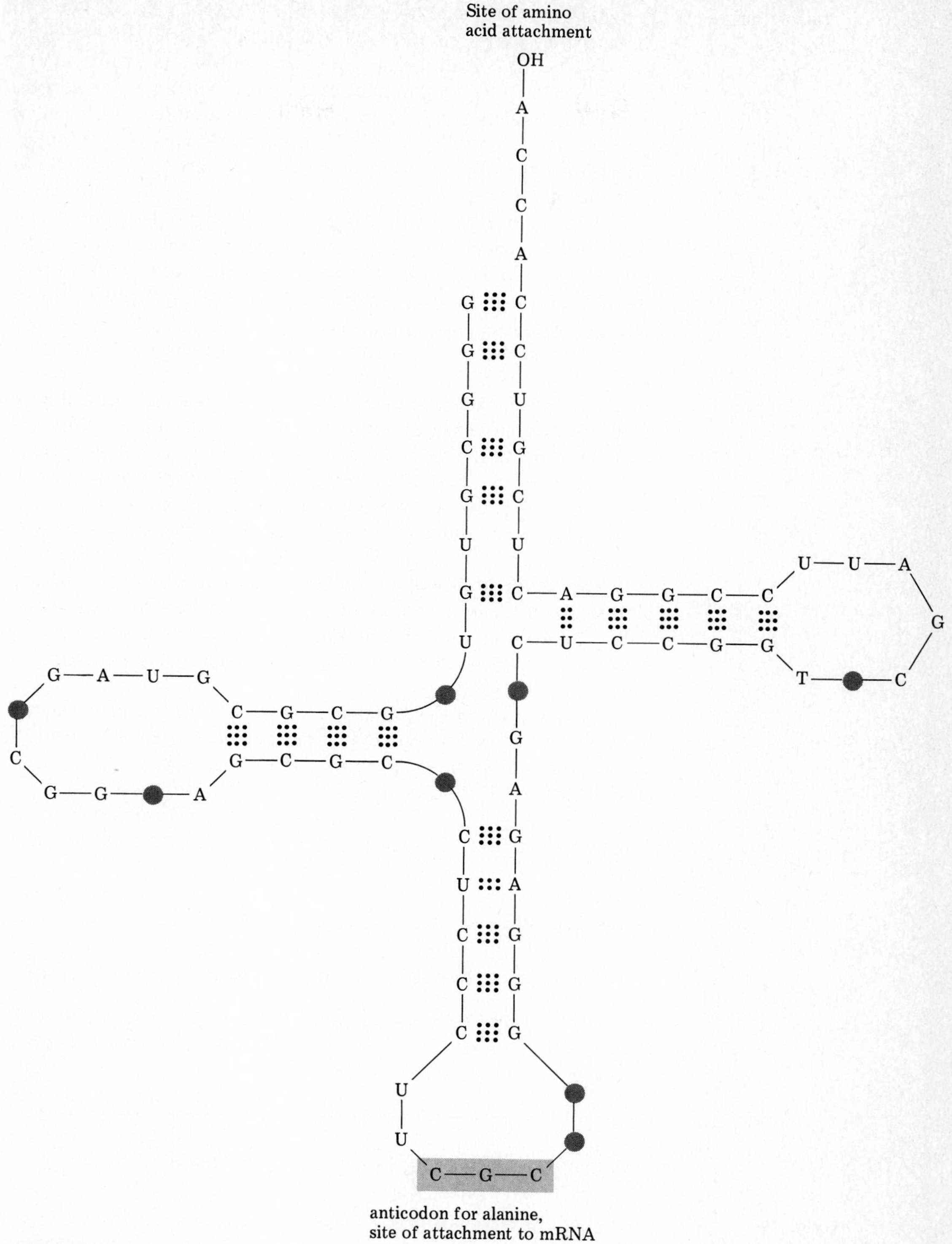

Figure 29.8
tRNA Structure
The dark circles represent modified bases that are present in small amounts in tRNA.

complements of the codons. Transfer RNA, which associates with messenger RNA in the ribosome, contains a base sequence that is an anticodon.

In 1962 the relationship between codons and amino acids began to be understood. *Escherichia coli* ribosomes were bound to a synthetic messenger RNA molecule consisting of only uridylic acid. The ribosomes associated only with the transfer RNA of phenylalanine. Therefore the base sequence U—U—U specifies phenylalanine, and messenger RNA containing only uridylic acid can produce only polyphenylalanine.

Gamow predicted that there might be more than one codon for each amino acid because there are 64 possible codons (Table 29.3). Thus a group of codons called synonyms can specify the same amino acid. The reason for having synonyms is still not entirely clear, but there must be advantages to the cell to have multiple ways of specifying its need for a particular amino acid.

There are some interesting relationships between codons specifying a common amino acid (Table 29.3). Amino acids specified by two or more synonyms occupy a single box in the table. For example, the neutral amino acids glycine, alanine, valine, threonine, and proline are specified by four codons that differ only in the third letter. Note also that these codons start only with C or G.

Arginine and leucine are each specified by both a quartet and a related set of two codons. Serine is unique in being specified by a codon quartet and an unrelated set of two codons. Methionine is the only neutral amino acid specified by a single codon.

The acidic amino acids, aspartic acid and glutamic acid, are specified by two codons each. However, these two sets of codons are related with the difference being in the third letter. The basic amino acids, lysine and histidine, have two codons each. However, in contrast to the acidic amino acids, the codons are not related.

Tyrosine and phenylalanine, which are structurally related aromatic amino acids, are specified by sets of two codons each that differ in the second letter. The third aromatic amino acid, tryptophan, has a single codon that is not related to the codons of the other aromatic amino acids.

The sequences UAA, UAG, and UGA are involved in chain termination of protein synthesis. This feature of the genetic code will be discussed in the next section.

Example 29.4

What are the codons for alanine? How are they related?

Solution. Table 29.3 shows the codons GCU, GCC, GCA, and GCG for alanine. These four codons are related by identical letters in the first two positions. Only the third letter differs.

[Additional examples may be found in **29.39–29.41** at the end of the chapter.]

29.9
Protein Synthesis and Translation

In order to translate the codons of messenger RNA into a sequence of amino acids, the following substances are required.

1. Amino acids.
2. ATP to activate the amino acid.

3. Transfer RNA.
4. Messenger RNA.
5. Ribosomal RNA.
6. Guanosine triphosphate (GTP).

Activation of Amino Acids

In order to prepare an amino acid for protein synthesis, ATP reacts with the carbonyl carbon atom to yield an activated product in which adenosine monophosphate is bonded to the amino acid. The product is called an aminoacyl adenosine monophosphate or an aminoacyl adenylate (Figure 29.9). This reaction is catalyzed by a group of enzymes known as aminoacyl synthetases, and a specific enzyme is required for activation of each different amino acid.

Esterification of tRNA

Aminoacyl adenosine monophosphates of the amino acids react with a molecule of transfer RNA. Although each transfer RNA has a unique structure, they all terminate in the sequence CCA. The carbonyl carbon of the amino acid forms an ester with a hydroxyl group of the terminal adenylic acid.

As discussed previously an anticodon in transfer RNA is responsible for its association with the codon of messenger RNA. The anticodon, which is present in a cloverleaf of the transfer RNA, is responsible for forming the proper aminoacyl RNA molecule and thus bringing the correct amino acid to the codon site in mRNA where base pairing occurs.

Site of Protein Synthesis

Messenger RNA and the ribosome associate starting at the 5′ end and the ribosome then traverses the messenger RNA chain to the 3′ end. After one ribosome starts its journey, other ribosomes may do the same thing. Thus several ribosomes simultaneously translate the same mRNA molecule. Each ribosome will be at a different position along the mRNA chain, and the several proteins will

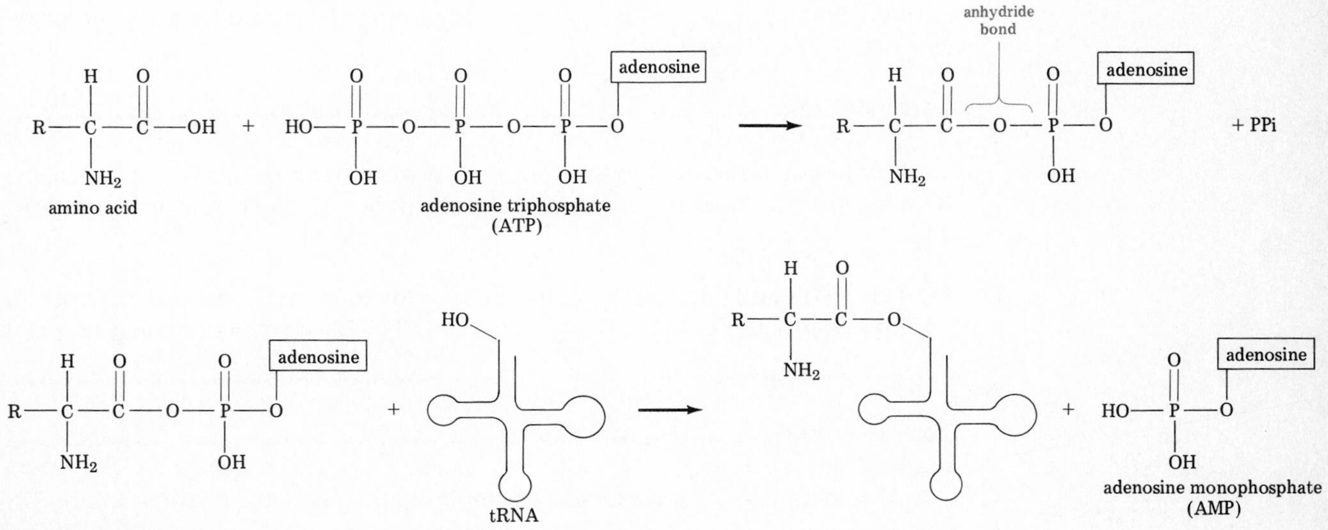

Figure 29.9
Formation of Activated Amino Acid and Activated tRNA.

beginnning of
protein

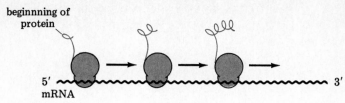

Figure 29.10
Ribosomes and mRNA
The ribosomes move
from the 5′ to the 3′
end of the mRNA and
function independ-
ently as they synthe-
size a protein chain.

be in various stages of growth. Eventually, identical proteins are formed because the same genetic message is being read (Figure 29.10).

Protein synthesis consists of three steps at the ribosome.

1. Binding of an aminoacyl tRNA by complementary base pairing.
2. Peptide bond formation.
3. Translocation.

The processes occur at two site in the ribosomes called the donor (P) site and the acceptor (A) site. During synthesis, a growing protein chain is transferred from the donor site to the amino acid located at the acceptor site (Figure 29.11). The protein chain is synthesized from the N-terminal toward the C-terminal amino acid.

Protein synthesis starts with formylmethionine (fMet) combined with a special tRNA. The fMet tRNA occupies a P site in (a). In (b) an aminoacyl tRNA specified by the codon in messenger RNA is shown in an A site. Peptide bond formation has occurred in (c) as fMet in the P site is transferred to arginine in the A site. In (d) the free tRNA for fMet has left the site, and the ribosome has moved to the next trinucleotide sequence to the right. This process is called translocation and is catalyzed by translocase with the expenditure of energy provided by GTP. The dipeptide is now located in a donor site. Another aminoacyl tRNA can now enter the A site, and the entire sequence is repeated until the protein is complete and growth of the peptide chain is terminated. Note that bonding is always from the carboxyl group of the peptide to the amino group of the next amino acid in the growing peptide chain.

Example 29.5

What anticodon must exist in the template recognition site of an aminoacyl tRNA in order to bind at the site having the codon sequence AGU in messenger RNA.

Solution. The anticodon must consist of complementary base pairs in order to bind aminoacyl tRNA to the codon sequence. The base sequence must be UCA.

Example 29.6

What peptide sequence is synthesized from the following codon sequence in mRNA?

—G—C—U—G—A—A—U—G—G—

Solution. Reading the letters as sequences of three-letter words, the codons are GCU, GAA, and UGG. These codons specify alanine, glutamic acid, and tryptophan, respectively. The peptide will be Ala-Glu-Trp.

[Additional examples may be found in **29.48–29.54** at the end of the chapter.]

Initiation and Termination

In the preceding discussion, two important features of protein synthesis, initiation and termination, were not considered. **Initiation** *means to start the growth of the protein chain.* **Termination** *means to conclude the synthesis of the protein chain and to release it from the ribosome.*

The process of initiation is quite complex. However, for a number of mRNA molecules, the first amino acid in the chain is a modified methionine called formylmethionine, and it is specified by AUG. In combination with other more complex factors, the formylmethionine signifies the N-terminal position of the growing peptide chain.

Termination of the protein synthesis is indicated by the termination codons UAA, UGA, or UAG. There are no tRNA molecules with the anticodon to pair with these trinucleotide sequences. The codons then bind some proteins called release factors. Binding of the factors somehow activates the enzyme needed to hydrolyze the peptide from the tRNA at the P site. The peptide chain is then free to leave the ribosome.

Protein Modification

Most proteins formed by translation of mRNA are not the actual proteins required by the cell. The proteins released from the ribosome are usually modified in some way. One of the most common modifications is the hydrolysis of the formyl group of formylmethionine or of methionine itself. In some cases the protein chains are cleaved, as in the conversion of proinsulin into insulin. Other modifications include joining the protein to another group such as a lipid or a carbohydrate.

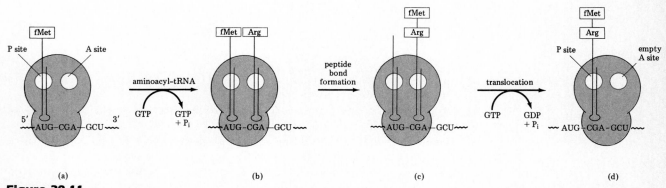

Figure 29.11
Synthesis of a Protein at Sites in a Ribosome.

Antibiotic Inhibitors

Antibiotics exert their effect on prokaryotes by inhibiting the formation of necessary proteins. Streptomycin interferes with the binding of formylmethionine tRNA and prevents the initiation of protein synthesis. Erythromycin inhibits protein synthesis by preventing translocation. Puromycin causes premature chain termination, which results in the release of an incomplete protein.

29.10
Gene Mutations and Genetic Disease

A **mutation** *of a gene is a change of one or more of the nucleotides in DNA.* Nucleotides may be substituted, added, or deleted. Before considering the consequences of mutation on protein synthesis, let's see what happens to a sentence when letters are substituted, added, or deleted. Since codons are three-letter words, our sentence is

THE BIG DOG SAW THE CAT

If substitution of one letter by another occurs, the message might or might not make sense. The degree of confusion depends on the letter substituted and the location in the word. For example, consider

THE BIG DOH SAW THE CAT
sense   sense

The first two words and the last three make sense, but DOH doesn't make sense. Some of the message is understood and there is little confusion.

Now let's consider what happens when H is substituted for the first letter of a word.

THE BIG HOG SAW THE CAT
sense

The meaning of the message is changed, but it does make sense. However, if you wanted to send information about a dog, with the word hog the sentence is seriously flawed.

Consider adding a letter to the sentence. If the words are still read three letters at a time, a major part of the sentence is nonsense. Let's add E after the G of DOG.

THE BIG DOG ESA WTH ECA T
nonsense

Deletion of a letter such as the G of DOG also gives nonsense.

THE BIG DOS AWT HEC AT
nonsense

Both additions and deletions are examples of frame shift errors, which cause a different series of words to be read.

Now let's consider the effect of substitution, addition, and deletion on codons. If DNA is mutated, then mRNA is mutated as well. As a consequence, protein synthesis is flawed. The seriousness of the effect that the defective protein has on biological function depends on the amino acid or acids that are changed.

The consequences of changing the third letter in a three-letter sequence may not be serious. Many amino acids have several codons that differ in the third letter. For example, valine is coded by GUA, GUG, GUU, and GUC. Interchange of the third letter will still cause the appropriate amino acid to be placed in the protein.

Mutations of even one letter can sometimes lead to serious biological damage. The improper amino acid may be placed at the active site of an enzyme and render it useless. Consider the codon UCG for serine. Mutation to give UUG results in leucine being substituted for serine. Since serine is located at active sites in trypsin and chymotrypsin, such a flawed enzyme would not hydrolyze proteins.

Although not necessarily involved in an active site, a single amino acid may change the conformation of a protein to alter biological function. In normal adult hemoglobin, glutamic acid is the sixth amino acid in the chain. The codons for glutamic acid are GAA and GAG. In sickle cell hemoglobin, valine is in the sixth position. The codons for valine are GUA, GUG, GUU, and GUC. Substitution of A, the second letter in the glutamic acid codon, by U gives codons for valine. The replacement of glutamic acid by valine changes the tertiary structure of hemoglobin. Glutamic acid is an acidic amino acid, but valine is a neutral amino acid.

There are serious consequences when a mutation changes the codon for an amino acid into a termination codon. One of the codons for serine is UCA. If G is substituted for C, the termination codon UGA results. As a consequence when this codon is reached the synthesis of the protein is stopped. Such a protein probably is useless.

Mutations caused by deletion or addition also have serious consequences. Beyond the point of change there is formed a completely unsuitable sequence of amino acids.

Example 29.6

What are the consequences of a genetic mutation that results in replacement of CGC in messenger RNA by CCC?

Solution. The change is in the second letter and is expected to result in the specification for a different amino acid in protein synthesis. The codon CGC specifies the basic amino acid arginine, whereas the codon CCC specifies the secondary amino acid proline. Since these amino acids differ substantially in structure and acid–base properties, the tertiary structure and function of the protein may be seriously altered.

[Additional examples may be found in **29.48–29.54** at the end of the chapter.]

29.11
Recombinant DNA

No pills, vaccination, or other current medical procedures can cure genetic diseases. Only intervention at the level of the nucleus of the cell in the DNA can correct a genetic abnormality. It is this awesome task that is being considered as the next advance of biochemistry.

At present the research on changing the gene structure is limited to the simplest of bacteria, *Escherichia coli* (*E. coli*). DNA chains are cleaved, a different piece of DNA is inserted, and the chain is reconnected. The process is called gene splicing, and the new DNA is called recombinant DNA.

Recombinant DNA has properties and functions characteristic of the original DNA as well as the inserted DNA. *E. coli* has been made to produce proteins that it normally does not make. For example, *E. coli* has been spliced with a DNA fragment responsible for producing the messenger RNA for human insulin. The *E. coli* then produces insulin, which it does not normally produce. Diabetics use insulin from animal sources, and these insulins differ from human insulin. Allergic reactions sometimes occur, and another animal insulin must be used. When bacterially produced human insulin becomes available, a long-standing medical problem will be solved since insulin should be free of any complications such as allergic reaction.

29.12
Viruses

Viruses, which cause a variety of diseases in plants and animals including chicken pox, measles, and influenza in man, are intermediate between living and nonliving material. They are composed of either DNA or RNA and protein, but do not have the necessary cellular components to support life. As a result, viruses cannot independently reproduce themselves.

Viruses are parasites of cells. They attack cells and take over the cell's chemical machinery. The viral chromosomes can be replicated in this host cell, and viral proteins can be formed by transcription and translation. Depending on the virus and the host cell, the viral multiplication can ultimately destroy the cell. In other cases the virus chromosome may be incorporated into the chromosome of the host cell, which now has new properties and functions.

Viral infection can be prevented by immunization with a specific vaccine. The vaccine contains an active or inactive form of a virus that stimulates the body's natural immune system to produce antibodies against the virus. These antibodies circulate in the body and inactivate invading active viruses before they can enter host cells. Viral infections can also be treated with drugs that inhibit nucleic acid synthesis.

Summary

The nucleic acids contain three types of components: phosphoric acid, a pentose, and a heterocyclic nitrogen base. Ribonucleic acids contain β-D-ribose, whereas deoxyribonucleic acids contain β-D-2-deoxyribose. The bases in ribonucleic acids are adenine, uracil, cytosine and guanine, whereas the bases in deoxyribonucleic acids are adenine, thymine, cytosine and guanine.

Nucleosides are *N*-glycosides of a pentose and one of the heterocyclic nitrogen bases. Nucleotides are phosphate esters of nucleosides. Polynucleotides contain diphosphate esters between the 5′ hydroxyl group of one pentose and the 3′ hydroxyl group of another pentose unit.

DNA consists of two polynucleotide chains coiled about an axis to form a right-handed double helix. The purine and pyrimidine bases, which are pointed to the interior of the helix, form hydrogen bonds. Cytosine forms three hydrogen bonds to guanine, and adenine forms two hydrogen bonds to thymine.

The double helix of DNA is duplicated by semi-conservative replication. The helix uncoils, and each chain serves as a template for the synthesis of its complementary chain. Each daughter DNA has one strand of the original DNA and a new strand.

RNA are single strand molecules which contain diphosphate esters from the 5′ hydroxyl group of ribose of one nucleotide unit to the 3′ hydroxyl group of another nucleotide. There are three types of RNA: ribosomal RNA, transfer RNA and messenger RNA.

DNA produces RNA by synthesis of a complementary strand by a process called transcription. Information coded in messenger RNA is used in a translation process to form proteins at a site in ribosomal RNA.

Learning Objectives

As a result of studying Chapter 29 you should be able to

- Name and draw the structures of the sugars and bases found in nucleic acids.

- Name the compounds represented by C, G, T, A, and U.
- Draw the structures of nucleosides and nucleotides and describe the bonding within these molecules.
- Explain the relationship between the composition of the bases in DNA.
- Outline the tenets of the central dogma of heredity.
- Define the terms replication, transcription, and translation.
- Outline the functions of mRNA, tRNA, and rRNA.
- Write an anticodon for a codon.
- Write an amino acid sequence of a peptide given the DNA or mRNA base sequence.
- Show how addition, deletion, or substitution of nucleotides leads to synthesis of a defective protein.
- Describe the action of viruses.
- Explain the action of antibiotics.

New Terms

The **anticodon** is a series of three bases of tRNA that base-pair with the codon of mRNA.

Antiparallel strands in DNA run in opposite directions when read from the 5' to the 3' end.

The **central dogma** is a series of theories about the transmission of hereditary information and protein synthesis.

Chromosomes are contained in somatic cells and possess genetic information.

A **codon** is a sequence of three bases found in messenger RNA.

Complementary base pairing is the matching of bases through hydrogen bonding.

Deoxyribonucleic acid is a polynucleotide containing deoxyribose, phosphate, and a mixture of adenine, thymine, guanine, and cytosine.

The **double helix** is an association of two polydeoxyribonucleotides that are antiparallel.

DNA polymerase is the enzyme that catalyzes the replication of DNA.

The **genetic code** in DNA contains all the information needed to maintain the identity of an organism.

Initiation means the start of growth of the protein chain.

Messenger RNA contains a series of codons required to synthesize proteins.

Mutation of a gene is a change in one or more nucleotides in DNA.

Nucleic acids are deoxyribonucleic and ribonucleic acids.

A **nucleoside** contains a pentose and a heterocyclic nitrogen base combined as a *N*-glycoside.

A **nucleotide** results from phosphorylating a nucleoside.

Phosphodiesters link nucleotides in polynucleotides.

Purine bases refer to adenine and guanine found in nucleic acids.

Pyrimidine bases refer to cytosine, thymine, and uracil found in nucleic acids.

Replication is a duplication process to generate DNA for cell division.

The **replication fork** is the site of replication of DNA.

Ribonucleic acids contain ribose, phosphate, and a mixture of adenine, uracil, guanine, and cytosine.

Ribosomal RNA is used to produce ribosomes, which are the site of protein synthesis.

The **ribosomes** consist of ribosomal RNA and protein.

RNA polymerase is an enzyme required to produce RNA by a transcription process.

A **template** is a pattern used to synthesize compounds with complementary bases.

Termination is the conclusion of synthesis of the protein chain.

Termination codons are codons that signal that the construction of a protein is complete.

Transcription is the means by which information for protein synthesis is passed from DNA to messenger RNA.

Transfer RNA molecules have an amino acid attachment site and a template recognition site needed to bring the proper amino acid to messenger RNA.

Translation is the transfer of information from a sequence of nucleotides to form a sequence of amino acids in a peptide chain.

Questions and Problems

Terminology

29.1 What are complementary bases and how do they associate?

29.2 Describe the differences between translation and transcription.

29.3 What are nucleosides and nucleotides?

29.4 What are codons and anticodons?

29.5 What is a frame shift?

29.6 To what process does translocation refer in protein synthesis?

DNA and RNA

29.7 What are the chemical differences in the structures of DNA and RNA?

29.8 What are the functional differences in DNA and RNA?

29.9 Name the three types of RNA and describe their structure and function.

29.10 What group bridges the sugars in DNA and RNA?

29.11 DNA and RNA are acidic materials. Explain why.

29.12 Describe the tenets of the central dogma.

Bases in DNA and RNA

29.13 Indicate whether each of the following is a purine or pyrimidine.
    (a) uracil   (b) cytosine   (c) guanine
    (d) adenine   (e) thymine

29.14 Which of the bases are found in DNA? Which are found in RNA?

29.15 What relationship exists between the bases in DNA?

29.16 What relationship exists between the bases in RNA?

29.17 What relationship exists between the percentage of the total purine bases and the total pyrimidine bases in DNA?

Nucleosides and Nucleotides

29.18 Write the structural formula for adenosine monophosphate.

29.19 Write a structural formula for a nucleoside containing thymine and deoxyribose.

29.20 What is the configuration of the N-glycoside bond of nucleotides?

29.21 What hydroxyl group of ribonucleosides is phosphorylated to give ribonucleotides?

29.22 How many free hydroxyl groups exist in a polynucleotide of deoxyribose?

29.23 In what direction is the structure of a polynucleotide read?

29.24 Cyclic AMP ($3',5'$-cyclic adenosine monophosphate) is a regulator of metabolic activity. It has a single phosphate that has esterified both the $3'$ and $5'$ hydroxyl groups of adenosine. Draw the structure of cyclic AMP.

The Double Helix

29.25 Describe the primary and secondary bonding in DNA.

29.26 The strands of DNA in the double helix are antiparallel. What does this mean?

29.27 Which of the following pairs do not occur in the DNA molecule? Explain why.
(a) G–A   (b) G–T   (c) C–U   (d) A–A
(e) C–T   (f) A–C

29.28 What forms the backbone structure of each strand of the DNA double helix?

29.29 Where are the base pairs located in the double helix?

Replication

29.30 Describe the replication process.

29.31 What is the replicating fork?

29.32 What is the complementary strand for a DNA strand with the sequence dGCATCAG?

29.33 What is the function of DNA polymerase?

Transcription

29.34 What mRNA would be formed from the base sequence in **29.32**?

29.35 What is the complementary base in mRNA for A in DNA? What base in mRNA is the complement of T in DNA?

29.36 What direction is used to describe the base sequence in mRNA?

29.37 How are the various types of RNA produced?

Translation

29.38 Describe the shape of mRNA.

29.39 What is a codon? What is an anticodon?

29.40 Does every base triplet code for an amino acid in protein synthesis?

29.41 What amino acid does each of the following codons in messenger RNA place in a protein chain?
(a) GUU   (b) CCC   (c) UUU   (d) ACG
(e) UGU   (f) GCA   (g) AGG   (h) CUA

29.42 How many nucleotides must be present in mRNA to synthesize a protein containing 250 amino acids?

29.43 What anticodon in tRNA will base-pair with the codons of **29.41**?

29.44 A portion of DNA has the sequence CCCTGTACACCT. What base sequence will form in mRNA? What peptide would be formed?

29.45 What base sequence in mRNA is necessary to form Val-Asp-Ala-Gly?

29.46 How are acceptor and donor sites in the ribosome used in protein synthesis?

29.47 What amino acid initiates protein synthesis?

Codons and Mutations

29.48 What are the genetic consequences of a frame shift?

29.49 Why are substitution mutations generally less serious than deletion or addition mutations?

29.50 Give two types of substitution mutations that will cause seriously defective proteins.

29.51 What aromatic amino acids have similar codons?

29.52 Which type of amino acids has the largest number of codons?

29.53 What differences are there between the codons for acidic amino acids and those for basic amino acids?

29.54 What is unique about the codons of tryptophan and methionine?

Selected Answers to Questions and Problems

Chapter 1

1.2 A theory is a concept, whereas a law is a statement of fact.

1.4 36%

1.6 43% C; 57% O

1.8 52% C; 13% H; 35% O

1.10

1.12 Scientists accept that there is a natural order that can be discovered.

1.14 Hypotheses are needed to test and to establish a theory.

1.16 So that they may be tested.

1.18 182 lb

1.20 5.0×10^5 barrels

1.22 10 s

1.24 25 mph

1.26 5.26×10^3 mL

1.28 7.1×10^2 mg

1.30 4 lb

1.32 2.5×10^3 mL

1.34 2.5×10^{10} cells

1.36 1×10^5 drops

1.38 1.00×10^{-4} nm/L

1.40 (a) .00156   (b) 4897   (c) 34680   (d) .0001987   (e) 502   (f) .00006059

1.42 (a) 3   (b) 4   (c) 3   (d) 2   (e) 4   (f) 3   (g) 3   (h) 2

Chapter 2

2.4 19.3 g/cc

2.6 1.1 g/mL

2.8 No; the density is too high.

2.10 $-35°C$

2.12 239 K

2.14 961°C; 1762°F

2.16 248°F

2.18 (a) 80 cal   (b) 10 cal   (c) 800 cal

2.20 2.2×10^7 cal

2.22 3×10^3 Cal

2.24 3.8 Cal

2.26 Their compositions are identical, and they are interconverted by physical changes.

2.28 No; the oxygen is released as a gas.

2.30 0.032 g

2.32 Matter is "chemicals."

2.34 A mixture of gases that is released by heating, and solids that are deposited after the water is boiled.

2.35 a, b, d, f, h, l, m, s, t are physical changes.

2.38 It is formed from chemical combination of elements.

2.40 X is a compound; Y and Z cannot be classified.

2.42 (a) Ca   (b) Cl   (c) Cu   (d) C   (e) Mg   (f) Hg   (g) P   (h) S   (i) Au   (j) Sn   (k) Br   (l) Ni

2.44 hydrogen

2.46 O and Si

2.48 O, C, and H

2.50 carbon dioxide + water $\longrightarrow$ glucose + oxygen

2.52 17.86 g of water is produced, and 3.14 g of oxygen does not react.

Chapter 3

3.2 Electrons surround a nucleus that consists of protons and neutrons.

3.4 The electron is much lighter than the proton.

3.6 (a) 24.0   (b) 17.9   (c) 235

3.8 1.21×10^{-8} cm

3.10 See Table 3.2.

3.12 The electron has a mass of 0.0005486 amu.

3.14 The radii are 10^{-13} and 10^{-8} cm.

3.16 It equals the number of protons; the subscript to the left of the symbol.

3.18 Given in the order electrons, protons, and neutrons:

(a) 3, 3, 4   (b) 15, 15, 16
(c) 47, 47, 60   (d) 27, 27, 32
(e) 31, 31, 38   (f) 80, 80, 121
(g) 80, 80, 119   (h) 55, 55, 84
(i) 11, 11, 11   (j) 8, 8, 9

3.20 88 protons; 88 electrons; 138 neutrons

3.22 (a) 14, 7, 7, 7, 7   (b) $^{14}_{6}C$, 6, 6, 8
(c) $^{40}_{20}Ca$, 40, 20, 20   (d) $^{79}_{35}Br$, 35, 35, 44
(e) $^{35}_{17}Cl$, 17, 17, 18

3.24 Each isotope contains one proton and one electron. The numbers of neutrons are 0, 1, and 2, respectively.

3.26 It has 4 additional neutrons.

3.28 (a) He   (b) H_2 (c) P_4 (d) N_2 (e) O_2 (f) S_8 (g) F_2 (h) Cl_2

3.30 $C_7H_5N_3O_6$

3.32 One molecule contains 10 carbon atoms, 14 hydrogen atoms, and 2 nitrogen atoms.

3.34 C_4H_{10}

3.36 Mn^{2+}; 23

3.38 Fe^{2+} has 24 electrons, whereas Fe^{3+} has 23 electrons.

3.40 (a) O^{2-} (b) S^{2-} (c) I^- (d) Cl^- (e) F^-
(f) Br^- (g) Na^+ (h) Ca^{2+} (i) K^+ (j) Mg^{2+}
(k) Zn^{2+} (l) Li^+

3.42 (a) oxide   (b) chloride   (c) sulfide
(d) lithium   (e) potassium   (f) calcium

3.44 (a) ferrous, ferric   (b) cuprous, cupric

3.46 (a) $FeCl_3$ (b) $NaOH$ (c) $Mg(OH)_2$
(d) CdS (e) KBr (f) Li_3N
(g) $Ba(NO_3)_2$ (h) $CsClO_4$

3.48 (a) $FeCl_2$ (b) Cu_2O (c) PbS (d) $HgBr_2$
(e) Fe_2O_3 (f) CuS (g) FeS (h) Cu_2O

3.50 (a) $CaCO_3$ (b) $NaHCO_3$ (c) $Mg(OH)_2$
(d) $Al(OH)_3$

Chapter 4

4.2 There are the same numbers of structural units given by the chemical formula.

4.4 They represent average weights of mixtures of isotopes.

4.8 50% of each isotope

4.10 28.1 reflecting the major isotope $^{28}_{14}Si$.

4.12 $12 + 2(16) = 44$

4.14 (a) 176   (b) 286   (c) 334   (d) 1351

4.16 (a) 58.4   (b) 119.0   (c) 62.0   (d) 45.9
(e) 56.1   (f) 151.3   (g) 126.7   (h) 297.1
(i) 40.3   (j) 319.2

4.18 (a) 106.0   (b) 132.1   (c) 399.9   (d) 187.6
(e) 158.0   (f) 146.3   (g) 310.2   (h) 58.3

4.20 (a) 75% C; 25% H
(b) 23.1% C; 73.1% F
(c) 5.9% H; 94.1% O
(d) 40.0% C; 6.7% H; 53.3% O
(e) 43.7% P; 56.3% O
(f) 36.1% Ca; 63.9% Cl
(g) 63.5% Ag; 8.2% N; 28.3% N
(h) 15.8% Al; 28.0% S; 56.2% O

4.22 47.4% C; 2.5% H; 50.1% Cl

4.24 (a) 15.3% Na; 84.7% I   (b) 74.2% Na; 25.8% O
(c) 71.5% Ca; 28.5% O   (d) 36.1% Ca; 63.9% Cl
(e) 52.9% Al; 47.1% O   (f) 60.3% Mg; 39.7% O
(g) 63.5% Fe; 36.5% S   (h) 55.9% S; 44.1% Fe

4.26 A mole contains Avogadro's number of formula units.

4.28 Divide by the atomic, molecular, or formula weight.

4.30 (a) 2.00   (b) 0.500   (c) 0.010   (d) 1.00

4.32 (a) 0.10   (b) 0.1000   (c) 0.000890   (d) 0.0050

4.34 (a) 0.40   (b) 8.0   (c) 1.1×10^2
(d) 80   (e) 18   (f) 1.1×10^3
(g) 0.3   (h) 3

4.36 (a) 6.02×10^{22} (b) 6.02×10^{24}
(c) 3.01×10^{23} (d) 6.02×10^{21}

4.38 (a) 0.50   (b) 0.00100   (c) 10.0

4.40 Coefficients are given in the order of appearance in the equations.
(a) 1, 3, 2, 3   (b) 1, 3, 1, 1
(c) 1, 6, 4   (d) 1, 3, 1, 3
(e) 1, 2, 1, 2   (f) 1, 3, 2, 3

4.42 Coefficients are given in the order of appearance in the equations.
(a) 1, 1, 1, 2   (b) 1, 1, 1, 1
(c) 2, 1, 2, 1   (d) 1, 2, 1, 2
(e) 3, 2, 1, 3   (f) 1, 1, 2, 1

4.44 1.0 mole

4.46 127 g

4.48 13 g

4.50 1271 g

Chapter 5

5.4 It has more elements.

5.6 (a) 2, VA   (b) 4, IIA   (c) 5, IIB
(d) 4, VIIA   (e) 2, IA   (f) 3, IIIA
(g) 5, IVA   (h) 3, VIA   (i) 5, 0
(j) 4, VIII   (k) 4, VIB   (l) 4, IB

5.8 Cs

5.10 Ar and K; Ni and Co

5.12 b, c, d, g, h and j are metals; a, e, f, and i are nonmetals.

5.14 francium

5.16 They have metallic properties and form cations.

5.18 (a) 10   (b) 10   (c) 18   (d) 18   (e) 10
(f) 18   (g) 2   (h) 10

5.20 No known element has a sufficient number of electrons to fill higher energy levels.

5.22 (a) 2 in first, 7 in second
(b) 2 in first, 8 in second, 8 in third, 1 in fourth
(c) 2 in first, 8 in second, 8 in third, 2 in fourth
(d) 2 in first, 8 in second, 6 in third
(e) 2 in first, 5 in second
(f) 2 in first, 8 in second, 2 in third

5.24 No; the mass number gives the number of protons plus neutrons. Only the number of electrons determines electron configuration.

5.26 (a) 2   (b) 4   (c) 5

5.28 (a) 1, $1s$ (b) 1, $3s$
(c) 7, $3s$ and $3p$ (d) 1, $2s$
(e) 7, $2s$ and $2p$ (f) 5, $2s$ and $2p$
(g) 6, $2s$ and $2p$ (h) 1, $4s$
(i) 6, $3s$ and $3p$ (j) 5, $3s$ and $3p$
(k) 3, $3s$ and $3p$ (l) 4, $3s$ and $3p$

5.30 (a) 2   (b) 2   (c) 2   (d) 2

5.32 (a) 2   (b) 1   (c) 1   (d) 2   (e) 1   (f) 2
(g) 3   (h) 2   (i) 1   (j) 1

5.34 (a) $:\ddot{O}:$ (b) $Na\cdot$ (c) $:\ddot{F}\cdot$ (d) $\dot{A}l\cdot$
(e) $\cdot\ddot{N}\cdot$ (f) $\cdot\ddot{P}\cdot$

5.36 7

5.38 (a) 4   (b) 6   (c) 5   (d) 8
(e) 1   (f) 5   (g) 2   (h) 3

5.40 2

5.42 (a) IVA   (b) VIIA   (c) 0   (d) VA   (e) VIIB

5.44 (a) 108°C   (b) 67°C   (c) 3.2 g/cm^3

5.46 It should be a member of Group IIB and have properties similar to mercury.

5.48 He

5.50 The ion should be larger.

5.52 He

5.54 (a) F (b) Li (c) Cl (d) S (e) Ge
(f) Mg

5.56 They have high ionization energies.

5.58 fluorine

5.60 The higher the electron negativity, the larger the tendency to form anions.

Chapter 6

6.2 The physical state, melting point, and boiling point of the compound.

6.4 +2 ions

6.6 The ionization energy is too high for C^{4+} formation, and the electronegativity too low for C^{4-}.

6.8 (b) $1s^2$ (d) $3s^0$ (f) $3s^0$
(h) $2s^2 2p^6$ (j) $4s^2 4p^6$ (l) $2s^2 2p^6$
(n) $2s^2 2p^6$

6.10 (a) NaCl (b) $CaCl_2$ (c) MgO (d) $MgBr_2$
(e) BaF_2 (f) $Ba_3 P_2$ (g) $Li_3 N$ (h) $AlCl_3$

6.12 Scandium loses the $4s^2$ electrons as well as the $3d^1$ electron.

6.14 Anions always have a larger radius than the atoms.

6.16 (a) Mg (b) K (c) Al (d) Br^- (e) S^{2-}
(f) N^{3-}

6.18 Only identical atoms may share electrons equally.

6.20 (a) H—S̈—H (with H below) (b) H—P—H (with H below)

(c) :F̈—C—F̈: (with :F̈: above and :F̈: below) (d) H—Së—H (with H below)

(e) :Br̈—C—Br̈: (with :Br̈: above and :Br̈: below) (f) :Cl̈—Si—Cl̈: (with :Cl̈: above and :Cl̈: below)

(g) H—Si—H (with H above and H below) (h) :Cl̈—P—Cl̈: (with :Cl̈: below)

(i) H—Sb̈—H (with H below)

6.22 Each atom, which has 7 valence electrons, shares one electron to form a single bond.

6.24 SnH_4 H—Sn—H (with H above and H below)

6.26 $SiCl_4$

6.28 The negative end of the polar bond is at the most electronegative atom.

6.30 (a) H (b) Br (c) H (d) O (e) Cl
(f) Si (g) H (h) N (i) P (j) P

6.32 (a) +1 (b) +3 (c) +3 (d) +5
(e) +3 (f) +5

6.34 (a) +6 (b) +5 (c) +3 (d) +2
(e) −2 (f) 0

6.36 (a) −3 (b) −2 (c) −1

6.38 (three resonance structures of NO_2^- type showing O, N, O arrangements)

6.40 :F̈ B :F̈ (with :F̈: below); 120°

6.42 The single bond and the triple bond are a maximum distance from each other at 180°.

6.44 (a) tetrahedral (b) planar
(c) tetrahedral (d) angular
(e) tetrahedral (f) angular

6.46 (a) tetrahedral (b) pyramidal
(c) angular (d) angular
(e) tetrahedral (f) pyramidal

6.48 All are sp^3.

6.50 d, e, g and h are sp^2; f is sp.

Chapter 7

7.2 A real gas has intermolecular attractive forces and occupies a fraction of the gaseous volume.

7.4 760

7.6 (a) 0.500 (b) 38.0 (c) 76.0

7.8 14.7 lb/in.^{2}

7.10 There is less air at this altitude.

7.12 No; the H_2 has a larger average velocity.

7.14 225 mL

7.16 4.2 L

7.18 1.4×10^3 L

7.20 375 K

7.22 750 K

7.24 2 atm

7.26 1.8 atm

7.28 3.7 atm

7.30 4 atm

7.32 507 torr

7.34 They are equal.

7.36 (a) 0.1 (b) 4.7 (c) 1

7.38 44

7.40 1.25 g/L

7.42 0.7

7.44 290 mL

7.46 62 torr

7.48 1.3 L

7.50 Individuals at higher altitudes have more hemoglobin.

Chapter 8

8.2 Boiling point may refer to any pressure, whereas normal boiling point refers to 1 atm.

8.4 humidity and wind

8.6 The vapor and liquid are established in equilibrium, whereas a room equilibrium is not established.

8.8 There is mercury vapor.

8.10 B

8.12 100°C, 120°C

8.14 The temperature of the grease is not affected by low atmospheric pressure because it is not boiling.

8.16 5.4×10^3 cal

8.18 There are different intermolecular attractive forces.

8.20 High pressure is produced as a result of the small area of the blade in contact with the ice.

8.22 8.0×10^4 cal

8.24 Higher pressure forces the molecules to remain in the liquid phase at the normal boiling point. As a result, a higher temperature is required to boil the liquid.

8.26 London forces, dipole–dipole forces, hydrogen bonds

8.28 Chlorine is more polarizable than fluorine, and its compounds have larger London forces.

8.30 Ethyl alcohol boils at a higher temperature because of hydrogen bonds.

8.32 ethane

8.34 neon $>$ helium

8.36 hydrogen bonds

8.38 The hydrogen bonds pull the molecules inward toward each other.

8.40 Ice is less dense because the solid state has a more open structure.

8.42 Lakes do not freeze from the bottom to the top.

Chapter 9

9.2 A saturated solution may contain only a small quantity of solute.

9.4 It is not homogeneous.

9.6 Some ionic and low-molecular-weight compounds are dissolved, whereas high-molecular-weight compounds are dispersed.

9.8 starch in pudding

9.10 Co

9.12 No

9.14 (a) 1.00　　(b) 4.00　　(c) 3.00　　(d) 0.080

9.16 (a) 50.0 g　　(b) 15.0 g　　(c) 1 g

9.18 1×10^{-3} L

9.20 parts per billion

9.22 volume/volume

9.24 (a) 0.20　　(b) 0.025　　(c) 0.5　　(d) 1.0

9.26 (a) 0.40 L　　(b) 10.0 L　　(c) 0.10 L　　(d) 0.200 L

9.28 (a) 10%　　(b) 0.1 M　　(c) 2 M

9.30 $H_2O + HBr \longrightarrow H_3O^+ + Br^-$

9.32 AgCl has a very low solubility.

9.34 unsaturated

9.36 Although nothing is observed, there is an equilibrium process occurring.

9.38 The solubility of a gas is less at a higher temperature.

9.40 There is an increased chance of gaseous molecules colliding with the surface of the liquid.

9.42 More solute would dissolve in the case of NaCl and KCl, but less would be soluble for $Ce_2(SO_4)_3$.

9.44 There are large attractive forces between ions in the solid.

9.48 Glycerol will dissolve in water and will hydrogen bond to water. The solubility in the nonpolar CCl_4 will be low.

9.50 The freezing point is lowered; the boiling point is increased.

9.52 There is a larger number of ions per formula unit.

9.54 (a) nothing　　(b) hemolysis　　(c) nothing
　　(d) crenation　　(e) crenation

9.56 There are equal numbers.

9.58 The liquid levels would remain unchanged.

9.60 3.7×10^4

Chapter 10

10.2 For loss of electrons by one substance there must be a gain of electrons by another substance.

10.4 b, g, and h

10.6 (a) AgI　　　　　　　　(b) gives $BaSO_4$
　　(c) gives no precipitate　　(d) gives $PbCl_2$
　　(e) gives HgS　　　　　　(f) gives $PbCrO_4$
　　(g) gives no precipitate　　(h) gives no precipitate

10.8 All are $Pb^{2+} + SO_4^{2-} \longrightarrow PbSO_4$

10.10 the reaction of an acid with a base

10.12 $HCO_3^- + H_3O^+ \longrightarrow H_2O + H_2CO_3$

10.14 (a) SO_2 is oxidized; O_2 is reduced.
　　(b) Cu is oxidized; Br_2 is reduced.
　　(c) Cl^- is oxidized; I_2 is reduced.
　　(d) H_2 is oxidized; WO_3 is reduced.

10.16 reducing agent

10.18 oxidizing agent

10.20 oxidation

10.22
$$\begin{array}{c} \quad\ \ H\ \ H \\ \quad\ \ |\ \ \ | \\ Br-C-C-Br \\ \quad\ \ |\ \ \ | \\ \quad\ \ H\ \ H \end{array}$$

10.24
$$\begin{array}{c} \quad\ \ H\ \ H \\ \quad\ \ |\ \ \ | \\ H-C-C-Br \\ \quad\ \ |\ \ \ | \\ \quad\ \ H\ \ H \end{array}$$

10.25 a and b are condensation; c and d are hydrolysis

10.27 exothermic

10.29 There is an increase in the order of the products

10.31 (a) no reaction
　　(b) Reaction may occur at a low temperature.
　　(c) Reaction may occur at a high temperature.
　　(d) Reaction occurs at all temperatures.

10.33 lower

10.35 Energy is released by chemical reactions

10.37 Reaction rates increase with higher concentration

10.39 Metabolic rates increase and consume mass

10.41 The rate of decomposition is slowed

10.43 One with 15 kcal/mole

10.45

10.47 (a) $K = \dfrac{[O_3]^2}{[O_2]^3}$ (b) $K = \dfrac{[NH_3]^2}{[N_2][H_2]^3}$

    (c) $K = \dfrac{[CH_3Cl][HCl]}{[CH_4][Cl_2]}$ (d) $K = \dfrac{[CO_2]^2}{[CO]^2[O_2]}$

10.49 The larger K indicates more product relative to reactant.

10.51 Conversion of glycogen into glucose occurs.

10.52 aid

10.54 (a) increase K (b) decrease K

Chapter 11

11.2 (a) A conjugate base is formed from an acid by loss of a proton.

    (b) A conjugate acid is formed from a base by gaining a proton.

11.4 Hydroxide ions are present in solution.

11.6 Acids turn litmus red; bases turn litmus blue.

11.8 These products contain base.

11.10 Test with litmus.

11.12 H_2SO_4; $NaOH$

11.14 Amphoteric substances, such as HCO_3^-, can.

11.16 $H_2CO_3 + H_2O \rightleftharpoons HCO_3^- + H_3O^+$
$HCO_3^- + H_2O \rightleftharpoons CO_3^{2-} + H_3O^+$

11.18 (a) $HClO_4$ (b) HCl (c) HNO_3 (d) HCO_3^-
    (e) CH_3CO_2H (f) H_2SO_4 (g) NH_4^+ (h) H_3O^+
    (i) H_2CO_3

11.20 $C_3H_3O_3^-$, -1

11.22 $C_4H_5NO_4^{2-}$

11.24 Strength refers to the degree of ionization, not the amount of the acid.

11.26 a, d, and f are weak.

11.28 No; HCl would be completely ionized.

11.30 $C_2H_2O_4 + H_2O \rightleftharpoons C_2HO_4^- + H_3O^+$
$C_2HO_4^- + H_2O \rightleftharpoons C_2O_4^{2-} + H_3O^+$

11.32 6.5×10^{-4}

11.34 1.4×10^{-14}

11.36 (a) 10^{-1}; 10^{-13} (b) 10^{-2}; 10^{-12}
    (c) 10^{-11}; 10^{-3} (d) 10^{-1}; 10^{-13}
    (e) 10^{-10}; 10^{-4} (f) 10^{-12}; 10^{-2}

11.38 It is decreased as a result of neutralization.

11.40 (a) 3.7   (b) 6.7   (c) 11.7   (d) 1.0

11.42 It can react with acid or base.

11.44 No; HCl is not a weak acid.

11.46 (a) An increase will decrease the pH.

    (b) An increase will increase the buffering capacity for base.

11.48 respiration and urination

11.50 Normality gives the concentration of hydronium ions.

11.52 (a) 0.20   (b) 0.1   (c) 0.20   (d) 1.0
    (e) 0.00020

11.54 15

11.56 .00416

Chapter 12

12.2 (a) $_0^1 n$ (b) $_1^1 H$ (c) $_2^4 He$ (d) $_{-1}^0 e$ (e) $_{-1}^0 e$

12.4 (a) 82; 124   (b) 92; 143   (c) 53; 74
    (d) 7; 6   (e) 11; 14   (f) 1; 2

12.6 Order of decreasing penetrating power is gamma > beta > alpha.

12.8 (a) $_{84}^{212} Po \longrightarrow \, _2^4 He + _{82}^{208} Pb$
    (b) $_{96}^{240} Cm \longrightarrow \, _2^4 He + _{94}^{236} Pu$
    (c) $_{99}^{252} Es \longrightarrow \, _2^4 He + _{97}^{248} Bk$

12.10 (a) $_{94}^{239} Pu$ (b) $_{84}^{218} Po$ (c) $_{92}^{230} U$

12.12 proton

12.14 (a) 50   (b) 25   (c) 12.5   (d) 6.2
    (e) 3.1   (f) 1.6

12.16 0.4%

12.18 2

12.20 22,280

12.22 Ionization of bonding electrons destroys molecules.

12.24 Ions are formed in the gas phase and conduct electricity.

12.26 3.7×10^{10}

12.28 The rem indicates the amount of biological damage.

12.30 beta

12.32 The cells are severely damaged by high dosage and cannot recover.

12.34 16 m

12.38 $_{16}^{32} S$

12.40 Energy is required for biological changes required in therapy.

12.42 It is incorporated as phosphate in bones and affects bone marrow.

12.44 $_{27}^{59} Co + _0^1 n \longrightarrow \, _{27}^{60} Co$

12.46 $_5^{10} B$

12.48 $_1^2 H$

12.50 (a) $_{94}^{237} Pu$ (b) $_{15}^{30} P$ (c) $_{53}^{131} I$

12.52 Material in nuclear reactors is of lower concentration.

12.54 (a) 5   (b) 2   (c) 4

12.56 Controlled high temperatures haven't been achieved in the laboratory.

Chapter 13

13.2 Isomers have different structures; conformers differ in spatial arrangement of atoms as a result of rotation about single bonds.

13.4 4 single bonds; 1 double bond and 2 single bonds; 1 triple bond and 1 single bond; 2 double bonds

13.6 (a) CH_4 (b) C_2H_6 (c) C_2H_4 (d) C_2H_2
    (e) C_3H_8 (f) C_4H_{10} (g) C_2H_6O (h) C_2H_6O

13.8 (a) C_4H_6 (b) C_5H_8 (c) C_5H_{10} (d) C_4H_8

13.10 (a) $C_3H_6Cl_2$ (b) $C_3H_6Cl_2$ (c) $C_2H_4Br_2$
(d) C_3H_6BrCl

13.12 (a) C_3H_6O (b) $C_5H_{10}O_2$ (c) $C_5H_{10}O_2$
(d) C_4H_8O

13.14 (a) CH_3CH_2Br (b) CH_3CH_2Br
(c) CH_2BrCH_2Br (d) $CH_3CH_2CH_2CH_2CH_3$
(e) $CH_3CH_2CH_2SH$ (f) $CH_3CH_2CH_2NH_2$
(g) $CH_3CH_2CH_2CH_2OH$ (h) $CH_3CH_2CH_2CH_2CH_2Cl$

13.16 (a), (b), (c), (d) [structural formulas]

13.18 (a), (b), (c), (d) [structural formulas]

13.20 (a), (b), (c), (d) [structural formulas]

(e), (f) [structural formulas]

13.22 c, d, f, g, h

13.24 (a) CH_3CBr_2Cl; $CH_2ClCHBr_2$; $CH_2ClCHBrCl$
(b) $CH_3CH_2CH_2OH$; $CH_3CH(OH)CH_3$; $CH_3CH_2OCH_3$
(c) $CH_3CH_2CH_2SH$; $CH_3CH(SH)CH_3$; $CH_3CH_2SCH_3$
(d) $CH_3CH_2CH_2CH_2CH_3$; $CH_3CH_2CH(CH_3)_2$; $C(CH_3)_4$

13.26 (a) ester   (b) ether   (c) aldehyde
(d) alkene   (e) alkyne   (f) alcohol
(g) mercaptan   (h) amine

13.28 The first is an alcohol, which will react with sodium. The second is an ether, which will not react with sodium.

13.30 Mercaptans smell bad.

Chapter 14

14.2 Branched alkanes contain some carbon atoms bonded to more than two other carbon atoms.

14.4 a, f, and h are saturated.

14.6 b, c, d

14.8 b, c, d

14.10 b and d are the same; a and f are the same.

14.12 (a) butyl   (b) *sec*-butyl   (c) t-butyl
(d) propyl   (e) isopropyl   (f) *sec*-butyl

14.14 (a) methylpropane   (b) propane
(c) methylbutane   (d) methylbutane
(e) 2-methylhexane   (f) 2,3-dimethylpentane
(g) 3-methylheptane   (h) 4-ethylheptane

14.16 (a) $CH_3CH_2CH(CH_3)CH_2CH_3$
(b) $CH_3CH_2CH(CH_3)CH(CH_3)CH_2CH_3$
(c) $(CH_3)_3CCH(CH_3)CH_2CH_3$
(d) $CH_3CH_2CH_2CHCH_2CH_2CH_3$
                CH_2CH_3

14.18 (a) 1,1-dichloroethane   (b) 1,3-dichloropropane
(c) 1,1-dichloropropane   (d) 1,1,1-trichlorobutane
(e) 1,1,4-trifluorobutane   (f) 2,2-dibromopropane

14.20 $CH_3CH_2CH_2CH_2CH_2CH_3$ $CH_3CH_2CH_2CH(CH_3)_2$
$CH_3CH_2CH(CH_3)CH_2CH_3$ $CH_3CH_2C(CH_3)_3$
$CH_3CH(CH_3)CH(CH_3)CH_3$

14.22 $CH_3CH_2CH_2CH_2Cl$ $CH_3CH_2CHClCH_3$
$CH_3CH(CH_3)CH_2Cl$

[structural formula]

14.24 (a) 2 primary and 3 secondary
(b) 4 primary, 1 secondary, 1 quaternary
(c) 3 primary, 1 secondary, 1 tertiary

14.26 (a) 2   (b) 2   (c) 4   (d) 3   (e) 3   (f) 4
(g) 6   (h) 5   (i) 4   (j) 5

14.28 (a) 0   (b) 0   (c) 0   (d) 1   (e) 1   (f) 0
(g) 0   (h) 1   (i) 2   (j) 3

14.30 CH_3Cl, CH_2Cl_2, $CHCl_3$, CCl_4

14.32 (a) 2 (b) 2 (c) 2 (d) 3 (e) 4 (f) 3
(g) 2

14.34 (a) [structure: cyclopropane with Cl] (b) [structure: cyclobutane with CH_3, CH_3] (c) [structure: cycloheptane]

14.36 (a) 1,1-dichlorocyclohexane
(b) *trans*-1,2-dichlorocyclohexane
(c) *cis*-1,3-dimethylcyclohexane
(d) 1,1,3-trichlorocyclohexane

14.38 (a) C_6H_{10} (b) $C_{10}H_{18}$ (c) $C_{10}H_{18}$ (d) $C_{10}H_{16}$

Chapter 15

15.2 Geometric isomers have the same bond sequence, but the atoms have different spatial arrangements.

15.4 Both have π bonds

15.6 a and c are alkanes, f and h are cycloalkanes, b and e are alkenes, d and i are alkynes, g is a cycloalkene

15.8 (a) C_6H_{10} (b) C_6H_8 (c) $C_{10}H_{16}$ (d) C_8H_{14}
(e) C_6H_6 (f) $C_{14}H_{20}$

15.10 $C_{12}H_{22}O$

15.12 [structure: cyclobutane] ; [structure: cyclopropane with CH_3]

15.14 $CH_3C{\equiv}CCH_3$ 2-butyne; $CH_3CH_2C{\equiv}CH$ 1-butyne

15.16 1-Butene has two hydrogen atoms bonded to the number 1 carbon atom.

15.18 (a) [structure: H, CH3 / C=C / H, CH2CH3] (b) [structure: H, H / C=C / Cl, Br]
(c) [structure: ClCH2 / C=C / CH2Br] (d) [structure: H, CH3 / C=C / H, Br]

15.20 *cis* for bond on left, *trans* for bond on right

15.22 (a) methylpropene
(b) 2,3-dimethyl-2-butene
(c) 2-methyl-2-pentene
(d) 2,5-dimethyl-2-hexene
(e) chloroethane
(f) 3-bromo-1-propene
(g) 2,5-dimethyl-2,4-hexadiene

15.24 (a) cyclohexene
(b) 2-methyl-4-ethylcyclohexene
(c) 3-ethylcyclopentene
(d) 2,4-dimethylcyclopentene

15.26 (a) [structure: H, CH3 / C=C / Cl, H] (b) [structure: Cl, Cl / C=C / CH3, CH3]
(c) [structure: H, H / C=C / H, CH2Cl] (d) [structure: CH3, H / C=C / CH3, Cl—CCH2CH3—CH3]

15.28 (a) 1-pentyne
(b) 2,2-dimethyl-3-hexyne
(c) 1-chloro-3-octyne
(d) 4,5-dibromo-2-hexyne
(e) 4-methyl-2-hexyne
(f) 2-chloro-6-methyl-3-octyne

15.30 The red color of Br_2 disappears. Cyclohexane will not react.

15.32 (a) $CH_3CH_2CHBrCH_2Br$ (b) $CH_3CH_2CHBrCH_3$
(c) $CH_3CH_2CH(OH)CH_3$ (d) [structure: cyclohexane with Br, Br, Br]

15.34 with propene: (a) $CH_3CHClCH_2Cl$ (b) $CH_3CH(OH)CH_3$
(c) $CH_3CHBrCH_3$ (d) $CH_3CHBrCH_2Br$ (e) $CH_3CHClCH_3$
with methylpropene: (a) $(CH_3)_2CClCH_2Cl$ (b) $(CH_3)_2C(OH)CH_3$
(c) $(CH_3)_2CBrCH_3$ (d) $(CH_3)_2CBrCH_2Br$ (e) $(CH_3)_2CClCH_3$

15.36 2

15.38 All react with 2 moles.

15.40 (a) ortho (b) para (c) meta

15.42 (a) ethylbenzene (b) isopropylbenzene
(c) *p*-diethylbenzene (d) 1,3,5-trimethylbenzene

15.44 (a) *m*-xylene (b) *o*-cresol (c) *p*-toluidine

15.46 (a) 1-phenyloctane (b) 1,3-diphenylpropane
(c) phenylcyclohexane (d) *trans*-1-phenyl-1-butene

15.48 1,2-benzanthracene, $C_{18}H_{12}$; 1,2,5,6-dibenzanthracene, $C_{22}H_{14}$; 3,4-benzpyrene, $C_{20}H_{12}$

15.50 (a) [structure: —CH2—CH(CH3)—CH2—CH(CH3)—]
(b) [structure: —CH2—CH(CN)—CH2—CH(CN)—]
(c) [structure: —CH2—CCl2—CH2—CCl2—]
(d) [structure: —CH2—C(CH3)2—CH2—C(CH3)2—]

Chapter 16

16.2 according to the number of alkyl groups attached to the carbon with the hydroxyl group

16.4 (a) $CH_3CH_2CH_2C(OH)(CH_3)_2$
(b) $CH_3CH_2CH(CH_3)CH_2OH$
(c) $CH_3CH(CH_3)CH(CH_3)CH_2OH$
(d) $CH_3CH_2CH(OH)CH(CH_3)_2$
(e) $CH_3CH_2C(OH)(CH_2CH_3)_2$
(f) $CH_3CH(CH_3)CH_2CH(OH)CH_3$

16.6 (a) [structure: cyclopentane with OH] (b) [structure: cyclobutane with OH, CH3]

(c) [structure: cyclohexane with OH and CH₃]

(d) [structure: cyclopentane with OH and CH₂CH₃]

16.8 (a) 2-methyl-2-pentanol (b) 3-methyl-3-heptanol
(c) 1,2-butanediol

16.10 (a) 1-butanol
(b) 2,4-dimethyl-1-pentanol
(c) 2-methyl-1-butanol
(d) 2,2-dimethyl-1-pentanol
(e) 3-ethyl-2-hexanol
(f) 2,3,4-trimethyl-1-pentanol

16.12 (a) 1,1-ethanediol (b) 1,3-propanediol
(c) 1,2,3-propanetriol (d) 1,2-propanediol

16.14 $CH_2(OH)CH(OH)CH_2OH$, 1,2,3-propanetriol

16.16 e and h are primary; a, c, f, and g are secondary; b and d are tertiary.

16.18 e is tertiary; all others are secondary.

16.20 There is a secondary alcohol on the top left ring. There is a primary alcohol on the top right ring.

16.22 a, b, and f are primary; c, d, and e are secondary; e and h are tertiary.

16.24 There are two hydroxyl groups per molecule, which can form hydrogen bonds with other molecules.

16.26 The compound resembles more closely a hydrocarbon.

16.28 $2\ CH_3CH_2OH + 2\ K \longrightarrow 2\ CH_3CH_2O^-K^+ + H_2$

16.30 sodium ethoxide

16.32 $CH_3CH_2OLi + HBr \longrightarrow CH_3CH_2OH + LiBr$

16.34 An oil is formed, which causes turbidity.

16.36 2-Methyl-2-butanol would react rapidly; 3-pentanol would react somewhat more slowly.

16.38 (a) fast (b) no reaction (c) slow

16.40 (a) $CH_2{=}CHCH_2CH_2CH_3$
(b) a + cis- and trans-$CH_3CH{=}CHCH_2CH_3$
(c) $CH_2{=}\overset{\overset{\displaystyle CH_3}{|}}{C}{-}CH_2CH_3$ and $(CH_3)_2C{=}CHCH_3$

16.42 (a) [cyclopentene structure] (b) [cyclohexene with CH₃ structures] and [cyclohexene with CH₃]
(c) [cyclopentane with =CH₂] and [cyclopentene with CH₃]
(d) [cyclopentene with CH₃] and [cyclopentene with CH₃]

16.44 secondary, [cyclohexanone structure with CH(CH₃)₂ and CH₃]

16.46 a and c

16.48 (a) [cyclopentanone structure] (b) [cyclohexanone with CH₃]
(c) no reaction (d) [cyclopentanone with CH₃]

16.50 [structure: HO–C(=O)–C(=O)–OH]

16.52 (a) $CH_3{-}ONO_2$ (b) $CH_3CH_2{-}OSO_3H$
(c) $(CH_3)_2CH{-}OSO_3H_2$

16.54 $CH_3{-}O{-}\overset{\overset{\displaystyle O}{\|}}{\underset{\underset{\displaystyle O{-}CH_3}{|}}{P}}{-}O{-}CH_3$

16.56 [benzene ring with OSO₃H group]

16.58 d and e

16.60 (a) diethyl ether (b) ethyl methyl ether
(c) phenyl propyl ether (d) methyl phenyl ether
(e) dicyclohexyl ether (f) diphenyl ether

16.62 They do not closely resemble water and do not have a O—H group to hydrogen bond with water.

16.64 There is no hydrogen bonded to oxygen.

16.66 $CH_3CH_2CH_2SH$ and $CH_3CH(SH)CH_3$

16.68 The thiol is volatile, but the ionic $CH_3CH_2CH_2S^-Na^+$ is not.

16.70 (a) CH_3SSCH_3 (b) $CH_3CH_2SSCH_2CH_3$
(c) [cyclopentane–S–S–cyclopentane] (d) [phenyl–S–phenyl]

Chapter 17

17.2 An acetal has a hydrogen atom bonded to the carbon that is bonded to two alkoxy groups.

17.4 Isomers that differ in the position of a proton.

17.6 There is a hydrogen bonded to the carbonyl carbon in the aldehyde.

17.8 b and f are aldehydes; a, c, and d are ketones; e is an ester.

17.10 (a) $CH_3CH_2CH(CH_3)CHO$
(b) $CH_3CH_2\overset{\overset{\displaystyle CH_2CH_3}{|}}{C}HCH_2CHO$
(c) $CH_3CH_2CH_2CH(Br)CHO$
(d) $CH_3(CH_2)_3CH(CH_3)CH(CH_3)CH_2CHO$

CH$_2$CH$_3$

(e) CH$_3$(CH$_2$)$_3$CHCHO

(f) CH$_3$CCl$_2$CHO

17.12 (a) [structure] (b) [structure with CH$_3$]

(c) [structure with Br] (d) [structure with CH$_3$ CH$_3$]

17.14 (a) ethanal   (b) pentanal
(c) butanal   (d) 2,2-dimethyl butanal
(e) methylpropanal   (f) 2-ethyl-3-methylpentanal
(g) 2-methylbutanal (h) 4-methylpentanal

17.16 (a) cyclobutanone
(b) 2-methylcyclopentanone
(c) 2-methylcyclohexanone
(d) 3-methylcyclohexanone

17.18 (a) α-chloropropionaldehyde
(b) γ-bromobutyraldehyde
(c) γ-chloro-β-methylbutyraldehyde
(d) α-chloro-β-phenylpropionaldehyde
(e) β-hydroxybutyraldehyde

17.20 cyclohexanol

17.22 There is no hydrogen bonded to an electronegative atom.

17.24 Water can hydrogen bond to both compounds.

17.26 (a) CH$_3$CH$_2$CO$_2$H   (b) CH$_3$CCH$_2$CH$_3$

(c) no reaction   (d) [cyclohexanone structure]

(e) [cyclohexane CO$_2$H structure]   (f) no reaction

17.28 A silver mirror is formed.

17.30 (a) 2-propanol
(b) 3-pentanol
(c) 1-hexanol
(d) cyclohexanol
(e) 4,4-dimethylcyclohexanol
(f) 3,3-dimethyl-1-pentanol

17.32 primary alcohol

17.34 d is a hemiacetal; e and h are hemiketals; a and f are acetals; b, c, and g are ketals.

17.36 a and f are hemiacetals; d is a hemiketal; b and e are acetals; c is a ketal.

17.38 CH$_3$CH=NCH$_3$

17.40 3-Pentanone will react with 2,4-dinitrophenylhydrazine.

17.42 The carbonyl carbon atom draws electrons away from the α C—H bond.

17.44 (a) [cyclopentene OH structure]   (b) CH$_3$CH$_2$CH=C with OH and H

(c) [cyclohexene OH structure]   (d) CH$_2$=C with OH and H

17.46 [two enol structures with H, OH, CH$_3$] and [second structure]

17.48 (a) CH$_3$CHO   (b) CH$_3$CH$_2$CHO   (c) (CH$_3$)$_2$CHCHO

17.50 The reaction occurs via an enediol intermediate involving the top two carbon atoms.

17.52 CH$_3$CH$_2$CH(OH)CH$_3$ and CH$_3$CH$_2$CCH$_3$

17.54 [cyclopentanone structure] [cyclobutanone with CH$_3$] [structure with CH$_3$]

[cyclopropyl CCH$_3$ structure] [structure with CH$_3$ CH$_3$] [structure with CH$_3$ CH$_3$]

[structure with CH$_3$ CH$_3$] [structure with CH$_3$CH$_2$]

Chapter 18

18.2 An acid has —OH attached to a carbonyl carbon, whereas an ester has —OR.

18.4 a molecule formed between acid molecules with the loss of water

18.6 (a) propanoic acid   (b) methanoic acid
(c) butanoic acid   (d) ethanoic acid

18.8 (a) pentanoic acid
(b) methylpropanoic acid
(c) 3-methylpentanoic acid
(d) 2,4-dimethylpentanoic acid
(e) 3-methylhexanoic acid
(f) 5,5-dimethyl-3-ethylhexanoic acid

18.10 (a) 3-chlorobutanoic acid
(b) 2-bromopropanoic acid

(c) 3-bromo-4-methylpentanoic acid

(d) 2-chloro-4,4-dimethylpentanoic acid

18.12 (a) 2-hydroxypropanoic acid

(b) 2-hydroxybutanoic acid

(c) 3-hydroxypropanoic acid

18.14 (a) glutaric acid (b) succinic acid

(c) pyruvic acid (d) oxaloacetic acid

(e) fumaric acid (f) lactic acid

18.16

$$CH_3C\underset{O-H\cdots O}{\overset{O\cdots H-O}{<...>}}C-CH_3$$

18.18 There are two carboxyl groups on adipic acid that hydrogen bond to water.

18.20 (a) chloroacetic acid (b) α,α-dichlorobutyric acid

18.22 The β-keto group draws electrons toward itself and away from the bonding electrons in the O—H group

18.24 formic acid

18.26 (a) $CH_3CO_2H + NaOH \longrightarrow CH_3CO_2Na + H_2O$

(b) $CH_3(CH_2)_{16}CO_2H + NaOH \longrightarrow$
$$CH_3(CH_2)_{16}CO_2Na + H_2O$$

18.28 (a) $CH_3(CH_3)_2CO_2Na + HCl \longrightarrow$
$$CH_3(CH_2)_2CO_2H + NaCl$$

(b) $CH_3(CH_2)_{16}CO_2K + HCl \longrightarrow$
$$CH_3(CH_2)_{16}CO_2H + KCl$$

(c) $(CH_3CH_2CO_2)_2Ca + 2\ HCl \longrightarrow$
$$2\ CH_3CH_2CO_2H + CaCl_2$$

18.30 (a) $CH_3CH_2CH_2CO_2Na$ (b) $CH_3CH_2CO_2K$

(c) $CH_3(CH_2)_{16}CO_2Li$ (d) $(CH_3CO_2)_2Ca$

18.32 $CH_3CH_2CH_2CO_2H$ and $(CH_3)_2CHCO_2H$

18.34 $CH_3CH_2CH_2CO_2CH_3$; $(CH_3)_2CHCO_2CH_3$;
$CH_3CH_2CO_2CH_2CH_3$; $CH_3CO_2CH_2CH_2CH_3$;
$CH_3CO_2CH(CH_3)_2$; $HCO_2CH_2CH_2CH_2CH_3$;
$HCO_2CH_2CH(CH_3)_2$; $HCO_2CH(CH_3)CH_2CH_3$;
$HCO_2C(CH_3)_3$

18.36 (a) ethyl formate (b) methyl butyrate

(c) octyl acetate (d) ethyl butyrate

(e) pentyl butyrate (f) pentyl acetate

18.38 (a) methyl benzoate (b) phenyl acetate

(c) butyl benzoate (d) phenyl propanoate

(e) phenyl benzoate (f) ethyl benzoate

18.40 (a) butyl butyrate (b) pentyl acetate

(c) octyl acetate (d) methyl butyrate

18.42 (a) acetyl (b) propionyl (c) formyl

(d) benzoyl (e) butyryl (f) formyl

(g) butyryl (h) acetyl

18.44 (a) $CH_3CH_2CH_2CO_2H$ and $CH_3CH_2CH_2CH_2OH$

(b) $CH_3CH_2CO_2H$ and $CH_3CH_2CH_2OH$

(c) CH_3CO_2H and CH_3CH_2OH

(d) CH_3CO_2H and $CH_3CH_2CH_2OH$

18.46 (a) $CH_3(CH_2)_2CH_2OH$ and $CH_3(CH_2)_2CO_2Na$

(b) $CH_3CH_2CH_2OH$ and $CH_3CH_2CO_2Na$

(c) CH_3CH_2OH and CH_3CO_2Na

(d) $CH_3CH_2CH_2OH$ and CH_3CO_2Na

(e) $(CH_3)_2CHCH_2OH$ and HCO_2Na

(f) $CH_3CH_2CH_2OH$ and CH_3CO_2Na

18.48 (a) $NaO_2CCH_2CH_2CO_2Na$ and CH_3OH

(b) $CH_3CH(OH)CO_2Na$ and CH_3OH

(c)

$$\underset{NaO_2C}{\overset{H}{<...>}}C=C\underset{H}{\overset{CO_2Na}{<...>}} \quad \text{and} \quad CH_3OH$$

(d) $NaO_2C(CH_2)_4CO_2Na$ and CH_3CH_2OH

18.50

$$\begin{aligned}
&CH-O-\overset{O}{\overset{\|}{C}}-(CH_2)_{16}CH_3\\
&CH-O-\overset{O}{\overset{\|}{C}}-(CH_2)_{16}CH_3 \quad + 3\ NaOH\\
&CH_2-O-\overset{O}{\overset{\|}{C}}-(CH_2)_{16}CH_3\\
&\qquad\qquad\downarrow\\
&CH_2OH\\
&CHOH \quad + 3\ CH_3(CH_2)_{16}CO_2Na\\
&CH_2OH
\end{aligned}$$

18.52 (a) $CH_3CO_2CH_2CH_3$ (b) $CH_3CH_2CO_2CH_2CH_2CH_3$

(c) ⟨phenyl⟩—CO_2CH_3 (d) CH_3CO_2—⟨phenyl⟩

18.54 Esters do not have a hydrogen bonded to an electronegative atom with which to form hydrogen bonds.

18.56 $c < a < b$

18.58 (a) $CH_3\overset{O}{\overset{\|}{C}}CH_2CO_2CH_3$ (b) $CH_3CH_3\overset{O}{\overset{\|}{C}}CHCO_2CH_2CH_3$
$$\qquad\qquad\qquad\qquad\qquad\qquad CH_3$$

(c) $CH_3\overset{O}{\overset{\|}{C}}CH_2CO_2CH_2CH_2CH_3$

18.60 Yes; all three hydroxyl groups can react with acids to form multiple esters.

18.62 glutaric acid and 1,4-butanediol

18.64

$$\begin{aligned}
&CH_2-\overset{O}{\overset{\|}{C}}\\
&\qquad\qquad\searrow O\\
&CH_2-\overset{}{\underset{\\ O}{C}}
\end{aligned}$$

Chapter 19

19.2 Classification of amines is based on carbon atoms bonded to nitrogen, whereas that of alcohols is based on carbon atoms bonded to the hydroxyl-bearing carbon.

19.4 primary near CO_2H, secondary in the ring

19.6 tertiary, secondary, primary

19.8 both tertiary

19.10 a and e are primary; b and d are secondary; c and f are tertiary.

19.12 b, c, and d are primary; a is secondary.

19.14 (a) 3-aminohexane
(b) 1-(N,N-dimethylamino)-butane
(c) 2-amino-4-methylpentane
(d) 2-aminobutane
(e) 3-(N-methylamino)-1-butanol
(f) 2-(N,N-dimethylamino)-butane

19.16 (a) 2-aminoethanoic acid
(b) 2-aminopropanoic acid
(c) 4-amino-2-chloropentanoic acid
(d) 3-(N,N-dimethylamino)-pentanal
(e) 3-amino-2-pentanol

19.18 $C_nH_{2n+4}N_2$

19.20 (a) $C_6H_{13}N$ (b) $C_6H_{13}N$ (c) $C_7H_{15}N$
(d) C_2H_5N (e) $C_5H_{11}N$ (f) $C_5H_{11}N$

19.22 4

19.24 Propylamine can form hydrogen bonds

19.26 There are two amino groups that form hydrogen bonds.

19.28 so that it is soluble

19.30 1-aminobutane, 2-aminobutane, 1-amino-2-methyl-propane, 2-amino-2-methylpropane

19.32 b, d, and e would form nitroso compounds; a, c, and f would not react.

19.34 triethylamine

19.36 The electron pair are delocalized with the π electrons.

19.38 (a) C_5H_5N (b) C_4H_5N (c) $C_4H_4N_2$ (d) C_8H_7N

19.40 (a) propionamide
(b) N-ethylpropionamide
(c) α-methylbutyramide
(d) N,N-dimethylbutyramide

19.42 (a) N,N-dimethylpropanamide
(b) methylethylmethanamide
(c) N-methyl-3-methylbutamide
(d) N-methylbenzamide

19.44 (a) $CH_3CH_2CH(CH_3)CONH_2$
(b) $CH_3CHClCH_2CONH_2$
(c) $BrCH_2CH_2CH_2CONHCH_3$
(d) $CH_3CH_2CH(C_2H_5)CON(CH_3)_2$

19.46 the carbonyl oxygen and the nitrogen atom

19.48 a > b > c

19.50 (a) $CH_2CO_2^-$ and $CH_3CH_2NH_2$
(b) $CH_3CH_2CO_2H$ and $CH_3NH_3^+$
(c) $C_6H_5CO_2^-$ and CH_3NH_2
(d) $C_6H_5CO_2H$ and NH_4^+

19.52 (a)

(b)

(c)

19.56 Morphine is a phenol; heroin is an ester.

19.58 secondary amine, tertiary amine, disubstituted amide

Chapter 20

20.2 The catabolic reactions release energy required for anabolic reactions.

20.4 Eukaryotes contain nuclei, whereas prokaryotes do not.

20.6 nitrogen, phosphorus, and sulfur

20.8 carbohydrates, lipids, and proteins

20.10 (a) carbohydrate (b) amino acid
(c) carbohydrate (d) amino acid
(e) lipid (f) carbohydrate

20.12 positive

20.14 negative

20.16 a, d, e, h

20.18 2

20.20 5

20.22 3

20.24 They designate configuration.

20.26 $-28.3°$

20.28 $+12°$

20.30 (a) (b)

20.32

20.34 8

20.36 The first two compounds are enantiomers, as are the second two compounds.

20.38 Biochemicals are asymmetric.

20.40 The L isomer will remain unreacted.

20.42 Only one member of any enantiomeric pair will react.

20.44 lipids and proteins

20.46 They exist to perform selective functions.

Chapter 21

21.2 acetal or ketal of a carbohydrate

21.4 No; it can be oxidized by certain oxidizing reagents, which then are reduced.

21.6 any of the aldopentoses shown in Figure 21.3

21.8 See Figure 21.2 D- and L-erythrose are enantiomers, as are D- and L-threose.

21.10 (a)

CHO
H—C—OH
H—C—OH
H—C—OH
CH₂OH

(b)

CH₂OH
C=O
HO—C—H
H—C—OH
H—C—OH
CH₂OH

(c)

CHO
H—C—OH
HO—C—H
H—C—OH
H—C—OH
CH₂OH

(d)

CHO
HO—C—H
HO—C—H
H—C—OH
H—C—OH
CH₂OH

21.12 It lacks a hydroxyl group at the number 2 carbon atom.

21.14 Glycogenolysis is the hydrolysis of glycogen; gluconeogenesis is the formation of glucose from pyruvic acid.

21.16 gluconeogenesis and glycogenolysis

21.18 4 g

21.20 (a) [structure with CH₃, O, OH, H] **(b)** [structure with CH₃CH₂, O, OH, H]

21.24 (a) galactose **(b)** mannose **(c)** ribose **(d)** ribulose **(e)** fructose **(f)** fructose

21.26 a, c, e, and f are α; b and d are β.

21.28 [structure with HO, CH₂OH, HO, O, HO, OH]

21.30 [structure with HO, CH₂OH, HO, O, HO, OH, H] more stable than [structure with HO, CH₂OH, HO, O, HO, OH]

21.32 yes

21.34 a, c, and e

21.36 [structure with CH₂OH, O, OH, HO, OCH₃, OH] and [structure with CH₂OH, O, OCH₃, OH, HO, OH]

21.38 [structure with HOCH₂, O, CH₂OH, HO, OCH₃, OH] and [structure with HOCH₂, O, OCH₃, HO, CH₂OH, OH]

21.40

H
R—C—OR
OR

and

R
R—C—OR
OR

21.42 (a) glucose and galactose **(b)** glucose **(c)** glucose and fructose **(d)** glucose

21.44 yes

21.46 Cellulose is β-linked, whereas the others are α-linked. Amylose is a linear polymer, whereas amylopectin and glycogen are branched polymers.

21.48 They do not have the necessary enzyme for β-linked glucose.

21.50 Glycogen undergoes glycogenolysis to produce glucose.

21.52 arabinose and ribulose

21.54

CHO
HO—C—H
HO—C—H
HO—C—H
H—C—OH
CH₂OH

CH₂OH
C=O
HO—C—H
HO—C—H
H—C—OH
CH₂OH

21.56 yes

21.58 It isomerizes to ribose via an enediol intermediate.

21.60 a, b, and c

21.62 It consists of an acetal and a ketal that are not in equilibrium with carbonyl functional groups.

21.64 an alduronic acid; mannose

21.66

CO₂H
HO—C—H
HO—C—H
H—C—OH
H—C—OH
CH₂OH

The carboxylic acid is at the 1 carbon atom instead of the 6 carbon atom.

Chapter 22

22.2 Facilitated diffusion does not require energy, whereas active transport does.

22.4 b and c

22.6 Oils are more unsaturated.

22.8 The former are derivatives of glycerol, whereas the latter are derivatives of sphingosine. Both are esters of phosphoric acid and contain long-chain fatty acids.

22.10 (a) monoterpene   (b) monoterpene
(c) sesquiterpene   (d) triterpene

22.12 only d

22.14 Estrone causes growth of lining of the uterus. Progesterone maintains the fertilized egg.

22.16 glucose and electrolyte balance

22.18 two ketones, primary alcohol, secondary alcohol, aldehyde, double bond

22.20 26-carbon acid and 30-carbon alcohol

22.22 long chains that can pack tightly

22.24 The wax can remain liquid at low temperatures.

22.26 16-carbon unsaturated acid

22.28 More double bonds lower the melting point.

22.30 $CH_3(CH_2)_7C\equiv C(CH_2)_7CO_2H$

22.32 The compound is isomeric with oleic acid, but has a *trans* double bond.

22.34 b

22.36 It can form a micelle with the ammonium ion on the surface.

22.38
$$\begin{array}{l} RCO_2-CH_2 \\ | \\ RCO_2-CH \\ | \\ RCO_2-CH_2 \end{array} + 3\,OH^- \longrightarrow 3\,RCO_2^- + \begin{array}{l} CH_2OH \\ | \\ CHOH \\ | \\ CH_2OH \end{array}$$

22.40 insulation and energy

22.42 palmitic, lauric, oleic

22.44 linoleic, oleic, linolenic

22.46 yes

22.48 3

22.50 ethanolamine, choline, serine, inositol

22.52 stearic acid, oleic acid, glycerol, phosphoric acid, choline

22.54 a phosphoric acid ester of glycerol; esterification with an alcohol

22.56 Lecithins contain choline; cephalins contain ethanolamine.

22.58 The former contain sphingosine, whereas the latter contain glycerol.

22.60 a hydrocarbon chain

22.62 They contain longer fatty acid chains.

22.64 The former contain monosaccharides, and the latter oligosaccharides.

22.66 base

22.68 peripheral, integral, transmembrane

22.70 hydrophobic bonding

22.72 The electrostatic bonds are broken by the detergent.

22.74 Increase in protein increases permeability.

22.76 (a) surface location
(b) contained within membrane
(c) reaches both sides of membrane

22.78 The polar portion cannot flip over and enter nonpolar interior.

22.80 The potassium must move from low concentration to high concentration.

Chapter 23

23.2 They are divisions of the total structure of proteins.

23.4 Proteins are higher molecular weight.

23.6 proteins that yield compounds in addition to amino acids when hydrolyzed

23.8 (a) chromoprotein   (b) glycoprotein
(c) phosphoprotein   (d) lipoprotein

23.10 Exercise increases HDL level, which lowers blood cholesterol.

23.12 See Table 23.1.

23.14 It designates configuration.

23.16 lysine

23.18 like L-glutamic acid, but with NH_2 group on the right

23.20 (a) glycine   (b) alanine
(c) phenylalanine   (d) proline
(e) serine   (f) cysteine

23.22 Asparagine has an amide group, whereas aspartic acid has a carboxylic acid group.

23.24 See Table 23.2.

23.26 It is not essential for adults.

23.28 Its biological value is the same as for beef.

23.30 no, because eggs have a high biological value

23.32
$$CH_3-\overset{\overset{\displaystyle NH_3^+}{|}}{CH}-CO_2^-$$

23.34 basic

23.36 near 7, less than 7, greater than 7

23.38 vasopressin, because of arginine

23.40 (a) 7   (b) less than 7   (c) greater than 7

23.42 It has basic amino acids.

23.44 It cannot have many acidic or basic amino acids.

23.46 Glycylserine has an amide of glycine; serylglycine has an amide of serine.

23.48 With amino acids represented as A, B, C, and D, six of them are ABCD, ABDC, ACBD, ACDB, ADBC, and ADCB.

23.50 zwitterion

23.52 (a) serine, glycine, alanine
(b) glycine, cysteine, valine
(c) aspartic acid, threonine, glycine

23.54 peptide, disulfide, ionic, hydrogen bonds, hydrophobic interaction; peptide and disulfide are covalent

23.56 serine, tyrosine, threonine

23.58 two chains bonded via disulfide bonds

23.60 It cannot form intramolecular hydrogen bonds.

23.62 a, b, e, and f

23.64 hemoglobin

23.66 Binding of each O_2 causes a conformational change that increases the capacity of hemoglobin for O_2.

23.68 loss of biological function

23.70 Disulfide bonds would be reduced.

23.72 Changes occur in secondary and tertiary structures.

23.74 Sulfide groups react with lead ion.

23.76 Hydrolysis occurs at N-terminal amino acid.

23.78 e gives Asp-Lys and Ser; f gives Lys and Tyr-Cys; h gives Arg and Glu-Ser

23.80 c gives Phe, Trp, and Ser; d gives Ser-Phe and Asp; f gives Lys-Tyr and Cys

23.82 Phe

Chapter 24

24.2 Competitive inhibitors bind at the active site.

24.4 an enzyme in a series of reactions that is inhibited by a final product

24.6 They are highly specific.

24.8 complementary shape and an active site

24.10 It has the improper configuration for the enzyme.

24.12 ase

24.14 (a) hydrolase   (b) isomerase
(c) ligase   (d) oxidoreductase
(e) hydrolase

24.16 (a) esters   (b) phosphates   (c) peptides

24.18 oxidoreductase

24.20 hydrolase

24.22 isomerase

24.24 ligase

24.26 shapes that fit together

24.28 rigid fit of substrate with enzyme

24.30 basic amino acids

24.32 Shape changes to adapt to the substrate.

24.34 the pK_a of the side chain is near 7

24.36 NAD^+ and $FADH_2$

24.38 Only one metal ion is required for each high-molecular-weight protein.

24.40 (a) increase until saturation
(b) increase
(c) maximum at optimum temperature
(d) a maximum at optimum pH

24.42 maximum velocity achieved at saturation

24.44 Its optimum pH is at a low value.

24.46 It indicates the relative efficiency of an enzyme-catalyzed reaction.

24.48 low

24.50 Changes in conformation occur.

24.52

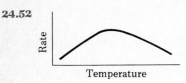

24.54 It is a dicarboxylic acid like succinic acid.

24.56 It can deactivate enzymes that are required only in low concentrations.

24.58 Such drugs replace a compound required by bacteria.

24.60 Lead will react with enzymes and cause denaturation.

24.62 There is a P—O—serine bond formed.

24.64 noncompetitive

24.66 A modulator combines with and changes the conformation of the enzyme.

24.68 Without the inhibitor, the pancreas would react and be destroyed.

24.70 The tertiary structure is changed.

24.72 The presence of the enzymes would cause clotting.

Chapter 25

25.2 Oxidative phosphorylation forms ATP in coupled reactions in which NADH and $FADH_2$ are oxidized.

25.4 negative

25.6 positive, negative

25.8 It cannot be further oxidized.

25.10 oxidation

25.12 (a) 4   (b) 3   (c) 2   (d) 4   (e) 3

25.14 exergonic

25.16 It is intermediate in a series of biological phosphates.

25.18 If the exergonic reaction has a sufficiently negative $\Delta G°$, the total $\Delta G°$ for the series of reaction can be negative.

25.20 The phosphate liberated is used to phosphorylate a substance in the coupled reaction.

25.22 Yes; its $\Delta G°$ is more negative than the $\Delta G°$ for ATP.

25.24 The $\Delta G°$ for hydrolysis of phosphoenolpyruvate is more negative than that for ATP.

25.26 oxidation of NADH and reduction of FP_1

25.28 oxidation of $FADH_2$ and reduction of FP_2

25.30 three

25.32 FP_1 and two cytochromes

25.34 reduction of oxygen

25.36 It is reduced.

25.38 two

25.40 acetyl, 2 moles carbon dioxide per acetyl group

25.42 four

25.44 The reduced coenzymes formed in the citric acid cycle are oxidized in the electron transport chain.

25.46 Step 4, succinyl thiokinase

25.48 It is an inhibitor of Step 1; its concentration affects Step 3.

25.50 to provide these compounds for the citric acid cycle, which occurs in the matrix.

25.52 the inner membrane

Chapter 26

26.2 glycolysis converts glucose into pyruvic acid; gluconeogenesis converts pyruvic acid into glucose.

26.4 mouth

26.6 Glucose and galactose must be converted into fructose.

26.8 glucose, galactose, and fructose

26.10 Glycogen hydrolyzes to glucose; glycogenesis and glycogenolysis

26.12 ribose 5-phosphate, and NADPH

26.14 NAD^+

26.16 to reduce pyruvic acid

26.18 2

26.20 formation of glucose 6-phosphate; hexokinase

26.22 (a) phosphoglucose isomerase
(b) phosphofructokinase
(c) phosphoglyceromutase

26.24 reduction

26.26 Lactic acid formed in muscles is oxidized in the liver to pyruvic acid and then glucose.

26.28 two; two

26.30 The NADH cannot cross the mitochondrial membrane.

26.32 yes

26.34 glycerol, fatty acids

26.36 oxaloacetic acid

26.38 4 ATP

26.40 CO_2

26.42 a three-carbon unit; an aldose

26.44 two

26.46 converted into glycolysis intermediates

26.48 kidney, intestine, liver

26.50 uridine diphosphate glucose

26.52 Cori's involves lack of debranching enzyme; Anderson's lacks enzyme to form branches.

26.54 conversion of fructose 1,5-diphosphate into fructose 6-phosphate

Chapter 27

27.2 release of fatty acids from adipose tissue

27.4 glycerol, fatty acids, monoglycerides, diglycerides

27.6 in micelles with plasma proteins

27.8 Tissue lipids are phosphoglycerides, phosphosphingolipids, and glycosphingolipids.

27.10 fatty acid mobilization, which is a hydrolysis reaction

27.12 Fatty acids easily cross the lipid bilayer.

27.14 Each cycle involves a different acid.

27.16 FAD

27.18 L

27.20 9, 8

27.22 7

27.24 electron transport chain

27.26 Oxidation occurs at the β position of the acid.

27.28 glycerol

27.30 acetone, β-hydroxybutyric acid, acetoacetic acid

27.32 starvation, diabetes, high fat diets

27.34 ketonemia, ketonuria, acetone breath

27.36 acyl carrier protein

27.38 liver

27.40 (1) carbon–carbon bond formation; (2) reduction; (3) dehydration; (4) reduction

27.42 3×7.3

Chapter 28

28.2 time to degrade and replace a protein

28.4 protein synthesis, biosynthetic reactions, produce energy

28.6 15%

28.8 to form proteins

28.10 oxidation to CO_2

28.12 10%

28.14 trypsin, chymotrypsin, and carboxypeptidases

28.16 pyruvic acid, α-ketoglutaric acid, succinyl coenzyme A, and oxaloacetic acid

28.18 leucine

28.20 conversion of amino acids into keto acids for the citric acid cycle

28.22 transamination reaction between aspartic acid and α-ketoglutaric acid

28.24 formation of glutamic acid

28.26 myocardial infarction

28.28 conversion of glutamic acid into α-ketoglutaric acid

28.30 NAD^+

28.32 ornithine, citrulline

28.34 elevated NH_4^+ in the blood

28.36 ornithine

28.38 energy equal to 4 ATP molecules

28.40 steps 3–5

28.42 oxidation, formation of an imine, reduction

28.44 3-phosphoglycerate

Chapter 29

29.2 Translation involves information processing from mRNA to form proteins; transcription involves information transfer from DNA to mRNA.

29.4 A codon is a series of three bases in messenger RNA; the anticodon is a series of three complementary bases.

29.6 movement of a growing protein chain from an A to a P site

29.8 DNA stores and transmits hereditary information; RNA is involved in protein synthesis.

29.10 diphosphate esters

29.12 See Section 29.4.

29.14 DNA has A, T, G, and C; RNA has A, U, G, and C.

29.16 No specific relationship exists.

29.18

29.20 β

29.22 none

29.24

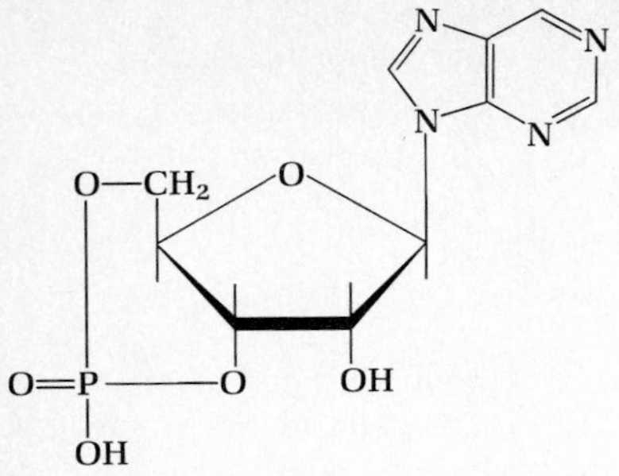

29.26 The sequences of bases run in the opposite directions.

29.28 deoxyribose and diphosphate linkages

29.30 DNA unravels to form a replicating fork, and new DNA strands are formed.

29.32 dC—G—T—A—G—T—C

29.34 C—G—U—A—G—U—C

29.36 5′ to 3′

29.38 "cloverleaf"

29.40 No; some terminate protein synthesis.

29.42 750

29.44 GGGACAUGUGGA, Gly-Thr-Cys-Gly

29.46 Growing protein is transferred from donor site to amino acid located at acceptor site.

29.48 Nonsense results, and an incorrect protein is formed.

29.50 substitution to give termination codons

29.52 Arginine and leucine each have six.

29.54 Each has one codon.

Index

Atomic Numbers and Atomic Weights of the Elements

Based on $^{12}_{6}C$. Numbers in parentheses are the mass numbers of the most stable isotopes of radioactive elemen

Element	Symbol	Atomic Number	Atomic Weight	Element	Symbol	Atomic Number	Atomic Weight
Actinium	Ac	89	227.0278	Europium	Eu	63	151.96
Aluminum	Al	13	26.98154	Fermium	Fm	100	(257)
Americium	Am	95	(243)	Fluorine	F	9	18.998403
Antimony	Sb	51	121.75	Francium	Fr	87	(223)
Argon	Ar	18	39.948	Gadolinium	Gd	64	157.25
Arsenic	As	33	74.9216	Gallium	Ga	31	69.72
Astatine	At	85	(210)	Germanium	Ge	32	72.59
Barium	Ba	56	137.33	Gold	Au	79	196.9665
Berkelium	Bk	97	(247)	Hafnium	Hf	72	178.49
Beryllium	Be	4	9.01218	Hahnium†	Ha	105	(262)
Bismuth	Bi	83	208.9804	Helium	He	2	4.00260
Boron	B	5	10.81	Holmium	Ho	67	164.9304
Bromine	Br	35	79.904	Hydrogen	H	1	1.0079
Cadmium	Cd	48	112.41	Indium	In	49	114.82
Calcium	Ca	20	40.08	Iodine	I	53	126.9045
Californium	Cf	98	(251)	Iridium	Ir	77	192.22
Carbon	C	6	12.011	Iron	Fe	26	55.847
Cerium	Ce	58	140.12	Krypton	Kr	36	83.80
Cesium	Cs	55	132.9054	Lanthanum	La	57	138.9055
Chlorine	Cl	17	35.453	Lawrencium	Lr	103	(260)
Chromium	Cr	24	51.996	Lead	Pb	82	207.2
Cobalt	Co	27	58.9332	Lithium	Li	3	6.941
Copper	Cu	29	63.546	Lutetium	Lu	71	174.967
Curium	Cm	96	(247)	Magnesium	Mg	12	24.305
Dysprosium	Dy	66	162.50	Manganese	Mn	25	54.9380
Einstenium	Es	99	(252)	Mendelevium	Md	101	(258)
Erbium	Er	68	167.26				

† Unofficial name